Brines and Evaporites

Brines and Evaporites

Peter Sonnenfeld

Department of Geology and Geological Engineering
University of Windsor
Windsor, Ontario, Canada

1984

ACADEMIC PRESS, INC.
(Harcourt Brace Jovanovich, Publishers)
Orlando San Diego New York London
Toronto Montreal Sydney Tokyo

ACADEMIC PRESS, INC.
Orlando, Florida 32887

United Kingdom Edition published by
ACADEMIC PRESS, INC. (LONDON) LTD.
24/28 Oval Road, London NW1 7DX

Library of Congress Cataloging in Publication Data

Sonnenfeld, Peter.
Brines and evaporites.

Bibliography: p.
Includes index.
1. Salt. 2. Evaporites. I. Title.
QE471.15.S2S65 1984 552'.5 84-6400
ISBN 0-12-654780-7 (alk. paper)

PRINTED IN THE UNITED STATES OF AMERICA

84 85 86 87 9 8 7 6 5 4 3 2 1

Contents

3. Water Masses

4. Precipitating Brines

PART II. PRIMARY PRECIPITATION

5. The Continental Environment

6. Evaporite-Related Marine Carbonates

7. Primary Marine Precipitates

PART III. ADMIXTURES TO THE PRECIPITATES

8. Substitutions and Accessory Precipitates

9. Behavior of Clastics in a Hypersaline Brine

10. Behavior of Base Metals

11. Organic Matter and Petroleum

PART IV. POSTDEPOSITIONAL ALTERATIONS

12. Syngenetic and Epigenetic Brine Movements

13. Secondary Sulfates

List of Illustrations

List of Tables

Preface

The purpose of this book is to present a synthesis of research findings in the field of naturally occurring concentrating brines and their precipitates, in order to encourage further research into problems never tackled nor solved and away from problems solved many times over.

To study the physical and chemical parameters inherent in hypersaline brine behavior makes one very much aware of how much has been done before and how much remains to be done. Many observations have been duplicated because previous publications were overlooked by subsequent investigators, and definitions remained vague. For example, a "deep-water" origin of evaporites was contrasted to a "shallow-water" origin, without the precise limits of the terms "deep" and "shallow" being defined. Consequently, both terms were often applied with equal enthusiasm to the same evaporite deposits, or even to the same depth ranges of brines. Using reasonable assumptions and available data on modern environments, many such controversies could easily be resolved.

Much work has gone into textural studies of evaporites and laboratory studies of the geochemical behavior of artificial brines, and useful information has thus been accumulated. However, reactions of artificial precipitates often differ from those from natural environments. We remain woefully ignorant of many aspects of the physics and chemistry of naturally occurring brines unaltered by human activities. For instance, no studies have been made of the Soret effect interference in concentrating brines, when cryophile ions try to slip by thermophile ones. Most data on the physical chemistry of brines have been collected from dilute solutions or from those with concentrations no greater than 3–4 moles.

We know little about the effects of organisms and their dissolved decomposition products in halite- and potash-precipitating brines. Although we have somehow assumed that no biota would be involved, organic decomposition products abound in all evaporites. The contribution of organisms, such as bacteria and algae, has hitherto been largely ignored in such studies. Neither geologists nor chemists have so far paid any attention to the effects on rates of crystallization and dissolution of dissolved organic materials, such as alcohols, protein lysates, or lipids. This has been of interest only to production foremen concerned with flotation techniques. All laboratory evaporation experiments, regardless of the methods employed, have ignored the sulfate reduction by bacteria, the consequent massive loss of hydrogen sulfide to the atmosphere or its recycling into oxygenated surface waters, and the resulting deficiency or common absence of magnesium and potassium sulfates in the precipitates. If organic compounds dissolved in the brine have a crucial (positive or negative) catalytic or trigger role to play in evaporite genesis, we also have to ask ourselves what happens to these compounds in a reducing medium. Massive quantities of such compounds can be generated in an evaporite environment, can complex, and can be discharged into subsurface aquifers along with brine encroachment. The intimate association of evaporite deposits with oil and gas fields in surrounding reservoir rocks is surely no accident.

Although the rapid concentration of metallic ions in brines is well known and the relative concentration of base metals in recent gypsum or in ancient anhydrite and halite is also known, the fate of metallo-organic and metal–chloride complexes in precipitating brines has not been investigated. It is easy to extend this list. More year-round studies of natural evaporite-precipitating lagoons are badly needed, from both a physicochemical and a biochemical point of view. Unfortunately, the little information available about such lagoons is based on very short term visits and not on continuous monitoring.

Over 4000 references, articles, papers, monographs, and textbooks were consulted in the preparation of this book. (In some cases, I had to rely on a translation of a text.) To keep the book to a reasonable size, some restrictions had to be placed on the content: The treatment of atmospheric conditions and oceanographic constraints affecting evaporite genesis had to be kept general, and the interested reader is referred to treatises on meteorology, climatology, and oceanography. The sections on polar, lacustrine, and groundwater-derived evaporites were also kept short, since the volume of such deposits is not large for the most part and these evaporites represent ephemeral deposits preserved in only very young sediments. In the same manner, Chapter 6 on marine carbonates is not all-encompassing, but is restricted to limestone and dolomite formation directly related to marine evaporite genesis. Again, there are many excellent texts available on open marine limestones of carbonate banks and reef growths. Chemical and physical properties of precipitating brines were discussed apart from syngenetic and epigenetic brines that alter the precipitates. Primary precipitates were dealt with in sequence inverse to their solubility, followed by accessory minerals. This includes not only cation or anion lattice substitutions but also siliciclastics, metal concentrations, and organic matter. Sulfatization was given special treatment, since many commercial potash deposits are derived from such metasomatism. Selected textural peculiarities were dealt with only as far as they affect our understanding of evaporite genesis. Chapter 15 summarizes the constraints and controls placed upon evaporite genesis by atmospheric, oceanographic, and topographic parameters that are evidenced by a comparison of several ancient evaporite basins that demonstrate similar developments in time. A brief discussion of each evaporite mineral was relegated to the final chapter, where composition and other data on the individual minerals were summarized.

The book leans heavily on the physics and chemistry of brines and precipitates. It will thus be of interest not only to sedimentologists dealing with salt sequences but also to low-temperature geochemists. Chapters on the peculiar properties of associated clays, metals, and organic matter should be of special interest to engineering geologists dealing with mixed-layer clays, to economic geologists searching for base metal deposits, and to petroleum geologists.

It is here that I have to express my gratitude for numerous suggestions by my colleagues at the University of Windsor, Dr. P. P. Hudec of the Department of Geology, Dr. D. T. Pillay of the Department of Biology, and Dr. A. Trenhaile of the Department of Geography. My thanks also go to Dr. C. G. van der Meer Mohr of the Geologisch-Mineralogisch Instituut der Rijksuniversiteit Leiden, The Netherlands, Dr. R. J. Hite of the U.S. Geological Survey, Dr. F. H. Fabricius, professor of marine geology at the Technical University in Munich, Dr. R. Kuehn, honorary professor of the University of Heidelberg, Dr. G. Busson, co-director of the Geological Laboratory, National Museum of Natural History in Paris, and Dr. M. Rouchy of the same institution, who critically read sections of an early rendition of the manuscript. Professor Dr. J. Lucas of the Geological Laboratory, Louis Pasteur University, Strasbourg, kindly enabled me to utilize his extensive library for some rarer source material.

Peter Sonnenfeld

1

Introduction

The term "evaporites" is restricted to rocks precipitated from a concentrated watery solution, i.e., hydrochemical precipitates from solutions concentrated by evaporation and formed either with or without biogenic influence. These rocks represent a distinct and fairly abundant group of economic deposits of sedimentary origin, and are closely interrelated in their environmental and geological conditions of sequential formation, geographic occurrence, physical and chemical properties, and industrial use.

Precipitation of massive evaporites is dependent on the continuous concentration of a brine to saturation, and to replenishment of waters extracted by evaporation. This requires environmental conditions to remain constant for a significant period of time, and thus time becomes one of the most important factors. This fact is not new; it was stressed a century ago by Ochsenius (1877). Occasional flash floods freshen the brine and contribute muds that are dispersed and altered upon settling. These muds serve as basin-wide time markers and are useful for detailed stratigraphic correlations. Massive evaporites accumulate only in basins starved for terrigenous influx.

Evaporites include chlorides and sulfates of sodium, potassium, calcium, and magnesium, carbonates of sodium and magnesium either with or without calcium, and nitrates of sodium and potassium. Subordinate in quantity to evaporitic rocks deposited from seawater or groundwater are admixtures of bromides, iodides, borates, and a few other complex salts, such as iron chlorides. Comparatively rare are authigenic secondary minerals derived from dissolved silica or from a fixation of sulfides. All these chemical admixtures are important indicators of depositional environments. However, they are volumetrically insignificant in any marine evaporite deposit and can become important only as a major component of some continental lake deposits. Other rocks that surround evaporites and are closely related to them are excluded from a detailed discussion. These include reef chains and other nonevaporitic carbonates, red beds accumulating on the shores of evaporite basins, and phosphorites forming in open waters outside evaporite basins in the same climatic belt and oceanographic environment.

Water masses are always moving, either as currents in open waters or through pore spaces as formation waters. Precipitated crystals are subject to repeated alteration by oxygenated or anoxic percolating brines, and they remain in equilibrium only with the last brine that passes by. It is important to remove the overprinting of subsequent brine movements in order to determine the original conditions of deposition. However, it is difficult to draw a precise boundary between paragenetic and diagenetic or epigenetic phenomena. Thus it is not always possible to exclude from the discussion all alterations to the precipitates after burial. Only alterations apparently produced by later tectonic deformation or volcanic intrusions are not considered.

Evaporitic rocks have been used since the dawn of human history. A Sumerian text, Ur III, speaks of part of a man's wages being paid in soda and gypsum (Levey, 1958). Salt has been used in all cultures, either mined or extracted from seawater. The Romans paid a "salarium" in halite, hence our modern word "salary" for wages, originally salt money. Knossos built his palace on the island of Crete out of big blocks, some of which were pure massive gypsum, preserved to this day. Methods of seawater extraction utilize brine concentration through natural evaporation by solar energy. This is the same process used in natural lagoons and lakes to precipitate sundry evaporite minerals. Instead of evaporating brines in artificial, roughly rectangular salt pans, another method practiced in Laos (Fig. 1-1) uses evaporation baskets. Only potash and magnesium salts had no large-scale practical use until the nineteenth century.

There are essentially five major groups of agents responsible for the ultimate composition of an evaporite sequence, as well as its location and morphology:

1. The intensity of the atmospheric elements in a semiarid climate, determined partially by solar activity and partially by latitudinal position.
2. The oceanographic conditions of current direction and density stratification, with resulting potential solar heating, anoxic bottom waters, and disappearance of bioturbation.
3. The syngenetic brine, open and exposed to the atmosphere, which concentrates beyond saturation and precipitates a series of progressively more soluble mineral suites while draining excess solute, first as bottom outflow and later increasingly as subsurface seepage.
4. The diagenetic or epigenetic brines, which may circulate either shortly after precipitation of a mineral or at any later time due to seepage from above, below, or laterally. Because the mineral always remains in equilibrium with

Fig. 1-1. Salt extraction in Laos (photo courtesy of Dr. R. P. Mendels).

passing brines, its mineral facies will reflect the composition of the latest contacting solution.

5. The original size and shape of a basin, restriction, length, and convolution of the entrance straits, direction of the opening with respect to the earth's rotation, and alteration of basin morphology by changes in relative rates of subsidence.

ENVIRONMENTAL SETTINGS

Individual crystals of any evaporite mineral can be found in a great variety of environments, ranging from lava caves and mine entrances to continental lakes and basins adjacent to open seas. Only large-scale accumulations are dealt with here. Special environments, such as caves, mines, weathering crusts, and gypsum as salt karst in arid lands, are not part of this discussion. The occurrence of evaporitic minerals in tunnels (Zednicek, 1954), in speleothems (Legrand, 1947; Kettner, 1948; Fischbeck and Mueller, 1971; Broughton, 1972; Hill, 1979), around fumaroles (di Franco, 1942; Mizutani, 1962), in lavas (Kawano, 1948; Taneda, 1949; Montoriol-Pouss, 1965), in metamorphic rocks (Nash, 1972), or in building stones (de Quervain, 1945; Gistl, 1940; Pochon *et al.*, 1949; Pochon and Jaton, 1967; Winkler and Wilhelm, 1970) is of great interest. However, in no case are these occurrences rock-forming and they remain beyond the scope of this discussion. Even more remote from the subject are epsomite, gypsum, halite, and sylvite crystals in lunar soils (Wood, 1977; Ashikhimina *et al.*, 1978), gypsum on Mars (Banin *et al.*, 1981), or the evaporite minerals reported from the moons of Jupiter.

There are four environments (Table 1-1) in which massive evaporite rocks can form: the polar environment, continental lakes in endorheic basins, groundwater precipitation, and subtropical marine bays and lagoons. Because of the predominance in volume of evaporites produced in the subtropical marine environment, these evaporites will be dealt with in greater detail.

Comments scattered throughout the literature suggest a correlation between peaks in volcanic activity and gypsum precipitation. No quantitative evaluation of such a correlation has ever been undertaken. It is true that some basic lavas are rich in SO_4^{2-} and that some basic volcanic eruptions are followed by a thick layer of halite that covers the countryside (Lotze, 1957). Fumarole encrustations around Central American volcanoes indeed precipitate halite, sylvite, gypsum, anhydrite, langbeinite, thenardite, and ralstonite (a hydrous sodium-magnesium-

TABLE 1-1

Environments of Evaporite Precipitation

Rock-forming evaporite minerals occur in:	Non-rock-forming evaporite minerals occur in:
Polar regions	Tunnels
Continental lakes	Speleothems
Groundwater precipitation	Fumaroles
Subtropical marine bays and lagoons	Mine efflorescences
	Weathered voids and fractures in:
	Building stones
	Lavas
	Metamorphic rocks

aluminum fluoride). These minerals are derived from exhalations of H_2O, SO_4^{2-}, CO_3^{2-}, HCl, HF, and their reactions with the atmosphere and wall rock (Stoiber and Rose, 1974). Sublimates of Mount Nyiragongo in Zaire include gypsum, thenardite, halite, sylvite, stalactites of aphthitalite, and a variety of exotic potassium fluorides (Herman *et al.*, 1961). Sublimates in Japan include halite, gypsum, anhydrite, epsomite, chloromagnesite ($MgCl_2$), sal ammoniac (NH_4Cl), and pickeringite ($MgA1_2(SO_4)_4 \cdot 24H_2O$). The alkali metals and alkaline earths are carried as chlorides and are deposited as sulfates. Aluminum and iron are carried as chlorides but deposited as oxides. However, silica is first carried as fluoride and is then deposited as silica (Osaka, 1965).

Whereas Ayrton (1974) theorized that some form of alkaline magmatism has its origin in the assimilation of evaporites and brines by rising basaltic magmas, Kalinko (1973) considered chlorine in volcanic emanations to be mostly recycled chlorine. Sozanskiy (1971) and Birina (1974) defended a plutonic origin of sedimentary salt sequences, denying that they are precipitated from formation waters trapped in volcanic processes. In any case, the amount of sodium chloride or calcium sulfate directly produced in conjunction with volcanic activity is relatively small, compared to the volumes found in even the smallest marine evaporite basin. A volcanic origin of massive evaporite deposits is thus not established. Otherwise, some evidence should have been found by now to suggest a gradual increase in ocean salinity, at least during geologic periods of no known halite or potash precipitation. Holland (1974) noted that the salinity of the world oceans does not appear to have increased markedly during the Phanerozoic time.

THE POLAR ENVIRONMENT

Freeze-drying is similar to evaporative processes in that it leads to concentration of brines. However, the polar environment does not produce layered marine evaporites, only groundwater-derived and lacustrine ones. Many of the minerals typical of the polar environment are considerably hydrated upon precipitation. This may be due to the structure of water. Liquid water forms clusters of H_2O molecules (Fig. 1-2) that decrease in size with rising temperature and at the same time reduce the amount of hydrogen bonding available to other cations from 52% at 6°C to 39% at 72°C (Choppin, 1965). The size of these clusters has an effect on precipitation rates, since hydrated evaporite minerals have very low solubility in cold water, i.e., water of large clusters. Although the number of water molecules with double hydrogen bonds (H_2O^{2-}) decrease progressively with temperature, it may be significant that single hydrogen bonds in water (H_2O^{-}) reach their highest frequency in the temperature range of maximum solubility of minerals hydrated with two water molecules (e.g., gypsum).

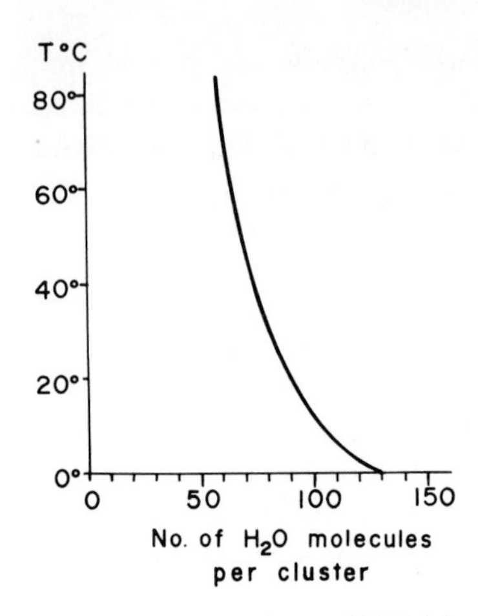

Fig. 1-2. Number of molecules in a cluster of H_2O (after Choppin, 1965).

THE LACUSTRINE ENVIRONMENT

Lacustrine evaporites accumulate only in basins with insufficient drainage, particularly closed or endorheic basins. Calcium and magnesium are mainly trapped in carbonates, whereas potassium is screened out by sorption and ion exchange. This results in a sodium enrichment of more than 100-fold compared to potassium (Eugster and Jones, 1979). Bacterial sulfate reduction is very moderate, being much less extensive than in marine lagoons. Sulfates and carbonates of sodium are therefore predominant in lacustrine environments, whereas they are nearly absent in marine precipitates.

The sequence of primary precipitates depends on the type of water supplied. Continental waters are enriched in bicarbonates and carbonates, and are well aerated. Upon concentration of the brine, the solubility of oxygen and carbonic acid decreases very rapidly; when stratified, the brine becomes anaerobic. Because of a deficiency of chlorine, continental water sources produce mainly sodium and magnesium carbonates and only minor amounts of calcium carbonates. Further concentration produces sodium sulfates and sodium-calcium sulfates, as well as minor amounts of calcium sulfates. Chloride ions generally enter lacustrine environments through leaching of older halite deposits in contact with groundwater.

Information on the chlorine content of rocks does not suggest that weathering and leaching processes could supply the quantities of chlorine found in surface waters and groundwater (Feth, 1981). Here the salts are mainly derived from leaching of older salt bodies or other rocks. However, there are doubts that chlorine can be considered to be the most conservative anion in this system.

GROUNDWATER PRECIPITATION

Precipitation of individual evaporite crystals by saturating groundwater occurs some distance from an open water surface. It leads to the development of crusts in soils, but does not generate bedded evaporite deposits.

THE SUBTROPICAL MARINE ENVIRONMENT

The composition of groundwater varies within certain ranges; seawater, however, displays a nearly constant ratio of its components. Only small variations in ratios are recorded: for example, the Na:Cl ratio in the world ocean is 0.86, and rises in the Mediterranean Sea to 0.88, in the Black Sea to 0.94, and in the Caspian Sea to 0.92, but is only 0.79 in typical formation waters (Tagaeva, 1935). The composition of seawater is given in Table 1-2 for a hypothetical chlorinity of 19 ppt. The putative salt content depends very much on which cations are joined to individual anions. Two such calculations are given for comparison. One has to keep in mind that marine deposits do not contain precipitates of NaF or $NaHCO_3$; nor do they contain primary magnesium sulfate,

TABLE 1-2

Composition of Seawater for a Hypothetical Chlorinity of 19 ppt[a]

Ion	kg/*t*	kg/m³	Compound	kg/*t*[b]	kg/m³[b]	kg/*t*[c]	kg/m³[c]
Na^+	10.5561	10.9930	NaCl	23.477	24.315	26.906	27.867
Mg^{2+}	1.2720	1.3174	$MgCl_2$	4.981	5.159	3.176	3.289
Ca^{2+}	0.4001	0.4144	Na_2SO_4	3.917	4.057	—	—
K^+	0.3800	0.3936	$CaCl_2$	1.102	1.141	—	—
Sr^{2+}	0.0133	0.0138	$MgSO_4$	—	—	2.252	2.332
Cl^-	18.9799	19.6575	$CaSO_4$	—	—	1.200	1.243
SO_4^{2-}	2.6486	2.7432	KCl	0.664	0.688	0.728	0.754
HCO_3^-	0.1397	0.1447	$NaHCO_3$	0.192	0.199	—	—
Br^-	0.0646	0.0669	$CaCO_3$	—	—	0.114	0.118
H_3BO_3	0.0260	0.0269	KBr	0.096	0.099	—	—
F^-	0.0013	0.0013	$MgBr_2$	—	—	0.086	0.089
	34.4816	35.7126	H_3BO_3	0.026	0.027	—	—
			$SrCl_2$	0.024	0.025	—	—
			$SrSO_4$	—	—	0.017	0.018
			NaF	0.003	0.003	0.003	0.003
			H_2O	965.518	999.986		

[a]A density of 1.0257, a chlorosity of 27.57, and a salinity of 49.79 ppt.
[b]Lyman and Fleming (1940).
[c]Fuechtbauer and Mueller (1970).

TABLE 1-3

Anion and Major Cation Ratios in Seawater and Continental Waters[a]

Ionic Ratio	Ratio in Ocean Water	Ratio in Continental Water
$Cl_2^{2-}:SO_4^{2-}$	7.19:1	0.468:1
$Cl_2^{2-}:CO_3^{2-}$	263.29:1	0.162:1
$SO_4^{2-}:CO_3^{2-}$	36.62:1	0.345:1
$Na_2^{2+}:Ca^{2-}$	12.75:1	0.142:1
$Na_2^{2+}:K_2^{2+}$	32.06:1	2.730:1
$Mg^{2+}:Ca^{2+}$	3.10:1	0.167:1

[a]After Niggli (1952).

sodium sulfate, or calcium chloride. The calculations by Fuechtbauer and Mueller (1970) better describe an evaporating marine brine.

The total amount of solute in the world ocean is on the order of 3.3×10^{16} tons, which could form about 28.0×10^{15} m^3 of precipitates that would fill all depressions in the world ocean deeper than 5650 m below sea level. Ocean waters contain chlorine as a primary anion and are enriched in magnesium, calcium, and potassium. Only part of the sulfate and carbonate/bicarbonate is present in ionic form; a significant percentage is tied to major cations as non-

TABLE 1-4

Ratio of Percentage by Weight of Dissolved Ions (Total Solute = 100) between Ocean Water and Averaged Continental Waters[a]

Ion	Ocean water: Continental waters
HCO_3^-, CO_3^{2-}	0.006
SO_4^{2-}	0.633
Cl_2^{2-}	9.734
Ca^{2+}	0.059
Mg^{2+}	1.091
Na_2^{2+}	5.283
K_2^{2+}	0.524

[a]After Niggli (1952).

ionized pairs. Upon becoming anaerobic, marine waters lose most of their carbonate and sulfate ion content as a result of bacterial degradation. There is a substantial difference in the ionic composition of marine and continental brines, either expressed as ionic ratios within each type of water (Table 1-3) or considered in comparison (Table 1-4). The relative importance of calcium, potassium, bicarbonate, and sulfate ions in continental waters is revealed in Table 1-4. Potassium is 165 times more concentrated in seawater than in river water (Livingstone, 1963) and thus plays a greater role in marine evaporites than in continental ones.

In contrast to continental settings, sodium and magnesium carbonates, as well as sodium sulfates, either are absent in evaporites from purely marine water sources or form local nests of secondary diagenetic alteration. The presence of significant amounts of sodium carbonates and sulfates always indicates a substantial component of terrigenous waters, i.e., drainage of a large river, or the later infusion of meteoric waters into a buried evaporite deposit.

If the brine remains anaerobic, rock salt is followed by sylvinite and carnallitite and eventually by bischofite. All other evaporite minerals either are precipitated under very special circumstances, such as those in the formation of tachyhydrite, or are products of secondary alteration, metasomatism, dehydration, or recrystallization. The primary mineral sequence has been the same in all Phanerozoic evaporite basins.

A cursory discussion of the major suites of evaporitic minerals is found in the appendix. The reader can refer to it to identify mineral names used in the text.

Part I
FORMATION OF THE BRINE

2

Air Masses and Climate

A dry climate, i.e., one with a negative water budget, is one of the key factors in the formation of evaporite deposits. Climate, in turn, is a function of moving air masses. Evaporites do not form in humid climates; they are mainly precipitated in areas that are only seasonally dry, as evidenced by plants swept into the deposits. Thus we must examine the conditions under which such seasonal dry spells are propitious for evaporite deposition. Dry climates occur in a subtropical and a subarctic zone in both hemispheres and are not necessarily dry year-round. However, the extent of the dry belt varies from year to year, and is particularly pronounced in periods of low solar irradiation, i.e., cool periods or periods of low transmission of excess latent heat from equatorial to polar regions.

CONTEMPORARY CLIMATOLOGICAL FACTORS

Concentration of a brine occurs either by evaporation or by freeze-drying. Both processes create a water loss, and both require contact of the brine with undersaturated air. If the water loss is not replenished concurrently, concentration sets in. Only undersaturated (i.e., dry) air can extract water vapor. The greater the undersaturation (expressed as relative humidity), the higher the rate of evaporation from a water surface. A brine saturated for halite needs winds with less than 76% relative humidity to continue evaporating (Stokes and Robinson, 1949; Wink and Sears, 1950; Usdowski and Usdowski, 1970; Kinsman, 1976). Fully saturated winds must then blow over hot land surfaces until their temperature rises by 3–5°C to lower their relative humidity sufficiently. This is achieved in a few kilometers and thus can be affected by a chain of windward islands. For a brine saturated with potash solute, the minimum relative humidity drops to 67% (Kinsman, 1976); precipitation of sylvite could then start. For carnallite, the relative humidity would have to drop to 46% (Eugster *et al.*, 1980); for bischofite, to 33% (Stokes and Robinson, 1949; Wink and Sears, 1950; Eugster *et al.*, 1980); and for antarcticite ($CaCl_2 \cdot 6H_2O$), to less than 29% (Arnold, 1981). Conversely, halite deliquesces at 75% relative humidity, sylvite at 80%, and carnallite at 53% (Holser, 1979b) (Fig. 2-1). Whereas the minimum relative humidity for evaporation of a bischofite solution is a continuous function of temperature, this is not the case for a halite-precipitating brine. The minimum relative humidity rises from 74.9% at 0°C to 75.6% at 30°C and drops again to 74.7% at 50°C (Wexler and Hasegawa, 1954).

However, extremely low relative humidities, which are attainable only in extreme continental climates, may not be necessary to precipitate hydrated salts. Most naturally occurring primary magnesium, potassium–magnesium, and calcium–magnesium chloride salts are hydrated. Their crystallization deprives the bitterns of even more water, averaging another 70–80% of the evaporative water loss per unit of time. This lessens the need for evaporative water extraction, which is attainable only with further reductions of relative humidity to values rarely achieved in the subtropics. In conjunction with the nighttime evaporative water loss, the density of the bitterns increases up to 1–2 ppt during crystalliza-

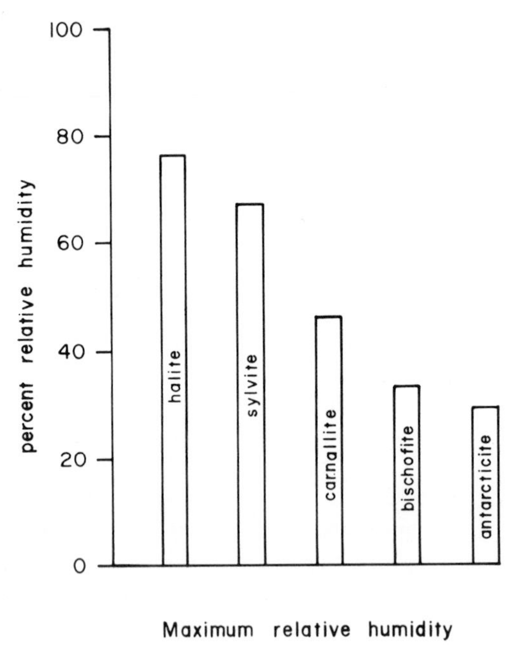

Fig. 2-1. Maximum relative humidity at which a brine, saturated for a given precipitate, ceases to evaporate.

tion. The warming up of the thickened brine during the daytime does not increase the solubilities sufficiently to redissolve much of the precipitate accumulated in the previous night. Pfennig (1962) recalculated the vapor pressure of saturated aqueous solutions of magnesium chloride at 15–60°C, since older values showed too much dispersion. However, a modest admixture of sodium chloride increases the vapor pressure over bischofite up to 10.5%. The vapor pressure again becomes identical to that of pure bischofite if more than 12.64% sodium chloride is added, because of the formation of $NaCl \cdot 2MgCl_2 \cdot nH_2O$ (Orekhova *et al.*, 1975). The influence of admixtures of various marine chlorides has not yet been adequately investigated.

Elton Lake, situated in the Caspian lowlands, experiences maximum evaporation in August at 46% relative humidity and minimum evaporation in October at 64% relative humidity. A secondary maximum occurs in May at 48% relative humidity. No evaporation takes place between November and March; on the contrary, at that time the brine absorbs some atmospheric moisture (Feigelson, 1940). Such absorption has also been noted in Tunisia by Perthuisot (1975), who observed an increase in brine volume during periods of high humidity. Some North African chotts are hypersaline lakes or rather swamps that are covered with a treacherous crust but do not desiccate completely; the air blowing over them is too humid, the substrate too impermeable.

Water Deficit

To concentrate a brine by evaporation or sublimation, the water losses per unit of time must exceed the sum of runoff and atmospheric precipitation. This is represented by the inequality

$$E/t > (R + P)/t \tag{2-1}$$

where E is the water loss through evaporation, R is the runoff (combined groundwater and river discharge), P is the atmospheric precipitation, and t is the unit of time. This inequality holds only in semiarid or arid lands. It represents a deficit in an area's water balance, which is one of the prime prerequisites for evaporite formation. A deficit in the water balance defines an arid or dry climate, whereas a surplus defines a humid or wet climate. Since true evaporation rates are very difficult to determine, various indirect methods have been sought to define an arid climate. As long as a close network of reliable evaporation data is not available, each of these methods has certain shortcomings. One of the simplest formulas is given by Koeppen (1931) in terms of total annual precipitation (P), mean temperature (T), and a constant (k); it was originally developed for a Mediterranean type of climate and still applies best to that climatic environment. Koeppen's definitions were as follows: arid climate: $P < (T + k)$; semiarid climate: $(T + k) < P < 2(T + k)$; humid climate: $P > 2(T + k)$. The constant k is equal to zero if the precipitation is concentrated in the winter and the dry season occurs in the summer (Mediterranean climate). The constant *k* equals 14 if the precipitation is concentrated in the summer and the dry season occurs in the winter (savannah climate). Finally, the constant *k* equals 7 if the precipitation is evenly distributed throughout the year and there is no specific dry season. The relationship of arid, semiarid, and humid climates at different temperatures and rates of precipitation is shown in Fig. 2-2.

The deficit in the water budget of an area is an index of its aridity, the consequences of which are manifold.

1. The amount of runoff is inversely related to the degree of aridity. As the aridity increases, the number of perennial streams decreases, and the number of wadis increases. With declining force of the streams, terrigenous clastics are reduced in size to fines and eventually cease to be delivered, resulting in a starved basin, where at first carbonate banks and eventually chemical precipitates are the predominant sediments.
2. The total atmospheric precipitation is also inversely related to the degree of aridity.
3. The frequency of downpours is inversely related to the degree of aridity, but their intensity is not. Flash floods are more destructive and possess more transportational force as aridity increases. This destructiveness is increased by the change from continuous vegetation cover in humid terrain to a discontinuous, patchy one in semiarid lands. Flash floods dump large amounts of mud into a marginal bay and cause clay intercalations in evaporite sequences.
4. The variability of annual rainfall values is nearly proportional to aridity. Semiarid regions possess an unstable climatic regime. The annual rainfall values may vary from the norm by 20 to 40%.
5. Morning dew is proportional to aridity, but only on land, affecting the desiccated margins of evaporite pans. Over water, the land breezes are chilled and fog arises. Saline waters have a lower vapor pressure than fresh waters; hence, a lower relative humidity is required (Walton, 1978) for overlying air to condense some fog or dew during cool nights. Perthuisot (1975) noted the abundant dew around the sebkha El Melah in Tunisia, which is a dry playa today. It is likely that the ample dew in drier terrain plays an important role in the redistribution of precipitated supratidal and playa precipitates (Shearman, 1970). It also affects recrystallization of salts derived from hygroscopic bitterns, which may adsorb atmospheric moisture and then dissolve in their own water.
6. More evaporation takes place from water surfaces outside equatorial latitudes in winter than in summer (Albrecht, 1949; Jacobs, 1951; Loewe, 1953). The lapse rate (or vertical temperature gradient) of water vapor from the warm sea to the cool air is greater in winter than in summer, when the temperature excess of the water is small or absent (Albrecht, 1951). Similarly, evaporation

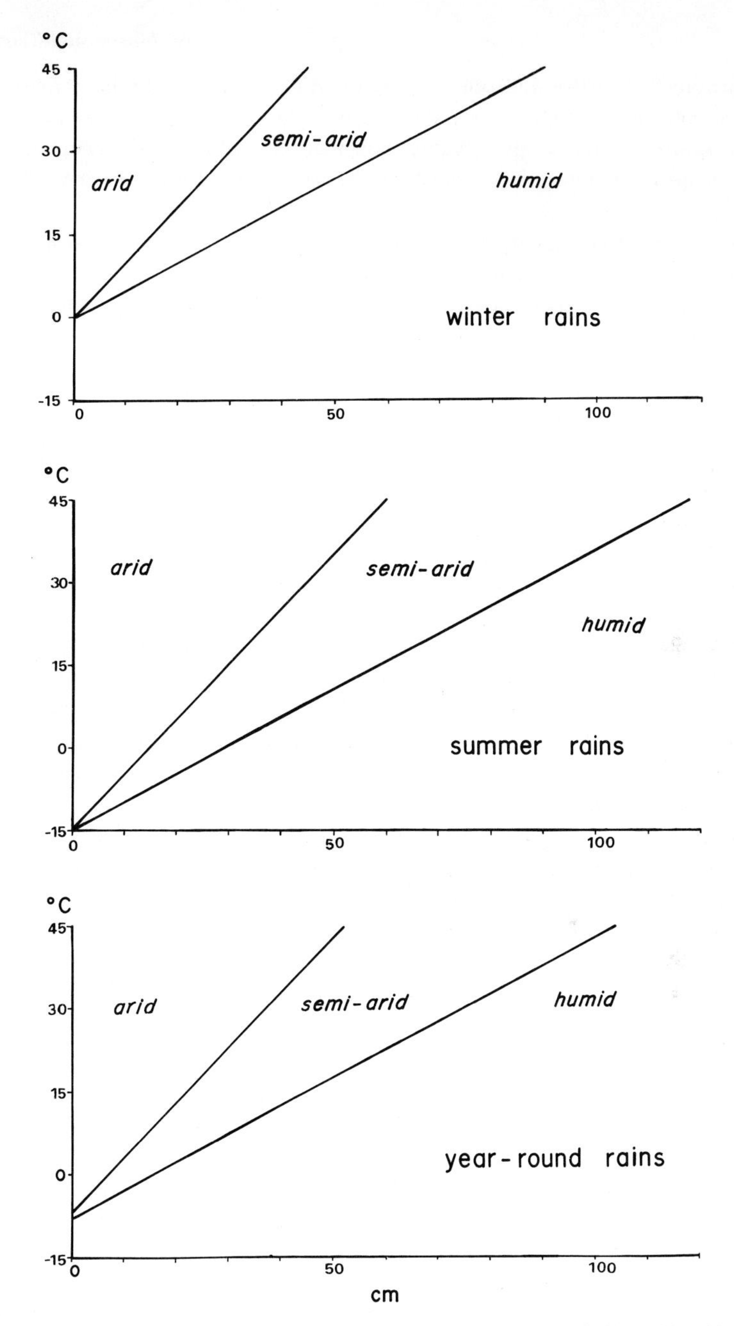

Fig. 2-2. Range of annual temperatures and precipitation in arid, semiarid, and humid climates (after Koeppen, 1931).

rates are highest in the subtropics at night, peaking before sunrise, when cold breezes from land warm up over the warmer sea. The Persian Gulf attains its highest salinity in the winter; evaporation reaches a minimum in the summer, when the air temperature is greater than the water temperature (Emery, 1956).

Evaporation losses in semiarid and arid regions are thus greatest in the winter season or in the predawn hours of a hot day. Since evaporation losses are enhanced by cold winds blowing from offshore and are reduced by warm winds blowing from land, the relative humidity is drastically reduced whenever strong continental winds are blowing, which may reverse the maxima of evaporation losses. If winds of high relative humidity blow over the water surface, further evaporation may cease. Hot continental winds blowing over large water expanses are cooled at the air–water interface, and their relative humidity rises. In contrast, saturated onshore winds blowing from the open oceans over a hypersaline lagoon warm up over the concentrated brine and are able to extract further amounts of water vapor (Walton, 1978).

The trade winds are most undersaturated when they originate from air descending in high-pressure cells on land, particularly on the eastern margin of such cells and in summertime. Winds heat up over land in the summer, rapidly reduce their relative humidity, and then cross marginal bays as warm, undersaturated winds of high water-carrying capacity. On an open, unprotected shoreline, the effect of the trade winds is thus reinforced along poleward shores. These winds can maximize their evaporation rates on the north shore of bays, arriving saturated at their south shores. This effect is mitigated along the equatorward shore on a coastline broken by small patches of land, peninsulas, and island chains. The water loss is increased again in embayments separated from the main body of water by even small land areas over which the winds blow. Whether the difference is large enough to alter significantly the surface water circulation in a marginal bay is not known. This system of local offshore/onshore breezes and regional wind systems can substantially alter the overall influence of global wind systems (Fig. 2-3). An example of such a displacement of peak evaporation rates by local wind systems is provided by the sebkha El Melah in Tunisia (Perthuisot, 1975). Summer winds and daytime breezes come from the sea, heat up, and thus decrease their relative humidity before blowing over the lagoon, whereas winter winds and night breezes are hot, dry winds from the Sahara. However, in this case, the data supplied by nearby weather stations are not necessarily applicable to an open lagoon. One cannot extrapolate from measurements of evaporation rates on land to those over a brine surface. Instrument readings that are not taken on land, i.e., that are taken in the center of an expanse of brine, are generally not available.

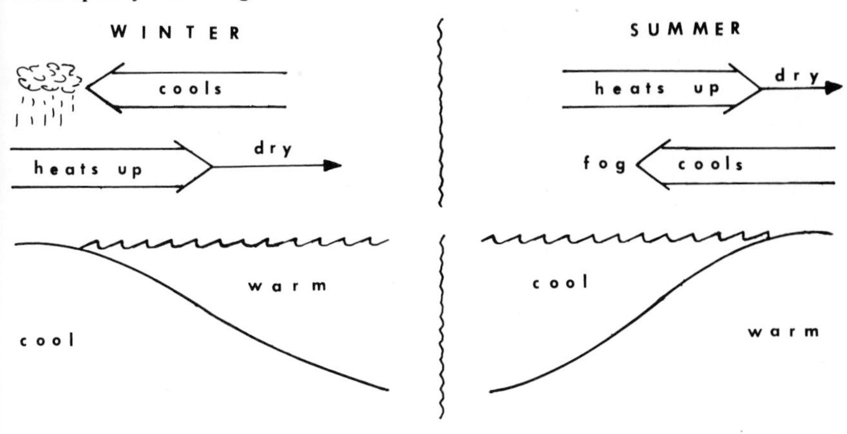

Fig. 2-3. Offshore and onshore winds in winter and summer.

Local wind systems can be generated even by very small lagoons and lakes. Dumont (1981) has studied a 76-m-deep hypersaline crater lake in Turkey, which has barely more than 0.5 ha of water surface. Summer heating of the barren desert soil around the lake produces an ascending air mass. The air above the lake is cooled by evaporation of the brine, is heavier, flows into the surrounding Low, and must be compensated by subsidence of air at some altitude above the lake center. Despite a windstill some distance from the lake, strong waves are generated by a local centrifugal wind originating over the lake. The surface brine, concentrated by evaporation, is driven against the shore and then sinks. Larger lagoons filled with hypersaline brine can be subjected to significant daily ponding along the shores.

Arid lands need not be dry year-round to produce evaporite sequences. They can have a short wet season, as long as the rainy season with its flash floods does not balance out the annual evaporation losses. Solution of evaporites can replace precipitation in such periods, causing minor diastems.

Overall, the rate of evaporation varies with several parameters:

1. The temperature of the air
2. The storm frequency
3. The humidity of the air
4. The wind speed
5. The wind direction (offshore or onshore)
6. The temperature of the brine
7. The concentration of the brine
8. The presence of surface crystals

Climate and Air Masses

Of all the climatic elements, the prevailing wind regime is the most important because it exchanges air masses in contact with the water surface, thus promoting continuing extraction of water vapor from the open water surface. Where winds move primarily equatorward, they warm up. Warm air rapidly increases its water vapor-carrying capacity, becomes undersaturated, and is capable of extracting more water from a brine (Fig. 2-4). However, increasing the concentration of the brine decreases its vapor pressure. The rate of water loss through evaporation is proportional to the difference between the partial vapor pressure in the air pass-

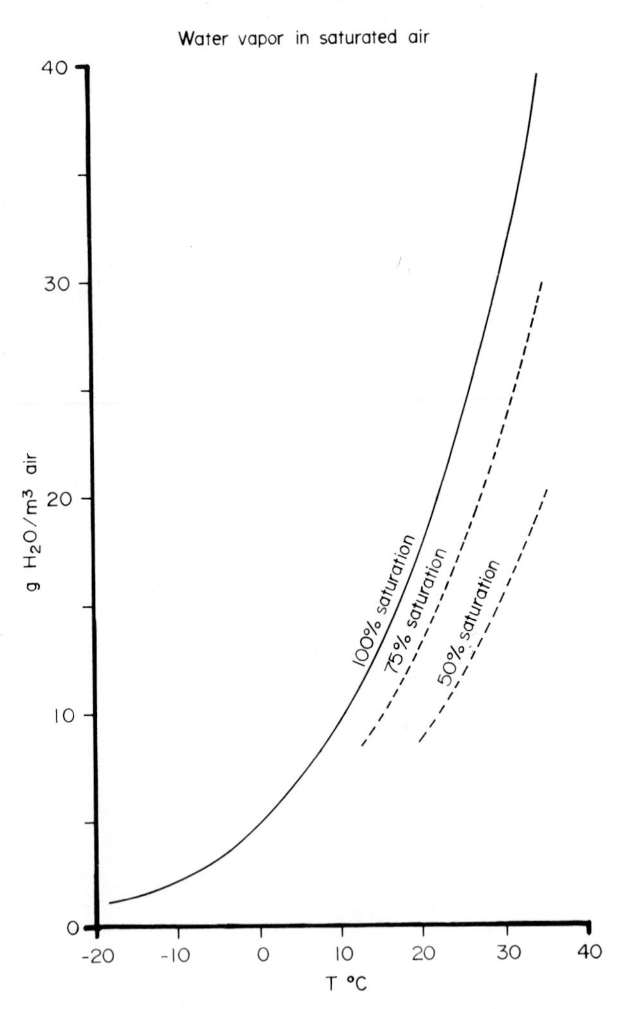

Fig. 2-4. Water vapor saturation of air.

ing over the water surface and the vapor pressure of water in the brine. Such data are not experimentally available for concentrating marine brines.

Winds moving equatorward are deflected westward by the Coriolis effect. These winds become prevailing easterlies. There are two sets of these winds in either hemisphere: The polar easterlies seasonally cross the Arctic and Antarctic Circles, and the subtropical easterlies or trade winds occur in lower latitudes. In both cases, the wind systems at the earth's surface have descended from high-altitude countercurrents. The air descending in the horse latitudes and moving poleward is also initially dry and rapidly desiccates any lagoon or land surfaces (Fig. 2-5). Descending air heats up adiabatically and lowers its relative humidity. Dry areas develop in wind regimes originating in descending air. There are thus two arid belts in each hemisphere: the subpolar belt, with low rates of atmospheric precipitation and evaporation, and the subtropical belt around the horse latitudes. The former is the realm of tundra vegetation and freeze-drying, e.g., in permafrost areas of Greenland and Antarctica. Within the polar easterlies, the relative humidity rises when the winds blow over antarctic ice masses or over Greenland ice, but it is not increased very much over arctic ice-free coastal plains. This polar belt produces only very small quantities of evaporite minerals in very localized areas. Tropical high-pressure belts are produced by descending air. The trade winds rise in the Intertropical Convergence Zone on the equatorial margins of Hadley cells of air circulation. Guided by the subtropical high-altitude jet stream, they return poleward from about 5° of latitude as the high-altitude antitrades, only to descend in the horse latitudes to form the trade winds moving equatorward and the prevailing westerlies in the opposite direction.

Orographic Effects

Rising saturated air cools adiabatically at a rate that is about 40% slower than the adiabatic temperature change in descending dry or undersaturated air, because condensation from cooling saturated air releases some heat. Saturated air that rises over a mountain chain cools at about 6°C per 1000 m elevation, but warms in the lee of the mountain at about 10–11°C per 1000 m descent, producing on the lee side a warmer regime at equal elevations. A decrease in the dew point temperature of 2°C per 1000 m also affects the final temperature and moisture saturation of the air.

Winds descending on the lee side of a mountain chain (foehn or chinook winds) warm up and lower their relative humidity. A rain shadow desert is the result (Fig. 2-6), frequently dotted with groundwater-fed lakes that precipitate evaporite minerals. The winds become most effective when they reinforce the general flow of air. Where a mountain chain strikes east–west along the windward side of a body of water and lies athwart the prevailing easterlies, the orographic effect reinforces the meridional one. The mistral, bora, tramontana,

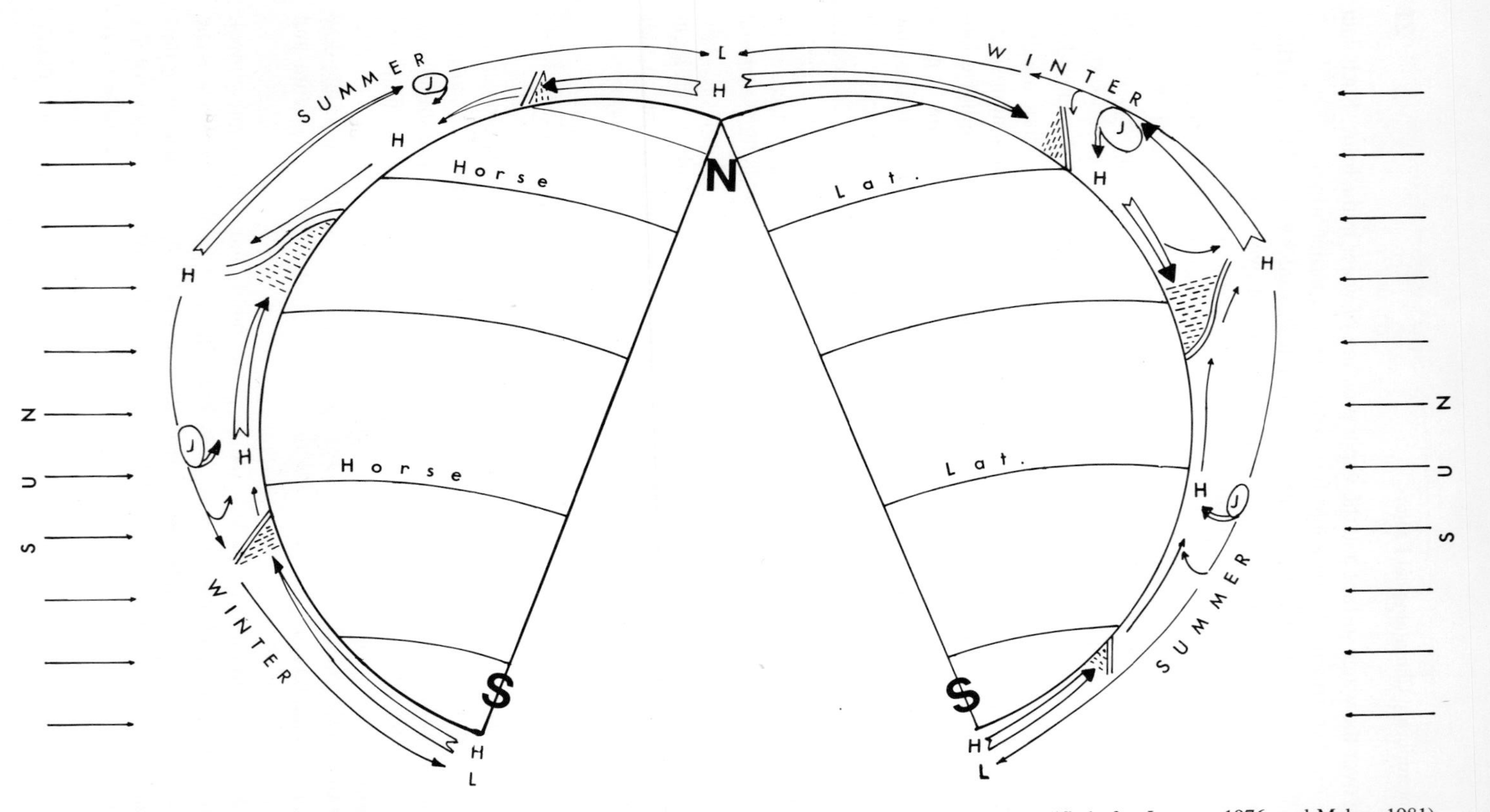

Fig. 2-5. Air circulation in summer and winter. H = high pressure; L = low pressure; J = jetstream (modified after Leroux, 1976; and Maley, 1981).

Fig. 2-6. Orographic wind shadow. The precipitation maximum lies on the windward side of rising air; the evaporation maximum lies on the leeward side of descending air.

and varadats or etesian are winds that blow down the lee of the mountains or through mountain gaps across the north shore of the Mediterranean Sea. Trade winds also blew over the Permian Cordillera onto evaporite basins in New Mexico and West Texas or, then in another hemisphere, across the Uralian land mass onto Permian evaporite basins now lying to the west of it.

THE MAIN EVAPORITE BELT

The width of the subtropical dry belt is not fixed. It is a function of incoming solar radiation in low latitudes and an energy deficit in high latitudes. The more solar energy is pumped into the tropical Hadley cells of the air circulation, the tighter the anticyclones (Svalgaard, 1973) and the narrower the dry belt (Fig. 2-7). The intake of solar radiation in the equatorial regions is exported, the energy transport being divided between atmosphere and oceans by a ratio of between 4:1 and 3:1 (Barrett, 1974). The total heat energy is converted into potential energy, sensible heat, and latent heat. With a reduction in intake of solar radiation energy, it is the amount of latent heat (stored as water vapor in the atmosphere) that is curtailed the most (Newell *et al.*, 1975). This is to say, rates of atmospheric precipitation in the subtropical regions are most severely affected.

Expansion of the dry belts in cooler periods has been linked to variability in the solar constant (Hammond, 1976a,b; Parkin, 1976). The less solar energy received at the earth's surface, the colder the equatorial sea and the weaker the Hadley cell (Wright, 1978). Less water vapor is then exported from the equatorial belt; the arms of the subtropical anticyclones are wider, and with them expand the subtropical arid belts. A reduction in the seasonal temperature gradient in the horse latitudes by cooling slows down the easterly migration of the high-pressure cells over dry land, where friction effects are more significant.

Air masses descending in the subtropical high-pressure belt (horse latitudes) from the upper troposphere produce an atmospheric high-pressure belt and de-

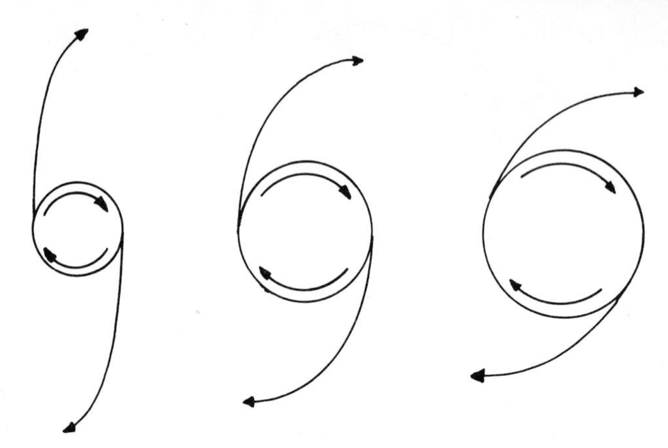

Fig. 2-7. Width of anticyclones. A tightly wound whirl results in a nearly meridional trajectory, whereas a loosely spinning vortex creates a latitudinal motion of lesser breadth.

press the water surface of the ocean underneath. These air masses warm up in their descent and become undersaturated. This belt is centered around 28°N and 27°S in the respective springtimes, and around 36°N and 27°S in the respective falls (with a somewhat greater amplitude over the South Pacific Ocean). Within the realm of the horse latitudes, the air has low relative humidity; hence, atmospheric precipitation is rare. Rivers entering such an area lose more water to evaporation than tributaries and groundwater supply, and thus progressively reduce their water volume. This is the main area of evaporite generation; past evaporite deposits also tended to center around similar paleolatitudes. The horse latitudes oscillate only slightly during the year, and the poleward and equatorward limits of the trade wind regime shift accordingly.

The descent of the air masses occurs in individual anticyclonal cells strung out along the horse latitudes. Descending air is compressed, and these cells are marked as high-pressure cells (highs) on meteorological charts. Anticyclonal cells of descending air are strongest and most persistent off the west coast of continents. Within each cell, the descending air dips eastward (Barrett, 1974), is more undersaturated on its eastern margin, and exerts there the greatest pressure at sea level. Consequently, we find the dry areas on western margins of continents slightly increasing their aridity eastward. As the air moves out of the horse latitudes, it turns poleward into the mid-latitude prevailing westerlies and cools along its path. Equatorward, it turns into the low-latitude trade winds, which are strongest on the east side of subtropical highs, i.e., off the west coast of the continents. Latent heat is transported both equatorward and poleward from latitude 25° (Barrett, 1974). The trade winds reach farthest north or south along west coasts and possess here a more pronounced meridional component. These are the areas of highest aridity, and evaporite basins are more common on the western margins of continents.

The anticyclonal cells are really giant cylinders of air. If the jet stream slows down due to reduced solar energy intake, it assumes meandering patterns around the globe, with wider amplitude. High-pressure cells get stuck in one place for months on end, prevent the more normal mixture of air masses, and result in weather extremes. Cold periods in one part of the globe then coincide with extremely dry periods in another part. As more energy is pumped in, the air accumulates in tubes of lessening diameter and increasing rates of spin. The horizontal convergence of the wind velocity approximately equals the fractional increase of its absolute vorticity with time (Barrett, 1974). At the same time, the descent of air must steepen in a cell of smaller diameter. This increases adiabatic heating. Individual high-pressure cells are then prevented from lingering in any one place, which reduces the probability of long dry spells. The shortening of the path of saturated air and the tightening of the vorticity of high-pressure cells reduce the width of arid regions. Especially affected are the semiarid regions of winter or summer rains on the margins of dry belts. Only the poleward areas get adequate winter rains; the equatorward limits receive adequate summer rains. In contrast, the more meridional paths of higher-energy westerlies bring the summer routes of midlatitude summer storms to Arctic Circle areas. The lush subarctic forests of the very warm Paleocene epoch can be explained by such meridional transport of latent energy into high latitudes.

The dry belt is not restricted to the poleward limit of the trade winds. The Mediterranean Sea is dominated today by prevailing westerly or northwesterly winds in both summer and winter. Only in the vicinity of the Strait of Gibraltar does the wind come from due west or southwest in the summer, and then blows in the same direction as any increased inflow of ocean waters moves in response to high rates of evaporation over the Mediterranean Sea.

Coriolis Effect on Air Masses

Both trades and antitrades start out as meridional winds. For a meridional motion of velocity v, we can write

$$v = \Delta \phi / \Delta t \qquad (2\text{-}2)$$

which is the rate of travel expressed in degrees of latitude (ϕ) per unit of time t. Velocities of air masses are commonly given in units of distance per hour. The Coriolis force (F_c) deflects the meridional motion to the right in the Northern Hemisphere, and the numerical value of the acceleration produced is expressed as

$$F_c = \omega v (2 \sin \phi) \qquad (2\text{-}3)$$

where ω is the angular rotational velocity of the earth, a constant through all latitudes at any given time, v is as defined above in Equation (2-2), and ϕ is the angle of latitude traversed.

This gives a rate of travel perpendicular to the meridional one (deflection) d_1 (see Fig. 2-8) equal to

$$d_1 = \omega \, \Delta t \, (\Delta\phi/\Delta t)(2 \sin \phi) \tag{2-4}$$

The ratio between deflection and meridional motion can be expressed as

$$d_1/v = \omega \, \Delta t \, (2 \sin \phi) = \tan \Theta \tag{2-5}$$

The Coriolis force also acts on the deflection d_2 by the same ratio between deflection and motion (d_2 in Fig. 2-8); thus

$$d_2/d_1 = \tan \Theta \tag{2-6}$$

Consequently, we obtain by substituting

$$d_2 \tan \Theta = v \tan^2 \Theta = v \, (\Delta\phi/\Delta t) \, \omega^2(\Delta t)^2(\Delta\phi/\Delta t)^2(2 \sin \phi)^2 \tag{2-7}$$

This leads to retrograde motion when $d_1 > v$ or $\tan \Theta > 1$, which can be expressed as

$$\omega = \Delta t \, (2 \sin \phi) \tag{2-8}$$

Since ω is constant throughout all latitudes and is unity by choice, forward motion in a meridional direction depends on the value of ($2 \sin \phi$), if we neglect friction effects, additional radiation energy input during motion, and heat losses. As the expression ($2 \sin \phi$) exceeds unity at angles greater than 30° of arc, *the meridional motion cannot be sustained beyond such a distance.*

Due to the Coriolis force, ocean currents do not sustain their meridional motion beyond about 30° of arc, unless deflected by a land mass. The Gulf Stream is deflected north and south by the coast of Europe well before it has traveled 30° of arc. The Kuroshiro Current reaches almost a full 30°, but only with the aid of an additional gyre between the North Equatorial and Japan currents. The same restriction applies to air masses. Equatorial air masses moving poleward also cannot sustain their meridional motion beyond about 30° of arc. An extension of the Hadley cells to the arctic regions at any time during the Phanerozoic eon (Predtechenskiy, 1950; Flohn, 1964) is thus not conceivable; only a more poleward deflection of the prevailing westerlies is possible. The subtropical belt of high-pressure cells has not changed its position very much in geological time in reference to the contemporary equator, since the trade winds and their high-altitude counterparts cannot blow over more than 30° of arc on the rotating earth's surface (Sonnenfeld, 1978). Consequently, Phanerozoic evaporite deposits likewise display a bimodal distribution, centering near 26° latitude

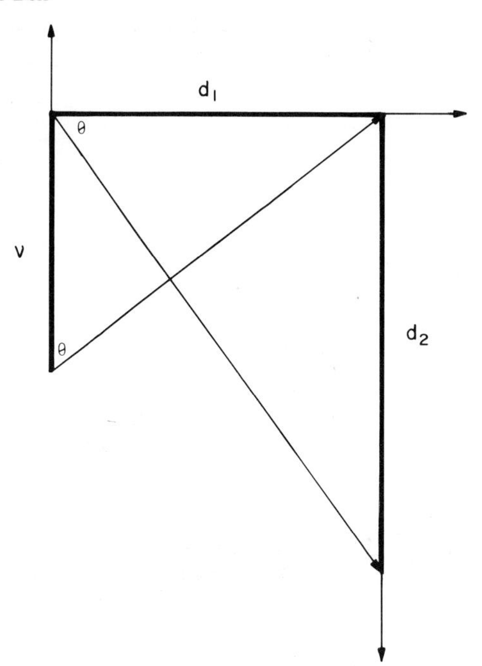

Fig. 2-8. Coriolis effect on a northward-moving mass (after Sonnenfeld, 1978). See text for explanation.

on the more continental Northern Hemisphere and near 19° latitude on the more oceanic Southern Hemisphere (Gordon, 1975).

Rate of Water Loss

The process of solute concentration is closely related to the activity of marine biota (Tarasov, 1939) and to evaporation losses. The former has not been studied quantitatively in hypersaline brines. The rate of evaporation is a function not only of the aridity of the area but also of the salt concentration of the brine. In concentrated brines, the rate of evaporation is reduced (Usdowski and Usdowski, 1970; Kinsman, 1976). The surface tension increases linearly with the molarity of the brine (Harkins and MacLaughlin, 1925). If plotted against the partial pressure of water vapor, the surface tension is the same straight line for sodium chloride, potassium chloride, and sodium sulfate (Tovbin and Boyevudskaya, 1956). The rate of evaporation rapidly decreases during halite precipitation (Chorower, 1941). A solution saturated for sodium chloride is about 30% of the rate of evaporation from a freshwater surface (Harding, 1949). This represents almost

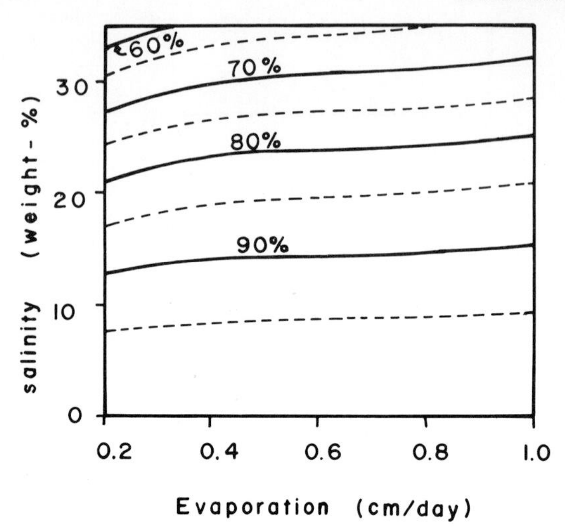

Fig. 2-9. Decrease of evaporative water loss with increasing salinity. The calculation is based on water and air temperatures at 15°C, relative humidity at 50%, and air pressure at 100 kPa. Albedo, latent heat of vaporization, and saturation water pressure are assumed to be constant (after Harbeck, 1955).

a 1% reduction in the rate of evaporation for each 1% increase in salt concentration. The water loss to be made up by inflow becomes smaller. Figure 2-9 shows the calculated decrease of evaporative water loss with increasing salinity, which is more pronounced in concentrated brines. Moreover, shallow basins contain warmer water, and thus evaporate faster than larger, deeper basins and concentrate faster.

Evaporation decreases at a constant temperature of the brine with increasing relative humidity of the air or increasing solute concentration of the brine. With increasing wind speed, evaporation increases only if it results in an increase in the difference in vapor pressures (Moore and Runkles, 1968). The evaporative water loss is intimately related to the vapor pressure of water over the brine. For saturated solutions of calcium chloride, magnesium chloride, potassium chloride, and sodium chloride, Acheson (1963) has calculated the vapor pressure of water in increments of 5°C up to 3.3 kPa, and Stoughton and Lietzke (1965) have calculated the osmotic coefficients and vapor pressure for concentrating sea salt solutions. Water vapor pressures and densities of seawater at rising temperatures have been calculated by Hara *et al.* (1932). Water vapor diffuses vertically and horizontally due to turbulent and molecular diffusion; the latter is negligibly small under conditions of fully turbulent flow (Weisman and Brutsaert, 1973). Figure 2-10 shows the decline in vapor pressure of concentrating seawater at constant temperature.

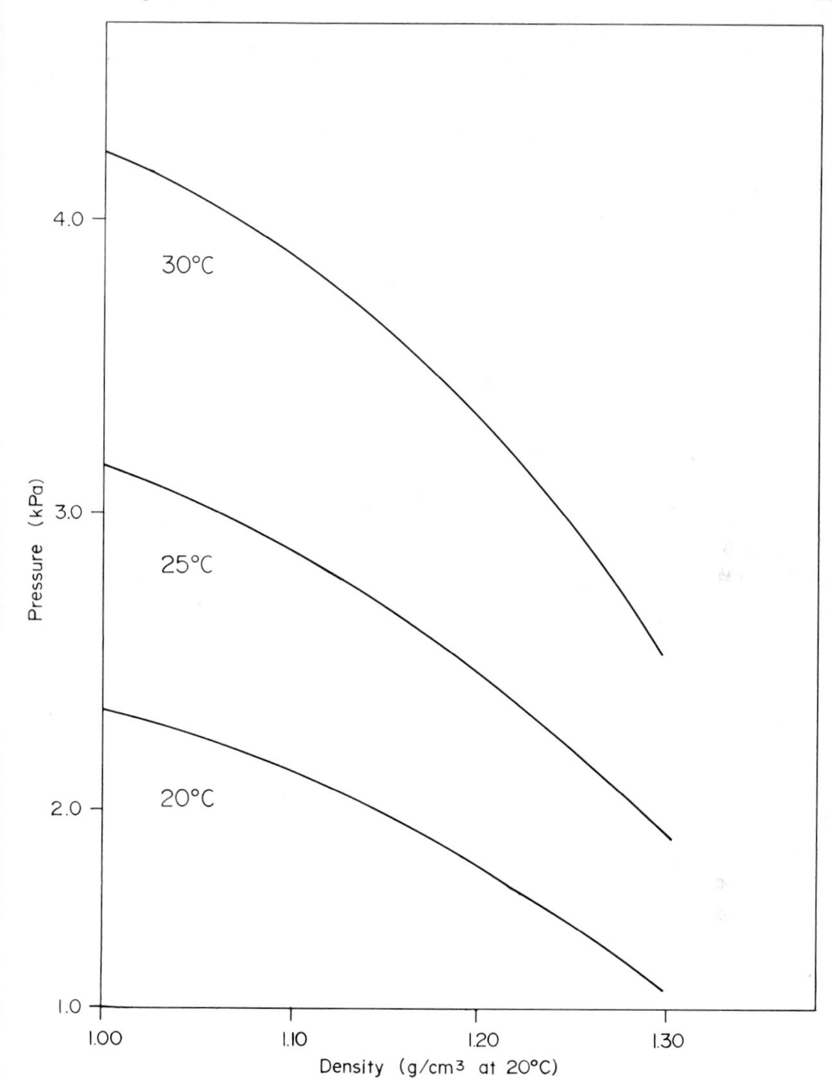

Fig. 2-10. Vapor pressure of seawater concentrates. [Reprinted with permission from Rothbaum (1958). Vapor pressure of sea water concentrates. *Ind. Eng. Chem., Chem. Eng. Data, Ser.* **3,** 50–52. Copyright 1958 American Chemical Society.]

RELATION OF DRY TO COOL PERIODS

Glacial stages of high latitudes have been related in the past to pluvial stages in the tropics in order to explain the excessive precipitation deemed necessary to generate large ice masses. Recent research has decisively proven as false the correlation that tropical "pluvial" (wet) periods accompanied the high-latitude ice ages (Nicholson and Flohn, 1980). Instead, they represent interglacials. Marked tropical aridity was synchronous with the peak of the last glacial stage. Sauthein (1978) found that sand dunes reached a maximum areal extent in the Sahara during the height of the last ice advance, and that such aridity can be correlated with the expanse of subtropical dry lands in both hemispheres. Because of longer fetch, the trade winds become more vigorous in dry periods (Parkin and Shackleton, 1973), as do the prevailing westerlies. The giant loess accumulations of the Pleistocene not only suggest strong winds but also indicate aridity. Both the emission stage of loess particles and the spreading phase must cover dry territories. Formation of dry, wind-blown loess soils in central Europe coincides with each of the coldest phases of the last million years (Kukla, 1970). Gypsum and halite content are increased in the top 30 m in loess deposits of the Ukraine (Burksev, 1954), and early Holocene loess deposits extend here to the Caspian coast of Iran and the Elburz Mountains (Bobek, 1937). In southern Iran, in the Zagros Mountains, and thence in the Taurus Mountains of southern Turkey, glaciation was restricted to the windward (southeastern) side exposed to storms from the Mediterranean Sea. The steep rise of the snow line inland indicates a sharp reduction of precipitation on the leeward (northern) side of these mountains (van Zeist and Wright, 1963, 1967) in the orographic rain shadow.

The arid conditions prevailing during the Pleistocene glaciations coincided with advective cooling of the ocean surface and suppressed evaporation (Flohn, 1978). Advective cooling is wind-driven and coincides with low evaporation, whereas convective cooling is by ascending air and usually coincides with high evaporation. The salt crystallization temperatures in the Pleistocene Lower Salt of California likewise prove a coincidence of salt precipitation with periods of substantially colder winters and cooler summers than today (Smith *et al.*, 1970), i.e., in the Wisconsin glacial stage. Pennsylvanian evaporites of Brazil appear to be coeval with red beds in Bolivia, which in turn interfinger with tillites (Helwig, 1972), another example of aridity being synchronous with a cold spell in the earth's climate. In Pleistocene eolian admixtures to deep-sea sediments, there is adequate documentation of a correlation between glacial stages and increased aridity, and of interglacial stages and increased humidity (Parmenter and Folger, 1974). The same holds for the Caribbean Sea and northern South America (Bakker, 1968; Zonneveld, 1968; Damuth and Fairbridge, 1970; van der Hammen, 1972, 1974). During the last advance of the Würm-Wisconsin glaciation, the climate was dry worldwide: in Australia (Walker, 1978) and in the Ethiopian highlands (Street and Grove, 1976), as well as in southern Europe (Florschutz *et al.*, 1971; Geyh and Rohde, 1972; Bonatti, 1966), whereas the Red Sea experienced a salinity maximum around 11,000–13,000 B.P. (Friedman, 1972). Approximately the same time interval is also responsible for the bulk of gypsum precipitation in the sebkha Mellala in Algeria (Boye *et al.*, 1978).

Comparative dryness with high rates of evaporation sets in somewhat later than cooling and then persists longer than the peak period of extreme cold temperatures. Aridity culminated toward the end of the Würm-Wisconsin stage in the southwestern Sahara, the northern Kalahari, and the Rajasthan deserts (Lamb, 1977), and also in central Turkey (Roberts *et al.*, 1979). The maximum cooling occurred around 18,000 B.P., but the deserts continued to expand, reaching a maximum around 12,000 B.P., when the glacial period had already started to wane. Desert area reached a minimum around 6000 B.P. The lush grazing grounds of North Africa that supported large herds of cattle during the postglacial temperature optimum all but disappeared in the last three millennia as the influence of the horse latitudes increased and the desert spread. Today the deserts are nearly equal in extent to the area occupied around 12,000 B.P. in the Sahara and the Middle East. This led Rognon (1980) to observe that dry periods follow glacial maxima and peak later. He opined that they occur when subtropical high-pressure cells draw equally on equatorial and polar air masses, i.e., when the high-altitude jet stream is realigned. The conclusion is therefore warranted that *high rates of evaporite accumulation correspond to periods of cold climate.*

The northern limit of evaporite deposition shifted within the same connected set of water bodies from the middle to the late Miocene in central Europe and the southern Ukraine south during a brief end-Miocene (Messinian) warm interval toward the Aegean Sea and the Po River valley. In Plio-Pleistocene times, the northern limit returned to the southern Ukraine (Sonnenfeld, 1974, 1977), and the southern limit came within 10° of the equator. Intense, sudden chilling of the Southern Hemisphere during the late Miocene caused a marked precipitation decrease in Australia (Kemp, 1978), because less water vapor was taken up into the atmosphere from colder seas. This chilling appears to have preceded the Messinian stage in the Northern Hemisphere. Even a slight cooling of the oceans in the face of reduced solar radiation results in a volume reduction. Not only are shorelines bared by a slight regression, but the cross-sectional areas of water flowing over sills are reduced. This can be critical for the water exchange (inflow/outflow ratio) of a basin.

Evaporite precipitation increases worldwide at given times in areas considerably distant from each other (Lotze, 1957). Often such peaks are preceded by continent-wide limestone formation. Peaks in coal formation thereby correlate with low points in evaporite production (Lotze, 1957). Surely, this must have a direct relation to the composition and quantity of incoming solar radiation.

Strike of Dry Belts

The dry belts (Sonnenfeld, 1978) currently possess a strike of about N25°E (or S25°E), which is only indirectly related to the inclination of the equatorial plane to the plane of the ecliptic. The strike of the dry belts is related more directly to such factors as ocean currents on the western margins of continents, prevailing wind directions, and air circulation patterns over land masses. These winds have higher friction coefficients and lower specific heat than the oceans. The dry belts extend inland from low-lying coasts in the trade wind regime, follow the outer margins of the Intertropical Convergence Zone, and veer poleward east-northeast from the Mojave Desert of California and the Sonora Desert of northern Mexico through Nevada and western Texas into Wyoming and southwestern Kansas, as well as from Morocco to the deserts of central Asia. They veer east-southeast in the Southern Hemisphere.

Plants and Animals

In an arid climate, a sebkha environment contains mangrove stands and various halophytes (Kendall and Skipwith, 1969). In a less rigid climate, one can expect to find more extensive vegetation, even if conditions are far from optimal for plant growth.

That massive evaporites are the product of a semiarid rather than an arid climate is suggested by the occurrence of fern pinnules and carbonized plant stems in shale intercalations of Pennsylvania evaporites in the Paradox Basin of Utah (Herman and Barkell, 1957). Gymnosperm fragments also occur in Permian salts in Kansas (Tasch, 1963). Coal seams and thin lenses of plant remains (Ivanov, 1933) give evidence of the existence of coniferous and cordaite forests, with undergrowth of tree ferns and scale trees (Abramova and Marchenko, 1964; Kudryavtsev, 1971) in the Permian Upper Kama region west of the Urals. Mississippian coal accumulations in a semiarid alluvial plain of New Brunswick, Canada, are not too distant from evaporites and have been compared to similar recent sedimentation in arid Australia (Legun and Rust, 1981). Paleobotanical evidence shows that the preevaporite marls were deposited into the Zechstein Sea in a dry climate not too different from the one prevailing during evaporite precipitation (Grebe, 1957; Daber, 1960). Halophytic plants, and even trees, are associated with gypsum nodules in the Jurassic evaporites of southwest England (West, 1975).

Broad-leaved trees and some conifers, including sequoia, were found in the Miocene Wieliczka salt deposits of Poland (Zablocki, 1928, 1930). Coniferous trees, apparently swept in by flash floods of freshwater streams, occur in both Zechstein halite and potash layers (Fulda, 1938), and similarly in potash and halite beds west of the Urals (Ivanov *et al.*, 1963). Tree trunks up to 7 m long were found that represent tropical or subtropical trees, based on an analysis of their solid wax content. Walther (1903) reported tree trunks, leaves, fruits, pine cones, crocodile and bird tracks, frogs, a variety of mammals from rodents to pachyderms, *Ptinus* beetles, dragonfly larvae, and freshwater fish from various Tertiary gypsum and salt deposits in Europe. Pinecones (Lotze, 1957), brushwood, amber, encrusted fruits and leaf imprints (Ochsenius, 1877), trees, and vertebrate footprints (Vyalov and Flerov, 1953) occur in mid-Miocene halite beds in the Carpathian foreland. Pontian coal seams with leaves of angiosperms adjacent to end-Miocene (Messinian) evaporite deposits (Sonnenfeld, 1974), or smaller animals, insects, and plant remains occur in these evaporites (Savelli and Wezel, 1978). Insects and plant remains have also been found in Oligocene potash deposits in the Alsace (Quievreux, 1935), suggesting an annual temperature not too different from that of present-day Algiers.

Since leaves do not travel very far in running water before being destroyed, Goergey (1912) reasoned that forests must have grown close to the Alsatian potash lagoon. This, in turn, means that the degree of aridity or the depth to the groundwater table were not excessive in lands upslope from halite- and carnallite-precipitating lagoons. Indeed, Oligocene coal beds up to 0.5 m thick occur close to the Oligocene evaporites in the Alsace (Harbort, 1913), with preserved trunks of palm trees and *Cinnammomum.* Such vegetation precludes the existence of an extremely arid climate nearby; the same can be inferred from the frequent intercalations of clays with brackish water faunas. In short, *evaporite deposition does not require an arid climate, but only a climate in which evaporation exceeds precipitation at least for some part of the year* (Schmalz, 1971). This is not to imply that evaporite precipitation cannot extend, in some instances, from semiarid regions into highly arid terrain.

Evaporite sequences also record the existence of warming trends with accompanying increases in precipitation. The sandy-clayey Gray Clay member of the Permian Zechstein evaporites is of fluviatile origin, delivered by a river coming from the south. Delivery of large water masses was possibly only if extensive atmospheric precipitation occurred in the hinterland. That is, the climate became more humid for a time, as is corroborated by the pollen content of the clay (Kosmahl, 1969).

Salt Steppes

Similarly, salt crusts are much more common in semiarid steppe regions than in true deserts. The drier the climate, the more salts are dissolved in a unit volume of groundwater, and the more likely is the formation of evaporitic crusts within the soil horizons. However, in true deserts, the groundwater table is often too low to permit evaporative drawdown; soluble salts are washed down through the soil horizons by rare flash floods and then move laterally. It has never been determined whether any ancient massive evaporites formed inside the main desert belt *sensu stricto* rather than in contemporaneous semiarid domains on the outer fringes of desert terrain.

SUMMARY OF CLIMATOLOGICAL CONDITIONS

Evaporite precipitation requires a climate with some, although not extreme, aridity. Semiarid terrain with wet seasons has produced the bulk of all evaporites, present or past. Such semiarid regions occur in the rain shadow of mountains, in subpolar latitudes, and in the realm of the horse latitudes. The anticyclonal belt of the horse latitudes has always hovered around 28° of latitude; it was narrower in warm periods and wider in cool periods of the earth's history. Consequently, cool periods are preferentially suited for widespread evaporite precipitation.

3

Water Masses

Evaporite precipitation from an open body of water is not only subject to atmospheric influences but is also conditioned by the circulation of surface and bottom waters and the resulting brine stratification. This stratification leads to warming of the waters along marginal shelves and results in migration of the solutes. A restriction of the indraft by narrows can be further enhanced by reef growth until the water exchange is nearly choked off. The inflow/outflow ratio and the underground seepage control the salt balance of the open body of water.

SUSTAINED WATER SUPPLY

A lagoon or lake requires a sustained water inflow to prevent ephemeral complete drying out and to introduce further volumes of solute. Only in this manner can layers of massive, bedded evaporites build up. The Mediterranean Sea today, shut off from all further water supplies, yields only 1.5–2.0 m of gypsum (= 90 − 120 cm of anhydrite) and 27–36 m of halite, in spite of its average water depth in excess of 1450 m. Thick gypsum (anhydrite) or halite sequences thus presuppose a continuous water supply during their deposition. Such a sustained water supply does not need to be uniform throughout the year; it is probably reduced to a trickle during the dry season.

Episodic flooding of a dried-out playa does not produce thick-layered evaporite sequences. It dissolves all previously precipitated halite and potash salts, except for irregular patches that were either not reached by a subsequent flood or were buried in a new sediment. Floodings result in irregular bedding, and all former layering is destroyed. A heavy desert dew also leaches exposed salts and redeposits them. Some of the redissolved salts are lost either through bottom outflow or as encroachment onto subsurface waters.

Sea-Level Fluctuations

If the entrance to a bay is very shallow, then even slight seasonal variations in sea level from a mean have a large effect on the water exchange. Oceanic sea levels fluctuate with changes in water temperature and consequent thermal expansion of the waters. The lowest sea level occurs near the spring equinox and the highest near the fall equinox. Sea levels fluctuate inversely with water levels in salt lakes or lakes in continental basins (Neev and Emery, 1967), which reach their lowest level at the end of the summer. Due to a variety of factors, such as solar tides, sea level can vary annually by 1 m or more from a mean value, an important factor when considering the shallowness of a bay entrance. A fast-flowing longshore current outside the lagoon entrance leads to a local lowering of sea level, which is called the "Bernoulli effect" (Wemelsfelder, 1970). Variations in the gravitational constant can amount to 0.16 mgal because of the moon and to 0.076 mgal because of the sun, or a total of 0.24 mgal. This may result in a weight change of only 0.24–0.26 g/m^3 of concentrating brine, depending on the attraction of the sun and the moon. However, the radial deformation of the earth has a total maximum amplitude of 78 cm, or an average of 20 cm in a 12-hr period (Nilsson, 1968). This can have an important impact on water exchange over a very shallow entrance sill.

Winds accentuate oscillations in sea level. A temporarily high sea level is mainly due to a consistent seasonal storm-driven water pile-up. The influence of wind stress is much greater than that of tides or of increased seasonal runoff. Along the poleward shores of a sea (the northern shore in the Northern Hemisphere), the trade winds drag the waters away and retard the maximum sea level by up to 10 weeks (Lisitsin, 1974). Where the trade winds blow onshore at the distal end of a bay, they push the surface waters toward the deeper parts. At the proximal end, the trade winds blow offshore and push the surface waters toward the shallower parts. Subsidiary shallower embayments at the proximal shore of such a bay are inundated if they open eastward and are drained if they open westward in either hemisphere.

Water Circulation

Vertical and horizontal gyres of ocean currents also tighten in warmer periods, in direct analogy to air circulation. Both the vorticity and the velocity of ocean currents increase with rising energy input. Reduced energy, on the other hand, increases the diameter of each gyre (Svalgaard, 1973).

The climate influences the circulation in open oceans as well as that in marginal bays. A bay level located in a semiarid belt is lowered by an evaporative water loss, inviting a hydraulic influx from the open sea to maintain a common sea level. As the influx spreads out over the bay waters, the combined head of a surface seawater layer and a more concentrated brine underneath is greater than an equivalent column of seawater outside the entrance to the bay. This hydrodynamic head sets up a bottom outflow (Scruton, 1953), i.e., an antiestuarine circulation (Fig. 3-1). The modern Mediterranean Sea is an example of such a circulation: Both at the Bosporus and at the Strait of Gilbraltar, the surface waters flow into the Mediterranean, and the bottom waters flow out. Estuarine surface outflow of the Black Sea thereby turns into antiestuarine surface inflow into the Mediterranean Sea. If the Black Sea were to develop a water deficit, both seas would have an antiestuarine circulation, and the Mediterranean Sea would act as preconcentrator to Black Sea brines. Similarly, a strait in the Suez area would deliver Mediterranean surface waters to the Red Sea, which has the greater water deficit; a bottom countercurrent would then concentrate Mediterranean waters. On a minuscule scale, the Suez Canal has precisely this circulation. *Antiestuarine circulation is driven by an evaporative water deficit that is not balanced by continental drainage. A lowering of sea level is prevented by surface inflow.*

The opposite circulation occurs in periods of cooler or wetter climate, when the ensuing water surplus in a bay pushes out surface waters and pulls in a bottom inflow. This is an estuarine circulation, and is found on a smaller scale in every fjord (Fig. 3-2). In the past, it applied to every pyritiferous black shale, e.g., the Devono-Mississippian Chattanooga–Bakken–Exshaw shales of North America and the Permian Kupferschiefer shales in Europe. Modern examples of marginal seas with surface inflow include the Black Sea, the Baltic Sea, the Hudson Bay, and the Sea of Okhotsk. In glacial periods of the Pleistocene epoch, the Mediterranean Sea reversed its flow into an estuarine fjord-type circulation (Mars, 1963; Vergnaud-Grazzini and Bartolini, 1970; Valette, 1972; Huang *et al.*, 1972; Huang and Stanley, 1973).

In a basin with a significant maximum depth below that of the entrance sill, a volume of dead water accumulates. The outflow is at an intermediate level; its interface to the dead body of water is tilted away from the sill crest in an

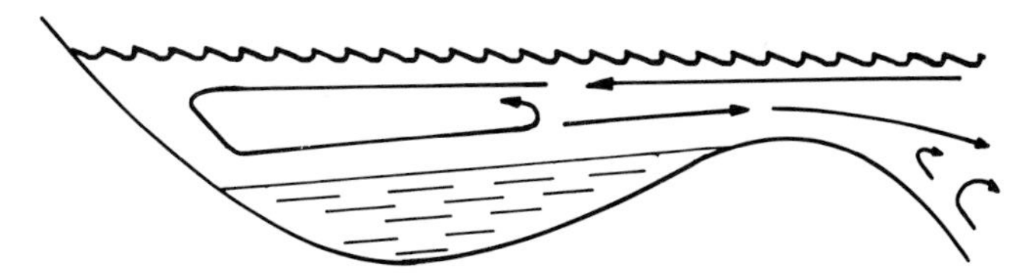

Fig. 3-1. Antiestuarine circulation in a deep basin in a semiarid climate. The interface to dead water dips away from the entrance sill.

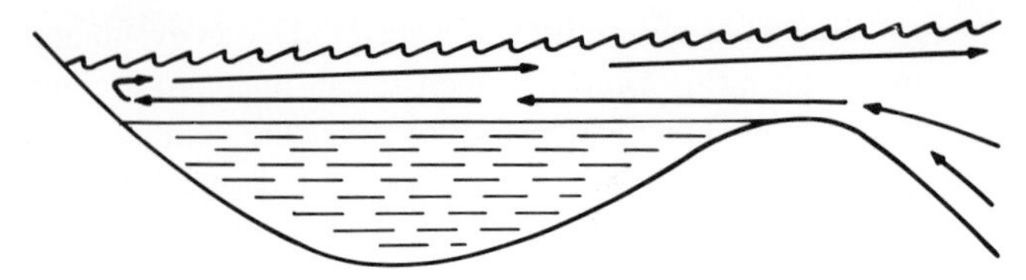

Fig. 3-2. Estuarine circulation in a deep basin in a humid climate. The interface to dead water slopes toward the outlet.

antiestuarine circulation (Fig. 3-1), but is tilted toward the crest in an estuarine circulation (Fig. 3-2). If the depression is shallow enough for the antiestuarine outflow to reach bottom (Fig. 3-3), much of that outflow is dragged back along the interface to inflow and increases the salinity of the bay. Such shallow depressions are flushed by an estuarine surface outflow (Fig. 3-4). *Estuarine circulation is driven by a water surplus generated by excessive runoff and* thus does not occur in dry areas, but *prevails only in humid climatic zones.*

Coriolis Effects on Water

There is a horizontal component to the Coriolis effect that deflects moving air or water to the right in the Northern Hemisphere and to the left in the Southern Hemisphere. Land masses in the path of these deflected currents will turn them into longshore currents. More important for the water exchange in a basin is the vertical component of the Coriolis effect, which tilts the interface between surface and bottom waters. The interface dips to the west and poleward, and this is where the surface layer is thickest. It is surely no accident that embayments with a water surplus, i.e., a surface outflow, open out either poleward or westward, e.g., the Sea of Okhotsk, the Black Sea, the Baltic Sea, the Hudson Bay, and even the North Sea. Seas with a water deficit, i.e., a surface inflow wedge, open to the east or equatorward, e.g., the Adriatic Sea, the Red Sea, and the Persian Gulf. For sills of equal height, a subsidiary embayment opening equatorward or eastward will not need as high an entrance sill as a bay opening in the opposite direction in order to keep most bottom waters from spilling out.

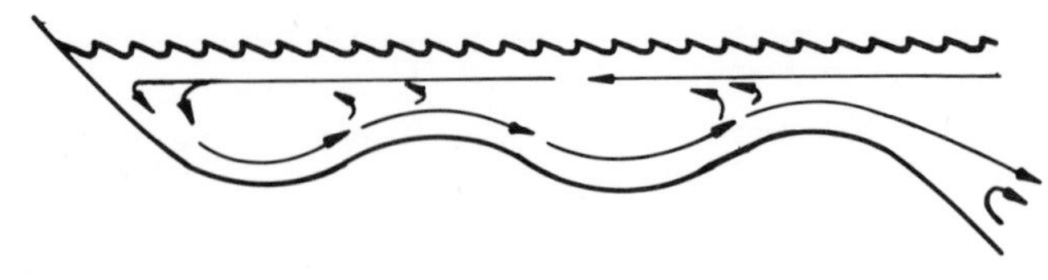

Fig. 3-3. Antiestuarine circulation in shallow depressions in a semiarid climate. The drag on the bottom outflow is directed into the lagoon.

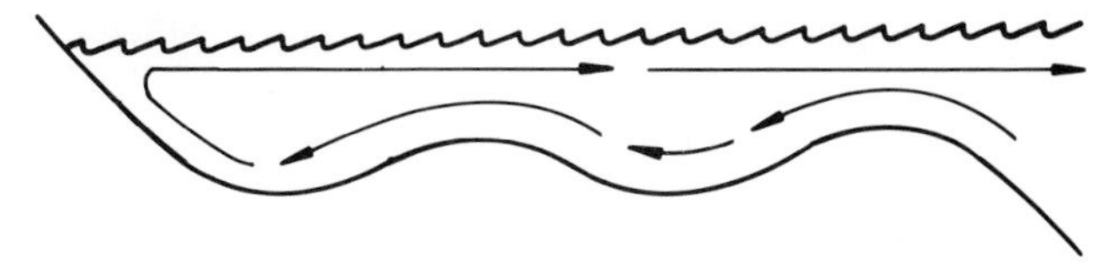

Fig. 3-4. Estuarine circulation in shallow depressions in a humid climate. The drag on the bottom inflow is directed outward.

The Devonian Elk Point, Permian West Texas, Zechstein, and Fore-Ural basins are but a few examples of ancient evaporite basins opening toward the paleoequator or in a former easterly direction. Most Middle and Upper Miocene evaporite basins in central and southern Europe also opened southward or eastward. At first sight, the Mediterranean Sea seems to be an exception. However, we have to bear in mind that the breakthrough at Gibraltar is mid-Pliocene in age, a time of considerable cooling and probably estuarine circulation, the reverse of the modern one. With a water surplus both draining from the Black Sea and generated in the Mediterranean Sea itself, the wedge of surface waters would have been thickest and most powerful in the west end. It is no accident that no major erosional canyon can be found on the Mediterranean side of Gibraltar, cut by inrushing Atlantic waters. An erosional canyon existing on the Atlantic side may be the telltale mark of the spillage of Mediterranean surface waters at a time when its sea level was somewhat lower due to water storage in ice masses. Once the breakthrough was established, Atlantic waters would have quickly mixed with Mediterranean waters; when a reversal of circulation occurred at the onset of the Holocene, Atlantic surface waters could easily wedge in.

Lower salinity values always occur on the right side of an observer looking in the direction of the current (Sonnenfeld, 1974), but such lateral salinity gradients are very small. They would not be stable otherwise. A salinity gradient occurs both in the Strait of Gibraltar (Lacombe and Tchernia, 1960) and the Strait of Hormuz (Emery, 1956), and even in the Dead Sea, where gypsum consistently precipitates at the beginning, followed by halite in lateral continuity (M. R. Bloch, in Valyashko, 1972). The direction of cross-bedding and the decrease in the size of Upper Miocene (Messinian) clastics are counterclockwise around the northern rim of the Mediterranean basins (Sonnenfeld, 1974, 1977; Sonnenfeld *et al.*, 1978). Counterclockwise flow has also been noted for Miocene evaporites in Transylvania, central Romania (Pauca, 1968). Since there is no modern bay or sea where current flow is not controlled by the Coriolis effect, a model of brine concentration that omits vertical and horizontal Coriolis effects is not realistic. Briggs and Pollack (1967) found that a computer model does not correspond to field data of evaporite distribution if it assumes a steady horizontal irrotational flow, i.e., if the inflow is assumed to spread out in all directions perpendicular to lines of equal salinity. The computer model approaches the field data for the roughly circular Michigan Basin only if a basinward flow from the surrounding shelves is assumed.

Since the currents flow in a horizontal plane, they divide the water column into distinct layers with different properties. Stagnant waters prevail in the wake of currents that bypass depressions without flushing them. If a deeper brine is in horizontal motion as a current, its density reaches a maximum within the highest velocity and declines toward the interface with surface waters, where frictional effects slow the motion. A surface current enters the Gulf of Kara Bogaz Gol from the Caspian Sea and moves counterclockwise along the shore. A countercurrent at depth moves clockwise toward the mouth (Dzens-Litovskiy, 1966). Presently, this countercurrent no longer exists but is dragged back in its entirety by inrushing surface waters. Lake Vanda in Antarctica is composed of four distinct layers. Water within each layer moves laterally. A horizontal current of 1 cm/s has been measured at a depth of 20 m (Ragotskie and Likens, 1964).

The Saturation Shelf

In order to remain within reasonable limits of annual evaporation rates, the brine surface area must be a multiple of the area of actual chloride precipitation. The vapor pressure of seawater is reduced by about 800 Pa before halite precipitates, and the rate of evaporation decreases proportionally. Unrealistic rates of evaporation have to be assumed if a column of normal seawater is to be evaporated to halite saturation in order to precipitate 10–40 mm of halite per year. However, if the brine surface exposed to evaporative water losses in the same

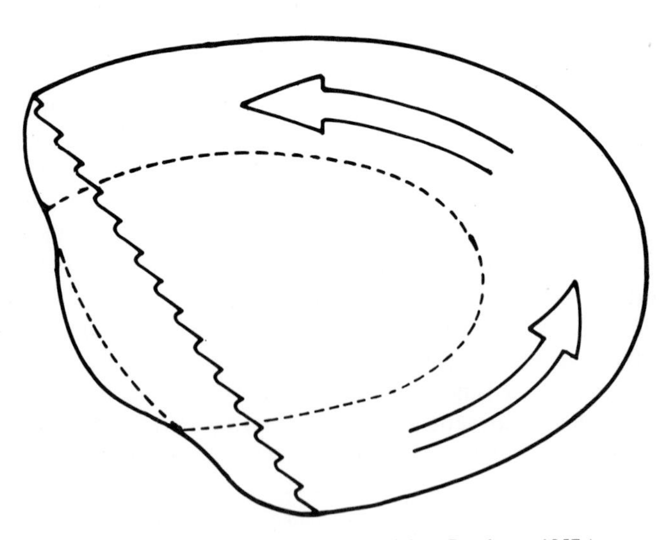

Fig. 3-5. The saturation shelf (after Richter-Bernburg, 1957a).

basin, or in a tandem set of connecting basins, is a multiple of the area of salt deposition, then evaporation rates suffice for a relatively small water deficit per unit water area. In that respect, the marginal carbonate banks and gypsum shelves, envisaged by Richter-Bernburg (1957a) under the term "saturation shelves" (Fig. 3-5) as part of any evaporite basin, become an essential prerequisite of halite or potash precipitation. Flowing over the shallow coastal shelves, the surface waters are warmer and evaporate faster as they encounter dry winds from land and receive more radiation per unit of depth. When they are sufficiently concentrated, the waters veer toward the deeper parts. The halite phase usually turns into a transgressive phase on the basin margins and is induced by the rapid subsidence of the basin axis. This subsidence then involves progressively wider zones of coastal areas. Rapid subsidence of the basin axis is thus also essential to halite precipitation, not the least because it also increases the brine surface area exposed to evaporation by submerging the flanks.

RESTRICTED WATER EXCHANGE WITH THE OCEAN (THE OCHSENIUS–KRULL MODEL)

Restriction of the outflow accelerates the concentration of resident bay waters. Both an entrance sill and a lateral constriction in the inflow straits are required to restrict the outflow effectively. Low-salinity influx sets up a barrier to outflow over a sill; constriction of the channel increases the effectiveness of this dynamic barrier (Scruton, 1953). Lucia (1972) estimated that the ratio between the surface area of the basin and the cross-sectional area of the inlet must encompass at least six orders of magnitude for the initiation of gypsum precipitation and at least eight orders of magnitude for the precipitation of rock salt (Fig. 3-6). Bays with an open entrance and an outward-dipping floor or deep-draft longshore currents promote oceanic exchange that tends to flush the bay. If these bays have a high water deficit, one can find on their margins typical sebkha environments of evaporite precipitation, proving that the climatic environment is conducive to massive evaporite deposition. Schlanger (1965) observed the same critical requirement of entrance restriction: Among Pacific island lagoons, only those with such a restriction precipitate evaporite minerals. All important evaporites accumulating today from waters of oceanic provenance occur upslope from an emergent but porous sill (Busson, 1980).

The barrier to the outflow of concentrated brines is most critical for the formation of salt deposits (Goldsmith, 1969); this has been known for more than a century. Ochsenius (1877, 1878, 1888) recognized the essential nature of the entrance bar; he thus elaborated on ideas previously expressed by G. Bischof, Hugh Miller, and Charles Lyell (Harbort, 1913). The model stemmed from an observation of the relationship between the Caspian Sea and the evaporite-pre-

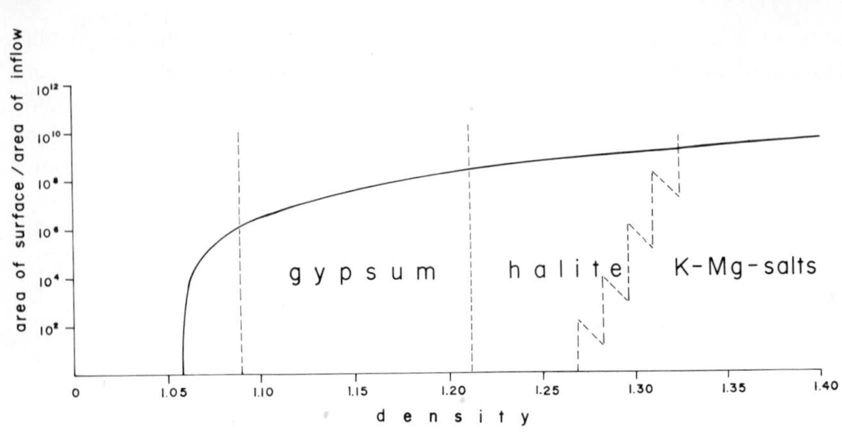

Fig. 3-6. Brine density in a basin plotted against its ratio between the surface area and the cross-sectional area of inflow (modified from Lucia, 1972).

cipitating Gulf of Kara Bogaz Gol (Ochsenius, 1876). Nonetheless, it is very unfortunate that Ochsenius did not state his concept of the functions of a bottom current very clearly. For the most part, he considered it chiefly as a destructive force breaching the entrance sill and terminating the evaporite deposition. Consequently, Riemann (1907), Everding (1907), and d'Ans (1947), in recapitulating Ochsenius' model, visualized only a continued inflow over the entrance bar as a necessary prerequisite for salt precipitation (Fig. 3-7).

The same is evident in Fulda's (Fig. 3-8) presentation of Ochsenius' model as one of inflow only. Fulda's (1930) concept is as follows: Seawater flows continuously into a bay, but only to compensate for evaporation losses. As the brine gradually concentrates, the biota are either expelled or die off. Eventually the stage is reached (Fig. 3-8.2) where calcium carbonate and iron oxide are precipitated. It should be noted that no such iron oxide layer is normally found underlying evaporite sequences. Further concentration of the brine beyond a density of 1.129 leads to gypsum precipitation. Once the brine density exceeds 1.218, sodium chloride begins to precipitate alongside gypsum. The residual brine constitutes a mother liquor that rests on the growing halite bed. This mother

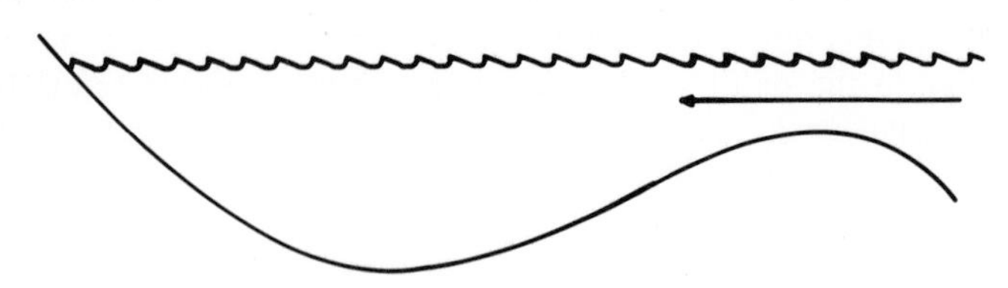

Fig. 3-7. The barred basin with marine inflow (after Ochsenius, 1877, 1893).

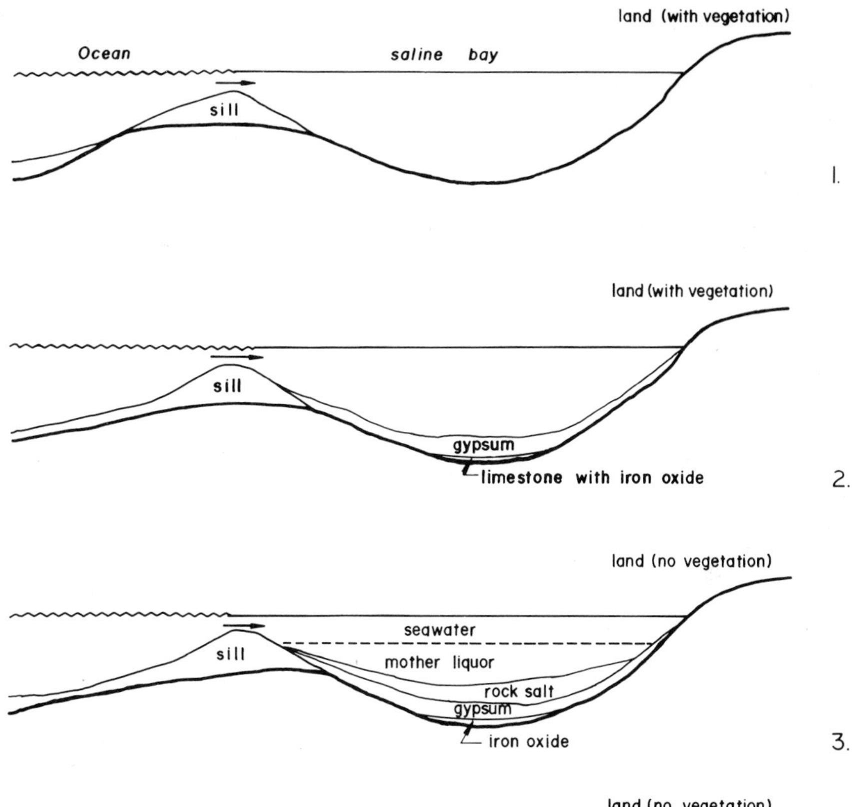

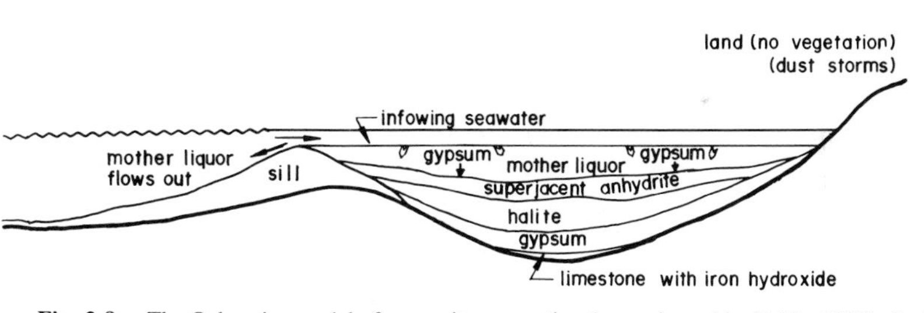

Fig. 3-8. The Ochsenius model of evaporite generation (as envisaged by Fulda, 1930). 1. A bar limits the water exchange with the ocean. All biota are either expelled or die off. 2. Limestone and iron oxides are precipitated, followed by gypsum. 3. Halite precipitates as a separate layer, overlapping on the margins. The thickest gypsum layer occurs under the thickest halite layer. A residual mother liquor contains more soluble compounds. 4. An outflow develops only when the residual brine (mother liquor) is diluted to produce the cover anhydrite. Calcium sulfate precipitates rain down from the density interface between mother liquor and seawater.

liquor contains very soluble salts, i.e., magnesium iodide, lithium chloride, and magnesium sulfate, in addition to some sodium chloride. At some stage (Fig. 3-8.4), the mother liquor breaks through the barrier and becomes a bottom effluent. The continuing seawater influx sets up a density interface, underneath which gypsum crystallizes out (presumably as hopper crystals) and descends through the mother liquor onto the halite as roof anhydrite. It is not explained how gypsum saturation can be maintained beneath the interface to normal seawater. The Zechstein embayment with its Stassfurt and other evaporite members can be explained (Fig. 3-9) by assuming an entirely closed barrier, with the mother liquor trapped in the bay. This mother liquor forms a layer that contains dissolved magnesium chloride with some magnesium bromide, calcium chloride, and other compounds. Magnesium sulfates enter laterally into the precipitating salts to precipitate a kieserite layer underneath the carnallite.

Although Lotze presented a slightly different diagram in an earlier edition of his standard volume on salt, his modification of it (1957, p. 224) was not too different from the interpretation presented by Fulda (cf. Fig. 3-8). Lang (1937), Hills (1942), Adams and Frenzel (1950), Landes (1963), and Sugden (1963) likewise interpreted Ochsenius' model as causing lateral juxtaposition of evaporite mineralogies by one-way flow. Ochsenius eventually clarified his ideas further by stating, in a not readily accessible publication (1893, p. 191), that halite precipitation requires a complete closure of the entrance barrier; that more concentrated bitterns then lie on the salt and stagnate until they are able to force a break through the barrier; that further inflow precipitates anhydrite by letting gypsum crystals descend through these hygroscopic bottom bitterns; and, finally, that this stage is followed by eolian dust clouds that settle on top as salt clay. Unfortunately, Ochsenius did not indicate what may have kept dust storms away until halite precipitation terminated, or what caused the residual mother liquor to attain the enormous force required to destroy the entrance barrier, which had withstood erosion by the much more voluminous earlier inflow.

The honor of being the first to recognize in the bottom outflow the essential

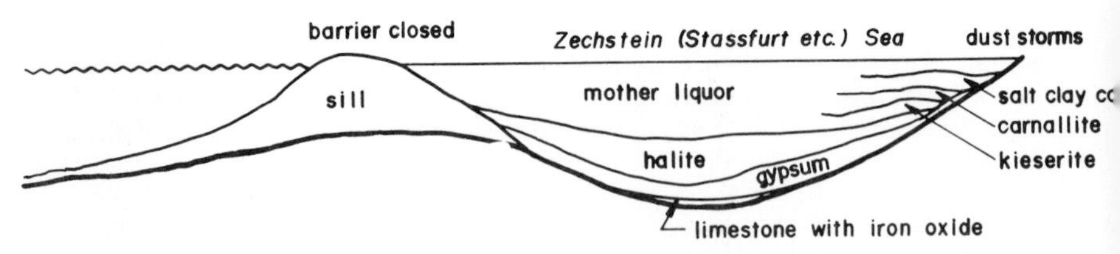

Fig. 3-9. The Ochsenius model as applied to the Zechstein basin (as envisaged by Fulda, 1930). A completely barren entrance prevents any brine exchange with the ocean. After precipitation of limestone with iron oxides, gypsum, and rock salt, the mother liquor precipitates carnallite at the distal end of the basin and infuses magnesium sulfate brine beneath the carnallite to produce a kieserite bed. The carnallite is capped by salt clay deposition.

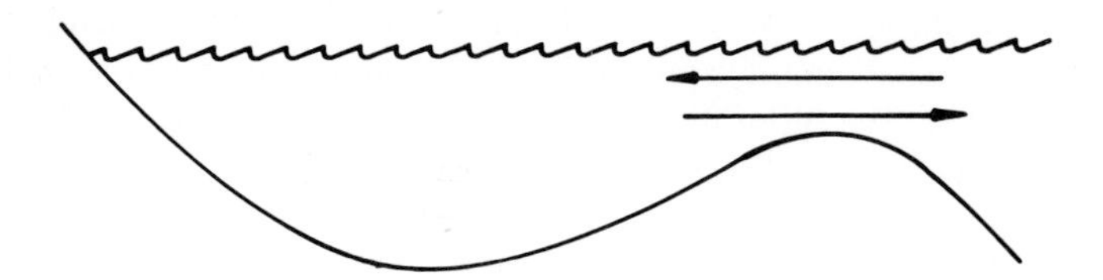

Fig. 3-10. The barred basin with two-way flow (after Krull, 1917).

part of a continuing water exchange during evaporite deposition must go to Krull (1917). Figure 3-10 shows a lagoon with continuous surface inflow and bottom outflow through a narrow inlet or strait. Originally, Krull based his modification of Ochsenius' model on his observation that there is a very significant deficit of halite compared to sulfate, and of magnesium salts compared to potassium salts. Unfortunately, his paper was quickly forgotten. Fulda (1930), King (1947), Sonnenfeld (1974), Holser (1979b), and Smith (1981) later arrived independently at the same modification of Ochsenius' bar theory. Barriers in the zone of the entrance strait, either above water as islands or below water as reefs and shoals, are essential to achieve a positive salt balance in evaporite basins. The greater the ratio between inflow and outflow in a basin with a water deficit, the faster the brine concentrates. *Barriers retard the removal of salt by the outward-flowing bottom current, and in so doing, cause the basin waters to become progressively saltier.*

The mechanics of the water exchange between bay and open sea are not very complicated. Only part of the total volume of seawater inflow is lost to evaporation. The remainder accumulates as a concentrated salt solution. Some salts precipitate eventually, but if it were not for the outflow, the remaining concentrate would eventually fill the bay, thereby setting up a density barrier to further inflow. A surface slope is induced by evaporation losses in the lagoon, and a pressure gradient develops toward the open sea. This shapes the inflow into a wedge-shaped body facing into the basin, and the outflow into a wedge-shaped body facing out. A depression of sea level near the bayhead causes an inflow. The rate of inflow necessary to counterbalance evaporation losses is several orders of magnitude greater than the mixing of inflow and resident brine by molecular diffusion. Inflow velocity is at a maximum just underneath the surface because of reduced friction at the air–water interface. This inflowing water of normal marine density has a pronounced tendency to override bay waters.

Tidal forces and the hydraulic head of the inflowing waters combine to drive the underlying brine toward the entrance, and thus sustain a two-way flow even over shallow margins. A slope of surfaces of equal pressure forms at depth due to excess densities, and the density interface stabilizes as the inflowing lighter waters override the brine. As friction falls off, the interfacial slope levels off and the interface becomes almost parallel to sea level, dipping slightly toward the bay outlet (Fig. 3-11). The tilt of the interface creates a potential for outflow of the basin which exceeds the hydrostatic head (Scruton, 1953), fostering a continuous inflow–outflow regime. This interface is strongly curved where the more saline waters exit from the bay entrance (Assaf and Hecht, 1974).

The level of the interface changes with the seasons; it drops in the summer and rises in the winter (Lisitsin, 1974), because inflow is augmented in the winter by increased surface drainage. The interface also oscillates with sea level. A slight increase in inflow lowers the interface and increases its surface area by making the wedges longer. Storm surges (up to a height of several meters) reinforce high tides and increase the hydrostatic head of surface waters in a bay. This accelerates the outflow. If the surges combine with maximum runoff during a rainy season, they can lead to episodic flushing of a shallow embayment. Conversely, negative storm surges can withhold enough inflow to bare the bottom waters to evaporation. Only topographic restraints, i.e., bottom friction, entrance restriction, and sinuosity of the entrance channel, can prevent a total flushing of the bay. *Evaporitic models assuming only a one-way flow are not feasible as long as a common sea level is maintained.*

The construction of a causeway across the Great Salt Lake produced a density difference between the two sides and a resulting two-phase flow through the causeway (Adams, 1964; Madison, 1970). Unsaturated brines flow north where halite precipitates, and saturated waters circulate back south at depth (Whelan, 1972). Only at a very steep gradient between the southern and northern lake levels does the flow become one-way (Lin and Lee, 1974). The same has been true of the Caspian Sea and the Gulf of Kara Bogaz Gol or the Black and Mediterranean Seas in the last centuries (Sonnenfeld, 1974). The progressive concentration of the Gulf of Kara Bogaz Gol from a gypsum-precipitating brine

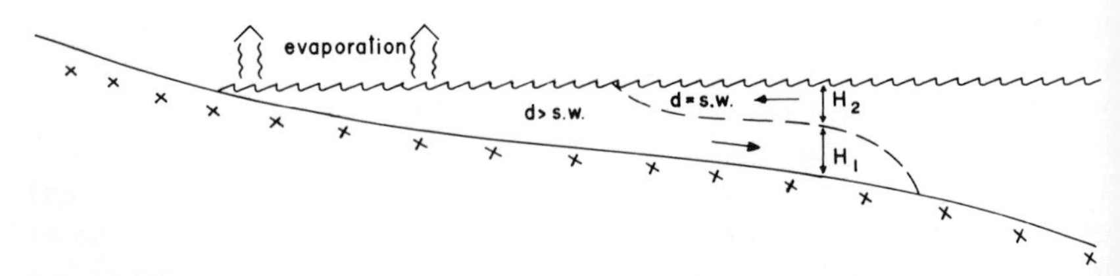

Fig. 3-11. The formation of an interface. Evaporation causes a depression of sea level in the warm, shallow lagoon and induces downhill inflow from the open sea. The lighter inflow overrides the brine concentrated by evaporation losses. The increased hydrostatic head at H_2 produces an instability; bottom waters tend to give way downslope, aided by gravity. The interface is tilted further until a two-way flow is established.

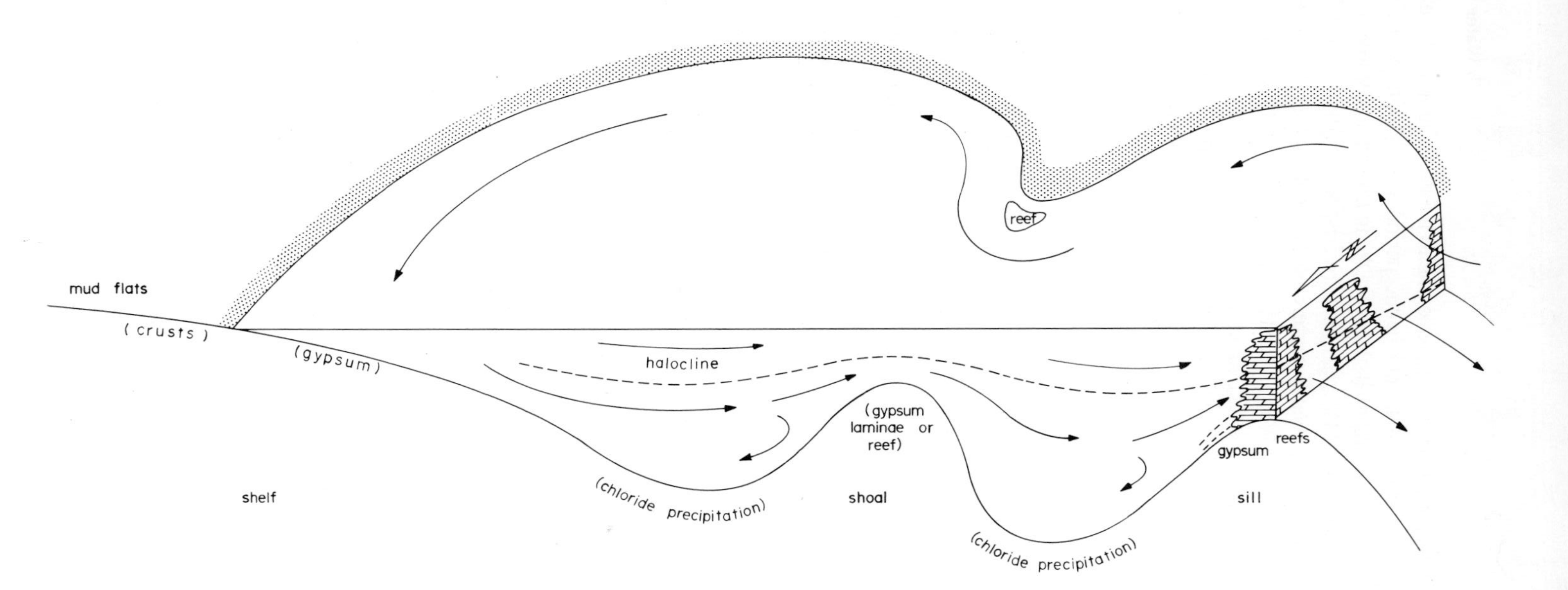

Fig. 3-12. The modified Ochsenius–Krull model of an evaporite basin with a tilted inflow/outflow interface, saturation shelf, and entrance restriction.

to a bittern with potash salts on its fringes is due to a 59% reduction in the cross-sectional area of the entrance channel in the period 1929–1956 (Sedel'nikov, 1958) under concurrent doubling of its length (Dzens-Litovskiy, 1966). In 1937–1941 alone, the inflow fell by 73% and has recovered only partially since (Fedin, 1979). No straits connecting water bodies of different salinity are currently without bottom counterflow. Waters made dense by evaporation sink to intermediate depths in the Persian Gulf, touch bottom in the Strait of Hormuz, and leave pockets of dead bottom waters in depressions (Emery, 1956). The water exchange across the Strait of Gibraltar is asymptotically approaching an equilibrium, also leaving pockets of dead water (locally with sapropelic sedimentation) in various Mediterranean deeps (Kullenberg, 1952). However, inflow through the winding Bosporus–Dardanelles straits blocks off the outflow into the Black Sea in some seasons (Sonnenfeld, 1974), i.e., when the Black Sea level rises in warm and wet decades.

We need only to expand Krull's two-dimensional picture (Fig. 3-10) into a three-dimensional illustration (Fig. 3-12) to incorporate current directions under horizontal and vertical Coriolis effects. The surface current moves counterclockwise along the shores, concentrating on shallow shelves. The interface between surface current and bottom brines is tilted: Entrance sills of equal depth allow flushing of marginal seas facing north or west, whereas marginal seas facing east or south are more restricted and outflow is hindered.

Current Velocity

As long as a connection remains between bay and open waters, the flow will try to maintain a common sea level. This requirement controls the velocity of the inflow. Only when the bay is filled with precipitate will the inflow cease. *The depth of the interface to the outflow is only a function of the ratio of the length of a strait to its total depth* (Anati *et al.*, 1977). The interface is inclined; thus, the longer the strait, the less fluid will pass through. Even small bumps on the floor of the entrance sill can cause interfacial waves, which have greater amplitudes the shorter the strait and which serve to choke the outflow (Anati *et al.*, 1977). Very small patch reefs represent such bumps and thus have a large impact on the water exchange. Each time a current is deflected, its velocity is sharply curtailed. Convoluted entrance straits thus inhibit the water exchange. Furthermore, the outflow is also depressed and impeded by uninodal standing waves (seiches) with a period of oscillation that is shorter the narrower the entrance. These waves originate from a nodal point present at every bay entrance (Lisitsin, 1974). Whereas inflow is dictated by the evaporative water loss, outflow is primarily determined by turbulence and friction. Friction against the walls of the entrance strait, drag along the interface between inflow and outflow, wind resistance, and internal viscosities limit inflow velocities. For that reason, *the size of the cross-*

sectional area of the entrance strait, or rather its width and configuration, is a most important factor in bay water concentration, second only to the basic postulate of a water deficit over the basin area per unit of time.

The outflow is slower than the inflow because it is retarded internally by the greater viscosity and externally on three sides (walls and floor) by friction against rocks in the entrance. The surface tension of the bottom brine is at least an order of magnitude smaller against the inflow than against air (Lerman, 1979). Furthermore, for the same flow rate, the stress between brine and air is about three orders of magnitude smaller than that between brine and bottom sediments (Assaf and Hecht, 1974). Increasing inflow velocities cause an exponentially rising drag of outflowing brines back into the basin, until all outflow is held back. Kruemmel (1911) estimated that a velocity increment caused by a 29-cm rise in the Black Sea level would be sufficient to block the 7-m-deep counterflow. That would quickly lead to a substantial reduction in Black Sea salinity.

Velocities of flowing salt water are rarely more than 1–2 m/s. Only exceptional cases of higher velocities have been recorded. The Saltstraumen of Saltfjord near Bodo in Norway carries, in a confined channel 152 m wide, over 14,000 m^3/s of water at a speed of 8 m/s. The Somali Current off the coast of East Africa attains a speed of 2.4 m/s in open waters, slightly faster than the maximum speed attained by inflow at Gibraltar. The maximum size of transported boulders in Upper Miocene (Messinian) conglomerates with Paratethys faunas that occur on the margins of eastern Mediterranean evaporites indicates water velocities in the 2.3–2.5 m/s range (Sonnenfeld, 1977). The inflow into the Gulf of Kara Bogaz Gol, an evaporite-precipitating bay off the Caspian Sea, sometimes reaches such velocities (Grabau, 1920), but in winter reaches only one-third to one-fifth of such values. Such velocities produce lateral erosion, which widens the channel and increases water delivery. Conglomerates were able to smother Messinian reefs growing in their path at the evaporite basin entrances near Adana, Turkey, and on Crete. The Strait of Gibraltar would have to reduce its cross-sectional area to one-quarter or one-fifth of its present size before the limits of inflow velocities would be reached that could deliver enough water to balance the current Mediterranean water deficit.

DENSITY STRATIFICATION

Once a density stratification develops between inflow and outflow, a pycnocline, a layer of rapid density changes, forms (Stewart, 1963a; Sloss, 1969). The steep vertical density gradient within this zone acts as a strong barrier to vertical fluid mixing. Separated from contact with the atmosphere, bottom brines quickly turn anaerobic.

A mixing zone gradually develops by diffusion. The molecular diffusion coefficient of an ion varies with the temperature of the water, the concentration of other ions, and its own concentration. It must also be a function of time and of depth (Erdey-Gruz, 1974). This mixing zone readily absorbs inflowing waters, but the interface between the mixing zone and the brine advances only very slowly (Fig. 3-13). In lagoons with episodic inflow, the interface can easily be overtaken by the evaporative lowering of the lagoon level. Conversely, each rainfall or influx sets up a new mixing zone and produces a multiplicity of descending interfaces.

A minimum density difference of 15 g/liter (Hudec and Sonnenfeld, 1980) appears to be necessary to maintain stratification in sulfatic lakes; a density difference of about 11 g/liter or less is required in chloride brines. The layering becomes more stable as this difference increases. Furthermore, the density difference across the interface decreases as the thickness of the upper layer increases. The two quantities are inversely proportional (Wu, 1977). Thus, large differences are possible only with a shallow interface between inflow and resident brine. For example, the density difference between brines immediately below the ice of Lake Vanda, Antarctica, and bottom brines 60 m down is 8.6 g/liter. The thermocline occurs at 45 m below the surface and marks a jump in density (Jones and Faure, 1967), i.e., the pycnocline. Furthermore, as lamellar currents develop in each of the layers, they tend to fortify the stratification: The maximum density occurs where the current velocity is the greatest, the minimum density is found along the steepest velocity gradient (Tollert, 1950a).

The differential surface tension of the brine is large enough to retard diffusion. Thus, a brine saturated for potash precipitation can be covered by brines saturated with halite (Valyashko, 1958) without mixing. Slow diffusion of a lighter, sodium-enriched brine into heavier magnesium bitterns will lead to halite precipitation (Raup, 1970), i.e., to a salting-out effect. In lagoons with continuous inflow, the interface stabilizes temporarily but is affected by periodic changes in rates of evaporation. A lowering of the water losses allows the mixing zone to descend faster.

In addition to an interface between inflow and a resident brine, other interfaces form. The dynamic viscosity of a brine increases with its concentration. For a sylvinite solution, it is 0.181 $N \cdot s/m^2$ at 20°C and gradually rises to 0.505 $N \cdot s/m^2$

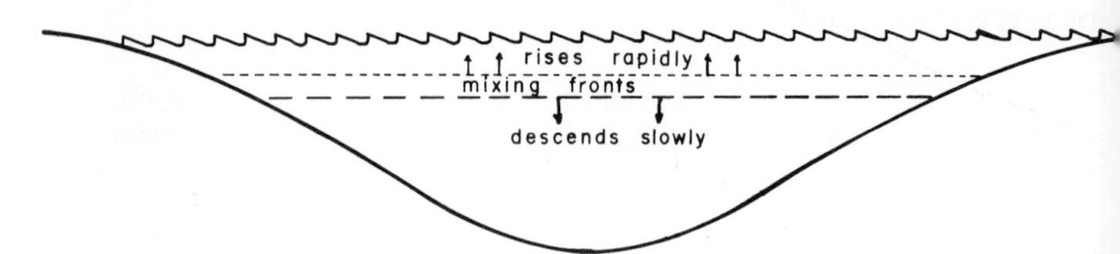

Fig. 3-13. Mixing fronts along an interface of a density-stratified brine.

in a bischofite solution that is essentially stripped of its potash and halite salts. The water in the brine, initially polymerized, depolymerizes in such highly concentrated solutions because of electrostriction of ions. Complexed molecules and other particles larger than the water molecules are then enriched in the zone of maximum viscosity; the water molecules are enriched in the zones of minimum viscosity (Tollert, 1950b). A secondary density stratification is thus created within the concentrating brine, particularly if the solute particles are enlarged by chloride complexing. The average thickness of the layers decreases with increasing salinity gradients and appears to be independent of the applied heat flux (Huppert and Linden, 1979).

A low-angle ray reflected from a shallow bottom is curved by multiple refraction into a parabola inside a saturated chloride brine which never reaches the interface to surface waters (Fig. 3-14) or is reflected back into the brine (Fig. 3-15). The multiple refractions are produced by a salinity gradient. A schematic diagram of the reflection and refraction of incoming solar radiation in a shallow, density-stratified body of brine is presented in Fig. 3-16. Only near-vertical

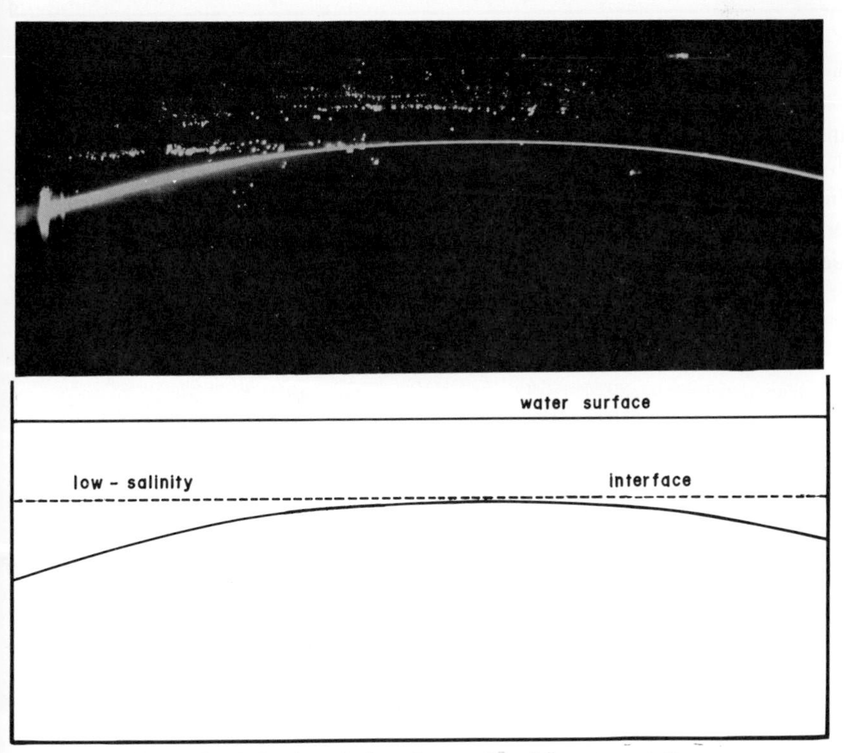

Fig. 3-14. Refraction of a laser beam reflected from the tank floor.

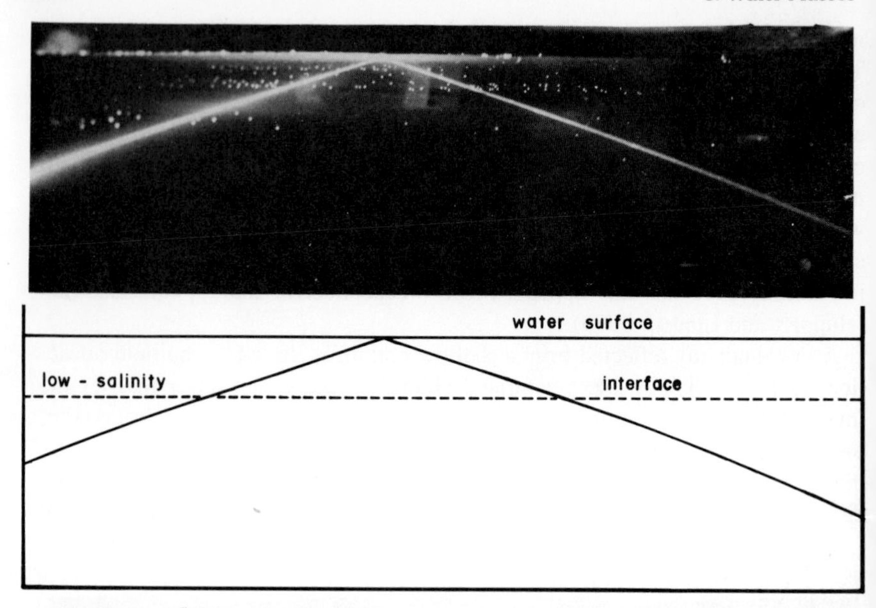

Fig. 3-15. Reflection of a laser beam from the air–water interface. The beam is curved in the hypersaline brine.

insolation is reflected out of the brine (Hudec and Sonnenfeld, 1974); reflections from the bottom at all other angles are either reflected back from the air–water interface or refracted by microstratifications. As absorption rates increase in concentrating brines, the solar radiation is quickly converted into heat. The development of multiple microstratifications is difficult to explain unless one invokes gravitational settling of chloride complexes. Ultra-high-speed centrifug-

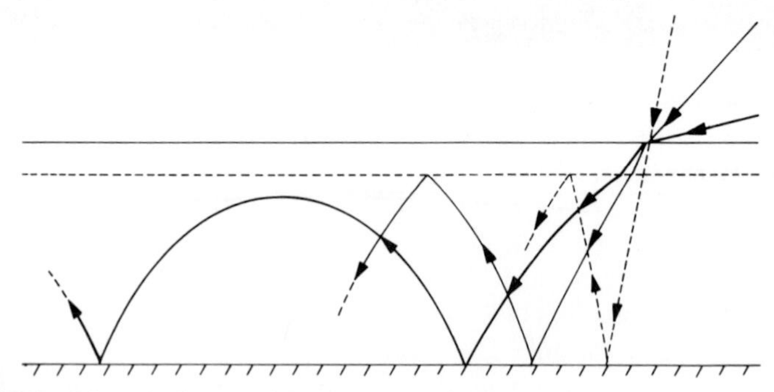

Fig. 3-16. Schematic diagram of the entrapment of solar radiation in a density-stratified brine. Rays reflected from a shallow floor are either refracted by microstratifications into a parabolic path or reflected back from the brine–air interface. Only nearly vertical reflections from the floor will escape.

ing of mixtures of chloride solutes (to model the gravitational effect on a compressed time scale) also causes faint gradations to develop. Concentration gradients develop even in homogeneous solutions and are related to temperature gradients (Tollert, 1950b).

The hydrostatic pressure gradient of overlying brines produces diffusion of salt in the direction of increasing salinity (Foronoff, 1962). This fortifies the tendency for salt to move in the direction of increasing dynamic viscosity. Consequently, most marginal seas and lagoons display a vertical density gradient, with bottom values substantially higher than mean values or near-surface determinations.

The former existence of such a stratification can occasionally be indicated by the bromide content of the salts. In undisturbed beds, the bromide content of halite should be one-tenth as high as that of the accompanying sylvite, but only one-seventh of that in carnallite. If it is less, this would show that the halite precipitated from a brine layer that was undersaturated for potassium chloride (Braitsch and Herrmann, 1964b). However, such differences could also be achieved by bromine migration, just as such migration must be assumed wherever the ratios are much smaller.

The interfaces are not horizontal. In large bodies of water, they are noticeably tilted by the Coriolis effect. In east–west striking straits, the interface between inflow and outflow is tilted toward the equator because of the earth's rotation, leaving a greater cross-sectional area for outflow along the poleward margin of the straits (Lacombe, 1961). If the high-salinity undercurrent attains higher levels on the poleward side of a bay, it also enters indentations of the coastline there more easily, and is more readily flushed out of indentations along the equatorward shore. Similarly, the undercurrent reaches to higher levels, resembling a vessel tilted simultaneously eastward and equatorward. The undercurrent is also shallowest in the embayments located farthest from the water intake (Sonnenfeld, 1974). Evaporitic lagoons are thus more common on the poleward side of shorelines, with straits opening toward either the equator or the east (in either hemisphere).

The interfaces sag over deeps and are subject to seiches even in small lagoons (Fig. 3-17A). These sags are ultimately caused by variations in barometric pressure and wind stress, or by a sudden discharge of flash floods. Their periods differ in closed and open basins, and are strongly influenced by the shape and length of the entrance channel. The depth varies from a little over a meter in small Caribbean lagoons or continental lakes, to a few tens of meters in antarctic lakes, to even greater magnitudes in larger bodies of water.

The wind is not strong enough to cause an overturn in a density-stratified hypersaline brine, as Maxim (1929, 1930, 1936) first observed and then underpinned experimentally in artificial, stratified brine pools. Turbulence near the surface of a water column can neither penetrate nor mix the water column if the

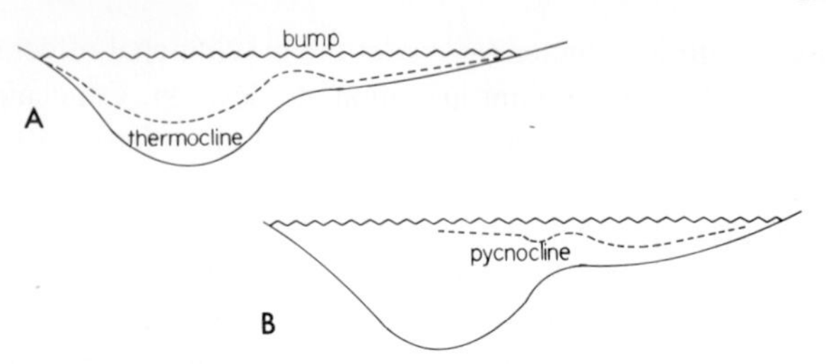

Fig. 3-17. Interface configurations. A. With a sufficient water supply to cover the whole lagoon. The thermocline sags over deeps and rises into the inflow. The inflow humps up the interface. Points above and below the interface that are of equal water depth must have equal density; the rising saline brine is far less dense because of thermal expansion. B. With an insufficient water supply. The inflow wedges out, leaving part of the brine uncovered. The wind rolls out the edges into a film of water, and evaporation causes the film to attain brine density. The inflow tongue wedges out both foreward and sideways, giving a triangular cross section. (Both configurations were observed at different times in Lago Pueblo, Los Roques Archipelago, Venezuelan Antilles, and Jan Thiel Bay, Curacao, Netherlands Antilles.)

column is mechanically stable. The turbulence generated by wind in surface waters is dissipated by work against buoyancy (Lerman, 1979). This author has observed the persistence of stratification under steady breezes of 20–22 knots and even during a hurricane. The interface is, of course, subject to internal waves and oscillates during strong winds; the oscillation period resonates with the wind period. During the strongest winds the surface waters are piled against the leeward shores, leaving bottom waters exposed to evaporation on the windward side. Winds increase the rate of evaporation by dislodging the local air–brine equilibrium. The evaporative downdraft (6–13 mm/day) then catches up faster with the lower interface of the mixing zone, which itself is moving downward by molecular diffusion (1–2 mm/day). Brine storage tanks in India that hold bitterns with a density of 1.330 are diluted by monsoon rains to a depth of 60 cm. However, they regain their original density within 15 days after a shower (Bhat and Oza, 1977).

Arkchangel'skaya and Grigor'yev (1960) have assumed that storms mix light surface waters with waters of the underlying brine layers and thus can average out the salinities to the depth of wave action, i.e., through the entire depth of a basin 30–50 m deep. This inference is not borne out by present-day oceanographic, limnological, or experimental data. First, the depth of the wave base in highly viscous, concentrated brines and bitterns is not known, but is definitely very small. Second, light surface waters resting on heavier brines constitute an extremely stable system. When Lake Eyre, South Australia, reflooded, a saturated brine layer formed over residual salt crusts, retarding further dissolution whenever the water depth exceeded one-half the maximum wavelength of surface waves (Dulhunty, 1976). Thus, *density stratification is absent, and mixing of the waters is complete only when the brine depth is appreciably less than half the wavelength of surface waves.* Salterns and salinas usually operate with very shallow water columns, and thus cannot serve as comparable examples.

In bays where the continuing inflow seasonally fails to cover the whole brine, the wedge of inflow fans out. The wind rolls it out to a film until it quickly evaporates and its salty residue diffuses (Fig. 3-17B). Effluent waters of the River Jordan cover only a very tiny portion of the Dead Sea before evaporating. Whenever the wind turns against the inflow, it piles up, and part of the lower layer remains exposed to evaporation, as Sloss (1969) documented in laboratory experiments. Strong winds enhance evaporation from seawater sprays; however, this is not an important factor in brines, as the greater surface tension impedes ruffling by the winds. The warm bottom brine appears to rise some distance into the inflow in Caribbean hypersaline lagoons (Hudec and Sonnenfeld, 1980). An analog occurs in river systems: During periods of low discharge of Mississippi River waters, a small amount of frictional stress is exerted by the superjacent fresh water on a wedge of ocean water traveling at least 220 m upstream (cf. Scruton, 1953). A high discharge of river waters expells the salt wedge.

To assure continuing concentration, at least some of the brine surface must be uncovered seasonally. The inflow has then weakened to form only a curved wedge, which is thickest near the center of the flow and tapers off on either side of the flow direction. This again requires a very restrictive entrance and a very shallow depth. Otherwise the inflow would enter without hindrance, quickly covering the total water surface. This phenomenon is observed in lakes with an inflow that is no longer sufficient to cover the whole lake surface. Jordan River waters veer toward the Jordanian coast of the Dead Sea, leaving most of the remaining lake exposed to evaporative drawdown.

Seepage and Encroachment

When the inflow is curtailed by entrance restrictions, the evaporative drawdown in the basin initiates a flow through the porous sediments of the entrance barrier into the lagoon. This occurs today in Lake Larnaca on the island of Cyprus, Lake Tekir-Ghoil in Romania, Lake Assal in Eritrea, Solar Lake on the Gulf of Aqaba, Sinai, and into a crater 1200 m below sea level on Ilha do Sale, Cape Verde Islands (Lecointre, 1960), to mention just a few instances. The flow can be so strong that it produces underwater springs, such as occur in the lagoons on Bonaire, in the Netherlands Antilles (Murray, 1969). Such underwater springs have been observed in many other instances. This subsurface flow occurs in response to a change in hydrostatic head between the outside sea level and the brine level (Sonnenfeld and Hudec, 1977, 1978). If the combined surface and groundwater inflow cannot counteract the evaporative drawdown, the basin starts drying out along the edges, particularly along bayhead shelves and shoals.

Grabau (1920), after observing brine concentration in salinas, made this type of occurrence the basis of his model of evaporite precipitation. If the inflow restriction is severe enough, a density-stratified brine can develop even in a equatorial humid climate. On the Palau Archipelago east of the Philippines, density-stratified anaerobic, hypersaline brines are formed by rainwater interfacing with brines full of hydrogen sulfide and ammonia (Hamner, 1981). Adjacent more open lagoons do not develop such stratification, because the tides mix up any incipient stratification.

In contrast, as the brine concentrates and becomes heavier, a progressively greater proportion of the brine escapes through the substrate, encroaching laterally on groundwater (Fig. 3-18). The lighter groundwater drainage has to override this encroachment and comes to the surface as pediment springs on the rim of salt flats surrounding the hypersaline brine (Sonnenfeld, 1979). These springs then evaporate before reaching the bay. Hypersaline bays usually have negligible surface runoff contributions, either from rare local flash floods or from rivers arriving over great distances from humid lands. Porous beds in the entrance sill can become the site of a two-way flow.

An analogy for such a flow is provided by seawater encroaching on less dense groundwater. A blunt-nosed wedge of seawater encroachment moves landward or retreats seaward, with oscillations in tides and in aquifer recharge. The dynamic forces causing these oscillations are responsible for dispersion of seawater into a zone of mixing near the surface and for ultimate discharge with the groundwater back to the sea (Cooper *et al.*, 1964). A negative temperature gradient down to 910 m in Florida also suggests a deeper cycle in which cold seawater enters, is warmed by the geothermal gradient, rises, and leaves with outflowing groundwater (Kohout, 1967). There is thus preliminary evidence that

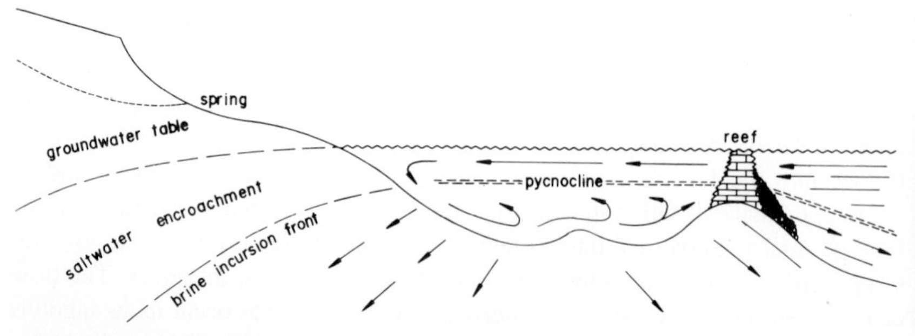

Fig. 3-18. Brine incursion fronts and reef growth. Reefal organisms settle at the bay entrance, where they feed on nutrients provided by the surface inflow. Dead reef trunks hinder the outflow of concentrated brine. Denser brines seep into the lagoon floor beyond the limits of the algal mat. The algal mat seals bottom sediments. The advancing incursion front of the bitterns forces groundwater to come to the surface at the knickpoint of the pediment, where it evaporates on the flats. A tesselated ground or a series of seasonal springs may develop here. (The depth of the initial basin and the reef height are vertically exaggerated.)

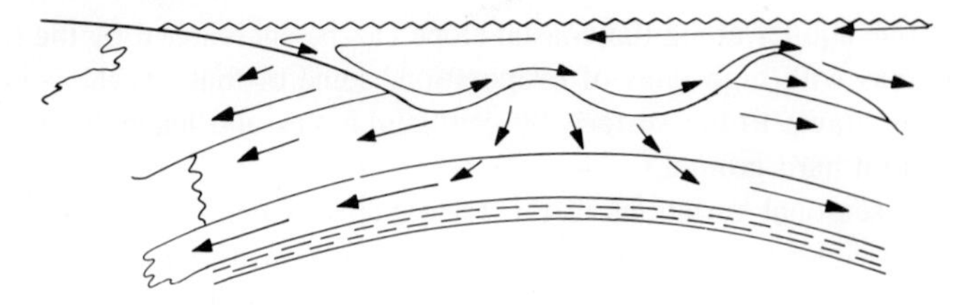

Winter: high evaporation, high rainfall rates

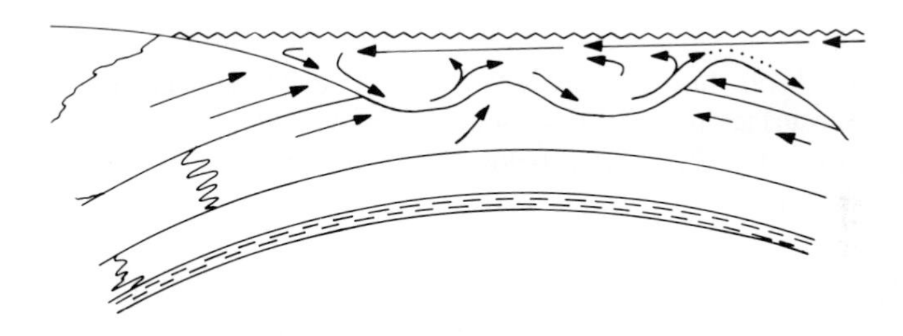

Spring and fall: high evaporation, low rainfall rates

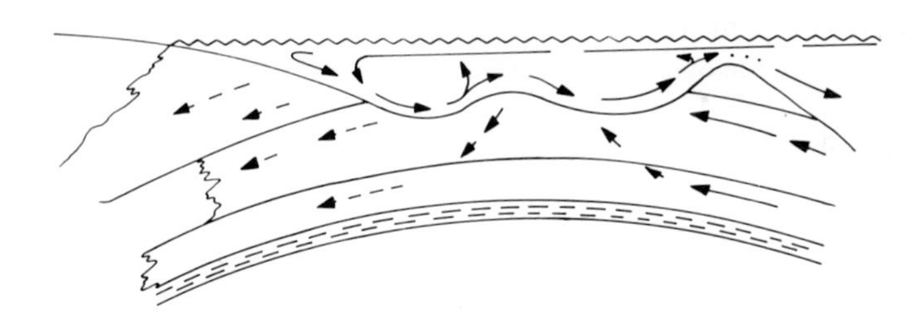

Summer: low evaporation, low rainfall rates

Fig. 3-19. Seasonal brine encroachment. The water level in the lagoon rises during winter rains; more brine encroaches on subsurface waters and also escapes through the entrance sill. In spring and fall, the hydrostatic head is lowered, and the intensity of the reflux depends on the effectiveness of the inflow restriction. Inflow continues during the summer through the permeable sill, but landward encroachment is at a minimum.

more than one aquifer along the coastal slope can be encroached by the heavier brine. In areas with high rates of evaporation, some of that seawater–groundwater mix is drawn to the surface by capillary forces and leaves its solute as desert roses on hard ground.

There are seasonal variations in the brine encroachment. For example, in a climate with winter rains, the rainy season coincides with high sea level, high rates of evaporation, and therefore slow inflow. Both the less impeded bottom outflow and the subsurface brine encroachment are at a maximum. This is also the season when storm frequency peaks and piles up waters at the distal end of the lagoon, increasing there the hydrostatic head on subsurface brines. Cyclonic storms (hurricanes) likewise experience a seasonal peak and affect vast water surfaces in subtropical latitudes. During the rest of the year, low rainfall coincides with high rates of inflow and reflux from the subsurface, but impeded outflow (Fig. 3-19). In a very shallow basin, the algal mat can spread and seal the floor. Subsurface reflux and seasonal pumping action are blocked. Only in the deeper parts of the basin, i.e., below the photic zone, can concentrated brines enter the substrate.

BIOLOGICAL INFLUENCES

Barren Outflow

A reversal from estuarine to antiestuarine circulation at the onset of a water deficit in a bay is preceded by a gradual deceleration in circulation. The reversal specifically affects the biota in the marginal basin. Bottom waters stop bringing in nutrients from the open ocean; instead, a bottom outflow starts that is largely depleted of nutrients and oxygen by ongoing consumption in the bay. Most profoundly affected is the benthonic fauna, which is suddenly deprived of waters well supplied with nutrients from outside the bay, and is thereafter exposed to waters that are deficient in nutrients, much warmer, much less oxygenated, and gradually concentrating. To some extent, such reversal also affects the nekton, but not the plankton.

As the resident brine concentrates, the inability of the benthos to maintain an osmotic equilibrium with progressively higher salinity concentrations leads to dehydration and ultimate extinction (Valentine, 1973, p. 130). The sedimentary record shows a sudden mass extinction in such instances, as is recorded in many preevaporite sequences, e.g., in the faunal eclipse in the first cyclothem of the British Zechstein sequence (Smith, 1980). This event is repeated after each freshening of the brine, as documented by dolomite or shale intercalations in evaporitic sequences. If these suddenly killed biota do not decay in the oxygen-deprived waters, they become the bituminous component of marls or shales associated with evaporite deposits. *Bituminous claystones, marls, or micrites are thus a fossil record of such ephemeral cessation of current movement prior to current reversal.* With the disappearance of bottom scavengers, burrowers, and grazers, the bioturbation of the bottom muds ceases, and laminated sediments can form. The number of species of the remaining fauna that can survive in the inhospitable environment is progressively reduced, but the few endemic species then produce large populations for lack of any natural enemies until the oxygen supply drops so low that only anaerobic bacteria and the facultative anaerobes among the blue-green algae can survive.

The outflowing hypersaline bottom current acts as a salinity barrier for bottom dwellers migrating from the less saline environments of the open ocean. *The seaward outflow of dense bottom brine seriously depletes the shelf in front of the entrance sill of nutrients by washing them down the continental slope* (Valentine, 1973, p. 115). This makes the barrier even more effective (Fig. 3-20). The warm effluent of an evaporating marginal sea spreads out into cooler waters of the ocean and covers the deeper waters like a blanket, preventing any upwelling nutrients from entering the bay. Without migration of additional forms, the surviving benthonic populations quickly become endemic. The effluent mixes with upwelling waters that are usually rich in phosphorus.

Hite (1978) suggested that there might be a genetic relationship between evaporites precipitating inside the marginal sea and the occurrence of phosphorites below the wave base in the adjacent open sea, with attendant iron deposits and dolomites in the phosphorites. In the last century, Ochsenius (1877) described such depletion by the outflow of the Gulf of Kara Bogaz Gol into the Caspian Sea; that outflow ceased early in this century (cf. Sonnenfeld, 1974). The Mediterranean Sea today is the most impoverished large body of water known. The waters are skimmed from a largely depleted North Atlantic water surface and have to support an indigenous food chain. The result is a very low phosphate content. In contrast, outflow from the Persian Gulf is high in nitrates

Fig. 3-20. Surface inflow brings in nutrients; the depleted antiestuarine bottom outflow washes down the entrance sill.

(Thompson and Gilson, 1937), possibly because the Persian Gulf, with its anti-estuarine flow direction, generates less organic matter than the adjacent open ocean (Emery, 1956).

Nutrient Inflow

The surface waters remain oxygenated and can take up some carbon dioxide from the atmosphere, and their organic carbon production is extremely high because they support a teaming life swept in from the inflow sources. Photosynthesizers among blue-green algae and bacteria are extremely euryhaline. In the Solar Lake on the east coast of the Sinai Peninsula, there are several bacterial plates stacked on top of each other (Cohen *et al.*, 1977). Desiccated fish in Upper Miocene evaporites in Italy or in Pleistocene ones in Israel are the corollary of swarms of minnows seen in the upper waters of modern evaporite-precipitating lagoons that feed on prolific plankton. Such schools of fish obviously represent the upper end of an extensive food chain. Similarly, the inflow into the Rann of Cutch, India, a giant natural salt pan, at times resembles a thick, brown, meat broth that later dries out to a film of organic matter and then rots away (G. Richter-Bernburg, personal communication, 1980).

Filter feeders such as large flocks of flamingos (*Phoenicopterus ruber ruber*) populate the salt pans of Bonaire, the Netherlands Antilles, indicating that an adequate supply of nourishment exists (Rooth, 1965). They are also found on salinas near Narbonne, France, and on the salt lakes of Kenya, and elsewhere. Other smaller birds also populate the brine ponds. Their droppings supply a large amount of the ammonia and phosphorus needed in the food chain of brine organisms. The hills around the salt pond of Gran Roque, Venezuela, are covered with a thick layer of guano. Curaçao, Bonaire, and Gran Roque have been commercial phosphate producers. Birds, however, are only a Cenozoic addition to the food chain of hypersaline brines.

Reef Growth

Reefs grow only in the photic zone in aerated waters of nearly normal salinity; they occur preferentially on the entrance sill facing incoming nutrients, or grow around river mouths for the same reason (Fig. 3-18). This is because the oxygen consumption of reefs tends to be extremely high (Verwey, 1931). While they grow, reefs produce more organic material than they consume (Peterson and Hite, 1969), generating a continuous influx of organic matter into the evaporite basin. For the onset of algal reef growth, the entrance sill must have been initially within the photic zone; most reefal organisms cannot survive in the anaerobic bottom effluent depleted in nutrients. A subrecent example is a lagoon on Carmen Island, California (Kirkland *et al.*, 1966), where reef growth gradually closed off the entrance, restricted the two-way water exchange, and thus induced first gypsum and then halite precipitation within the lagoon (Fig. 3-21).

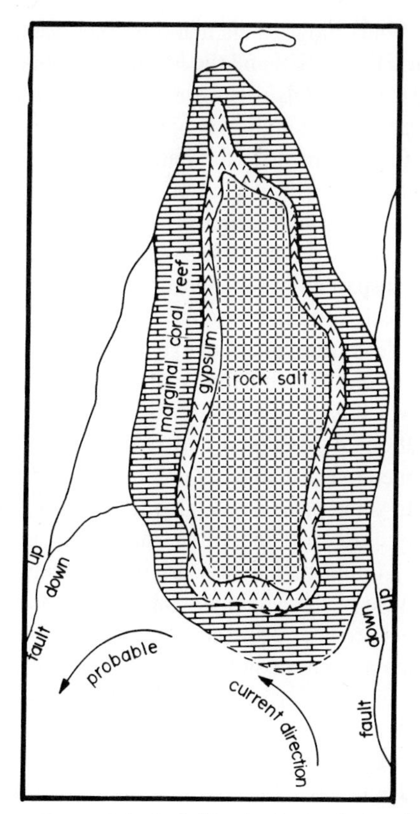

Fig. 3-21. Lagoon on Carmen Island, California. A marginal reef gradually grew across the mouth of a previously open embayment, restricting the water exchange with the open sea. The bay converted thereafter into a hypersaline, halite-precipitating brine pool (after Kirkland *et al.*, 1966).

At the entrance sill, the reefs serve to restrict the water exchange with open marine waters by reducing the effective cross-sectional area of the entrance. It is very common to find ancient reef chains facing open marine conditions on one side, while evaporite accumulations interfinger with the back reef. For example, the Upper Devonian Hardisty barrier reef in central Alberta faced open seas on one side; massive calcium sulfate precipitation took place in the back reef areas. Barrier reefs are following the periphery of the Permian Basin of West Texas (King, 1948) and the Zechstein coastline in East Germany and Poland (Poborski, 1970). The Zechstein reefs of Germany are restricted to sills, shoals, and island rims; stagnating waters with bituminous shales and carbonates are typical of the deeper parts of the basin prior to the onset of evaporite deposition (Richter-Bernburg, 1957a). The same holds true for Middle and Upper Devonian reefs of Alberta and Saskatchewan. Reefs settle at the entrance of a confined basin, not only because of reduced competition by other biota but also because of a continuous supply of inflowing nutrients near the water surface, i.e., at the level where the reef organisms are flourishing. Reefs mark the entrance to the Pennsylvanian Paradox Basin in Utah (Hite, 1970) and the Middle Devonian Elk Point Basin in northwestern Alberta, Canada. *Lush reef growth is thus an essential by-product of antiestuarine circulation,* although it is not confined to such localities.

The growing reefs further constrict the entrance and further impede the exchange of water with the open sea. As the brine becomes hypersaline, the reefs become populated only by a very impoverished fauna, but an extraordinary wealth of stromatolites exists (Paul, 1980). At the same time, the presence of reefs implies clean, transparent waters, an absence of murkiness, and thus a dearth of terrigenous clastic influx. While the reef grows upward, bottom brines may become denser. A dead trunk of a reef extending into bottom brines does not affect life on top of the reef. Eventually, even halite may precipitate in vugs while the top of the reef is still flourishing. Such salt-filled vugs occur in the Silurian Salina Formation of Michigan and Ontario. Reef growth around the Michigan Basin continued over chemically precipitated rocks into both the inlets and the basin proper. In some places, reef growth was snuffed out repeatedly; in others, reefs eventually closed off individual inlets (Droste and Shaver, 1982). Gill and Briggs (1970) found reef debris originating from a growing reef top beneath the first salt cycle of the Michigan Basin. This proves that onset of the growth of marginal reefs predates brine concentration; the debris does not, however, necessarily originate from erosion of all of the final reef top. The Middle Devonian Rainbow reefs of Alberta grew only at shallow depths in waters of near-normal salinity (Klingspor, 1969), although halite accumulated in their back reef area against dead reef trunks.

Busson (1978) wondered why reefs did not grow outside the constricted basin, but populated both the inlet and sills within the basin. Laminites usually terminate against the reef base. He ascribed the reef growth to the reduction in competition by progressive elimination of benthic life and the heaping up of mud mounds by wave action prior to stagnation of an initially very shallow pool. Indeed, patch reefs on the eastern margin of the Silurian Michigan Basin grew preferentially on mounds of crinoidal debris. The progressive scarcity of reef builders and their ultimate absence in favor of algal growth in the reef top signals the drowning of the reef. The upward growth of the reef has not been able to keep pace with the rise of the hypersaline brine (Busson, 1978).

That life continued to flourish on top of the reef is shown even more clearly by the Middle Devonian Winnipegosis reefs of Saskatchewan, which grew to a height of over 100 m above the halite base (Wardlaw and Reinson, 1971). These reefs must have grown during halite precipitation (as evidenced by lateral growth of the Quill Lake carbonate member over anhydrite and halite; Fig. 3-22). The Winnipegosis reefs, growing during Whitkow salt deposition, represented a shoal in the photic zone. They were skirted by gypsum (now altered to anhydrite). Toward the end of halite precipitation, the reefs became exposed. The

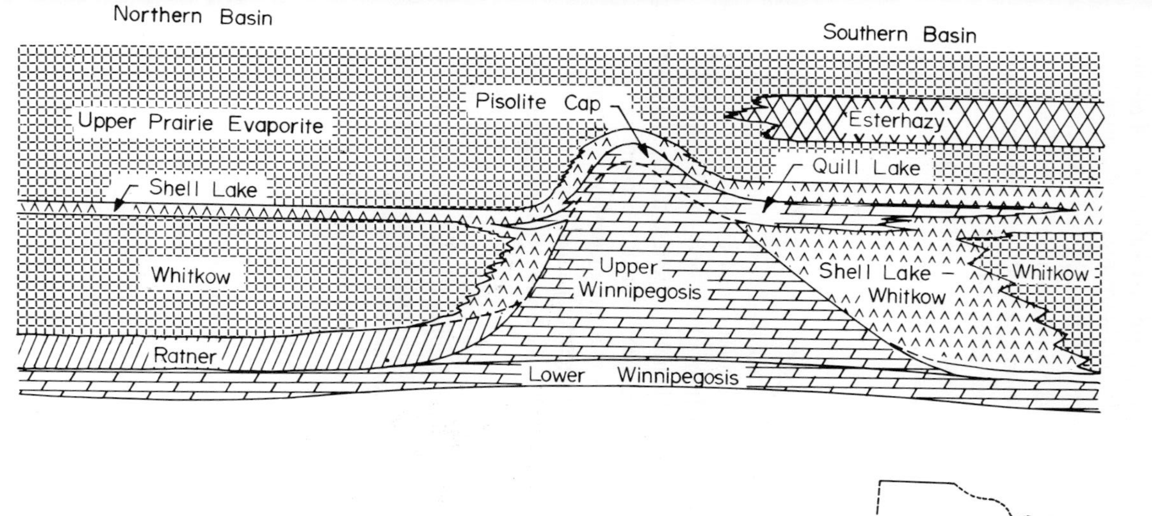

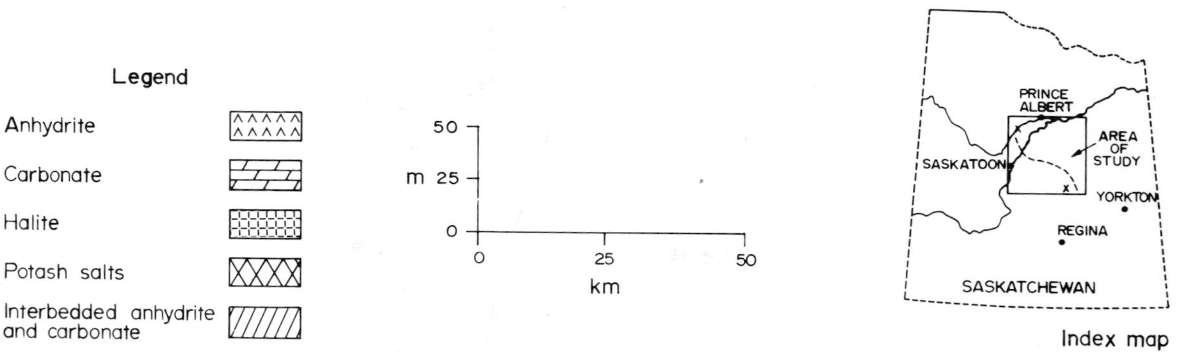

Fig. 3-22. Devonian Elk Point reef in Saskatchewan, Canada (after Wardlaw and Reinson, 1971).

pisolitic Quill Lake Member formed during a period of lowered water level. The overlying anhydrite bank bears witness to the progressive subsidence of the shoal and the decreasing wave energy.

Several such reefs are scattered throughout the distal parts (= Saskatchewan) of the Elk Point Basin. Consequently, surface brines could not have been saline enough during Ratner and Whitkow deposition to kill all reef growth. None of these reefs survived when the brine became saturated for carnallite. The red carnallites and sylvites capped by corrosion surfaces suggest that by then density stratification had ceased, the brine had oxygenated, and the carnallite had frequently dried out (Baar, 1974).

Productivity in Evaporite Basins

Seawater flowing into a hypersaline lagoon is rich in plankton and other organisms including crustaceans, shellfish, diatoms, and schools of fish. It also contains a substantial amount of exudates from phyto- and zooplankton. Whatever organisms or organic compounds accidentally penetrate the pycnocline are exposed to the hygroscopic brine, desiccate, and eventually become part of the food chain for halophilic bacteria. Algal mats that harbor various protozoa (particularly ciliates), bacteria, pseudocoelomate Aschelminthes (Rotifera and Nematoda), and small shellfish effectively seal the bottom of a moderately saline basin in its shallow parts and, with the dark color of the mats, increase solar absorption and evaporation losses. The environment is characterized by high temperatures, high light intensity, and low solubility of gases, particularly oxygen and carbon dioxide. The organic content depends not only on salinity and temperature, but also on the availability of oxygen: North African sebkhas, low in oxygen, are generally rich in organic matter and hydrogen sulfide, but are odorless. However, sebkhas along the Persian Gulf are more oxygenated and contain less organic material (Perthuisot, 1977).

The widely held view that evaporite basins are devoid of life and thus of organic matter should definitely be put to rest. The Gulf of Kara Bogaz Gol contains 59.6–67.1 mg/liter of organic carbon, mainly derived from nine species of algae and two species of bacteria; the adjacent Caspian Sea contains only 6 mg/liter of C_{org} (Bordovskiy, 1964). Dissolved organic material concentrates in salt basins, but not as rapidly as the salt, because a part of the organic matter is oxidized.

Even the Dead Sea is alive (Wilkansky, 1936). Its surface layer contains 8.9×10^6 red halophilic bacteria and 4×10^4 *Dunaliella* algae (red algae) per cubic centimeter (Kaplan and Friedman, 1970). The Dead Sea furthermore contains up to 8.3 mg/liter of dissolved organic carbon (Neev and Emery, 1967). The northern, more hypersaline part of the Great Salt Lake, Utah, contains 300 g/m^3 of biomass (Post, 1977), twice the amount present in the less saline southern part of the lake (Stephens and Gillespie, 1976). Continental lakes contain different assemblages of biota than marine hypersaline lagoons. Nonetheless, lakes with meromictic density stratification are the most fertile in accumulating organic matter. Lake Faro in Sicily produces 44 g/m^3/year of organic matter by photosynthesis; halotrophic bacteria produce 29 g/m^3/year. The wet biomass of ciliates alone amounts to 1000 g/m^3 within the red layer of the water (Sorokin and Donats, 1975). The highest production rate of organic matter recorded in a saline lake amounts to 58.16 g C/m^3/day or 2601 g C/m^2/year (Hammer, 1981). Halophile flagellates, amoebae, bacteria, and fungi populate underground lakes in the Wieliczka salt mine in Poland and are capable of resisting an osmotic pressure of 2.16 MPa (Namyslowski, 1913a,b).

In hypersaline lagoons, the quantities of organic matter can be even larger. Peirce (1914) was the first to investigate quantitatively the biota in California salinas; later, Carpelan (1957) found the rate of carbon assimilation in certain California salinas to be 20–30 times that of open ocean waters. Marine salinas in Puerto Rico contain 100 g/m^3 organic carbon, which is equivalent to about 200 g/m^3 biomass in a system kept artificially aerobic. The density-stratified Solar Lake on the Gulf of Aqaba, Sinai, produces in the lower part of its still-aerobic epilimnion a primary biomass of 36 g/m^3/year; in contrast, 4 m inside the anoxic hypolimnion, the biomass production is two orders of magnitude higher (Hirsch, 1980), with a maximum at 4960 mg C/m^3/day (Krumbein and Cohen, 1964). This is the highest productivity ever recorded in nonpolluted natural waters; it is exceeded only by that of sewage ponds. The lake has a chemo-organotrophic bacterial production of 16.9 g C/m^3/day (Hirsch, 1980), which, if maintained even throughout only part of the year, would represent an enormous productivity. Thus, chemo-organotrophic production by heterotrophic bacterial activity far exceeds primary production. It should be compared to a productivity of only 100 g/m^2/year of open surface in the world oceans (Combaz, 1974), where lack of nutrients, especially nitrates, is the limiting factor. In an anoxic hypersaline basin, LaRock *et al.* (1979) found a 5-fold increase in biomass above the density interface, compared to that at corresponding depths in open oxygenated waters; a 15-fold increase prevailed in the anoxic zone below. Microbial activity at the pycnoclines in the Sargasso Sea generates large amounts of dissolved free amino acids and carbohydrates (Liebezeit *et al.*, 1980). It was well known in the last century that birds around the shores of the Gulf of Kara Bogaz Gol were feeding only on fish eyes, as the amount of organisms swept in from the Caspian Sea was excessive. The waters in this embayment contain up to 67 g/m^3 fatty and fulvic acids, and up to 30 g/m^3 basic, phenolic, and acidic organic compounds (Niyazov and Dzhumayev, 1983).

There is a correlation between the total salinity and ionic composition of the brine and the types and numbers of species able to survive. The higher the salinity, the more difficult it becomes to survive in the environment (Ferroniere,

1901) for both stenohaline and stenoionic organisms. Although the number of species that can adjust to the rising osmotic pressure differential in concentrating brines rapidly decreases, the number of individua that have made this adjustment increases. These individua are no longer threatened by their natural enemies or by grazing and browsing animals. Halophilic bacteria that grow in saturated solutions build up a high sodium chloride content in their cells (Golikova, 1930), and many strains of bacteria are able to adapt if the sodium chloride concentration increases gradually (Anderson, 1945); for example, *Vibrio costicolus* responds to increments in sodium chloride concentration in a regular and measurable manner. Sodium bromide, potassium, or magnesium chloride can substitute for sodium chloride. Cations thereby exert a greater stimulating effect than anions on the response of the organism (Flannery *et al.*, 1952). Both sodium and chloride ions act favorably on bacterial growth; only calcium or magnesium chloride may be substituted for sodium chloride, but not potassium, ammonium, or lithium chloride or sodium bromide (Zaslavskiy and Harzestein, 1930). The relative toxicity of the cation concentration gradually increases from $(Ca + Mg) < Sr < Ba < Be < Zn < Cd < Hg$ (Koulumies, 1946) or $Na < K < NH_4 < Li < Cs$ and $Cu < Au < Ag$ (Alha, 1946). The microflora of hypersaline lakes is not well known; it includes eukaryotes (such as red and green *Dunaliella*), heterotrophic halobacteria (such as *Halobacterium halobium* and *Halococcus*), green and purple anaerobic photosynthetic sulfur bacteria (such as *Ectothiorhodospira*), cyanobacteria (such as *Phormidium*), facultative anaerobic photosynthesizing algae (such as *Oscillatoria* and *Microcoleus*), and several bacteriophages (Borowitzka, 1981).

Stromatolites, which are excellent source rocks for bitumina (Friedman, 1980a), persist well past halite saturation. Nitrogen-fixing blue-green algae, diatoms, various protista, salt fungi, brine shrimp, and larvae and pupae of brine flies can also persist well beyond gypsum and even halite saturation. The chlorophytes *Dunaliella salina* (Dunal) Teodoresco and *Stephanoptera gracilis* (Atari) G. M. Smith are dominant among the algae still thriving during gypsum precipitation and up to saturation with sodium chloride (Carpelan, 1957). Dehydration of evaporite minerals is, of course, lethal to organisms. Halotolerant *Ectothiorhodospira* survives in mirabilite and assorted brines, but it does not survive in thenardite (Tew, 1980).

Absence of nitrogen fixers limits biomass production (Larsen, 1980), although many heliothermal continental lakes possess waters with a decided ammonia taste (Brecht-Bergen, 1908). The large quantities of ammonia observed in a Recent density-stratified hypersaline brine pond on the Palau Archipelago east of the Philippines (Hamner, 1981) have an ancient corollary in the high ammonia content in carnallites in the German Zechstein, which decreases upward in the section (Biltz and Marcus, 1908). Most gases trapped in halites, sylvites, and carnallites show a preponderance of nitrogen, presumably derived from the decomposition of ammonia. Even the ammonia substituting in the crystal lattice of carnallite and sylvite ultimately has an organic origin.

Bacteria digest proteins even in very concentrated brines. If grown in a 25% sodium chloride solution, they produce a glycerol extract with proteolytic properties (Stuart and Swenson, 1934). Amino acids as decomposition products derived from proteins have been found in Silurian salts in New York State, in Permian Wellington salt of Kansas and Oklahoma, and in a Recent salt on Bonaire, Netherlands Antilles (Brunskill and Vallentyne, 1966; Tasch, 1963, 1970); they are also the main nitrogen compounds in bitumina (Sokolova *et al.*, 1962). At extremely high sodium chloride concentrations, *Sarcinia littoralis* forms no pigment, grows only at a relatively low redox potential (Eh), and has a marked reducive metabolism (Stuart and James, 1937). Aliphatic hydrocarbons are generated by a number of algae, bacteria, and fungi (Weete, 1972); for example, alkanes or naphthenic hydrocarbons are produced by marine sulfate-reducing bacteria such as *Desulfovibrio desulfuricans* and *Desulforistella hydrocarboniblastica* (Seliber, 1950; Sisler and ZoBell, 1951; Hvid-Hansen, 1951a; Davis, 1968). They are symbiotic with anaerobic *Actinomyces* that produce organic compounds from bicarbonates (Hvid-Hansen, 1951b).

Overall, the procaryotes as a group can be considered as major contributors to alkane production (Bird and Lynch, 1974). *Chlorella vulgaris,* an alga that is heterotrophic in darkness and autotrophic in light, produces 1–2 ppt of saturated hydrocarbons, and normal alkanes in the C_{17-36} range (Patterson, 1967). However, Winters *et al.* (1969) could not produce alkanes higher than C_{18} by cultivating blue-green algae. Yet alkanes with a maximum in the C_{15-17} range and a secondary maximum in the C_{20-30} range occur in green, brown, and benthic algae (Youngblood *et al.*, 1971). Gelpi *et al.* (1970) found a similar bimodal distribution of alkanes and aliphatic hydrocarbons in 12 species of blue-green algae. Even diatoms contain 2–15% hydrocarbons (Lee and Loeblich, 1971). Bacteria are important not only for the generation of hydrocarbons, but also in further modification, as they decrease the lighter fraction and thus alter the density in the presence of traces of ammonium chloride. Bacteria are useful in cracking and opening rings, reducing the presence of cyclic hydrocarbons or the amount of naphthenic acid, changing saturated hydrocarbons to unsaturated ones, and increasing the bromide index of crude oil (Imelik, 1948). Halogenated aliphatic compounds are readily transformed by denitrifying bacteria (Bouwer and McCarty, 1983).

Algae have been found in the Main Anhydrite Member of the Zechstein evaporites (Kosmahl, 1969), as inclusions and interstitial films in granular and fibrous halite, or associated with the hematite in halite (Tilden, 1930; Jones, 1965), or even in potash salts (Smith, 1973). Putrefactive anaerobes are present in nearly all samples of mined rock salt (Yesair, 1930; Bien and Schwartz, 1965), but none are alive. Bacteria and bacterial spores were found encased in crystal centers in Permian, Mississippian, Devonian, Silurian, and Lower Cambrian halites (Mueller and Schwartz, 1953; Dombrowski, 1966). Osmophile bacteria have been found in Silurian-to-Tertiary salts in various localities (Reiser and Tasch, 1960); iron bacteria have been found in the Yorkshire salt deposits (Strong, 1956). The organic matter in mud partings of Searles Lake, California, is black in dried form and orange in dilute form, contains a large amount of molybdenum, has pronounced fungicidal properties, and is probably derived from green and red bacteria (Flint and Gale, 1958). Finally, there are up to 1 million bacteria and fungi in each gram or milliliter of crude oil (Schwartz, 1972).

Coorongite, an organic compound accumulating along lakes in South Australia, the Balkhash Lake in Siberia, and along lakes in East Africa, is produced by the algal bloom of one organism, *Botryococcus braunii,* which is also present in torbanites and oil shales (Cane, 1977). It is caused by aerobic polymerization of unsaturated lipids and anaerobic production of hydrocarbons (Stadnikoff, 1930; Breger, 1970). Earlandite (calcium citrate tetrahydrate) is found in deposits of the Weddell Sea (Bannister and Hey, 1936); even less common organic compounds are preserved in subpolar areas.

Murkiness

The color of microorganisms and floating algal debris adds to the murkiness and turbidity of the water and thus increases solar absorption. Eventually, the only organisms that can survive during halite crystallization are certain cyanophyta and red halophilic bacteria. The absence of such microorganisms reduces salt production in salinas, yet their presence yields contaminated salt and promotes food spoilage by halophilic bacteria. This is the reason why sterile fossil salt is preferred for pickling and conservation purposes. The red bacteria depend on a steady supply of organic material, since they are not photosynthetic. They not only raise the temperature of the pond by increasing solar absorption but also seal the pond floor with organic material. Consequently, the bright red crystallizer pans in salinas never leak (Davis, 1974). However, flourishing bacterial colonies seal pond floor permeability, whereas decaying colonies cannot do so. If the algal and bacterial community is forced to go into a seasonal state of dormancy in natural brines, accumulated organic oozes can leak out.

Red Brine Color

The accumulation of a dense bacterial plate immediately below the density interface gives the brine a reddish to pinkish color. The strong tendency to form red or pink pigments is a remarkable characteristic of strict halophiles among bacteria and algae (Zajic, 1969). Their pigments provide resistance to sunlight

(Horovitz-Vlassowa, 1931), whereas the development of colorless varieties is severely hampered in bright sunlight (Dundas and Larsen, 1962).

RADIATION EFFECTS

Adsorption of Radiation

Density-stratified hypersaline brines absorb even diffuse sunlight and turn it into heat. The temperature of the brine is inversely proportional to light transmittance (Crutzen *et al.*, 1970). Incoming solar radiation is partially absorbed by the surface layer, and some of this is returned at night as back radiation. Absorption varies with different wavelengths of incoming radiation, and the mean coefficient of extinction rapidly decreases with depth. The albedo (or reflectivity) of a water surface is independent of the water temperature and of dissolved solids at low concentrations, but is reduced in density-stratified hypersaline brines, where only a fraction of incoming solar radiation is reflected back into the atmosphere. Most of the radiation leaves as evaporative energy. However, as turbulent convectional heat transport is prevented by the density stratification, high temperatures will occur inside the water body (Emden, 1940).

Concentrated brines generally have lower specific heats than more dilute seawater solutions. In the case of sodium, calcium, and magnesium chloride solutions, the specific heat is a function of both temperature and concentration (Anghelescu, 1953). However, this varies from solute to solute. A concentrated potassium chloride solution has a lower specific heat than a corresponding sodium chloride solution (Bogorodskiy and Dezideriev, 1935). As the brine concentrates, its index of refraction rises and a smaller percentage of rays is capable of escaping. More heat is produced, since the specific heat of the brine decreases with rising concentration. Simultaneously, the rate of evaporation decreases, and with it the heat loss. A steep chemocline within the photic zone makes a lagoon surface a solar collector, but the lower this interface is located, the less radiation can be trapped. Consequently, precipitating hypersaline brines gradually need less energy to maintain their temperature regime. The same solar radiation then produces correspondingly higher temperatures (Borchert, 1969). Figure 3-23 shows the calculated temperature difference of hypersaline brines compared to freshwater bodies. With increasing evaporation losses, the temperature rises more rapidly at higher salinities. Moreover, the greater the density difference between resident brine and incoming waters, the greater the energy released when the two solutions of different salt concentration mix. For every 1 m^3/s of inflowing seawater, that energy can amount initially to a few megawatts, eventually rising with increasing concentration of the resident brine to about 25 MW.

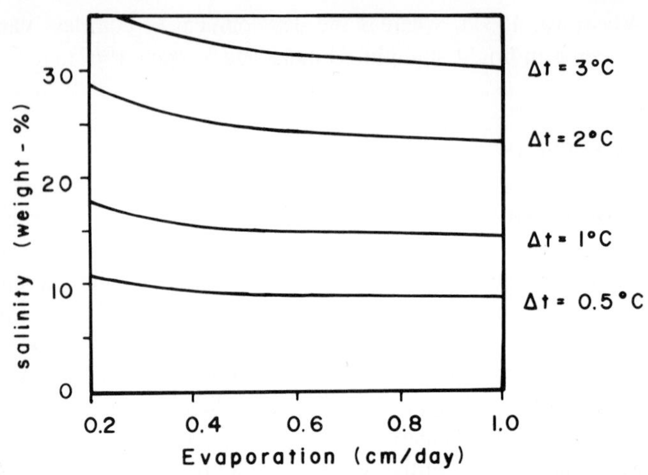

Fig. 3-23. Temperature difference between hypersaline brine and fresh water with an increase in salinity (after Harbeck, 1955). The calculation is based on water and air temperatures at 15°C, relative humidity at 50%, and air pressure at 100 kPa. The albedo, latent heat of vaporization, and saturation vapor pressure were assumed to be constant over the temperature range involved.

Where the radiation penetrates the interface between surface waves and brine, it is easily trapped. The greater density of the brine increases the scattering of the rays and lengthens their path; thus, their chances for absorption are enhanced. Whenever the rays hit a shallow bottom they are reflected, but only a small portion of the reflected rays can escape through the interface (Hudec and Sonnenfeld, 1974); the rest is reflected back into the brine. The interface acts as a one-way mirror for about 89% of incoming radiation. Kenat (1966) estimated the amount of solar radiation absorbed and transmitted to the bottom to be 94% in brine ponds crystallizing halite. Cole *et al.* (1967) found that 75% of the radiation entering two saline lakes in Arizona was absorbed in 1.25 m of water from surface to thermocline. The next 25 cm absorbed 23.5%, so that only 1.5% remained at a depth of 1.5 m. When a lake is unstratified, half of the sunlight reaches its bottom at 2.0–2.5 m. When a lake is stratified, 99% is absorbed above the bottom.

Heating Effects

Intensity of evaporation decreases as concentration increases, and solar radiation is then used mainly to raise the temperature (Borchert, 1969). Shallow marginal seas are always warmer than adjacent open ocean waters (Debenedetti, 1982), because the shallower the basin, the less brine must be heated by solar radiation entering per unit area. The specific heat of a concentrated brine is also lower than that of a dilute one; furthermore, less heat is lost through evaporation as the concentration rises. A brine so heated can attain a temperature about 20–40°C higher than that of the surface waters (R. G. Wetzel, 1975). A small side effect is the influence on the pH. Hydrogen solubility in pure water reaches a minimum at 37°C.

The rebound of radiation in extensive shallow regions appears to be essential to the heating process, with hot brines spreading out along the interface to surface waters. Of two Siberian hypersaline lakes 150 m apart, only the one with extensive shallows heats up; the one with vertical walls dissipates the incoming radiation (Fig. 3-24). The former produces scalding temperatures. The latter encounters summer temperatures above freezing only in the surface layer; on the bottom it permanently precipitates hydrohalite ($NaCl \cdot 2H_2O$), which is stable only at freezing temperatures (Dzens-Litovskiy, 1968). Saturation shelves (Fig. 3-5; Richter-Bernburg, 1957a) heat up faster than deeps because the surface area is larger per unit volume. Stratified brines heated up by solar radiation occur in all latitudes from Ellesmere Island in the Arctic to Antarctica (Sonnenfeld and Hudec, 1980); such lakes are referred to as *heliothermal lakes* if their heat is derived entirely from solar radiation. Those lakes that freeze over in winter remain hot only if a freshwater lens intervenes between the ice and the hypersaline brine. An exceptionally dry fall in 1975 left brine lakes in British Columbia unstratified, and no hot brine developed under the ice that winter. Fall rains in other years have assured stratification under the ice and a perennial hot water lens.

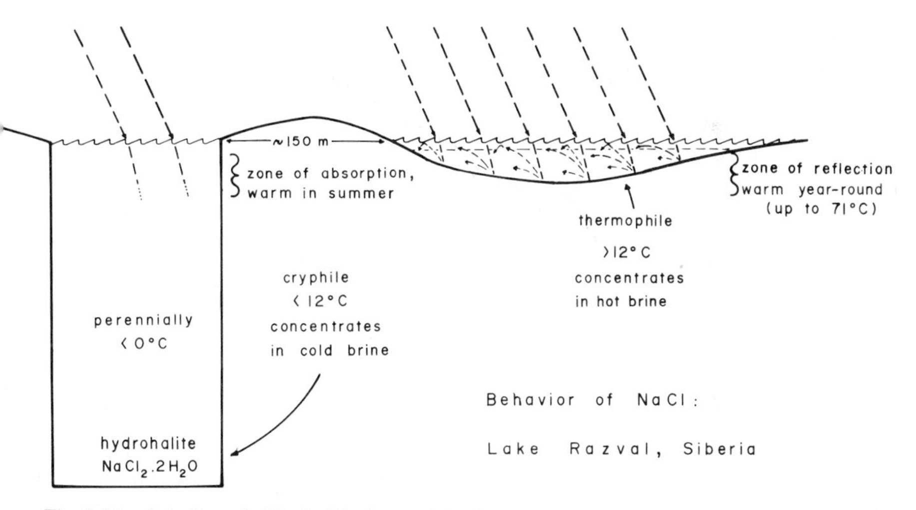

Fig. 3-24. Lake Razval, Siberia. The bottom brine in the deeper lake with vertical walls remains cold, whereas an adjacent pool heats up because of light reflection from the shallow bottom (after Dzens-Litovskiy, 1968).

The surface water heats up at the air–water interface in the late afternoon and then cools again by evaporation. This layer is termed the "epilimnion" in lakes and the "epithalasson" in the sea (Fig. 3-25). The temperature difference between surface waters cooled by evaporative heat loss and brine heated by absorbed radiation is called a "thermocline." It separates the surface waters from the hypersaline brine. Heating from below by radiation absorbed beneath the thermocline and concurrent cooling from above by evaporation losses stabilizes the salinity interface, or "halocline," and prevent mixing (Lewis *et al.*, 1982). Whereas the thermocline sits on top of a mixing zone, the "pycnocline" (density interface), "halocline" (salinity interface), or "chemocline" (concentration interface) is often slightly below the thermocline, separating the anoxic monimolimnion from the oxygenated mixolimnion. The hot lens cools at its base in distinct steps (Hudec and Sonnenfeld, 1980); the densities also increase in steplike fashion, although the steps may be extremely small. The effective thermal conductivity is directly proportional to the molecular thermal diffusivity and inversely proportional to the ratio between the salinity gradient and the heat flux (Huppert and Linden, 1979). Since the heat flux is about four orders of magnitude faster than the salt flux, heat is transported much more rapidly than salt (Gregg and Cox, 1972). In the absence of a continuous supply of heat both during the day and at night, this disparity will increase until the temperature step is inadequate to lift the salt. A remnant salinity step forms, and a stable interface between adjacent brine layers is the result.

The hot brine rises toward the inflow (Hudec and Sonnenfeld, 1980) and is preferentially discharged (cf. the bump in the thermocline in Fig. 3-17 moving

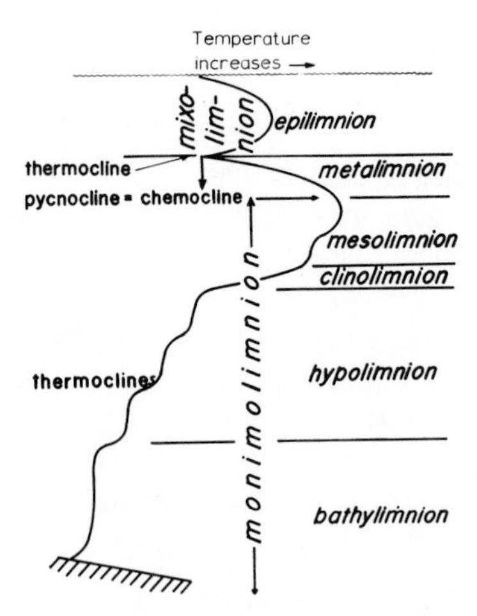

Fig. 3-25. Nomenclature of a hypersaline brine pool. The suffix "-limnion" pertains to lakes and should be replaced by the suffix "-thalasson" when referring to layering in the sea.

toward the inlet). The hot brine also spreads out over deeper waters as a hot water lens with a sharply defined upper transition to cooler surface waters. As the thermocline (Fig. 3-17A) dips into the basin, waters of different temperature become laterally juxtaposed to each other. This system is stable only if waters of equal depth have an equal density, or in other words, if the hotter brines have a higher salinity. A very small horizontal concentration gradient is the result. In an elongated lagoon system, such as that of the Ojo de Liebre lagoon of Baja, California, the salinity increasing toward the bayhead is counteracted by an increase in the temperature of the brine. In midsummer, the decrease in density due to the rising temperatures can match the density increase due to a rising salinity. Excess brine then moves seaward as bottom flow in one tide and as surface flow in the next tide (Postma, 1965).

The surface brine loses heat to evaporation and thus remains relatively cool. The bottom brine, heated by trapped sunlight, uses some of that energy to pack more tightly the ensemble of molecular space cells of all solutes contained in seawater. The solute entities (ions and molecules) need a specific volume to acquire complete freedom of motion. The critical concentration is attained when all available molecular space is used. Seawater is very close to this value, containing about 0.6 mol of solute entities (Fujiwara, 1979), and is close to the value at which the heat of infinite dilution (ΔH) changes from a positive to a negative value.

Rising brine temperatures cause a progressive dehydration of hydrated solute species; falling temperatures have the opposite effect. For example, magnesium chloride has the following stability fields (Strakhov, 1962):

$MgCl_2 \cdot 12H_2O$	−33.6°C to −16.4°C
$MgCl_2 \cdot 8H_2O$	−40.0°C to −3.4°C
$MgCl_2 \cdot 6H_2O$	−15.0°C to +116°C
$MgCl_2 \cdot 4H_2O$	+116°C to +118.5°C
$MgCl_2 \cdot 2H_2O$	+181.5°C to +300°C

Hexahydrate bischofite is thus the common mineral in subtropical marine evaporites; the octohydrate and the dodecahydrate occur only in polar environments. The tetrahydrate and the dihydrate have not been found in sedimentary environments.

The higher the salinity of a lagoon, the lower its freezing temperature and the longer it evaporates from an open water surface. Lagoons stay open year-round, even in Antarctica, if they contain over 100 g cations per kilogram of chloride ions (Matsubaya *et al.*, 1979) or about 135 g cations per kilogram of sulfate ions. This places a limit on freeze-drying and thus on the volume of precipitates possible under polar conditions. However, if lagoons are covered by low-salinity runoff in the late fall, they freeze over. A low-salinity water layer beneath the ice separates it from the hot brine. If fall runoff is omitted in any year, no heating takes place.

Thermal Diffusivity

Thermal diffusion of halides occurs in physicochemical systems containing more than one component if the temperature distribution is not uniform. There is a large disparity between the relaxation time for concentrations and that for temperatures. Thermal diffusion results in a separation; some of the components accumulate in the warmer regions, and others in the cooler regions of the system (Chanu, 1967). Ion migration across the chemocline thus fortifies the salinity difference, since most of the major ions in seawater are thermophile. The heating from below reinforces the thermocline (Stern, 1980) and thus retards diffusion. Thermal convection can be eliminated by linear shear (Ramser, 1957) of currents in plane parallel layering. The solar heating of density-stratified brines has been known for nearly a century; likewise, the high stability of a hot brine layer covered by cooler surface waters (Ziegler, 1898; von Kaleczinsky, 1901; Rozsa, 1911, 1913a, 1915; Liesegang, 1916; Maxim, 1929, 1930, 1936; Sonnenfeld and Hudec, 1980).

When a brine maintains a temperature gradient for some time, a concentration gradient is set up. This was discovered by Ludwig in Vienna in 1856 and rediscovered by Soret in Paris in 1879 (Chipman, 1926). The Soret effect (or thermal diffusivity) of solute migration along a temperature gradient can produce a stabilizing contribution to the density gradient (Hurle and Jakeman, 1971). A concentration gradient of up to 1% can be maintained in a potassium chloride solution by differential heating of a long trough (Borchert, 1933). The Soret effect is very sensitive to small temperature variations and shows a distinct minimum as a function of the concentration of individual chlorides. It increases with temperature but decreases with concentration in sodium chloride solutions (Tanner, 1927, 1952; Alexander, 1951; Chanu and Lenoble, 1955, 1956; Longworth, 1957); in potassium bromide solutions, however, it increases with concentration (Chanu and Mousselin, 1961). The Soret coefficient is additive for the ions involved (Murin and Popov, 1953), but anionic entropies of transfer are not additive in concentrated solutions (Chanu, 1958). A plot of the Soret effect against temperature and concentration is hyperbolic in shape for sodium and potassium chloride and sodium bromide (Lerman, 1979); the maximum is near 235 ppt for sodium chloride, 412 ppt for sodium bromide (Tanner, 1927), and nearly 185 ppt for potassium chloride (Eilert, 1914). For equal concentrations, the values of the Soret coefficient increase for chloride anions with the ionic radii of alkaline earth cations, but decrease with the ionic radii of alkali metal cations (Ikeda, 1959; Ikeda *et al.*, 1968). For calcium and magnesium chloride solutions, the Soret coefficient is very small because the ionization factor compensates for the Soret effect (Porter, 1927). The values of the coefficients of self-diffusion for H_2O and anions are parallel for a wide range of concentrations; toward the greater concentration of the brine, the values for cations and anions converge as the hydration of the cations diminishes (Salvinien and Brun, 1964). The self-diffusion is inversely proportional to the viscosity for an invariable kinetic entity and remarkably constant for the partial pressure of chlorine (Salvinien and Brun, 1964).

Cryophile and Thermophile Solutes

The terms "thermophile" and "cryophile" are here used in the sense defined by Borchert (1933, 1940) or Borchert and Muir (1964). Accordingly, thermophile solutes wander into warm brines and precipitate upon cooling. Cryophile solutes wander into colder brines and precipitate in warm brines. Unaware of these definitions, Strakhov (1962, p. 235) called the former "cold-loving" or "cryophile" and the latter "warm-loving" or "thermophile." In effect, thermophile solutes increase their solubility with temperature, whereas cryophile solutes decrease their solubility with temperature. In other words, thermophile salts have a negative temperature coefficient (e.g., all potassium and magnesium solutes other than kainite), cryophile salts a positive one (e.g., kainite, iron or zinc chloride).

At a temperature of 49.5°C, the Soret coefficient of sulfates is much larger than that of corresponding chlorides. The time required to reach saturation is shorter at that temperature for potassium sulfate than for magnesium or sodium sulfate. The change in the Soret coefficient with changing temperature is also greater in sulfates than in chlorides. In contrast, the ratio of Soret coefficients of two different solutes with common anions rises in the following order: Mg $<$ K $<$ Na $<$ Ca $<$ Ba (Hirota, 1942; Schott, 1973).

There are some compounds that change the sign of their temperature coefficient, and with it the direction of migration at some point within the normal range of evaporating brines. Sodium chloride has a solubility minimum at 12°C, and gypsum a solubility maximum at 37.5°C. The presence of other solutes alters these solubility minima. In a saturated potassium chloride solution, sodium chloride is decidedly cryophile, i.e., has a positive temperature coefficient (Blasdale, 1918); the same holds for a magnesium chloride solution (Langauer, 1932). In a concentrating brine, halite is cryophile below 12°C (Caldwell, 1973b), particularly in the presence of magnesium chloride. It thus wanders into cold bottom layers of saline heliothermal antarctic lakes. Above 12°C it is thermophile and wanders into warmer waters. It therefore crystallizes in subtropical lagoons in the daytime, when the surface waters heat up and concentrate by evaporation. Heat flux direction itself also changes with concentration. A concentrating calcium sulfate solution changes from endothermal to exothermal at about 15.5 mmoles/liter (Hulett, 1901), which is close to saturation at 25°C.

Cryophile solutes reinforce the thermal density gradient, but by weakening the stability of the fluid layer, they eventually destroy the concentration gradient.

Instability occurs when the Rayleigh number (the product of the gravitational constant, thermal expansion coefficient, adverse temperature gradient, and the fourth power of the thickness of the layer divided by the thermal diffusivity coefficient and the kinematic viscosity) exceeds 1708 in pure liquid layers. In brines with predominantly thermophile solutes, the critical Rayleigh number is increased substantially; in brines with cryophile solutes, it is decreased (Legros *et al.*, 1972). If a concentration gradient can be dissipated by diffusion compared with the rate of dissipation of thermal gradients, it may lead to overstability; concurrent sedimentation can enhance this overstability (Hurle and Jakeman, 1969). When the Soret effect is negative and the heat flux is from below, thermally diffusive salt movement stabilizes the solution. Convection can begin only when the Rayleigh number is much larger than the pure fluid value of 1708 (Caldwell, 1973a).

In the open ocean, the thermally driven salt diffusion is roughly half of the diffusion driven by concentration (Gregg and Cox, 1972). It can approach equality in warm hypersaline brines, since the Soret coefficient is dependent on the concentration of the solution (Snowdon and Turner, 1960). The rate of heat propagation decreases very slightly with increasing concentration concurrent with a lowering of the specific heat; in contrast, the thermal diffusivity of salt decreases more rapidly with increasing concentration, but increases with temperature. The inversion of the sign of the Soret effect, known also from calcium and ammonium chloride, has not been looked for in other types of natural solutes. Likewise, the Soret effect has not been adequately investigated in complex hypersaline brines, where the movement of thermophile solutes interferes with the movement of cryophile solutes. Such an inverse effect has been observed in some solutions, in which a heavy constituent migrates against the direction normally expected (Gillespie and Breck, 1941; Hirota, 1941). At best, all of these values are known only qualitatively for natural hypersaline brines.

Heat of Solution, Dilution, and Crystallization

Influx of more seawater or rainwash dilutes the brine and dissolves some of the precipitated salts along the fringes of the embayment. The amount of heat so generated or consumed depends on the concentration of the brine. For example, the alkali halides go through a maximum heat of solution at 0.18 moles per mole of H_2O and through a minimum heat of dilution at 0.14 moles per mole of H_2O (Wuest and Lange, 1925).

Once the brine is saturated and starts precipitating various salts, heat is also released. Typically, a precipitating salt pan contains a warmer brine than an undersaturated salina. Per cubic meter of precipitate, halite releases five times more heat than gypsum, sylvite releases three times more heat than halite, and hydrated potassium–magnesium salts release a multiple of the heat generated in sylvite crystallization.

From fluid inclusion studies, Kityk *et al.* (1980) found that in the Pripyat Basin of Byelorussia and the Dneper-Donets basins of the Ukraine, Upper Devonian halites precipitated at a brine temperature at or above 35°C; potash minerals formed when the brine temperature reached 60–65°C during the day. Above a density of 1.275, natural evaporation is very small because a thick floating crust forms; the bittern then attains temperatures in the range of 70–80°C (Elschner, 1923). This occurs well before potash salts start precipitating.

As Table 3-1 shows, polyhalite, kainite, and langbeinite consume heat when crystallizing, whereas halite and sylvite increase the water temperature. In carnallite, the rise in temperature due to potassium chloride crystallization outweighs the drop due to $MgCl_2 \cdot 6H_2O$ crystallization, producing a net warming effect. Anhydrite alone shows no effect, possibly because of its low specific heat (0.1753). However, a mixture of anhydrite and sylvite shows greater cooling upon solution than does sylvite alone (Richardson and Wells, 1931). Gypsum absorbs a very small amount of heat when dissolving up to 37.5°C; above that temperature, it evolves heat (van t'Hoff *et al.*, 1903). Thermodynamic properties of moderately concentrated aqueous chloride, bromide, and sulfate solutions were calculated by Wicke and Eigen (1953); those of saturated solutions, and of mixtures of concentrated solutions are unknown.

TABLE 3-1

Heat of Crystallization from a Watery Solution

Mineral	Chemical formula	kJ/cm^3
Gypsum	$CaSO_4 \cdot 2H_2O$	+2.33
Halite	$NaCl$	+11.64
Sylvite	KCl	+36.91
Carnallite	$KCl \cdot MgCl_2 \cdot 6H_2O$	+20.76
Tachyhydrite	$CaCl_2 \cdot MgCl_2 \cdot 6H_2O$	+32.10
Arcanite	K_2SO_4	+72.96
Picromerite	$K_2SO_4 \cdot MgSO_4 \cdot 6H_2O$	+90.09
Epsomite	$MgSO_4 \cdot 7H_2O$	+26.51
Mirabilite	$NaSO_4 \cdot 10H_2O$	+115.824
Anhydrite	$CaSO_4$	−65.17
Bischofite	$MgCl_2 \cdot 6H_2O$	−19.19
Langbeinite	$K_2SO_4 \cdot MgSO_4$	−124.34
Thenardite	Na_2SO_4	−6.21

BRINE CONCENTRATION

Salinity of Source Waters

There is no minimum salinity requirement for the influx into an evaporite basin. It is a common fallacy to assume that thick evaporite deposits must be derived from normal marine salinities of seawater. Undoubtedly, most ancient massive evaporite deposits were derived from such marine brines, because no groundwater drainage or surface runoff could supply the required volumes of salt. However, brackish or even fresh waters can also lead to hypersaline brine formation on the scale of continental lakes. For equal volumes of water the only requirement is for more time, as the inflow salinity controls the amount of salt entering in solution per unit of time. The length of time required to reach saturation also affects the isotopic composition: The more time a brine needs to concentrate, the heavier it gets isotopically.

In order to assess the possibility of halite precipitation from freshwater sources, we can consider a hypothetical lake with no outlet, whose water level is maintained by river waters (density $\rho = 1.0025$). At typical evaporation rates of 2000 to 5000 mm/year of fresh water in desert or semidesert terrain, it would take at least 146 years per meter of average lake depth to reach halite saturation. In nature, such lakes lose some brine as subsurface encroachment on groundwater; this increases the time required to reach a given concentration. Eventually, an equilibrium is established between inflowing dissolved solids, precipitation of salts, and brine leakage. A corollary is found in interconnected lake chains in semiarid lands, which often grade by progressive concentration from pure carbonate lakes to sulfate and chloride lakes (Hudec and Sonnenfeld, 1980).

There are two examples of water bodies currently precipitating salts, but fed by fresh or brackish waters. The Dead Sea is fed by the River Jordan and by some salt springs that redissolve older salts. Its salinity accumulated over a mere few thousand years (Neev and Emery, 1967). Now it precipitates halite at the south end on its shallow shelf. The Gulf of Kara Bogaz Gol has always been fed by the dilute waters of the Caspian Sea; in addition to mirabilite and glauberite, it precipitates mainly calcium salts (Kolosov *et al.*, 1974) and some halite (Fedin, 1979). Potash minerals in minute quantities occur along its distal supratidal zone.

There are many more examples postulated for fossil evaporites. A freshwater source has been assumed for Oligocene potash and halite deposits in the Rhine valley (Wilser, 1926). Both Oligocene and Miocene evaporite deposits in the Middle East seem to have had brackish rather than normal marine water sources (Stoecklin, 1968; Sonnenfeld, i981). Another large continental body of water, the Paratethys Sea, was connected to the Tethys Sea throughout Phanerozoic time but was separated from it by land masses of varying areal extent (Sonnenfeld, 1981). It extended at times from central Europe to central Asia; the Black, Caspian, and Aral seas are its remnants. It appears to be unique in that the various basins of the Paratethys Sea consistently alternated between hypersaline evaporitic precipitates and normal marine or even freshwater deposits. No other major body of water has oscillated in that manner for such a long time. Miocene salt horizons follow Oligocene, Eocene, Cretaceous, Jurassic, and older salt bodies. Almost everywhere, Sarmatian freshwater deposits overlie these older Miocene salts on either side of the Carpathian and Caucasus mountains. In places they are overlain by younger evaporite deposits.

Salt Balance

As long as more salt is entering into the basin than is flushed out per unit of time, the concentration of the brine will increase. Saturation is then only a matter of time, as long as other factors remain constant. The volume of influx increases in response to an increasing water deficit through evaporation losses, and the concentration of the outflow also increases. At any moment, the following equation holds

$$V\rho_I + I\rho_I + R\rho_R + P\rho_P = A\epsilon + V\rho_O + O\rho_O \tag{3-1}$$

where I is inflow, O outflow, R runoff, and P atmospheric precipitation, all per unit of time respectively, A the surface area, V the volume of the bay, ϵ the rate of evaporation, ρ_I, ρ_R, ρ_P, and ρ_O the densities of inflow, runoff, atmospheric precipitation, and outflow/bay waters, respectively.

The total evaporative water loss (E) is, of course, a function of the surface area of the marginal bay. However, the configuration of the shoreline affects the water loss in that the more convoluted it is, the larger is the water loss of a constant area; further, the greater the circumference of the bay, the smaller is the area for a given circumference (Lotze, 1957). A change in climate alters both the circumference and the surface area by either baring or drowning extensive coastal flats.

Overall, the progressive water exchange in a bay can be expressed as follows:

$$\begin{aligned} &V(1 + s_I) + (I + dI/dt)\,(1 + s_I) - (E + dE/dt) \\ &= V(1 + s_I + ds/dt) + (O + dO/dt)\,(1 + s_I + ds/dt) \end{aligned} \tag{3-2}$$

where E is the evaporative water loss and s_I is the salt content of the inflow, defined as the density ρ of the brine less the density of water. In other words, the volume of water in the bay (V) that initially had the same density ($1 + s_I$) as inflowing waters, plus an increasing amount of inflow ($I + dI/dt$), reduced by evaporative water losses ($E + dE/dt$), equals a bay water volume (V) of increased density ($1 + s_I + ds/dt$) and an outflow ($O + dO/dt$) of corresponding concentration.

A steady state requires that the evaporation losses equal the difference between inflow and outflow, or

$$E = I - O \tag{3-3}$$

and

$$dE/dt = dI/dt - dO/dt \tag{3-4}$$

Thus, we can attempt to solve Equation (3-2) by differentiation:

$$d[s - (I/O)s_I]/dt = (O/V)[s - (I/O)s_I] \tag{3-5}$$

and then integrate

$$\int d\frac{[s - (I/O)s_I]}{[s - (I/O)s_I]} = -(O/V)\ dt \tag{3-6}$$

From this, we can determine the salinity s of the basin (again expressed as density ρ less the density of water) after any period of time as

$$s = s_I[I - (I/O)e^{-(O/V)t}] \tag{3-7}$$

Conversely, the time t required to reach a given salinity (Sonnenfeld, 1980) is then expressed as

$$t = (V/O)\ln\frac{I/O}{(I/O)s_I - s} \tag{3-8}$$

With Equations (3-7) and (3-8), one can easily prove the following:

1. The more restricted the outflow O per unit of time, the faster the bay waters reach saturation. Inflow and outflow have a tendency to strike a balance, so that the terminal salinity in the embayment is a function of the ratio between inflow and outflow. From Equation (3-3), we can see that an outflow equaling evaporation losses, i.e., that amounts to half the inflow, needs only to contain twice the solute content of the inflow to balance the salt budget of the basin.

For gypsum saturation, the inflow needs to be at least 4.8 times the outflow; for halite saturation, it is close to 11 times the outflow. Once the basin actually starts precipitating halite, the inflow of normal seawater must be 8.3–10.7 times the outflow volume of brine in order to maintain the brine in the halite saturation field. A greater inflow leads to dilution of the brine. A lesser inflow leads to further concentration and eventually to the precipitation of salts of higher solubility. Potash salts start precipitating when the outflow and seepage combined amount to less than 11.56% of inflow or more than 88.44% of inflow evaporates. Every meter of a seawater column has to be reduced below a height of 115.6 mm.

2. The greater the volume of the basin, the longer the time required for each salinity increment. Based on calculations analogous to those presented here,

Valyaev (1970) concluded that the speed of salination and salt accumulation is inversely proportional to the depth of a basin.

3. Saturation is reached basinwide for gypsum. It is therefore possible to correlate individual gypsum, anhydrite, and halite beds over tens of kilometers in the same depression provided this depression has not been tectonically disturbed. However, this does not apply to gypsum beds of adjacent basins. The conspicuous gypsum bed in one basin may belong to the same paleontological substage as a similar gypsum bed in an adjacent marginal bay. They are certainly not contemporaneous in the strict sense of the word if the volumes of water in the two bays differ in size. Correlation between one bay and another must take this slight time differential into consideration. Bays that are in direct communication with each other without an intervening barrier, of course, reach saturation simultaneously.

4. Even in the small-scale environment of a laboratory experiment, the deeper the saturated brine, the slower the growth of crystals and the smaller their size. For that matter, the growth of crystals is also slower, and their size is smaller, the more permeable the substrate, another source of water loss (Arthurton, 1973). Rapid growth of large crystals is thus possible only in very shallow environments.

5. The longer the time required to bring a basin to saturation, the less probable it is that the rates of evaporation and/or subsidence remain constant. This makes saturation of very voluminous basins more difficult to achieve and thus less probable. Most thick evaporite sequences show evidence of synsedimentary rapid subsidence.

6. The long period of progressive concentration from the salinity of inflowing waters to the salinity of a gypsum-precipitating brine is usually marked by thick marl deposits of increasingly impoverished fauna and a reduction in the number of surviving species. It takes more time to reach saturation for gypsum than it does thereafter for each of the progressively more soluble salts. Prematurely terminated preevaporite sequences of marlstones are more common in the sedimentary record than actual evaporites.

7. *If the cross-sectional area of the entrance strait is too large, then the bay strikes a balance of water exchange even through a very shallow entrance strait,* i.e., the denominator in Equation (3-8) goes to zero, and time t to infinity. The greater the inflow velocity (itself a combined function of entrance cross-sectional area and water deficit or surplus of the bay), the more likely is the development of an intermediate, progressively thinning countercurrent, leaving deep waters stagnant (Dietrich *et al.*, 1975) (Fig. 3-26).

The Gulf of Cariaco, Venezuela, maintains such an equilibrium over its entrance sill for 10 months each year; then, after a short rainy season, it flushes out the accumulated dead water. Most fjords also have an inflow/outflow regime that leaves a volume of dead water behind the entrance sill, anaerobic and full of biogenic hydrogen sulfides, but neither freshening nor concentrating.

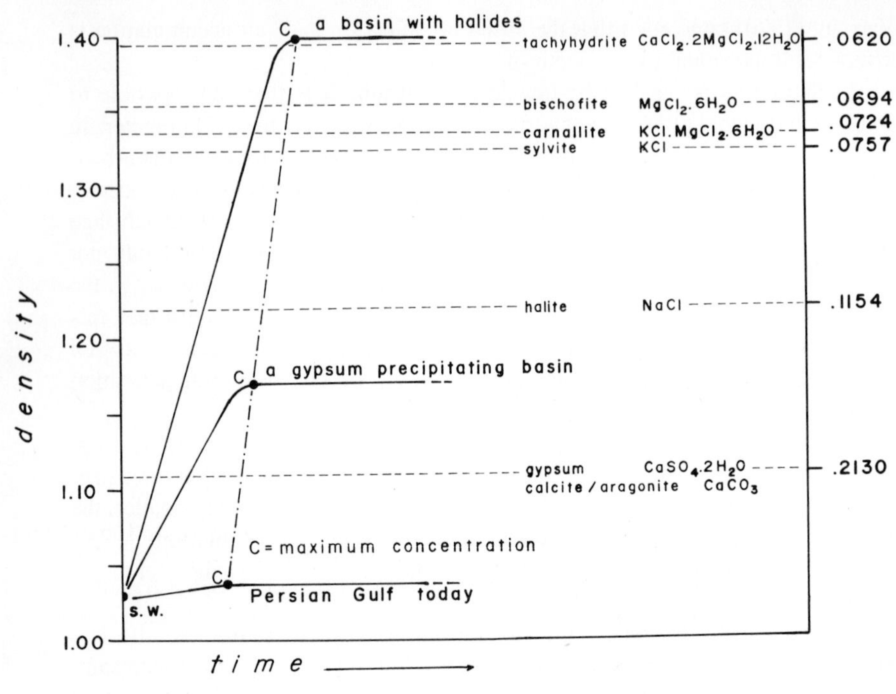

Fig. 3-26. Brine density related to inflow/outflow ratio. Equilibrium at some finite density is obtained unless basin volume, entrance configuration, or evaporation losses change.

8. Major evaporite basins, such as the Zechstein, Elk Point, West Texas, Hormuz, Upper Kama, East Siberian, and Fars basins cover an area on the order of $n \times 10^5$ km^2. At an annual rate of evaporation of 2–3 m, the waters required to replenish evaporation losses, outflow, and seepage total about 10^5 m^3/s. With a 10- to 100-m-deep surface inflow and an inflow velocity of 1–2 m/s, it can be shown that the width of the inflow must not exceed about 10 km. If more water enters the basin, there is no progressive concentration of the brine. The reason we do not have present-day examples of large evaporite basins is simply that none of the subtropical marginal seas have such a narrow inlet. The only sea with a narrow enough inlet is the Black Sea, but it does not have a water deficit.

9. *The salinity in the upper layer increases due to turbulent mixing.* The salinity of Black Sea waters entering the Mediterranean Sea increases 3 ppt in the 60-km-long Dardanelles but only 2 ppt in the 30-km-long but narrower Bosporus (Assaf and Hecht, 1974). Overall, however, the turbulent energy loss due to narrowing of the channel plays a minor role (Stommel and Farmer, 1952).

10. Within the entrance strait, there is a direct relationship between in-

flow/outflow regime and salinities. Assaf and Hecht (1974) found the empirical proportionality

$$(1 - s_U/s_L)^3 \propto \frac{q^2 L W_a}{D^4} \quad (3\text{-}9)$$

where s_U and s_L are the salinities of inflow and outflow, q represents the sum of the throughputs $u_1 \cdot A$ and $u_2 \cdot A$ (velocity times cross-sectional area of each flow direction), and L the length of the channel. W_a is then the average width equal to one-half $(W_U + W_L)$ measured at the surface and the bottom, and D is the thickness of the two layers, equaling the sum of the upper and lower thicknesses, D_U and D_L.

According to Knudsen's hydrographic equation (Dietrich *et al.*, 1975), the following relationship holds in a channel open to one side, in which two fluids flow in opposite directions:

$$E = I(1 - s_U/s_L) \quad (3\text{-}10)$$

where E is the water deficit and I the inflow per unit of time, and s_U and s_L are the inflow and outflow salinities. Of course, this relationship does not take into consideration hydrodynamic flow equations, friction, or drag along the interfaces.

Moreover, as was discussed in the previous chapter, the increasing salinity of the resident brine reduces its rate of evaporation; consequently, the amount of inflow and its drag on the outflow are also reduced. Increasing the brine volume by subsidence then becomes even more important in keeping inflow/outflow from reaching a balance. However, exact measurements are then necessary and are hard to make. In applying Knudsen's hydrographic equation to the Strait of Gibraltar, presently the entrance to a set of basins with a water deficit, we obtain from an inflow of 55,200 km^3/year, with a salinity of 36.25 ppt and an outflow salinity of 37.75 ppt, a water deficit of 881 mm/year over the area of 2,507,000 km^2. However, the total evaporative water loss reduced by rain and runoff is 965 mm/year (all figures from Dietrich *et al.*, 1975), indicating a 10% discrepancy in the data base. There are not enough long-term measurements of the throughput at Gibraltar in both directions to establish with certainty whether or not a balanced solute exchange exists.

OTHER CONCEPTUAL MODELS

The Model of Wind-Driven Upwelling

This model, proposed by Brongersma-Sanders (1971), requires prevailing winds to blow oceanward through the inlet into a bay; a salinity gradient inside

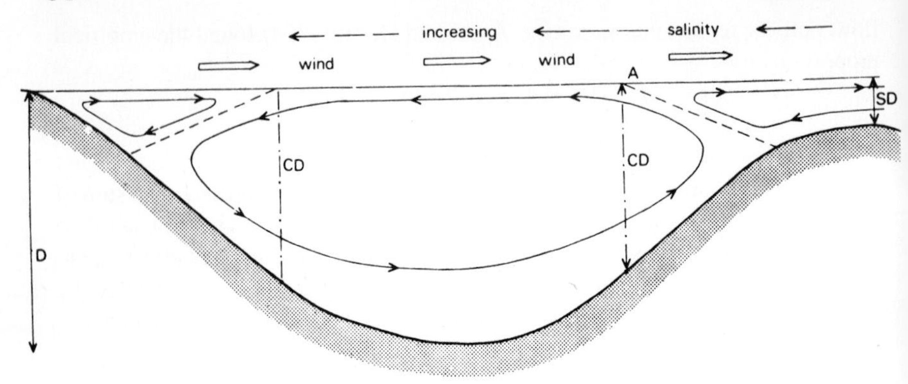

Fig. 3-27. The wind-driven model of upwelling. At a critical depth, the wind pulls up bottom brines and prevents further exchange with the ocean (after Brongersma-Sanders, 1971).

the basin eventually causes convectional upwelling of bottom brines at some critical depth near the entrance (Fig. 3-27). The upwelling brines inside the basin prevent inflowing bottom waters from entering from outside; these ocean waters are dragged up and then deflected into a surface outflow.

Internal circulation in the basin leads to concentration and eventual saturation. There is no known example of this type of circulation in present seas. Embayments with an evaporative water deficit always show an antiestuarine circulation of surface inflow and bottom outflow (Dietrich *et al.*, 1975). Although local winds in the Strait of Gibraltar often blow oceanward, the prevailing wind direction is into the Alboran Sea (Perthuisot, 1975), tempered by local onshore/offshore regimes. Winds also blow into the evaporite-precipitating Gulf of Kara Bogaz Gol east of the Caspian Sea; this was also true of winds blowing into the Permian Basin of West Texas. Moreover, with rising concentration, the brine becomes progressively more viscous. It is extremely doubtful whether winds can overturn a gravitationally stable system of dense brines subjected to normal marine inflow and induce mixing, let alone induce upwelling of viscous hypersaline brines into less dense surface waters.

The Open Shelf Model

A long reach of shallow water on an open shelf (Deicha, 1942b; Arkhangel'skaya and Grigor'yev, 1960) is insufficient to induce saturation of a brine to gypsum precipitation. The partially saturated brine would escape as undertow, as can be seen today in the Strait of Hormuz, the entrance to the Persian Gulf. Here waters gradually concentrate toward the shores and inlets, but the heavier brines are flushed and drain into outward bottom currents. Salt crusts form only where gulf waters have been trapped in barred bays and have been prevented from refluxing by seasonal tide effects. Braitsch (1971) concluded that the onset of an unhindered reflux of concentrated solutions would prevent a stable increase in salt concentration along a saturation shelf without a barrier.

Friedman (1980a) has resuscitated the ideas of Arkhangel'skaya and Grigor'yev (1960), previously also espoused by others, and has postulated that neither density stratification nor an entrance sill is necessary. He suggested that in an epeiric sea, in which the deep ocean is hundreds or thousands of kilometers away, a rate of evaporation exceeding that of precipitation by nearly 3 m indicates that much of the water cover must have vanished as it evaporated to dryness. Thus, evaporites accumulated in large stretches of the epeiric sea floor. Along with subsidence of such continental masses, sheets of seawater from the distant deep ocean provided periodic influx of water for the precipitation and accumulation of evaporites. Thus, thick strata of evaporites spread across epeiric seas. As stated by Friedman (1980a), epeiric sea depths of 30 m or less favor the deposition of widespread sheets of evaporites. We have to assume that these epeiric seas shared a common sea level with the world ocean. Evaporation does not suddenly dry out an open sea in contact with the oceans, and influx is continuous. The influx does not become periodic if no barrier exists between ocean and sea. Seasonal reversals of flow, as known from the Red Sea and from the Gulf of Cariaco, Venezuela, only serve to flush the embayment.

We can imagine an epeiric sea shelf with an inflowing current as envisaged by Arkhangel'skaya and Grigor'yev (1960) or Friedman (1980a) and calculate its water loss. Let us take a slice of the epeiric sea shelf that is 1 km wide and 1000 km long (Fig. 3-28). At an assumed evaporation rate of 3 m/year the total evaporative water loss is 95 m^3/s. For a water column 30 m deep, an inflow velocity of 3.2 mm/s or 11.4 m/hr would suffice to replenish the evaporative water loss. This flow is too slow to impart enough friction to stall the undertow, and it would not allow the brine to saturate. Every ocean current known today is faster than that. We can compare this inflow with a lacustrine situation: The

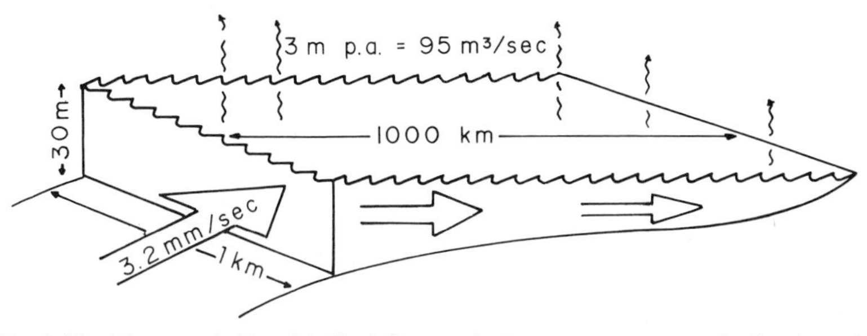

Fig. 3-28. The open shelf model. The inflow required to counteract evaporation losses would be infinitesimal.

current velocity at a depth of 20 m in the small Lake Vanda, Antarctica, is more than three times that of this required current velocity. Moreover, complete drying out would produce an isochemical precipitate, i.e., one in which the compounds are in the same proportions as in the solute. Such a deposit is not known from any ancient marine evaporite deposit.

The "Deep Basin" Model

In this model, a stationary body of water of oceanic depths (Fig. 3-29) stagnates, whereas surficial circulation eventually saturates. A body of dead anoxic water fills the deep and initially creates a sapropel facies in the sediments. Most evaporite basins do not have a euxinic or sapropel facies at their base; some have it intercalated in the evaporite sequence as an interval of freshening. This is not to say, however, that rapidly subsiding basins may not have contained local troughs reaching well below the photic zone at the onset of evaporite precipitation. Eventually, hopper crystals of various precipitates start dropping (Schmalz, 1969). However, there is no way envisaged for hopper crystals of gypsum, halite, and other minerals to separate laterally, as their formation is deemed to be strictly a function of the concentration of the whole stagnant body of water in the basin.

Support for this model was drawn from an assumption that the Zechstein evaporites are 1200 m thick (Schmalz, 1969). However Fiege (1939) had earlier calculated the median cumulative thickness of all five cyclothems of Zechstein rocks to be 840 m. Lotze (1957, p. 222) allocated to the same sequence a

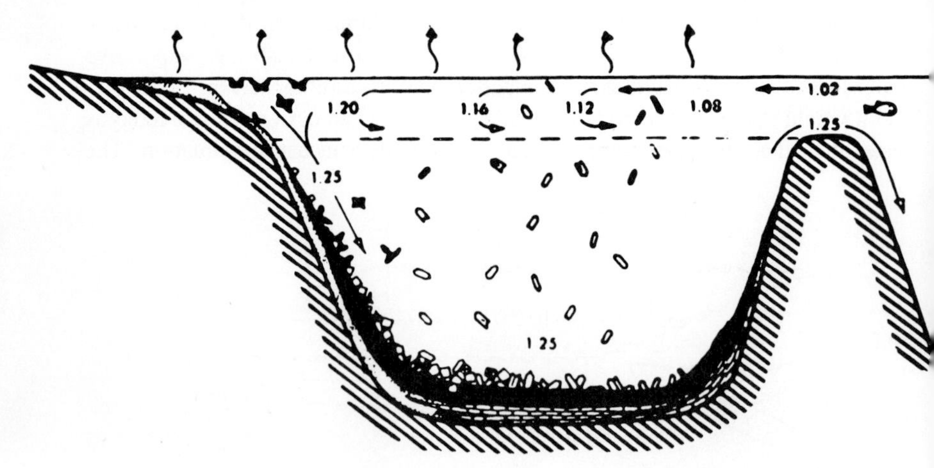

Fig. 3-29. The deep evaporite basin model. Halite and gypsum precipitate at the surface and are preserved at depth. Sapropel facies with anaerobic benthos underlies evaporites throughout the basin (after Schmalz, 1969).

maximum of 500 m, whereas Baar (1952) gave a cumulative "normal" thickness of 360.4 m, including even preevaporitic Kupferschiefer beds and a basal conglomerate. Moreover, there is no evidence in the carbonates and clastics separating the individual evaporite cyclothems that they might not have covered a reasonably flat and shallow sea floor. If all five evaporite sequences were merely filling a preexisting basin, the beds separating each of the evaporite cyclothems should show a very significant basinward slope and facies changes from the shore to below the wave base, and ultimately to below the limits of the photic zone. This has not been documented.

A forerunner of this model is the concept of Bischof (1864) that the Zechstein sea must have been 18,600 m deep, based on the potential compound thickness of all evaporite-related sediments. Many of the older estimates of the depth of evaporite basins tended to be on the high side, underestimating concurrent rates of subsidence. Schmalz (1969) also postulated 600 m as the depth of the Midland Basin of Texas and 330 m for the Elk Point Basin of Alberta. Adams (1936) had suggested a depth of over 700 m for the Permian evaporite basin of West Texas, probably in the range of 1800–2400 m.

The model may explain a single transgressive sequence of rocks, but it fails to account for the lateral facies changes in a basin, the regressive phases, and the common multiplicity of cyclothems. There is no evaporite basin that went to halite saturation or beyond without producing repeated cycles of vertical and horizontal zonation of evaporite minerals. Kuehn (1954) has examined the bromide curves of the individual units of the Zechstein evaporites. Assuming no leaching or epigenetic redistribution of the original bromide content, he arrived at different depths for the individual units, effectively supporting synsedimentary subsidence. At certain times, the subsidence exceeded precipitation; at other times, it lagged behind. A very serious objection to this model is the apparent high ratio between areas of the water surface and of halite precipitation, which demands an unreasonably high annual rate of evaporation to bring the brine to saturation. The required surface area of a basin will be treated in greater detail in Chapter 15.

The "Deep Dry Basin" Model

In this model, a body of water of oceanic depths dries out, but it is periodically replenished. The model was supposed to apply to the Upper Miocene Mediterranean Sea, the Devonian Elk Point Basin of western Canada, and the Silurian Michigan Basin (Hsu, 1972). The basic assumption of this model is that the present configuration of the sea floor is inherited from preevaporite times—in the specific case of the Mediterranean Sea, from pre-Messinian (at least mid-Miocene, if not Oligocene) times. The individual basins of the Mediterranean Sea were periodically inundated by Atlantic waters. The Strait of Gibraltar operated as a grand gate valve, closing whenever the basin had filled up with water and opening whenever the basin had dried out. The evaporites of Sicily thus represented an uplifted part of the Balearic Sea floor (Hsu *et al.*, 1973a). The assumption of repeated floodings was necessary, since the evaporation of a quantity of water equal to the volume of the Mediterranean Sea could not account for the volume of evaporites drilled into or delineated on seismic sections. The brine depth in the Upper Miocene Mediterranean Sea was deemed to have been at a high water level in excess of 2000 m, perhaps as much as 4000 m below the level of the open ocean (Hsu *et al.*, 1973a,b). The Balearic Basin desiccated and filled up again numerous times, perhaps 8–10 times (Hsu, 1973), or 11 times (Hsu *et al.*, 1978), by a repeated collapse of the Strait of Gibraltar. That tectonic movements raising and lowering the entrance sill control the flow into an evaporite basin had first been suggested by Fiege (1939). This model represents a rejuvenation of Wilfahrth's (1933) giant flood hypothesis (Fig. 3-30), which requires the evaporating surface of the salt sea to lie below sea level, separated from the ocean by a land bridge that is inundated periodically by megatides. In turn, such a brine surface lower than sea level postulates the existence of an old depression, completely blocked off, with a depth at least equal to the salt thickness (Braitsch, 1971). Fiege (1939) objected to any hypothesis involving giant floods on the grounds that such sudden floods should also have occurred in nonevaporitic environments, preceding or following the evaporite deposition. No sedimentological record of such floods can be found. Furthermore, this model does not explain the occurrence of commercial potash deposits sandwiched between under- and overlying halites in Sicily that are juxtaposed to gypsum sequences. Atlantic waters, entering and eventually concentrating to precipitate overlying halites, would first have dissolved any potash salts left behind when the preceding inundation had dried out. They would not have covered precipitated potash salts with a new layer of halite.

The presence of brackish-water faunas, initially reported only from eastern Mediterranean sites, seemed to necessitate the postulate of a connection to the Paratethys, which at that time represented a freshwater reservoir. Cita (1973, Fig. 3) and Cita *et al.* (1978, Fig. 12) have suggested that the Atlantic sea level was higher than the Paratethys level from the late Miocene to the earliest Pliocene (cf. Fig. 3-31), whereas Hsu *et al.* (1978) assumed equal levels. In effect, the model represents an oversized endorheic playa basin. It was conceived to account for the occurrence of undoubtedly shallow-water features now situated beneath the Balearic Sea floor. There are stromatolitic growths that have become totally gypsified (Nesteroff, 1973a,b; Rouchy and Monty, 1981), algal mats that call for deposition in the photic zone, desiccation cracks and other evidence of subaerial exposure (Schreiber, 1973) or wave activity, evidence from bromide studies (Kuehn and Hsu, 1974), and other shallow-water features (Ruggieri and Sproveri, 1978).

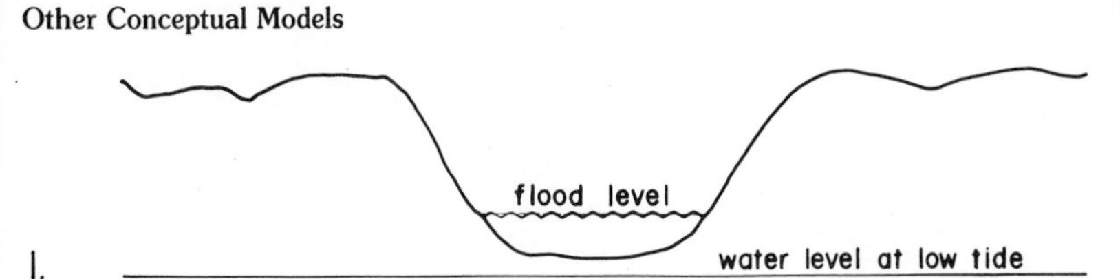

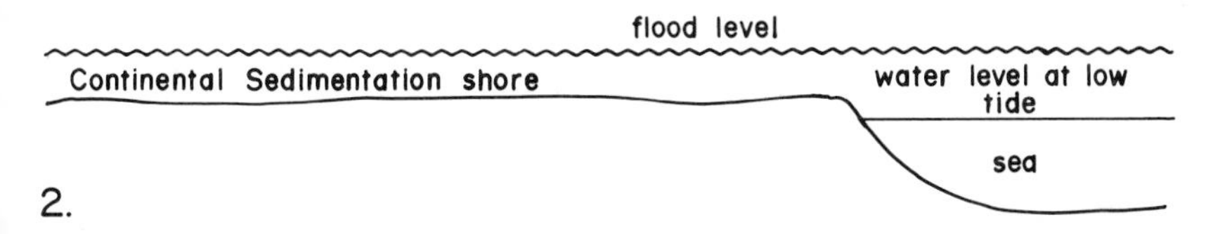

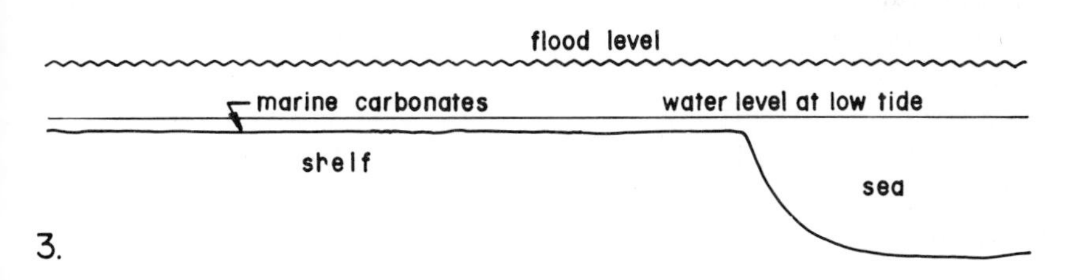

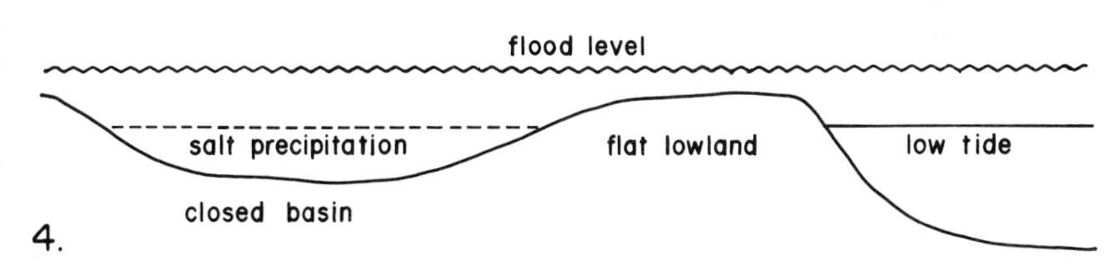

Fig. 3-30. The formation of different facies under the influence of megatides [Wilfarth's giant flood hypothesis at interpreted by Fulda (1937)]. 1. Elongated valleys are inundated in the flood stage and bared at low tide. The breakers wear down the valley walls and distribute the resulting debris. Exposure to atmospheric oxygen at low tide produces red clastics. 2. Inundation of lowlands: Only fine-grained material is spread out. Marine and continental influences alternate. 3. Inundation of a shallow shelf: The shelf is covered by seawater even at low tide, and a carbonate bank forms. 4. Inundation of a deep depression, separated from the ocean by a lowland plain. The plain is flooded at high tide. At low tide it dries out, but the water in the depression is trapped and evaporates in a hot climate.

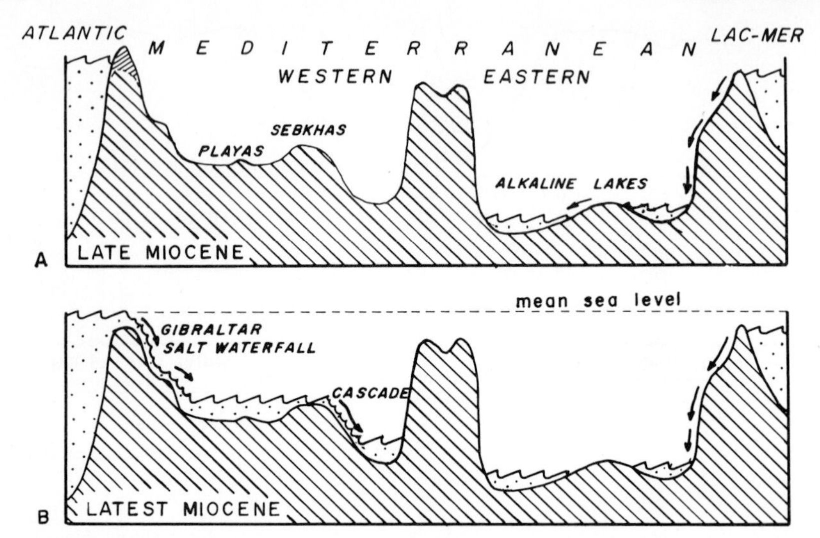

Fig. 3-31. Model of the deep dry basin. A. The dry stage. Playas and sebkhas in the western and alkali lakes in the eastern Mediterranean area. B. The refilling stage, with a waterfall at Gibraltar in the west and inflow from the Paratethys (or Lac-Mer) in the east (after Cita, 1973).

No evidence is available in any of the basins to which the model was applied that the wall rock is significantly older than the precipitates in the basin, that wall rock members do not interfinger with evaporites, or that correlation is not possible between boreholes through wall rock and boreholes into the salt sequence. In basins where adequate grids of radioactivity logs exist, correlations carry from shore to shore without such old depressions in evidence. For that reason, the application of the model to the Silurian Michigan and Devonian Elk Point basins (Hsu, 1972) is not feasible; in neither case do the evaporites fill a preexisting depression whose walls are older than the evaporites. In the Elk Point Basin there is enough well control to correlate marker beds across the basin into onshore deposits, using hundreds if not thousands of well logs, and it is easy to demonstrate the synchroneity of evaporites with circum-evaporite sedimentation.

If Atlantic waters were shut out, Paratethys waters would still have filled any depression to the level of the Paratethys Sea, regardless of the depth of the depression. Conversely, a breakdown of the sill at Gibraltar would have led to a substantial inundation of the Paratethys by saltwater encroachment. Free access of waters from the Atlantic Ocean to the Paratethys would have led in time to normal marine salinity in the Paratethys, even without an elevation difference between the two water levels. Instead, the Paratethys remained a brackish-to-fresh body of water throughout the Messinian stage. Thus, there could not have been any meeting of Atlantic and Paratethys waters. Dolomitic marl intercalations at Deep Sea Drilling Project, Leg XIII, site 124 in the center of the Balearic Sea contain brackish-water planktonic faunas covered by several tens of meters of massive anhydrite. They also underlie black (sapropelic?) gypsum at site 374 in the Ionian Sea (Schrader and Gersonde, 1978). The same fauna is found in Messinian outcrops in Tuscany, Sicily, Algeria, and Spain (Hsu, 1973). Fuchs (1874) was the first to point to eastern European affinities of faunas in Miocene outcrops near Syracuse, Sicily. This fauna could easily have been swept in from surrounding estuaries; we know very little about Messinian drainage now covered by Saharan sands.

Others have also invoked deep, dry depressions as receptacles of marine evaporites. A 670- to 730-m-deep lagoon was envisaged for the Delaware Basin, a relatively small basin in the fifth cyclothem of Permian salt deposition in West Texas, whereas all seven cycles of halite precipitation were deposited from continuously inflowing brines no more than 1 m deep (Adams, 1963). The 2400 m of Permian evaporite deposits under the Caspian Lowlands were deemed to have been deposited in a preexisting deep basin (Oshakpayev, 1977), although several authors had commented on the shallow-water features preserved in these evaporites. Oshakpayev (1977) sought to explain both the multiplicity of potash-, magnesium-, and borate-rich horizons, and the source of the halites and other evaporite minerals, by assuming hydrothermal brines seeping from great depths through this deep-sea floor. Only in this respect does his explanation differ from the assumption of repetitive sluice gate activity postulated for an entrance sill in the Strait of Gibraltar [as Hsu (1973) aptly described this event: ‘‘the flood-gate . . . swung open and shut repeatedly . . . and then was irreparably crushed’’].

In assuming that Balearic, Sicilian, and Levantine evaporites were deposited several kilometers below the present sea level, the model ignores the presence of coeval gypsum and anhydrite beds now uplifted to 1500 m in central Italy, to 2800 m in the Taurus Mountains of eastern Turkey (Blumenthal, 1955), or to 3000 m in Tunisia (Burollet, 1971), as well as onshore gypsum deposits in southern France, Calabria, the Ionian Islands, Albania, and the Mediterranean coast of Egypt and Israel (Sonnenfeld, 1975), and halite in several areas of west-central Italy (Roda, 1970; Martina, 1974) (Fig. 3-32). The Miocene evaporite beds of Romania, slightly older but derived from a communicating body of water, are also found 4 km below sea level in the Transylvanian Basin and 1300 m above sea level on its rim (Pauca, 1968). Moreover, the presence of salt domes under the western Balearic and eastern Levantine seas suggests that the evaporites have been subjected to severe postdepositional tectonic stresses. The overburden is still not very thick, and the salt has not been mobilized in the thickest sections with the greatest depth of overlying water masses. Some salt has, no doubt, been redissolved in Plio-Quaternary time, giving rise to the present distribution as shown in Fig. 3-33.

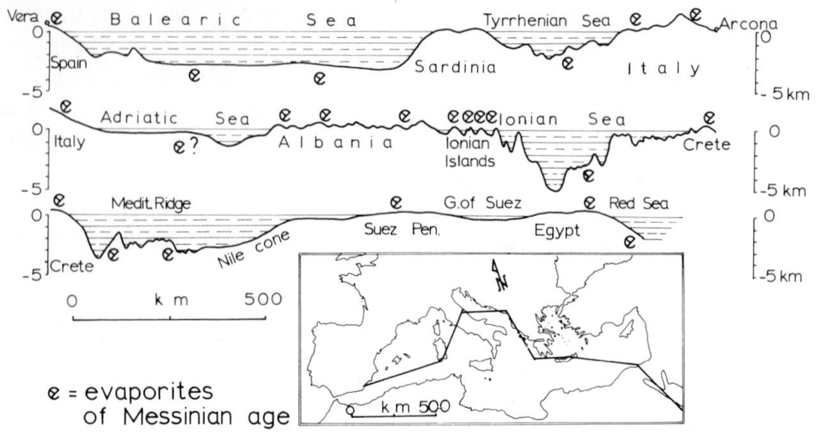

Fig. 3-32. Continuous west–east topographic profiles throughout the Mediterranean region. Messinian evaporites now beneath the sea floor are indicated below the line of section. The profile crosses a depression in the Mediterranean Ridge (after Sonnenfeld, 1975 *J. Geol.* **83,** 287–311. Copyright 1975 University of Chicago).

A growing number of local studies indicates that the individual basins of the Mediterranean Sea collapsed in Plio-Quaternary times. There are Plio-Quaternary normal faults with throws of up to 2000 m in the Tyrrhenian Sea, nearly always rejuvenating preexisting tectonic trends (Fabbri and Curzi, 1979), and the Tyrrhenian Sea is a product of Pliocene subsidence (Selli and Fabbri, 1971; Fabbri and Selli, 1973). The Tyrrhenian margin underwent Plio-Quaternary

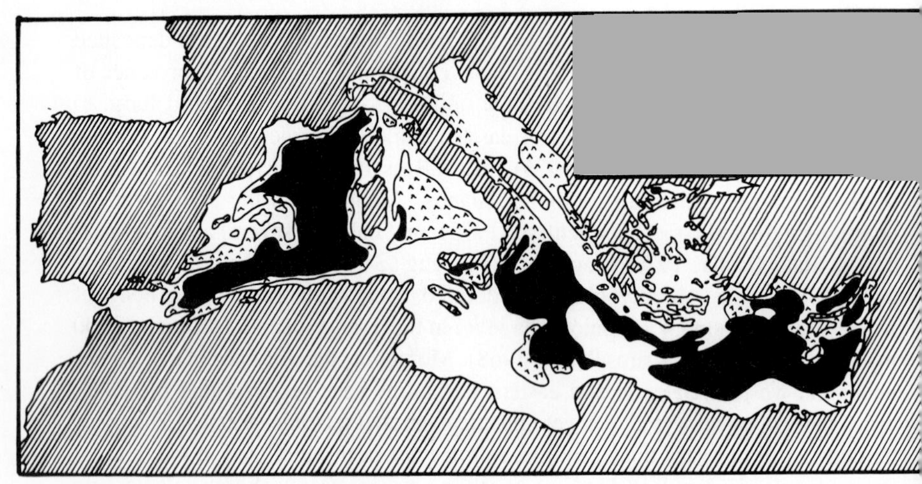

Fig. 3-33. Distribution of Messinian evaporites in the Mediterranean region, as known at present. Neither the current Mediterranean sea floor morphology nor the coastline configuration resembles that of the Messinian sea (after Rouchy, 1980).

block faulting, submergence, tilting, and finally, inversion of the tilt movements, resulting in canyon rejuvenation (Wezel *et al.*, 1981). Winnock (1981) indicated vertical throws of up to 3000 m in post-Messinian times on the sides of the Pelagian block south of the Caltanissetta Basin of Sicily. The Ionian Sea collapsed in early Quaternary times (Fabricius and Hieke, 1977), and later the Aegean Sea (Cvijić, 1908) and the Levantine Sea (Sonnenfeld, 1975; Neev *et al.*, 1976; Gvirtzman and Buchbinder, 1978) collapsed also. Portions of the Levantine Sea are very young indeed, judging by the occurrence of mid-Pleistocene Nile River sediments on the north slope of the Mediterranean Ridge (Hsu *et al.*, 1973a). They are now separated from the delta by the Nile abyssal plain. Overall, rapid subsidence continued longer in the eastern Mediterranean Sea, since evaporite thicknesses under abyssal water depths are greater than anywhere else in the Mediterranean region. The onset of this subsidence can probably be placed in the Upper Miocene interval of rapid chemical precipitation. Van der Zwaan (1982) collected data on transgressive onlap in Messinian and pre-Messinian (Tortonian) sediments, together with an absence of cold-water faunas that would have been present in deep basins.

Deeply incised submarine canyons are cited as support for the existence of a deep, dry depression. Such submarine canyons are not restricted to the interior of the Mediterranean Sea, but occur also on the shelves of open oceans, such as on the west coast of Africa and along eastern North America, Indonesia, and elsewhere. Typical is the Ceduna Terrace, a bathymetric terrace, 500 km long by 200 km wide, off South Australia. Its surface is cut by numerous submarine valleys and canyons now submerged 500–2500 m (Fraser and Tilbury, 1979). Moreover, the distribution of canyons does not follow the present outline of Mediterranean evaporites, but instead follows the steps in the slope of the individual depressions in this sea (Mikhailov, 1965; Sonnenfeld, 1975). In France, canyons on the Mediterranean coast (Var, Durance, Rhône rivers) are evidently related to vertical movements of the continent, since such canyons are also known from the Atlantic coast (e.g., near Cape Finisterre or the Loire River mouth), possibly caused by changes in the buoyancy of the rigid crust.

Running water is needed to cut canyons. Canyons in the Mediterranean Sea floor have thus at least three major causes:

1. Erosional channels of increased submarine groundwater discharge in humid periods (cf. Johnson, 1939). Most formation waters are significantly more dense than seawater.
2. Preexisting depressions (river beds, mud flow channels, etc.) foresteepened by the rapid subsidence of basin flanks. The canyons cut into marginal slopes of deeper ledges within the Mediterranean Sea are possibly in this category, since these canyons do not extend to the present shelf margin.
3. Uplift and later subsidence of the continental crust, in part initially as a

lateral compensation and later as a more moderate expression of the rapid basin subsidence and consequent drowning or burial of river channels.

The canyons in the Mediterranean Sea floor are not of the same age. A canyon buried in southern Israel (Gvirtzman, 1969; Sonnenfeld, 1975) is a continental river channel of early Miocene age or older, with a thin tongue of Messinian evaporites in its mouth. Its burial depth can be partly accounted for by the isostatic loading of overlying rocks. Nesteroff (1973b) found a Miocene channel parallel to the present French coast, formed when there could not have been any steep slope at right angles to its course. He thought that the numerous canyons along the Riviera represented a rejuvenated Middle Miocene system. The same age may be assigned to some block faulted grabens and canyons in Egypt and along the Lebanese coast. The floor of a valley in the East Mediterranean Ridge exposed Middle Miocene marls throughout Pliocene times (Zemmels and Cook, 1973). It is important that no canyon has been found in the Alboran Sea that shows erosional evidence of waters cascading repeatedly from the Gibraltar sill into a deep, dry Balearic Sea; nor is there such a canyon at the mouth of any other pre-Messinian passage to the Atlantic Ocean. Even major seepage through the dry sill wall would have left behind an erosional signature.

A barrier at Gibraltar formed a land bridge for migrating mammals (de Bruijn, 1973; Azzaroli and Guazzone, 1979). A significant elevational difference between this barrier and a dry Mediterranean depression could have been maintained only if the uppermost 3000 m of the barrier contained no fractured or permeable beds. Otherwise, Atlantic waters would have seeped in as seawater seeps into many depressions along semiarid or arid coasts today. Furthermore, the hydrostatic head of the groundwater level in the Alps, Dinarides, and Taurides would have produced substantial artesian flows.

The Messinian gypsum formations show no evidence of large-scale dehydration followed by rehydration, as an analysis of the salt budget shows that Mediterranean evaporites were precipitated in a brine and were not the product of repeated evaporation to total desiccation (Debenedetti, 1976). The bromide content of Messinian evaporites in the Mediterranean Sea suggests deposition near sea level (Kuehn and Hsu, 1974). The occurrence of gypsified algal laminites (Nesteroff, 1973b) supports this conclusion. If the present Balearic Sea were to be deprived of all inflow and were to dry out, gypsum would start precipitating in 835 m of water, provided that the bottom waters remained well oxygenated. This is nearly 1800 m below the present sea level and thus would leave unexplained the presence of Messinian gypsum beds encountered on stable blocks of the Pelagian Sea off Tunisia in 300–500 m of water. These blocks may have subsided slightly since the Messinian, but they certainly show no evidence of uplift. Halite would follow when the water level had dropped to 265 m, i.e., well below the photic zone. The land bridge at a closed Gibraltar Strait (12,800 m across at its narrowest point) is not entirely composed of impervious rocks. At an average permeability of 1 μm^2, a very low average permeability for clean sandstones, the walls of this land bridge would generate an annual throughput of 3.15×10^5 m^3/m^2 of Atlantic waters at a velocity of 1 cm/s. Aquifers with a combined thickness of 600 m would suffice to counteract net evaporation losses over the present Mediterranean Sea, which are on the order of 2200–2400 km^3/year. The Gibraltar sill is more than twice that height, measured from the bottom of the Alboran Sea. Typical hydrocarbon-bearing sandstone reservoirs in Texas have permeabilities up to 4.6 μm^2; friable, uncompacted sandstones have one to two orders of magnitude more. The average permeability of a compacted but uncemented sandstone would drop the required aquifer thickness to under 150 m. Continuous percolation of seawater through a land bridge into an evaporite basin has been observed on satellite images of Laguna Mormona in Baja California (Vonder Haar, 1976).

However, runoff and rain then did not cease, as evidenced by a large influx of meteoric waters into both eastern and western Mediterranean basins (Longinelli and Ricchiuto, 1979). Consequently, a balance between inflow of continental waters and drawndown by evaporation would have been reached well before halite saturation (Debenedetti, 1982). With declining water activities in a concentrating brine and rapidly declining vapor pressures with a concurrent increase in temperature, it is doubtful whether the brine can reach saturation for halite under increased atmospheric pressure in a very deep depression. Since any marginal sea decreases its surface area substantially whenever sea level is lowered, the evaporative water loss decreases more rapidly than the volume because the concentration rises. Salinas maximize the surface area by leading their brines through a series of shallow condenser pans before allowing them to enter precipitating pans. Broad shelves and shoals have the same function in natural basins. A sea decreases its brine surface and thus its total water loss per unit of time when shelf areas begin to dry out and sea level drops along the inclined continental slope. The Messinian evaporites are overlain by the latest Miocene brackish-to-freshwater deposits in coastal Spain, Italy, Tunisia, and Cyprus. There is no way in which floodings by seawater can produce freshwater deposits; only a brackish-to-freshwater source can produce such a freshening. These latter deposits are the best evidence that the supply of continental waters did not cease when seawater was choked off.

However, no terminal Miocene deposits were found on top of Messinian evaporites in the Balearic Sea (Nesteroff, 1973b), suggesting that this part of the evaporite basin system had dried up and turned into a huge salt flat. Desiccation cracks and other evidence of subaerial exposure found in Balearic cores (Schreiber, 1973) support the contention that the western part of the Mediterranean Sea dried up prior to the Pliocene transgression. In southern Spain, Orti Cabo and Shearman (1977) reported a southwestern migration of Upper Miocene evaporite facies toward the mouth of the local subbasin.

The various intratidal and supratidal facies recognized by Schreiber (1973), Schreiber and Decima (1976), Schreiber and Friedman (1976), and Schreiber *et al.* (1976) suggest flood and ebb levels. Tides occur only in a marginal bay connected to the world ocean. Even large endorheic basins do not produce a tidal regime (Debenedetti, 1982).

SUMMARY OF OCEANOGRAPHIC CONDITIONS

A deficit in the water budget of a marginal sea induces an antiestuarine circulation that moves inside the sea in a counterclockwise direction in the Northern Hemisphere. All the brine is flushed out again unless an entrance bar coupled with a narrow indraft prevents a complete water exchange. The inflow sets up a density stratification of oxygenated open marine waters overlying anoxic, more concentrated brines; the interface becomes a trap for solar radiation. The bottom outflow is depleted of oxygen and nutrients and thus prevents the immigration of bottom dwellers; part of the bottom brine displaces formation fluids, since a dense brine near the earth's surface is gravitationally unstable. Eventually, the open outflow of concentrated brines ceases, and brine loss continues only through seepage. Reefs that extend above the interface between inflow and outflow can continue to grow, as only their crown is alive and there is a rich plankton supplied by the inflow; but nekton and bottom dwellers first become endemic in the hypersaline brine and then die out.

4

Precipitating Brines

Both surface and subsurface brines are involved in evaporite precipitation. Surface brines are mainly derived from seawater or from groundwater springs. Subsurface brines are derived from surface brines soaking into the subsurface, migrating through aquifers, and reacting with the country rock. Concentration is induced by evaporation or membrane filtration, but continued sediment–brine interaction then alters the solute. Dissolution, precipitation, ion exchange, biological activity, and brine mixing are involved (Lerman, 1970). As the brine concentrates, its physical properties change, first slowly and later more rapidly once precipitation of the least soluble salts has commenced. Its chemical composition also changes, partly through absorption and partly through biological activity. Changing salinity and temperature gradients and different biological activities in oxygenated surface waters and anoxic bottom brines create anomalous pH regimes. Consequently, an isochemical and isothermal evaporation of a brine in oxygenated laboratory environments does not yield results comparable to those of natural conditions.

Oceanographic literature reports brine saturation in salinity [weight percent (wt %) of solute, often reported as parts per thousand (ppt)], chlorinity (weight of halogen recalculated as equivalent chloride per unit weight of brine, also often reported as ppt), or chlorosity (chlorinity per liter). Once the crystallization point of halite is reached, the chloride content of the solution stays approximately constant. Consequently, it is best to differentiate concentrating hypersaline brines and bitterns by density, adjusted to standard temperature (20°C). The latter is also the easiest to measure in the field with reasonable accuracy.

We can isolate four sources of brines that precipitate evaporite rocks:

1. The water of the oceans.
2. Old salt bodies that have supplied salt, e.g., to Lake El'ton in the Caspian depression or the Great Salt Lake, Utah.
3. Leaching of older nonevaporitic rocks of marine origin and redeposition of groundwater solutes in endorheic basins.
4. Chlorine of volcanic emanations that reacts with sodium weathered out of other rocks. The amount of juvenile and remobilized chlorine added annually by volcanic activity to the oceans is not known, but is likely very small.

Three factors determine the nature of precipitating salts: temperature, salinity, and interference by other solutes. Both evaporating and freezing solutions precipitate individual compounds sequentially. However, mere saturation of a brine does not lead to precipitation. Precipitation commences only when a certain degree of supersaturation is reached and then starts over a wide area. The required degree of supersaturation appears to be temperature dependent and is lower with rising pH or dropping Eh values.

Once nucleation has started, precipitation proceeds very rapidly until a temporary equilibrium is reestablished with the brine concentration. Further extraction of water vapor again has to bring about supersaturation before precipitation resumes. This process can lead to fine laminations, as found, e.g., in ice cubes. The brine stripped of most of its sodium chloride content is referred to as "bittern" because of the bitter taste of a magnesium chloride solution. The residual bittern left to soak into the precipitate is sometimes referred to as "mother liquor."

PHYSICAL PROPERTIES OF BRINES

In the process of losing water, the seawater influx gradually turns into a hypersaline brine with reduced volume (Fig. 4-1) and somewhat altered physical and chemical properties. The changing activity of water during evaporation of the concentrating undersaturated brine leads to an inversion of the isotope enrichment (Gonfiantini, 1965), superimposed on bacterial selection of isotopes.

Eutonic and Eutectic Points

As the concentration of a brine by evaporation is increased, a point is finally reached at which no further increase in concentration occurs and the solution dries out. All the salts present are then precipitated simultaneously, with no change in composition (Table 4-1). This is the eutonic point, which is reached in nature at a salinity of about 35–40% and an air temperature of 32°C. Whereas evaporating solutions progressively concentrate, freezing solutions are gradually

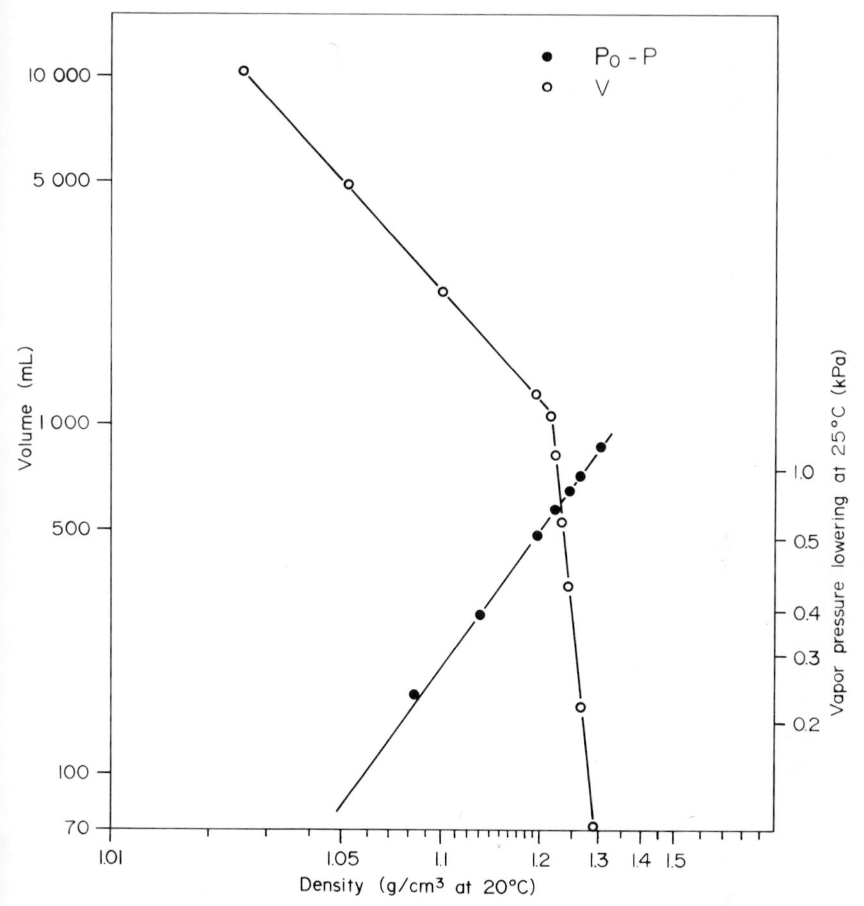

Fig. 4-1. Relation of the volume and vapor pressure lowering of seawater concentrates to the density of a solution. Open circles refer to volume, full circles to vapor pressure lowering in kPa. [Reprinted with permission from Rothbaum (1958). Vapor pressure of sea water concentrates. *Ind. Eng. Chem., Chem. Eng. Data, Ser.* **3,** 50–52. Copyright 1958 American Chemical Society.]

diluted by precipitation of salts. This is due to the lower solubility of most salts in cooler brines. Eventually, however, all the salts and the water go into a solid phase simultaneously, without changing their composition. This is the eutectic point, which is reached in natural brines between −21° and −35°C (Strakhov, 1962). In the polar environment, potassium and magnesium chloride begin to crystallize out at about −36°C (Burton, 1981), and $CaCl_2 \cdot 6H_2O$ forms at −54°C (Dort and Dort, 1970). It follows that all the compositional changes of brines are restricted to the temperature range of about −55° to +35°C, a range of 90°C.

TABLE 4-1

Decrease in the Amount of Water in the Brine with Increasing Concentration of Seawater

Degree of saturation	Water evaporated (kg/m^3)	Water remaining in the brine (%)
Seawater	0	100.00
Calcite	707.85	28.44
Gypsum	795.09	19.62
Halite	897.08	9.31
Sylvite	950.49	3.91
Carnallite	958.31	3.12
Bischofite	961.97	2.75
Tachyhydrite	970.67	1.87
Eutonic point	989.17	0.00

With increasing temperature, the precipitates change toward less hydrated forms and favor the formation of double and triple salts.

Increasing salinity of the brine raises its hygroscopy and thus *lowers the temperature range of the stability of anhydrous minerals* and partially hydrated compounds. Complexing occurs with progressively fewer water molecules as the temperature of the brine rises. Hydration also decreases with increasing molecular volume (Van Ruyen, 1953). Initially, seawater contains 92 water molecules per ion. This is reduced to 4.5 water molecules per ion before halite starts precipitating, and to 3 molecules per ion at the eutonic point. Salt precipitation is thus the result of a deficiency of free water molecules to act as a solvent (Kostenko, 1982). The energy of hydration increases with increasing molecular weight, but is about three to four times as high for alkaline earth cations as for alkali metal cations. The halide anions require the same amount of energy for hydration as alkali cations (Aleksandrovich and Pavlyuchenko, 1966). Sodium, potassium, rubidium, cesium, and magnesium chlorides or bromides complex with slightly more than 10 water molecules for each monovalent ion and 15 for each bivalent ion at 0–20°C (Euken and Herzberg, 1950; Darmois, 1957). The 10–15 molecules of water occupy two layers in hydrated ions (Eigen and Wicke, 1951). The chlorides also complex with each other. $KCl \cdot MgCl_2$ complexes form at ambient temperature (Ermolenko and Levitman, 1950); $KCl \cdot CaCl_2$ double salts form only above 37.7°C (Assarsson, 1950). In concentrated solutions, magnesium chloride can complex with sodium chloride into Na_2MgCl_4 or with ammonium chloride into $(NH_4)_2MgCl_4$ or with zinc chloride into $ZnMgCl_4$. There is also a faint complexing with iron chlorides, but none with aluminum, potassium, calcium, or strontium chlorides (Titov, 1949).

As the concentration increases, the ions tend to interfere more and more with the degree of hydration of neighboring ions. This affects the migration of water molecules from the surface layer into deeper parts of the brine, and vice versa. The surface tension then becomes indirectly a function of the concentration (Yashkichev, 1963). Interference of other solutes can greatly alter the stability fields. For instance, increasing concentrations of magnesium ions lower the solubility of gypsum and halite or favor the precipitation of aragonite over calcite. Even undersaturated solutions can mix to yield a supersaturated solution of one of the components (Wigley and Plummer, 1976). Centrifuging a sodium chloride solution can yield a concentrate with a density of 1.196. Similarly, gravity acting over long periods of time should be able to accumulate concentrated complexed solutions at the bay bottom. If two basins of different morphology concentrate their heavier solutes, the one with the greater surface area will produce more concentrates that would eventually spill over any barrier to a smaller, deeper basin. It has not been investigated whether gravitational effects on concentrated complexed solutions can lead to precipitation of evaporite minerals.

The mixing of solutions with different compositions, molalities, and activities of individual ions in the mixture are nonlinear functions of their end-member values. Partial pressure, mean molal activity coefficients, and activities of solute have been calculated for supersaturated solutions of alkaline chlorides at 25°C by Hidalgo and Orr (1968). As the salinity increases, the specific volume of the brine decreases, and so does its compressibility. The coefficient of thermal expansion becomes smaller, but the boiling point rises and the rate of evaporation decreases.

The specific heat (*Cp*) decreases with rising salinity (*s*, in percent) of natural brines according to the following formula (Ponizovskiy *et al.*, 1953):

$$Cp = 1 - 0.01307s \tag{4-1}$$

The same amount of solar irradiation thus produces higher temperatures in more concentrated brines. The molar heat of chloride solutions has a maximum at 70°C and is directly proportional to the square root of the concentration (Eigen and Wicke, 1951), but the heat of solution of sodium chloride changes signs at about 60°C; precipitation above that temperature then releases further amounts of heat (Liu and Lindsay, 1972).

The surface tension σ of natural brines increases with salinity according to the following formula (Danil'chenko *et al.*, 1953):

$$\sigma = 3.815s + 726 \text{ J/cm}^2 \tag{4-2}$$

Surface active substances can reduce the surface tension by about 3×10^{-7} J/cm^2.

The viscosity is related to the hydration potential. For instance, magnesium sulfate is heavily hydrated in aqueous solution and profoundly alters the water structure in its immediate vicinity. As the temperature of the brine rises, more of the water molecules are truncated and the viscosity drops. *The viscosity of a hypersaline brine is a complex function of temperature and the nature and concentration of the constituent solutes. Chloride brines are more viscous than sulfate brines; the latter, in turn, are more viscous than carbonate or nitrate brines.* The viscosity increases with decreasing atomic weight of alkali earth cations; however, it is greater with sodium than with ammonia or potassium as the cation. It also increases with rising concentrations of sodium, magnesium, calcium, and iron chlorides, or sodium bromides, sodium, potassium, ammonium, and magnesium sulfates (Sato and Hayashi, 1961), but drops with increasing concentrations of potassium, ammonium, rubidium, and cesium chlorides and bromides (Rutskov, 1961). A rise in temperature in each case counteracts the change. Potassium chlorides and bromides have a negative viscosity at all concentrations, i.e., one smaller than water, and it decreases with increasing concentration (Chesnokov, 1962). The fraction representing change in viscosity per change in temperature increases to saturation in potassium chlorides and sulfates, but decreases in magnesium chlorides and sulfates or in sodium sulfates. Only for sodium chloride is there a maximum at 2–3 g/liter, and then the viscosity decreases to saturation (Ezrokhi, 1952). Overall, a marine brine increases its viscosity by at least a factor of 10 in the process of progressive concentration.

Onset of Precipitation

Calcite starts precipitating from Aral Sea water at a density of 1.100 or a 3-fold concentration. Gypsum follows at 1.115 or a 4.5-fold concentration (Lepeshkov and Bodaleva, 1952). In contrast, Caribbean seawater, which is about 10% more concentrated than standard seawater, according to Baseggio (1974), should commence precipitating gypsum at a density of 1.0897 and halite at 1.2185 if little or no carbonate is present. However, the effect of supersaturation rather than mere saturation was not taken into account in his calculations. Whereas halite starts precipitating at a density of 1.212, corresponding to a salinity of 35% (Clarke, 1924), precipitation of potash salts from marine brines begins when the brine reaches a density of 1.3235 (Strakhov, 1962). Braitsch and Hermann (1964a) estimated a somewhat lower density of $1.241 + 0.0006924t$ (t = °C) for the onset of sylvite precipitation from seawater stripped of its magnesium sulfate. Bischofite starts forming at a density of 1.353 (Sedel'nikov, 1958), and tachyhydrite follows at a density of 1.395. Evidently, all evaporites precipitate from brines with densities below 1.4 (Fig. 4-2). Bhat *et al.* (1979) brought seawater by solar evaporation to a density of 1.349 when the residual bittern contained 35.82% magnesium chloride, 6.60% magnesium sulfate, 1.00% po-

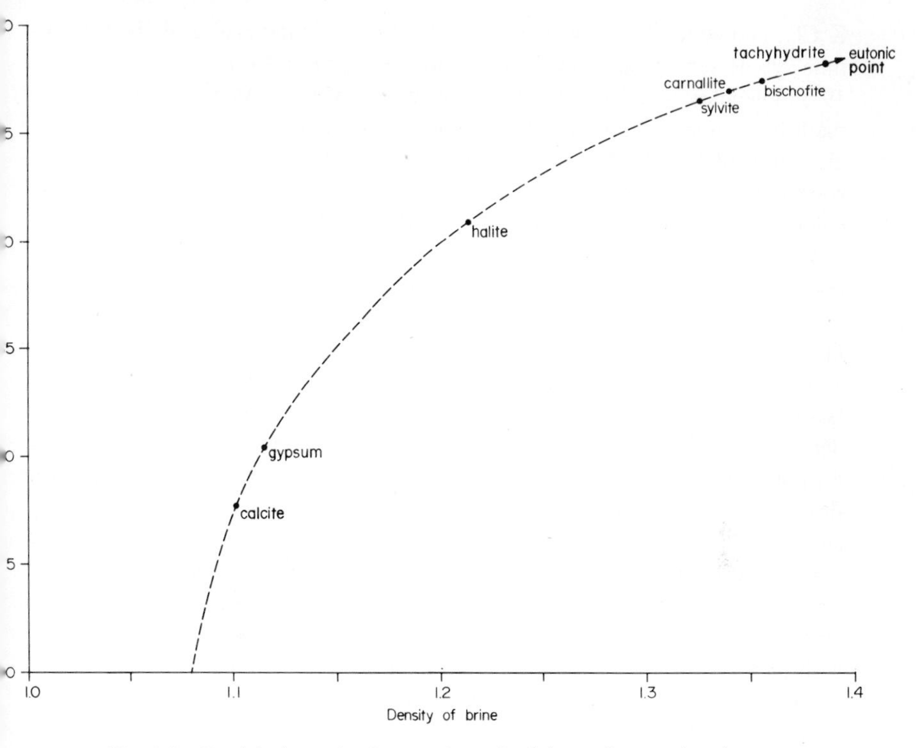

Fig. 4-2. Precipitation path of an anoxic marine brine to the eutonic point.

tassium chloride, and 2.30% sodium chloride. When the same bittern had reached a density of 1.368, about 40% of the magnesium chloride had precipitated as bischofite (of 95.76% purity).

The combined outflow and seepage into the subsurface must be reduced to a fraction of the outflow that can be calculated from salt balance equations. For seawater with a salinity of 34.84 and a density of 1.0245, the outflow must be sufficiently constrained so as not to exceed 21.30% of the inflow, per unit of time, for the commencement of gypsum precipitation, 11.56% for halite, 7.57% for potash salts, 6.94% for bischofite, and 6.20% for tachyhydrite.

Seawater contains about 4.24 moles of Na_2Cl_2, 0.088 mole of K_2Cl_2, and 0.633 mole of $MgCl_2$ per 1000 moles of H_2O. Gypsum-precipitating brines are mainly sodium chloride brines low in magnesium chloride. Once halite starts precipitating, the proportions of magnesium, calcium, and potassium chlorides begin to rise. By the time the first potash salts start precipitating, the $MgCl_2$:NaCl ratio exceeds 10:1 (Hartwig, 1955). At 25°C, a marine brine saturated for sodium chloride contains 55.5 molecules of Na_2Cl_2, 1.1 molecules of

K_2Cl_2, and about 1.1 molecules of $MgCl_2$ per 1000 molecules of H_2O. By the time sylvite saturation is reached, the brine contains 4.45 molecules of Na_2Cl_2, 19.5 molecules of K_2Cl_2, and 19.9 molecules of $MgCl_2$. At carnallite saturation, the brine contains 2.0 molecules of Na_2Cl_2, 5.5 molecules of K_2Cl_2, and 70.5 molecules of $MgCl_2$. By the time bischofite starts precipitating, the brine contains 1.0 molecule of Na_2Cl_2, less than 0.5 molecule of K_2Cl_2, and 106 molecules of $MgCl_2$ per 1000 molecules of H_2O (Hintze, 1915).

When the brine reaches a density of 1.32, i.e., shortly before the onset of potash precipitation, the volume of precipitated salts equals the volume of the remaining brine (Valyashko, 1951). Precipitated potassium and magnesium salts initially will be impregnated with up to 40–50% brine, much of which eventually seeps out. Subsurface leakage increases with rising density of the brine; it overtakes outflow over the entrance sill by volume when the brine has reached a density of about 1.25, i.e., still within the field of halite precipitation. Saturation for potash salts is thus reached only well after subsurface leakage exceeds outflow over the entrance sill by a wide margin, or after surface outflow has stopped altogether.

In artificial environments, the onset of precipitation is somewhat delayed. Calcium carbonate precipitates only when the brine is reduced to 53% of its original volume, gypsum at 19%, and halite at 9.5%, coprecipitating not only with gypsum but also with magnesium chloride and sulfates. Sodium bromide follows at 3.9% of the volume, and potassium and magnesium salts precipitate only after the volume has dropped to less than 1.66% (Gardner *et al.*, 1967). Neither coprecipitation nor sodium bromide crystal formation is observed in ancient marine evaporites.

COMPOSITIONAL CHANGES

Potash Deficiency

A potassium deficiency develops in the early phases of marine brine concentration if it is measured against magnesium (Busson and Perthuisot, 1977; Levy, 1980; Nadler and Magaritz, 1980). This happens mainly by adsorption of some potassium onto active surfaces in clays, silica gels, and various detrital minerals in the sediments. However, this is followed by a preferential potassium enrichment in the most concentrated brines due to sodium losses by precipitation (Eugster, 1980a). By the time the brine is very concentrated, precipitated calcium carbonates and sulfates have effectively separated the brine from direct contact with subjacent terrigenous clastics, unless the subsidence of the basin was very asymmetrical.

Sulfate Deficiency

The sulfate–sulfide boundary is lifted in natural hypersaline brines from beneath the sediment–water interface to near the base of the interface with inflowing waters. Consequently, there is no consumption of organic matter by anaerobic sulfide producers beneath the sediment–brine interface, as one finds in fjords or at the bottom of the Black Sea. The consumption takes place in the water column. Organic matter settling onto the lagoon floor can thus be preserved. In some modern salinas, the workers artifically induce anaerobic conditions into the unstratified brines to prevent the precipitation of iron oxides (Herrmann *et al.*, 1973).

The main difference between density stratification in fjords and in evaporite pans is the location of sulfate reduction. Sulfate scavengers reside in bottom sediments of fjords, and much of the generated hydrogen sulfide is then consumed in iron fixation, first in the form of hydrotroilites and later as pyrites. In hypersaline brines (Fig. 4-3), sulfate reduction occurs mainly in the water column above the sediment–water interface (Romanenk *et al.*, 1976). Hydrogen sulfide escapes and does not attack dissolved ferrous hydroxides. There is therefore no hydrogen sulfide available in the bottom sediments, which explains the scarcity of pyrite in evaporite sequences.

Where the interface is shallow enough for bottom waters to reach into the photic zone, a pink horizon composed of prokaryote bacteria, called a "bacterial plate," develops beneath the mixing zone between oxygenated surface waters and bottom brines. Only prokaryote cells can utilize inorganic energy sources directly, because the electron-transport system lies in plasma membranes. Eukaryotes have a more complex cell structure that houses their electron-transport system in organelles inside the cell (Ehrlich, 1978); therefore, they need electron donors in solution. Sulfate-reducing bacteria are normally heterotrophs that can

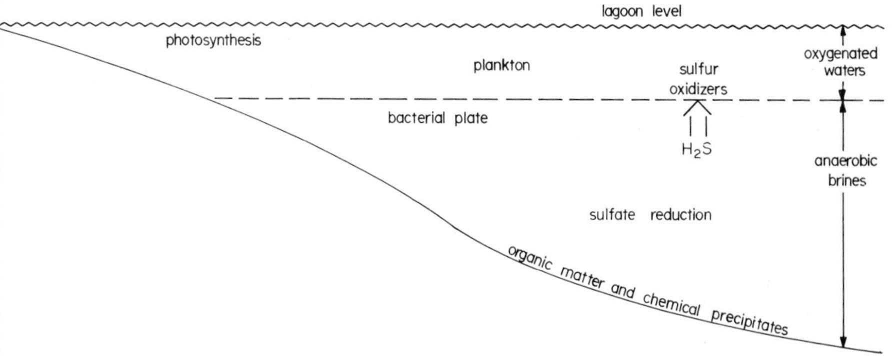

Fig. 4-3. Sulfur reduction in a density-stratified hypersaline brine.

become facultative autotrophs, since they can also utilize inorganic carbonate, provided there is a souce of molecular hydrogen, as is supplied by the oxidation of iron (Butlin and Adams, 1947). As heterotrophs, the sulfate reducers prefer to use the available dissolved organic matter in their metabolism. Consequently, the contamination of water bodies with organic matter strongly accelerates the process of sulfate reduction (Tezuka *et al.*, 1963), since about half of the organic matter in marine sediments may be mineralized by anaerobic sulfate respiration (Fenchel and Joergenson, 1977). Sulfate-reducing bacteria thrive on volatile fatty acids as electron donors. However, even slight amounts of MoO_4^{2-} completely inhibit the SO_4^{2-}-reducing activity (Soerensen *et al.*, 1981). With the exception of saturated alkanes, all other fresh organic matter, in contrast to fossil matter, can be sulfurized at ambient temperatures; however, only sulfur from hydrogen sulfide can sulfurize in this manner (Bestougeff and Combaz, 1974). Such reactions occur soon after deposition; the volume of elemental sulfur decreases with depth (Palacas *et al.*, 1968). Nonetheless, only a very small fraction of the hydrogen sulfide is involved in generating organo-sulfide complexes or hydrous ammonium sulfide. Any hydrogen sulfide reacting with organic matter (Casagrande *et al.*, 1979) is then deposited with it.

Dissolution of gypsum results in both charged and uncharged calcium and sulfate species in solution. The reaction can be written as follows:

$$CaSO_4 \cdot 2H_2O = xCa^{2+} + xSO_4^{2-} + (1 - x)CaSO_4^0 + 2H_2O \quad (4\text{-}3)$$

where $x < 1$ and the notation $CaSO_4^0$ represents an ion pair which is a soluble but uncharged species. As sulfate reducers dispose of the sulfate ions, the resulting excess of cations raises the pH. Precipitation of gypsum allows the pH to drop. The same effect is obtained simply by heating a calcium sulfate solution. At temperatures above 62.5°C, the calcium and sulfate ions cease to be present in equal amounts. In brines heated to 99°C, the CaO/SO_3 ratio is 0.5066/0.3414, or 1.4839 (Arth and Chrétien, 1906), representing a loss of one-third of the sulfate.

Additional seawater brings in more dissolved sulfate, and should therefore progressively increase the hydrogen sulfide concentration of the brine. This is not the case. The water usually smells only very faintly of hydrogen sulfide, if at all. The seawater-fed Solar Lake on the coast of the Gulf of Aqaba, Red Sea (Eckstein, 1970), contains only a fraction of the hydrogen sulfide found in anaerobic bottom waters of continental meromictic lakes (Sorokin, 1975). *Substantial quantities of hydrogen sulfide thus escape to the atmosphere.* The sulfate deficiency in density-stratified marine lagoons increases with rising brine concentration, since the solubility of hydrogen sulfide decreases rather rapidly with both rising temperature and rising brine concentration (Rasmussen, 1974). The rate of hydrogen sulfide escape to the atmosphere peaks whenever the concentrated brine reaches a seasonal temperature maximum, i.e., when the hot water lens develops under seasonal density stratification. Since the density stratification itself is best developed during seasonal high water inflow, the two events coincide in time. The original precipitation of marine salts thus occurs in an environment that is both anaerobic and grossly deficient in sulfates. Such primary sulfate deficiency has been noted by many, including d'Ans (1947), Borchert (1967).

Analogous to density-stratified lagoons is the situation in meromictic continental lakes. Here also, the bulk of the hydrogen sulfide forms in the water column and not in the sediment (Sorokin and Donats, 1975). This production of hydrogen sulfide can be very substantial. In Ain-ez-Zauia, a Libyan lake, the water contains 20 g/m^3 of H_2S at the surface and 108 g/m^3 near the bottom, with an annual production of 53.68 kg of sulfur per cubic meter of water (Butlin and Postgate, 1955). Similar hydrogen sulfide concentrations have also been reported from Lake Faro in Sicily (Genovese, 1962). Saline brines generate two to three times less hydrogen sulfide than even sulfate-poor freshwater lakes (Sorokin, 1975). The prevailing atmospheric temperature regime is also very important. The Aral Sea is located in a dry belt, but well outside the subtropics. It is very rich in potassium sulfate and magnesium sulfate (Nikolayev and Fradkina, 1949), but is deficient in sodium chloride and magnesium chloride on account of its continentality. The lacustrine bacterial population here is evidently not sufficiently numerous to digest all the accumulating sulfate ions.

Sulfate oxidizers live symbiotically with photosynthesizers in the bacterial plate. They take advantage of hydrogen sulfide trapped on top of the stagnant waters and increase the absorption of radiation at this level (Bradbury, 1971). Such bacteria oxidize hydrogen sulfide in two stages (Sorokin, 1970). The energy contained in consumed organic matter is thereby released through the sulfur cycle. More energy is required to oxidize hydrogen sulfide than to oxidize an equivalent amount of organic matter. Biological oxidation of hydrogen sulfide by autotrophic bacteria does not lose all the energy in the form of heat, as does chemical oxidation. Up to 25% of the energy is utilized for chemosynthesis and is thus included in the biological turnover (Sorokin, 1972). Thus, not all hydrogen sulfide escapes to the atmosphere; photosynthetic sulfur bacteria utilize about 40% of the hydrogen sulfide produced (Romanenk *et al.*, 1976), which leads to a second generation of gypsum (Baas-Becking and Kaplan, 1956; Bonython and King, 1956).

The concentrated layer of euryhaline photosynthesizers (mostly *Cyanophytes*) will somewhat counteract the progressive oxygen depletion in the uppermost levels of the bottom brine. Bubbles of free oxygen are frequently seen along the density interface on sunny afternoons, when algal photosynthesis reaches its maximum. Even where sulfides are absent, the oxygen is still produced but not consumed. Gas domes under salt crusts of the Great Salt Lake, Utah, contain primarily oxygen and nitrogen, but no carbon dioxide (Post, 1980). Primary sulfate minerals are therefore restricted to bay floors in the photic zone, i.e., shelves, shoals, and ephemerally drying-out shoreline facies.

If the sulfur cycle represented a closed system, one would expect all reduced hydrogen sulfide to be reoxidized by the bacterial plate of photosynthesizers or by dissolved oxygen in surface brines. Since initially there are more sulfate ions than calcium ions in seawater (28.26 versus 10.23 moles per cubic meter), major quantities of magnesium sulfate would have to precipitate. This in fact occurs in the closed system of a laboratory beaker, as Usiglio (1849) has shown. In his experiments, magnesium sulfate started to precipitate after gypsum and continued to precipitate in diminishing amounts during halite crystallization. In fossil evaporite sequences, one normally does not find a layer of epsomite or kieserite sandwiched between anhydrite and halite. Kieserite occurs only in sulfatized potassium horizons where additional sulfate has been supplied later. Consequently, one must assume an open system in nature, in which some of the generated hydrogen sulfide escapes and is not available for precipitation. The amount of hydrogen sulfide that escapes may be sufficient to keep some of the calcium ions in solution (Sonnenfeld *et al.*, 1976). These ions would ultimately lead to tachyhydrite precipitation or drainage into formation waters rich in calcium chloride. In modern sediments, there are diffusional losses into the brine that cause an underestimation of sulfate reduction by a factor of about 10; a count of sulfate-reducing colonies yields an underestimate by about three orders of magnitude (Joergenson, 1978).

It is, however, possible to assess the amount of sulfur lost. Magnesium stays as a conservative element in natural evaporite pans well into the period of halite precipitation. During and immediately after gypsum precipitation, the magnesium does not precipitate in any minerals, nor does it substitute as yet in any clays or carbonates. The Mg/Ca ratio determined in the early stages of halite precipitation is thus a measure of the calcium preserved in the brine. The brine then no longer contains any significant amount of reactive sulfate ions. The difference between the amount of sulfate ions entering the basin with fresh seawater and the amount precipitated in gypsum represents the quantity of sulfate ions utilized by anaerobic bacteria and discharged as hydrogen sulfide:

$$\mathrm{Ca}_{\mathrm{inflow}} - \mathrm{Ca}_{\mathrm{in\ brine}} = \mathrm{Ca}_{\mathrm{precipitated}}$$

$$\mathrm{Ca}_{\mathrm{precipitated}} = \mathrm{sulfate}_{\mathrm{precipitated}}$$

$$\mathrm{sulfate}_{\mathrm{inflow}} - \mathrm{sulfate}_{\mathrm{precipitated}} = \mathrm{sulfate}_{\mathrm{metabolized}}$$

The amount of sulfate precipitated and not remobilized will vary, but will rarely exceed 15% of the inflowing sulfates. Because of the constant inflow of sulfate ions in solution, the anaerobic bacteria have no need to attack precipitated sulfates. They do so only when the supply in solution becomes insufficient.

Osichkina (1980) estimated that chloride brines had lost up to two-thirds of their sulfate ions when they precipitated only halite, sylvinite, and carnallite in Upper Jurassic evaporites of central Asia. The actual loss was probably much higher. In antarctic waters, the oxygen-producing photosynthesizers are subdued or absent, and thus colloidal sulfur accumulates at the interface with suface waters. Nonetheless, even here, 76% of the sulfur is lost compared to seawater (Burton and Barker, 1979).

Borchert (1967) had pointed out that a sulfate deficit compared to seawater is characteristic only of primary brines. To explain the presence of potassium and magnesium sulfate minerals in an evaporite sequence, one must postulate the early epigenetic seepage of oxygenated brines, exchanging their less soluble sulfate ions for more soluble chloride ions in the precipitate. The common occurrence of minerals with less hydration than the original precipitate indicates that the process of sulfatization occurred at slightly elevated temperatures, i.e., above about 60°C.

The observation that sulfatic potash ores are generally absent from pre-Permian, Mesozoic, and Paleogene evaporites suggests that sulfatization is not obligatory. Only descending epigenetic (metamorphic) brines contain an excess of sulfate ions, compared with seawater, as deeper formation waters are also anoxic. It is thus necessary to find a source of these oxygenated sulfatic brines. Such a source is found only in the well-oxygenated surface waters, which easily become turbulent near the shoals and shores and are thereby ventilated (Fig. 4-4). On islands, they can underrun the Dupuit–Ghyben–Herzberg freshwater lens of lower density; in coastal regions, the groundwater flow occupies the same position and forces encroaching seawater to deflect into subsurface aquifers. Surges in groundwater discharge occur after each major rainfall. Delivery of

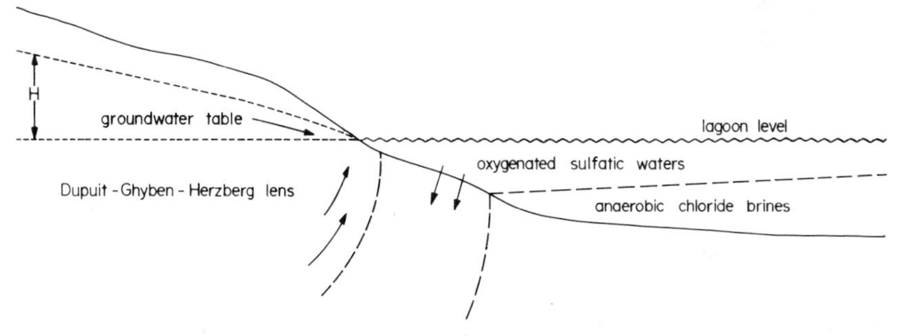

Fig. 4-4. Types of water in a density-stratified lagoon: Anoxic chloride brines are in contact with the lagoon floor; oxygenated sulfatic surface waters touch ground only in shoals and along shoreline shallows. If there are no broad coastal flats, artesian mixing with fresh waters can occur as long as the groundwater table elevation (H) is high enough to drive out waters from the Dupuit-Ghyben-Herzberg lens. As the density difference to the hypersaline brine increases, the freshwater lens decreases in volume and thickness.

solute from groundwater sources and its mixing with only weakly concentrated surficial seawater are at a maximum during the rainy season, when the density stratification is best developed in the open brine. A thick layer of oxygenated surface waters rich in sulfate ions then becomes available to seep into shelf sediments.

Chloride Deficiency?

It is always assumed that only sulfates are lost from the brine and that chlorides are the conservative anions. Comparable data on chlorine losses of hot brines do not exist, but concentrated hydrochloric acid solutions lose almost half of their chlorine upon heating. Since chlorine gas rapidly escapes from chlorinated sewage ponds unless they are covered, it is probable that some chlorine also escapes in time from stratified hypersaline brines, thus increasing the anion deficiency. After Owens Lake, California, lost its main tributary due to diversion of the river around the time of World War I, it has gradually turned into a brine lake that has lost nearly half of its solute in the last 70 years (Alderman, 1984). Brine seepage downward into the subsurface is an obvious explanation. However, nearly three and a half times more chlorine than sodium was lost, so that the lake gradually turned into a soda lake. This has likely happened by loss of chlorine to the atmosphere. The same process seems to operate in turning the sodium chloride lakes of northern Egypt into natron lakes. However, continental lakes contain a finite amount of chlorides; thus, this process cannot occur on a scale similar to that of marine embayments that are continuously fed by inflowing seawater: None of the Phanerozoic marine evaporite deposits show an upward transition to soda lake mineralogy. So far, potential chlorine losses from exposed hypersaline brines have not been investigated. Such chlorine losses would obviously also affect the Br/Cl ratio, a very important tool in the interpretation of the depositional history of evaporites. Furthermore, the cation surplus in the brine is increased even further.

Ion Imbalance

Table 4-2 gives the molar concentration of major ions in seawater that have to be accounted for in progressive precipitation of dissolved mineral species. Carbon dioxide dissolved in a concentrating marine brine is consumed in calcium carbonate precipitation, and about half of the remaining calcium stays in solution even after gypsum precipitation. As the sulfate–sulfide boundary moves up from beneath the sediment–water interface into the water column of a hypersaline brine, SO_4^{2-} is broken up. More and more sulfate anions are bacterially converted to hydrogen sulfide, which is ultimately released to the atmosphere. The constant influx of new quantities of solute provides new food for the sulfate

TABLE 4-2

Molar Concentration of Major Ions in Seawater of Density 1.025

Ion	Moles/m^3
Na^+	470.86
K^+	9.76
Ca^+	10.23
Mg^{2+}	53.64
Cl^-	548.74
SO_4^{2-}	28.26
Br^-	0.83
HCO_3^-	2.39

reducers. Most sulfate ions in inflowing ocean waters are tied to magnesium cations and are not dissociated in the brine. The exact amount of magnesium complexed as magnesium sulfate is not known (Horne, 1969). Heavily hydrated magnesium sulfate can dissociate in steps:

$$MgSO_4 = Mg(OH_2)\cdot SO_4 = Mg(OH_2)_2\cdot SO_4 = Mg^{2+} + SO_4^{2-} \quad (4\text{-}4)$$

and

$$MgSO_4 + 2H_2O = H_2S + Mg(OH)_2 + 2O_2 \quad (4\text{-}5)$$

The sulfate removal leads to a deficiency of anions and produces an excess of cations, since the bacteria do not replace scavenged anions with new ones. A corresponding quantity of cations, specifically magnesium, is forced to form soluble hydroxides or organometallic complexes (Fig. 4-5). This tends to raise the pH. The process is not entirely speculative; at a pH of 8.4, some Australian halite-crystallizing salinas cloud with magnesium hydroxide that began coming out of solution at a pH of 8.2 (N. Sammy, personal communication, 1983). This happens well after whitings of calcium carbonate and sulfate spicules have occurred in the concentrating basins. As long as any bacterial activity of anaerobic sulfate scavengers continues, more and more magnesium ions are driven into hydroxide bonds and as such alter the crystal lattice of swept-in clay minerals to mixed-layer varieties with brucite pillars. An estimated 36% of magnesium is thus available to produce brucite pillars in clays or enter subsurface aquifers as hydroxides or organo-metallic complexes. The amount of such reactive magnesium in 1 m^3 of seawater would be sufficient to dolomitize the aragonite precipitated from 55 m^3 of seawater. Since no other primary sulfates have been found in meaningful quantities in ancient marine evaporite deposits, we must assume that all of the remaining sulfate ions are decomposed to hydrogen sulfide, i.e., about

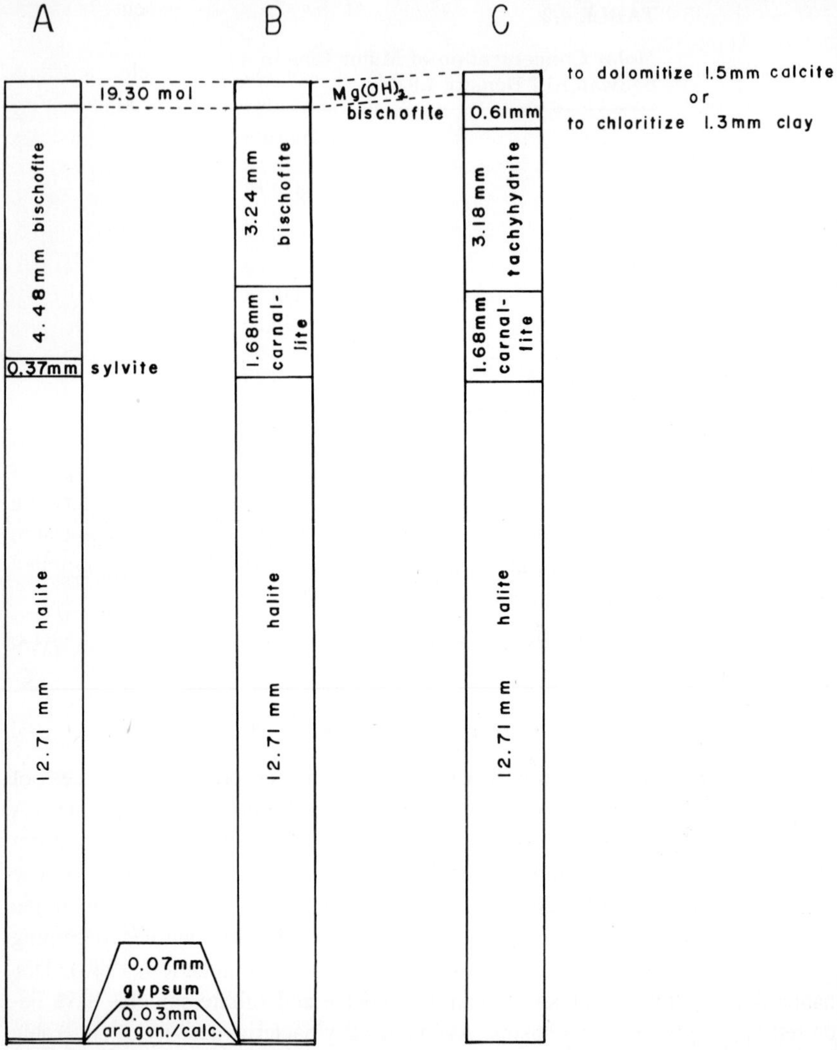

Fig. 4-5. Volume of precipitates per cubic meter of seawater in a basin with a sulfate deficiency. A. Precipitation under density stratification. B. Precipitation of exposed hypersaline brine (A and B: all calcium as carbonate or sulfate). C. Precipitation with all the sulfate converted to hydrogen sulfide.

86% of the incoming sulfate ions. Once all of the sulfates are spent, bacterial activity diminishes.

Another source of hydroxides is bischofite. It breaks down at elevated temperatures, progressively loses water molecules, and eventually produces some Mg(OH)Cl; hardly any hydroxychloride is produced from carnallite (Grube and Brauning, 1938). As long as there is sufficient sodium chloride present in the brine, it prevents the hydrolysis of bischofite (Orekhova *et al.*, 1974) and strongly suppresses the formation of magnesium hydroxides. They would thus begin to form only at a brine density of about 1.35. Indeed, Pasztor and Snover (1983) found that a freshly mixed calcium chloride brine of density 1.39 contained about 25 mmol/liter of hydroxyl ions. In contrast, magnesium sulfate solutions produce brucite from cement paste, i.e., a mixture of clay and limestone (Greschuchna, 1976).

Isothermal versus Solar Evaporation

Much experimental work has been done to establish the stability fields of the individual precipitating minerals and their parageneses at given temperatures. The pioneering work of Boeke, van t'Hoff, Jaenicke, d'Ans, Kurnakov, and their disciples was thereby of the greatest importance. However, there is a substantial difference between results, specifically solubility diagrams, obtained under controlled isothermal conditions, and results obtained by natural solar evaporation even from an unstratified, oxygenated brine. Natural brines are exposed to a range of seasonal temperatures and thus are polythermal. Consequently, both Borchert in Germany and Valyashko in the USSR later turned toward solar evaporation models. Up to a density of 1.29, both isothermal and polythermal evaporation models give reasonably similar results. In more concentrated brines, magnesium chloride is enriched by solar evaporation compared to isothermal values, whereas potassium chloride and magnesium sulfate are found in lesser quantities and limited to a maximum value (Oka and Inagaki, 1942a,b; Oka and Hishikari, 1944). The results can be readily applied to clay-lined salinas and salterns. However, one must not forget that the chemistry of natural marine brine pools is severely affected by the metabolism of anaerobic bacteria and various photosynthesizers, in addition to the polythermal conditions of solar evaporation.

In their investigation of the brines of Lake Qarun (southwest of Cairo, Egypt) Estefan *et al.* (1980) found that laboratory evaporation unduly increases the presence of sulfates in the brine. Under conditions of solar evaporation at 30°C, sulfate ions reach a maximum at 87 g/liter, whereas gypsum, halite, sylvinite, carnallitite, and bischofite precipitate one after the other. If solar evaporation is forced to continue, it produces composite sulfates such as aphthitalite and blœdite. Such terminal bitterns of a marine embayment tend to infiltrate into any porous substrate, inducing early diagenetic alterations and transformations.

Nitrogen Content

The solubility of oxygen decreases sharply in concentrating seawater-fed lagoons. At a sevenfold brine concentration, oxygen solubility drops to one-third of its original value (MacArthur, 1916). The growing deficiency of free oxygen beneath the pycnocline of density-stratified brines encourages anaerobic decomposition of part of the accumulating organic matter, utilizing available nitrates, nitrites, and eventually sulfates. The anaerobic hypersaline environment is not strictly comparable to the anoxic environment of fjords, as anoxic waters of normal salinity contain only two to three times the concentration of both soluble and particulate organic carbon compared to open ocean waters. If greater quantities of organic matter are found in the bottom sediments, they may indicate faster rates of accumulation rather than slower rates of decomposition in an anaerobic environment (Richards, 1970).

Some of the algae in hypersaline brine ponds, notably the Cyanophyta, are active nitrogen fixers. The nitrogen content of salinas (determined by the Kjeldahl method) rises more rapidly than the salinity in concentrating brines, and is up to two orders of magnitude greater than in inflowing ocean waters. There is a good correlation between the location of the chemocline to anoxic waters and the top of a layer of nitrite enrichment in anoxic ocean waters (Horne, 1969). All the nitrate and nitrite ions must have been consumed before sulfate reduction begins, which happens after oxygen level has fallen below 0.11 ml/liter. Nitrogen gas very commonly occurs in halite and potash salts, mainly as inclusions and along crystal faces, always under high pressure. The nitrogen gas occurrences are particularly noted in Permian salts that were partially sulfatized.

Horne (1969) observed an impressive correlation between hydrogen sulfide and ammonia concentrations, and illustrated the sequence of chemical reactions with a hypothetical organic material, $(CH_2O)_{106}{\cdot}(NH_3)_{16}{\cdot}H_3PO_4$, that decomposes by depleting available oxygen first, then nitrates and nitrites, and finally sulfates. The normal oxidation in surface waters can be represented by the following equation:

$$(CH_2O)_{106}{\cdot}(NH_3)_{16}{\cdot}H_3PO_4 + 138O_2 \rightarrow$$
$$106CO_2 + 122H_2O + 16HNO_3 + H_3PO_4 \qquad (4\text{-}6)$$

Denitrification sets in when oxygen is exhausted:

$$(CH_2O)_{106}{\cdot}(NH_3)_{16}{\cdot}H_3PO_4 + 84.8HNO_3 \rightarrow$$
$$106CO_2 + 42.4N_2 + 148.4H_2O + NH_3 + H_3PO_4 \qquad (4\text{-}7)$$

or

$$(CH_2O)_{106}{\cdot}(NH_3)_{16}{\cdot}H_3PO_4 + 94.4HNO_3 \rightarrow$$

$$106CO_2 + 55.2N_2 + 167.2H_2O + H_3PO_4 \tag{4-8}$$

Eventually, sulfate reduction takes over:

$$(CH_2O)_{106}{\cdot}(NH_3)_{16}{\cdot}H_3PO_4 + 53SO_4^{2-} \rightarrow$$

$$106CO_2 + 53S^{2-} + 16NH_3 + 106H_2O + H_3PO_4 \tag{4-9}$$

The denitrification of protein derivatives yields both ammonia and phosphate, which become available for algal and bacterial metabolism. The former is the principal form of available nitrogen; the nitrate content decreases with increasing salinity (Carpelan, 1957). Ammonia reaches its maximum in carnallite (up to 77 ppm); there is none in tachyhydrite (Biltz and Marcus, 1908). Amides and amines are the most likely sources of such biogenic nitrogen. The bulk of the nitrogen found in evaporites is indeed biogenic in origin (Osichkina, 1978a). This explains why most of the nitrogen found in vacuole inclusions of evaporites appears to be of organic origin.

NATURE OF SEDIMENTS

Waters entering an embayment from the open sea are clean, i.e., they have a minimal admixture of terrigenous clastics. Their concentration leads to precipitation of essentially clean evaporites free of argillaceous content. Since inflowing surface seawater harbors a substantial amount of biota, the precipitates are contaminated with organic decomposition products.

Terrigenous clastics enter an embayment from two sources:

1. Increasing velocities of inflow increase the rates of erosion along the banks of the inflow channel. The amount of clastics is a function of the length of these straits, their degree of convolution, and the velocity of the flow. However, only suspended material reaches the basin. Any sands and gravels dislodged by the rip current move outward with the bottom countercurrent. If the entrance strait is lined by carbonate banks and reefs, some of the fine debris goes into solution or is scavenged by a variety of organisms dwelling inside the basin. Overall, the contribution of erosion by inflow remains small until the open outflow over the entrance sill ceases, and all further outflow is confined to seepage into the bay floor. Once the inflow has suppressed the open outflow, its velocity is great enough to erode the channel bed. The Gulf of Kara Bogaz Gol is an example of this process today. Ever since the open outflow ceased in this century, the inflow began to spread an apron of black bituminous shales around the channel orifice (Kara = black; Boghaz = channel, strait, estuary; Gol = any body of water: lake, river).

2. Rainwash and runoff occur in semiarid lands with irregular frequency but gargantuan force. They sweep accumulated products of tropical weathering into depressions, including dry playas and open lagoons. In a lagoon filled with hypersaline brine, such rainwash invariably produces an ephemeral surface layer of low-salinity waters and an interstratal delta of flocculating clays that spread out widely before slowly settling onto the lagoon floor. Fresh supplies of bicarbonates, primarily of calcium and sodium, are also added to the system. *The source of clay beds intercalated with evaporite sequences is therefore not the open sea, but the surrounding shores.* The clays indicate a temporary increase in the rates of atmospheric precipitation and a consequent reduction in the water deficit. The inflow then slows down; drag is reduced, which results in increased outflow. The reduction in the inflow/outflow ratio brings about an overall freshening of the brine, not merely the superposition of a freshwater layer derived from the runoff. In the German Zechstein sequence, the salt clays are mutually exclusive with carnallite and thus indicate temporary freshening (Langbein, 1964). Virtually all clays deposited in marine evaporite sequences are then heavily altered. Bodine and Standaert (1977) introduced the term ''hyperhalmyrolysis'' to cover mineral reactions in hypersaline brines.

Marine chemical sediments always display the same primary sequence of calcium carbonates and sulfates, followed by a variety of chlorides. In laboratory experiments, the first salt precipitating after halite is epsomite ($MgSO_4 \cdot 7H_2O$). An admixture of epsomite to halite was indeed reported by Strakhov (1962) in the contemporary halite of the Gulf of Kara Bogaz Gol, beyond the areas of mirabilite and glauberite precipitation. This sequence of precipitates is typical of a continental lake deposit, but is not found in evaporite sequences of marine origin. Its absence is due to the scarcity of sulfate ions in the brine. Furthermore, epsomite is unstable above 28°C (Braitsch, 1964), a temperature normally exceeded by that of subtropical hypersaline brines. Seawater evaporation experiments closed to air or open to carbon dioxide initially produce gypsum, followed by mirabilite, but they do not produce magnesium salts (Simpson, 1965).

Solar evaporation of inland bitterns from the Rann of Cutch area of India, which are low in sulfates, produces mixed potassium–magnesium salts at low densities at the same relative concentration of potassium and magnesium chlorides, irrespective of variations in sulfate content. Bischofite starts precipitating when potassium is depleted (Choudhari, 1969). In marine evaporites, halite is followed by either sylvinite or carnallitite, and these, in turn, are followed by bischofite or tachyhydrite. Very occasionally, an intercalation of primary polyhalite can be found; otherwise potassium and magnesium sulfates are secondary diagenetic alterations of precipitated minerals. The potassium–magnesium salts often precipitate only in the innermost subsidiary bays, where preconcentration has stripped the brine of less soluble salts.

The sequence of potash and magnesium minerals occurring in natural evaporite basins may be determined by three factors:

1. The deficiency of sulfate ions due to microbial degradation, which prevents the early precipitation of magnesium sulfates. This sulfate deficiency was absolute in the Oligocene Upper Rhine Valley evaporites, where not even kieserite is found. It was also extreme in the Ronnenberg seam in the third cyclothem of Zechstein salts in Germany (Siemens, 1961). There are many other Devonian, Jurassic, Cretaceous, and Tertiary deposits nearly devoid of sulfates other than anhydrite. Massive magnesium and potassium sulfates are consequently the product of epigenetic percolation of oxygenated sulfate brines.
2. The rate of crystallization of different minerals. Chloride compounds, such as halite, sylvite, and carnallite, crystallize more rapidly than the brine concentration can be altered by evaporation from saturation of one mineral to saturation of the next. Sulfate minerals such as kieserite, kainite, langbeinite, and other double salts precipitate slowly and lag behind the rate of evaporation of the solution. Thus, they are not formed directly by solar evaporation, but only via intermediate precursors (Strakhov, 1962). *In a concentrating brine, sulfate precipitation is thus overtaken by chloride precipitation.*
3. The sulfur deficiency of marine hypersaline brines. Because of this deficiency, magnesium does not crystallize as a sulfate in natural primary precipitates of marine environments, but only as a chloride. There are 964.46 moles of chlorine and 1.46 moles of bromine entering with surface inflow with every 1 million moles of ocean water. This translates into 827.58 moles of halite, 17.51 moles of carnallite (and/or sylvite), and 34.61 moles of bischofite (and/or tachyhydrite). After these salts are precipitated, all available halides are consumed. At that point, 42.16 moles of magnesium are still present in the residual bittern, presumably as hydroxides, since no other anions are available, unless the magnesium was used for building brucite pillars into clay minerals or for dolomitization (cf. Fig. 4-5). Chemical analyses of halites and potash rocks frequently report a small excess of magnesium oxide (e.g., Kuehn and Hsu, 1978), which is ascribed to analytical discrepancies.

THE pH OF BRINES AND BITTERNS

The pH of a marine brine does not seem to rise until chlorides start precipitating. The alteration of siliciclastic material is very minute in gypsum deposits but is extensive in halites. Red halite is rare, indicating an inability of bivalent iron to convert to the trivalent variety, another sign of low Eh and high pH values. Red halite becomes more frequent in carnallitites or sylvinites. In contrast to modern salinas, fossil primary evaporite deposits provide ample evidence of high pH values. Contrary evidence has come only from recrystallized and metasomatosed deposits that have reacted with percolating younger brines.

Published pH values of hypersaline brines obtained by direct emersion of an electrode can be erroneous, as Pasztor and Snover (1983) have pointed out. Both the salt and water can act as solvating species in concentrated brines and lead to errors once the concentration exceeds 0.2 molal. A 50% caustic solution yields a pH measurement with a glass electrode of about 7, although the solution is strongly basic. As distilled water is added, the pH rises to 14, giving the impression that a 4% solution (1 molal) is more basic than a 50% caustic solution (25 molal). Diluting the brine 10:1 gives better results but does not eliminate the error. Dilution causes hydrolysis of the salts in solution and reduces their proton-donating power. Moreover, many salts, especially hydroxides, are more soluble in concentrated chloride solutions than in water. Consequently, accurate titration is the only reliable method of determining the pH of hypersaline brines. This comment has to be kept in mind in evaluating the pH values mentioned in this chapter.

The addition of neutral salts and organic solvents to concentrated brines changes the pH. Forces of ionic attraction become prominent in concentrated electrolyte solutions; thus, the apparent concentration of ions is lower than the actual one (Bliefert, 1978). The addition of chloride ions reduces the apparent pH, and the addition of sulfate ions increases it (Schwabe and Ferse, 1962). Hydration of ions is responsible for the diminution of ionic activity; consequently, the measured pH value decreases in saline solutions with increasing concentration (Krumgalz, 1980), the amount of decrease varying with the mixture of dissolved salts involved. Moreover, the elevated temperature of hot hypersaline brines slightly lowers below 7 the neutral point on the pH scale (Helgeson, 1964) (Fig. 4-6). In chloride brines or mixtures of brines, the pH decreases slightly with rising temperature because the temperature coefficient of the ionization constant of water is negative (Zubakhina and Gerasimenko, 1974).

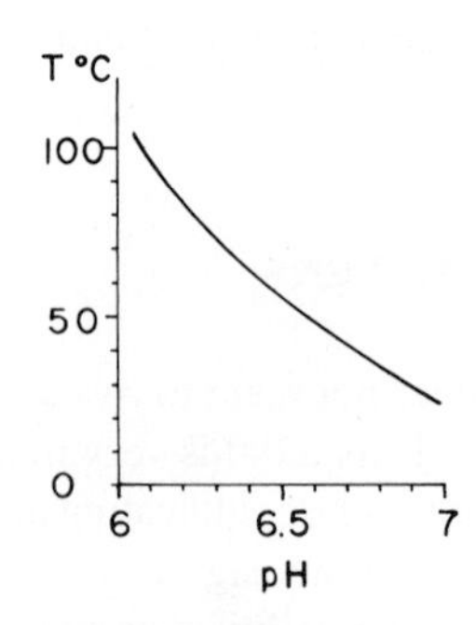

Fig. 4-6. Changes in the neutral point of pH measurements with temperature (after Helgeson, 1964).

Formation waters have a higher pH than distilled water at equal CO_2 pressure. With increasing CO_2, the pH decreases. However, this decrease is buffered by the presence of a variety of salts (Kuznetsov and Kutovaya, 1977), which can even overcome the negative temperature coefficient of the ionization constant of water.

Very dilute solutions are neutral at pH 7.0, and are acidic at lower values. As the brine concentrates, the value of the neutral point becomes fractionally lower. Thus, a pH of 8 in a very concentrated brine is said to be almost one order of magnitude more basic than the same pH value measured in ordinary lake water. Normal glass electrodes introduce errors when used in concentrated brines and indicate a pH value at least 4–5% too low (Baumann, 1973). Perthuisot (1975) noted that samples consistently furnished a pH reading one-half to one order of magnitude lower in the field than in the laboratory. He ascribed this to the presence of greater amounts of hydrogen sulfide in freshly collected samples. Pasztor and Snover (1983) found a similar discrepancy of 0.5 in reading the pH of freshly prepared concentrated halide solutions and taking another reading 10 days later. Halogen acids completely dissociate in concentrated solutions into alkali or alkaline earth ions and acid radicals, and the latter progressively dissociate further (Rao and Rao, 1936). A solution of analytical-grade sodium chloride has a constant pH at molar concentrations of 3 to 5. Commercial grades ionize because of impurities and become either more acidic or more alkaline.

The solubility in seawater of both calcite and aragonite drops drastically with increasing pH. At a water temperature of 30°C, it drops down at a pH of 8.5 to only about 6% of the solubility at a pH of 7; the drop is even greater at lower temperatures (Wattenberg and Timmermann, 1936). As the biota on the bottom of the lagoon precipitate magnesian calcite or aragonite, the Mg/Ca ratio increases in the open waters above and may reach 13:1. At the same time, the oxygen supply vanishes and the aragonite becomes saturated with hydrogen sulfide. The algal mats then act as a semipermeable membrane to carbon dioxide, thereby reducing the pH in the open waters to 6.5 (Bauld *et al.*, 1978). Beneath the mat, however, methane forms, an indicator of a high pH.

The oxygenated surface layer of stratified brines has a positive Eh and a pH ranging from 5.6 to nearly 8.0, lowered from an open marine value of 8.2–8.5 by plankton and by the absorption of atmospheric carbon dioxide. In this environment aragonite becomes unstable and is converted to gypsum; the algae continue to thrive (Grabau, 1920; Nesteroff, 1973a,b; Sonnenfeld *et al.*, 1976, 1977), supplying much of the oxygen for the conversion.

The pH rises as the brine concentrates; oxygen and bicarbonate solubilities then become negligible. As the salinity increases, the pH and alkali content of the brines passes from the control of the organic mass to the salt mass. If the pH is initially basic in carbonate lakes, the gradual concentration reinforces the progressive elevation of the pH, and with it the alkalinity. The same is true of lakes concentrating in an open, i.e., drained system (Strakhov, 1962; Maglione, 1976). In aerobic microbial decomposition of dead algal matter, part of the developing carbon dioxide should lower the pH. Instead, the pH stabilizes near 9 regardless of whether the initial pH was 5, 7, or 9 (Otsuki and Hanya, 1972). In two saline lakes in the Tibesti area of Chad, the pH rises slowly to a maximum of 9.6 as the brine precipitates first huntite, dolomite, pirssonite, and gaylussite, and later trona, natron, thenardite, and eventually sylvite (Gac *et al.*, 1979). Phototrophic bacteria and algae remove carbon dioxide and raise the pH beyond 9 in many wind-mixed freshwater lakes. The alkalinity decreases concurrently due to the precipitation of aragonite and calcite (Tschoertner, 1969). The lower solubility of carbon dioxide in brines and the competition for algal oxygen drive up the pH from around 7.5 in open brines and artificial salinas to 9.0+ beneath the density interface in stratified brines. For phototrophic bacteria in Wadi Natrun, Egypt, the optimum is a pH of 8.5–9.0 (Imhoff *et al.*, 1978a).

The pH drops when bacteria attack organic matter or hydrocarbons (Nicol, 1942). Lowering of the pH is then due to a combination of geochemical and biological factors, generating a complex relationship among temperature, solute concentration, pH, alkaline reserve, and absorption of visible radiation (Tarasov, 1939). From values near those for normal seawater at the surface of Solar Lake, a density-stratified hypersaline playa lake on the coast of the Gulf of Aqaba, Red Sea, the pH declines to barely alkaline values at the bottom, where only 30 ppm hydrogen sulfide are found (Eckstein, 1970). This may be an indication of the level of bacterial activity. In experimental evaporation of normal seawater, one observes a drop in pH with increasing concentration (Rieke and Chilingar, 1962), caused by the absence of biogenic sulfate reduction in such experimental evaporation. In artificial salinas as well, a drop in pH is noted as soon as the concentration of calcium ions begins to decline (Herrmann *et al.*, 1973). Calcium removal should increase the pH (Bodine, 1976). The Eh declines only after the brine has concentrated further.

The anaerobic bottom brines carry calcium ions and limited amounts of hydrogen sulfide. Here the bacteria scavenge oxygen and decompose gypsum to calcite (Friedman, 1978), thereby liberating HS^-. Sulfate reduction by bacteria in itself produces a high pH (d'Ans, 1967), as CaS is highly unstable and is hydrolyzed to HS^-, Ca^{2+}, and OH^-. With increasing initial pH and increasing sodium chloride concentration, the Eh decreases in bacterial growth media, and thus bacterial growth rates increase (Stuart and James, 1938; Stuart, 1940).

Addition of salts to either acidic or basic solutions leads to higher H^+ activity, because part of the water is bound in the hydrated form. Organic solvents cause the pH to vary with the dielectric constants of these solvents (Schwabe, 1964). Most metal chloride complexes are slightly acidic. They deform the water shell of the ions, so that hydrogen atoms are loosened and become hydrogen ions. There is then potential food for bacteria.

Among several hypersaline lakes in the Toledo region of Spain, those with the highest magnesium content show pH values above 8 (Bustillo *et al.*, 1978). These lakes are slightly more acidic only in the spring, when chloride as the primary anion is diluted by sulfates due to algal decay at the end of the rainy season. Bodine (1976) reasoned that the decline in pH is a function of increasing magnesium and chloride ion activities and is probably caused by the formation of one of the oxychlorides. To depress the pH, only 27 ppm of magnesium hydroxychloride tetrahydrate [$Mg_2(OH)_3Cl \cdot 4H_2O$] are required during gypsum precipitation; and only 70 ppb are required after the brine saturates for halite. Since this compound breaks down very easily, it is destroyed early in the diagenetic aging process. However, although it has been produced in the laboratory, it has yet to be identified in nature. The apparent source of carbon dioxide explosions in German potash mines is believed to be a magnesium chlorocarbonate ($MgCl_2 \cdot MgCO_3 \cdot 7H_2O$) that occurs in magnesium-rich samples but is absent in potash ores. This compound is very unstable and decomposes into a variety of hydroxychlorides (Schmidt, 1960). Chlorates are much more soluble than chlorides, whereas bromates and iodates are much less soluble than corresponding bromides and iodides.

The voltage differential between a positive Eh of oxygenated surface waters and a negative Eh of anoxic bottom waters is about 400–800 mV (Fig. 4-7). The cations are thus attracted to the bottom brines, whereas the anions should move toward the positive surface layer. This would facilitate the expulsion of CO_3^{2-}, HS^-, and possibly even Cl^-, and the concentration of metal hydroxides in the bottom waters. The electrochemical potential of this system has not been adequately investigated. In this context, Kostenko (1982) observed that the concentration of heavy metals decreased in surface waters and increased in the hypersaline bottom brines that were in contact with precipitating salts. Both surface runoff and discharging groundwater carry dissolved products of chemical weathering, mainly in form of carbonates and bicarbonates. Since their anions have only very limited solubility in concentrated chloride brines, upon contact with the brine they either precipitate as carbonate banks or the carbon dioxide is expelled, leaving an excess of cations.

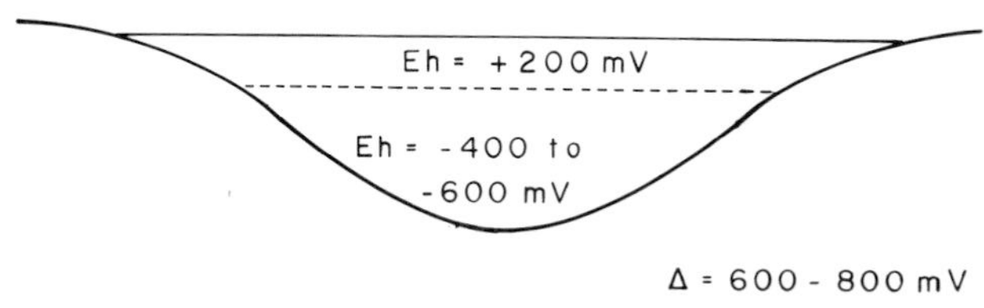

Fig. 4-7. Redox potential difference in a density-stratified brine.

pH in Bottom Brines

The redox potential (Eh) of the bottom layer of hypersaline brines is usually strongly negative (−400 to −500 mV), and the pH hovers around 5.0–9.5. In natural stratified lagoons in the Venezuelan and Dutch Antilles, the author measured a constant pH of 8.4 at the interface and up to 9.6 farther down in the anaerobic chloride brine. Presumably in near-surface waters, Ward *et al.* (1970) found in similar hypersaline lakes in Yucatan a pH of 8.4–8.6 with a salt content of 13.8–16.6%. In sulfate lakes of British Columbia, Canada, the pH rises from 8.3 in surface waters to 9.4 in bottom brines, with the Eh being sometimes slightly positive in surface waters and up to −250 mV in bottom brines. A high pH and a low Eh are due to the high content of organic matter that is trapped when it sinks through the interface from oxygenated surface waters that are teeming with life. This organic matter cannot be converted to carbon dioxide, because of a lack of oxygen. Virtually all organic additives increase the pH to 5.0–9.0 (Kitano and Hood, 1965). A small addition of biogenic nitrogen drives up the pH very rapidly. Even a solution of 10^{-4} ammonia (less than 50 mg of ammonia per cubic meter of brine) produces a pH of 9.1 (Osichkina, 1978a). In an alkaline medium, the equation

$$NH_3 + H_2O = NH_4^+ + OH^- \tag{4-10}$$

shifts to the left and the ammonia escapes to the atmosphere. High NH_4^+ values are associated only with acid formation waters (Bogomolov *et al.*, 1970). NH_4^+ substitutes for K^+ in illites in increasing amounts in proximity to areas of stratiform mineralization in Alaska (Sterne *et al.*, 1982). This can be explained only by an increasing acidity of the pore waters in the vicinity of such mineralization.

A high pH restricts animal life because of larval tolerance limits, and plant life because of the limited carbon source for photosynthesis. At a pH of 9.0, the concentration of carbon dioxide is practically zero, and HCO_3^- is less abundant than CO_3^{2-} (Moberg *et al.*, 1934). Algae unable to utilize HCO_3^- will not photosynthesize in hyperalkaline ponds (Oesterlind, 1950). Plants that are able to assimilate HCO_3^- are limited by the extremely low concentration of that ion in waters with a pH between 9.5 and 10.0 and find the concentration of hydroxyl ions too toxic (Carpelan, 1957).

pH of Brine Encroachment

Seawater overriding saline groundwater in a sebkha on the west coast of the Suez Peninsula yields pH values in excess of 8.5 in the subsoil near the shores of the sea or near the shores of open lagoons, but it produces lower pH readings farther inland (Gavish, 1975). A similar decrease in pH values inland was observed along the Trucial coast, Persian Gulf (Curtis *et al.*, 1963; Butler, 1969). However, if concentrating brine is diluted with distilled water, an increase in pH is the result (Krumgalz, 1980). In open salinas in California, the highest pH values occur in midpond regions (up to 9.8), dropping off toward the shores to 8.1 (Carpelan, 1957), and reach a daily peak in the afternoon. Salt marshes in semidesert environments have a pH of about 8.6 at the surface, dropping to 8.2 at a depth of 25 cm (Chapman, 1974).

LONG-TERM CHANGES

The sequence of primary minerals found in marine evaporite sequences is identical in all Phanerozoic occurrences. This suggests three things: First, the laws of physics and chemistry governing evaporite precipitation were constant over that time interval. Second, where evaporite precipitation is governed by biogenic factors, these also acted at all times during this period. And finally, the ionic composition of the initial brine, whatever its total salinity may have been, did not deviate too far from that of modern brines. Deposition of major volumes of evaporites, e.g., the Permian ones in Europe and North America, must have had an impact on the salinity of the world ocean. The Zechstein evaporites contain about 10% of the sulfate now dissolved in the world ocean (cf. Holser, 1979b); if we add the sulfates in the Permian basins west and south of the Urals, the amount is even larger. The volume of Neogene sulfates and halides precipitated in various parts of Eurasia must have extracted enough salts to reduce the open marine salinity by at least 15–20%.

Virtually all cations dissolved in seawater can be derived from the weathering of preexisting Archean rocks. However, this is not the case with anions, specifically sulfate and chloride ions. They must have been present in the primordial ocean, if not in the atmosphere, since weathering would not produce the quantities now dissolved in the oceans and stored in evaporite deposits (Lotze, 1957). If the cations were added by weathering processes, the primordial ocean should have been excessively acidic; however, this was not the case for at least the mid-Archean ocean, judging by the occurrence of carbonate banks and algal precipitates.

Valyashko and Lavrova (1976) determined the Br/Cl ratio of many halites and their fluid inclusions, extracted in alcohol. Alcohol was used because halite is almost insoluble in it. The authors then plotted the data against geological time. For all Cambrian to Recent halite deposits, the values can be plotted on a single curve, i.e., they are all of the same generic type, derived from the concentration of seawater. Both bromide and rubidium have been used for an interpretation of a variety of environmental parameters prevailing at the time of deposition, such as brine temperature, concentration, distance from shore, depth of the lagoon, and others. The use of bromide and rubidium for such interpretations of ancient marine evaporite sequences is based on the premise that they were deposited from concentrating seawater not too different from present-day oceanic waters in its initial Br/Cl and Rb/K ratios. No evidence has yet been uncovered to dispute this contention. Yet it is obvious that volcanic activity contributes quantities of water vapor, sulfur, and chlorine to surface waters; it is unlikely that the proportions of ions going into solution to exhalated juvenile water have been constant through time. Consequently, one must assume that the presence of major buffering agents, such as evaporite precipitation, clay conversion, dolomitization, silicification, or hydration of silicate minerals in weathering, to name but a few, is sufficient to hold seawater composition to within narrow limits.

Nevertheless, long-term changes are observable in evaporite composition and mode of deposition. Middle Proterozoic and older gypsum pseudomorphs (McClay and Carlyle, 1978; Golding and Walter, 1979) indicate that evaporitic conditions and the partial pressure of oxygen required to produce sulfate precipitates have been available for the last 1.5–3.0 billion years. These are essentially pseudomorphs in what are the first red beds, and may thus be equated with groundwater-derived similar pseudomorphs in younger sediments. They are not derived from evaporation of an exposed brine surface.

The first bedded anhydrites, evidently precipitated in an open body of water and preserved and not replaced by other mineral suites, are Middle-to-Late Proterozoic in age; bedded barite is preserved in even older, Archean horizons (Heinrichs and Reimer, 1977; Walker *et al.*, 1977; Barley *et al.*, 1979). These occurrences indicate a significant oxygen level in the local environment and predate the appearance of the first metazoa (Guillou, 1972) or the end-Proterozoic substantial increase in the $\delta^{34}S$ value of sulfur deposits (Claypool *et al.*, 1980). The cause of this increase may be related to an increased presence of anaerobic sulfate scavengers in addition to a spread of photosynthesizers in hypersaline brines and ocean waters. The sharp increase in $\delta^{34}S$ values (Fig. 4-8) takes in almost the whole Ediacarian period, but essentially reached its peak in the Lower Cambrian (J. Chen *et al.*, 1981). The cause of this increase may be related to an increased oxygen supply in the atmosphere due to a rapid spread of photosynthesizers in hypersaline brines and ocean waters. All Paleozoic salt basins are deficient in halite compared to calcium sulfates, but this deficiency becomes progressively smaller toward the end of the era. At the same time, potash salts become more frequent, reaching a worldwide peak in the Permian period. Calcium salts are the only sulfates present in pre-Permian salt basins. Both bischofite and kieserite are absent, and indeed, no massive potassium and magnesium sulfates are known from pre-Permian evaporites (Zharkov, 1978; Zharkov *et al.*, 1979).

The $\delta^{34}S$ value reached a minimum at the end of the Paleozoic era, but has been rising slowly ever since without ever again reaching Paleozoic values (Schidlowski *et al.*, 1977). A more detailed investigation reveals a subordinate low in the Lower Cretaceous, and high values in the Cambrian and Upper

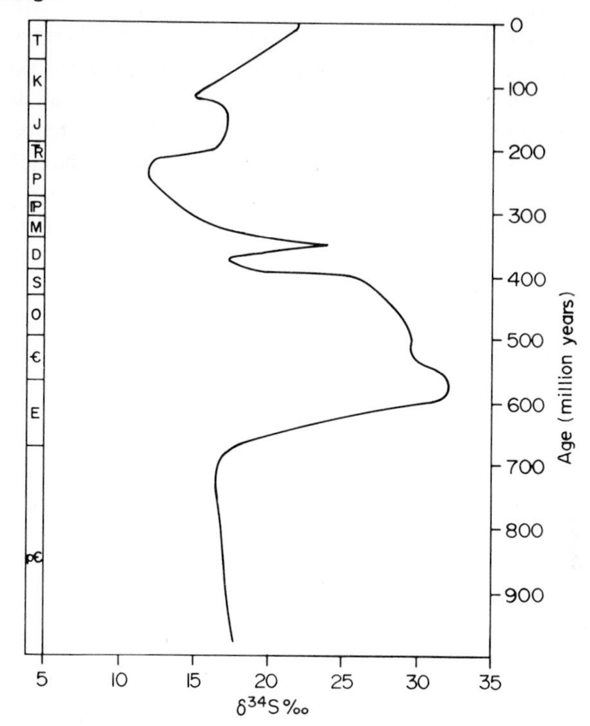

Fig. 4-8. Summary sulfur isotope curve for sulfates (after Claypool *et al.*, 1980).

Devonian (Claypool *et al.*, 1980). Only major oscillations in the $\delta^{34}S$ curve should be taken as significant, as $\delta^{34}S$ values are not uniform; they can vary by more than 1.5% within a single bed less than 40 m thick or in (six to eight) consecutive beds. There can be a variation of 2% in contemporary occurrences 130 km apart or in beds belonging to the same cyclothem (J. Chen *et al.*, 1981). Tokyo Bay today registers a $\delta^{34}S$ value 6 ppt lower than the open ocean (Sakai, 1957), whereas the algal bloom in Lake Erie causes a swing in the opposite direction, to 6 ppt more than in the ocean. Nonetheless, the variation in the isotopic composition of coeval sulfates is confined to a very small range.

The Phanerozoic minimum of $\delta^{34}S$ values, which occurred in the Permian, coincides with a minimum in the $\delta^{18}O$ value of marine sulfates (Sakai, 1972; Claypool *et al.*, 1980), and with the appearance of massive secondary potassium and magnesium sulfate minerals. One potential explanation can be offered for the extremely high $\delta^{34}S$ values in pre-Permian sulfate deposits. If sulfate-reducing bacterial strains were particularly effective in Paleozoic hypersaline brines, the heavy sulfur would preferentially accumulate in precipitating gypsum and the lighter sulfur isotope would eventually escape into the atmosphere. If the bacterial and algal populations were decimated toward the end of the Permian period in marine environments, as L. R. Moore (1969) has suggested, then Late Permian seawater percolating into as yet poorly compacted Permian evaporites would allow sulfate ions to react unhindered with precipitated potassium and magnesium chlorides. To this day, the $\delta^{34}S$ values have not attained the high Paleozoic values; one could therefore surmise that the sulfate reducers have not recovered completely from their end-Permian decimation. The absence of sulfates other than calcium salts from pre-Permian evaporites is then explained by the greater efficiency of sulfate reduction in Paleozoic periods (Fig. 4-9), rather than their complete dissolution. A greater consumption of sulfate ions by bacteria left, of course, more magnesium available for dolomitization. This could explain the widespread dolomitization of Paleozoic carbonates.

The increasing utilization of sulfur in the Paleozoic goes hand in hand with progressively increasing C_{org} storage, which declined thereafter. The Palezoic transfer of oxygen from carbon dioxide to sulfate amounted to several times the total free oxygen content of the Holocene atmosphere (Garrels and Lerman, 1981). This could be interpreted to mean a larger rate of accumulation of bacterial and algal organic matter in the Paleozoic than thereafter. The $\delta^{13}C$ values show a gradual enrichment of about 2.8 ppt in the Permian (Magaritz and Turner, 1982), which represents about double that amount in organic matter buried in sediments. Holser (1977) has noted that drastic increases in the $\delta^{34}S$ values coincide with sudden increases in the $^{87}Sr/^{86}Sr$ ratio, which is raised by 2×10^{-4} in the Late Proterozoic, by 1.3×10^{-4} at the end of the Middle Devonian, and by 1.0×10^{-4} in the early Triassic, only to slide back to lower values after each increase. The very low values in the Permian (Fig. 4-10) and again in the Jurassic (Veizer and Compton, 1974; Burke *et al.*, 1982), however, do not seem to correlate; they could very well be related to salinity changes in the world ocean. The overall decrease in the $^{87}Sr/^{86}Sr$ ratio from the Late Proterozoic to the Jurassic or Early Cretaceous and the gradual increase thereafter to present values comparable only to Early Cambrian ones, cannot be caused by water temperatures: The Ordovician and Permian glaciations correspond to low values, the Pleistocene one to high values. The heavy oxygen values ($\delta^{18}O$) also change; in evaporitic carbonates, $\delta^{18}O$ increases systematically from Cambrian to Tertiary samples (Pilot *et al.*, 1972).

High salinity, high relative humidity of the overlying atmosphere, and high initial isotope composition each restrain isotopic enrichment. Nevertheless, the $\delta^{18}O$ enrichment in closed drainage basins not connected to the open sea will rise above 6 ppt (Lloyd, 1966) and may then even exceed 30 ppt (Fontes and Gonfiantini, 1967b). Crystallization leads to an enrichment of heavy oxygen in gypsum (Gonfiantini and Fontes, 1963), and quite possibly also of heavy hydro-

---

Fig. 4-9. Distribution of evaporite deposits in time. Bedded deposits are restricted to post-Ediacarian formations, potassium-magnesium sulfates to Permian and Neogene formations.

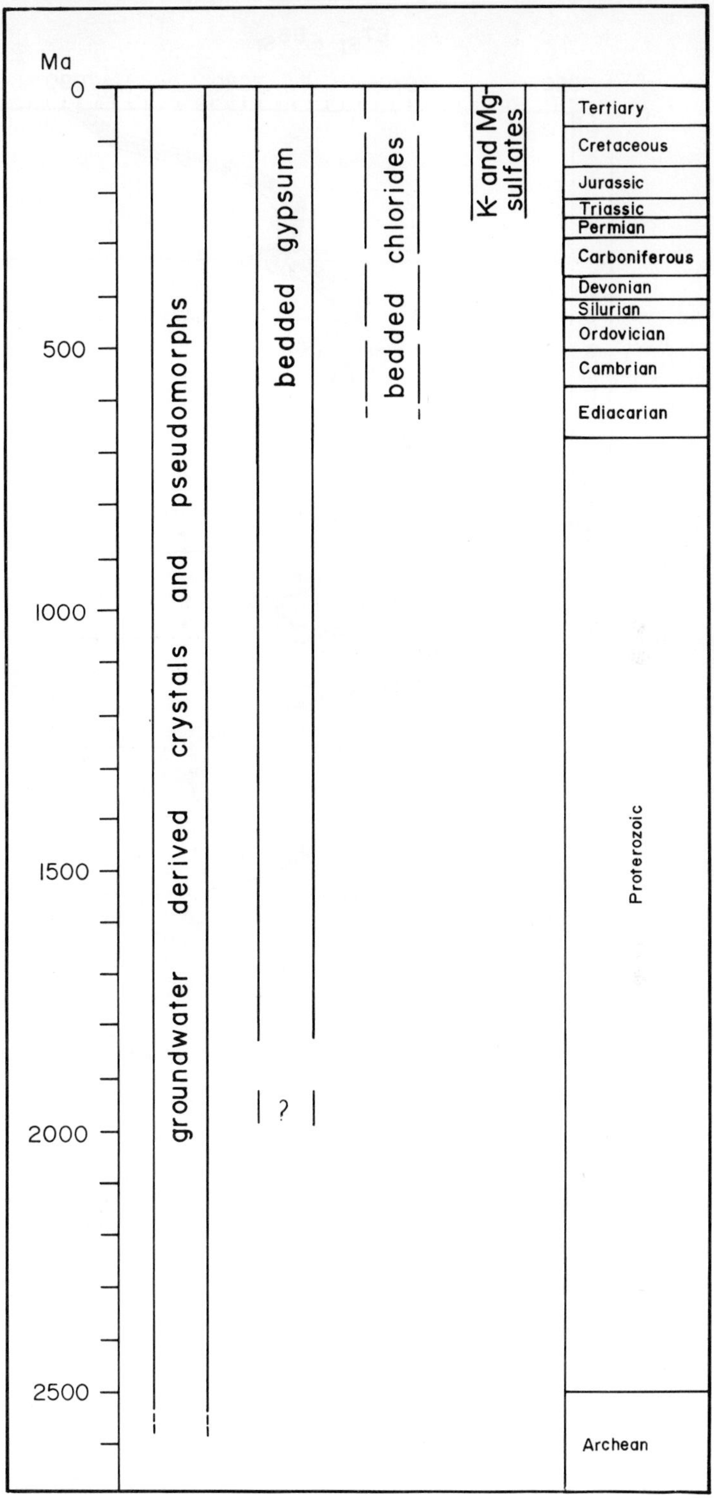

Ma
0
500
1000
1500
2000
2500

groundwater derived crystals and pseudomorphs

bedded gypsum

?

bedded chlorides

K- and Mg-sulfates

Tertiary
Cretaceous
Jurassic
Triassic
Permian
Carboniferous
Devonian
Silurian
Ordovician
Cambrian
Ediacarian
Proterozoic
Archean

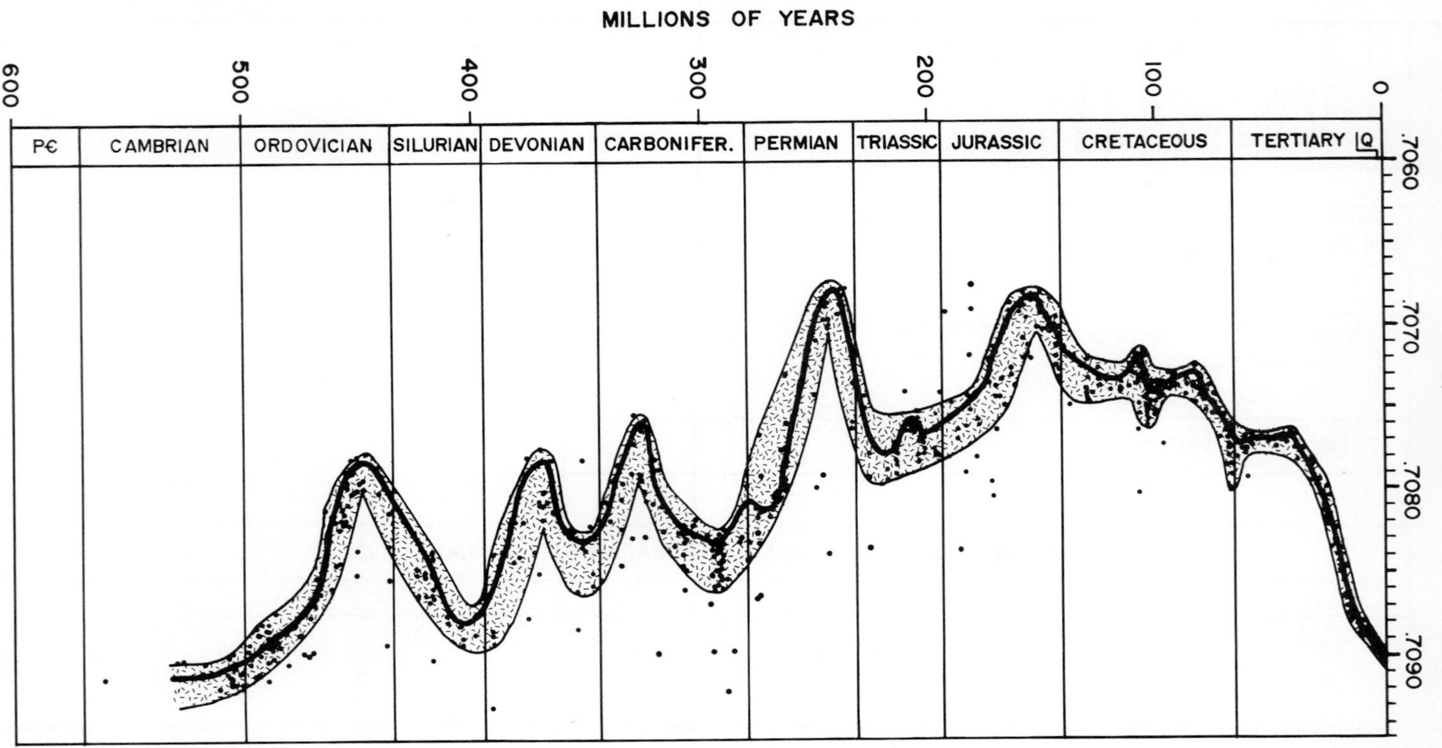
MILLIONS OF YEARS
600
500
400
300
200
100
0
P€
CAMBRIAN
ORDOVICIAN
SILURIAN
DEVONIAN
CARBONIFER.
PERMIAN
TRIASSIC
JURASSIC
CRETACEOUS
TERTIARY
Q
.7060
.7070
.7080
.7090

87Sr / 86Sr

gen, as initially they are left in the water but are later preferentially incorporated from more concentrated brines. In contrast, hydrogen incorporated into lacustrine carbonate and borate minerals discriminates in favor of the lighter isotope. Further study is suggested, however (Holser, 1979c). Under certain circumstances, deuterium and ^{18}O values in hypersaline brine deposits can approach those of continental freshwater deposits (Jauzein, 1982). If one plots oxygen isotope values against carbon isotope values of evaporitic limestones, one must also know the paleotemperature of the brines in which the limestones were formed. If the brine contained a hot water lens, as most density-stratified brines do, then on a plot of carbon versus oxygen isotopes, the limestone precipitated in a hot brine plots in the freshwater field and simulates strong continental isotopic affinities (A. Jauzein, personal communication, 1982), as the salinity effect overwhelms the temperature effect of oxygen isotope partitioning. Consequently, many limestones precipitated in hot hypersaline brines can be misinterpreted as freshwater derived if only isotope ratios are considered. On the other hand, if we keep in mind that most of the limestone accretion occurs in surface waters, which can represent a brackish-to-fresh influx, this accretion would likely contain a fauna with calcified skeletal remains. When the limestone subsides below the thermocline, it is exposed to hot brines, but isotopic changes may result only if suitable ions are available. Moreover, since most limestone banks represent nearshore deposits, the impact of seasonal flash floods on both carbonates and sulfates is likely to be considerable.

In contrast to sulfur, oxygen, and carbon isotopes, the chlorine isotopes do not fractionate. The ratio $^{35}Cl/^{37}Cl = 3$ is the same in samples from various sources (Hoering and Parker, 1961). Whether the magnesium isotopes fractionate has not been adequately investigated because of some analytical difficulties with separation. There seems to be an indication that the heavier magnesium isotope does not travel as readily through an aquifer as the lighter one when limestone aquifers are being dolomitized.

Massive primary tachyhydrite occurs only in Lower Cretaceous basins now located in very low latitudes. This cannot be explained by assuming progressively greater isolation of evaporite-precipitating basins alone; such isolated basins must have existed in other periods. The mineral is stable only above 21.95°C (van t'Hoff *et al.*, 1912), and this minimum temperature is raised by 1°C for every 2300 m of overburden. Salts buried several hundred meters maintain such a temperature on a year-round basis, but the minimum burial needs to be deeper as the latitude becomes higher. Exposed salts become subject to seasonal cooling and potential decomposition.

There is no halite or potash deposition for a 35 to 40-million-year period

---

Fig. 4-10. Phanerozoic strontium isotope ratios in seawater (after Burke *et al.*, 1982, *Geology* **10,** 516–519, but thereafter corrected by the authors).

straddling the Cretaceous–Tertiary boundary (Sonnenfeld, 1978). There are similar periods devoid of halite deposits in the Lower Ordovician (Tremadoc, Arenig, and Llanvirn) (Lefond, 1969; Zharkov, 1978) and parts of the Pennsylvanian (Zharkov, 1978). The latter author also lists a Lower Silurian (Llandoverian and Wenlockian) time interval. In all other periods of Phanerozoic earth history, there occur marine evaporites yielding massive rock salt and potash deposits in some parts of the world.

Although one could derive a galactic periodicity, there is a much simpler explanation available for time intervals lacking chloride deposits. This explanation is based on periods of excessive rainfall and high solar radiation, as depicted by rapid plant development in each of the time intervals devoid of salt. The Lower Ordovician is likely the time of the first appearance of lichens, the time when other nonvascular plants became the first occupants of terrestrial environments, of soggy soils in river flats. The early Silurian marks the first appearance of club mosses as the first vascular plants spreading across the continents. The Lower Pennsylvanian marks the rapid worldwide expanse of the coal forests. The end-Cretaceous/early Tertiary is not only the time of progressive decimation of primarily large animals but also the time of rapid diversification of flowering plants. Considerable evidence has been presented that the last of these intervals, around the Mesozoic–Cenozoic boundary, is caused by a breakdown of the ozone layer and consequent unhindered influx of ultraviolet radiation (Sonnenfeld, 1978). If that conclusion has merit, then this absence of an ozone layer and consequent high solar radiation might also be the cause of the absence of bedded chloride deposits in Precambrian strata (Fig. 4-9).

Sodium, sodium-calcium, and hydrous magnesium carbonates are known only from Cenozoic rock suites. This is not to say that they did not form in earlier periods. Any buried carbonate rocks would have come in contact with hygroscopic formation waters enriched in calcium chloride, and would thus be altered to calcium or calcium-magnesium carbonates in their presence.

It has been proposed that early Paleozoic seas were low in their potassium content and that potash deposits could not form until excessive land vegetation preconcentrated potassium and released it to the rivers (Rutherford, 1936), or that there could not be any Devonian potash deposits (Ochsenius, 1897). Such ideas are no longer tenable in view of the discovery of Devonian potash deposits in Saskatchewan, in Byelorussia, and in the Ukraine, Silurian ones in Michigan, and Cambrian ones in Siberia. We must also guard against reaching hasty conclusions about the hitherto apparent absence of any other rock type from Paleozoic or early Mesozoic evaporites. A hypothesis deriving thick evaporite deposits from large-scale evaporation produced by the approach of giant stars or stellar systems (Omori, 1960) cannot be sustained. Evaporite-free time intervals are in the minority, and thick deposits are too frequent, spread over all continents (Kozary *et al.*, 1968). No other effects (earth tides, etc.) that would result from such galactic encounters are evident. Similarly, a hypothesis about the submarine volcanic origin of rock salt deposits (Rode, 1944) is not borne out by any field observation of evaporite beds.

SUMMARY OF BRINE CHARACTERISTICS

All evaporites form in the range of 218–308°K. As the brine concentrates, its increasing salinity raises the hygroscopy, lowers the temperature range of the stability of anhydrous minerals, and changes all of the physical properties of the brine, such as viscosity, specific heat, compressibility, specific volume, boiling point, rate of evaporation, and thermal expansion. It also changes its chemical properties through partial precipitation, adsorption of cations of active clay and silica gel surfaces, and the biological activity of algae and anaerobic bacteria. With the removal of major quantities of hydrogen sulfide, a percentage of the cations must enter into hydroxide bonds. Sulfate precipitation is thus soon overtaken by chloride precipitation because of a scarcity of sulfate ions. The pH of the brine remains near that of seawater in the surface layer but rises in the bottom brines due to silicate hydrolysis and ammonia dissolution. The redox potential is positive in the surface waters exposed to the atmosphere but becomes negative beneath the interface with bottom brines, giving rise to a voltage differential on the order of 400–800 mV.

Since the sequence of primary minerals is identical in all Phanerozoic occurrences, the composition of seawater could have varied only within narrow limits. However, there are variations in the sulfatization of evaporite minerals that can be explained by varying the intensity of sulfate destruction by anaerobic sulfur bacteria.

Part II
PRIMARY PRECIPITATION

5

The Continental Environment

Evaporites form on the continent in hypersaline lakes or in the soil as groundwater precipitates. Their chemistry is specific for each particular drainage environment and differs from that of marine evaporite sequences. Sodium and magnesium carbonates and sulfates are prominent, but there is a considerable overlap of individual evaporite mineral parageneses. We can distinguish a polar regime in which bacterial activity is minimal and hydrated minerals dominate, because their stability field lies within cold temperatures. A second environment is offered by open lakes in endorheic basins that are either groundwater fed or replenished by surface runoff. In both cases, the waters often deliver redissolved older salts. Finally, groundwater percolating into shallow soil horizons evaporates to form individual crystals or widespread crusts.

THE POLAR REGIME

Melting ice readily produces a surplus in the water budget, and with it an estuarine water circulation. The rates of evaporation from open water or sublimation from ice surfaces are too small to cause a reversal to an antiestuarine water circulation. Hypersaline lakes and lagoons maintain an open water surface longer in the polar environment due to the depression of the freezing point (up to 21°C) by solute content. Lagoons open to the sea, even where icebound, do not develop brine concentrations conducive to evaporite precipitation. No such hypersaline lagoons are known from either the Greenland or antarctic coasts. Massive marine evaporites form only in subtropical lagoons with high rates of evaporation. It is thus most unlikely that the Permian halites and potash deposits of West Texas and New Mexico are derived from the cooling of waters flowing under an ice cover, as was suggested by Udden (1924).

Evaporite minerals in the polar environment form either from drying of sea spray or from evaporative concentration of dissolved weathering products. Torii *et al.* (1981) maintained that snow and ice are the main sources of salt in the Dry Valley lakes of Antarctica. Keys and Williams (1981) observed that chloride minerals decrease in frequency landward with increasing distance from antarctic coasts, which is what occurs at coasts in other climatic zones. In contrast, sulfates and nitrates increase away from the coast. As the contribution of marine sodium declines, the weathered-out calcium and magnesium become more important cations in the polar continental environment; they are transported primarily by groundwater. To summarize, recystallization processes that result in the separation of compounds by fractionation are the prime cause of salt distribution, followed by migration through deflation and groundwater movement. The proximity of marine sources and the presence of suitable weathering substrates do not seem to be as important. Anaerobic sulfate reduction is subdued, but not absent; photosynthetic and heterotrophic bacteria are common along density interfaces; and algal mats can produce an effervescence of oxygen bubbles. Below the interface, hydrogen sulfide, methane, and ammonia are produced from a rain of organic compounds (Burton, 1981). The contribution of biota to the lake chemistry is thus not negligible even in the extreme rigors of a polar environment.

Sodium Sulfates

Seawater concentrated by evaporation becomes saturated first with calcium carbonate and then with gypsum at a 4.5-fold concentration. However, seawater concentrated by freeze-drying is saturated with mirabilite ($Na_2SO_4 \cdot 10H_2O$) (Matsubaya *et al.*, 1979) at a 4-fold concentration, due to the colder temperature regime and to a concomitant decrease in biological sulfur consumption (Fig. 5-1). Overall, primary sodium sulfate and sodium–calcium sulfate precipitation is restricted to waters exposed to very low temperatures or to those of continental origin. In each case, they show the presence of an excess in sulfate ions, i.e., an oxidizing environment. Gypsum does not precipitate in the presence of sodium bicarbonate; calcium carbonate and sodium sulfate form instead (Delecourt, 1946). At 0°C, a brine saturated for sodium and magnesium chlorides can contain only 5000 ppm sodium sulfate (Grokhovskiy and Grokhovskaya, 1980). Mirabilite is restricted by climate to higher latitudes (Strakhov, 1962). Thus, sodium sulfates in marine lagoons are a measure of seasonal cooling.

Mirabilite precipitates in the shallower horizons of antarctic lakes or along the coast. Continental runoff precipitates mirabilite onto any available substrate,

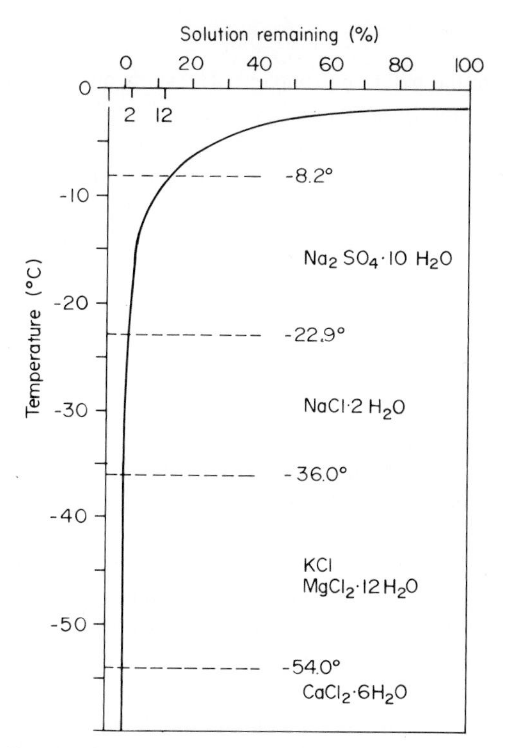

Fig. 5-1. Crystallization of salts caused by freezing of seawater (from Dort and Dort, 1970).

even the Ross ice sheet (Brady *et al.*, 1979). The upper temperature limit of mirabilite is 13°C (Krull, 1919) in a seawater-derived brine, but it remains stable up to 32.3°C in fresh water. In lacustrine brines, mirabilite crystallization begins at 14–16°C (Vlasov and Gorodkova, 1961). The extent to which an algal mat covering the mirabilite beds is involved in the precipitation has not been investigated. The mirabilite precipitation in supraglacial and periglacial meltwater lakes of Antarctica is derived from sulfate-enriched glacial waters. These waters redissolved thenardite left behind in evaporation of capillary-drawn water above the permafrost (Bowser *et al.*, 1970). Thenardite (Na_2SO_4) is also formed by subaerial dehydration (Dort and Dort, 1970) and is thus often a secondary mineral.

Also found in this environment are epsomite, nitrates from weathering [such as darapskite ($NaNO_3 \cdot Na_2SO_4 \cdot H_2O$) and soda niter ($NaNO_3$)], iodate-enriched soils (Gibson, 1962), and sundry hydrous iron, copper, and aluminum sulfate efflorescences (Vennum, 1979). However, compared to lacustrine environments, sodium carbonates play a very subordinate role.

Gypsum

Gypsum is rare in the polar environment. The snow on the islands of Svalbard (Spitsbergen) is covered by a layer of gypsum, believed to be derived from salt spray and windblown seawater from a freezing ocean (Corbel *et al.*, 1970). Antarctic gypsum, unlike its subtropical counterpart, contains significant quantities of iodates (Johannesson and Gibson, 1962). However, because there is little gypsum precipitation and no large-scale biogenic aragonite extraction, much of the calcium remains dissolved in cold brines, despite the increased solubility of atmospheric carbon dioxide in cold waters, and ends up in groundwater enriched with calcium chloride.

Sodium Chloride

At an eightfold concentration, freeze-drying saturates the brine for hydrohalite ($NaCl \cdot 2H_2O$) (Matsubaya *et al.*, 1979), a mineral unstable at temperatures above freezing. Hydrohalite does not begin to precipitate from seawater brines above −23°C (Thompson and Nelson, 1956), but at a slightly higher temperature in Siberian lakes (Dzens-Litovskiy, 1968), where interference by magnesium ions is lacking. At the bottom of these deep lakes, hydrohalite crystallizes out where the deeper waters do not heat up. Where steep shores do not allow warm waters to form on shoals and to slide thence into deeper parts, the bottom brines remain perennially cold. Hydrohalite precipitates in the deep, cold waters of antarctic ocean-fed lakes, even where a solar hot water lens has formed at shallower depths (Craig *et al.*, 1974). Part of the deposit is covered by a porous algal mat at depths below 28 m, indicating sufficient exposure to sunlight. Ultimately, the sunlight will convert the hydrohalite to halite during the brief southern summer.

Precipitates of calcium and magnesium chlorides increase their water content as the brine's temperature goes down. The same can be assumed to be true of sodium chlorides. Indeed, Dravert' (1908) reported minerals from Lake Eskulap, Siberia, with the chemical formulas $2NaCl \cdot 5H_2O$, $NaCl \cdot 2H_2O$, and $NaCl \cdot 10H_2O$, but these have not been confirmed.

Potassium Salts

Potassium sulfates are less soluble than sodium sulfates at temperatures below 11°C (Delecourt, 1946), and thus should precipitate in the subpolar environment wherever potassium becomes available. They are rarely found because potassium ions are easily taken up by clay minerals. Another solid mineral at freezing temperatures is carnallite ($KCl \cdot MgCl_2 \cdot 6H_2O$) (Igelsurd and Thompson, 1936), which can be precipitated when potassium is supplied. In Siberian lakes, bloedite always forms before halite saturation is reached (Valyashko, 1962). However, some of the halite may occasionally contain a tiny fraction of sylvinite (Rubanov, 1966).

Magnesium Salts

Sakiite (magnesium sulfate hexahydrate) exists down to 3.7°C (Pel'sh, 1953), not just down to 13°C, as was previously assumed. The dodecahydrate and the octohydrate of magnesium chloride ($MgCl_2 \cdot 12H_2O$ and $MgCl_2 \cdot 8H_2O$), as well as the dodecahydrate of magnesium sulfate ($MgSO_4 \cdot 12H_2O$), are stable only near freezing temperatures (van t'Hoff *et al.*, 1912; Strakhov, 1962). Similarly, carnallite disintegrates if the temperature drops to −12°C (van t'Hoff *et al.*, 1912).

Calcium Chlorides

Calcium eventually precipitates as tachyhydrite [$CaCl_2 \cdot (MgCl_2)_2 \cdot 12H_2O$] in the summer or as antarcticite ($CaCl_2 \cdot 6H_2O$) in the winter, together with ikaite (hydrocalcite) ($CaCO_3 \cdot H_2O_{1-6}$). Ikaite is also found as a stable mineral at temperatures between +7°C and −4°C at Ika Fjord in Greenland (Pauly, 1963a,b). Whereas antarcticite precipitates only in very cold brines, the minimum temperature for tachyhydrite precipitation is 19.5°C (d'Ans, 1933) or 21.95°C (van t'Hoff *et al.*, 1912; d'Ans, 1961). The antarcticite is derived from $CaCl_2$ brines supplied by continental groundwater. Don Juan Pond in Antarctica loses per year two to three times its volume of water to evaporation and sublimation without going dry (Harris *et al.*, 1979).

Hypersaline lakes and salt efflorescences in Arctic environments have been investigated much less than occurrences in Antarctica, although they have been frequently reported as containing mainly sodium and magnesium sulfates leached out of bedrock (Jensen, 1889; Nordenskjoeld, 1914; Boecher, 1949; Fristrup, 1953; Soerensen *et al.*, 1970a,b). Only highly alkaline polar lakes and soils accumulate salts. Lakes with acid or neutral pH are very low in salt content (Boecher, 1949).

SUBTROPICAL LAKES

Continental lakes that precipitate evaporite minerals occur only in hydrologically closed basins, i.e., basins in which evaporation exceeds inflow. Such basins are called "endorheic," because they are fed by inflow, but have no significant outflow and often are also fed by groundwater influx. Rainwater facilitates weathering reactions that become the chief sources of solutes. The greater the net evaporation, the smaller is the area of a closed lake. In turn, the smaller the lake area, the more numerous are the topographic opportunities for closed lakes to form (W. B. Langbein, 1961). Small closed lakes are thus far more frequent than large ones. Most lakes that precipitate evaporite minerals occur in areas of less than 500 mm/year atmospheric precipitation and a ratio of net evaporation (in millimeters) to temperature (in degrees Celsius) between 50 and 100. Many such closed endorheic basins occur in the lee of mountain chains, in their rain shadow. Where there is no adequate supply of inflow to maintain an open water surface, such a lake soon turns into a salt marsh and eventually dries out completely. Most closed lakes are thus transitory. Eugster and Hardie (1973) and Eugster (1980a) have given excellent reviews of the environmental conditions of such lakes.

Both vertical and horizontal stacking of evaporite minerals are common in closed lakes. Hunt (1960) observed in the salt pan of Death Valley, California, that chlorides occur in the center and on top, sulfides around or below the chlorides, and finally, carbonates around and below the sulfates. In Saline Valley, California, the halite in the center is surrounded by glauberite, the latter by gypsum, and the rim is occupied by alkali earth carbonates (Hardie, 1968). A similar concentric zonation of precipitates has been observed in the Chad Basin in Africa (Maglione, 1974b), where again the most soluble minerals occur in the center. This lateral zonation of evaporite minerals is the consequence of differences in solubility that produce a fractionation of anions (Eugster, 1980a). The most soluble compounds travel the farthest.

Sodium Salts

Simple and compound sodium carbonates are common among early-stage precipitates of lacustrine water concentration. Primary sodium carbonates are specific to the subtropical lacustrine environment; they are found in neither the polar environment nor marine lagoonal precipitates. The saturation of carbonate or bicarbonate waters requires a much smaller degree of concentration than the saturation of sulfate or chloride brines. *The progressive concentration of brines in alkaline lakes leads to a preferential precipitation of sodium carbonates followed by sulfates and chlorides or of corresponding magnesium compounds.*

Sodium sulfates and carbonates rarely survive bacterial diagenesis after burial, and thus are exceptional in ancient sediments. For example, lacustrine marls of Miocene age contain glauberite ($CaSO_4 \cdot Na_2SO_4$) and thenardite (Na_2SO_4) at various localities in northern Spain (Rios, 1968). Another fossil example is furnished by trona beds ($NaHCO_3 \cdot Na_2CO_3 \cdot H_2O$) and shortite ($Na_2CO_3 \cdot 2CaCO_3$)-bearing shales in the Eocene Green River Formation of Wyoming (Deardorff, 1963). Considerable replacement is common: Shortite is altered to northupite directly or via gaylussite or pirssonite, or even a trona–dolomite mixture as intermediary, or it is altered to bradleyite (Fahey and Mrose, 1962;

Maglione, 1974a). The waters seem to have been deficient in sulfates, magnesium, and potassium. Desiccation was preceded in each cycle by precipitation of halite in trona crystal interstices in the central or lowest part of the lake, giving the facies distribution a bull's-eye pattern. Similar halite precipitation in trona interstices without the occurrence of sulfates has been reported from Lake Magadi, Kenya (B. H. Baker, 1958). It is likely that in both instances sulfates are scavenged by bacteria in the anaerobic environment which preserved the bitumina of the Green River shale.

According to Strakhov (1962), soda lakes form mainly in terrain with exposed crystalline rocks or with polymict feldspathic or argillaceous sandstones. On the periphery of soda lakes are sodium-magnesium lakes. Cation exchange with clays produces calcite or gypsum precipitation, and the lakes are gradually enriched with magnesium sulfates and chlorides, while being depleted of carbonates. They eventually turn into sulfate lakes. Introduction of groundwater enriched in calcium chloride then turns sulfate lakes into chloride lakes under continued gypsum precipitation.

Sodium sulfates are common among lacustrine precipitates, in part again because of the lower Ca/Na ratio in lakes. They precipitate around Great Salt Lake, Utah, and around many lakes in arid and semiarid regions. Gaylussite, gypsum, and mirabilite can grow interstitially in mud flats soaked in brine (Eugster, 1980a). The solubility of thenardite decreases with rising temperature, but that of mirabilite increases with temperature. At a concentration of 205–260 ppm sodium chloride, the transition temperature is 25°C, but it is lowered in brines with a higher sodium chloride concentration (Block *et al.*, 1968). Bacterial reduction of sodium sulfates leads to sodium carbonates, e.g., in Wadi Natrun, Egypt (Gubin and Tsekhomskaya, 1930; Verner and Orlovskiy, 1948). The alkalinity is related quantitatively to the sulfate reduced, but is significantly higher than theoretically calculated (Abd El-Malek and Rizk, 1963; Imhoff *et al.*, 1978b).

Primary sodium carbonates and sulfates do not occur in subtropical marine evaporite sequences, because of the higher Ca/Na ratio in normal seawater compared to most continental brines. A late-stage epsomite precipitation recrystallizes in the presence of halite and water to mirabilite, leaving magnesium chloride in solution; this reaction is easily reversible. Other sodium sulfates occur as nests in evaporite sequences only when a late-stage sulfate concentration has brought the brine into equilibrium with epsomite or when meteoric groundwaters have gained access to the precipitated evaporite series. However, there are also lacustrine salt deposits composed of gypsum, sodium, and calcium carbonates, and sodium and potassium chlorides with a total absence of magnesium (Lopez and Borrell, 1944).

Lakes with Cl^- as the principal anion precipitate halite and other minerals akin to those found in subtropical marine evaporites. However, the source of the chlorine is often an older leached halite deposit. Congruent dissolution of soluble minerals is an important mechanism for recycling evaporites (Eugster, 1980a). Lake Ushtagan and adjacent lakes in central Kazakhstan precipitate halite and concentrate calcium chloride in the brine as the dominant solute (Posokhov, 1949). The deficiency in sulfates and magnesium can only be explained by groundwater supplying a calcium chloride/sodium chloride brine.

Compound Sodium Salts

Tertiary glauberite in valleys of the Tien Shan Mountains decomposes in the weathering zone into mirabilite, thenardite, and gypsum (Shcherbina, 1948, 1949). The groundwater saturated with sodium sulfate forms in winter mirabilite ice bodies that are up to 7.5 m thick (Dzens-Litovskiy, 1962). However, glauberite also forms with gypsum as precursor (Eugster and Hardie, 1973; Kolosov *et al.*, 1974); both burkeite and aphthitalite may form by recrystallization of an earlier precipitate. Friedel (1976) suggested that soil encrustations are also formed by the metastable $Na_2Mg(SO_4)_2 \cdot 5H_2O$, which then alters to bloedite. The latter is frequently found in small quantities in continental playas; a specimen from the Konya Basin in Turkey led to its designation as "konyaite" (van Doesburg *et al.*, 1982) in lieu of the previously used term "protoastrakhanite." Some salt lakes in central Asia develop two horizons of halite–bloedite deposits, as a density-stratified brine is supersaturated for one compound and reaches saturation for another precipitate that requires crystallization centers (Poyarkov, 1957). The crystal habit of the bloedite is then a function of the concentration of the brine (Bokiy and Kachalov, 1963).

Calcium Salts

Calcium compounds are generally subordinate to sodium compounds in continental environments. This is due to the property of waters rich in sodium carbonates to precipitate calcium carbonate and magnesium carbonate very early in the concentration cycle. Calcium carbonate crusts form on the floor of some lakes in the dry western region of Alberta, Canada; calcium sulfate occurs in several closed continental basins of the Chilean Altiplano (Hurlbert *et al.*, 1976; Risacher, 1978). Monohydrocalcite has been reported from both Lake Kivu (Africa) and Solar Lake (Sinai) even in the presence of aragonite (Stoffers and Fishbeck, 1974).

In a mixture of dissolved calcium and sodium sulfates, gypsum precipitates only when calcium sulfate exceeds 1.6–2.0 ppt (Hill and Wills, 1938). Mirabilite or glauberite forms at lower calcium sulfate concentrations. Once sulfate and chloride take over as the principal anions, sodium carbonates are no longer precipitated. Any remaining small amount of carbonate and bicarbonate ions is tied to a portion of the available calcium and magnesium. From rather limited, mainly nearshore sampling, the Gulf of Kara Bogaz Gol was interpreted to have a roughly circular distribution of calcium-sodium sulfates and halite in the relatively dry period of the 1950s (Fig. 5-2) (cf. Hsu, 1972). A much more detailed survey in 1971 (Kolosov *et al.*, 1974) could not confirm the presence of the bull's-eye pattern of sediment distribution commonly cited in the literature. Instead, the bay floor can be subdivided into subparallel zones of sediment facies that trend roughly north-northwest–south-southeast.

The Gulf of Kara Bogaz Gol is thus a very poor example for any model of a bull's-eye distribution of evaporite sedimentation. On either side of a belt of exposed, largely fossil gypsum, there appears to be a roughly symmetrical zone of decomposition of gypsum to glauberite. This turns into a zone of hydromagnesite and calcite generation from the decomposition of both glauberite and gypsum in the entrance area, an area thus characterized as a carbonate-gypsum region. A mirabilite region in the Kurguzul inlet along the southwestern shore, often reported in previous surveys, is really secondary, leached out of glauberite

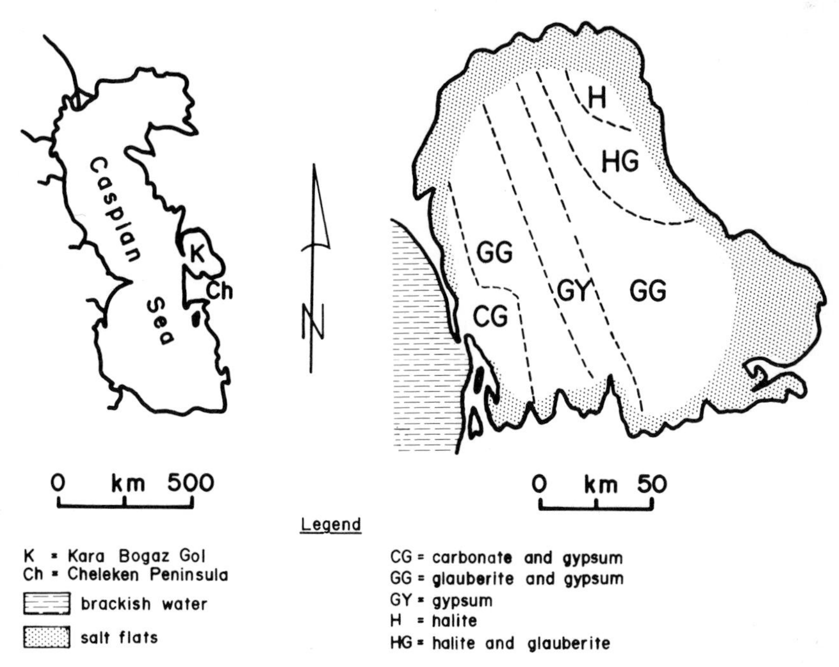

Fig. 5-2. The Gulf of Kara Bogaz Gol. The sediment facies in 1971 (after Kolosov *et al.*, 1974) shows facies boundaries at right angles to the inflow, symmetrically arranged on either side of an ancient gypsum exposure. Minerals of higher solubility precipitate only in the distal regions, where glauberite precipitation is outstripped by that of halite. Along the distal northeastern shores, some individual crystals of potassium minerals have also been found.

beds (Kurilenko and Frolovskii, 1982), that will turn into mirabilite upon subaerial exposure. Halite, as the principal precipitate, appears to be relegated to the distal part, where in drier years a variety of potassium sulfate minerals have been identified in small quantities. In the sebkha Oum el Krilate in southern Tunisia, a similar sequence is found: Gypsum is here being leached by runoff that is producing calcium-enriched sulfatic brines. These brines reprecipitate gypsum in the center, and mirabilite and thenardite around the rim (Perthuisot, 1976).

Potassium Salts

Potassium chlorides and sulfates are generally absent from continental evaporite sequences. Potassium ions liberated in the weathering of alkaline basalts and related rocks are removed from the groundwater by absorption onto clay wafers. However, potassium-rich interstitial brines form in some lakes, producing various potash salt crystals either on the desiccating margins or on polygon crests. Polyhalite occurrences underneath a Turkish salt lake are antithetic to gypsum and thus represent a diagenetic alteration by percolating brines (Irion and Mueller, 1968; Mueller and Irion, 1969a).

Alunite [$KAl_3(SO_4)_2 \cdot (OH)_6$] and jarosite [$KFe_3(SO_4)_2 \cdot (OH)_6$] occur in potash lakes in Western Australia (Cairns, 1948).

Potassic brines are known especially from China. Sodium sulfates and chlorides precipitate in the Basin Q in Hubei Province, central China, under concurrent enrichment of brines in potassium, magnesium, and bromine (B. Wu *et al.*, 1980). Since precipitated salts remain in equilibrium with changing intercrystalline brines, Xu (1982) calculated the phase transformation effect of intercrystalline brines at Chaerhan Salt Lake in Qinghai, China. He determined experimentally both the diffusion and dissolution coefficients in the dynamic equilibrium between sodium-potassium-magnesium chloride brines and salts in the lake. A local addition of minor sulfate ions occurs on the south shore; calcium and magnesium ions are supplied on the north shore. Eventually, the brine produces a carnallitite. The sulfates are used up to produce kainite on the south shore, evidently due to a lack of calcium ions at this end of the lake. Another lake precipitates a mixture of carnallite and halite, in its distal areas away from the river's mouth, from brines with a magnesium chloride/magnesium sulfate ratio of about 35 (Sun, 1974). Carnallite is also precipitated into interstices between halite within 15 cm of the surface in a series of lakes in the Tsaidam Basin of south-central China (Strakhov, 1962). Primary carnallite thereby changes to kainite by diagenesis.

Magnesium Salts

Magnesium carbonates and hydrous carbonates occur only where carbonated groundwater is in contact with basic igneous rocks and then exudes into lakes in a semiarid environment. This occurs in the Coorongs of South Australia, in the carbonate lakes of British Columbia, Canada, and in the East African rift valley. Sodium-magnesium carbonate-sulfates, carbonate-chlorides, and even carbonate-phosphates are known from such environments as more complex but rarer minerals. The exact mineral species precipitated depends upon the partial pressure of carbon dioxide prevailing in brine and sediment. In contrast, primary magnesium carbonate (magnesite) is absent from marine evaporite sequences.

Endorheic Accumulations

Posepny (1877) hypothesized that all evaporite deposits were formed either in endorheic closed basins, by fluviatile concentration and evaporation, or by eolian transport from a nearby shore. In that respect, his ideas predate similar ones of Walther (1903, 1924). Following Walther, Holland and Christie (1909) and Holland (1912) suggested that salt in Rajasthan, India, is windblown from the Rann of Cutch. He used this concept as a model for the accumulation of Keuper salts in England. Pratt (1962) revived the idea that Rajasthan salt is windblown, derived from halite efflorescences in the Rann of Cutch. However, most of the sediments in the Rann of Cutch are river derived, with only relatively small amounts of halite and gypsum in the sand, precipitated from annual floodings of the sea. The volume of trapped seawater does not seem to be sufficient to account for all the salt precipitation in inland lakes. Again, taking the Rann as an example, the groundwater beneath the Great Rann of Cutch is enriched in potassium and magnesium, containing up to 40 g/liter potassium chloride, 410 g/liter magnesium chloride, and 52.4 g/liter magnesium sulfate (Kachhara, 1977), with the Mg/Ca ratio rising from 3 at the coast to 240 inland, due to gypsum crystallization (Glennie and Evans, 1976).

Saxena and Seshadri (1956) later proved that Rajasthan salt is derived from local groundwater that loses potassium, magnesium, and bromide ions en route, and picks up sodium carbonates and sulfates. The hypersaline lakes in the area all have silty bottoms; only Lake Luhkaransar contains a salt crust in the bottom silt and gypsum in the clays 110–127 cm below the sediment–water interface. The gypsum and salt crusts thin toward the lake margins (Singh *et al.*, 1972). Gypsum precipitation strips the brine of all calcium ions. A sodium chloride solution remains; it is very undersaturated for sodium sulfate and carbonate. Finally, Paliwal (1977) showed that some of these lakes derive their salt from dissolution and dry season reprecipitation of salts in salt domes outcropping in topographic depressions.

Maglione (1976) differentiated endorheic basins with little or no relief and feeble rates of subsidence (e.g., the Chad Basin) from tectonic basins with rapid subsidence and great relief. As an example of the latter group, he classified lakes in the East African rift valley, such as Lake Magadi, Kenya.

Leaching of Older Salts

Many salt lakes derive their solute from recycled earlier salt deposits. The desert rivers of southern Iran today leach emergent plugs of Cambrian salt. Lake Zuni in New Mexico (Bradbury, 1971) and the saline lakes south of the Urals (Dzens-Litovskiy, 1968) derive their salt content from the dissolution of Permian salt bodies. Great Salt Lake, Utah, was charged in Pleistocene times with eroding Jurassic salts (A. R. Pratt *et al.*, 1966). Todate, gypsum precipitation in the latter is negligible because of a calcium deficiency, with calcium carbonate precipitating at the inflow (Eardley, 1938).

Examples of ancient evaporite deposits deemed to be derived from the leaching of older salts are not uncommon. Lower Cambrian salts leached by groundwater in Siberia appear to have been reprecipitated into Jurassic endorheic depression in the Sayan region north of Mongolia (Antsiferov, 1979). Koester (1972) postulated a marine embayment extending deep into Arizona in Pliocene times, as there are many small salt bodies in the Basin and Range Province of the southwestern United States, and in central Arizona, that are presumably Tertiary in age. They are extremely low in bromide content, and have no apparent connection either to an ancestral sea or to an older salt body nearby (Eaton *et al.*, 1972). The prevalent interpretation of their origin is a derivation from a solution of older salts now gone. Pliocene salts with glauberite beds in Nevada weather into fibrous gypsum. Gypsum and anhydrite are here almost completely absent; magnesium, potassium, and sulfate contents are very low (Mannion, 1963). The source of this salt appears to be evaporites exposed at that time on the uplifted Colorado Plateau and the rims of the Paradox Basin. In later stages, these outcrops supplied growing quantities of calcium sulfate. Other isolated salt bodies, in the vicinity also have exceedingly low bromide contents which is suggestive of reprecipitation (Eaton *et al.*, 1972). Analogous Upper Oligocene evaporites in the basin fill of the Sevier Desert, western Utah, were deposited in a closed endorheic basin (Lindsay *et al.*, 1981).

The Tertiary halite and gypsum of southeastern France are similarly said to be derived from leaching of Triassic salts surrounding this deposit (Collot, 1880; Truc, 1980). Lamanon de Paul (1782) and Finaton (1934) were earlier advocates of the concept that Eocene gypsum in the Paris Basin had been deposited in a lake originally filled with fresh water (Fontes *et al.*, 1963; Fontes, 1968) and then saturated with leached Permian and Triassic evaporites (Fontes *et al.*, 1966; Fontes and Letolle, 1976). However, blue marls overlying the gypsum of the Paris Basin are marine. Thus, the sea must have entered shortly afterward, and there may have been other occasional entries of the sea (Renard, 1975). These seawater incursions altered the Br/Cl ratio of interstitial brines, so that today the liquid inclusions of gypsum crystals, on average, have a somewhat higher Br/Cl ratio than seawater, which in turn is substantially higher than in continental waters. The inclusions thus indicate regular salinity oscillations within the basin (Sabouraud, 1974). Nury (1968) therefore thought that the Tertiary gypsum at Aix-en-Provence, France, is not an endorheic deposit derived from draining Triassic evaporites. He interpreted the gypsum as a deposit of a marine embayment of variable salinity which became the burial ground for a variety of insects, mollusks, and fishes. Similarly, some intermittent marine invasions into the Williston Basin have been conjectured as possible additional sources of brine (Gallup and Hamilton, 1953).

The source of Tertiary sulfates (gypsum with minor thenardite, mirabilite, and epsomite) in the Ebro Basin of northeastern Spain appears to be leaching of Triassic sulfates (Birnbaum and Coleman, 1979). This origin is also claimed for Mesozoic salts in central Europe (Trusheim, 1971) and England (Brunstrom and Kent, 1967), as well as for those in the Williston Basin in North Dakota. In Britain they occur in places with carbonates and anhydrites which are not accounted for by mere salt solution. In the Williston Basin, they occur without carbonates at the base of the salt, but with secondary cementation along the flanks of the salt basin (Zieglar, 1963; Reed, 1963). The extreme position in that respect was taken by Everding (1907), who interpreted all the upper cyclothems of the Zechstein evaporites as having been produced by resedimentation, partly in the solid state and partly in solution. This hypothesis was severely criticized by Borchert and Muir (1964). However, a salt solution can also add salinity to open ocean waters. Neogene plugs composed of early Cambrian salt are presently forming salt hills in the Persian Gulf that are subject to solution (Emery, 1956). Similarly, the solution of Bittern Lake salt beds adds to the salinity of the Suez Canal, the Gulf of Suez, and the northern Red Sea (Wuest, 1934).

GROUNDWATER PRECIPITATION

Groundwater not only dissolves evaporitic minerals in its path but is also charged with the products of chemical weathering. Precipitation from groundwater in semiarid lands depends on the position of the permanent water table. If that water table is very shallow, the evaporation exerts suction. The water drawn up from the groundwater table evaporates within the aerated zone or at the surface, leaving dissolved salts behind (Petrov, 1976). These salts precipitate at the surface as desert roses or beneath the surface as crusts where the suction power of evaporation is inadequate to bring the brine to the surface. If these precipitates are hygroscopic, the desert soil may remain perennially moist (Wetzel, 1928). Concentration by capillary evaporation above the water table is one method of bringing about gypsum or halite precipitation. Pouget (1968) pointed to a second method in regard to some of the crusts in southern Tunisia: Dilution of the groundwater reduces its chloride content and causes a drop in the degree of salinity; if this solution is saturated in respect to gypsum, the reduced solubility product of gypsum causes its precipitation.

Groundwater discharge in semiarid and arid regions leads to substantial encrustations due to evaporative water loss. Williams (1968) observed that sites of surface accumulation of nonmarine evaporite minerals usually coincide with discharge zones for regional groundwater flow systems of high mineralization. For example, the Great Salt Lake in Utah is a discharge zone for a groundwater flow system for which the distant Wasatch Range is the major recharge area. More than 13 kg/m^2 of solute are delivered annually to the areas of crust formation. The increase in groundwater mineralization is a function of the length of the flow path, its route, and the velocity of the flow. As a major transport medium of dissolved solids, saline groundwater discharges toward the ocean floor were believed to be a potential supply of solute for marine evaporites (Rutten, 1954; Williams, 1968). However, generally speaking, most groundwaters are enriched in calcium, whereas marine brines are enriched in magnesium. Moreover, formation waters are mainly carbonated, and normally carry only negligible amounts of sulfate and chloride ions in comparison to marine brines.

If the water table is so deep that evaporative drawdown no longer plays a significant role in the aerated zone, periodic downpours soaking into the soil cause leaching and redeposition in the form of a cementing phase of crystallization, a crust, or a series of individual euhedral crystals. The effect of crystallizing evaporites, their dehydration and rehydration on weathering in the soil, has been discussed by Mabbutt (1977). The crystallization pressure is directly proportional to the logarithm of the supersaturation ratio. In a sodium chloride solution supersaturated by a factor of 2.0, halite exerts a pressure of 66 MPa (Winkler and Singer, 1972).

Subsurface rock salt dissolution leads only to reprecipitation in semiarid lands, where endorheic basins or lagoons with severe entrance restrictions would generate evaporite precipitation in any case. Dissolution of buried salts in humid climates merely leads to a slightly increased solute load in both groundwater and surface drainage. An example would be the alluvial aquifers in central Kansas that are contaminated by sodium chloride leached from the buried Permian Wellington Salt member (Whittemore *et al.*, 1981). Another example would be the salty creeks of Wood Buffalo National Park, Northwest Territories, Canada, that leach Middle Devonian evaporites. A third example would be the saline soils along the river banks and valley floors of the upper Angara River region in Siberia, where Cambrian salts are being leached out (Khismatullin, 1961). In contrast, dissolution of gypsum or anhydrite leads to reprecipitation even in humid terrain. An example is the redeposited Cenozoic gypsum in Switzerland derived from dissolved Triassic occurrences (Schneider, 1965).

Nitrates and Iodates

Nitrates accumulate only where annual rainfall remains below 10 mm; iodides oxidize here to iodates and become immobile (German Mueller, 1968). Both compounds are thus found only in areas of extreme aridity. They accumulate through capillary migration of nitrogen-bearing groundwater (George Mueller, 1958) and form a nitrate caliche, i.e., a saline-cemented regolith (Ericksen and Mrose, 1972). In addition to halite, gypsum, glauberite, and bloedite, the caliche always contains a wide variety of rare minerals, such as niter (KNO_3), nitratine or soda niter ($NaNO_3$), and darapskite [$Na_3(NO_3)SO_4 \cdot H_2O$], which form only between 13.5 and 50°C. Dietzeite [$CaCrO_4 \cdot Ca(IO_3)_2$], lautarite [$Ca(IO_3)_2$], sodium perchlorate, potassium dichromate (tarapacaite), sodium dichromate, and sodium iodate have also been identified. Potassium nitrate is especially known from Bolivia. Small nitrate deposits are known from almost all deserts, e.g., from the continental sebkha Ulad Mamud in the Sahara, from Namibia, and from Death Valley, California. Chilean nitrate deposits, by far the largest nitrate deposits, are composed of lentils and tongues within a salt debris, covered by a fine dust. Underlying silty layers are perennially moist from rising groundwater. Halite within the nitratine-enriched caliche layer is commonly blue (Wetzel, 1932). The origin of Chilean nitrate efflorescences is deemed to be ultimately related to volcanic emanations from Mount Chachani (Ericksen, 1961; Fiestas y Contreras, 1966) that were picked up by the groundwater; however, they could also have an exogenic explanation.

Using solar energy for photoreduction, titanium minerals such as rutile, which is common in desert sands, act as catalysts in the nitrogen cycle wherever soils are unprotected by vegetation. Atmospheric nitrogen combines with chemisorbed water to form ammonia and lesser amounts of N_2H_4. Although microorganisms may have a part to play, the reaction proceeds in desert sands even under sterile conditions (Schrauzer *et al.*, 1983). Fog and dew are then prominent agents that redistribute the highly deliquescent efflorescences. Desert air has a high absolute water vapor content but a low relative humidity. This leads to the generation of static electricity and to a high ionization potential. Atmospheric nitrogen is oxidized; the oxides then dissolve in atmospheric moisture, rains, fog, and dew. Such nitrogen oxides are also common elsewhere in atmospheric precipitation and increase at least 96-fold from polar to tropical rains. In Chile the frequent morning fog seeps into the ground, but it cannot penetrate far, because the underlying beds are cemented with halite. This halite crust thus becomes essential for nitrate accumulation. Downdip, the water reaches the surface later in the day when the fog has disappeared, and it evaporates. In this fashion, the nitrates migrate daily to low points in the topography (Wetzel, 1932). It should be noted that buried ancient nitrate deposits are not known. The thus represent an ephemeral evaporite of desert lands that dissolves upon burial below the water table of formation waters.

Gypsum and Halite Crusts

Gypsum crusts are the most common type of crusts. They are found in all deserts and semidesert rims: in Australia; Somalia; and the Sahara, the Namib, Thar, Gobi, Kara-Kum, and Atacama deserts; in Syria and Bahrein; and from Turkey to Afghanistan and Kazakhstan (Auden, 1952; Dregne, 1968; Watson, 1979). A high gypsum content makes the soil impervious, gives it a high moisture-holding capacity, and requires a greater hydrostatic potential and more time to flush waters through (Rabochev, 1949). Frequently, the underlying waters are under considerable artesian pressure and may tesselate the soil. Piccard in 1868 was the first to connect the formation of gypsum crusts with rising groundwater, although actually a drop in the water table by increased evaporation is often responsible for gypsum crust formation. An extensive overview of gypsum crust development was published by Watson (1979).

The source of the dissolved calcium sulfate need not be dissolution of older anhydrite or gypsum deposits, or seawater encroachment. Weathering of sulfides can supply the sulfate anions. Pyrite and macasite can be oxidized, leading to the growth of large gypsum crystals (Churinov, 1945). Pliocene pyritiferous calcareous clays weather into gypsiferous clays in the Ukraine (Yurk and Lebedeva, 1960) and Bulgaria (Zafirov, 1959; Kuleliev and Kostov, 1960). Weathering in Katanga, Zaire, of sulfide-bearing schists supplies sulfate ions that interact with limestone to form a gypsum crust to a depth of 200 mm (Raucq, 1954). Similarly, a gypsum crust, deeply embedded in the soil, is underlain by basalt in Soviet Georgia (Klopotovskiy, 1949), and gypsum occurs even as an alteration product after feldspar in lava vesicles (Haegele, 1939). In the sandy desert of Kara-Kum, gypsum is found only in zones of recent tectonic subsidence, but calcite occurs in zones of recent tectonic uplift (Tolchel'nikov, 1962). A corollary is the occurrence of gypsum in depressions between sand dunes in Rajasthan, India (Sastri, 1962). However, gypsum crusts in Algeria occur in the higher ground surrounding playas (or "chotts"), where groundwater evaporates in sandy soils.

Position of Groundwater Precipitates

The sequence of evaporitic precipitates in the soil horizon depends on the provenance of the waters. It is different in a purely continental groundwater regime or in one influenced by brine encroachment from the sea. Krupkin (1963) noticed that chloride ions move faster than sulfate ions, and that calcium or magnesium ions move faster than sodium ions. He suggested that the velocity of migration of ions depends on their capacity to interact with soil electrolytes and to form new compounds of lesser solubility.

Where secondary leaching through intermittent wetting by dew and very sporadic rainfall causes downward migration, Yaalon (1965) showed that *downward leaching causes sulfates to accumulate in shallower horizons and chlorides in deeper horizons.* The chlorides move down nearly at the speed of percolating waters; sulfate ions attain only two-thirds of that speed. This sequence also occurs if the brine is affected by seawater encroachment onto coastal sand flats, where oxygen is carried in by the seawater. In that case, gypsum precipitates first above the groundwater table, followed by halite above the gypsum (Fig. 5-3A). This has been observed in sand flats on the west coast of the Sinai Peninsula (Gavish, 1975, 1980) and in mud flats of Laguna Madre, Texas (Masson, 1955). The same reversed sequence is found even in coastal sebkhas, where the major solute supply appears to be through groundwater that mixes with encroaching seawater, such as in Libya (Rouse and Sherif, 1980) and Egypt (West *et al.*, 1979). Sandy gypsum crystals are known from raised beaches in Somalia (MacFadyen, 1950) and inland from the high-tide mark on the Trucial coast, Persian Gulf (G. Evans and Shearman, 1964). Again, this can be explained by Yaalon's (1965) and German Mueller's (1968) experiments, since halite will have traveled a greater distance from the source than gypsum.

Where atmospheric precipitation is the only source of water, a gypsum crust can develop within the soil by capillary upward migration of water during dry spells. The height of the capillary rise of electrolyte solutions is proportional to the increasing concentration; it also changes with the type of electrolyte, with $NaCl > NH_4Cl > KCl > MgCl_2 > CaCl_2$ (Bottini, 1933). These electrolyte solutions are younger than the calcareous caliche (Kulke, 1974). *Chloride ions migrate farthest from the water-logged zone* (German Mueller, 1968), *but gypsum marks the depth of wetting in a leached soil horizon* (Krupkin, 1963) (Fig. 5-3B). However, a dilute sodium chloride solution does not travel as fast as a saturated calcium sulfate solution (Belkin, 1940). In turn, gypsum is found farther than silica. This is evidenced by gypsum accumulations at the end of opal stalactites in weathering amphibolites (Antonyuk, 1959). Rising groundwater thus accumulates chlorides above sulfates. German Mueller (1968) found experimentally that sodium chlorides move at more than five times the speed of sulfates and leapfrog under conditions of high humidity. *The vertical sequence of precipitates from halite below through gypsum to calcite above* (Linck, 1946) *indicates* three things in continental basins: first, *that calcite accumulates in the region of an adequate carbon dioxide supply;* second, *that gypsum is found at the base of the aerated zone and its oxygen supply;* and finally, *that halite forms in the anoxic waters with a negative redox potential, or immediately above the permanent water table.*

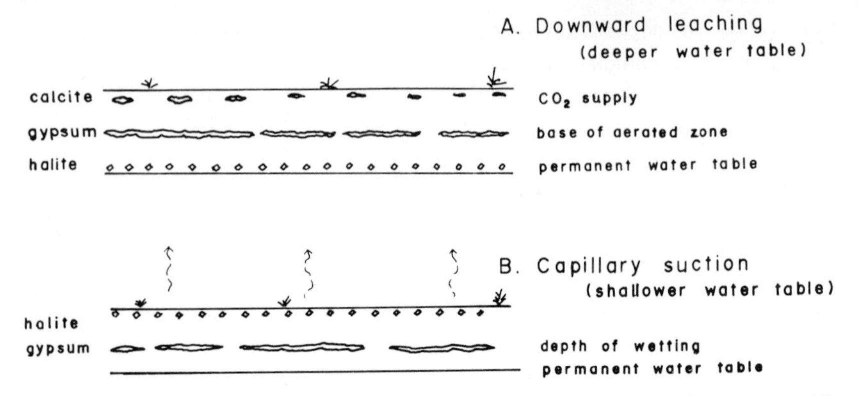

Fig. 5-3. Precipitated crusts in semiarid soils. A. Downward-leaching to a deeper water table. B. Capillary suction from a shallower water table.

If the capillary water movement is entirely upward from a shallow water table, a salt crust will develop at or near the surface, underlain by gypsum crystallizing in the aerated zone above the water table. A shallow water table under dry mud flats promotes evaporative pumping (Hsu and Siegenthaler, 1969); the interstitial growth of evaporite minerals is favored in that part of the mud flats which is soaked in saturated brine (Eugster, 1980a,b). A halite crust, 2–3 mm thick, on top of the Chemchana playa in Mauritania, corresponds to a gypsum crust found 1.5–2.0 m beneath the surface (Chamard, 1973) in slightly alkaline waters. Gypsum beneath a thin crust of halite is also found around lakes in Western Australia (Bettenay, 1962). Along the Laguna Madre, Texas, thin laminae of gypsum form within the soil, due to ascending capillary-transported groundwater with a salinity of 200 ppt—higher than that of Laguna Madre (Fisk, 1959). The same holds for Broad Sound, Queensland, Australia (Cook, 1973), where interstitial pore waters with a salinity of 120 ppt precipitate flat discoid crystals in the supratidal soil; the halite crystallizing at the surface is blown away by the wind. In Antarctica, soil profiles in coastal Enderby Land show gypsum at 0.40–0.50 m below the surface. Halite effloresces seasonally at the surface and is removed by katabatic and cyclonic winds of the fall season (MacNamara and Usselman, 1972).

A vertical sequence of minerals corresponds to a horizontal separation of mineral facies. The solutions percolating through a soil horizon that is subject to repeated wetting and drying are only gradually depleted in SO_4^{2-}, Mg^{2+}, and SiO_2, because sodium and chloride ions are more mobile and migrate first (Drever and Smith, 1978). This is a strict analogy to the lateral zonation in hypersaline lakes and lagoons.

Two Horizons of Precipitation

Two halite horizons are possible, depending on the depth of the water table. Calcite-bearing silts in Death Valley, California, are encrusted with halite above a layer with gypsum and borate nodules. Chlorides occur again underneath the nodules, on top of the capillary fringe of the water table (Hunt *et al.*, 1966). Two chloride horizons have also been found within saline soils in central Czechoslovakia (Pelisek, 1941) and in Tadzhikistan, USSR (Vaksman and Onishchenko, 1972), at the surface and again near the groundwater table. On occasion, two gypsum horizons occur. On the high plateau of South Oran in Algeria is the Chott Ech Chergui, a playa that contains two layers: Capillary evaporation produces lenticular gypsum crystals and beige-to-reddish rosettes in the aerated zone. Gray prismatic and tabular gypsum crystals have formed underneath the water table in a brine of 155 g/liter salinity (Bertrand and Jelisejeff, 1974). Since the level of the water table oscillated in many regions during glacial and interglacial times, it is surprising that two levels of precipitates are not found more often. This would suggest that fossil crusts are rapidly scavenged.

Euhedral Crystals

Some of the crystals grow to large sizes, displacing surrounding sediment. Gypsum crystals up to 50 mm long have been reported 0.60–1.00 m below the surface of a playa margin in Tunisia (Sassi, 1969). A giant halite pseudomorph has been described from Upper Triassic sandstones of northeastern Bavaria (Heller, 1976). In the Laguna Madre, Texas, the gypsum crystals and rosettes of intergrown discoid crystals increase in size from less than 0.5 mm at 5–8 m in depth to 80–100 mm at 0.5–1.0 m, and to 500 mm at 5 m (Kerr and Thompson, 1963). Another example is the Clayton Playa in Nevada, which is covered by a thin halite crust. Gypsum precipitates upon drying of the sediment, immediately, forming crystals up to 5 mm in diameter, which grow to a diameter of 75 mm within 11 months (Moiola and Glover, 1965) at a very slow rate of precipitation. Eventually, the gypsum crystals are dehydrated to bassanite and anhydrite in the silty clay saturated with sodium chloride. Vertical gypsum twins 3–7 m in length have been described from Upper Miocene evaporite beds in Spain (Dronkert, 1976; Shearman and Orti Cabo, 1978). They are also known from many other deposits, both of gypsum and of halite. Such growths are invariably the product of slow or intermittent movements of brines, not necessarily saturated, but subject to capillary evaporation.

Selenite crystals 50–150 mm long occur in clays around the Great Salt Lake, Utah; all of them are vertically oriented at right angles to the bedding planes. Eardley and Stringham (1952) suspected the involvement of bacteria in selenite growth. Bacterial activity is indicated in a contemporary example: A firm, gritty black sand in New Caledonia covered by a veil of dried bacteria is underlain by orange, more fluid mud containing translucent gypsum crystals, up to 80 mm long, with elongate mud inclusions. The crystals display phases of dissolution alternating with growth, and occur at the base of iron sulfide oxidation (Avias, 1958). Selenitic crystals, twinned with fishtail fins pointing up, can also occur as a clean precipitate; they are then interpreted to have grown into an open brine column (Holser, 1979b). They may be simply the subaqueous exit points of slowly moving interstitial brines, similar to those that produce tepee structures (see Chapter 14) in shoals. Vertical movement of brine is also indicated wherever the gypsum is arranged in nearly vertical crystals resembling a cauliflower.

In a system containing crystals in equilibrium with a saturated solution, large crystals will grow at the expense of smaller ones. Any system tends to reduce the area of its interfaces to a minimum, because the chemical free energy of a given mass of solid increases with its surface area. If larger crystals grow inside a pore space, they first try to fill the void and then continue to grow against the constraint imposed by the walls of the pore. They will expand the pore and, in extreme cases, can fracture the rock. A pore will be enlarged if

$$S < \sigma \, dA/dV \tag{5-1}$$

where S is the mechanical strength of the rock, dA/dV is the increment in area over increment in volume, and σ is the interfacial tension between the crystallization force and the saturated solution. For rocks of equal mechanical strength, those with large pores separated from each other by regions of very small pores will be most liable to failure (Wellmann and Wilson, 1963). This is an important factor in salt weathering, which is common in dry regions whenever efflorescences are found.

Large euhedral or imperfect crystals form only when the concentration of the solution proceeds slowly and does not differ too much from the rate of diffusion. Large crystals therefore grow in very dilute, grossly undersaturated solutions (Linck, 1946). The large halite crystals required commercially for curing fish are grown in slowly concentrating solutions over a period of 2 weeks or more. Small crystals to be ground into table salt are produced by evaporating the brine quickly (Sherlock, 1921). A sodium chloride solution, preconcentrated to 290 g/liter, will produce cubic halite crystals less than 1 mm across if evaporated at 50°C. However, if the same solution is packed into dry forest soil or such a soil suspension is added, the crystals will grow to 5–7 mm (Koide and Nakamura, 1943). The rates of evaporation in warm climates are too high to keep exposed waters at constant undersaturation. A crystal thus grows very slowly in groundwater only at some distance from an open evaporite basin. The crystals are younger than the covering strata, but their growth is completed before their cover compacts. Since halite crystals occur even in bottom sediments of antarctic lakes, groundwater-derived crystal growth can be either subtropical or subpolar.

Dendritic crystallization, common in soils, requires a slow rate of crystallization and a limited supply of solution. It is caused by growth predominantly at the corners of crystals. The presence of gelatin and related products facilitates dendritic crystallization, whereas iron chloride suppresses it. To produce dendritic crystals, 665 g/liter of $FeCl_3$ is needed in a potassium chloride solution, whereas 1.62 g/liter is enough in a sodium chloride solution (Til'mans, 1951, 1952). A small amount of ferrocyanide has the same effect (Shuman, 1966). If the upward movement of water reaches the surface, desert rosettes form, giving sandy, impure crystal aggregates of gypsum, barite, and other minerals (Traveria Cros *et al.*, 1971). Associated with them may be corroded small anhydrite crystals (Bellair, 1954). A slow growth of gypsum crystals at artesian groundwater discharge points is thought to be responsible for mounds of gypsum covered by a veneer of very fine-grained clastics in the Laguna de las Barrancas, Argentina (Teruggi *et al.*, 1974). Desert rosettes are very often peripheral to playa deposits (Watson, 1979). The type of host rock is unimportant, since gypsum forms even in peats of semiarid zones, as a precipitate above the groundwater level. In humid woodland areas, gypsum is leached below the groundwater level and is replaced by pyrite (Tarakanova, 1973). Gypsum precipitated in the soil eventually erodes out and forms dunes, such as the famous gypsum dunes at White Sands, New Mexico (Herrick, 1900). Other gypsum dunes occur around the Great Salt Lake of Utah (Eardley, 1962), in southern Tunisia (Trichet, 1963), and on the leeward side of saline lakes in Western Australia (Bettenay, 1962). Recent gypsum in southern Spain is found in 10 different crystal habits, all formed by a capillary rise of groundwater. The morphology and size of the crystals vary with the porosity and permeability of the substrate, brine salinity, strontium content, trace element concentration, and the presence of clay or salt intercalations (Carenas and Marfil, 1979). Halite characteristically forms fibrous aggregates that are curved or bent (Wetzel, 1928). Such crystal habits have been referred to as "capillarites" (Reimer and Utter, 1979). Their growth rate is 0.5 mm/day, but actual growth is very intermittent.

Gypsum and Halite Pseudomorphs

In the Namibian Desert, gypsum overlies halite (Gevers and van der Westhuyzen, 1932; Cagle and Craft, 1970). Intrasedimentary crystallization often produces a vertical stacking of calcite crusts near the surface, sand gypsum crystals underneath, and finally sandy halite at depth. This sandy halite, which is impure, may contain up to 25% potassium chloride (Linck, 1946; Kaiser and Neumaier, 1932; Scholz, 1968). The solution did not saturate by evaporation at the soil surface, but underwent a very slow concentration inside the sedimentary rock (Linck, 1948). Both gypsum and halite crystals preserve the original micro-stratification of sands and silts.

The gypsum and halite are later removed, leaving large (up to 100 mm long) pseudomorphs in fine-grained sands, silts, and clays. The crystal molds create impressions in both the substrate and the covering layers. The red color of the sediment indicates an increased sodium chloride content, since sodium chloride mobilizes iron (Scholz, 1968). In a similar Triassic example, the leached void after dissolution of a crystal is eventually filled from above (Haude, 1970). Sand-filled pseudomorphs are frequently found in soils surrounding a saline playa lake (Courel, 1962; Plaziat and Desprairies, 1969).

Pseudomorphs after euhedral gypsum and halite crystals with the original textural features preserved have been reported from red beds of Cambrian or even Proterozoic age (Henderson and Southgate, 1980; Marzela, 1981). These pseudomorphs should be distinguished from halite cube impressions in limestone laminites that are pseudomorphs after hopper crystals. Such salt impressions have been found in Maryland, Virginia, and West Virginia in Upper Silurian limestones (Ludlum, 1959). Dolomite-filled halite casts are known from France (Dropsy, 1978) and Oklahoma (Merritt, 1936).

Other Alterations

Another mineral sequence is also very frequent. The sebkha Oum el Krilate in southern Tunisia contains an argillaceous gypsum layer contaminated upward with mirabilite, bloedite, thenardite, and glauberite. Only the surficial layers also contain halite (Perthuisot, 1976). Bloedite may thereby be produced from mirabilite and thenardite under the influence of magnesium sulfate solutions at temperatures above 20.6°C (Kuehn, 1952a). In contrast, hydration of minerals drawn to the surface of the weathering zone by exposure to atmospheric humidity (Mortensen, 1933), efflorescences in mines due to pyrite oxidation, and weathering of minerals to gypsum due to acid rain (Winkler and Wilhelm, 1970) are not caused by groundwater.

If circulating fluids are charged with calcium sulfate, they lead to dedolomitization according to the reaction

$$CaCO_3 \cdot MgCO_3 + CaSO_4\,(aq) \rightarrow 2CaCO_3 \downarrow + MgSO_4\,(aq) \tag{5-2}$$

The magnesium sulfate then generates an epsomite efflorescence, leaving behind a coarse crystalline cavernous limestone (Tatarskiy, 1949). The absence of such dedolomitization speaks against the migration through dolomites of brines saturated with calcium sulfate that may have originated in anhydritization of gypsum beds.

Sebkhas

A special case of groundwater-derived evaporites is represented by modern and ancient sebkha environments, in which seawater encroaches on a shore and mixes with the groundwater. Concentration of the brine leads to precipitation of gypsum and moderate-to-small quantities of halite. Commercial quantities of potash or magnesium salts are absent. Both gypsum and halite can be transported by wind in the dry season, piled up into dunes or blown away. Carbonates usually show mud cracks and other evidence of intermittent subaerial exposure. A sebkha development as a strandline facies behind skeletal and oolitic limestones occurs along the Trucial coast, Persian Gulf (Kinsman, 1969), and is widespread along the margins of the Mississippian Windsor Basin of Nova Scotia and its subbasins (Schenk, 1969), or in each of the Zechstein horizons of Britain, based on oolites in the carbonates and ripple marks and desiccation cracks in associated clays (Stewart, 1954). In the Permian Basin of West Texas, an example is offered by Guadelupian evaporites (Jacka and Franco, 1974), which underlie far more voluminous Ochoan evaporites.

The term ''sebkha'' comes from the Arabic. The spelling is based on that preferred in the American Geological Institute's ''Glossary of Geology'' and ''Webster's Third New International Dictionary.'' It depends largely on whether the researcher did the field work in North Africa (sebkha) or along the Persian Gulf (sabkha). Hsu (1973) even uses the spelling ''subkha.'' The term is also transcribed as ''sebkra,'' ''sabkhah,'' ''sebkhan,'' ''sebchet,'' ''sebjet,'' or ''sebcha.'' It is more or less equivalent to the Spanish ''playa,'' meaning a lake or lagoon floor that dries out at least seasonally. A sebkha is strictly coastal in Arabia, where an interior playa is called ''mamlahah;'' in central Asia the term ''sebkha'' applies to interior playas, locally pronounced ''sabkhet,'' similar to the Egyptian and Iraqui ''sebkhat.'' Such an interior playa is referred to as ''nor'' in Mongolia, ''shott,'' ''chott,'' ''zahrez,'' or ''garaet'' (garaa) in North Africa. As a clay pan, it is called ''tikir'' in Siberia, ''khabra'' in Saudi Arabia, and ''qu'' in Jordan. As a saline pan, it is called ''kavir'' in Iran, ''salar'' in Chile, and ''tsuka'' in Mongolia (Cooke and Warren, 1973). Many of these terms are found in local place names. Because the North African transcription ''sebkha'' has precedence in dictionaries over the Arabian ''sabkha,'' the former is used here in analogy to the more common transliteration ''Mekka'' instead of the Saudi Arabian ''Makkah.'' North American geological usage is ambivalent: The editor of the *Journal of Sedimentary Petrology* allows both ''sebka'' (Rooney and French, 1968) and ''sabkha'' (Butler, 1969).

According to Perthuisot (1975), the term ''sebkha'' (correctly pronounced in Arabic as ''sebkhat'' because it ends in a vowel) designates a depression of greater or lesser extent that is either closed or open to the sea, is subject to inundations, and has a flat floor on which saline soils (with or without a salt crust) prevent the growth of any vegetation. Only the rim is populated by halophilic xerophytes. The term "sebkha" is thus purely descriptive, covering many diverse geomorphological features.

Sebkhas can be divided into continental and paralic types. Continental sebkhas are situated in the center of closed endorheic depressions without communication to the sea, and are mainly groundwater fed. They occur in the following settings:

1. On the floor of dried-out lakes that are intermittently inundated by flash floods, so that the evaporation of these waters leaves behind a salt-encrusted sediment.

2. Where the groundwater table approaches the soil surface. This often happens in dried-out lakes. Capillary evaporation or efflorescence causes crystallization from the groundwater. This may lead to salt crusts or to salt-cemented terrigeneous clastics.

Paralic sebkhas include the following:

1. Supratidal plains with detrital or biodetrital sedimentation that gradually advances seaward. The morphology is comparable to that of sebkhas along the Persian Gulf.

2. Enclosed coastal lagoons fed in part by wadis (arroyos) and in part by storm tides. The resulting sheet of brine is hypersaline; summer evaporation suffices to dry it up, leaving a salt crust. If they do not dry out completely, these lagoons are nourished during the dry season by a supply of artesian seawater or groundwater.

3. Lagoons now totally surrounded by dry land, except for a neck of varying length that furnishes a connection to the sea. They represent former lagoons now filled with sediment but also subject to inundation, possibly due to renewed subsidence. The salinity of the waters filling such embayments depends largely on the degree of constriction of the neck.

Because of the great interest in sebkhas of the Persian Gulf type, it would be best to restrict this term in the geological literature to supratidal plains; continental sebkhas are better known as "endorheic basins." Lagoons fed only by storm tides are, in effect, ephemeral or seasonal coastal lakes. The term "lagoon" should be reserved for bodies of water with a perennial but limited connection to the sea. The size of the lagoon is thus of no concern, although some authors have hesitated to use this term for bodies of water of the size of the Zechstein or Elk Point seas.

Sebkhas in the Gulf of California and along the Persian Gulf experience shoreline regression totalling 1–2 m/year (Kinsman, 1969). Any such retreat of the sea by 1–2 km/1000 years makes a sebkha environment a geologically ephemeral event. The whole southern rim of the Persian Gulf represents an emerging coastline, produced by sedimentary offlap and a relative fall in sea level. This resulted in the development of broad, gently seaward-sloping planar areas. The water table beneath these areas always slopes seaward and is never more than 1.0–1.5 m below the surface. The sebkha surface has thus attained a deflationary equilibrium. Strong onshore winds drive thin sheets of seawater several kilometers inland. This seawater is concentrated by evaporation and then slowly infiltrates into the ground. Magnesite, dolomite, and gypsum form when this seawater brine reacts with onshore sediments. At the same time, continental groundwater moves seaward and eventually mixes with seawater brines. Capillary evaporation of this groundwater produces gypsum and anhydrite (Patterson, 1972). A 3- to 5-km-wide belt is flooded at monthly intervals along the Trucial coast, and an additional 8 km at 4- to 5-year intervals. The former is marked by the occurrence of up to 30 cm of gypsum mush that changes landward to anhydrite. The latter contains highly dolomitized carbonates and anhydrite that is recrystallized in a characteristic chicken wire pattern (Butler, 1969).

Perthuisot (1980) noted that sebkhas are characterized by the following:

1. A net dominance of detrital elements and, rarely, thick evaporites.
2. A relatively meager contribution of calcium and magnesium carbonates.
3. The coexistence of salts precipitated in open water with salts of capillary evaporation in a concentric arrangement of evaporite mineral species.
4. Often, a peculiar geochemistry of the sediments, both of clay components and of chemical precipitates.
5. In the modern sebkha, gypsum is not precipitated from open lagoonal waters, and is therefore never found on the floor of the lagoon (Cuff, 1969; Butler, 1973). The first evidence of this mineral is found just beneath the surface algal mat, in the intertidal sediments, some 100 m from the low-water level. From there on, landward for some 5 km, gypsum is common in the top 2 m of the soil (Bush, 1973).
6. In contrast to ancient anhydrites, sebkha anhydrites show lattice imperfections, electron hole centers caused by the replacement of sodium and chloride ions, and a higher strontium content. Further, all crystals enclose foreign mineral grains (aragonite, dolomite, celestite) (Cuff, 1969).

A variation on the sebkha environment is provided by wind-driven precipitation. Fog along the coast of Namibia deposits halite (Boss, 1941), and the wind carries inland large quantities of hydrogen sulfide emanating from an anaerobic sea bottom. En route, they oxidize to sulfuric acid. Experimentally, Perrier (1915) found that limestone kept in a sulfate solution converts to gypsum only very slowly over a period of years. Nonetheless, Martin (1963), Scholz (1968), and Goudie (1971) explained the surface gypsification of limestones located even several hundred kilometers inland by their exposure to acid atmospheric precipitation. Components of atmospheric sea salt do not travel equally well; sodium chloride decreases inland with increasing distance from the coast, whereas

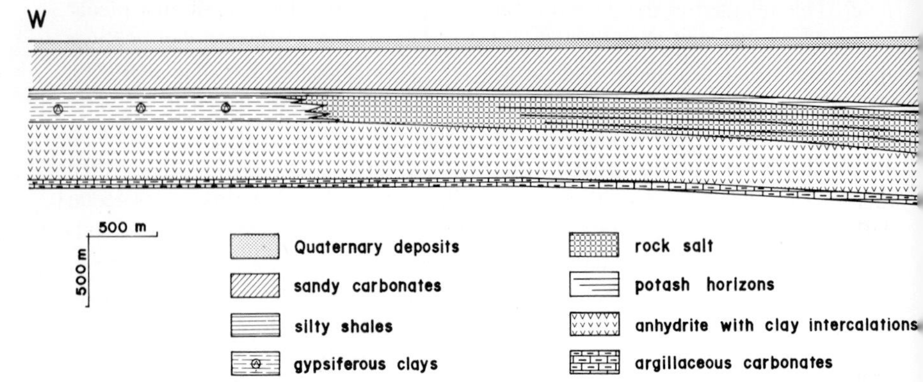

Fig. 5-4. Schematic cross section through the western margin of the Upper Pechora evaporite basin. Note the sebkha-style gypsiferous coastal clays receding and the potash horizons encroaching onto the shelf (modified from Ivanov and Levitskiy, 1960).

calcium sulfate, calcium carbonate, and magnesium carbonate increase (Koyama and Sugawara, 1953). Saharan aerosols over the tropical North Atlantic Ocean also contain gypsum that is derived from conversion of calcite particles by atmospheric sulfur species (Glaccum and Prospero, 1980).

The coastal areas of any major evaporite basin are exposed to episodic flooding by an oscillating sea level, and thus will develop a sebkha-style broad coastal flat with nodular anhydrites and halite crystals growing in a silty clay. An example is provided by the western margin of the Permian Upper Pechora Basin west of the Urals, where coastal clays interfinger with offshore halite beds that contain potash horizons on the shelf (Fig. 5-4).

SUMMARY OF CONTINENTAL EVAPORITES

In polar dry areas, the evaporite minerals are mainly derived from evaporative concentration of weathering products. Sodium sulfates predominate over calcium sulfate, but other sulfates are rare. Sodium, magnesium, and calcium chlorides form hydrated mineral species because of the cold temperatures; potassium salts are extremely rare.

In subtropical lakes, likewise, the chief source of brine mineralization is weathering reactions. Both surface runoff and groundwater leach and recycle very substantial quantities of ancient evaporites. Sodium carbonates and sulfates predominate over corresponding calcium salts. Potassium salts are rare; magnesium salts occur only where carbonated groundwater is in contact with basic igneous rocks.

Groundwater raised to the surface by evaporation creates desert roses of a variety of minerals. If the drawing power of evaporation is not large enough to bring groundwater to the surface, the water evaporates in void spaces and produces a crust. Gypsum crusts mark the depth of wetting at the base of the aerated zone; halite crusts form above them if the groundwater rises, but beneath them if the accumulation is due to downward leaching. Slow concentration of groundwater leads to the formation of large euhedral crystals that may later be replaced by other minerals or leave casts filled with sand.

6

Evaporite-Related Marine Carbonates

A complete discussion of the conditions of precipitation, depositional environments, and diagenetic alterations of carbonate rocks is beyond the scope of this book. Nonetheless, since massive calcium and magnesium carbonates are commonly found around marine evaporite sequences as the first solutes to precipitate out of an evaporating brine or as diagenetic or epigenetic successors to evaporite minerals, a very brief discussion of aspects pertaining to evaporite genesis is in order. Some carbonates around evaporite deposits are bedded, some are reefal buildups, and some are alterations due to bacterial activity or percolating brines. Most, but not all, of them are altered either to dolomite or to magnesite. Biogenic carbonate precipitation is extensive outside the evaporite belt, but the subtropical warm waters, which frequently carry rich planktonic fauna and flora, are the habitat of many sessile carbonate-secreting organisms that are both euryhaline and euryionic. Since the surface waters remain at moderate salinity until very advanced stages of brine concentration are reached, some of these settle preferentially into indrafts, being protected from relatively stenohaline scavengers by the incremental salinity.

Many parts of the subtropical and tropical oceans are today supersaturated for calcium carbonate (Dietrich *et al.*, 1975), but only in their shallow horizons. However, spontaneous precipitation remains an exceptional event. Moreover, the calcium carbonate precipitation from a bicarbonate solution depends on the magnesium concentration in the brine (Vetter, 1911). In contrast, the solubility of carbonates (and, for that matter, borates) increases as the sodium chloride concentration rises (Hunt *et al.*, 1966). When evaporating Mediterranean waters in his laboratory under well-oxygenated conditions, Usiglio (1849) found hematite, rather than calcium carbonate, to be the first precipitate. This is not the case in nature.

As the brine concentrates, the calcium carbonate comes out of solution first, provided there are enough anions available. No bicarbonate can remain in solution at a three- to fourfold concentration of a brine derived from seawater (Wattenberg, 1936). The solubility of carbonate ions is likewise severely diminished (Fig. 6-1). The waters contain very little free carbonate ions by the time the brine is saturated for gypsum, the precipitate of the next higher solubility (Hardie and Eugster, 1970). The carbonate ions required for ongoing calcium carbonate precipitation thus must be generated by organic reactions. Carbonate ions can also be brought in by runoff, i.e., surface inflow, but then they must precipitate before the surface layer evaporates.

Carbon dioxide decreases in concentrating brines in five ways:

1. It is taken up by plants.
2. It is lost by heating the water.
3. It is consumed in bacterial activity, which rises with increasing salinity.
4. It is lost in increasing salinity.
5. By variations in the hydrostatic head (Bradbury, 1971).

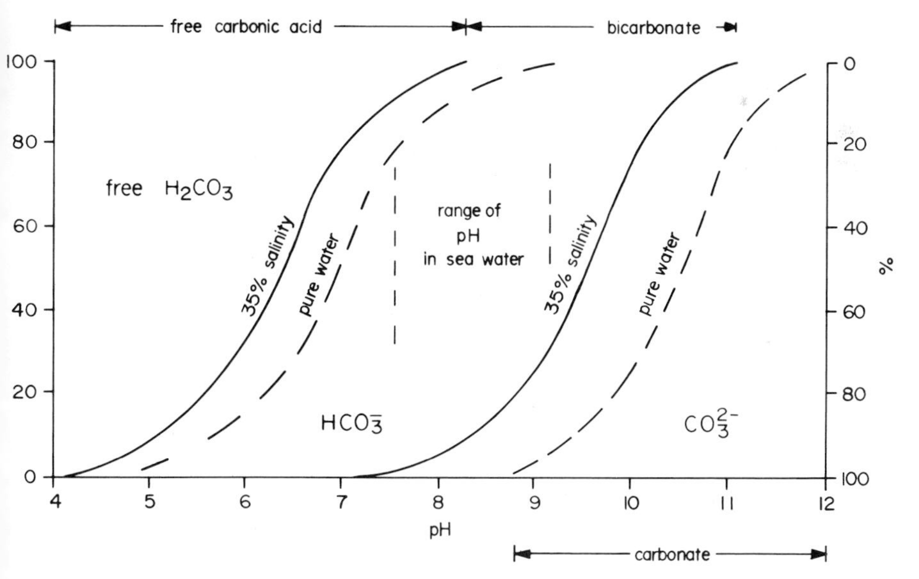

Fig. 6-1. Carbonate-bicarbonate solubilities as a function of pH and salinity (after Wattenberg, 1936).

ARAGONITE

Aragonite precipitation reaches a maximum at a molar Mg/Ca ratio of 5.0 (the present ratio in seawater is 5.05). It is thermodynamically unstable at surface temperatures and pressures, but is the only phase precipitated once the magnesium concentration exceeds 243 ppm. Magnesium ions in solution inhibit the precipitation of both magnesium calcite and aragonite, but their effect on the former is more profound (Chen *et al.*, 1979). These ions subdue not only the precipitation of calcite but also that of vaterite, small amounts of which may form in solutions containing no magnesium. Increasing amounts of sodium chloride in solution somewhat negate the influence of the magnesium ions. With rising water temperature, the required minimum level of magnesium concentration decreases; the concentration of magnesium is then accompanied by a decrease in free CO_3^{2-} (Lippmann, 1973). This facilitates aragonite precipitation.

The inorganic precipitation of aragonite needles is well documented, notably in the form of "whitings" (Wells and Illing, 1964). The sudden precipitation of aragonite crystals in a large body of seawater is indicative of considerable preexisting supersaturation. The spontaneous precipitation may be an indication of increased residence time of the surface waters. Under continuing evaporation, horizontal water movement is temporarily slowed, causing the ion activity product to drop. Aragonite (and vaterite) form under rapid rates of precipitation; calcite forms under slow rates (Kitano and Hood, 1965). However, aragonite precipitation in an aquarium is always bacterial and does not occur in a sterile environment (Billy *et al.*, 1976).

Various ions present in seawater affect the stability of calcium carbonate polymorphs; adsorption of foreign cations onto crystal surfaces fosters preferential crystallization of one mineral species or inhibits the recrystallization of another. The presence of $CoCl_2$ in minute quantities favors the precipitation of calcite and vaterite, but above 390 ppm it favors that of aragonite (Barber *et al.*, 1975). Magnesium ions in the water inhibit calcite formation and favor aragonite precipitation (Pytkowicz, 1965; Kitano *et al.*, 1979a), whereas barium ions, for instance, have the opposite effect. The solubility of calcite increases with the rising magnesium content of the brine until aragonite becomes the stable phase above 8.5 mole % (Berner, 1975). However, Chen *et al.* (1979) maintain that in lacustrine environments the precipitation of magnesian calcite is favored over that of aragonite with an Mg/Ca ratio greater than 10, high concentrations of sodium chloride, and temperatures of 6–35°C. Aragonite, together with small quantities of monohydrocalcite ($CaCO_3 \cdot H_2O$) predominates at lower temperatures. A slight decrease in the dissolved calcium due to aragonite precipitation might be sufficient to lower the molarity of calcium in seawater (401 ppm), so that radial fibrous growth takes place (Lippmann, 1973). Magnesium ions selectively poison the sideways growth of calcite; thus, calcium carbonate prefers to crystallize as micritic or fibrous aragonite. Calcitic microspar forms only in the absence of magnesium ions in subsurface brines or in meteoric waters (Folk, 1974). Aragonite allows much less magnesium than calcite to enter the crystal lattice; later diagenetic alteration of aragonite to calcite generates a low-magnesium limestone. Isochemical recrystallization to calcite thus decreases the originally available void space by about 6%. Other cations also have an important role to play in calcium carbonate stability.

Whereas copper adsorption is greater in salt water than in fresh water, the opposite is true of zinc adsorption. Even total adsorption of all copper in solution does not inhibit the transformation of aragonite to calcite. However, zinc adsorption inhibits this transformation very strongly; more than 90% of zinc in solution is eventually adsorbed onto aragonite (Kitano *et al.*, 1976). The evidence on anions is not yet unequivocal. Whereas Bischoff (1968) found that sulfate ions strongly inhibit aragonite formation, Kitano *et al.* (1975) found the opposite: The presence of sulfate ions favors aragonite precipitation, and borates inhibit aragonite and favor calcite crystallization (Kitano *et al.*, 1979b).

The solubility of calcium carbonate in seawater drops sharply with rising pH (Fig. 6-2). Addition of sodium carbonate to seawater raises its pH and promotes some aragonite precipitation (Chave and Suess, 1970), which increases as the amount of dissolved matter rises. Coprecipitation of sodium or potassium is also a function of pH and temperature, and causes a variety of crystal defects (A. F. White, 1977). It is not the total concentration of carbonate ions which determines

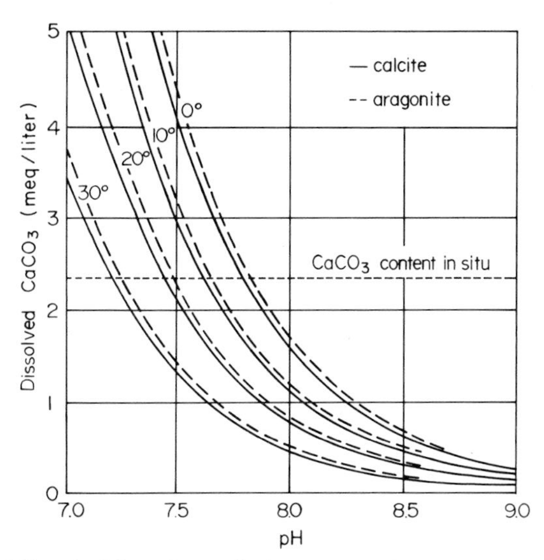

Fig. 6-2. Solubility of calcite and aragonite in seawater (19 ppt chlorine) as a function of pH (after Wattenberg and Timmermann, 1936).

the supersaturation, but the effective concentration, i.e., the activity. This is low because of the tendency of carbonate ions to form complexes with calcium, magnesium, and sodium ions. Two-thirds of the total CO_3^{2-} is neutralized as magnesium carbonate. Evaporation reduces the amount of free CO_3^{2-} (Lippmann, 1973). Aqueous anion activities suggest that carbonate is the only anion species in the solid solution (White, 1977).

The bulk of all marine carbonates is biogenic in origin, no matter how much they have since been altered diagenetically. The amount of dissolved organic matter and other biogenic parameters control the rate of precipitation and determine whether aragonite is precipitated, whether magnesian calcite is formed, and whether the two form intergrowths (Alexandersson, 1972). The mineral surfaces are easily saturated with a monomolecular layer of organic compounds that inhibit or reduce the reaction rates of inorganic carbonate equilibria (Suess, 1970). Organic matter with a weak affinity for calcium ions gives faster rates of precipitation and thus leads to aragonite crystallization. Organic matter with a strong affinity for calcium ions fosters the formation of complexing, reduces the rate of calcium carbonate precipitation, and therefore leads to calcite crystallization (Kitano and Hood, 1965).

Spherulitic aragonite grains in Great Salt Lake, Utah, and elsewhere appear to be directly related to the presence of proteins or their decomposition products (Fabricius, 1977). Ammonium carbonate, produced by denitrifying bacteria in the putrifying organic matter on the sea shelf, reacts with calcium in solution to produce spherulitic aragonite. If any sodium carbonate were generated, it would yield calcite (Grabau, 1920). The same occurs in the presence of minute amounts of sodium hexametaphosphate (Reitemeyer and Buehrer, 1940) or of sodium salts of many organic acids. The nature of the body fluids of many euryhaline organisms may thus influence the choice of aragonite or magnesian calcite as skeletal building material (Lippmann, 1973). Many phyla produce aragonite shells in a marine environment rich in magnesium; this also leads to comminuted aragonitic skeletal sands, silts, and muds. Even on the level of bacteria and algae, aragonite and magnesian calcite producers exist (Krumbein and Cohen, 1964).

The effect of sodium carbonate solutions percolating through a sequence of evaporite minerals has so far not been adequately investigated. Tertiary volcanism produced large quantities of carbon dioxide that percolated through Zechstein halites and filled vacuoles, yet very little sodium carbonate was produced. Waters rich in alkali carbonate react with gypsum to form vaterite rather than calcite, an unstable modification of calcium carbonate (Floerke and Floerke, 1961). Circulating waters that contain sodium carbonate transform primary gypsum into calcium carbonate: In the 30–40°C range, a high concentration yields vaterite, a low concentration calcite, and a still lower concentration aragonite. The presence of sodium chloride thereby supports the formation of calcite

(Wilde, 1961). However, seepage of crude oil into the oxidized zone in Kirkuk, Iraq, transforms pure white anhydrite into coarse, greasy-looking, fibrous aragonite (Al-Hashimi, 1972).

Freshwater encroachment removes the magnesium barrier and allows precipitated aragonite to convert to calcite. Subsurface brines essentially stripped of their magnesium content by dolomitization or by sorption effects on clays may play the same role for buried aragonites. Even in these cases, dissolved organic matter influences the process. Basic and neutral amino acids accelerate the recrystallization of aragonite to calcite, whereas acidic amino acids inhibit the process (Jackson and Bischoff, 1971).

MAGNESIAN CALCITE

Phyla that build a calcite skeleton commonly use magnesian calcite, i.e., high-magnesium calcite, which later loses its magnesium content diagenetically. Fossil high-magnesium calcites are not known. It has hitherto not been documented whether, in recent sediments, the magnesium content of such calcite particles converts to tiny dolomite crystals or is swept away in solution. Ancient analogs seem to allow the former to happen. Ancient limestones are composed of low-magnesium calcite; modern calcite cements in contact with seawater are of the high-magnesium variety. Otherwise modern skeletal limestones are composed of either high-magnesium calcites or aragonite. Deep-water, low-magnesium calcites do not seem to be derived from recrystallization of either aragonite or high-magnesium calcite; Friedman (1972) suggested that they may be the product of anaerobic degradation of gypsum. High-magnesium calcite is precipitated by phytoplankton in freshwater lakes, concurrent with inorganic aragonite coating of leaves (German Mueller, 1971). Bischoff (1968) saw in high-magnesium calcite a metastable intermediate phase during the seawater transformation of aragonite to calcite. Nonetheless, direct precipitation of high-magnesium calcite in the skeletal parts of some phyla seems to be well established. Eventually, the magnesian calcite changes to ordinary calcite in calcareous algae, where no magnesium ion is present. Magnesium ions in the solution prevent this transformation, since the magnesium carbonate content increases due to dissolution of calcium carbonate (Ohde and Kitano, 1981).

MAGNESITE

Magnesite is often found as an alteration product after basal or roof anhydrites or in nests. Putrefying organic matter favors the formation of magnesite (C. L. Baker, 1929), because bacteria can then decompose the original precipitate and generate the required carbon dioxide. Magnesite as an evaporite deposit is often a form of impure mesitite, i.e., the iron–magnesium carbonate containing 50–70% magnesium carbonate and 30–50% iron carbonate, with a minor admixture of both manganese and calcium carbonates. An increasing concentration of the brine causes an increase in the density of the crystal faces, changing from an isometric to a more platy habit. In rock salt, magnesite is rhombohedral, with small amounts of a pseudo-octahedral habit; in primary sylvite, it is rare. In carnallite and in sylvite after carnallite, the habit becomes prismatic or platy (Stewart, 1949; Yarzhemskiy, 1955; Sedletskiy and Mel'nikova, 1970). Magnesite forms either with nesquehonite as precursor (Leitmeir, 1953; Leitmeier and Siegel, 1954) or from hydromagnesite. Artificial hydromagnesite and aragonite can be produced by adding sodium carbonate to seawater (Alderman, 1965).

The magnesite that occurs at the expense of gypsum underneath halites of the sebkha El Melah in Tunisia (Perthuisot *et al.*, 1972) may have a two-step origin. Gypsum, converted first to calcite by anaerobic sulfur bacteria, was subjected to percolating magnesium chloride solutions that completed the alteration. The chronic sulfate deficit in the sedimentary sequence of this sebkha (Perthuisot, 1974; Perthuisot *et al.*, 1972) is proof that sulfur bacteria were involved in the sulfate depletion, since sulfates are still present in an aerated nearshore belt. Magnesium-rich bitterns stripped of their potash by precipitation produce magnesite as the prime carbonate, even when diluted by runoff. When entering previously dolomitized limestones, these bitterns can alter these rocks to magnesites liberating calcium chloride (Johannes, 1966). This appears to have happened in perisaline dolomites of the Triassic salt basin in the Atlas Mountains (Kulke, 1978), and in the Permian Wellington halite of Kansas, where the thin layers of magnesite contain a coquina (Jones, 1965).

Descending meteoric waters enriched in clay-derived sodium produce nests of sodium sulfate minerals in buried marine evaporite sequences, such as aphthitalite or bloedite. If the waters are enriched not only in sodium but also in bicarbonate ions, they become responsible for the occasional magnesite nests in buried evaporites. Magnesite, together with pyrite, has been found in Zechstein halites and carnallites in Germany, in halites at Zipaquira, Colombia (Buecking, 1911), and elsewhere. Dolomite (at times fossiliferous) often predominates in Zechstein evaporites on the basin margins and on sills, whereas in the interior deeper parts of the basin, it is replaced by magnesite within anhydrite rocks or by bituminous limestone (Kuehn, 1968). Magnesite then occurs in nodular form, mainly in clay horizons close to carnallite beds (Koch *et al.*, 1968) or below anhydrite above a gray clay horizon (Langbein, 1964).

A variety of hydrous magnesium chlorocarbonates ($MgCl_2 \cdot MgCO_3 \cdot 6–7H_2O$), stable under anaerobic conditions, change to a basic chlorocarbonate [$MgCl_2 \cdot 2MgCO_3 \cdot Mg(OH)_2 \cdot 6H_2O$] if exposed to air (Walter-Levy and Maarten de Wolff, 1949). There are also a variety of basic magnesium chlorides

$[MgCl_2 \cdot 1.9Mg(OH)_2 \cdot 1.8H_2O]$ that are stable at 25°C (Bianco, 1951). These compounds have been recognized in the aging of concrete, but none of them have so far been identified as metastable compounds in an evaporite basin.

DOLOMITE

Dolomite becomes more stable than calcite at pH values above 7.8 (Fig. 6-3). Thus, it is the thermodynamically stable phase in seawater with a normal salinity and Mg/Ca ratio (Stumm and Morgan, 1970). Yet it does not precipitate spontaneously without the aid of organic activity. In a straightforward analogy to high- and low-magnesium calcite, there are also two dolomites within the temperature range of subtropical lagoons. One of these dolomites contains an almost equal amount of calcium and magnesium carbonates; the other has about 10 mole % excess calcium carbonate (Reeder and Wenk, 1979). South Australian dolomites are enriched in ^{18}O and depleted in ^{13}C relative to calcite (Clayton *et al.*, 1968). The observation that evaporite-related dolomites are ^{18}O enriched indicates that they were produced by hypersaline brines (Clark, 1980b).

Recent dolomites often seem to have a precursor in huntite (Kinsman, 1967; Irion and Mueller, 1968; Perthuisot, 1974). However, in Lake Keilambete, Victoria, Australia, recent dolomite occurs in association with calcite, and no aragonite is present (Bowler, 1970). This may indicate that here the aragonite has been a precursor. Experiments at elevated temperatures showed that aragonite converts to dolomite four times faster than does calcite (Katz and Matthews, 1977).

Dolomites in the Coorong lakes of South Australia were thought to be groundwater derived (Sonnenfeld, 1964). This has been confirmed (von der Borch *et*

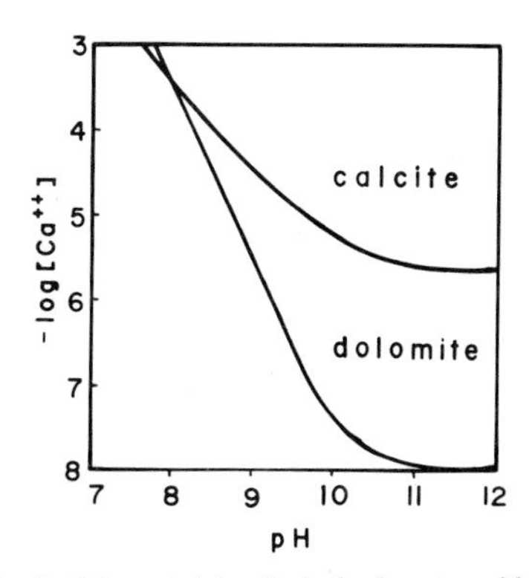

Fig. 6-3. Stability fields of calcite and dolomite in fresh water with a concentration of 0.0005 *M* magnesium ions and 0.002 *M* total carbonate carbon (after Stumm and Morgan, 1970, copyright John Wiley & Sons, Inc.).

al., 1975). Both Badiozamani (1973) and Folk and Land (1975) proposed that mixing of groundwater and seawater is critical for dolomite precipitation; Gaudefroy (1960) observed coarse dolomite crystals in the vicinity of an ancient stream discharge. One has to bear in mind that many metals are more soluble in concentrated brines, and precipitate upon dilution by formation or meteoric waters (Pasztor and Snover, 1983).

Dolomitization in the Florida aquifer occurs in brackish waters, i.e., in waters less saline than seawater (Hanshaw *et al.*, 1971), with a strontium content of less than 250 ppm (Randazzo and Hickey, 1978). Authigenic dolomite along the southwest coast of Florida was also interpreted as the result of the interaction of organic matter and hypersaline waters with a carbonate substrate (Taft, 1961). Dolomite nodules around burrows of crustaceans in recent offshore sediments along the west coast of Scotland are produced by the interaction of carbon dioxide ions and ubiquitous magnesium ions with an initial calcium carbonate phase. The carbon dioxide ions are produced by bacterial sulfate reduction, which causes increased alkalinity and is fueled by the decaying organic matter (Brown and Farrow, 1978). Organic matter also contributed a portion of the carbon involved in Recent dolomization on the island of Naxos in the Aegean Sea (Koppenol *et al.*, 1977). In the Solar Lake, Gulf of Aqaba, Red Sea, the aragonite is being dolomitized during the winter never less than some 30 cm below the sediment–water interface (Aharon *et al.*, 1977). Winter there is the season of a high water level, density stratification and eventual mixing, a 37°C temperature difference between surface and bottom waters, and high production rates of organic matter.

The kinetics of dolomitization of either aragonite or calcite are strongly dependent on temperature. The presence or absence of other cations or anions, particularly strongly hydrated ions, seems to be the most important condition for dolomite formation. Certain amino acids and soluble proteins inhibit the reaction (Gaines, 1980), whereas strongly hygrophilic elements (e.g., lithium) assist in catalytic dolomite formation (Gaines, 1974). Most secondary dolomites contain a small amount of iron. Increasing amounts of ferroan dolomites are produced under stagnant reducing conditions. Since ferroan calcites form at near-neutral pH, the presence of ferroan dolomites may also indicate reduced pH values during formation. Patterson and Kinsman (1982) found that dolomitization along the Persian Gulf proceeded at an Mg/Ca molar ratio of 6, an ambient temperature, a pH of 6.3–6.9, a brine saturated with respect to gypsum, and a chloride concentration of 115–130 ppt with hydrogen sulfide present. However, Maunaye *et al.* (1981) found a molar ratio of 1.2 adequate and the sodium chloride concentration irrelevant in the presence of urea and elevated temperatures. Under elevated temperatures, all of the calcium in calcite lattices becomes exchangeable and results in nonstoichiometric phases of dolomite that are deficient in anions if calcite is exposed to magnesium chloride brines (Kiss, 1981).

Pierre *et al.* (1981) studied the sublittoral formation of dolomite in quartz–feldspar sandstones of the Ojo de Liebre lagoon in Baja California (Mexico) that occur in and under a gypsum lens. They envisaged the dolomite to be a product of marine magnesium sulfate brines mixing with continental groundwater charged with calcium bicarbonate. Similarly, magnesite in alpine salt deposits is interpreted as a diagenetic reaction product of limestone with magnesium sulfate solutions (Schroll, 1961). However, these assumptions may not be valid, as dolomite is not stable in sulfatic brines. The interaction of dolomite with sulfatic groundwater produces epsomite efflorescences in southern Israel (Mazor and Mantel, 1966), and anhydrite precipitates in Recent sediments in the Savikamsk lakes, USSR, wherever calcium bicarbonate waters meet magnesium sulfate brines (Rubanov *et al.*, 1964).

The elimination of calcium sulfate from seawater seems to be an essential preliminary step in dolomite formation (Liebermann, 1967). If the sulfate is removed from the brine and the solubility of bicarbonate ions remains negligible, then even the slightest amount of oxidation of organic matter must result in instant reactions with any available calcium carbonate and the ultimate utilization of magnesium ions to precipitate such carbonic acid. As soon as magnesium enters clay minerals in the form of brucite bridges and converts the clays into mixed-layer varieties, it is also available to react with any products of organic metabolism, i.e., oxidation of carbon compounds. Authigenic dolomite on marble surfaces occurs in the presence of algae, fungi, and lichens, but never together with gypsum (Del Monte and Sabbioni, 1980).

Nikol'skaya and Gordeeva (1967) synthesized dolomite at 50°C and a Mg/Ca ratio of 2:1 from gypsum in a solution of 250 ppt magnesium carbonate and an excess amount of 40 ppt sodium bicarbonate. Baker and Kastner (1981) synthesized dolomite at 200°C, and also found that reduction of the sulfate ion concentration, either by the action of microbes or by mixing with fresh water, exerted the principal control on dolomite formation. They further ascertained that the Mg/Ca ratio need be only slightly above unity if the ionic force of the solution equals the ionic force of ocean waters. In the strip between diurnal and storm tides on the Arabian coast of the Persian Gulf, where the salinity reaches 275 ppt, dolomite occurs in sediments together with gypsum. The higher parts of the intertidal zone are occupied by gypsum alone; below the diurnal tide level, aragonite occurs (Wells, 1962).

Davies *et al.* (1975) suggested that some molecular fractions of humic acids inhibit aragonite formation, whereas for dolomite production high alkalinity is more important than a high Mg/Ca ratio, coupled with a brine concentration of 6- to 8-fold seawater salinity and a consequently high CO_3^{2-}/HCO_3^- ratio. This latter ratio is produced by decaying organic material which releases methane; the methane is utilized to reduce gypsum, with concomitant generation of excess carbon dioxide. Dolomitization appears to occur here in close proximity to the strand line, soon after the deposition of the sediment. Preferential areas are sites with excess calcium, i.e., with increases away from drainage channels. Dolomitization does not increase inland (Patterson and Kinsman, 1982), as McKenzie *et al.* (1980) postulated.

Relatively high concentrations of phosphates tend to favor sulfate synthesis (Zajic, 1969, p. 68) and thus gypsum production. It is surely no accident that most recent dolomite occurrences are reported from lagoons near abandoned phosphate mines or near active guano accumulations (Bonaire, Curaçao, Los Roques, Jarvis Island, Crane Key, etc.). Dolomite on Sugarloaf Key, Florida, is restricted to the shoreline strip of red mangrove growth, and is absent from black or white mangroves farther inland. The pH of surface waters here is 7.8–8.0; that of interstitial waters is 7.3–8.3 (Atwood and Bubb, 1970). Crane Key, with its bird population, and Sugarloaf Key are the only islands in the 260-km-long chain of Florida Keys where dolomite has been found.

Whereas dolomite formation requires high alkalinity, a reduction in alkali bases produces in the same medium rosettes of gypsum; several such pathological instances in equine or human urine were reported by Lapinski (1906), who suspected that ammonia plays a crucial role. Mansfield (1979) also suggested an organic origin of dolomite, i.e., by urease-producing bacteria. A dolomitic urolith of a Dalmatian dog was deemed to be due to an infection caused either by anaerobic urease-producing bacteria or by anaerobic uric acid-fermenting bacteria (Mansfield, 1980). Linck (1937) noted earlier that ammonium chloride is a good catalyst in dolomitization. Possibly this is due to the ability of ammonium salts to increase the solubility of calcium sulfate. Matsuoka *et al.* (1978) produced dolomite at slightly elevated temperatures, using ammonium carbonate as the source of carbon dioxide and ammonia. They found ethylene diamine and ethyl urea to be organic compounds very effective in catalyzing dolomite, even at 180°C. Van Tassel (1965) synthesized dolomite in the presence of urea, albeit at 200°C, a temperature that is much higher than can be expected to occur in nature in near-surface environments. Ammonium chloride has been reported from evaporite sequences that contain intercalated, highly dolomitized limestones. Clay intercalations and brines in German Zechstein deposits contain ammonium chloride (Biltz and Marcus, 1908; Linck, 1942), as do potash salts in Permian Upper Kama deposits west of the Urals (Apollonov, 1976). These observations can be expanded. Nitrogen-fixing sulfate reducers, such as strains of *Desulfovibrio* and *Desulfotomaculum*, are common in oil field brines (Nazina *et al.*, 1979). If they digest the sulfate ions of infiltrating surface waters, seawater, or hypersaline brines, they produce carbonate ions. A dolomitizing reaction can then proceed as follows:

$$CaCO_3 + CO_2^{2-} + H_2O \rightarrow Ca(HCO_3)_2 \quad 6\text{-}1)$$

which is the leaching stage recognized in most epigenetically dolomitized sequences, followed by

$$Ca(HCO_3)_2 + Mg^{2+} + \text{organic matter} + SO_4^{2-} \rightarrow$$
$$H_2S + CaMg(CO_3)_2 + \text{bacterial metabolism} \quad (6\text{-}2)$$

which is the dolomite crystallization stage. This may be one of the paths of secondary dolomitization in an aquifer resulting in sulfuretted brines. Such brines, deficient in magnesium and sulfate ions but containing calcium chlorides and hydrogen sulfides, are common in medium-to-coarsely crystalline dolomite rocks. The local rate of metabolism may then dictate the uniformity of crystal sizes observed in most secondary dolomites. The data on dolomites and dolomitization were summarized by Sonnenfeld (1964) and by Chilingar *et al.* (1979).

Dedolomitization can proceed in the presence of sulfates as follows (Braddock and Bowles, 1963; Perel'man, 1967):

$$CaSO_4\ (aq) + CaCO_3{\cdot}MgCO_3 \rightarrow 2CaCO_3 + MgSO_4\ (aq) \quad (6\text{-}3)$$

Dedolomitization is likewise linked to arid lands, as the resulting calcium carbonate dissolves in a humid climate. Diluted circulating waters lead to calcian dolomites and eventually dedolomitization (Katz, 1971), especially if the waters are fresh waters charged with carbon dioxide in the vadose zone (Sha *et al.*, 1979). These waters should have a very low Mg/Ca ratio, a carbon dioxide partial pressure considerably below 50 kPa, and a temperature below 50°C (de Groot, 1967). Dedolomitization leads to extreme strontium impoverishment in the resulting calcite (Shearman and Shirmohammadi, 1969), which can then be subjected to redolomitization (Clark, 1980b).

Subrecent dolomitization of limestones along exposures carved out by Pleistocene glaciers (cf. Sonnenfeld, 1964) is produced by rainwater diluting groundwater. Lacustrine dolomite crystals in subhumid to semiarid terrains (Sonnenfeld, 1964) are likewise not related to marine evaporite formation.

BASAL CARBONATE UNIT

In marine sequences, it is common to find a basal limestone unit, often secondarily dolomitized. Occasionally the carbonate bank may not cover the whole basin floor, and evaporite minerals are encased in nearshore clays. This is, for instance, the case in Mississippian evaporites of Nova Scotia (R. Evans, 1970). As in the Netherlands and north Germany, the British Zechstein evaporite cyclothems start with oolitic or algal limestones and stromatolites laterally grading into laminated dolomites on the shelf margin, with reefal carbonates toward the basin entrance (Smith, 1970).

As the brine concentration increases, the carbonates progressively retreat to the margins, to the vicinity of the bay entrance, or even to the shelf area outside the entrance threshold. If antechamber basins succeed in concentrating the brine beyond calcium carbonate saturation, the inner basins may start an evaporite sequence with calcium sulfates. The ratio between the solubility products of calcium sulfates and carbonates is 4700 (Stieglitz, 1908). One could thus expect to find 4700 times as much limestone in and around an evaporite basin as gypsum or anhydrite. Quite commonly, the limestones and secondarily dolomitized limestones represent a multiple of the volume of anhydrites, but no one has really taken up Stieglitz's challenge to estimate relative volumes for a given basin, even where subsurface control is excellent.

STRONTIANITE AND WITHERITE

Strontianite, the strontium carbonate, has a solubility commensurate with that of calcium carbonate, and strontium substitutes for calcium in the calcite lattice. Aragonite takes up twice as much strontium as does calcite. Strontium removal seems to go hand in hand with the conversion of aragonite to calcite (Siegel, 1958, 1960; Friedman and Brenner, 1977). In contrast, witherite, the barium carbonate, which occurs in limestone vugs, is rarely a primary mineral. It is mostly an alteration product after barite.

SUMMARY OF EVAPORITIC CARBONATES

Aragonite is the first mineral to precipitate from concentrating seawater. However, most aragonite and calcite is extracted from seawater by biogenic processes. In contrast, calcite and magnesite are frequently found as alteration products of gypsum or anhydrite due to microbial activity in an anoxic environment. Dolomites are likewise often the alteration product of limestones exposed to concentrated magnesium-rich brines that have been stripped of their calcium sulfate content. Barium carbonates precipitate into limestones early in the concentration cycle, whereas strontianite is often secondary. The strontium, originally substituting for calcium, was expelled from the aragonite or calcite lattice during recrystallization.

7

Primary Marine Precipitates

GYPSUM

The principal primary marine sulfate is gypsum. Anhydrite and bassanite, the other two forms of calcium sulfate in nature, are usually produced by secondary alteration of already precipitated gypsum, i.e., its partial or complete dehydration. Several detailed textural studies of gypsum and anhydrite have appeared, the first of these by Hammerschmidt (1883).

Calcium sulfate precipitation can occur in the following ways (after Corbel *et al.*, 1970, and Shearman, in Holliday, 1968):

1. Direct precipitation of gypsum in marine hypersaline brines.
2. Direct precipitation of gypsum in saline lakes.
3. Direct precipitation of anhydrite to form, e.g., nodular anhydrite.
4. Hydration of anhydrite by surface waters to form gypsum.
5. Dehydration of gypsum to form bassanite and ultimately anhydrite.
6. Dehydration of gypsum to form bassanite and rehydration to form gypsum flour.
7. Dissolution and reprecipitation of gypsum.
8. Dissolution and reprecipitation of anhydrite.
9. Dissolution of gypsum and reprecipitation of anhydrite.
10. Dissolution of anhydrite and reprecipitation of gypsum.
11. Pseudomorphs of polyhalite after gypsum.
12. Oxidation of sulfur deposits.
13. Reaction of sulfuric acid present in certain naturally charged waters with limestone.
14. Low-temperature hydrothermal synthesis. The dehydration/rehydration effects will be discussed in Chapter 13.

Gypsum Saturation

The partial pressure of carbon dioxide in the ocean has always been smaller than $10^{-2.9}$, the point of dolomite decomposition, and greater than $10^{-3.7}$, the point of gypsum formation (Stumm and Morgan, 1970). In hypersaline brines, this partial pressure of carbon dioxide rapidly drops below the latter value. When the brine begins to precipitate gypsum, it has passed calcium carbonate saturation. Neither carbonate nor bicarbonate ions are then available in significant quantities in the solution to bind calcium ions. If decarbonated seawater evaporates at 35–60°C, it first forms gypsum. It is the removal of carbon dioxide in the brine that aids gypsum precipitation. The molal solubility of 1.57×10^{-3} is independent of temperature in natural brine pools (Shaffer, 1966). Adding calcium chloride to seawater supersaturated to calcium carbonate produces no increase in pH but induces a slow precipitation of gypsum (Chave and Suess, 1970). It is possible that groundwater, often enriched in calcium chloride, has the same effect when mixing with encroaching seawater, and thus plays a role in forming gypsum crusts in intertidal and supratidal environments, especially in coastal sebkhas.

Inside a marine bay, gypsum precipitation commences with a sharp basal contact, but with an occasional transition to overlying laminae. This indicates that the brine had attained at least 4.5 times the concentration of seawater. Since calcium sulfate can remain in solution up to a certain degree of supersaturation, precipitation often starts only after about 80% of the seawater has evaporated. By that time, the salinity exceeds 117 ppt (Elderfield, 1976). As a calcium sulfate solution concentrates, it reaches saturation at 15.33 mmoles at 25°C, if the dissolved particles were about 2 μm in diameter. However, saturation is reached only at 18.2 mmoles with particles 0.3 μm in diameter. A significant increase in solubility is attained by merely shaking a saturated solution with comminuted gypsum particles (Hulett, 1901). In nature, wave action can accomplish this. Gypsum crystallization changes both the ionic and the isotopic composition of the residual brine. It depletes the brine of about 15 ppt deuterium but enriches it by 4 ppt heavy oxygen (Fontes and Gonfiantini, 1967a,b). Water trapped in fluid inclusions is also enriched in heavy oxygen (Fontes, 1966).

The shape of the gypsum crystals is a function of pH and of foreign cations. Gypsum crystallizes best at a pH of 7–8; the higher the pH, the stubbier are the crystals. Acicular crystals are restricted to acidic environments (Barta *et al.*, 1971). However, foreign cations also produce acicular crystals (Forestier and Kremer, 1952), e.g., large quantities of dissolved silica (Murat, 1977).

Oxygen Supply

In a density-stratified lagoon, the surface brine remains oxygenated because of its lower salinity and consequent higher oxygen solubility, and because of continuing gas exchange with the atmosphere. Based on laboratory experiments, Sloss (1969) concluded that calcium sulfates precipitated above the interface between oxygenated surface waters and brine. This is the case in many coastal lagoons, where gypsum grows after the rainy season. Raup (1982) proved that oxygenated seawater brines of different concentration can precipitate gypsum upon mixing. However, he excluded from his experiments the important contribution of life to the process by first eliminating all biota from his seawater samples. Nature does not use aseptic samples. The experiments therefore shed light only on gypsum precipitation in ephemeral, extremely shallow standing bodies of water where density stratification does not develop and all waters remain oxygenated.

Gypsum precipitation is a function not only of the concentration beyond saturation but also of the oxygen supply in anoxic, density-stratified brines, together with a sufficient supply of sulfate ions. Progressively more concentrated brines carry less and less oxygen in solution (Fig. 7-1). By the time halite starts precipitating, the solubility of the oxygen in the brine is less than one-third of that in seawater (MacArthur, 1916; Sozanskiy, 1973), and the brine is functionally anaerobic even if exposed to atmospheric oxygen (Kinsman *et al.*, 1974) (Fig. 7-2). Consequently, any sulfate ions are prone to be stripped of their oxygen if anaerobic bacteria are present.

It is *the lack of oxygen* that *restricts gypsum precipitation largely to the photic zone* (Sonnenfeld, 1979), where algae and photosynthesizing purple bacteria generate oxygen, and sulfur bacteria oxidize hydrogen sulfide to create a new supply of sulfate ions. Algal and bacterial photosynthesizers are often found in distinct layers within modern gypsum laminites. They thrive on ultraviolet radiation, for which gypsum laminites are transparent (Hausmann and Krumpel, 1930; Tagaki, 1931). Absorption through calcium sulfate in solution controls the transparency of seawater for ultraviolet radiation and reduces it rapidly with increasing wavelength (Hurlburt, 1928). The volume of gypsum precipitation thus depends on the algal oxygen production beneath the interface between aerated but undersaturated surface waters and anaerobic but supersaturated brines. Lagoons in the Dutch and Venezuelan Antilles precipitate gypsum when the brine is density stratified; in seasons when the surface water has evaporated, the gypsum decays partly by dissolution and partly by decomposition. It reconstitutes a fresh surface after the next rainfall or seawater influx, when the algal population is also regenerated. That this process remains incomplete is proven by the presence of significant residual amounts of calcium in the brine (Sonnenfeld *et al.*, 1976).

Although most of the oxygen produced by algae escapes as gas bubbles, enough oxygen remains dissolved during the day for the waters to become supersaturated with it: The O_2 evolved by photosynthesizers is not lost to the atmosphere as rapidly as it is produced (Carpelan, 1957). It is available to

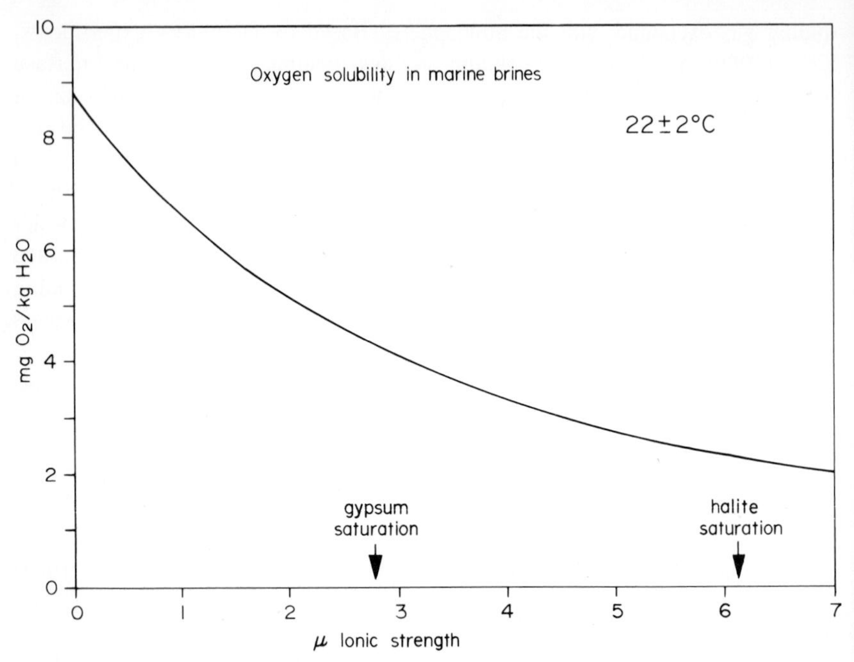

Fig. 7-1. Solubility of molecular oxygen in an artificial marine brine at 22° ± 2°C (after Kinsman *et al.*, 1974).

convert some rising HS^- ions back to SO_4^{2-} ions. In contrast, liberation of carbon dioxide and concurrent depletion of O_2 have been reported (van der Meer Mohr, 1972) within a series of ponds in a salina. These actions occurred primarily in gypsum-precipitating ponds. The source of this carbon dioxide is from the corrosion of algal aragonite by sulfate ions. Carbon dioxide is then again consumed during the day by blue-green algae that replenish the O_2. Consequently, gypsum can be seen as a product of osmotic phenomena inside the algal mat (Perthuisot and Jauzein, 1978). Anhydrite (which possibly was gypsum before ashing) is found as a precipitate inside the cells of blue-green algae (Leung, 1981). On the other hand, another alga, *Apanothece halophylica* (also known as *Coccochlorus*), produces mats that prevent gypsum precipitation (N. Sammy, personal communication, 1983).

Pyrite and calcite taken together out of bituminous shales produced gypsum within 6 days when exposed to humid air. In contrast, pyrite and dolomite produced gypsum in humid air overnight (Dean and Ross, 1976), as dolomite is more vulnerable to sulfatic waters. Gypsum pseudomorphs after rostra of belemnites southeast of Kiev, Ukraine, were deemed to be derived from the interaction

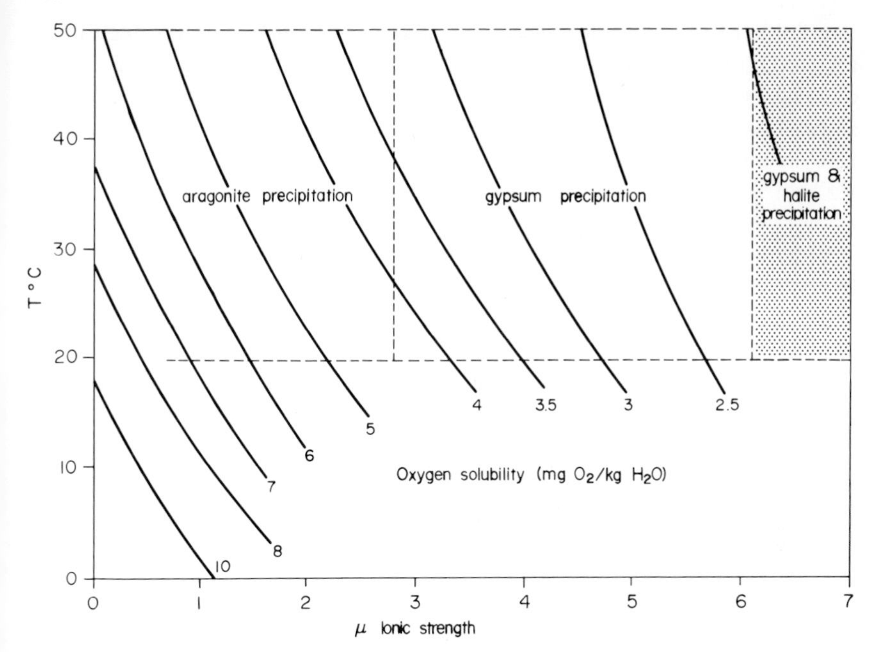

Fig. 7-2. Molecular oxygen solubility in parts per million, plotted against the temperature and ionic strength of the brine. The majority of marine evaporites crystallizes in the range of 20°–50°C. Once gypsum starts to precipitate, oxygen solubility has fallen to less than 4 ppm. When halite starts to precipitate, oxygen solubility has dropped to less than 2 ppm (after Kinsman *et al.*, 1974).

of the calcite shell with sulfuric acid produced by the oxidation of pyrite (V. N. Chirvinskiy, 1928).

Whereas gypsum precipitation is limited by the oxygen supply in the photic zone, the zone of algal oxygen production, anhydrite may be inhibited by its high nucleation energy (Braitsch, 1971) and becomes mainly a product of gypsum dehydration by percolating hygroscopic brines. Truly primary anhydrite is thus very rare, as is primary polyhalite. Both are restricted to the outer fringes of some modern lagoons, where they occur in very small amounts.

Sulfate Supply

If all the calcium ions in inflowing seawater were captured by sulfate ions, only about 36.2% of the available sulfur would be utilized. Additional calcium ions are introduced by ion exchange with episodic input of clay minerals; some calcium forms carbonates that are not altered into gypsum, and some calcium remains in solution. Nonetheless, more than two-thirds of the incoming sulfate ions fail to be precipitated. Beneath the interface with surface waters, the sulfate ions are stripped of their oxygen in the anaerobic bottom brines. The rate of removal of sulfate ions by reduction to sulfide is roughly proportional to the sulfate concentration in seawater. This concentration of sulfate ions approaches zero in marine sediments rich in decomposable organic matter (Lasaga and Holland, 1976). The autotrophic sulfate reducers consume more than three times as much H_2 as is required for hydrogen sulfide production.

Sisler and ZoBell (1951) found that any available carbon dioxide was reduced by the same strains to lipids and members of the alkane series. This represents extreme competition for any available carbon dioxide liberated in gypsification of algal stromatolites. In supratidal flats, such as in the Abu Dhabi sebkha, over 90% of the incoming sulfate ions have been either precipitated or metabolized when the brine is at, or near to, halite saturation (Butler *et al.*, 1973). Braitsch (1971) suggested that an excess of sulfate ions is precipitated with calcium ions derived from dolomitization of shelf carbonates. He supported this argument by the observation that the sulfate deficiency reaches a minimum where the Zechstein Main Dolomite is laterally replaced by a thin bituminous shale. This is not possible, since dolomitizing solutions spread out farther and farther into an aquifer, and there is no way for liberated calcium ions to reflux for tens or hundreds of kilometers back into the evaporite basin.

A sulfur deficiency found in ancient evaporites (Borchert, 1959) apparently occurs very early in the concentration cycle when hydrogen sulfide is formed and escapes. This author has produced a gypsum precipitate from anoxic lagoonal bottom waters by simple addition of sulfuric acid and even traces of hydrogen peroxide. Neither calcium nor sulfur nor sufficient concentration of the brine has apparently been lacking—only oxygen. The same conclusion may be reached about formation waters from the observation of calcium sulfate scaling in the production casing of many oil wells. Isolated gypsum crystals intersecting halite crystal boundaries in the Salina salt of Michigan may have been formed as late as the time of drilling (Dellwig, 1955). Olausson (1961) and Arrhenius (1963) suspected that some of the gypsum found in deep-sea cores was formed by oxidation in storage, as the first description of gypsum in Mediterranean deep-sea cores stemmed from Kullenberg (1952), who had examined cores stored for more than a year and a half.

Any inflowing current hugs the shore of a basin and presents a slight density gradient away from the coast. Entering brines therefore reach saturation for gypsum first on the shelves of nearshore regions. Photosynthesizers can flourish, and thus gypsum precipitation can build up on these shelves and steepen the slope to any deeper parts of the lagoon. At the same time, however, the coastal areas are also the sites of the greatest influence of continental drainage. Kushnir (1982) has tried to show that the Upper Miocene (Messinian) gypsum and anhydrite occurrences in the Mediterranean area appear to have been marine only at sites a significant distance from ancient shores, whereas nearshore sites were dominated by continental influences.

Gypsum Stability

Gypsum is stable only in waters saturated with oxygen, i.e., well-aerated waters, so that the activity of sulfate-reducing anaerobic bacteria is prevented. As a deep-water sediment, one can find gypsum forming only in a standing body of water where the aggregate inflow is less than the losses by evaporation. Consequently, the water masses can overturn; convection currents then bring bottom waters to the surface and oxygen-rich waters to the bottom. All sulfate-consuming anaerobes are killed in such aerated waters; in addition to calcium sulfates, the additional undestroyed sulfate ions precipitate as sodium and magnesium sulfates.

Bacterial destruction of gypsum leads either to low-magnesium calcite (Miropol'skiy, 1935; Krotov, 1935; Neev and Emery, 1967; Friedman, 1972), which is ^{12}C enriched (Sokolov, 1965), or to magnesite with huntite as the intermediary (Busson and Perthuisot, 1977). The situation is comparable to that of fjords, in which extraordinarily high sulfate reduction rates can be attributed largely to the shallowness of the overlying oxygenated layer that allows a high flux of nutritious organic matter into the stratified bottom water (Orr and Gaines, 1974). That the conversion of anhydrite or gypsum to calcite is significant is shown by the calculation that the Permian Castile Formation of Texas and New Mexico contains in its carbonate–anhydrite couplets about five times as much carbonate as is theoretically expected (Briggs, 1957). Upper Jurassic evaporites in Sussex, England, show the same bacterial replacement of gypsum by calcite under stagnant reducing conditions (Holliday and Shephard-Thorn, 1974).

Lattman and Lauffenburger (1974) proved that gypsum is a precursor to low-magnesium carbonate: They altered gypsum experimentally to calcite in the presence of suitable organic matter, with the concurrent development of hydrogen sulfide at a pH of 7.5–8.5 and a redox potential (Eh) of -280 to -500 mV. The pH rises rapidly during the anaerobic incubation of bacteria and the Eh falls quickly, but calcite crystallization follows more slowly (Morita, 1980). A first step in the breakdown of gypsum is probably its comminution. Where an algal mat or a bacterial layer settles at the bottom of a gypsum crust, the gypsum crystals crumble into a gypsum sand. Geisler (1981) observed this in southern France, and this author in Caribbean lagoons.

The prerequisite to gypsum decomposition is the presence of bitumina (Pawlowska, 1962). Many anaerobic methane-producing bacteria begin an almost symbiotic relationship with strains of *Desulfovibrio,* which then produce the following reaction:

$$n\mathrm{CaSO_4} + n\text{-alkane} \rightarrow n\mathrm{CaCO_3} + n\mathrm{H_2O} + n\mathrm{H_2} + n\mathrm{S} + \text{pore space} \qquad (7\text{-}1)$$

A primary sink for anaerobically generated methane is therefore sulfate reduction, not aerobic oxidation (Barnes and Goldberg, 1976).

Gypsum crystals growing off the Trucial coast, Persian Gulf, contain poikilitic carbonate inclusions often arranged in herringbone fashion (Wood and Wolfe, 1969). Analogous to this oxidation of hydrogen sulfide is the gypsum formation by sulfur-oxidizing bacteria in building stone (Pochon *et al.*, 1949; Pochon and Jaton, 1967); the hydrogen sulfide here is ultimately derived from acid rain. The oxidation of sulfur and the synthesis of gypsum in this case liberate substantial quantities of heat (Karavaiko *et al.*, 1962). Building stones that are partially gypsified by bacteria may also contain black streaks caused by blue-green algae (Cyanophyceae) (Gistl, 1940).

Solubility Controls

Gypsum and halite precipitation is not particularly temperature sensitive. Whereas the solubility of halite is nearly constant over the range of temperatures encountered in a precipitating lagoon, that of gypsum in water reaches a maximum at 37.5°C. At 0°C it has about the same solubility as it does at 100°C (van t'Hoff *et al.*, 1903). The velocity of a gypsum solution varies from one crystal face to another (Tolloczko, 1910) at a ratio of (010):(110):(111) = 1:1.76:1.88. Saturation of the brine for gypsum or halite is achieved primarily by high rates of evaporation, not by temperature changes. However, gypsum precipitates faster in the wet season, when algae are thriving, or at the peak of algal activity during the daytime, when photosynthesizers supply oxygen for conversion of hydrogen sulfide to sulfate ions. Halite precipitates in the dry season, when evaporative concentration is at a maximum, and at night (or in the early morning hours), at the peak of evaporative water loss. This is the case today in gypsum-precipitating lagoons on the Los Roques archipelago and other islands off the Venezuelan coast. It has also been postulated for the Middle Devonian Prairie Evaporite Formation of Saskatchewan, Canada (Wardlaw and Schwerdtner, 1966).

The solubility product increases with brine concentration and with temperature (Shaffer, 1966). The oxidation of H_2S to H_2SO_4 can cause gypsum precipitation to begin at certain calcium ion concentrations, even if the solubility product is not reached (Marr, 1959). This occurs because the calculated activity coefficients for brines and other natural electrolytic media do not correct completely for all interactions among ions in aqueous solutions (Bennett and Adams, 1972). Precipitation then continues, although brine concentration may vary. Calculated activity coefficients must be corrected for specific salt effects in mixed solutions, for hydration, and for ionic strength (Robinson and Stokes, 1949; Stokes and Robinson, 1949; Glueckauf, 1955; J. R. Wood, 1975). If a term representing Van der Waals forces is added to the fundamental equation of the Debye-Hueckel theory, it remains negligible in dilute solutions (Oka, 1938). In any case, a calculation of activity coefficients may not be valid in solutions containing a significant amount of chlorides, i.e., over 3.4 moles NaCl, 0.9 moles $MgCl_2$, or 1.0 mole $CaCl_2$ (Bates *et al.*, 1970), because of competition for water of hydration.

Gypsum solubility initially increases as the brine concentrates (Ostroff and Metler, 1966). However, it decreases with an increasing excess of calcium ions or in the presence of excess sulfate ions (Metler and Ostroff, 1967). It also decreases in the presence of sodium sulfates (Morozova and Firsova, 1956). Solubility data on concentrated sodium chloride solutions have always been based on the assumption that the amount of sulfate ions present in the solution equals the amount of calcium ions determined. This is not the case in warm brines (Arth and Chrétien, 1906; Glew and Hames, 1970).

Within the temperature range of 0–60°C, the chloride concentration of the brine has a more pronounced effect on gypsum solubility than does a change in temperature. Denman (1961) found a significant increase in the solubility of gypsum with a rising sodium chloride concentration in the brine. However, Cameron (1901, 1907), Hara *et al.* (1934), Foret (1943), Shternina (1948, 1949, 1960), and Marshall and Slusher (1966) found (at variance with data by Posnjak, 1940, and Kruell, 1933) that there is a decided maximum solubility at a given temperature (Fig. 7-3): Increasing the sodium chloride concentration beyond that point decreases gypsum solubility. A maximum is reached at about 40 moles of NaCl per 1000 moles of H_2O (d'Ans, 1965). Gypsum solubility decreases in sodium chloride solutions above 3.0 moles and it becomes unstable in sodium

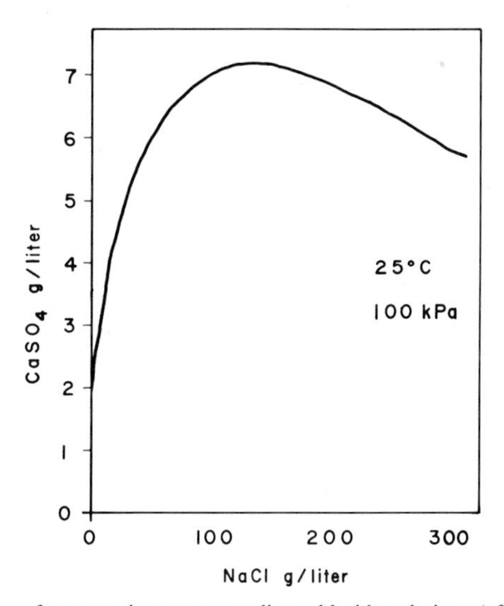

Fig. 7-3. Solubility of gypsum in aqueous sodium chloride solutions (after Shternina, 1960).

chloride solutions above 3.68 moles. Geologic interpretations based on the concept that saturated calcium sulfate solutions become undersaturated by dissolution of halites (Schreiber and Schreiber, 1977) are valid only if the supply of halite is extremely limited.

The same pattern is observed in bitterns. The solubility of gypsum reaches a maximum in about 10 wt% of magnesium or calcium chloride in the brine; in contrast, that of bassanite has no maximum and decreases continuously with rising concentration of chlorides (Ponizovskiy *et al.*, 1974). In a potassium chloride solution, gypsum solubility decreases with temperature (Kruell, 1933). A compound salt ($Na_2SO_4 \cdot 5CaSO_4 \cdot 3H_2O$), isomorphous with bassanite, can form in warm brines. The higher the temperature, the more the formation of disodium pentacalcium sulfate is enhanced; the higher the sodium chloride concentration, the lower is its solubility and that of gypsum. At 50°C this compound does form, but it decomposes in 30 min to gypsum and thenardite (Rogozovskaya *et al.*, 1977). Both gypsum and disodium pentacalcium sulfate will eventually be transformed into anhydrite (Glew and Hames, 1970) and, therefore, have not yet been observed in nature.

Organic Compounds

The precipitation of gypsum is affected by the concentration, amount, and pH of proteolytic solutions and the composition of amino acids present (Nagai *et al.*, 1952). Gypsum crusts stabilized by algae contain up to 256 ppm of organic nitrogen, and a majority of the samples show evidence of nitrifying activity. However, no evidence of bacterial nitrogen fixation was found in these samples (Shields *et al.*, 1957). It is suggested that the breakdown of proteins by denitrifying bacteria liberates sulfate radicals from amino acids such as methionine ($C_5H_{11}O_2NS$) or cystine ($C_6H_{12}O_4N_2S_2$), which can react with dissolved calcium hydroxide before being stripped of their oxygen. Both of these amino acids and several others have been identified in Silurian salt of New York State (Brunskill and Vallentyne, 1966). Several species of bacteria are capable of producing hydrogen sulfide from cystine (Matsui and Fukawa, 1955). Even in concentrations as low as a few parts per million, cystine or other organic compounds have a profound effect on the crystal form, producing substantal deviations from normal crystal habits (Ploss, 1964). Organic carbon isotope ratios, then, resemble those produced in anaerobic environments by bacterial decarboxylation (Hahn-Weinheimer *et al.*, 1978). However, a high pH is a prerequisite, since organic amides that normally prevent agglomeration of the gypsum precipitate are ineffective in very acidic environments (Rogozovskaya *et al.*, 1979).

It was well known for over half a century that the presence of proteins delays the precipitation of gypsum (Budnikoff, 1928a,b; d'Ans *et al.*, 1955; Kuntze, 1966; d'Ans, 1967, 1969). For this reason, organic compounds are used commercially to prevent scaling in crude oil production equipment and to retard the transformation of plaster of paris into gypsum. The best retarders are protein molecules with at least two carboxyl groups that have two- or five-carbon atoms between them (Kuntze, 1966). Naphthenic acids also increase gypsum solubility (Zelizna and Roskosh, 1973). Resins, sugars, xylan, and other vegetable compounds likewise stabilize calcium sulfate solutions and diminish the transformation of bassanite into gypsum. Xylan, at a concentration of 250 ppm, is capable of preventing gypsum precipitation (Korol'kov and Tyagunova, 1956). Gypsum can adsorb up to 3 ppt of xylan; the stability of any supersaturated solution is enhanced by substances that can be adsorbed in the solid phase (Tovbin and Krasnova, 1955).

There has recently been a revival of studies using various organic macromolecules to inhibit crystallization by adsorption onto calcium in gypsum lattices (Wied and Syrojezkina, 1965; Combe and Smith, 1964, 1965, 1966; Liu and Nancollas, 1973, 1975; Liu and Griffith, 1979). Although polymers (polyelectrolytes) are useless, esters are quite effective in inhibiting gypsum crystallization. Even 2 ppm polyphosphonate is capable of preventing gypsum precipitation from a solution concentrated to more than three times the normal saturation (Vetter, 1972), bringing it into the range of halite precipitation. Fifty parts per million of sodium polyphosphates can keep five times more gypsum in solution than would be contained in a normally saturated brine (Burcik, 1954). Gypsum also becomes more soluble if traces of ammonium chloride are present (Seidel, 1917; Kruell, 1933), as the introduction of ammonium salts accelerates gypsum dissolution. The nitrogen content of evaporite deposits in almost entirely of organic origin; ammonium chloride is present in fluid inclusions in sylvite (Cifrulak and Cohen, 1969) and thus was also available in the original brine. Farmers in Britain have always stored manure, altering it with gypsum layers; the mixture then loses only an insignificant amount of its ammonia (Sherlock *et al.*, 1938). A 50% urea solution increases the solubility of gypsum fivefold compared to plain water (W. Sakai, 1943).

Longer-chain alkanes and fatty acids are more readily adsorbed and foster the formation of more tabular, equidimensional crystals (Barcelona and Atwood, 1978). Humic acids and other organic compounds, as well as a variety of inorganic cations that are present in the solution, influence the relative growth rate by forming epitaxial deposits on the specific habit faces of gypsum (Chepelevetskiy and Evzlina, 1937; Edinger, 1973). Elongate prismatic crystals grow in low-pH environments, with or without the presence of organic matter. Organic acids added to a calcium sulfate solution produce smaller crystals than inorganically precipitated gypsum (Hiyama and Fukui, 1952). Discoid (lenticular) crystals grow only under alkaline conditions in the presence of dissolved organic matter, especially phenolic compounds; this inhibits nucleation and thus prevents growth of prismatic crystals (Cody, 1979). That may explain the prevalence of discoid and tabular gypsum crystals in desert soils. In short, *the addition of inorganic compounds to the solution fosters acicular crystals, whereas organic compounds foster stubby crystal growth* (van Rosmalen *et al.*, 1976).

Gypsum often serves as a nucleus for salts of higher solubility. Recent gypsum encased in polyhalite or halite has been described from southern Tunisia (Busson and Perthuisot, 1977). The former may result from the incomplete alteration of the original calcium sulfate by potash-bearing bitterns.

CELESTITE

Strontium sulfate solubility (0.114 kg/m^3 at 30°C) is considerably lower than calcium sulfate solubility (2.09 kg/m^3 at 30°C). Evaporating seawater is saturated for strontium sulfate at one-quarter of the original volume (German Mueller, 1962). Strontium therefore prefers to crystallize out as celestite ($SrSO_4$) druses in limestone. Conversion of aragonite to calcite releases strontium to pore solutions (German Mueller, 1962), as does aragonite replacement by dolomite (Evans and Shearman, 1964). As strontium-rich aragonite turns into strontium-poor calcite, celestite can precipitate, as it did in Permo-Triassic deposits in Dalmatia, Yugoslavia (Scavnicar and Scavnicar, 1980). In the Main Dolomite Member of the Zechstein evaporites in northern Germany, celestite occurs as fracture fillings and as cement of void spaces (Theilig and Pensold, 1964), but in the Alps it replaces aragonitic skeletal material in a calcite matrix. To this day, very little of the surrounding limestone has been altered to dolomite (Scherreiks, 1970).

Along the Trucial coast of the Persian Gulf, one finds celestite in small quantities as individual crystals or aggregates in the uppermost gypsum layers. Celestite occurs more abundantly inland, where strontium-rich aragonite has been converted to strontium-poor dolomite (Kinsman, 1966), inland of the high-tide mark, either around 20-cm-long sand-filled gypsum crystals or poikilitically enclosed in them (Evans *et al.*, 1969). Celestite is also found as encrustations on fossil fragments (Jayaraman, 1940). A very small amount of strontium substitution occurs in the gypsum crystal lattice. Gerhard Mueller (1964) has shown that an influx of continental bicarbonate waters produces the same results as a higher initial strontium content: Celestite precipitation commences earlier.

Strontium sulfate solubility decreases with increasing temperatures. As sunlit shelf areas heat up, they become preferential sites of celestite precipitation. This action is a direct analogy to the behavior of strontium in aragonite and calcite. The more rapid the rate of evaporation, the higher is the percentage of strontium built into a gypsum crystal lattice (Watson, 1979). Primary gypsum and anhydrite have a similar strontium content (R. Langbein, 1968). Celestite is isomorphous with anhydrite and polyhalite (Herrmann, 1961b), and always forms very small crystals in the former (Muegge, 1913; Link and Otteman,

1968). However, in the gypsification of anhydrite, strontium is liberated and precipitates as euhedral celestite. A solution containing 100 ppm strontium, which comes into contact with anhydrite, partially precipitates celestite without any strontium sorption. Equilibrium is reached when the solution contains about 20–35 ppm strontium (May *et al.*, 1961).

Some evaporite basins show a maximum in celestite precipitation when a brine has concentrated up to the onset of gypsum precipitation and again when calcium sulfates terminate in the roof of an evaporite deposit (Strakhov, 1962) during the progressive freshening of the brine. In others, celestite precipitation increases with brine concentration until a maximum is reached just short of the onset of halite precipitation and even the halite still contains some celestite (R. E. Taylor, 1937). However, at some time during the precipitation of halite, all residual strontium is stripped from the brine (Zherebtsova and Volkova, 1966), because the solubility of strontium chloride is reduced as the sodium chloride concentration increases (Chou and Haas, 1981).

In paleogeographic terms, the occurrence of celestite is tied to shoals and precipitates there together with carbonates prior to gypsum sedimentation (Fuechtbauer and Mueller, 1970). An upward increase in strontium content in the Stassfurt rock salt member in the German Zechstein Basin, and from the center of the basin to its margin (Braitsch, 1971), can be related to the configuration of brine currents. These directions of increase in strontium are identical to the directions of increasing bromide content in the halite (Baar, 1952).

Nests of celestite with anhydrite in rock salt are either a product of diagenetic alteration by residual brines or are epigenetic minerals (Voronova, 1968). Such secondary celestite is often marked by large crystals to which anhydrite is still attached (Rudolph *et al.*, 1966). Finely dispersed strontium sulfate (Zaritskiy, 1960) has recrystallized into these large crystals. The Upper Keuper marl in Britain contains in its upper evaporite horizon celestite nodules and granular celestite boulders up to 1 m in diameter. They are the lateral equivalent of the upper two gypsum beds in the formation. Inside, there are cubic crystals that are pseudomorphs after halite. Pseudomorphs after calcite are also found toward the center of the boulder, whereas toward the rim the calcite crystals are leached and replaced by cavities. Some of the celestite nodules have overgrowths of calcite and sphalerite (Curtis, 1974).

BARITE

Barium is eliminated from the brine very early, because both its carbonate and its sulfate have a very low solubility (22 g/m^3 of carbonate and 2.22 g/m^3 of sulfate in water at 18°C). They precipitate as nodules and nests within the limestone framework. Barium in solid solution in anhydrite is expelled upon recrystallization. Some barite may have a biogenic origin, as several organisms concentrate barium during their life cycle.

Barite is associated with evaporite basins, occurring as concentric concretions in soils formed through repetitive swelling and shrinking (Stoops and Zaveleta, 1978), as nodular concretions in clays (Barbier, 1976) mimicking a sebkha environment, or as basin margin beds (Macquar, 1978; Lagny, 1980). The barite, then, is found only in the area of brine mixing, where fresh water encroaches on seawater (Fuchs, 1978) or seawater overruns lakes (Suess, 1982). It is often a very early replacement of gypsum in beds that were evidently precipitated in shallow waters (Lambert *et al.*, 1978). The sulfate appears to be altered by bacterial action in deposits as old as those of the early Proterozoic (E. M. Cameron, 1982). The question remains whether the barium is due to local hydrothermal emanations, which then should have carried with them other ions, notably lead. Precipitation of barium and strontium goes hand in hand with the calcium depletion of the brine, and a continuous series from barite to celestite exists. Even though the sulfate of barium is less soluble than that of calcium, ion association can alter the precipitation sequence. A deficiency of sulfate ions and a disproportionate concentration of calcium ions could allow calcium sulfate to precede barium sulfate precipitation; the progressive depletion of calcium in the brine would then allow barium to replace calcium in the precipitate.

MARINE SODIUM CHLORIDES

Further concentration of gypsum-precipitating brines gives rise to halite precipitation. Whereas hydrohalite begins to precipitate at an 8-fold saturation in cold brines, a 10-fold concentration is required for halite precipitation in subtropical brines (Krull, 1917; Borchert and Muir, 1964). However, if they are stratified or agitated by storms, the onset of precipitation is delayed even further, to about an 11-fold concentration, since no nucleation occurs at the evaporative surface. The precipitation of rock salt in nature takes place in the temperature range 30–35°C (d'Ans, 1947; Braitsch, 1964). The total amount of sodium chloride preserved in evaporite formations is only a small fraction of the amount presently dissolved in the waters of the world ocean. Individual authors have suggested different ratios, ranging from a 1:6 to 1:2 (Grokhovskiy and Grokhovskaya, 1980).

Slow evaporation of a brine produces (100) faces in halite, sylvite, or even potassium bromide. Once a critical rate of crystallization is exceeded, the octahedral habit takes over (Kern, 1952a). This critical rate of evaporation is smaller for sodium chloride than for potassium chloride or potassium bromide. The rate of formation of (111) faces increases with the dielectric constant of the solution (Kern, 1952b). The octahedral habit is also produced in the presence of carboxymethyl cellulose and other organic compounds (Shuman, 1966). A summary of the physical and chemical data on rock salt has been published by Gevantman *et al.* (1981).

Contribution of Biota

The presence of a variety of alcohols decreases halite solubility (Emons, 1967). *Although organic compounds retard gypsum precipitation, they accelerate halite crystallization.* A change from a cubic to an octahedral habit occurs whenever organic matter is present in adequate amounts, because it is adsorbed onto faces with the greatest surface area and then inhibits further crystallization (Koshurnikov and Mokievskiy, 1948). In the laboratory, even simple organic compounds, e.g., acetic acid or ethyl alcohol, are capable of doing so (Hintze, 1915).

Some of the red halophilic bacteria require at least 15% sodium chloride for growth and often grow in saturated solutions. The ability to tolerate salt is thereby definitely related to the ability to accumulate potassium within the cell. That stromatolites persist well into brines supersaturated to sodium chloride is documented by the occurrence of specimens totally replaced by halite in both Permian salts from the North Sea and Silurian salts from Michigan (Friedman, 1980a). In brine storage ponds and crystallizers, a heavy rain which lowers the salinity might kill large numbers of red halophilic bacteria, causing the brine to become clear (J. S. Davis, 1974). The bacteria are evidently ill equipped to handle a sudden change in osmotic pressure. The rate of halite precipitation is slowed down, and a very clear salt is produced. The slower crystallization of the halite also reduces its bromide content (Wardlaw, 1970). Clearance of the brine reduces the rate of absorption of solar rays and the rate of evaporation. Therefore, if no bacterial plate forms, small amounts of dye are added to solar ponds filled with sodium or magnesium chloride brines to accelerate evaporation by 10–20% (Kane and Kulkarni, 1949). Periodic massive killing of the brine biota occurs in contemporary lagoons in Patagonia, Argentina, where halite layers alternate with slime layers; the slime represents the bacteria killed in the rainy season. The precipitate hardens with depth (Brodtkorb, 1980). This may be a recent analog of fossil varve couplets composed of halite and organic matter, such as laminae of dark opaque material in Upper Silurian salts of New York State that contain 0.44–2.14 weight-percent of organic carbon (Treesh and Friedman, 1974) and a concurrent fetid smell of H_2S.

Impurities

Freshening by rainwash that dumps a layer of clay into the brine will affect precipitated salts. Halite stratification in Triassic evaporites of the Lorraine,

France, has been intensely disturbed at the base of clay intercalations (Langier, 1959). In the Silurian Salina salt in Detroit, near the margin of the Michigan Basin, one finds discordant secondary halite masses with inclusions of dolomite and anhydrite. These dolomites and anhydrites have virtually horizontal bedding and seem to have been washed into a temporary cavity in semilithified halite. Bedding is, on occasion, continuous from the primary into the secondary halite. Water apparently moved through cavities and channelways in the salt after at least partial lithification (Dellwig and Evans, 1969). Contaminations of halite are relatively rare, with the exception of occasionally coprecipitated minor amounts of anhydrite and sylvite. The amounts of secondary quartz, pyrite, and heavy minerals are usually negligible. In addition to solid impurities, halite contains liquid and gaseous inclusions in tiny pores and interstitial voids. Hydrogen sulfide predominates; it can be detected by crushing or cutting. Air and brine inclusions give halite a milky white color (Reimer and Utter, 1979).

In compact rock salt beds, halite grains usually have irregular borders and are small, ranging from a fraction of a millimeter to several millimeters in diameter. Halite precipitates initially as a slush. As the concentration of the brine increases, the amount of water adsorbed to halite crystals drops sharply until it becomes a monomolecular film (Harkins and McLaughlin, 1925). The eventual recrystallization of this watery slush produces unusually large halite nests of quite large, absolutely transparent crystals. The impurities are thereby transposed to the periphery of the crystals. The growth of such ''saltspar'' in otherwise granular rock salt requires calm waters (Murzayev, 1941).

In pure form, halite is colorless and transparent, with a glassy luster. Green discoloration of salt occurs in the Hallstadt mine in Austria; the coloring agent has been identified as atacamite [$Cu_2Cl(OH)_3$](Cornu, 1907). Green atacamite forms elsewhere in the weathering zone of copper ores under the influence of saline waters. However, at this locality, the situation may be reversed: The copper comes to the salt. Green salt is formed only where Celtic miners forgot their copper tools. Such green discoloration is different from that of green salt at Wieliczka, Poland, where green clay particles fill spaces between halite crystals (Manecki and Pawlikowski, 1975); these clay particles are alteration products of an andesitic tuff. Occasionally, halite is smoky, pink, light blue, dark blue, lavender, violet, or purple and greenish in a sylvinite zone (Cornu, 1908; Aprodov, 1945). Organic matter causes a dark brown discoloration. Inorganic (clayey) impurities color the salt gray to dark gray. Yellow, pink, and red colors can be produced by traces of iron oxides. Small quantities of sylvite can give the salt a yellowish hue. Close to an overlying potash horizon, the halite in Byelorussia is yellow-orange (Lupinovich and Kislik, 1965). Usually yellow salt is either contaminated by bitumina or has been exposed to a slight amount of radioactivity. Yellow salt is often an intermediary to blue salt (Przibram, 1936), as is violet rock salt (Przibram, 1950). Black halite contains large platelets of hematite (Richter, 1964).

There does not seem to be a uniform cause for the blue discoloration of rock salt (or the violet color of sylvite). Some of the causes are discussed below.

Lopez-Rubio and de la Rubia-Pacheco (1950) found that some salt has a peppery taste. The blue variety has a density of 2.262, and the transparent variety a density of 2.2779, compared to an ordinary density of 2.164. The authors interpreted this as an admixture of Na—Cl = Cl—Na (density, 4.328) in a ratio of 1:18. Blue salt is harder than ordinary salt, has a conchoidal fracture, and dissolves more rapidly in water. It also has a lower index of refraction (Pustyl'nikov, 1975). Blue halite samples may give an alkaline reaction upon dissolution (Cornu, 1908; Guthrie, 1929), but some do not (Spezia, 1909). Vinokurov (1958) found a higher pH in several blue salt samples compared to colorless specimens. He found a lesser magnetic susceptibility and a higher rate of dissolution in blue rock salt.

Blue rock salt requires an increase in the number of color centers per unit of volume and can be produced by adding sodium vapor or vaporized sodium containing mother liquor, or by subtracting Cl^- by ionization (Siedentopf, 1905). Ionization can be achieved by cathode rays, x-rays, ultraviolet light, β and γ radiation, or even electric sparks. Irradiated halite is light sensitive and will change color when exposed to sunlight. The discoloration disappears if the rock salt is heated to 275°C. An artificial discoloration induced by heating in sodium vapor is maintained even beyond 400°C (Spezia, 1909; Przibram, 1929), particularly if the specimen has been first exposed to lateral compression (Ivanov, 1953). However, some samples, initially discolored, in time show a return of the color (Doelter, 1925). Colorless salt next to blue salt thereby discolors faster than colorless salt from elsewhere (Ludewig and Reuther, 1923). The blue color in halite can also be produced by artificial x-ray irradiation, whereby the colors range from blue-black closest to the radiation source to pale blue and purple farther from it. Such discoloration decreases after irradiation ceases (W. Borchert, 1948) even if the sample is kept in the dark (Doelter, 1925). Bombardment with β-particles also changes the color by adding an electron to Na^+ and neutralizing it (Przibram, 1929) in irregular surroundings (Przibram, 1931). Originally, this was explained as a consequence of radiation producing the sodium subchloride Na_2Cl (Pieszczek, 1905, disputed by Siedentopf, 1905) or splitting off chlorine to leave a sodium-enriched sodium chloride complex (Hoffmann, 1934). It is easier to separate Cl^- and leave a sodium enrichment than to achieve the opposite. The exact hue and color intensity depend on the particle size and the sodium enrichment (Przibram, 1927c) of submicroscopic Na_2 crystals (Kahanowicz, 1932). Cathode rays produce a metallic sodium coating on halite (Siedentopf, 1908), which is sputtered cathode material always located where the electron strikes (Frisch, 1927). Most artificial radiation sources utilized to induce a color change are several orders of magnitude stronger than the naturally occurring radiation in evaporite beds. The first to suggest ^{40}K radiation was Boeke in 1909 (Kirchheimer, 1976). However, most of the ^{40}K radiation attenuates in a few millimeters of salt, since there is only 119 ppm of ^{40}K present in the potassium of evaporites.

Sylvite turns violet if subjected to radiation, but the color is not permanent at room temperature and fades rapidly in sunlight (Ludewig and Reuther, 1923). An admixture of 5.4 ppt potassium or rubidium chloride in rock salt is sufficient to induce color changes (Aprodov, 1945). The loss of radiation-induced color over time in either halite or sylvite is due to gradual recrystallization (Przibram, 1929). Colored halite can sometimes be related to the potash content of the evaporite sequence: In some occurrences, colored salt is restricted to the contact zone between halite and sylvinite (Przibram, 1929). In other instances, it never occurs below potash salts (Chirvinskiy, 1943), and it is not associated with sylvite (Gawel, 1947). Yet it is related to the potash salt sites (Chirvinskiy, 1945). Nonetheless, blue salt in Oligocene evaporites of the Upper Rhine Valley is not tied to sylvite, but occurs in the nodular anhydrite–halite region, together with variegated marls or in clays more than 15 cm above a sylvite horizon. Blue halite never occurs adjacent to red sylvite, because the hematite absorbs all radiation emanating from the radioactive potassium isotope (Pustyl'nikov, 1975). Fluid or sylvite inclusions in blue halite are always surrounded by a halo of colorless halite approximately the thickness of any radiation influences (Kirchheimer, 1976). That halo should present the deepest blue colors if color is introduced by radiation. Kirchheimer (1976) mentioned that M. R. Bloch succeeded in producing blue rock salt in about 10% of all crystallization runs from a solution saturated for both potassium and sodium chloride and containing 320 g/liter of magnesium chloride.

Nonetheless, if one does not restrict oneself to studying only laboratory specimens divorced from their sampling sites, one quickly concludes that some blue salt occurrences obtained no more radiation than adjacent undiscolored crystals, and grew in the same environment. Consequently, after an exhaustive evaluation of all the data pertaining to light-sensitive blue and purple halite, Kirchheimer (1976, 1978) cautioned that radioactivity may not be the exclusive explanation for the phenomenon. But even the formation of color centers by proton irradiation is more complex than was previously reported: It is a function of the dosage, rate, temperature, exposure to light, strain, and the distribution of any impurities (Bird *et al.*, 1981).

Some blue salt may contain colloidal metallic sodium (Savostyanova, 1930; Kahanowicz, 1932; Liermann and Rexer, 1932; Przibram, 1947). Purple salt contains small, dispersed sodium particles; blue salt contains large sodium particles (Przibram, 1936; Holdoway, 1974). The stoichiometric excess of alkali metal by itself leads to no color change. The visible color is due to electrons trapped in the vacant sites. Darkening of the crystal by x-ray irradiation is caused by ionic diffusion (Seitz, 1946). A primary carnallite in the Werra and Fulda deposits of the German portion of Zechstein evaporites was decomposed to sylvite and blue halite with open channels that now contain mobile intercrystalline carbon dioxide (Hartwig, 1954). The gas is probably derived from a basaltic intrusion.

The color seems to be associated with defects in the crystal lattice, which are produced either by metallic sodium (Cornu, 1910) or by other foreign cations; these lattice defects act as color centers. Rapid growth of crystals results in a deep blue color, and slower growth in a clearer, more violet color (Przibram, 1929), suggesting that the degree of disarrangement of the crystal lattice determines the shade and hue of the color. Solid solutions of chlorides with bromides have a greater number of lattice defects than pure compounds (Wollam and Wallace, 1956). The variety of crystal defect structures in halite crystals has been summarized and tabulated by Shlichta (1968). A fast rate of crystal growth fosters the occurrence of lattice errors. There is an optimum number of structural disturbances, an optimum concentration of impurities (Przibram, 1951). The most intense discoloration occurs at points of strongly distrubed crystal structure (Blank and Urbach, 1927). In a solid solution of sodium and potassium chloride at 25°C, the heat of formation reaches a maximum in an equimolal solution, but lattice spacings and Schottky defects reach a maximum near a 30.1% concentration of potassium chloride (Barrett and Wallace, 1954).

Blue halite that is powdered and compressed again becomes yellow but regains its blue color after exposure to light (Przibram, 1927b). In the same way, if rock salt is subjected to pressure, it quickly turns yellow under the influence of β radiation (Ludewig and Reuther, 1923; Przibram and Belar, 1924; Zekert, 1927; Przibram, 1927a) as a function of stress (Wieninger, 1950); however, it then changes to blue after exposure to light or heat (Wieninger and Adler, 1950). Preheating the sample to 150°C inhibits the discoloration (Przibram, 1926). Application of a pressure of 980 MPa arrests any recrystallization by radiation (Przibram, 1932). That grinding or heating makes the color disappear has been known since 1775 and 1864, respectively (Kirchheimer, 1976). If the sample becomes deformed under pressure, it eventually turns black (Przibram, 1927a; Tammann, 1931). The plasticized portions show a lasting, more intense coloration due to an increase of sodium sites where the lattice was loosened (Smekal, 1927). The greater the degree of deformation, the more rapidly the color disappears (Haberfeld, 1933). Frequently, the discoloration of natural samples is younger than the mechanical stress to which the salt may have been exposed, judging by the orientation of the blue color (Andrée, 1912).

Certain blue salt samples cloud over upon heating (Yamamoto, 1938) because hydrated cations may have been built into precipitating halite (Truesdell and

Jones, 1969). Some samples evolve a gas upon solution (Cornu, 1908); others do not, but discharge organic liquids upon heating (Prinz, 1908). Some samples of blue salt are enriched in helium (Valentinov, 1916), but the helium content is about four times smaller than that of syvite and slightly smaller than that of carnallite or kieserite (Strutt, 1909). Some blue salt has a slightly higher energy content and offers a distinct glow at 350°C that is absent in colorless salt (Guthrie, 1929); some instances of blue color are an attendant phenomenon of spark discharge through halite crystals (Steinmetz, 1932).

Doelter (1910, 1920, 1925, 1929) maintained that the shades of color are due to colloidal pigments, the individual shades being caused by different degrees of dispersion. Lead chloride built into weak spots of halite increases the elastic limit of rock salt about threefold (Blank and Smekal, 1930). Together with helium, the lead may have radium as a progenitor and is a cause of some blue discoloration (Born, 1934). Serra (1949) found that blue halite from Calascibetta, Sicily, contained copper, calcium, barium, magnesium, iron, and manganese, and that bivalent copper was a particularly important ingredient. Halite can derive its color from finely dispersed iron (Gawel, 1947). Bivalent alkali earth ions likewise wander into color centers (Seitz, 1951). Copper and zinc enrichment produces ruby red and pale green aragonite, the variety called ''mossotite'' (Friend and Allchin, 1940a), but colloidal gold (Friend and Allchin, 1940b) causes discoloration of halite and anhydrite. Repshe (1936) had earlier found colloidal gold in sylvite samples, and Popov (1975) noticed that carnallite contains three times as much gold as anhydrite; halite with a significant admixture of clay contains about twice as much gold as pure halite. The bromine concentration thus plays an important role, because gold complexes are stable and because bromide catalyzes dissolution of gold and inhibits its precipitation (Miller and Fisher, 1974; Popov, 1975).

The first investigator to derive some of the blue halite colors from colloidal metals such as gold, silver, and copper was Siedentopf (1905). This type of discoloration is akin to the gold enrichment of pale blue and deep red celestite lining channels where water had penetrated the mineral (Friend and Allchin, 1939), the reddish-violet of sylvite, or the blue and green of potassium bromide due to colloidal gold (Blank and Urbach, 1927). Even fine globules of sulfur may be involved, as some crystals discharge sulfur upon heating (Prinz, 1908). The exact color is then a function of the diameter of the impurity (Doelter, 1920): Particles 80–90 nm in size discolor to bluish-violet; 90–110 nm, blue; 110–120 nm, green; 130–150 nm, yellow; and 150–180 nm, orange. Absorption at wavelengths typical of sulfur ions found only in natural crystals indicates the presence of colloidal sulfur and an increase in the number of color centers (Wieninger, 1951), confirming Doelter's (1920) earlier observation.

Kennard *et al.* (1937) took issue with the idea of coloration by pigments: Although blue salt seems to contain a variety of elements in trace amounts (aluminum, lithium, potassium, silicon, titanium, magnesium, strontium, barium, etc.), they maintained that the color merely has structural causes and is not induced by any particular pigment. This conclusion is in accord with that of Andrée (1912), who had maintained that no coloring matter is present. All in all, the causes of blue discoloration in rock salt and purple discoloration in sylvite require further study of samples with determined stratigraphic positions collected in different mines.

PRIMARY MARINE POTASH AND MAGNESIUM SALTS

In past centuries, German miners had grouped the salts of higher solubility under the term "Abraumsalze" ("trash salts" or "rubbish salts"). Since these salts were of no apparent use, they had to be removed (to remove = "abräumen" in German) to trash heaps before the halite could be mined. Nowadays the successors of these miners prefer the term "Edelsalze" ("noble salts") in view of their high commercial value; these salts have been exploited for their potash content since 1861. There are no equivalent terms in the English language, because potash mining in the United States and Britain commenced well after the "cinderella salts" had turned into "noble salts"—after World War I in the former country and after World War II in the latter. The optical and crystallographic constants for alkali and alkaline earth chlorides, as well as their hydrates and mixed crystals, were summarized by Assarsson (1957).

Potash Precipitation

It is noteworthy that primary pure sylvite or carnallite deposits are very rare; almost all of them constitute intergrowths with halite. Such halite commonly has a granular texture, indicating the presence of agitated waters (Dubinina, 1951a). The intergrowths of halite with sylvite and halite with carnallite are called "sylvinite" and "carnallitite," respectively. Crystal inclusions suggest that many sylvite- or carnallite-precipitating bitterns are either enriched with gases or in contact with the atmosphere (Dubinina, 1951b). Sodium chloride solubility in saturated potassium chloride solutions decreases with increasing temperature, reaching a minimum at 105°C (Cornec and Krombach, 1932). A saturated solution thus crystallizes only sylvite upon cooling (Blasdale, 1918). D'Ans and Kuehn (1938) described cloudy sylvinite as a coprecipitate of sodium and potassium chloride from a brine containing magnesium chloride.

An important distinction has been observed in the only large inland commercial potash lake in the world, the Chaerhan salt lake in China: Carnallite deposits from open brines are distinctly stratified; carnallite beds alternate with halite intercalations as couplets that are not necessarily annual. However, precipitation of carnallite from interstitial brines produces a disseminated structure with carnallite occurring in the interstices of a halite crystal matrix (Yang, 1982). The halite in sylvinites often seems to have already existed as mush when potash precipitation commenced. In the Devonian Prairie Evaporite of Saskatchewan, Canada, the individual potash beds can be traced laterally into an equivalent halite bed by correlating clay markers (Worsley and Fuzesy, 1979). The potash layers therefore seem to be the product of interstitial precipitation in halite slush in a depression. From intergrowths of sylvite and halite in Poland, Rozen (1926) deduced that the potassium chloride solutions entered the halite along translation planes. Kuehn (1951) mentioned that the bromide content of koenenite matches that of associated halite but is much smaller than that of associated sylvite. This might indicate that the potash minerals precipitated much later out of a more concentrated brine into an already formed halite slush, or that there was a difference in the bromide content of the bitterns.

A mere daytime accumulation of such bitterns in depressions in open waters or underneath an interface in the brine can lead to precipitation of potassium or magnesium chlorides, but not of sulfates. Several small sub-Pyrenean potash troughs separated by gypsum or anhydrite were fed by an Eocene arm of the Mediterranean Sea into northern Spain (Rios, 1968). They are composed of sylvinite and carnallitite, but with the exception of gypsum they totally lack sulfates, particularly magnesium sulfates. To precipitate sulfates, a brine must first be oxygenated by direct exposure to the atmosphere. This requires a very shallow water column. Nevertheless, salinities are then too high for significant carbon dioxide or oxygen solubility and diffusivity in the brine or for the survival of even extremely euryhaline algae: Both light and sound are quickly attenuated in bitterns rich in magnesium salts, and the increasing surface tension drastically curtails the diffusion of atmospheric gases. In the Miocene evaporites of Sicily, the horizons with potassium–magnesium sulfate lenses are truncated by an unconformity covered by gypsiferous calcarenites and gypsarenites, followed by several gypsum cycles (Decima and Wezel, 1971), and show evidence of contamination by meteoric, i.e., oxygenated, waters. Bedded potassium and magnesium sulfates are generally secondary alteration products of precipitated chloride minerals. They will be discussed in greater detail in Chapter 13.

There are no known potash deposits that were not originally encased in halite. In contrasts, *all major halite deposits contain some potash deposits,* albeit sometimes only in small quantities (Lotze, 1957). All potash and magnesium salts are very temperature sensitive; they share this property with sodium sulfates. Except for kainite, they diffuse from cooler toward warmer brines, increasing their concentration in warmer brines during the daytime. That is, the Soret effect, also referred to as the "thermal diffusion coefficient," has a positive value only for kainite and a negative value for all other solutes in an evaporite basin. They all precipitate at night or in the rainy, cloudy season, and can redissolve on sunny days unless covered by halite. Once supersaturation is reached, cooling by evening offshore breezes stops further evaporation but induces precipitation. The rate of nucleation is a function of the rate of cooling. A sodium chloride solution saturated at 50°C will reach critical supersaturation when cooled to 0–1°C. A potassium chloride solution saturated at the same temperature reaches its critical supersaturation at 35.1°C, whereas a potassium bromide solution reaches it at 42°C.

The critical temperature of supersaturation remains constant at rapid cooling rates, i.e., those faster than 0.36°C/s. At slower rates, the critical temperature rises slightly (Hirano, 1954). In tropical lagoons with insufficient inflow to cover all of the water surface, the shallow near-entrance areas heat up in the daytime, but only by about 2–3°C, because of lateral thermal diffusion. This temperature difference is sufficient for temperature-sensitive solutions to wander in and saturate the brine. A nighttime temperature drop then causes precipitation. As the temperature of the bitterns reaches a daily minimum before sunrise, the rates of evaporation reach a peak at about the same time. Consequently, the coolest temperatures are found near the surface, and further heat loss is possible only by slow conduction.

If the potassium chloride content of a sodium chloride brine exceeds about 35%, *the brine reduces its density by cooling* (Cornec and Krombach, 1932) (Fig. 7-4). For equal concentrations, sodium chloride solutions are heavier than potassium chloride solutions, and potassium–magnesium chloride solutions are even lighter (Heykes, 1924). If a floating crust develops, it can be broken up only by windstorms and then drifts toward the leeward shores. This explains the common occurrence of sylvite deposits near the margins of depressions or on slopes as the saturated bitterns start to rise upon cooling, e.g., the Miocene potash deposits of Iran (Stoecklin, 1968), the Permian potash deposits of New Mexico (Jones and Madsen, 1968), the recent mirabilite sedimentation in the Gulf of Kara Bogaz Gol (Kolosov *et al.*, 1974; Strakhov, 1962; Sonnenfeld, 1974), the Silurian sylvinite deposits in the Michigan Basin (Elowski, 1980), and the Jurassic ones in central Asia.

Strakhov (1962) suggested that potash deposits can form only where differential tectonic subsidence drains interstitial brines enriched in potassium from a larger halite-covered area into depressions. This is another format of saturation shelves, as suggested earlier by Richter-Bernburg (1957a). Potash beds would then mark regions of maximum rates of subsidence during late stages of salt deposition. Indeed, the larger the area of evaporating brine, the more likely it is that this brine will reach potash saturation. However, such a model applies only to a brine basin that seasonally contracts much of its surface area, so that basin margin precipitates can be washed into the central depression by dew, fog, or

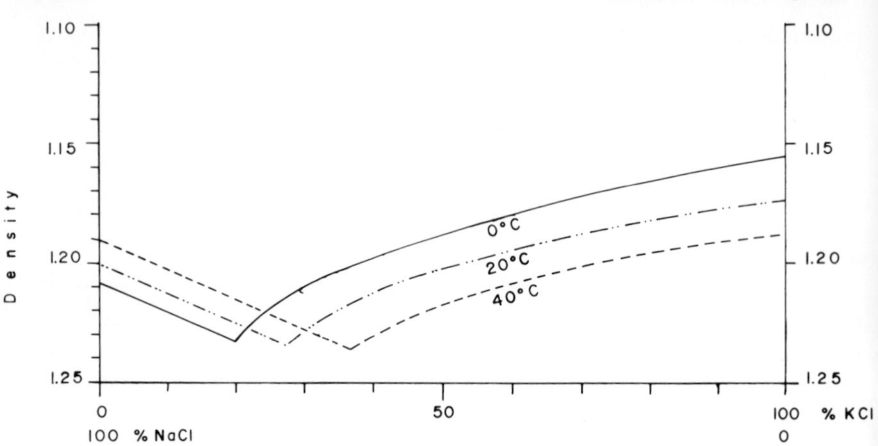

Fig. 7-4. Density of various mixtures of KCl and NaCl in solution at various temperatures (after Cornec and Krombach, 1932).

eventually rain. Strakhov's model has been observed on Lake Eyre, Australia (Bonython, 1956). A saucer-shaped halite precipitate is washed into the center of the depression by successive rains, and re-forms there as a thicker deposit of reduced areal extent.

Potassium Chloride Solubility

Initially, concentrating seawater precipitates halite first as chloride salt and then continues to do so until the NaCl:KCl ratio is somewhere between 2.28:1 and 2.00:1, depending on the temperature of the brine (Goergey, 1912). Thereafter sylvite starts coprecipitating. If the brine reaches an NaCl:KCl:$MgCl_2$ ratio of between 0.364:1:6.41 and 0.266:1:4.60, carnallite becomes the preferred precipitate (Goergey, 1912). To prevent sylvite from precipitating before the brine reaches saturation for carnallite, the brine must maintain an $MgCl_2$:KCl ratio of greater than 10:1 and an $MgSO_4$:$MgCl_2$ ratio of smaller than 0.4:1 (Tseng Cheng *et al.*, 1965). Carnallite deposition thus corroborates a continuing sulfate deficiency of the brine, that d'Ans (1947) and Borchert (1967) have observed. As the magnesium chloride concentration in the brine increases, it is more apt to alter swept-in mica flakes to corrensite under liberation of potassium (Lippmann and Savascin, 1969). This in turn decreases the $MgCl_2$:K_2Cl_2 ratio. The presence of increasing quantities of magnesium chloride in solution also reduces the solubility of halite and, even more rapidly, that of sylvite in proportion to the increasing ionic strength of the brine (Rozsa, 1920; Akerlof, 1934; Zdanovskiy, 1946; Ezrokhi, 1949; Wardlaw, 1968; Wood, 1975). This retrograde solubility of potassium chloride in complex chloride brines at a temperature of 35°C has also been observed by Lightfoot and Prutton (1946, 1947, 1948, 1949). Similarly, a small amount of calcium chloride greatly depresses the solubility of potassium chloride in the brine (Lee and Egerton, 1923). However, although sylvite and carnallite are extremely insoluble in cooling saturated brines rich in magnesium chloride, the drop in solubility is slightly less pronounced when a little calcium chloride is present in the brine rich in magnesium chloride (Igelsurd and Thompson, 1936). Increasing calcium chloride content in the brine at a constant magnesium chloride concentration increases the amount of halite in carnallitite and eventually leads to conversion of sylvite to carnallite, thus decreasing the magnesium chloride concentration in the brine (Savinkova *et al.*, 1976). Sylvites in Devonian evaporites of Byelorussia are replaced by halite and carnallite; in turn, carnallite is at times replaced by halite (Varlamov *et al.*, 1974). Sylvinite is the main constituent of potash salts in the Michigan Basin (Matthews and Egleson, 1974).

In order to explain these sylvinite occurrences, Nurmi and Friedman (1977) suggested that the brine is first stripped of its magnesium chloride content by dolomitization. However, they offer no model to explain how a potassium–magnesium chloride bittern, generated in the open evaporite basin, is filtered through its shelf carbonates and, stripped of magnesium chloride, is returned to the pool and then starts to precipitate potassium chloride, a compound that is much less soluble than magnesium chloride. It is more likely that the brine first lost its potassium chloride by slow precipitation, and that the residual magnesium chloride bittern then seeped into the substrate.

The metal content of the concentrating brine can have a crucial significance. Minute quantities of tin, and even smaller amounts of lead, have a stabilizing effect on supersaturated potassium chloride solutions and eventually lead to precipitation of transparent octahedral crystals in lieu of the normal cubic habit. Small quantities of trivalent iron cause the crystal habit to be deformed and the transparency to disappear (Schnerb and Bloch, 1958). Potassium chloride crystals take up either calcium or strontium up to an ionic ratio of 1:4000 (Pick and Weber, 1950). Strontium thereby raises the density, and calcium lowers it. Each bivalent ion replaces two potassium ions but occupies only one lattice position, leaving a hole in the cation lattice.

Since potassium chloride is less soluble than magnesium chloride and decreases its solubility in cooling magnesium chloride brines, it reaches saturation first. Where saturation is achieved slowly, i.e., by lateral daytime diffusion to warmer bitterns, one would expect sylvite (or sylvinite) to precipitate at night. If saturation is rapid, i.e., by high rates of evaporation of an exposed brine, potassium chloride and magnesium chloride coprecipitate as carnallite, provided the temperature remains above −12°C. In sterile laboratory experiments, it is difficult to precipitate sylvite from concentrating seawater at ambient temperature. Its primary nature has therefore been questioned at times. However, the presence of ammonium chloride decreases the solubility of potassium chloride, until this solubility virtually vanishes at a concentration of 273.5 g/liter ammonium chloride (Seidel, 1917).

Complex cyanides reduce the solubility of sylvite (Steinike, 1962), and so do many organic compounds. The solubility of potassium chloride also decreases with increasing presence of alcohols in the brine. Urea complexes magnesium chloride and increases its solubility (Emons, 1967), thus hindering its coprecipitation. In effect, organic decomposition products that act as surfactants to reduce the surface tension accelerate the deposition of potash minerals. Even at this brine concentration, the presence of dissolved organic compounds exerts a profound influence on the types of precipitates formed: In an analogy to halite, *sylvite and carnallite precipitation is also accelerated by the presence of organic compounds.*

Harvie *et al.* (1982) have calculated the quaternary invariant points at 25°C for different mixtures of sodium, potassium, magnesium, and calcium and the corresponding water activities. However, it is necessary to confirm these with experimental data, since mixtures of solutes may deviate considerably from calculated values. For instance, Fournier *et al.* (1982) found that the solubility of quartz in sodium chloride solutions of varying strength cannot be calculated using a simple model in which the activity of water is directly related to decreasing water vapor pressure as salinity decreases. In this case, the calculated data represented minimum values, which were always smaller than the experimental determinations.

PRIMARY CARNALLITE ($KCl \cdot MgCl_2 \cdot 6H_2O$)

Carnallite is frequently the next potash mineral deposited in quantity after halite, and it often precipitates concurrently with clear sylvite in other parts of the basin. Carnallitite is by far the most common primary potash deposit. Storck (1964), among others, therefore considered it to be the only primary potash facies. It is rarely a potash ore because of its low potassium content (14% in carnallite, and usually closer to 5% in carnallitites or carnallite–halite mixtures). Carnallite precipitation requires a daytime brine temperature of 41°–47°C (d'Ans, 1947; Braitsch, 1964). This temperature range is quite commonly reached by hypersaline brine pools exposed to solar radiation, even if the radiation is diffused under cloud cover.

Carnallite is also frequently a secondary mineral. Alterations of carnallite and secondary carnallite are discussed in context with diagenetic and epigenetic brines.

PRIMARY SYLVITE (KCl)

A thick sylvinite deposit, particularly a layered one, reflects uniform hydrological conditions of concentration, rates of influx, evaporation, and subsidence (Zen, 1960). Quaternary sylvite (KCl) occurs as cubic crystals in halite in the Danakil depression of Ethiopia. The sylvite contains microinclusions of anhydrite and occasionally thin anhydrite coatings, and is thus evidently primary (Augusthitis, 1980). Kainite occurs only on the margins, in the vicinity of sulfur springs. Because sylvite precipitation lies outside the stability field for the whole range of temperatures encountered in isothermal evaporation in sterile laboratory experiments, Holwerda and Hutchinson (1968) earlier interpreted this sylvite deposit as an alteration: Freshly precipitated carnallite was deemed to have been converted promptly each season to sylvite by residual bitterns seeping into the ground, a form of preburial leaching. Local occurrences of polyhalite and kieserite lenses indicated multiple phases of secondary dissolution and redeposition that complicated matters (Holloway, 1945). However, under natural conditions of solar evaporation, sylvite can precipitate at 25°C, and even at lower temperatures in cooling brines (Valyashko and Solov'eva, 1949, 1953). Thus, the deposit can very well represent a primary occurrence. Seasonal drying out and flooding by new inflow from the nearby Red Sea could not build up a thick sylvinite layer. Each new influx would redissolve some if not all of the precipitated salts, but it would also cover the remainder with some anhydrite and halite laminae. Seasonal drying out can lead to additional potash precipitation only if the returning brine is already saturated for potassium chloride (Fig. 7-5).

Other sylvinite deposits have also been interpreted as primary precipitates. According to Labonova (1949) and Dubinina (1951b), the Miocene sylvite in the Carpathian region and the Permian sylvite in the Upper Kama River area west of the Urals have textures compatible only with a primary origin (Ivanov and Voronova, 1963, 1972). On the basis of solar experiments, Valyashko and Solov'eva (1953) concurred that primary sylvite could indeed occur. Primary sylvinite has been recognized in Oligocene deposits in the Rhine Valley (Baar and Kuehn, 1962; Braitsch, 1966). Braitsch (1967) observed that sylvite crystals often appear as geopetal infill with smooth tops. He suggested that the crystals grow below the thermocline. In sinking, they partially dissolve, and consequently, halite precipitates in the form of rosettes. The existence of a density and thermal stratification here can also be inferred from the bromide and rubidium content of the sylvinites and halites (Braitsch, 1966).

Mottled and Variegated Sylvites

The origin of mottled sylvite differs from that of variegated sylvite. Variegated sylvites seem to be secondary in all cases; their contacts with carnallites are

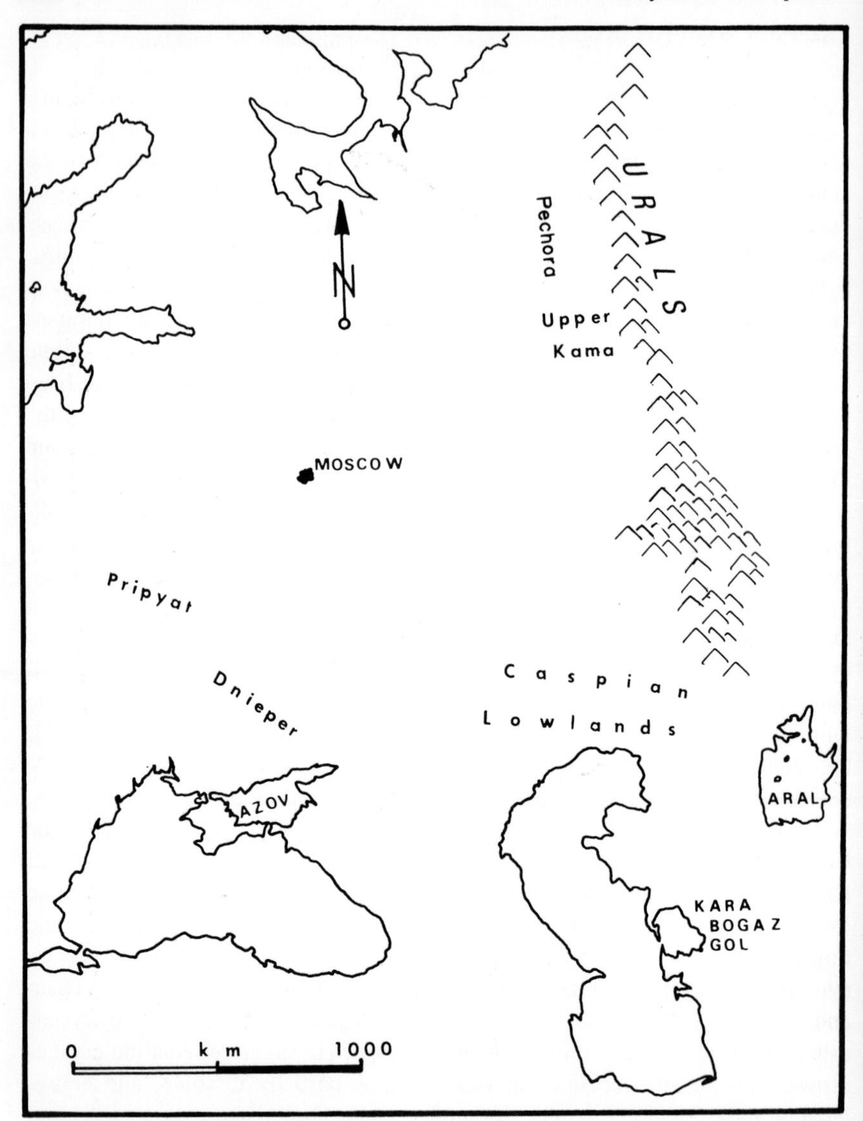

Fig. 7-5. Location map of major evaporite deposits on the Russian Platform. Devonian: Pripyat and Dnieper Basins. Permian: Pechora Basin, Upper Kama Basin, and Caspian Lowlands. Recent: Kara Bogaz Gol, Aral Sea, Sea of Azov.

usually destroyed (Vakhremeyova, 1954). Kopnin and Maloshtanova (1980) referred to variegated sylvinites in the same deposit as "diagenetic alterations." The calcium sulfate content of these sylvinites is derived from continental sources and is two to three times smaller than that in overlying halite or in banded sylvites where the calcium sulfate is of marine origin (Rayevskiy, 1967). Mottled sylvite has also often been taken as a secondary crystallization after carnallite. However, Ivanov (1963) has shown that in the Upper Permian evaporites on the west flank of the Ural Mountains, the mottled sylvite is a primary facies laterally contiguous with carnallites, often alternating with them (Fig. 7-6). The mottled sylvite is characterized by coarse crystals, a very small amount of hematite inclusions, a massive structure of the annual layers, and high gas saturation. The main minerals are milky white sylvite and light-to-dark-blue halite. Banded sylvinites form in the last stages of sylvite precipitation just before the brine saturates for magnesium chloride and starts forming carnallitites (Vakhremeyova, 1954).

The gas content of microinclusions can shed some light on the origin of the precipitate. Ivanov (1963) reported that the gas content of carnallite and red sylvite is here high in noble gases (argon, krypton, xenon) and methane, hydrogen, and hydrogen sulfide, in addition to some carbon dioxide. In mottled and banded sylvites there is more gas, mainly N_2, and small amounts of carbon dioxide and hydrogen sulfide. Associated halites have 70 times less gas in their inclusions. The absence of neon and ^{36}Ar is evidence that there was no contact with the atmosphere. The difference in the amount and type of gas content between sylvites (except for red secondary sylvites) and carnallites proves that conditions were very specific during crystallization of each of these facies.

A basic feature seems to be the deposition of mottled sylvite under anaerobic conditions of hydrogen sulfide contamination. The mottled sylvite is situated near the ancient shores that received all the runoff from the Uralian land mass. Mottled sylvites formed beneath the cover of inflow, even though the inflow was insufficient to cover all the exposed brine surface. In contrast, carnallites precipitated beyond the density stratification of the brine, where bottom waters were in contact with the atmosphere and oxidized any iron compounds. Supporting evidence is drawn from the observation that to this day, sylvinites give alkaline pH readings of 8.2–8.5, whereas the carnallites offer an acidic pH of 6.7 (Vovk, 1979). Fresh waters supply the calcium sulfate in mottled sylvites (Rayevskiy, 1967). Indeed, Urazov (1930, 1932) had come to the same conclusion much earlier, namely, that the annual runoff came from the east and thus must have overridden the concentrated brine in that part of the lagoon. The existence of inflow of waters from the Uralian land mass is further documented by the occurrence of carbonate and clay intercalations along the eastern shores of the brine basin. They are absent along the other shores.

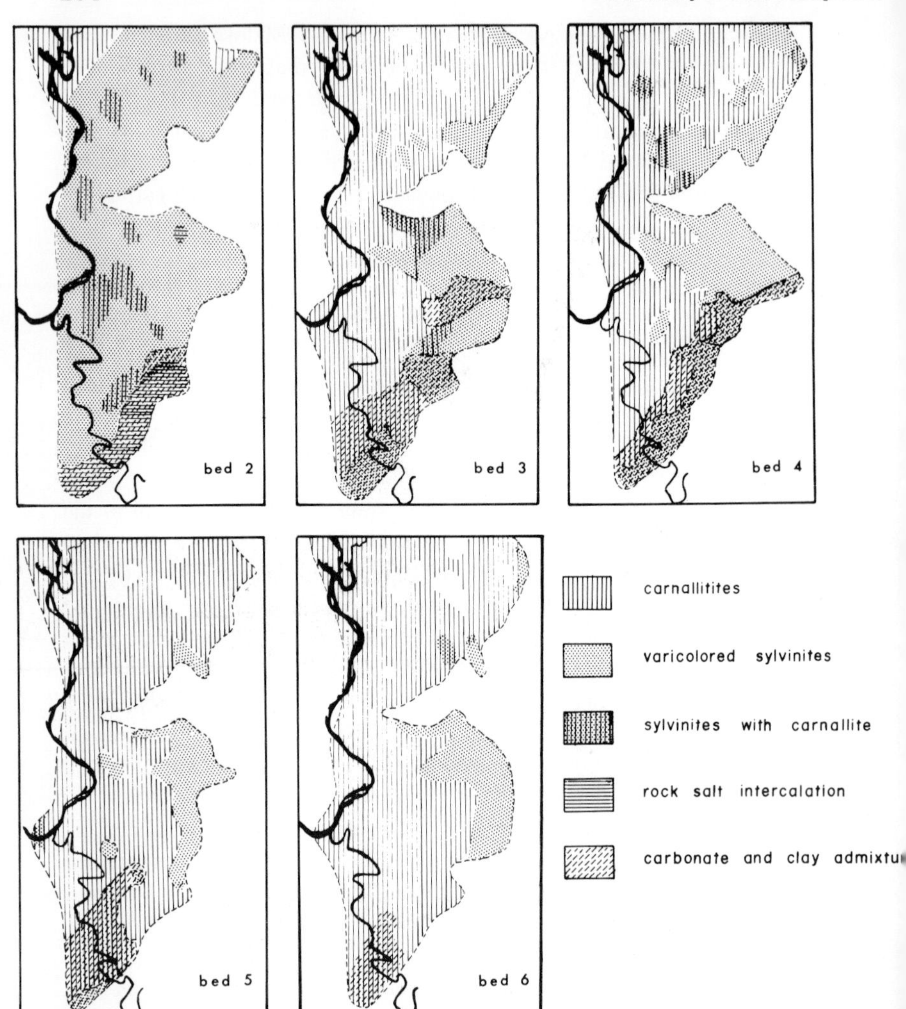

Fig. 7-6. The Permian Upper Kama evaporite deposits west of the Urals. Varicolored sylvites hug the eastern shores, where runoff from the ancestral Ural Mountains would have created a density stratification. Carbonates and clays accumulated at the southeastern end of the basin (redrawn from Ivanov and Voronova, 1975).

Convection currents form from denser solutions surrounding precipitated carnallite to less dense solutions surrounding precipitated sylvinite (Bloch *et al.*, 1952). When the inflow-covered sylvite-precipitating brine cools at night to reach saturation, the uncovered carnallite-precipitating brine slowly tends to equilibrate the small salinity differential. A daytime heating up of the covered brine creates a much larger density differential and thus higher current velocities in the undersaturated warmer brines. Sylvite precipitation requires a smaller concentration of a natural brine than does carnallite crystallization. Up to a concentration of 60 moles $MgCl_2$ per 1000 moles H_2O, the stable precipitate is sylvite with halite, i.e., sylvinite, at 20°C. At a greater concentration, carnallite is stable with halite, whereas sylvite becomes metastable (Karsten, 1954) (Fig. 7-7).

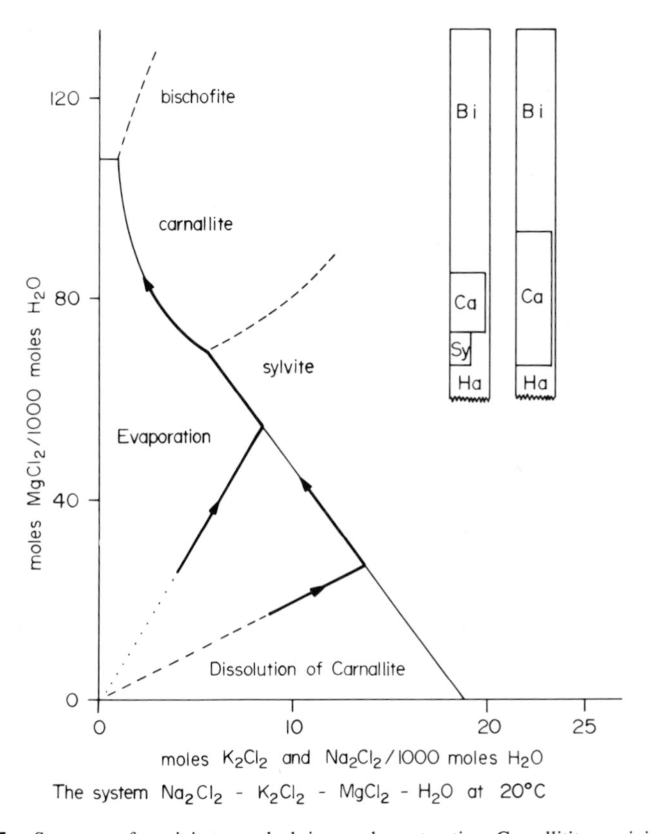

Fig. 7-7. Sequence of precipitates as the brine reaches saturation. Carnallitite precipitates above halite if the brine concentrates very slowly (upper concentration path and right-hand stratigraphic column). Evaporation losses of an exposed hypersaline bittern are small per unit of time, particularly if a floating crust has formed. An influx of waters undersaturated with respect to carnallitite dissolve it; the bittern precipitates sylvite after saturating for KCl (lower concentration path and left-hand stratigraphic column). The presence of calcium chloride in the bittern favors carnallitite over sylvinite (after R. Evans, 1970).

RINNEITE ($FeCl_2 \cdot 3KCl \cdot NaCl$)

Rinneite crystallizes above 26.4°C; below that temperature, it coprecipitates with sylvite or forms an isomorphous replacement in carnallite. Since rinneite is unstable in the presence of oxygen, its occurrence proves that the evaporite deposit was not subjected to any oxygenated (i.e., sulfatic) brines circulating from either above or below. The absence of rinneite in carnallite seams is easy to understand if carnallite is the product of precipitating homogenized bitterns exposed to the atmosphere, and sylvite formed in a density-stratified medium. However, rinneite has a very narrow stability field even at 55°C, which is little altered by an increasing magnesium chloride concentration (d'Ans and Freund, 1954). Its massive occurrence therefore always points to very special circumstances during its precipitation.

Seawater is deficient in iron, and the concentration of a volume of seawater to carnallite saturation would still contain only negligible amounts of iron. If we consider the ability of organisms to concentrate iron hydoxides in the brine, we can explain the sporadically reported high iron values in Recent marine gypsum deposits or abandoned salinas. Iron chlorides form even today. Heavy storms in the winter of 1951–52 drove seawater into exposed pyrite deposits on the island of Elba, Italy. The brine changed the iron sulfide to copiapite, a hydrated iron sulfate, and to orange-red $FeCl_3 \cdot 6H_2O$, with some halite and gypsum (Garavelli, 1958). The solubility of iron increases in concentrating magnesium chloride solutions (Whitman *et al.*, 1925). Nonetheless, the origin of iron in evaporite deposits, either as chloride or as hematite, requires further investigation.

Rinneite is not uncommon in some evaporite deposits. It occurs in Jurassic potash deposits in Uzbekistan, Triassic ones in Tunisia, and Devonian ones in the Tuva region north of Mongolia. Although common in Permian evaporites in Britain, rinneite is absent from Permian evaporites in New Mexico (Stewart, 1954). This reflects a different postdepositional history of the two basins. Rinneite forms clusters within milky white sylvinite that is totally devoid of hematite in the uppermost evaporite cyclothem of the Zechstein in Yorkshire. Extensive recrystallization and replacement occur in attendant halite and anhydrite beds (Stewart and Vincent, 1951). Originally, Stewart (1951) considered it here to be a primary precipitate, which would also make its host sylvinite a primary deposit. Later he changed his mind (Stewart, 1956) when he observed that the rinneite occurs after carnallite, in conjunction with halite and sylvite.

The presence of rinneite instead of hematite in the uppermost evaporite cyclothem in the Zechstein of eastern Yorkshire (Stewart, 1951; Stewart and Vincent, 1951) and the concurrent absence of iron oxides in the surrounding sylvinites may be caused by an absence of bacteria that was probably due to a low pH. The existence of a low pH is here corroborated by the admixture of a detrital quartz silt with the halite that is related to an erosional surface. $FeCl_2$

readily substitutes for $MgCl_2$ in the carnallite lattice because of similar ionic radii (d'Ans and Freund, 1954; S. S. Adams, 1969). The Fe^{2+} activities necessary for $FeCl_2$ (lawrencite) or $FeCl_2 \cdot 3KCl \cdot NaCl$ (rinneite) precipitation occur in normal seawater only at pH values lower than 4.6 (Marr, 1957a), which are difficult to attain in saturating bitterns. However, such values are common in epigenetic brine movements. Rinneite in the Stassfurt Member of German Zechstein evaporites occurs in nests and small intercalations along joints or contacts with the clay member; thus, here it is also an epigenetic relocation (Kuehn, 1966).

The parageneses of rinneite can shed some light on its mode of origin. In the western part of the Cambrian evaporite basin in Siberia, rinneite covers at least 5000 km^2 near the distal bayhead. Neither carbonates nor anhydrites are present, and the rinneite occurs within the potash horizon, either as independent crystals in ''annual'' partings or overgrown with sylvite. It occurs together with sylvite–carnallite and occasionally with sylvite, but never with pure carnallite. It has inclusions of anhydrite, carnallite, gas, or brine, or is itself an inclusion in sylvite (Kolosov *et al.*, 1968; Kolosov and Pustyl'nikov, 1970b), and occasionally shows pseudomorphs after carnallite. Rinneite here marks a region of very low partial pressure of oxygen and a markedly increased complexing strength of both potassium and sodium.

In each case, rinneite seems to occur in marginal parts of the basin that are most remote from access to the ocean (Kolosov and Pustyl'nikov, 1970a). It may be produced by a very early desiccation of iron-bearing carnallite. Erythrosiderite and douglasite are presumed to be alteration products after rinneite (Kolosov and Pustyl'nikov, 1967, 1970b). Rinneite in the Ronnenberg seam of the German Zechstein evaporites contains carnallite, anhydrite, or rounded halite inclusions, but is also overgrown by sylvite (Siemens, 1961).

BISCHOFITE ($MgCl_2 \cdot 6H_2O$)

Further concentration of the brine causes bischofite precipitation, but only if there is no calcium chloride available to form tachyhydrite. The bittern must also be totally devoid of sulfate ions (Azizov, 1979), and it is usually restricted to small regions within the carnallite areas (Korenevskiy, 1963). These regions represent residual pools of bitterns after much of the area has temporarily dried out. Bischofite layers can attain a thickness of several tens of meters (de Ruyter, 1979). They precipitate very rapidly, since they reach saturation very close to the eutonic point of the bittern. Permian bischofites west of the Urals are encased in carnallite, but their bromide content is higher than that of the latter (Kazantsev *et al.*, 1974; Valyashko *et al.*, 1979). However, they are not the terminal deposits in a dried-out lake or in sebkha-style coastal salt flats. They form only if there is still some mother liquor left, or at least if some interstitial solution is being discharged under a hydraulic gradient directed upward (Azizov, 1979). This requirement of a brine cover is also shared by primary tachyhydrite, suggesting that the drying out of a basin would leave only potash deposits as terminal facies.

Bischofite here is accompanied by clays, quartz, borate minerals, or anhydrite. At first sight, this could suggest contamination during redeposition of potassium–magnesium salts (Yermakov and Grebennikov, 1977), even though the deposit covers a very large area (Grebennikov and Yermakov, 1980). Since bischofite occurs here together with carnallite and halite and does not occur in any other subbasin closer to the seawater intake (Valyashko *et al.*, 1976b), it probably represents part of the original sequence of precipitates. The associated carnallite was also not precipitated in a dried-up playa, but under a cover of brine.

Secondary bischofite forms by alteration of carnallite in water (Dana, 1951) undersaturated for potassium chloride. It occurs frequently in potassium sulfate deposits, because magnesium chloride is less soluble in a sulfate brine than in a chloride brine (Yarzhemskiy, 1966). The resulting bischofite then contains inclusions of anhydrite and kieserite (Yarzhemskiy, 1967).

Bischofites are rare not only because bitterns do not often go to bischofite saturation. Exposed to even a moderate geothermal gradient, bischofites are easily subjected to thermal hydrolysis, leading to many basic reaction products. The presence of potassium chloride, and to a lesser extent sodium chloride, reduces this hydrolysis (Tittel, 1959). Circulating undersaturated magnesium chloride solutions can bring about the dissolution of bischofite, but saturated sodium chloride solutions are much more effective in this respect. The salting out of halite during the dissolution of bischofite decreases the sodium chloride concentration and with it the rate of dissolution. This rate is 2.6 times as fast along a horizontal sheet of bischofite as it is along a vertical face (Reznikov and Bel'dy, 1974), a measure of the difference between the horizontal and vertical permeabilities.

TACHYHYDRITE ($CaCl_2 \cdot 2MgCl_2 \cdot 12H_2O$)

Within the photic zone, calcium remains in solution where bacterial sulfate reduction is rampant, and it precipitates gypsum where oxidation of hydrogen sulfide occurs. Since photosynthesizers are nightly oxygen consumers, they die out as oxygen solubility decreases in the concentrating brine beyond a threshold minimum. Calcium depletion is then delayed below the chemocline until highly soluble calcium chlorides can precipitate. The temperature of the concentrating brine rises as its specific heat decreases. With increasing temperature, the stability field of tachyhydrite grows at the expense of both bischofite and calcium chloride hydrates. Similarly, baeumlerite ($KCaCl_3$) increases its field at the expense of carnallite and calcium chloride hydrates (Assarsson and Balder, 1955).

Massive tachyhydrite beds occur in a Cretaceous basin of Brazil (Wardlaw, 1972a,b; Szatmari *et al.*, 1979), in the Gabon and Congo basins (Lambert, 1967; de Ruyter, 1979), and in Thailand (Hite and Japakasetr, 1979). All three sites are noteworthy for a lack of carbonates and sulfates of either calcium or magnesium. The associated halite is frequently sapphire blue, possibly indicating the presence of excess sodium ions in the halite crystal lattice, i.e., a deficiency of chloride.

Two possible modes of origin have to be considered:

1. Each occurrence could be a secondary deposit. Borchert (1940, p. 108), unaware of massive tachyhydrite deposits in Thailand, Gabon, and Brazil, considered primary tachyhydrite to be impossible. He had seen tachyhydrite–sylvite intergrowths only in the Zechstein evaporites, and these appeared to be secondary (Braitsch and Herrmann, 1963; Kuehn, 1969) on the basis of a departure in the bromine ratio from theoretical values. However, Hintze (1915) had noted that tachyhydrite occurs on occasion with kainite but never with secondary sylvite; with carnallite it forms angular intergrowths several tens of millimeters thick. This means that the K_2/Mg ratio in the brine was in the range of 3–9 (cf. Chapter 13).

Bischofite in contact with precipitated calcium chloride hexahydrate produces tachyhydrite at temperatures between 21.95° and 25.15°C (Kuehn, 1952a). The minimum temperature for tachyhydrite formation is 19.5°C (d'Ans, 1933). A source of circulating calcium chloride solutions is not difficult to find. Both inland brines in evaporite basins and sebkhas (Levy, 1980) and subsurface brines on continents (e.g., Fritz and Frape, 1982a,b) tend to be enriched in calcium chloride. The percolation of such brines is suggested by the presence of up to 35% calcium chloride in Miocene sylvinites in the Carpathian belt (Fiveg and Rayevskiy, 1971). Similarly, in Byelorussian potash deposits, banded sylvite gives way laterally to sylvite, with widespread dissolution, recrystallization, and replacement features and higher magnesium chloride, calcium chloride, and bromine contents (Lupinovich and Kislik, 1971). Secondary tachyhydrite may also have formed in the Stassfurt Member of the German Zechstein sequence by the interaction of warm calcium chloride solutions with carnallite (Kuehn, 1955) or by infiltration of dolomitizing solutions from above (d'Ans, 1961). This produced a strontium enrichment.

Jaenecke (1915) interpreted tachyhydrite in the German Zechstein as a coprecipitate with sylvite after destruction of carnallite by a $CaCl_2 \cdot 2H_2O$-carrying current, since the deposit decreases in size westward. However, Braitsch (1971) considered a tachyhydrite–sylvite paragenesis to be unstable at any temperature. In contrast, Kling (1915) assumed a secondary origin by the interaction of magnesium chloride brines with anhydrite. He suggested that this could explain the intergrowth of secondary kieserite and carnallite with tachyhydrite. A horizon of extraordinarily large anhydrite crystals found above the tachyhydrite layer represented to him a precipitate of excess sulfate.

D'Ans (1961) also considered tachyhydrite to be only a secondary mineral, formed by double decomposition of carnallite or bischofite, with or without evaporation of the mother liquor. Entering calcium carbonate solutions would leave dolomite and magnesite as by-products, and would either precipitate tachyhydrite in place or penetrate deeper into the substrate, precipitating tachyhydrite there upon cooling. Entering gypsiferous brines could form tachyhydrite, in conjunction with kieserite and sylvite, by decomposing carnallite or sylvite mixed with bischofite. The problem with this explanation is that one must account for the presence and later disposition of calcium carbonate or calcium sulfate in solution in brines so extremely concentrated that they are normally stripped of such compounds.

2. Tachyhydrite is occasionally a primary precipitate. Calcium chloride must thus be retained in the brine in very significant quantities to produce massive tachyhydrite deposits. Where most of the calcium was depleted by early gypsum precipitation, the terminal bittern produced bischofite instead (Fig. 7-8).

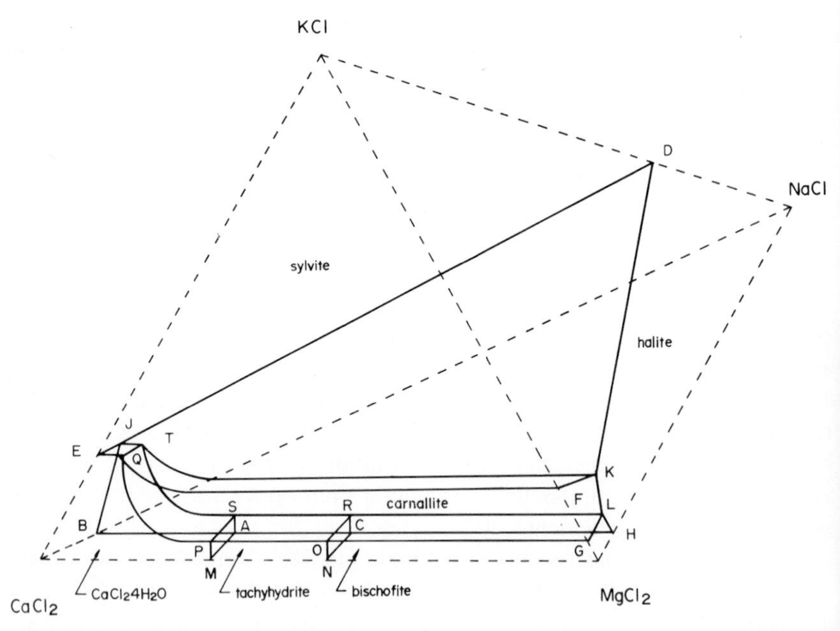

Fig. 7-8. The quinary system $CaCl_2$–$MgCl_2$–KCl–NaCl–H_2O at 35°C. [Reprinted with permission from Meyer *et al.* (1949). Equilibria in saturated solutions. V. The quinary system $CaCl_2$–$MgCl_2$–KCl–NaCl–H_2O at 35°C. *J. Am. Chem. Soc.* **71**(4), 1236–1237. Copyright 1949 American Chemical Society.]

Tachyhydrite occurs sandwiched into carnallite in the Sergipe Basin of Brasil (Benavides, 1968; Coutinho and Fernandez, 1973, Borchert, 1977), as well as in the continuation of this tachyhydrite sequence in Gabon, West Africa (Belmonte *et al.*, 1965), and in the Congo Basin (Lambert, 1967). Associated sylvite is not primary, but is derived from carnallite (Wardlaw, 1972a,b) on account of its rubidium enrichment (38 ppm). At Sergipe, Brazil, the tachyhydrite deposit shows a bromide and strontium content appropriate to a brine derived from oceanic sources, but already preconcentrated en route. The tachyhydrite averages 1800 ppm strontium and 3000–3700 ppm bromide; in contrast, the carnallite contains 4728 ppm bromide and the halite 360 ppm bromide (Wardlaw, 1972a,b). In German Zechstein evaporites, tachyhydrite never occurs together with sylvite. For that reason, Marr (1957b, 1959) considered this tachyhydrite rather than the sylvite to be primary, citing support by the late E. Fulda. It occurs together with kieserite (Zwanzig, 1958; Kuehn, 1968), which would have been altered to anhydrite by any extraneous calcium chloride influx, unless the kieserite was the younger salt. Fulda (1928) considered tachyhydrite to be primary because it occurred in conjunction with carnallite for considerable distances. The tachyhydrite zone is restricted to the same stratigraphic horizon and is uniformly 1 m thick.

The tachyhydrite sequence in Brazil is locally interspersed with one or two carnallitite and halite sequences, showing the sequence of a basin progressively saturating from halite, through potash and calcium salts, and back to mere halite saturation, and suggesting at least two partial freshenings by temporary decreases in inflow/outflow ratios before being covered by minor amounts of anhydrite and marls. However, the constancy of the strontium content through the upper tachyhydrite zone and the very regular upward increase in bromide content could be achieved only where crystallization was uninterrupted by any sudden freshening of the brine, i.e., where outflow had ceased and further inflow matched evaporation and seepage losses. Even then, a preconcentration basin strongly suggests that a sulfate and carbonate deficiency is created in the inflowing brine, and it is partially stripped of halite and potash salts.

A primary deposit of tachyhydrite must have formed in an embayment separated from the open sea by a long, narrow, and convoluted channel in which the brine was gradually concentrated. The channel would have had to be long enough to move the brine out of a regime of a maritime climate of high humidity into a continental climate of low humidity (Hite and Japakasetr, 1979). The relative humidity required to evaporate further quantities of water from a bittern concentrated to tachyhydrite saturation (which is 310 times that of seawater) is so low that it is attained only in continental interiors, not in paralic environments. Such a long channel would normally fill up quickly with precipitating carbonates and sulfates. This model therefore requires a further postulate. The channel must have been deep enough so that the bottom would remain below the photic zone, yet narrow enough to prevent significant reflux of bottom brines. In that case, algae and other photosynthesizers would be unable to settle on the bottom. Neither reefs nor gypsum crusts could then accrete. The anaerobic sulfur bacteria, converting SO_4^{2-} to hydrogen sulfide, would have allowed the sulfate to escape. Essentially, only a chloride brine would then be allowed to pass through the straits. In that scenario, the bottom of the straits could remain muddy and thus adsorb some of the dissolved calcium. At no time could the sequence have dried out, since tachyhydrite is not stable when exposed even briefly to the atmosphere (Wardlaw, 1972a,b). The absence of red hematite staining would also preclude exposure to oxygenation. A playa or sebkha model thus does not apply. It is significant that the tachyhydrite beds in Brazil and Gabon are always underlain by carnallite; its crystallization strips the brine of further volumes of water without requiring any evaporation losses.

For the evaporites of Gabon and the Congo, some of the corresponding gypsum may well have been deposited in a preconcentrator basin, as Lower Cretaceous gypsum occurs in Angola (Mascarehas Neto, 1961). Tachyhydrite precipitation can thus be seen as a function of the effectiveness of bacteria to strip the brine of sulfate ions and to expel the resulting hydrogen sulfide gas without reoxidation within the surface waters, while keeping most of the calcium and magnesium in solution. Calcium sulfates would then be absent if there were no near-shore shallow shelves for photosynthesizers to settle, as both calcium and magnesium sulfates form well before tachyhydrite saturation if the sulfate anion is available. Furthermore, for every unit volume of tachyhydrite, previous precipitation of half a unit of carnallite underneath or in an antechamber is necessary to eliminate the dissolved potassium.

SUMMARY OF PRIMARY PRECIPITATION

By the time the brine reaches saturation of calcium sulfate, it has lost almost all of its bicarbonate and carbonate ions, and the solubility of atmospheric oxygen has become extremely low. Since anaerobic bacteria actively convert all incoming sulfate ions into hydrogen sulfide, gypsum can precipitate only in the photic zone, where algal and bacterial photosynthesizers generate new supplies of oxygen. Gypsum remains stable in the oxygenated environment of shallow shelves; gypsum crystals rolling into deeper waters become subject to decomposition into calcite. Gypsum precipitation is profoundly affected by the presence of organic matter and ammonium chloride, which increase its solubility and inhibit its crystallization. Likewise, crystal shape is affected by the presence of organic or inorganic compounds in solution. Strontium and barium are built into the gypsum lattice, but are easily moved into crystals of their own, particularly in the warmer waters of shoals.

In contrast to gypsum, the precipitation of halite and other chlorides is aided by the presence of organic matter, which reduces the solubility of sodium and potassium chlorides. Blue discoloration of halite has many causes that can be related either to the presence of impurities or to a nonstoichiometric Na/Cl ratio.

Primary potassium minerals are sylvite and carnallite, which are always encased in halite. Both minerals precipitate on basin slopes, since potash brines saturate quickly upon cooling and at the same time reduce their density. Sylvinite appears to be deposited under anoxic conditions in a density-stratified medium, whereas carnallite precipitates in an exposed brine. Very early dehydration of carnallite leads to crystallization of rinneite. Where most of the calcium was depleted in the brine by early gypsum precipitation, the final product of chloride crystallization is bischofite. Where significant quantities of calcium choride are retained in the brine or added from formation waters, the final crystallization leads to tachyhydrite. The warmer the brine, the greater the stabilitiy field of both tachyhydrite and baeumlerite. *Both have to be precipitated under brine cover, as they disintegrate upon total desiccation of the brine pool.* Both are so hygroscopic that they adsorb atmospheric moisture and then dissolve in their own crystal water. *Dried-out* playa lake stages of evaporite basins terminate with sylvite, carnallite, or their later alteration products, but do not preserve either bischofite or tachyhydrite.

Part III

ADMIXTURES TO THE PRECIPITATES

8

Substitutions and Accessory Precipitates

Most evaporite rocks contain both mechanical and chemical admixtures. Some cations, such as rubidium, cesium, thallium, and ammonia, substitute in the crystal lattice of carnallite. Others form discrete minerals, such as manganese, iron sulfides, or oxides. Accessory neoformation of quartz, feldspar, and clay minerals will be dealt with in the discussion of terrigenous clastics. Borates form a great variety of accessory minerals, particularly in evaporite basins influenced by volcanic emanations. Phosphates are extremely rare in evaporite sequences, and so are iodides. Fluorides occur mainly as fluorite in magnesium carbonates and as sellaite in rocks containing compound magnesium sulfates. Bromides do not form independent minerals, but substitute for chlorides in increasing amounts with rising brine concentration. They are thus important indicators of the original conditions of concentration and precipitation. Distribution coefficients between brine and precipitate have been published by Holser (1979c) for various trace elements substituting in evaporite minerals.

RUBIDIUM AND CESIUM

Rubidium remains in solution in the brine until potash deposits form (Fig. 8-1). A small amount is coprecipitated with sylvite, and the bulk with carnallite. Rubidium has a distribution coefficient greater than unity in carnallite; precipitation of the latter impoverishes the brine in rubidium (Holser, 1979c). Some is adsorbed into clays: Vermiculite fixes some rubidium, cesium, and potassium (Amphlett, 1964). By the time bischofite starts precipitating, the bittern has

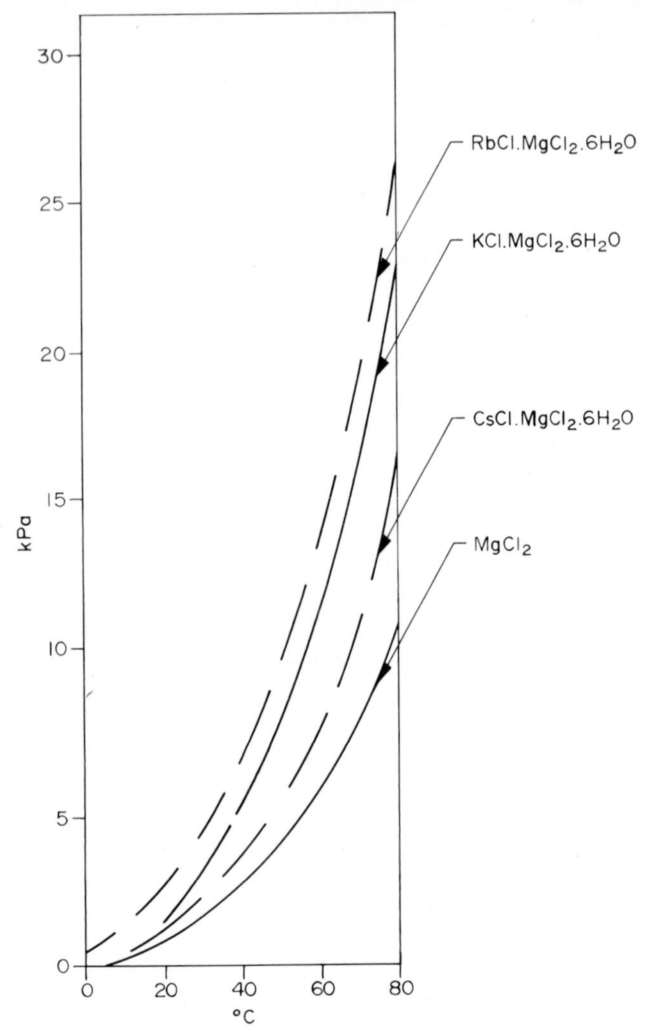

Fig. 8-1. Separation of alkali chlorides as a function of pressure and temperature (after Busch, 1959).

already been stripped of all of its rubidium (Zherebtsova and Volkova, 1966). Carnallite precipitation commences with a rubidium content of 347 ppm. The ratio of rubidium in carnallite and in solution is about 17:1 in the presence of ammonium chloride (Kuehn, 1963). In carnallite, both rubidium and cesium replacement of potassium decrease with temperature and with a decreasing K/Mg ratio in the residual bittern (McIntire, 1968; Schock and Puchelt, 1971).

In the sylvite lattice, the amount of incorporated rubidium is also dependent on both the concentration and temperature of the bittern. In primary sylvites the overall K/Rb ratio is 968, whereas in carnallite it is 405, dropping to 40 by repeated recrystallization (Petrova, 1973). Overall, the rubidium content is high in carnallite and slightly less in primary sylvite, but a rubidium enrichment then occurs in secondary sylvite (Kuehn, 1972).

The partitioning coefficient of cesium between brine and sylvite is not a function of temperature, but decreases with the addition of sodium chloride. In contrast, the partitioning coefficient of rubidium is independent of the presence of sodium chloride at 25°C. It increases with rising temperature if sodium chloride is absent, but decreases if it is present (Mumindzhanova and Osichkina, 1982). Cesium is built into the carnallite lattice up to about 160 ppb, as $CsCl_2 \cdot MgCl_2 \cdot 6H_2O$, but it does not occur in either sylvite or halite. In yet smaller quantities, it is found in clay particles within rock salt or sylvinite (Mormil', 1974).

The host minerals vary from one deposit to another. Rubidium is present in both sylvinite and langbeinite in German Zechstein deposits, but none can be found in kainite or polyhalite (Heyne, 1913). Nevertheless, in the Ukraine, rubidium occurs only in kainite and carnallite and is never found in sylvite (Petrichenko and Slivko, 1967). Rubidium and cesium are both associated with carnallites in Miocene evaporites of the Carpathian foreland and in Jurassic evaporites in central Asia. Sylvinites are enriched in either of these metals only if they are inherited from a carnallite precursor that was decomposed by continental runoff and was recrystallized (Osichkina, 1978b). This indicates a different history of the basins, particularly of the nature of postdepositional percolation of fluids. Synsedimentary resedimentation of salts washed in from basin margins (Braitsch, 1967) can alter the rubidium and cesium distribution. In the Oligocene salts of the Upper Rhine Valley, the rubidium content decreases upward, then increases and is very much enriched. The amounts of rubidium and cesium present in these carnallites are more than double those of the Permian Zechstein carnallites (Schock and Puchelt, 1970).

THALLIUM

The thallium atom has the same diameter as the rubidium atom. Traces of thallium, therefore, occur in all potash deposits. The Tl/K ratio appears to be almost constant near 8×10^{-7}, but is independent of the Rb/K ratio. The absolute values show the influence of periodic dilution of the brine (Malikova, 1967). In Miocene carnallite–sylvite beds in the Carpathian region, rubidium is preferentially concentrated in carnallites and thallium in sylvites (Slivko *et al.*, 1973). This is to be expected, as the distribution coefficient is greater than unity for carnallites (Holser, 1979c).

AMMONIA

Ammonium chloride substitutes in carnallite. In Permian evaporites of the Upper Kama area west of the Urals, carnallites contain 600 ppm but sylvites only 7–136 ppm (Apollonov, 1976), and the ammonia content varies directly with the rubidium content (Apollonov, 1980). The release of ammonia from carnallites decomposing to sylvites contributes to mine gases.

LITHIUM AND BORON

Lithium and boron continue to concentrate in the brine even beyond bischofite precipitation (Zherebtsova and Volkova, 1966). Both are concentrated in evaporitic clay minerals rich in magnesium, but not in associated chemical precipitates (Tardy *et al.*, 1972; Trauth, 1977). Lithium occurs in halite in water-soluble form (presumably as a chloride) in concentrations up to 0.2 ppm, but in polyhalite up to 10 ppm (Lepeshkov *et al.*, 1970), in carnallites usually 8–9 ppm and in sylvites up to 16 ppm (Kropachev *et al.*, 1972). Although the lithium content increases sharply at the top of halite, it decreases gradually upward in carnallite. It also enters spaces in the crystal lattice of borates. Lithium, beryllium, scandium, gallium, germanium, selenium, zircon, niobium, indium, tellurium, the lanthanides (rare earth elements), tantalium, and rhenium do not occur in pure evaporites in more than a few parts per billion. They are found only in association with terrigenous clay intercalations (Boyko, 1966).

Boron has been linked to basic volcanic emanations, but an underrated source of boron is sedimentary muscovite. It is of paramount importance for boron transport because of its ability to accommodate this element in its crystal lattice (Harder, 1959). In Mississippian evaporites of southern New Brunswick, Canada, eight different borate minerals were identified in halites, sylvites, anhydrites, and red mudstones; the boron was apparently derived from contemporaneous volcanic exhalations in the area (Roulston and Waugh, 1981).

Boron is primarily attached to clay minerals, and is liberated during their decomposition in a concentrating brine. Boron substitutes for aluminum in the clay silica tetrahedron or occupies empty spaces in the crystal lattice (Harder, 1959, 1961; Hingston, 1964; Fleet, 1965). When the clays are altered in hypersaline bitterns to mixed-layer varieties rich in magnesium, the original lattices are destroyed and boron is liberated. Thereafter, it either forms independent minerals or enters sulfates such as anhydrite (Ham *et al.*, 1961). Although in some localities borates are associated with clay laminations and with anhydrite (Milne *et al.*, 1977; Anderle *et al.*, 1979), they can also occur antithetic to anhydrite (Hite and Japakasetr, 1979). The latter condition suggests some early gypsum replacement by borates derived from clay influx. Boron liberated in the hydration of anhydrite forms new minerals, but such minerals are usually not found in the anhydrite adjacent to secondary gypsum (Ham *et al.*, 1961). Mueller (1969) found ulexite nodules in the zone of incomplete gypsification of anhydrite. The clay attached to the nodules contains glauberite that effloresces into mirabilite upon exposure to air. Boron mobilization and subsequent ulexite generation are apparently tied to the entry of oxidized sodium brines.

Open marine clays, especially illites, soak up some of the available boron, but illites are not available in hypersaline brines; consequently, boron fixation does not occur in sepiolites or related minerals (Fontes *et al.*, 1967) typical of rock salt and potash sequences. Borates remain in solution in brines and bitterns; a little is also taken up by newly formed secondary tourmaline. Triassic salts of the Atlas Mountains do not contain sepiolites; hence, some boron is fixed in tourmaline crystals that often have a detrital core (Kulke, 1978). In Germany they appear to be restricted to the carnallite or polyhalite–kieserite–sylvite region of the evaporite sequence and to adjacent halite beds. In Thailand, Britain, and Nova Scotia, Canada, the borates are enriched in the uppermost halite as disseminated crystals, clusters or laminations, constituting up to 8.5% of the rock (Milne *et al.*, 1977; Anderle *et al.*, 1979; Hite and Japakasetr, 1979). Apparently, the borates remain in solution even during precipitation of potash minerals and subsequent dilution of the brine to halite saturation. Clay incursions at this late stage in the evaporite cycle find a sufficient borate concentration in the brine to act as a catalyst for precipitation of borate crystals. Oligocene evaporites in the Alsace, France, contain only a few parts per million of boron in sylvites, rising to 15–20 ppm near clay contacts, but the clay intercalations themselves contain 50–500 ppm (Bertrand, 1937).

Borates are absent in those marine evaporite deposits that are devoid of potassium–magnesium salts. This suggests that hydrated borates (which can complex in the marine environment) precipitate only during the terminal desiccation process, when underlying salts have hardened and prevent syngenetic downward filtration of residual bitterns. Almost all boron is lost if only 2000 ppm of the originally entering seawater can escape (Osichkina, 1978a). In Britain, borate minerals do not occur outside the potash field (Stewart, 1965). Likewise, in Byelorussia, the borate content is highest in sylvinites (1200–1400 ppm), lowest in halite (20–30 ppm), and intermediate in carnallite (60–70 ppm) (Pavlyuchenko *et al.*, 1961a). If terminal bischofite solutions dry out, boron-bearing evaporite deposits can be subject to leaching, leading to boron enrichment in a weathering crust (Strakhov, 1962; Anderle *et al.*, 1979). Borates are prone to reach extreme supersaturation and thus precipitate very slowly. No experimental documentation exists for boron precipitation in brine environments. Boracites can vary laterally, being pure magnesian boracites in one part of the basin and changing to iron–manganese–magnesium boracites elsewhere (Heide *et al.*, 1980). The relationship of individual borate minerals to the rim (borates enriched in calcium) or the center (borates in sodium) of lakes (Bowser and Dickson, 1966) has never been investigated in borate laminations in ancient evaporite basins.

MANGANESE

Manganese is primarily built into reefs and other carbonates along the entrance to an evaporite basin as manganocalcite. It also occurs in significant quantities in gypsum, but decreases rapidly in chlorides. Clay intercalations adsorb manganese, so that there is an inverse relationship between the amount of clay in an evaporite section and the amount of manganese in evaporite minerals (Korenevskiy, 1970). Very rarely, one finds halite of a dark pink color caused by manganese. An oriented growth of cryptomelane in the form of black hairs has been found in Permian sylvites in Carlsbad, New Mexico (Sun, 1962). Where the basal Zechstein copper slate is overlain by limestone, the latter is altered to brown iron ore containing up to 30% hematite and psilomelane. This ore was mined in Thuringia from the eighteenth century to the end of World War I, primarily for its manganese content. In the Cambrian evaporite basin of Siberia, rinneite contains 2.1% $MnCl_2$, a much higher value than the fractional percentages reported from other areas (Kolosov and Pustyl'nikov, 1967, 1970b).

IRON

Although iron precipitates before calcite in laboratory seawater evaporation experiments, most of it normally stays in solution in nature. If the brine becomes oxygenated during the precipitation of rock salt or potassium–magnesium salts, goethite or hematite stains the precipitate brownish-red. Only 2 ppb of trivalent iron suffice to give halite a yellowish color (Elschner, 1923; Lotze, 1957), and a few more parts per billion turn the salt red. Lepidocrocite, a ruby red to blood red variety of FeO(OH), has also been found (Ivanov and Voronova, 1975). Very rarely is the color due to red Fe^{3+} salts [erythrosiderite ($KCl \cdot FeCl_3 \cdot H_2O$] or blackish-brown iron salts [molysite ($FeCl_3$)].

Trivalent iron is reduced by bacteria in an anaerobic medium according to the reaction

$$2Fe(OH)_3 + H_2S = S + 2Fe(OH)_2 + H_2O \qquad (8\text{-}1)$$

(Gusseva and Faingersh, 1974). In contrast, in an oxygenated environment, bivalent iron is converted to trivalent hydroxide by chemolithotrophic bacteria (Ehrlich, 1978) according to the reaction

$$Fe(OH)_2 \rightarrow FeO(OH) + H^+ \qquad (8\text{-}2)$$

The ferric/ferrous iron ratio increases in the brine with rising salinity (Osichkina, 1978a). In the presence of reducing substances, $Fe(OH)_3$ is reduced in concentrating magnesium chloride solutions (d'Ans and Freund, 1954). Otherwise, coagulation of $Fe(OH)_3$ colloids is delayed in the presence of chlorides, more so with sodium chloride than with magnesium chloride (Charmandarian and Andronikova, 1952). Conversion of clays in very concentrated bitterns to high-magnesium varieties also releases some FeO(OH) or Fe_2O_3 to the brine (Caroll and Starkey, 1960). In highly hygroscopic brines, such as a saturated sodium chloride solution, the trivalent iron precipitates initially as brownish goethite [FeO(OH)], which is the solid phase in brines. It would exist in that form only for a very short time interval before being dehydrated to hematite, provided that the redox potential (Eh) values are high enough to stabilize Fe_2O_3 (Berner, 1970a). Both sylvite and carnallite grow epitaxially with hematite (Leonhardt and Tiemeyer, 1938; Johnsen, 1909). Marginal zones of sylvite are often completely impregnated with hematite. This might imply that the enrichment with iron did not start until about the end of sylvite crystallization (Kuehn, 1968), i.e., after the redox potential had turned positive. Most hematite-bearing evaporites probably originated as goethite- or hydrogoethite-bearing deposits that were subsequently stripped of their hydroxyl radicals by hygroscopic brines. Goethite fibers are occasionally recognized as inclusions, e.g., in Devonian carnallites in Saskatchewan (Wardlaw, 1968).

Iron sulfides can also be hydrogen donors (Schwartz, 1972):

$$FeS_{(hydrotroilite)} + H_2S = FeS_{2(pyrite)} + H_2 \tag{8-3}$$

However, pyrite is found only in extremely minute quantities in anhydrites and in some halite sequences. For the most part, hydrotroilite is utilized as

$$2FeS + 3\ H_2O = Fe_2O_3 + 2H_2S + 2H^+ \tag{8-4}$$

The bacterial conversion of iron sulfides to iron oxides was first investigated by Winogradsky (1887, 1888). It is possible only in the relative absence of decomposable organic matter, as only then can an Eh be maintained that is high enough to stabilize hematite (Berner, 1970a).

To foster faster evaporation, a stratification is not allowed to develop in artificial salinas. The pH then remains below open sea values. A reduction in algal growth in artificial salinas further depresses the pH and produces a positive Eh. The precipitating halite is then stained red by ferric iron unless the Eh is dropped artificially to slightly negative values by periodically churning up of anaerobic bottom muds (Herrmann *et al.*, 1973; Schneider, 1979). The hydrogen sulfide and the ammonium sulfide contained in the mud clear the salt of any traces of iron (Elschner, 1923). A reddish-brown, iron-rich ferric layer accumulates in salinas above the discontinuity between churned and oxygenated precipitates above and anaerobic sediments below (Schneider, 1979). Salinas evidently represent an environment different from that of natural evaporite basins, since such a layer is not observed in natural stratified lagoons where the oxygenated zone does not reach the lagoon floor. Any dissolved iron here retains its bivalent form. In the Permian Hutchinson salt of Kansas, there are red precipitates in syneresis cracks of shale intercalations, concentrated in a series of bands parallel to the salt–shale contact; they indicate a periodic accumulation of red hematite over thick shale layers (Dellwig, 1968). These shales cover a smooth salt contact, which represents an erosional salt solution surface produced by a freshening of the brine.

Where fossil halite is stained red, an interlude of higher oxidation potential is indicated. Since there is no surface affinity of halite to hematite (Storck, 1964), pinkish hematite is common in facies of leaching. Ferric hydroxides have settled on halite crystal borders in the Permian Donbass deposits of the Ukraine (Bobrov *et al.*, 1968). In New Mexico, red halite is restricted primarily to the entrance area (Jones and Madsen, 1968), due to exposure of the entrance sill to atmospheric oxygen. The red halite in the lowermost Zechstein salts is probably due to formation in very shallow, oxygenated waters.

The presence of bacteria and related organic forms (Dellwig, 1963, 1968), or of blue-green algae (*Phormidium antiquum*) (Tilden, 1930), in the Permian salt of Kansas suggests an organic origin for the hematite. Bacteria were also found in Permian red salt in Britain and Germany (Mueller and Schwartz, 1953; Tasch, 1963). Sheaths of *Leptotrix* bacteria, which accelerate direct hematite crystallization, are found pseudomorphed to koenenite, and occur worldwide in hematite-rich carnallites and sylvites (Kuehn, 1961, 1968). Cyanophytes are responsible, according to Monty (1980), for the deposition of remobilized iron in sylvite and carnallite. Hematite can thereby also be derived from the iron carbonate as an exothermal reaction (Singh, 1972):

$$2FeCO_3 + O \rightarrow Fe_2O_3 + 2CO_2 \tag{8-5}$$

In contrast to these bacteria-bearing salts, Triassic red fissure salts in Germany were found to be sterile (Klaus, 1953).

Naturally precipitated halite beds are usually white or gray, betraying a negative redox potential of their brine. That is, the zero Eh surface occurs inside the water column, at the interface of density stratification, as water in contact with the atmosphere obtains a positive Eh. Permian halite layers in Stassfurt, Germany, are mainly gray and show evidence of a negative Eh (Storck, 1964). They contain fossils and pyrite, whereas the red clays do not (Lotze, 1957). Oxidation of iron is generally restricted to shoals, whereas gray chemical sediments prevail offshore, in somewhat deeper waters (Jurgan, 1969). Gray salts usually contain some bitumina or hydrogen sulfide, reddish salts rarely. Occasionally, the presence of hydrocarbon gases can reduce red salt to white (Hartwig, 1936). Because of the association of halites with pyrites and bitumina, it is wrong to place their natural environment of precipitation above the sulfate–sulfide fence into an only mildly negative or even positive redox potential field, as Krumbein and Garrels (1952) have done.

The Silurian Salina salt is white in the salt mines on the rim of the Michigan Basin. In the center of the basin, red halite without detectable organic material alternates with anhydrite–dolomite laminae, rich inorganic material, and pyrite (Dellwig, 1955). The change in the redox potential during halite precipitation turned the brine into an aerobic one; all the organic material was destroyed, and the ferrous iron was converted into the ferric oxide. One possible explanation is the analogy of the Jordan River, which brings oxygenated waters into the Dead Sea but spreads out in arcuate fashion over only a very small portion of this sea. Similarly, inflow into the Michigan Basin was probably insufficient to cover all of the brine surface.

Where an excess of red lateritic clays is swept into the basin, it may exhaust the capacity of the brine to reduce all the iron. The Triassic salts of the Atlas Mountains are red only in the clay-rich parts (Kulke, 1978). In Permian salts of Kansas, the red hematite is restricted to veins in shale intercalations, suggesting that the iron was primarily derived from the shale (Dellwig, 1963). Hematite-stained halite is much less common than hematite-stained carnallite. Oligocene evaporites in the Upper Rhine Valley of Alsace are composed of gray halite layers alternating with red carnallite layers (Rozsa, 1920). This relationship is very commonly observed in other deposits. The explanation can be found in a periodic exposure of the brine to atmospheric oxygen during carnallite precipitation, and in periodic covering by new inflow during halite deposition. Carnallite and secondary sylvite, as well as kainite, glaserite, syngenite, and polyhalite, are quite commonly stained red; tachyhydrite, langbeinite, and magnesite are very rarely stained; anhydrite, kieserite, loewite, vant'hoffite, and primary sylvite are never red (Richter, 1962). Similarly, halite is never red, but always gray, if it occurs below or alongside carnallite; it then emits the odor of hydrogen sulfide if crushed. In contrast, halite is commonly red if it occurs beneath leached potash horizons, such as clayey anhydritic or sylvitic leaching zones, or in langbeinite–halite mixtures (Richter, 1964). In that case, the Fe_2O_3 content of halite is greater than that of carnallite.

A small fraction of the total iron is also extracted by precipitation in the gypsum lattice (Sonnenfeld *et al.*, 1976). Some of it percolates into the bay floor with terminal bitterns.

PHOSPHATES

Isokite, wagnerite, and apatite are three alkali earth phosphate fluorides that occur in evaporites in addition to the soroborate lueneburgite, another compound phosphate (Braitsch, 1960; Kuehn, 1968). Lueneburgite has been identified in Zechstein carnallite and hard salt horizons, in marls overlying Upper Miocene evaporites underneath the Ionian Sea (Mueller and Fabricius, 1978), and in cores from sylvinite horizons in New Mexico and Texas (Schaller and Henderson, 1932). No alkali metal phosphates have been found in evaporite sequences, either as simple salts or as compound phosphate-halides. The four minerals mentioned are known only from recrystallized evaporite deposits altered by epigenetic fluids. It is noteworthy that the phosphorus/fluorine ratio in Zechstein evaporites is similar to that of nearby oil field waters (Kuehn, 1968). A phosphate-bearing halite bed, 5 m thick, has been described from a small structure in the Ukraine (Khrushchov and Stroev, 1973), but this is unusual. Also reported from here was hamlinite, a hydrous strontium–aluminum phosphate. A series of Paleogene cyclothems in Tunisia end either with gypsum or with apatite in separate cycles, but the two minerals never coexist in the same cycle (Lucas *et al.*, 1979).

IODIDES

Iodides are easily converted into iodates in an aqueous solution. Sodium iodate is about 220 times more soluble than sodium chloride, and thus remains in the residual solution. There is 50–84 mg iodine in a cubic meter of seawater. Precipitated iodates are found only in trace amounts in polar environments or in extreme tropical deserts, and in both cases are associated with nitrite deposits.

The iodine content in halite and potash minerals varies considerably from one deposit to another. Rock salt from various localities in Spain yielded 577–631 ppb iodine (Cavayé Hazen and Hoyos Ruiz, 1953). Sylvinites and sylvites have yielded 0–5.9 ppb (Geilmann and Bartlingek, 1942). Rock salt, potassium sulfates, and sylvite of the Stassfurt Member, German Zechstein evaporites, contain much less iodine than does sea salt. Schnepfe (1972) detected no iodine in his samples of German Zechstein halite. However, halite samples from Polish Miocene evaporites yielded double the amount present in sea salt; kainite yielded 10.5 times this amount (Erdmann, 1907, 1910b).

FLUORIDES

Sea salt, when dried, contains up to 36.73 mg/kg fluorine in raw salt but only 8.74 mg/kg in refined salt, in comparison to 3.75 mg/kg in fossil rock salt (Okamura and Matsushita, 1967).

Fluorides are much more frequent as traces in limestones surrounding the evaporite pan than in evaporites precipitated within that pan. In a neutral sodium fluoride solution, gypsum will be altered to fluorite at room temperature (Duff, 1972). However, such solutions are not available in evaporitic environments, and fluorite precipitates directly from the brine instead. Sellaite and fluorite dissolve congruently; the fluorite solubility decreases as the sellaite mole fraction increases in the solution. In addition to fluorite (CaF_2) and sellaite (MgF_2), a whole series of sodium, potassium, and compound magnesium fluorides have been described, mainly in the immediate vicinity of volcanoes; there is some doubt, though, that the latter two occur in evaporite sequences. Lake Borzynskoye near the Mongolian border precipitates mirabilite and halite with less than 3% of an admixture of sodium and calcium carbonates. Fluorine varies from traces to 26.6 mg/kg in the air-dried salt. It is found in the intercrystalline brine, because it cannot precipitate for lack of calcium or magnesium (Filippova *et al.*, 1969). This makes it dubious that villiaumite (NaF), neighborite ($NaMgF_3$), or carobbite (KF) can be found in evaporite sequences, since alkali metal fluorides are three to four orders of magnitude more soluble in brines than are alkaline earth fluorides.

In evaporitic sediments, magnesium fluoride predominates over calcium fluorite, particularly in anhydrites, because CaF_2 is unstable in the presence of magnesium sulfate at all temperatures (Sahama, 1945). Stable salts are those for which the sum of the heats of formation is larger than it is for the alternate pair of possible combinations (Lebedev, 1952). Calcium fluoride (fluorite) is stable with magnesium carbonate (magnesite), but not with magnesium sulfates. In German Zechstein sediments, fluorine is most commonly found in fluorite contained in dolomites (Koritnig, 1951; Fuechtbauer, 1958; Siemeister, 1961). Magnesium fluoride (sellaite) is stable with calcium sulfate (anhydrite), but not with calcium carbonate. This is due to the increasing solubility of fluorite with rising magnesium or sodium concentration, and to its decreasing solubility with rising calcium concentration (Kazakov and Sokolova, 1950). Several patents were issued as early as 1887 to A. Feldmann to produce MgF_2 from CaF_2 by exposure to a concentrated magnesium chloride solution or to dissolved carnallite (Leonhardt and Berdesinski, 1953).

The presence of sellaite in veinlike form (Heidorn, 1931) or together with rinneite (Siemeister, 1961) speaks for percolating oxygenated magnesium sulfate solutions but against ascending calcium solutions from a late diagenetic or epigenetic gypsum–anhydrite transformation of basal beds (Kuehn, 1968). However, Maus *et al.* (1979) described an occurrence of sellaite embedded in fluorite. The intergrowth suggests simultaneous precipitation.

BROMIDES

Bromide Content

Bromine, iodine, and fluorine continue to concentrate in the brine until the eutonic point is reached (Zherebtsova and Volkova, 1966). Partial molal volumes (e.g., of potassium chloride and potassium bromide) vary in the brine linearly with the volume ionic strength, even to very high concentrations; sulfates of potassium deviate from this linearity at higher concentrations (Wirth, 1937). The transition from $NaBr \cdot 2H_2O$ to NaBr occurs at 50.7°C in a neutral solution, but at lower temperatures in a composite solute (Nikolayev and Ravich, 1931). Solubility gaps in free energy curves indicate a critical solution temperature of 50°C for NaCl–NaBr, KCl–KBr, and KCl–RbCl (Hovi and Hyvoenen, 1951). However, bromine does not form separate and distinct minerals in an evaporite sequence. Instead, it substitutes for chlorine in solid solution in the various chlorides, with the partitioning coefficient being different for each mineral (Boeke, 1908; Chirkov, 1935; Chirkov and Shnee, 1937; Holser, 1979c). The Br/Cl distribution coefficient between precipitate and solution is dependent on both the temperature and the concentration of the solution. It amounts to 0.014×10^{-3} for halite at 40°C, 0.17×10^{-3} for sylvite at 40°C, and 0.32×10^{-3} for carnallite at 21°C. Halites containing more than 30 ppm bromide can thus be primary (Puchelt *et al.*, 1972). These figures are lower than the values calculated earlier (Valyashko, 1956a,b; Kuehn, 1968; Zherebtsova, 1970; Herrmann *et al.*, 1973). Examples of carnallites with such low bromide content occur in salt deposits east of Saratov underneath the Caspian lowlands (Shlezinger *et al.*, 1940a) and in the Michigan Basin (Kunasz, 1970). Seawater contains 65 ppm bromine, and brine saturated for halite contains 510 ppm; rock salt precipitation should commence with a bromide content of 65–75 ppm (Holser, 1979c). However, Great Kalkaman Lake in northeastern Kazakhstan precipitates rock salt that has a bromide content of 100–200 ppm at the onset of halite crystallization (Kostenko, 1974).

The bromide content in an evaporite sequence is not constant, because several factors are known to affect the Br/Cl ratio (Table 8-1). The following are some causes of bromide variation:

1. The more bromine there is in the brine, the more the chlorides will initially be enriched in their Br/Cl ratio. By a gradual increase in its bromide content, an undisturbed salt sequence ought to depict faithfully the progressive concentration of the crystallizing brine (Fig. 8-2). The invariant point of KCl–carnallite is reached at higher concentrations of potassium, magnesium, and chloride ions, the higher the original bromine content (Boeke, 1908; Bloch and Schnerb, 1953). However, the concentration of brines in modern salterns does not lead to a change in the Br/Cl ratio of fluid inclusions in halite. The latter remains the same as that of the inflowing seawater (Sabouraud-Rosset, 1974).

2. Precipitating salts are usually altered to varying degrees, showing secondary and, at times, even tertiary recrystallization, often accompanied by metasomatism. In part, this alteration is almost concurrent with precipitation; in part, it occurs after a substantial time lag. Thus, the present bromide content need not

ABLE 8-1

anging Ion Ratios in Seawater Concentrating to the Eutonic Point[a]

Ion ratios	Ocean	Onset of halite precipitation	Onset of sylvite precipitation	Carnallite precipitate	Eutonic point
Density	1.023	1.215–1.227	1.283–1.290	1.305–1.344	1.325–1.359
$10^3 \cdot Br^- / Cl^-$	3.4	4.7–5.5	17.5–21.0	20.1–23.8	21.9–24.4
Cl^- / Br^-	300	326	57.0	—	41.5
Mg^{2+} / Cl^-	0.11	0.16–9.17	0.75–0.77	0.86–0.96	0.97–0.98
Na^+ / Cl^-	0.87	0.79–0.82	0.15–0.16	0.03–0.09	0.03–0.04
$10^3 \cdot K^+ / Cl^-$	18.6	20.2–30.5	87.5–115.7	9.9–73.9	4.2–6.5
$10^3 \cdot K^+ / \Sigma$ ions	10.2	11.4–16.9	50.6–63.1	6.2–42.6	1.9–4.1
K^+ / Br^-	5.4	4.7–5.6	4.8–5.4	0.4–3.2	0.1–0.3

[a] After Ivanov and Voronova (1972).

always depict the original bromide distribution at the time of the initial crystallization; it would, indeed, be very unusual if an original bromide ratio remained unaffected by recrystallization and anion substitution. Only recrystallization in the dry state will fail to alter the initial bromide content (Wardlaw, 1970). This must have happened in the German Muschelkalk salt, as recrystallization apparently took place without loss of bromides (Dellwig and Kuehn, 1980).

Recrystallization in the absence of the original mother liquor will normally lead to impoverishment of bromide in the chlorides (Valyashko, 1956, 1959). The process of redissolution *in situ* may explain the very low bromide content of

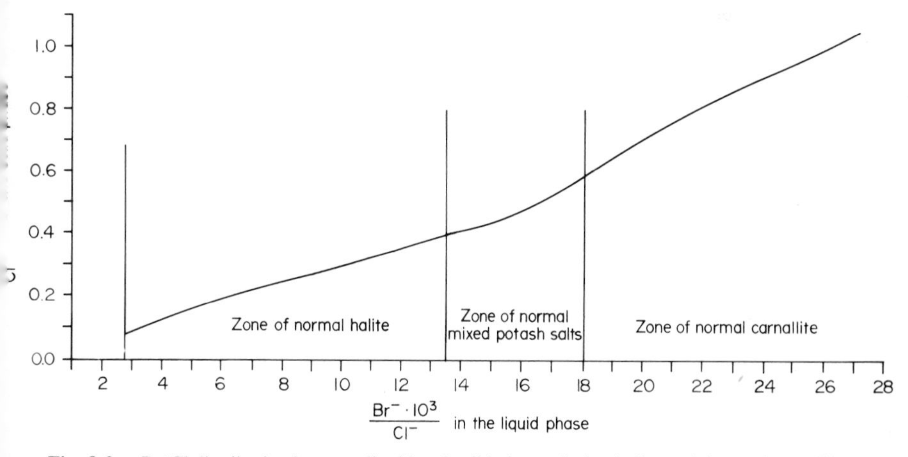

Fig. 8-2. Br/Cl distribution between liquid and solid phases during halite, sylvite, and carnallite precipitation (after Valyashko, 1962).

some apparently primary carnallites along a basin margin. The low bromide content of Permian evaporites in New Mexico is a sign that the deposit is a second-generation mineral assemblage (Combs, 1975). However, certain formation waters adjacent to ancient evaporite basins contain Br/Cl ratios far in excess of those derived from concentrating seawater (Wardlaw and Watson, 1966). This suggests an as yet unidentified source of bromine enrichment of subsurface waters, as organic bromine concentration is not applicable. Such brines would raise the Br/Cl ratio in recrystallizing chlorides. Consequently, there need not be any changes in the potassium values concurrent with increments in the Br/Cl ratio, making the ratio an inadequate tool for potash exploration (Trofimuk, 1970).

Sylvites in the Ronnenberg seam within the Leine (third) cyclothem of German Zechstein evaporites are considered to be secondary after carnallites because of the behavior of the bromide content. The bromide content gradually increases upward underneath the potash seam, but it is low in halites at carnallite–sylvite boundaries, in sylvites, or in salts above the potash seam. Furthermore, the bromide content in halites associated with carnallites differs from that in sylvinites (Reichenbach and Boehm, 1976). Similarly, the Br/K ratio in evaporites in the eastern Alps decreases in halites with the increasing presence of potassium. Whereas potassium is leached from recrystallized rock salt, bromide is here not reduced in the secondary halite (Reinold, 1965). However, the Miocene halites of the Ciscarpathian foothills contain a $10^3 \times$ Br/Cl ratio of 0.4–0.5 where primary, but a ratio of only 0.02–0.2 where recrystallized (Petrichenko *et al.*, 1974). Recrystallized beds have essentially an unchanged amount of potassium, magnesium, and sulfate, but an increased calcium content compared to primary sections.

3. All calculations of bromide content are based on the assumption that the bromide content of seawater has been constant throughout the Phanerozoic eon in all seas and oceans. Presently, the Br/Cl ratio varies slightly from open oceans to the various marginal seas. The bromide content of Upper Jurassic evaporites of central Asia, nonetheless, is 3–10 times higher than that of Lower Cretaceous evaporites in the same area, with a Br/Cl ratio of 0.103 versus 0.035 in thicker Lower Cretaceous beds, and 0.008 toward the margin of the basin (Sedletskiy and Derevyagin, 1969). The difference increases with increasing potassium chloride content. It does not indicate a different primary brine composition from epoch to epoch. Rather, the difference is probably due to a postdepositional downward translation of bromide by Cretaceous brine seepage and consequent exposure of lower evaporites to brines highly enriched in bromine.

4. Plastic deformation leads to migration of bromine out of the crystal lattice (Popov and Sadykov, 1970) and consequent bromide impoverishment.

5. Rapid cooling of a brine decreases the average bromine coefficient of distribution for the brine/precipitate (Chirkov and Shnee, 1936). It is the increas-

ing rate of crystallization that facilitates the more rapid transition of bromine to the solid phase, not the type of salt or its concentration (Chirkov and Shnee, 1939).

6. The partitioning coefficient, i.e., the equilibrium concentration in crystal/brine, depends on the pressure and temperature of the brine and on the concentration of the other ions present. With slow desorption of ions from seasonally swept-in clastics, with an episodic dilution of brine by continental and marine waters, and with ongoing brine seepage losses, the concentration of trace ions is continuously changing and must have interference effects. Superimposed daily and seasonal temperature variations complicate the variability.

A high magnesium chloride content of the brine increases the absorptive power of halite for Br^- about 1.5 times. The bromine in a brine is initially attached to highly hydrated magnesium chloride complexes (Tollert, 1956), preventing it from entering into a solid solution with halite in amounts equaling those in a solution containing no magnesium chloride. As the concentrating brine strips the magnesium chloride of its water molecules, more bromine becomes available to enter into a solid solution with halite. In contrast, dissolved magnesium sulfate has no effect on the bromine partitioning coefficient (Valyashko *et al.*, 1976a). The precipitate of potassium chloride also contains more bromide in the presence of magnesium ions than in their absence (Boeke, 1908), because potassium chloride saturation of a brine presupposes the onset of stripping of magnesium chloride complexes of some attached water molecules. In potassium chloride precipitates, there is a minimum in the mole ratios [mol Br^-/(mol Br^- + mol Cl^-)] at about 20–30 moles Mg^{2-} in the brine at various bromine concentrations. However, at low initial bromine concentrations this does not hold, and neither a minimum nor an increase with rising magnesium concentration can be observed (Bloch and Schnerb, 1953).

7. Influx of only small amounts of magnesium or cesium chloride lowers the coefficient of distribution of bromine (Chirkov and Shnee, 1936). In the German Zechstein evaporites, ratios were assumed of $Br_{(rock\ salt)}$:$Br_{(kainite)}$:$Br_{(sylvinite)}$ = 1:1.9:4.7, and $Br_{(halite)}$:$Br_{(kainite)}$:$Br_{(sylvite)}$:$Br_{(carnallite)}$ = 1:6:10:13.3, but these ratios become inaccurate if traces of magnesium chloride were present in the carnallite (d'Ans and Kuehn, 1940, 1944). However, in nature, the theoretical ratio of 1:13 between halite and carnallite is hardly ever found (Kuehn, 1968). Valyashko (1956, 1959) published a ratio for $Br_{(halite)}$:$Br_{(sylvite)}$:$Br_{(carnallite)}$:-$Br_{(bischofite)}$ = 1:5:9:13, and Zherebtsova (1970) merely reduced the bischofite value to 11.5. Braitsch and Herrmann (1963) then reduced the ratio further to $Br_{(halite)}$:$Br_{(sylvite)}$:$Br_{(carnallite)}$:$Br_{(bischofite)}$ = 1:10 ± 1:7 ± 1:9 ± 1.

In the Permian Upper Kama deposits west of the Urals, Myagkov and Burmistrov (1964) determined a ratio of $Br_{(halite)}$:$Br_{(carnallite)}$ of 1:5–10 and $Br_{(sylvite)}$:$Br_{(carnallite)}$ of 1:2–3. Kolosov and Pustyl'nikov (1970b) even reported a $Br_{(NaCl)}$:$Br_{(sylvite)}$ ratio of 1:17 in Cambrian salts of Siberia. The German

Zechstein evaporites show a ratio of $Br_{(halite)}:Br_{(carnallite)} = 1:1.2$ to $1:4.4$ with the concurrent appearance of secondary polyhalite, kieserite, and kainite (Loeffler, 1960b).

8. The solution of a carnallite bed produces a brine of high bromine concentration. Halite recrystallizing in equilibrium with such a brine then has a high bromide content (Baar, 1963). Experiments have shown that precipitated carnallite equilibrates with new brines rather rapidly, sylvite equilibrates with them more slowly, and halite equilibrates extremely slowly. It becomes increasingly difficult for bromide ions to enter a crystal through surface zones that have already achieved a bromide equilibrium. Neither stirring nor diminution of the crystal size of halite seems to have an appreciable effect on the rate of exchange (Kuehn, 1968).

9. The bromide content also varies with decreasing concentration of the brine. A significant reduction in the bromide content is observed immediately underneath anhydrite intercalations (Kuehn, 1954), as a temporary freshening of the brine then leaches some of the precipitated bromides. Dilution of the brine by fresh water reduces the bromide content sharply. However, windblown halite is likewise deficient in bromide (Holland and Christie, 1909). If a primary carnallite containing admixed halite is obtained by evaporation of Dead Sea water and is then decomposed with fresh water, a precipitate of sylvite and halite is formed. Evaporating the remaining solution leads to precipitation of a secondary carnallite and admixed halite, but with a more than halved Br/Cl ratio. Even a tertiary carnallite can thus be produced with a Br/Cl ratio that is reduced further (Bloch and Schnerb, 1953). Influx of less dense waters into a lagoon sets up a density interface. As the upper layer of brine concentrates to halite saturation or develops a mixing zone, halite crystals fall through the interface (Raup, 1970). They are initially not in equilibrium with the bromine content of the lower, more concentrated brine. Similarly, the influx of groundwater under artesian pressure can decrease the Br/Cl ratio, since groundwater has a Br/Cl ratio of <1.

Sylvinites can also be stripped of their bromide. The Lower Permian sylvinites near the Urals are characterized by lower Br/Cl values than those either calculated or experimentally determined. However, the variegated sylvinites, the milky white sylvites, and the red laminated sylvites contain more bromide than the halite (Shleimovich, 1976); this is evidently the effect of redistribution by circulating waters. A bromide enrichment found near halite inclusions is apparently due to bromide diffusion from the halite crystal lattice into that of sylvite (Apollonov *et al.*, 1977). The Upper Devonian sylvinites in the Pripyat Basin have even lower Br/Cl ratios. In both cases, the Br/Cl ratio is normal in the accompanying halites (Kislik *et al.*, 1970). Furthermore, this process also explains the observation (Boeke, 1908) that bromine concentrates more in the center of the basin than on its margins, where rainwash has a major effect.

10. Patterson and Kinsman (1977) noticed that the zone of mixed continental

and marine brines moves seaward at a rate of 0.3–1.0 m/s along the Trucial coast. Continental waters (Br/Cl $<10^{-7}$) are thus overriding denser marine waters (Br/Cl $>10^{-6}$) via a mixing zone. A vertical profile would give a progressive reduction upward in the Br/Cl ratio in such sebkha evaporites.

11. Both bromide content and liquid inclusions have been used for paleotemperature determinations, with inconclusive results (Dreyer *et al.*, 1949; Dellwig, 1955; Wardlaw and Hartzell, 1963; Braitsch and Herrmann, 1964a,b). If we could be certain that the $MgCl_2/K_2Cl_2$ and $MgCl_2/K_2Br_2$ ratios have not been altered by circulating brines, then the bromide distribution could give us an indication of the temperature of evaporite formation. Herrman (1976,1977) assumed such absence of alteration in the Oligocene evaporites of the Upper Rhine Valley and determined that the brine temperature increased from 20°C in the first bed to 50° ± 10°C in the third bed, with a horizontal temperature gradient of 10°C toward the lagoon center.

While recrystallization alters liquid inclusions (Peach, 1949), bromides frequently migrate into percolating brines during such recrystallization processes (Holser *et al.*, 1972). High bromide values may indicate elevated temperatures of formation, or may simply represent an enrichment due to the conversion of carnallite to sylvite (Kolosov and Pustyl'nikov, 1970b). However, Bloch and Schnerb (1953) found practically no change in Br/Cl ratios with temperature in either the brine or the precipitate. Only the position of the invariant points changed for sylvite–carnallite and carnallite–bischofite. In any case, such temperature determinations are bound to be of little value, since they represent at best only one point on a daily or seasonal temperature curve.

12. Bromine is taken up by some algae (Levy, 1977), and soil bacteria can also achieve bromide depletion (M. R. Bloch in Holser *et al.*, 1972); kerogens fix bromine in direct proportion to their carbon content (Lange, 1970). Furthermore, sorption by clays depletes the bromine content of the brine (Osichkina, 1978c).

Fluid inclusions in halites precipitating in mangrove swamps of New Caledonia are enriched in bromine and contain almost double the amount normal for halites. Sabouraud-Rosset (1974) thought that the abundant organic matter in the swamps might in some way be responsible for this increase. There is a residual enrichment of brominated organic matter during the decay of biota; the Br/C_{org} ratio then correlates with the $\delta^{13}C$ values and rises with the increasing presence of macroalgae (Mayer *et al.*, 1981). As organic matter absorbs bromine, the Br/Cl ratio drops (Rozen, 1970). Thus, large variations in Br/Cl ratios in samples with the same particle size are a function of variations in the content of clay and organic matter (Kostenko, 1973). This is particularly true in the environs of reefs, where sundry algae absorb bromine from surface waters but release it later in the hypersaline brine during the decay of dead organic matter.

13. The clay content of the evaporite deposit also plays an important role

(Kostenko, 1973). In Jurassic salts of the Gissar range in central Asia, both halites and sylvites contain two to four times more bromide where the insoluble residue exceeds 10% (Popov and Osichkina, 1973) than where no clay inclusions exist. The same is true of the Devonian potash horizons of Byelorussia, where the primary carnallite contains a considerable amount of insoluble residue. The increased bromide content (<2030 ppm) has a linear relationship to the clay content (Pavlyuchenko *et al.*, 1961b). The bromide content of the Permian Upper Kama evaporites west of the Urals is likewise entirely controlled by the lithologic composition and the stratigraphic level (Mokhnach, 1978).

14. The bromide content varies laterally (Baar, 1952; Ogienko, 1959; Schulze, 1960b; Raup *et al.*, 1970) in various basins and decreases toward the basin margin (Schulze, 1958; Braitsch and Herrmann, 1964a,b; Braitsch, 1967; Sedletskiy and Derevyagin, 1969; Kunasz, 1970; Hemmann, 1972). It is higher in shallow subsidiary basins than in the main basin (Braitsch, 1967), because it rises with the increasing depth of the overlying brine (Kuehn, 1953). After attaining a maximum, it may decline again in the deepest parts of the main basin (Ogienko, 1959). Lateral variations can be substantial even over short distances. Bromide values vary widely within any one bed of the Stassfurt Member in the second cyclothem of German Zechstein evaporites, when followed through the workings of a single mine. Simon (1972) therefore assumed the presence of two populations of potash crystals. A lateral doubling of the bromide content in the lower salt of the Khorat evaporites of Thailand between two bore holes could thus indicate a salinity gradient at the time of deposition (Hite and Japakasetr, 1979), a shoreward location of one of the bore holes, or both.

There are reasons for a bromide variation other than the distance from the basin margin. The dark infill of polygons in rock salt contains more bromide than the well-bedded halite deposits around them (Richter-Bernberg, 1980). Neither the depth of the water nor the distance from shore is materially different there; the concentration of the brine had probably changed. Triassic salts in Newfoundland contain less bromide than coeval and originally probably contiguous evaporites in France (Jansa *et al.*, 1980). There are likewise great lateral differences in absolute bromide content in beds of equal age and stratigraphic position in the central part of the Zechstein Basin, as Haltendorf and Hofrichter (1972) found in the third cyclothem in the Hannover area. As yet, there are not enough bromide determinations across any basin to explain satisfactorily such observed lateral variations.

15. In the evaporation of Dead Sea water, halite starts precipitating at a Br/Cl ratio of $0.5–0.3 \times 10^{-3}$. As soon as carnallite starts precipitating, the Br/Cl ratio increases sharply to about 10^{-3}. A further sharp increase is noted when bischofite starts precipitating ($20–25 \times 10^{-3}$) (Bloch and Schnerb, 1953). Such sharp discontinuities in the Br/Cl distribution curve have not hitherto been considered in evaluating the bromide content of ancient potash deposits. At the onset

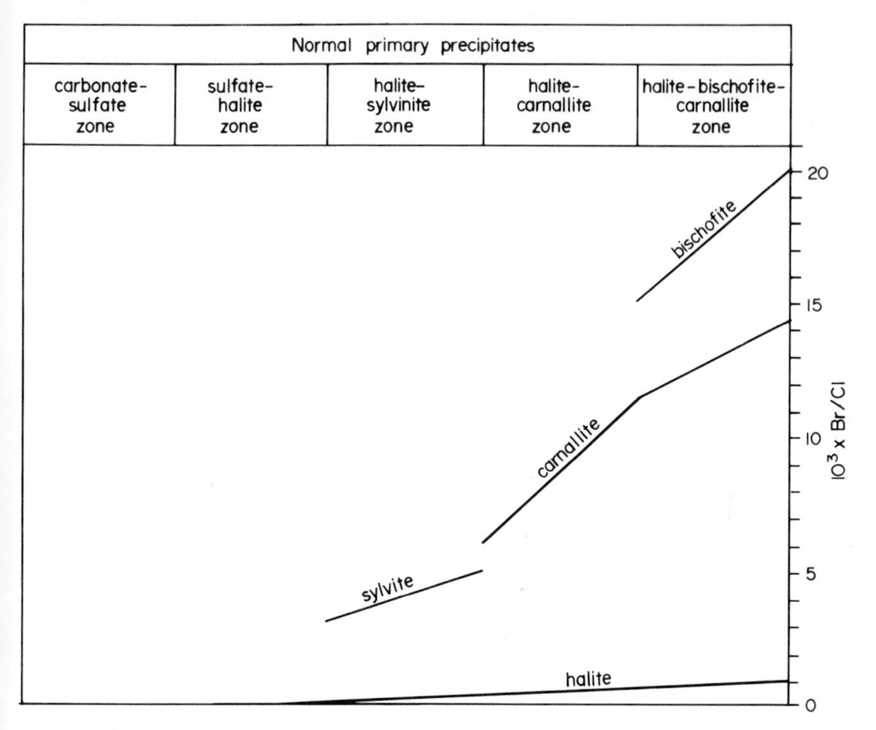

Fig. 8-3. A $10^3 \times$ Br/Cl ratio of primary evaporites (after Valyashko, 1962).

of precipitation of the next more soluble salt, the Br/Cl ratio suddenly rises as a result of even a slight admixture of such a salt of higher solubility (Fig. 8-3).

On the whole, one cannot take raw Br/Cl values and hope to shed light on the salinity and temperature of the original brine, the rate of precipitation, or the distance from shore. A thorough understanding of the lithostratigraphic relationships within the basin, and the consequent hydrogeological history, is a prerequisite to such an undertaking.

Bromide Distribution Curves

In a salt sequence produced from a finite amount of concentrating brines, the bromide distribution curve would show a gradual increase in the bromide content of halite (Fig. 8-4) and a more rapid increase, once the potash salts began to precipitate. Such a simple curve is almost never found in nature.

At best, one finds a curve that generally zigzags to higher values, suggesting repeated dilution of an overall concentrating brine (Fig. 8-5). The dilutions are

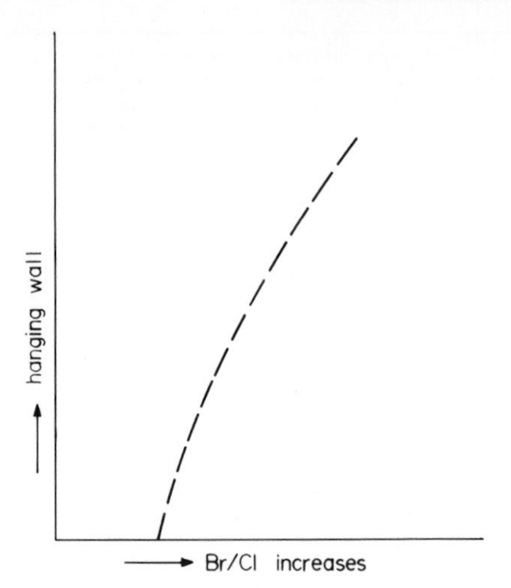

Fig. 8-4. Normal bromide distribution curve.

never large enough to reverse the trend. Kendall (in Kuehn and Hsu, 1974) has proposed an ingenious explanation for such zigzag curves: Dense, warm brines concentrating on evaporative flats along the basin periphery slide into the deeper basin already saturated with respect to halite. The descending brine would either tumble to the bottom or would extend as a submarine delta along a surface of equal density. When the warm brine cools to surrounding temperatures, halite precipitates with a high bromide content. As the brine flow gradually mixes with

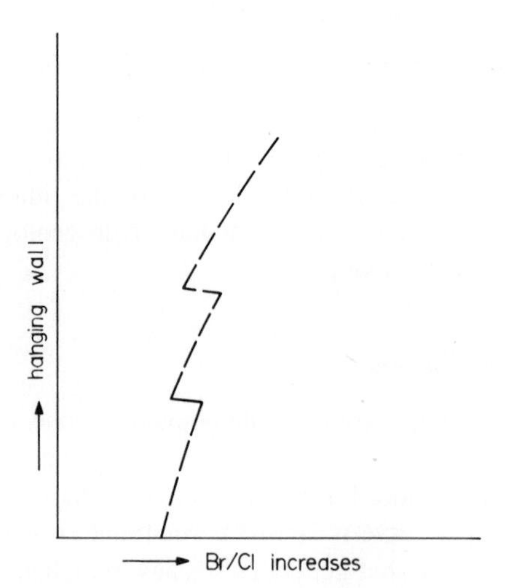

Fig. 8-5. Bromide curve with dilutions.

the basin brine, its bromine content decreases until it reaches the background level of the brine basin. Another amount of hot brine cascading down the shelf slope repeats the sequence. The broader the shelf areas, the larger is the volume of concentrated brine that is periodically dumped into the deeper halite-precipitating depressions and the more frequent are such incursions. This can produce very erratic bromide profiles.

Often, a marked increase in bromide is found at the base of the salt sequence (Fig. 8-6). All halites that are part of a second or higher cyclothem of evaporite deposition show such a higher bromide content in their basal part. This is explicable only by the leaching of bromide from the underlying salt (Hite and Japakasetr, 1979) during the deposition of this basal portion. In addition to reworking salts precipitated in an earlier cyclothem that went to desiccation, the brine could have picked up the residual brine of that cyclothem in a central depression and spread out the bromides dissolved therein. An example of such a sequence is provided by the Miocene salts of Wieliczka (Poland), where the bromide content oscillates between 20 and 221 ppm (Garlicki and Wieworka, 1981). Schulze (1960b) provided an alternative interpretation: The waters preconcentrate on shelves and then move into the deeper parts and enrich precipitating salts. Higher salt levels are no longer subject to such lateral translation and produce a more normal bromide distribution.

Many salts are grossly deficient in bromide (Fig. 8-7), often so much so that the bromide content approaches values one to two orders of magnitude smaller than those calculated from an ideal curve. This is taken to mean that the salts recrystallized in the presence of meteoric waters. The phreatic waters were quickly saturated for sodium chloride and thus did not continue to dissolve sodium chlorides. They only leached their bromide and carried it away. This is

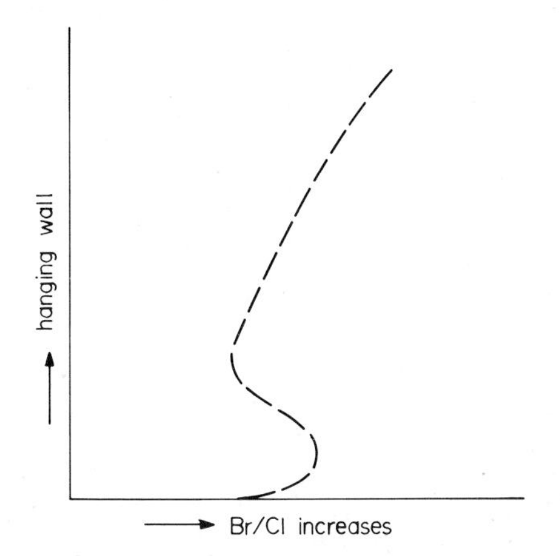

Fig. 8-6. Bromide curve that increases at the base.

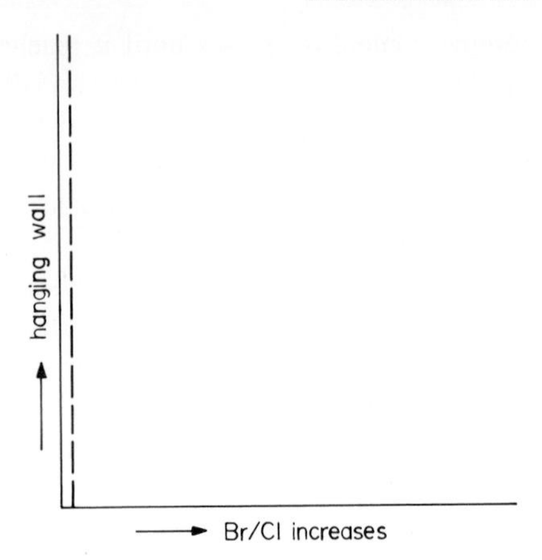

Fig. 8-7. Bromide curve with a uniformly low value.

the cause of the common enrichment of ancint formation waters in bromine. For instance, in the potash deposit of Tyubegatan in Turkestan, the Br/Cl ratio is nearly constant and uniformly small, regardless of the NaCl/KCl ratio (Nabiev and Osichkina, 1965). Jurassic halites of central Asia have a bromide content of 0–110 ppm, with a Br/Cl ratio of $0.018–0.299 \times 10^{-3}$ wherever any bromide is present (Popov, 1977). Similarly, the bromide values of evaporites in Catalonia, Spain, are much too low and are vested mainly in the potash minerals (Pueyo Mur, 1977). The late Lower Permian Flowerpot and Blaine salts of Kansas are coarsely crystalline, embedded in a matrix of silty mudstone (Holdoway, 1978). The low bromide content (<5 ppm) suggests recrystallization in the presence of undersaturated waters; the coarse crystallinity indicates that the major portion of the salt has been dissolved away. The crystal size indicates very slow growth in a stable environment; the texture suggests displacive recrystallization.

A bromide curve that first shows an increase and then stabilization at some higher value (Fig. 8-8) indicates that inflow and evaporation plus seepage losses had reached an equilibrium, and that the brine was not concentrating any further. Such a curve has so far not been found, but then not every salt section has been analyzed for its bromide content. However, if found, it would be an unusual case. It would mean that the forces driving the basin to concentrate its brine have been completely stalemated.

If percolating brines are enriched in bromine, they will increase or homogenize the bromide content of precipitates. The bromide content increases in proportion to the length of the path of ascending brines soaking through an evaporite deposit (Koch *et al.*, 1968), but depicts only the equilibrium of precipitates with the last brine circulation with which they have been in contact. A uniformly high bromide content throughout the section (Fig. 8-9) indicates that the bromide was

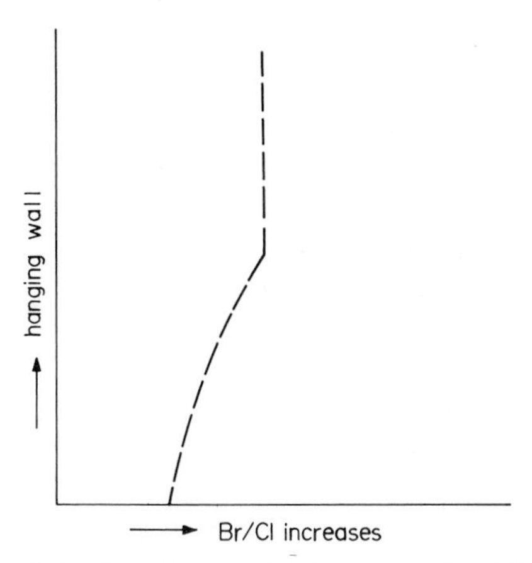

Fig. 8-8. Bromide curve that increases and stabilizes.

homogenized after the precipitation of the salts by percolating younger brines. The Lower Cambrian evaporites of the Irkutsk amphitheater in southern Siberia are strongly enriched in bromide (Popov and Osichkina, 1973), averaging 0.6–0.65% bromide in sylvites (Kolosov and Pustyl'nikov, 1972). In Lower Cretaceous evaporite deposits of central Asia, halite occurs only with sylvite, but with a high admixture of clastics. Only the underlying Upper Jurassic evaporites also contain carnallite. Sulfate-bearing solutions did not percolate through the deposit, as sulfates are lacking. The bromide content is constant both vertically and horizontally in the Jurassic evaporites (Sedletskiy and Derevyagin, 1971). Since it should vary both with distance from shore and with changes in concentration, as it does in the overlying Cretaceous evaporites, the bromide content

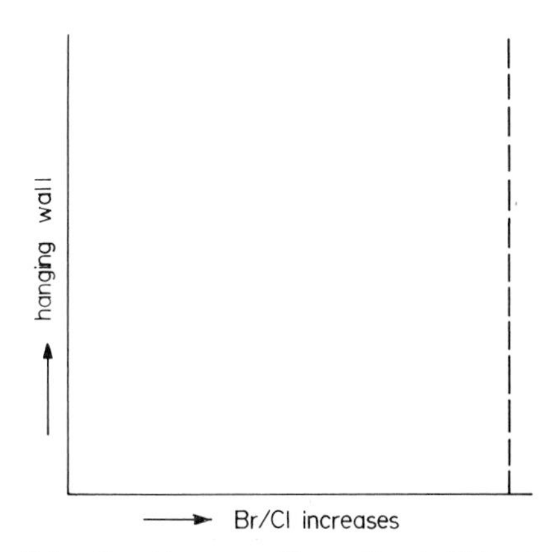

Fig. 8-9. Bromide curve with a uniformly high value.

of these deposits must have been homogenized by percolating chloride solutions. If residual bitterns with a high bromide content soak into underlying precipitates in the terminal stages of evaporite deposition, they will homogenize the bromide content of all contacted crystals, even if the bittern was first slightly diluted by rainwash.

Decima (1976) reports a bromide content of about 400 ppm in salts immediately above a basal anhydrite of Miocene evaporites in Sicily (Fig. 8-10). This value remains constant for all of the lower salt member. The middle salt member, which contains kainite and polyhalite beds of commercial-grade ore, is again nearly uniform in bromide content, but only at about 60% of the value attained by the basal member. Since 400 ppm bromide represents the bromide content in salts nearly saturated for potassium chlorides, it is unlikely that such a concentration was achieved instantly in the first halite crystallizing after gypsum precipitation had ceased.

The reduction of the bromide content to almost nil in the upper salt, the impoverishment of the potash horizon, and the bromide enrichment of the lower salt in this case point to a downward translation of bromide by percolating waters. The lowermost salt member could not have derived its high bromide content directly from a freshwater contribution. The presence in this deposit of potash minerals almost solely in the form of sulfates supports the contention that the reduced bromide content in the middle salt member is due to sulfate-bearing solutions affecting the unit. In this case, the solutions could well represent a Pliocene percolation of overlying seawater into an as yet unindurated and uncompacted salt sequence. The $\delta^{18}O$ of the water of crystallization is consistently

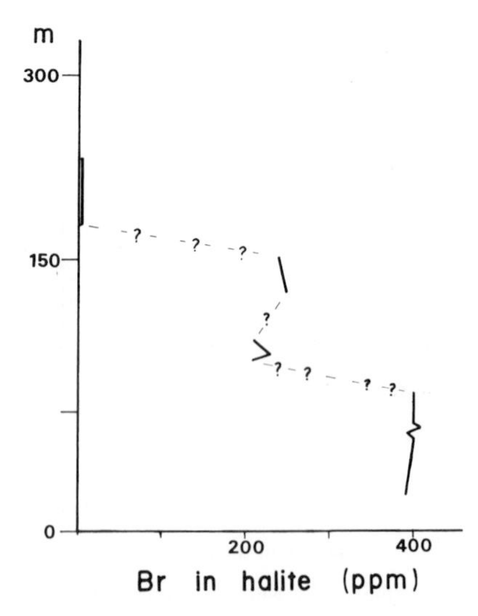

Fig. 8-10. Bromide distribution in a Sicilian potash mine (after Decima, 1976).

more negative than in modern salinas (Longinelli *et al.*, 1978). This suggests a large contribution of fresh water in the lower part. This interpretation is texturally corroborated by the occurrence of chaotic gypsum brecciation. The stable isotopes of ^{18}O, ^{13}C, and ^{2}H in carbonate and sulfate laminae in the basal as well as the upper units also suggest strong influences of continental waters (Pierre and Catalano, 1976). The upper nonselenitic gypsum appears to be resedimented gypsum that dissolved in a basin filled with meteoric waters over a long time. Biological metabolism caused the striking enrichment in light sulfur (Longinelli *et al.*, 1978). The same conclusion was reached after an investigation of the clay sediments in the basal portion (Coradossi and Corazza, 1976).

Bromine can be liberated in dehydration and recrystallization by ascending brines (Borchert and Muir, 1964; Holser *et al.*, 1972; Baar, 1974). Often the basal salt cyclothems are depleted of bromide. Ivanov and Voronova (1968) concluded that the salts of the Permian Upper Pechora Basin are not resedimented but primary, despite their very low bromide content. The Permian Salado Formation of New Mexico is also an evaporite deposit low or depleted of its bromide content (Lindberg, 1946; Adams, 1969). However, it is not necessarily a recycled salt deposit, merely a leached one. There is ample evidence of brines entering both from above and from below (Adams, 1969). Since there is a pronounced antipathy of bromides to sulfates, this leads to an impoverishment of bromide toward sulfate beds.

An interesting situation pertains to the two oldest cyclothems of evaporite deposition in the Devonian Elk Point Basin of Alberta, Canada. A basal sequence of red beds is overlain by the Lotsberg and Cold Lake halites separated from each other and from the overlying Prairie Evaporite Formation by dolomitic and marly shales and marly dolomites and limestones. Red halite, exhibiting a fibrous habit typical of evaporite minerals in fracture fillings, is found in the basal red beds, within the Lotsberg salt, and within the shales capping the Cold Lake salt. The younger Prairie Evaporite and the halite fracture fillings in the shales between Prairie Evaporite and Cold Lake salt have a much higher bromide content than the fracture fillings and the salt beds farther down. Halite in the lower two salts, as crystals in the shale intercalations between the lower two salts, or in the basal red beds has an extraordinarily low bromide content. This can be explained by recrystallization in percolating fresh water (Wardlaw and Watson, 1966), but the higher bromide content of the fracture fillings indicates that more concentrated brines were able to move through at a later date along distinct paths. Prior to the Prairie Evaporite transgression, the shales capping the lower two salts probably formed an inland depression. Two scenarios are possible:

1. Meteoric waters washed into the lower part of an endorheic depression (i.e., one without outward drainage) from surrounding slopes with less than a 1° angle. They soaked through the salt and the shale intercalations, quickly saturat-

ing for sodium chloride but not for sodium bromide and then moving out mainly laterally while continuing to leach bromides.

2. Groundwater, which was saturated for sodium chloride, entered laterally, leached bromides, and then discharged into the depression as artesian springs. Eventually, these discharges must drain into the sea or evaporate. Groundwater discharge produces salt crusts in many semiarid soils. Later, transgressive seas would have dissolved any accumulated residual crusts. The basal Prairie Evaporite would then show an abnormally high bromide content. This is not the case. In the former scenario, the exiting brines would enrich formation waters; this has been documented (Wardlaw and Watson, 1966).

Either way, the assumption of water movements through the salts and through enveloping shales demands the presence of a primary permeability and an initial ability to react with percolating waters. This precludes a timing of the movements after the evaporites have lost their permeability to water by compaction. It is also important when one considers the possibility of flushing out organic compounds originally trapped in evaporite sequences or of later introducing carbonated or sulfate-bearing waters to alter chlorides into secondary sulfates and carbonates.

The Lower Permian Rotliegend evaporites beneath the Zechstein horizons of northwestern Europe are also very deficient in bromide (Haslam *et al.*, 1950) and appear to be devoid of sulfates and carbonates. Brunstrom and Warmsley (1969), Glennie (1972), and Holser (1979b) therefore suggested a continental origin, possibly derived from the dissolution of Devonian evaporites. Within only 1000–2000 km of the Lower Permian Rotliegend evaporite there are Devonian evaporites (Holser, 1979a). The nearest evaporites to the Elk Point Basin of Alberta (with a similar bromide deficiency) are the Silurian ones in the Michigan Basin, also 1500–2000 km away. Surface transport of salt over such distances is not very likely in view of the continuous ion exchange of rivers with groundwater and riverbed linings.

It is rather unusual that both the Prairie and Zechstein evaporite sequences should be underlain by thick halites of continental provenance. There is no other corroborating evidence to prove that such major evaporite basins start out as depositories of large volumes of continental redissolved halite. Although calcium salts (sulfates and carbonates) are indicative of a marine origin, sodium salts are usually present in a continental regime. However, the sodium sulfates and carbonates are also absent. If ascending formation waters were squeezed through these salt sequences under artesian pressure, one would expect the basal salts to be depleted of bromide, whereas higher stratigraphic units would remain unaffected. Moreover, such waters initially undersaturated in calcium sulfate would pick up some of this solute in passage through the basal anhydrite, whether it was still gypsum at that time or already dehydrated.

9

Behavior of Clastics in a Hypersaline Brine

The volume of terrigenous clastics is very small in evaporite basins. One must bear in mind that density-stratified waters in evaporitic basins are generally of low energy, i.e., they cannot transport coarser clastics. Their rate of clastic sedimentation is therefore reduced by a factor of nine or more, compared to basins with homogenized waters (Busson, 1978). If it were not for massive chemical precipitation, such depressions would be extremely starved basins. Because of the bottom outflow from basins with a deficit in their water budget, no marine clastics are swept in. All sands and clays are washed in by episodic flash floods from surrounding lands. The hypersaline brines dissolve much of the silica and alter compound aluminum silicates into magnesium-rich varieties of mixed-layer clays. Consequently, there are only thin intercalations of clays or of gypsiferous sandstones that mark brief interludes in the regime of chemical precipitation.

SANDS AND SILTS

There is a decided fining of clastics toward any evaporite basin (Hunt *et al.*, 1966), so that onshore sediments are frequently only red clays. The preponderance of clays and silt-sized particles and the near absence of coarser clastics inside the halide facies of evaporites suggest delivery by waters moving slowly over gentle gradients. If the aggregate water deficit of the brine is large, the influx has a high velocity and thus a substantial carrying capacity for clastics. Any discharge of the few rivers emptying into an evaporite basin is deflected and dragged along by such a forceful current. Unlike a clay slurry, sand or gravel carried by the discharge is too dense to override a hypersaline brine. It is dumped at the river mouth. Only the fines can be washed along the coast and move along the shallow shelves. Statistically, it should be possible to discern the heavy mineral assemblage in such a contribution of fines to the bedload of any longshore current. Gypsiferous sandstones are common along the margins of evaporite basins, and gypsum is the only evaporitic rock sequence that may have a significant admixture of terrigenous coarse clastics. Usually, these gypsiferous sandstones are fairly coarse-grained or conglomeratic. Messinian gypsiferous sandstones along the northern margin of Upper Miocene Mediterranean evaporites suggest initial current velocities of 2.2–2.4 m/s, decreasing westward to 0.5 m/s (Sonnenfeld *et al.*, 1983).

Deeper, halite-precipitating depressions are virtually free of silt, sand, or pebbles. The energy levels of brine currents during halite or carnallite precipitation are not high enough to transport coarse clastics. Sandstorms and dust storms occur in all areas of discontinuous vegetation cover and must have affected every evaporite basin. Perthuisot (1975) observed that in exposed crusts, all crystals facing the wind are yellow with dust. Gypsiferous sandstones and siltstones are evidence of pH values remaining only slightly alkaline during gypsum precipitation. However, sandy halites and sandy carnallites are unknown in salt basins; a few very thin sandstone intercalations are invariably associated with a freshening of the total brine column. This can be explained only by the dissolution of blown-in quartz grains in brines of high pH. In other words, the pH of natural brines must rise during halite precipitation in order to solubilize silica. Halite-bearing sandstones are rare and almost always require two generations of deposition, such as small brine puddles dried out on a sandy beach or sand bar, or deltaic fans encroached on by brines. Sandstone intercalations in halite represent ephemeral sharp reductions in salinity, are fine-grained, coincide with corrosion or dissolution surfaces, and are almost always sheathed in anhydrite grains. Only a few such distinct sandstone beds are known from rock salt sequences; no silty or sandy sylvinite sequence has been recorded. The explanation is relatively simple: In an oxygenated, acidic environment, both quartz and aluminum silicates are stable; in an anaerobic, alkaline environment, they are not.

Fragments of quartz sandstone are interbedded in salt in Colombia; the sandstone clasts, are, however, embedded in a matrix of black clay (McLaughlin, 1972). A detailed study of the Permian Salado Formation in New Mexico uncovered individual bands of halitic fine-grained sandstone close to several corrosion surfaces, often associated with clayey halite (Jones, 1973). The major sandstone intercalation, the 3-m-thick Vaca Triste Member, is sheathed in anhydrite. Its bromide content is extremely low (Cheeseman, 1978), suggesting a sudden reworking in a very diluted brine. The sandy members occur at the end of cyclic sedimentation of detrital, anhydritic, and halite layers. The overlying

Rustler Formation has in its basal part a halitic sandstone overlain by a salt-free sandstone capped by anhydrite (Powers *et al.*, 1978).

In the Donbass deposit in the Ukraine, there is a gradual facies change from halite in the basin center to clayey halite, to halite with progressively coarser siltstone, and eventually to sandstone near the basin margin. Even a fine-grained conglomerate with a gypsum cement was encountered in one locality. Both halites and anhydrites are strongly recrystallized (Bobrov, 1973).

Clay and sand intercalations in the Solotvin evaporite deposit are enriched in carbonates, anhydrite, and barite. Both the carbonate and halite cements of the quartz grains are usually mixed with gypsum (Korenevskiy, 1960).

Toward the top of the fourth halite cyclothem of Miocene evaporites in Poland, there occurs on the north rim a thin intercalation of very fine-grained quartzose sandstone (Garlicki, 1979) mixed with anhydrite grains and carrying macerated plant material. Brines infiltrating the substrate from a Miocene evaporite basin soaked through the Carpathian flysch and precipitated halite into the void spaces (Krayushkin and Osadchiy, 1965).

In the Guadelupan evaporite sequence of the Permian Delaware Basin of West Texas, sandstones were also cemented by halite (Waldschmidt, 1958; Jacka and Franco, 1974). This happened before they acquired overgrowths such as those commonly found in overlying sands cemented by dolomite or anhydrite. The sequence here represents a sebkha environment with small brine pans (Jacka and Franco, 1974), a precursor to the Ochoan massive halite and potash sequences in the same basin. An alternative interpretation deriving the anhydrite and halite cements of these sandstones from groundwater movements would be analogous to the modern halite-cemented sandstones of South African salt vleys (Grabau, 1920). Halite cements of gravels and sands have been found in Namibia, Chile, the Gobi Desert, Egypt, and south of the Urals (Lotze, 1957).

Siltstone is even rarer than sandstone. A case of silty halite documented from a marine setting occurs in the final stage of the last evaporite cyclothem in the Permian Zechstein of Yorkshire, where the halite contains some detrital quartz silt, talc, and chlorite (Smith, 1974). The halite nucleated and grew slowly within the sediment, in many respects akin to groundwater precipitation. The largest halite crystals are found where the halite/clastics ratio is high. Accompanying clays were reworked by repeated solution and reprecipitation due to inundations and retreats of the brine from the distal part of a mature basin-margin plain. The oscillations in brine level were caused by periodic large-scale expansion and contraction of an extensive basin-center playa (Smith, 1971). The halite on top of the carnallite has been reworked, i.e., the silt is a secondary admixture (Smith, 1974). A nonmarine example is the Pliocene lacustrine salt deposits in Nevada (Mannion, 1963), which are derived from the leaching of older salt occurrences. Dominance of silt and fine sand rather than clay in the insoluble residue of this salt suggests strong periodic flooding from swollen rivers.

Ash falls are not uncommon in the vicinity of ancient evaporite basins. Tuffs and tuffaceous coarse and fine clastics are reported from Cambrian evaporites in Siberia (Sokolov, 1969). Similarly, dacitic tuff beds intercalate with Miocene anhydrites at the base of halites in southern Poland. Halite has saturated dark gray tuffs, but is present only sparingly as cement in white tuff (Kamienski and Glinska, 1965). Two sandstone intercalations in the upper part of the oldest salt unit at Wieliczka in Poland are composed of quartz sand and andesitic tuff, which are cemented by anhydrite and epigenetic halite (Pawlikowski, 1978). The quartz is considerably recrystallized, and the tuff devitrified. Both features are a measure of halmyrolysis, or solution by brine. In the Azgir Upland northwest of the Caspian Sea, halites are intercalated with periodic rhyolitic tuffites, up to 2 m thick, that contain hillebrandite [$CaSiO_3 \cdot Ca(OH)_2$], barite, searlesite, unaltered hornblendes, and micas (Lobanova, 1960). Alkaline hypersaline lakes alter the more basic tephra to zeolites and eventually to feldspar (Surdam and Parker, 1972).

SILICIFICATION

Silicification or chertification of rocks situated downdip from some evaporite deposits is frequent (Hite, 1970). An intermediate stage in the conversion of quartz sand to colloidal chert may be cristobalite. Deltaic sandstones in the Neogene Tajo Basin of Spain correlate with layers composed of 60% cristobalite intercalated in thenardite–mirabilite–halite beds (Ordonez *et al.*, 1976). The presence of sulfates promotes the formation of fibrous varieties of chalcedony (Arbey, 1980).

Whenever the access to atmospheric carbon dioxide becomes limited in a stratified brine, silicate hydrolysis proceeds more rapidly than carbon dioxide uptake; for this reason, the pH also rises. Silicate hydrolysis proceeds even more readily when oxidation of hydrogen sulfide produces sulfate ions and ultimately causes gypsum precipitation (Jones, 1966). Silica solution in seawater causes a slight drop in both pH and magnesium content, according to the reaction (Hirano and Oki, 1978)

$$2Mg^{2+} + H_4SiO_4 = Mg_2SiO_4^0\,(aq) + 4H^+ \qquad (9\text{-}1)$$

However, as the silica saturates the brine, the pH rises (Helgeson, 1968; Helgeson *et al.*, 1969). Silicates are combinations of weak acids with strong bases. Hydrolysis reactions of lithic fragments are thus alkaline, because they consume hydrogen ions. If water percolates through a vertical tube filled with granite powder, a pH of 9.6 can be measured in the collected solution (Millot, 1970). Springs issuing from sites where ultramafic rocks are altered to serpentine frequently give a pH reading approaching and even exceeding 12 (Pfeifer, 1977;

Feth, 1981). Decomposition of some terrigenous dust in brines therefore increases the pH values.

At a pH of 9+, silica converts to ionized orthosilicic acid (H_2SiO_4), which becomes a true solution rather than a colloidal gel. The solubility of silica becomes enhanced (Fig. 9-1) because of the formation of monomeric and multimeric or polymeric silicates by the progressive reaction of multiple silica molecules with several hydroxide radicals (Stumm and Morgan, 1970). As long as the activity of orthosilicic acid remains smaller than its solubility product, SiO_2 can continue to dissolve. However, the addition of small quantities of magnesium or aluminum ions to the brine drastically reduces the solubility of silica (Weaver and Pollard, 1973). The presence of organic matter exerts a profound influence on silica solubility. Basic amines released during the breakdown of proteins aid the solubilization of silica and the dissolution of silicates. Organic materials stabilize silica hydrosols by forming protective colloids. Carbon dioxide production later destroys the organic matter, lowers the pH, and leads to precipitation of silica (W. E. Evans, 1964; Zajic, 1969).

Seawater is grossly undersaturated in silica, and its inflow into an evaporite basin dilutes the silica concentration in the resident brine. Groundwater discharge from crystalline terrain, however, is frequently enriched in silica, containing more than 1% silica by weight. The impact of this groundwater-dissolved silica is thus likely to be more pronounced in the bottom brines of the distal parts of a basin where the effect of fresh influent is weakest. Moreover, the groundwater discharge surges after each major rainfall, and the delivery of solute is thus

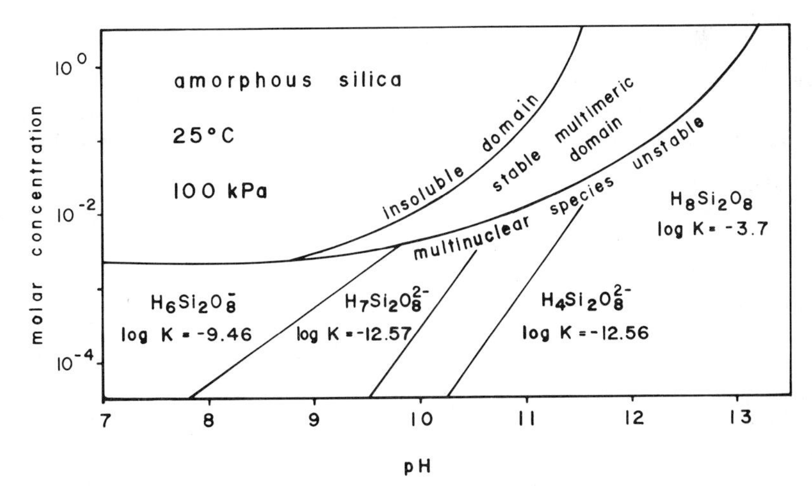

Fig. 9-1. Solubility of hydrated silica species in equilibrium with amorphous silica. K = equilibrium constant (modified after Stumm and Morgan, 1970, and others).

at a maximum during the rainy season, when density stratification is best developed over the whole basin and both stenoionic and stenohaline biota that utilize some silica can find a brief period of bloom.

SECONDARY QUARTZ

Secondary quartz is produced in mangrove swamps under the influence of organic matter (Avias, 1949). Both colloidal and dissolved silica are drained with the bitterns into permeable rock underneath. This silica precipitates later in voids as authigenic quartz and feldspar crystals. The secondary quartz crystals are probably derived from cristobalite and porcellanite that recrystallized at prevailing temperatures after burial (Yarzhemskiy, 1949a; Ernst and Calvert, 1969; Mizutani, 1977). A near-total absence of any opaline silica confirms this process. Pelikan (1900) reported an instance of pseudomorphs of precious opal after gypsum, but silica replacement of evaporite minerals is otherwise rare in Phanerozoic sequences. At least part of the opal and quartz in Miocene evaporites of Spain represents syngenetic silicification; the quartz crystals either replace gypsum crystals or replace calcite after gypsum (Bustillo Revuelta, 1976).

Epigenetic quartz, often in a rosette-like arrangement and with clay inclusions, has been reported from the Permian salt of the Donets Basin and the Miocene salt of the Carpathian region (Yarzhemskiy, 1949a), from the Mississippian halite and potash deposits of Nova Scotia (Bancroft, 1957; R. Evans, 1970), the Triassic salt in the Atlas Mountains (Kulke, 1978), and the Permian Zechstein evaporites of Germany (Braitsch, 1958). Euhedral crystals and rosettes of quartz in the Silurian Salina salt of Michigan are associated with anhydrite–dolomite laminae (Dellwig, 1955), and thus with brine freshening. Quartz concretions in gypsum and anhydrite can be considered syngenetic (Brownell, 1942) only if one includes the time span until the end of the particular evaporitic cyclothem. Even then, the recrystallization from a silica gel to a quartz crystal probably occurred decidedly later.

Some salts are decidedly enriched in authigenic quartz; others are not. That is probably a function of the supply of detrital silica to the original brine. The Cambrian salts in southern Siberia contain 8800 ppm authigenic quartz, an amount far in excess of a normal marine silica contribution to an evaporite basin (Sokolov, 1971). The quartz must be derived from reprecipitation of dissolved terrigenous silica. Furthermore, anhydrites of the upper suite contain silica nodules up to 5 cm in diameter. In contrast, the Jurassic evaporite sequences in central Asia contain an extremely low terrigenous admixture, mainly in the form of euhedral authigenic quartz crystals with hydrocarbon inclusions in vacuoles. The pseudo-dipyramidal quartz crystals contain inclusions of hematite, anhydrite, and carbonates. Neither opal nor chalcedony has been found (Sedletskiy *et al.*, 1971).

Permian and Triassic salts in Germany contain euhedral crystal aggregates of quartz and potash feldspar, both often etched and with clay inclusions in larger crystals (Brasack, 1867; Braitsch, 1958), with calcite inclusions or coal chips (Schettler, 1972), or with anhydrite inclusions (Dreizler, 1962; Braitsch, 1971). Corroded quartz crystals can even be found as inclusions in other euhedral quartz crystals (Koch *et al.*, 1968). The quartz crystals are evidently of very early secondary origin, since a difference in shape exists between crystals grown in shelf and offshore localities (Nachsel, 1966) and in different petrographic provinces of the salts (Koch *et al.*, 1968). Films of minute quartz crystals coat anhydrite crystals that occur inside halite-forming pseudomorphs after gypsum (Zimmermann, 1907). Fibrous silica may substitute for quartz and calcite as a coating on quartz crystals with anhydrite inclusions, or with calcite intergrowths (Demangeon, 1966). On the margins of an anhydrite basin, the dissolved silica may eventually reach supersaturation in a bittern that is undersaturated in relation to anhydrite. End-Paleozoic ore-bearing beds in France are associated with such quartz pseudomorphs after anhydrite (Arnold and Guillon, 1981). The assumption of pressures in excess of 500 kPa and temperatures of 150°C for halite and sylvite inclusions in quartz of Triassic evaporites in northern Tunisia (Perthuisot *et al.*, 1978) may well be on the high side.

Whereas Grimm (1962a,b), Nachsel (1966), Franz (1967), and Schettler (1972) used secondary quartz as a marker mineral and facies indicator in Permian evaporites in Germany, Sedletskiy *et al.* (1971) found no relationship of quartz shapes or quantities to hosting rock types in Upper Jurassic evaporites of central Asia. No correlation exists here between the amount of silica and either the total salt precipitated or its major components.

Celestite nodules in the British Keuper marls contain quartz grains which must have iron as a necessary minor constituent to yield an amethystine color (Curtis, 1974). Similarly, Tarr (1929) found doubly terminated quartz crystals in gypsum of New Mexico, colored by residual hematite.

GRAINS OF OTHER SILICATES

Potash feldspars are known from Late Triassic gypsum in southwestern Germany (Lippmann and Savascin, 1969), or from clay intercalations in Devonian evaporites of Saskatchewan, Canada (Mossman *et al.*, 1982), whereas basal anhydrites of the Zechstein carry authigenic sodic feldspar (albite) (Dreizler, 1962). The precise reason for the difference in feldspar chemistry is not known, but is probably related to the magnitude of the sodium concentration and activity in the circulating brine. Although feldspar neoformation occurs in gypsum and anhydrite beds, secondary feldspars are not found within the potash precipitates. The pH of the bitterns must have been too high.

Other mineral species are very rare. Neoformation of tourmaline (Kuehn, 1968) and serpentine grains has been reported on rare occasions. Rutile needles have also been found sporadically in carnallites. Zircon, tourmaline, garnet, and epidote grains in carnallite probably represent blown-in material (Muegge, 1913; Braitsch, 1958; Mormil', 1974), as do amphibole (Wetzel, 1939) and staurolite (Kuehn, 1968). Carbonate lakes in Africa have yielded a series of minerals that are essentially sodium silicates with various degrees of hydration, such as magadiite, kenyaite, kanemite, and others (Maglione, 1976). It is doubtful whether these minerals are stable enough to survive for geologically significant time intervals, particularly in the presence of hygroscopic brines.

CLAYS

Carbonates grade oceanward into open marine shales. Sulfates and halites, however,do not grade laterally into clayey rocks (Langbein, 1978). For the duration of evaporite precipitation, clastics of marine provenance are effectively shut out of the basin by the outflowing undercurrent. Only clastics washed in from surrounding mud flats or slopes will enter episodically. Consequently, Kopnin (1966) established the ratio of anhydrite to insoluble residue as a significant environmental indicator: Under the influence of rainwash and drainage, this ratio is smaller than 0.5 in nearshore areas. It is greater than unity in parts of the evaporite basin supplied by feeders of marine waters.

Intercalations

Shale intercalations in evaporite sequences are invariably the mark of ephemeral periods of freshening. This is as true of the red and gray "Salt-Clay" facies of the Permian Zechstein Basin (Hollingworth, 1942; Richter-Bernburg, 1957a) as it is of the Permian West Texas Basin (Adams, 1969), the Pennsylvanian Paradox Basin of Utah (Hite, 1970), or the Ukrainian deposits (Bobrov *et al.*, 1968). Although the clays themselves are derived from surrounding lands, the freshening of the brine by runoff often allows a restricted marine fauna to regain a temporary foothold. An example is furnished by marine fossils in clay intercalations of the Zechstein evaporites (Zimmermann, 1907).

Clays and finer silts do not form classical turbidity currents in brines, but disperse along the pycnocline as submarine deltas or interflows (Fig. 9-2). When they finally sink into the brine, they give rise to widespread time markers,

Fig. 9-2. Clay slurry spreading out along a density interface.

provided they are only altered and not dissolved. Banding in these clay intercalations survives only without wave action or vertical pumping of water. The sinking clay particles form perfect Bénard cells (Bénard, 1901), simulating convective motion on top of the interface to surface waters, but these cells form only in the horizontal plane (Fig. 9-3). The cells have a concentric structure similar to the one described by Dauzère (1912). Within the hypersaline brine, the descending clay particles are slowed at subsidiary interfaces that must form by gravity acting on chloride complexes. In the tank experiment shown in Fig. 9-4, the hypersaline brine was thoroughly mixed before a clay/water slurry was slowly poured on top of the brine.

Whether the surface drainage enters the basin as ephemeral wadis or as floods of a perennial stream, it can carry with it shore-derived pollen, spores, and even leaves for some distance without destroying them. Leaves are easily macerated and do not travel very far in rivers and creeks, but display greater longevity along a brine interface. From river mouths, the flood waters can also transport planktonic and even loosened benthonic organisms, again moving them along the interface without significant abrasion. Some of these organisms may even continue to live for a while in the surface waters, until they slip through the interface

Fig. 9-3. Bénard cell produced by feeding a clay-water slurry into a density-stratified brine (vertical view). The clay drops quickly to the interface between the fresh water and the concentrated brine, and then slowly sinks into the brine in a series of rotating cells.

into deadly hypersalinity. Evaporative drawdown is faster than the advance of a mixing zone into a concentrated brine, particularly where the wind rolls out surface waters to form a film. It is important to realize that the presence of brackish-water faunas in clay intercalations of halites does not necessarily indicate a freshening of the whole brine column.

The settling velocity of the clays is inversely related to the temperature of the brine; clays settle out faster in warm brines than in cold ones, and in hypersaline brines faster than in bitterns enriched in potassium or magnesium (Ramsay, 1876). Consequently, clay intercalations are thicker outside the potash zone and thinner where carnallite or secondary minerals after carnallite prevail (Langbein, 1964). The settling of the clay may proceed without any noticeable reduction in the salinity of bottom brines. In New Mexico, clay intercalations cover euhedral crystals of halite that would have been beveled or corroded by any freshening (Jones, 1974). The extremely slow downward advance of the interface between freshened surface waters and bottom brines is arrested by horizontal currents

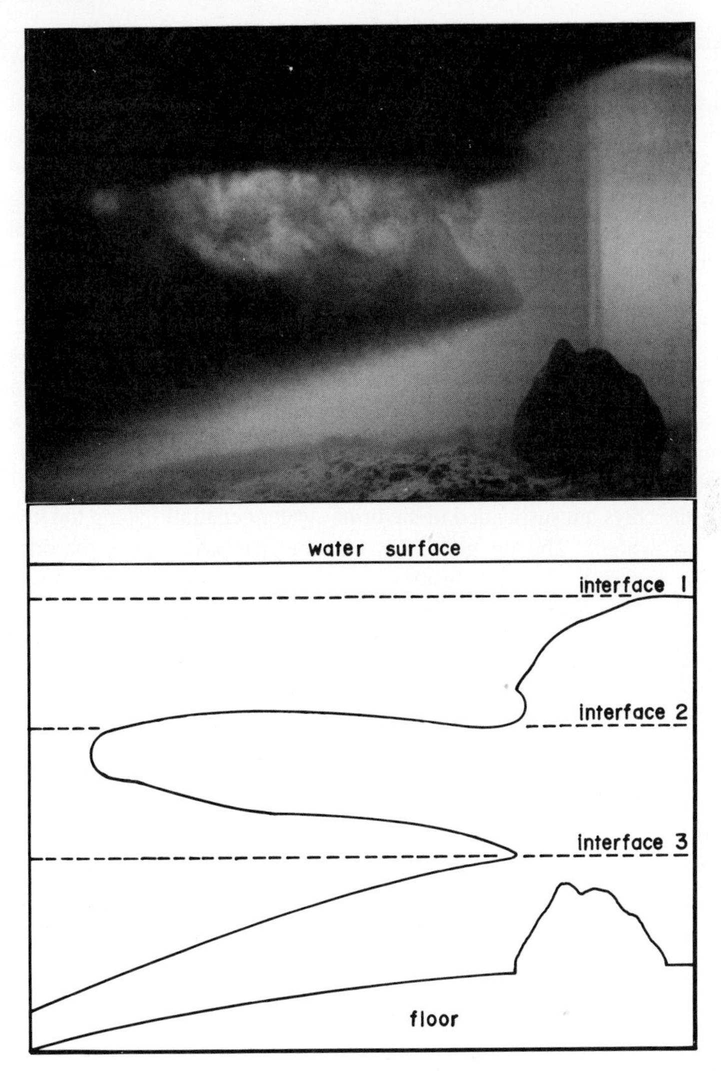

Fig. 9-4. Clays descending through a hypersaline brine are arrested momentarily at interfaces of the microstratification.

unless an amount of fresh water is supplied that is excessive if measured in terms of basin volume.

Clays blown into artificial salinas enter an unnatural environment. The brine is constantly kept at a negative redox potential, i.e., it continues to be reducing, but is also slightly acidic. A lowering of the pH due to increasing salinity, then, does not cause any notable changes in the clays (Quaide, 1958).

Ion Exchange

Freshwater clays delivered by runoff release hydrogen upon contact with the brine, which is taken up by biota or by dead organic matter. Conversion of dissolved silica to a hydrous magnesium silicate likewise liberates hydrogen ions. This not only reduces the solubility of carbon dioxide (Hardie and Eugster, 1970) but also serves to hydrogenate macerated organic material commonly suspended in the brine. The clay mineral type is thereby a function of evaporite facies. Clay intercalations in halites most often contain more authigenic than detrital minerals, reconstituted after the original clay minerals had disintegrated (Braitsch, 1958; Petrov and Chistyakov, 1964; Pastukhova, 1965; Kudryavtsev, 1971). The frequency of these secondary minerals decreases rapidly in salts of higher solubility. They are absent in langbeinite layers (Braitsch, 1958), which are themselves diagenetically altered from hydrated precipitates. Blown-in dust dissolved and was not reconstituted in very concentrated brines.

While the clays are suspended in the brine, water penetrates along the layering of the clay wafers. The integral diffusion coefficient is higher for diffusion within the wafers than for diffusion through the clay wafers, as diffusion is slowed at the clay–solute interface. This occurs possibly because anions find it difficult to enter through the charged clay interface, whereas cations experience difficulty in existing from it (Dutt and Low, 1962). An ion exchange then occurs between brine and clays: Cations are exchanged or adsorbed with a preference for multivalent cations of higher atomic weight. In low sodium chloride concentrations, the ion exchange predominates, but at high concentrations adsorption is dominant (LeDred and Wey, 1965). The clays first take up sodium, and a decrease in the electric repulsion occurs between the clay wafers (Rosenqvist, 1955, 1966). Sodium ions reduce the water enrichment in the interior of the clay particle and reduce its swelling ability. By creating an unstable, loose pore structure, they cause a higher permeability, easier drying, and low strength in the dry state. The diffusion coefficient of sodium chloride decreases with increasing concentration, due to a salt-induced change in the fractions of cations and anions in the more viscous adsorbed water. Freshwater clays release not only hydrogen but also calcium ions in exchange for sodium. The sodium depletion of the brine then decreases the calcium sulfate solubility (Strakhov, 1962) and promotes gypsum precipitation.

At the same total concentration, a mixed system (sodium plus magnesium) has more adsorbed ions than a homoionic system (Helmy, 1963). Cation mobility follows the sequence $Fe^{3+} < Al^{3+} < Co^{2+} < Cd^{2+} < Ca^{2+} < Mg^{2+} < Rb^{2+} < K^{+} < Na^{+} < NH_4^{+} < Li^{+}$ (Strakhov, 1962). Lack of cation mobility can be overcome by a sufficient concentration of the brine, which also fosters anion leakage. Anion leakage is in part compensated for at high salinity by the deswelling of the clays and by a decrease in anion mobility in the clay phase relative to cation mobility. The desorption due to univalent anions equals about one-third of that due to the bivalent cations present in the same concentration (Helmy, 1963). Sorption and desorption of organic molecules render some of the cations in interlayers nonexchangeable, and thus cause montmorillonites and other clay minerals to chloritize (Heller-Kallai *et al.*, 1973). Organic acids dissolve framework cations (Si, Al, Fe, Mg) more readily than the brine. The cations then form metallo-organic complexes or chelates. The dissolution of aluminum with respect to silica is controlled by the complexing capacity of the solvent acids (Huang and Keller, 1972).

No terrestrial clay minerals survive unaltered in a marine hypersaline brine. At low pH values, the clays remain poor ion exchangers, because of the strong association of the exchange sites with hydrogen. At moderately high pH values, they become very efficient exchangers by converting to the sodium form. However, this is an intermediate stage. A very subordinate role or the complete absence of sodic clays is characteristic of all marine evaporite sequences, suggesting the prevalence of high pH values in bottom brines.

Flocculation

Freshwater clays flocculate upon contact with brine. The flocculation of sodium montmorillonites is initially induced by an abrupt decrease in repulsion energy between particles, whereas the attraction energy remains the same. Further addition of electrolytes decreases the repulsion energy more gradually. The electrokinetic potential of clays thus decreases as the concentration of sodium or potassium chloride and other solutes increases (Vardukadze, 1959). At the same time, the pH rises to 8.5–9 during hydrolysis, but drops if the system comes in contact with the atmosphere (Schramm and Kwak, 1984). The presence of less than 1 ppt NaOH peptizes an aqueous solution of palygorskite by replacing exchangeable cations with sodium. A strong association between particles is thus achieved by the first increments of salt in the solution. Chlorides of monovalent elements decrease the swelling of clays linearly with brine concentration, and chlorides of bivalent metals decrease it exponentially (Prikryl, 1966). A 5% calcium chloride solution induces the same flocculation as a 25% sodium chloride solution (Ovchenko and Tretinnik, 1962). Of all the clays, the montmorillonites are particularly slow in sinking through the pycnocline. Upon contact with the interface, they produce greater flocs that gradually float to the bottom, bouncing on every interface of the microstratification (Fig. 9-5). Figure 9-6 shows the semiliquid nature of the settling flocs.

More concentrated NaOH causes intensive dispersion of the solid phase by forming flocculating thixotropic structures. The higher the solute content, the larger the floccule size and the higher the permeability of the sediment; once the sodium chloride concentration exceeds 20 ppt, the clay forms a sediment the size

Fig. 9-5. Montmorillonite flocs descending from the fresh water–brine interface. The size of the flocs ranges up to 8 cm in diameter.

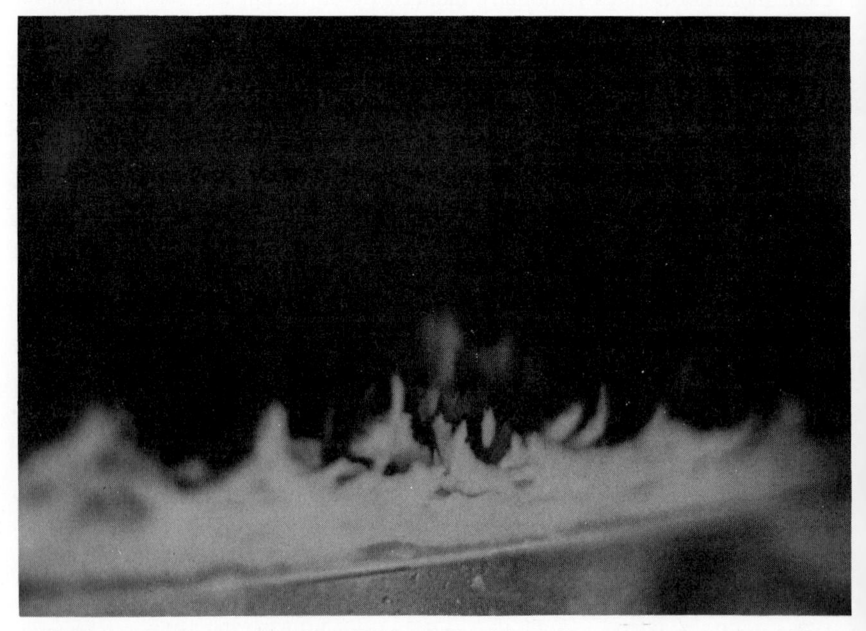

Fig. 9-6. Clay flocs settling through a hypersaline brine arrive at the bottom in a semiliquid state.

of coarse silt or sand (Little-Gadow, 1974). The higher the concentration of the solute, the greater the stability of floccules against tremors. A plateau of increasing flocculation is achieved at about 40 mmoles per liter of sodium chloride, at 0.25 mmole per liter of calcium chloride, but at only 0.08 mmole per liter of aluminum chloride (Ferreiro and Helmy, 1974). Aluminum hydrolyzes easily as

$$Al^{3+} + H_2O \leftrightharpoons (AlOH)^{2+} + H^+ \tag{9-2}$$

Therefore, only a small fraction of the aluminum ions is available, and the flocculation value of aluminum is reduced. Both phosphates and organic materials also reduce the flocculation effect of the elctrolyte (Rosenqvist, 1955).

Forces of repulsion and attenuation are more important than gravity and shear stress (Hahn and Stumm, 1970). If an undisturbed sample of clay or silt is subjected to a shear or compression test and is then remolded with its natural water content, the two stress–strain curves are different (Skempton, 1953).

Syneresis

Syneresis is the spontaneous separation of liquids from a gel or colloidal suspension during aging, resulting in shrinkage cracks. Such cracks, if induced in dry air, are referred to as "desiccation cracks"; they also occur underwater due the influence of hygroscopic brines, such as concentrated sodium chloride brines or potash-precipitating brines. Syneresis cracks in shale intercalations in the Permian Hutchinson salt of Kansas terminate by pinching out, or are abruptly truncated by salt or anhydrite (Dellwig, 1968). Efflorescences and crusts of halite–epsomite–bloedite, mirabilite–thenardite, or bloedite seal a soil surface, reduce evaporation, and prevent such cracking of the soil (Driessen and Schoorl, 1973). Brine storage tanks for bitterns with a density of 1.330 quickly develop cracks when built of concrete, but not if built of bricks or equipped with linings (Bhat and Oza, 1977). There have been many costly but avoidable leakage problems in brine storage tanks lined with untreated freshwater clays that developed desiccation cracks under brine cover. Several soil mechanics texts mistakenly follow Richter's (1941) concept that subaqueous shrinkage cracks do not occur, but that such cracks must in each case represent subaerial mud cracks or even frost cracks.

If stored underwater, even in standing water, clay at its liquid limit develops syneresis features. Electrochemical forces between clay particles are sufficiently strong to cause a realignment of these particles. Only clays with a very low plasticity index will not develop such fissures (Skempton, 1953). Both syneresis and subaqueous shrinkage appear to occur and create vertical fissures during the osmotic transfer of fresher water from a very fine-grained sediment to the more saline open brine (Turk, 1974). Swelling clays develop cracks when the salinity changes; nonswelling clays crack only in saline solutions (Plummer and Gostin, 1981).

Warm hygroscopic brines strip down the adsorbed water layers. The clay platelets flocculate, leaving voids in card-house fashion, or as submarine desiccation cracks mimicking frost damage (White, 1961; Pusch, 1973), or they simulate raindrops (Juengst, 1934). The greater the propensity of cations to polarize in a brine acting as an electrolyte, the greater the tendency for the clay to produce fissures (Rosenqvist, 1955). The card-house texture produced by flocculated clay wafers is particularly resistant to compaction in an evaporite sequence (Kulke, 1979). Figure 9-7 shows the remnant porosity in clays that descended through a hypersaline brine. Precipitating salt preserves the spacious texture, and a later drop in the permeability of the salt prevents the evacuation of trapped interstitial fluids. The pore space is greatest in original montmorillonites rich in sodium, or in illitic clays, and is least in kaolinites (Stadtbaeumer, 1976, 1977), since concentrated sodium solutions are able to maximize dehydration. In argillaceous siltstones inundated by hypersaline brines, the clay structure deswells, causing the entire rock to shrink (Bernstein, 1960) and to cease to be an aquiclude. However, pure clay layers remain conduits for percolating fluids long after burial.

Fig. 9-7. Remanant porosity in clays that descended through a hypersaline brine. Interstitial fluids are trapped when salt precipitation resumes above the clay marker. They prevent the collapse of the void spaces. Clay markers can later serve as conduits for circulating epigenetic brines.

Mixed-Layer Clays

Chloritic mixed-layer minerals dominate in hypersaline brines (Millot, 1970). There is a wide variety of mixing conditions, concentrations, and other physical and chemical parameters suitable to produce chlorite-like structures in montmorillonite. This suggests an adsorption process on the unit layer level. The most complete alteration results in six $Mg(OH)_2$ molecules per unit cell. Completely reacted magnesian chlorites are stable in water; only partially reacted ones are still expandable (Slaughter and Milne, 1958). Magnesian clay minerals that dominate in evaporite sequences have an Al_2O_3:SiO_2 ratio of less than 1:4 and greater than 0:4. The MgO:SiO_2 ratio is commonly less than unity. Magnesian montmorillonites (smectite and saponite) are not as common as the generally prevalent mixed-layer clays derived from them. Magnesian septechlorite found in clay intercalations of the Devonian Prairie Evaporite Formation in Saskatchewan are absent in the surrounding red beds (Mossman *et al.*, 1982), and thus prove alteration processes in the brine environment.

The conversion of almost all clays in halite and carnallite sequences to mixed-layer, high-magnesium varieties is possible only at a high pH value of the waters, which at times might have approached or exceeded 10.0, at least beneath the brine–sediment interface. Brucite pillars could not form in the clay lattices at a lowered pH and are the best evidence for high pH values developing under density stratification. With rising pH, aluminosilicates rapidly increase their uptake of magnesium (Perrott, 1981). The conversion of blown-in dust to magnesian chlorites and brucite-containing mixed-layer clays in a magnesium-rich brine liberates free hydrogen and raises the pH (Hardie and Eugster, 1970). However, as the magnesium chloride concentration increases, the pH values are reduced due to hydrolysis. The addition of $Mg(OH)_2$ to the solution will then not increase the pH, but dilution will do so (d'Ans and Katz, 1941). Nonetheless, at temperatures above 25°C, i.e., temperatures common in subtropical evaporite pans, the pH does not even have to reach 8 before equilibrium relations of aluminosilicates and seawater are firmly within the magnesian chlorite field at or below quartz saturation (Helgeson and MacKenzie, 1970).

Cement paste, when exposed to magnesium sulfate solutions, readily allows brucite to form, while producing only gypsum, calcite, and compound calcium–aluminum salts when corroded by solutions of sodium and potassium sulfates (Greschuchna, 1976). Carbonate–clay intercalations in evaporites are thus prone to generate brucite when exposed to magnesium sulfate solutions. Chemical analyses often assign available magnesium to dolomitic clays instead of checking for brucite pillars in mixed-layer clay minerals (G. Richter-Bernburg, personal communication, 1980). Excess magnesium that is dissolved in a magnesium chloride brine complexes to $Mg_2(OH)_3Cl \cdot 4H_2O$ (d'Ans *et al.*, 1966), and this compound is the likely converter of intercalated clays to brucite-rich varieties.

However, mixed-layer varieties are also produced in stagnant seawater of normal concentration in the presence of organic matter (Sigl *et al.*, 1978), such as in the sapropel cores retrieved from the bottom of the Mediterranean Sea. The crystallinity of clays apparently increases in the presence of sodium chloride or calcium sulfate as a catalyst. Chloride and sulfate anions break down the water structure, presumably increasing the mobility of the small clay-building cations, and facilitate their organization. However, the exact role of these electrolytes is not understood (Weaver and Pollard, 1973). High pH values thereby foster continuous silica layers and discontinuous hydroxyl layers.

Overall, magnesian chlorites are common, and many names have been applied to the evaporitic varieties. Vermiculite, a mixed-layer clay, may in part be derived from biotites that were stripped of their potassium. Palygorskite and sepiolite are fibrous, and contain structural hydroxyls, zeolitic, and bound water. Sepiolite, a lath-shaped magnesian clay, is not stable at a pH lower than 8–9 (Velde, 1977); palygorskites and smectites form only at a pH higher than 8.5 (Lomova, 1979). The Al_2O_3:MgO ratio is drastically reduced from that common in illites and montmorillonites (Weaver and Pollard, 1973), indicating removal of Al_2O_3. Sepiolite occurs in evaporitic environments more alkaline than necessary for attapulgite formation, where the aluminum concentration is low and the magnesium concentration high—in other words, farther from shore. A conversion of clays to sepiolite takes place in continental lakes that are precipitating thenardite–mirabilite–halite associations, in which the transitional zone contains palygorskite, chlorite, attapulgite, and sepiolite. This has been observed on the margins of the Neogene Tajo Basin in Spain (Carames Lorite *et al.*, 1973). Both sepiolite and sodium-rich hydromicas are common in soda-rich semiarid soils (Strakhov, 1962) (Table 9-1).

TABLE 9-1

Typical Clay Minerals Specific to Evaporite Basins, with Representative Chemical Composition (Substitutions of Iron, Magnesium, and Aluminum Common)

Mineral	Composition
Magnesium-montmorillonite	$3MgO \cdot 3Al_2O_3 \cdot 4Al(OH)_3 \cdot 24SiO_2 \cdot nH_2O$
Attapulgite	$3MgO \cdot Al_2O_3 \cdot Al(OH)_3 \cdot 12SiO_2 \cdot 12H_2O$
Palygorskite	$MgO \cdot Al_2O_3 \cdot 2Mg(OH)_2 \cdot 8SiO_2 \cdot 8H_2O$
Sepiolite	$3MgO \cdot$ — $\cdot Mg(OH)_2 \cdot 6SiO_2 \cdot 6H_2O$
Talc	$2MgO \cdot$ — $\cdot Mg(OH)_2 \cdot 4SiO_2$
Antigorite	$MgO \cdot$ — $\cdot 2Mg(OH)_2 \cdot 2SiO_2$
Vermiculite	$Fe_2O_3 \cdot$ — $\cdot 9Al_2O_3 \cdot 6Mg(OH)_2 \cdot 12SiO_2$
Corrensite	$Fe_2O_3 \cdot MgO \cdot 5Al_2O_3 \cdot 7Mg(OH)_2 \cdot 9SiO_2$
Chlorite	$Fe_2O_3 \cdot 2MgO \cdot Al_2O_3 \cdot 8Mg(OH)_2 \cdot 6SiO_2$
Chlorite variants: amesite, aphrosiderite, bertherine, clinochlore, corundophilite, penninite, prochlorite, pseudochlorite, ripidolite, xylotile	

Halloysites, kaolinites, and allophanes have not been reported from evaporitic clay sequences. Some chlorites were often confused with kaolinite in older literature (Braitsch, 1971). Kaolinite changes in magnesium solutions of low concentration to tosudite, a clay mineral composed of alternating layers of montmorillonite and chlorite. The degree of alteration depends on the availability of magnesium. At higher magnesium chloride concentrations, kaolinite is altered to chlorite and serpentine (Kotel'nikova *et al.*, 1979). Serpentine grains have been reported only very occasionally from evaporite sequences.

Clay minerals in shale intercalations that are derived from a temporary freshening of the brine are commonly altered to magnesium-rich varieties, because at higher pH values the sodium cations are forced out by magnesium cations; this accounts for the near absence of sodium clays and the predominance of magnesium clays in evaporites. The brine, rich in magnesium chloride, encroaches on the clays, and the hygroscopic bittern front leaches OH ions and crystal waters, as well as pore waters, from shales, including adsorbed watery films (Berner, 1971). Despite the high concentration of sodium in the brine, the sodium montmorillonite eventually turns into one of the clay minerals rich in magnesium, since magnesium ions from a marine brine preferentially replace hydrogen ions in clay minerals (Caroll and Starkey, 1960).

Even in marine basins precipitating gypsum, the clays either convert to mixed-layer varieties rich in magnesium (palygorskite, attapulgite, sepiolite, talc, vermiculite, corrensite, or swelling pseudochlorite and ferruginous chlorites) that adsorb potassium ions or they eventually convert to illites rich in potassium. Such a decomposition of montmorillonite or illite does not occur in magnesium-poor continental lakes that precipitate sodium sulfates or calcium–sodium carbonates (Droste, 1961), probably because of their potassium deficiency. However, even in open ocean environments, the montmorillonite decreases at the expense of illite and chlorite (Johns, 1963).

Where potassium-rich bitterns percolate upward from underlying potash deposits, a layer of bentonitic montmorillonite is locally altered to potassium-bearing illite in Colorado (Keller, 1963). Illite has also been reported as a major constituent of clay partings in other North American Paleozoic marine evaporites (Droste, 1963; Lounsbury, 1963; Mossman *et al.*, 1982) and in the upper parts of the Gray Clay intercalation in the German Zechstein evaporites (Niemann, 1960). However, illite–saponite is restricted to polyhalite or adjacent anhydrite, and is absent from pure clay beds, halite, or polyhalite-free sections (Powers *et al.*, 1978). In contrast, barred-basin marine evaporites in the Upper Proterozoic of the Amadeus Basin of central Australia, primarily composed of gypsum and anhydrite, contain diagenetic chlorite (Stewart, 1979). Conversion to illite and consequent collapse of the lattice spacing of montmorillonite to structures with very low expandabilities are dependent on the availability of potassium ions. The presence of ions of greater hydration energy, i.e., sodium, calcium, and magnesium, slows the reaction and may halt it altogether at lower temperatures (Johnston, 1983). It remains to be established whether illite represents a shoal and shoreline facies to magnesian chlorites, akin to the anhydrite–rock salt relationship.

Corrensite increases toward barren zones at the expense of penninite in the first cyclothem of Zechstein evaporites in Germany. The barren zones are evidently produced by later epigenetic brine movements; changes in clay composition occur around the conduits over the smallest distances. In this instance, the changes can document only secondary soaking processes (Weber, 1961). Corrensite is also more common than illite in Upper Triassic gypsum in southwestern Germany, where it may have formed from illite or kaolinite (Schlenker, 1971), or after mica (Lippmann and Savascin, 1969) under the influence of solutions rich in magnesium. Some of the secondary quartz in the gypsum may also be due to corrensite formation. Thus, quartz dissolution is not the principal mechanism for charging brines with H_4SiO_4 (Eugster, 1980a). In palygorskite beds of the Permian Salado Formation in New Mexico, corrensite, or rather its less regularly interstratified clinochlore–saponite analogue, predominates. The Mg/Al ratio is reduced to nearly half that of corrensite. Relatively impermeable argillaceous seams contain a more mature assemblage of corrensite, illite, and clinochlore, whereas adjacent evaporitic beds contain disseminated clinochlore–saponite, talc, and other immature mixed-layer clays. Bodine (1978) and Bodine and Laskin (1979) postulated migrating pore fluids with high $Mg^{2+}/(H^+)_2$ activity ratios and a pH sufficiently low to allow the aluminum concentration to greatly exceed the H_4SiO_4 concentration.

Magnesian clays in the German Zechstein anhydrites are contaminated with magnesite. The growth normal to bedding planes with columnar and spherulitic intergrowths suggests *in situ* formation (Langbein, 1978). There is an upward gradation in the diagenetic chlorites from sheridanite through clinochlore to penninite (Niemann, 1960), with a concurrent reduction in illite content. Clays associated with Neogene anhydrites in eastern Turkey are enriched in magnesium, iron, and potassium (Baysal and Ataman, 1980). The Triassic evaporites in Britain are embedded in a clay, enriched in silica and magnesium, composed of illite, corrensite, palygorskite, and sepiolite, with some montmorillonite and expanded chlorite (Warrington, 1974). Silica enrichment can be interpreted as aluminum impoverishment, but aluminum compounds, such as aluminum chlorides, must then remain in solution, because they are rare in evaporite sequences.

A whole range of hydrous aluminum sulfates and aluminum sulfate–carbonates exists in arid soils where sulfides are weathering out. Sulfuric acid solubilizes aluminum and decomposes clays. The absence of such hydrous aluminum sulfate minerals inside a marine evaporite sequence confirms the contention that sulfate ions were not available in the brine. Yet aluminum must go into solution, since there is a marked deficiency of it in salt clays (Braitsch, 1971).

Aluminum hydroxide is more soluble in a potassium chloride solution than in plain water, and its solubility increases with rising pH (Monzhalei, 1938). Aluminum probably remains in solution as aluminum chloride. As such, it plays an important role in determining the fate of accumulating organic matter. The chloride members of the Zechstein (halite, sylvite, carnallite) are enriched in aluminum (Herrmann, 1958; Elert, 1963), but the secondary potassium–magnesium sulfates are not.

Koenenite

The dissolution of clay particles blown into halite-precipitating brines is confirmed by the frequent occurrence of koenenite, a light yellowish or reddish mineral with a silky luster and a density of 1.98. It is flexible, weakly bitter, and soluble in a hot magnesium chloride solution. The chemical formula was given as $4MgCl_{2.5}Mg(OH)_2\cdot 4Al(OH)_3$ by Kuehn (1951) and revised to $4NaCl\text{-}\cdot 4(Mg,Ca)Cl_{2.5}Mg(OH)_2\cdot 4Al(OH)_3$ by Allmann *et al.* (1963) and Lohse (1963). Some of the koenenite may be due to misidentification: The koenenite inclusions reported from carnallite in western Canada (Wardlaw and Schwerdtner, 1963) later proved to be fibrous goethite (Wardlaw, 1968). Other fibrous koenenite might similarly have been misidentified.

The Gray Salt Clay Member of the German Zechstein evaporites is mainly composed of koenenite and quartz that formed concurrently. The formation is split by a clayey magnesite member which becomes anhydritic on top and below (Schulze and Greulich, 1962). Magnesite is a product of the same magnesium brines that altered the clay intercalations to koenenite. The magnesite itself is probably an alteration product of calcite after gypsum, as indicated by the incomplete degypsification toward the top of the magnesite. Kuehn (1961, 1968) tried to distinguish finely dispersed koenenite in the Main Anhydrite and Gray Salt Clay Members of the German Zechstein evaporites as primary, and thus distinct from the secondary fibrous koenenite in the fissures and cavities of these members. The two types may, however, be nearly synchronous in origin.

The presence of koenenite and chlorite is mutually exclusive (Feuchtbauer and Goldschmidt, 1959). The Oligocene evaporite minerals of the Rhine Valley are noted for the common presence of koenenite and euhedral potash feldspars and, consequently, for their lack of chlorites (Hiller and Keller, 1965). Koenenite formation increases with magnesium chloride concentration but is small in the presence of sulfate brines (Berdesinski, 1952). Dissolution of clay particles in the latter leads to the formation of alunite, $KAl_3[(OH)_6|(SO_4)_2]$. A rarer derivative of clays is white-to-brown zirklerite $[4Al(OH)_3\cdot 4MgCl_{2.5}Mg(OH)_{2.7}H_2O]$, also found in several evaporite sequences, with calcium and iron substituting for magnesium, $\rho = 2.6$.

Light yellow loewigite (or ignatievite) $(K_2O\cdot 3Al_2O_{3.4}SO_{3.9}H_2O)$, $\rho = 2.58$, sometimes forms a residue after carnallite dissolution. The fine fraction of the clays is also the depository of various phosphatic minerals. Wagnerite [$Mg_2(F|PO_4)_2$], apatite [$Ca_5(F,Cl|PO_4)_3$], and isokite [$CaMg(F|PO_4)$] have been identified (Weber, 1961).

Facies Distribution

Clays of the same horizon vary in composition as the adjacent salts change facies (Weber, 1961). Velde (1977) produced a facies distribution scheme using data by Millot (1970) and Fontes *et al.* (1967). Kaolinite and montmorillonite with minor chlorite appear to be restricted to the rims of evaporite basins, where detrital material is dumped and then quickly buried. Some magnesian montmorillonite and chlorite may be swept out into nearshore sites, eventually to give way to palygorskite and farther away to sepiolite in offshore areas. Sepiolite occurs in the area of calcium depletion of the brine under concurrent silica instability; it is not stable below a pH of 8–9. Muscovite and chlorite are interbedded only in nearshore sites of Zechstein anhydrites. Offshore they are replaced by talc (Braitsch, 1958, 1971). In the wall rock of the Mississippian carnallite–sylvite deposit of Nova Scotia, Canada, the clays are also altered to talc (R. Evans, 1970), which forms talc pseudomorphs after various evaporite minerals but does not form inclusions.

Palygorskite is often the dominant mineral in the carbonate zone of arid soils, where groundwater moves vertically by capillary action and a pH greater than 7.8 is sufficient to precipitate calcite. Sepiolite then occurs above or below the carbonate zone, and montmorillonite is present beyond these zones (Velde, 1977).

SUMMARY OF TERRIGENOUS ADMIXTURES

Sands and conglomerates are commonly associated with gypsum deposits but are rare as intercalations in halites and are not known from potash deposits. Coastal flats around an evaporite basin are usually composed of very fine-grained clastics, because runoff does not possess a great force even during flash floods. Hydrolysis of silicate minerals and quartz increases the pH of the brine and leads to secondary quartz and feldspar crystals and to recrystallization of clays into mixed-layer varieties rich in magnesium.

10

Behavior of Base Metals

Commercial deposits of sulfides of base metals occur frequently in ancient evaporite basins. Within a concentrating brine, the biota accumulate some metals in their bodies and release them upon decay. Such metals initially enter the crystal lattice of evaporite minerals but are removed by recrystallization and then travel into the subsurface with migrating hypersaline brines.

CONCENTRATION OF METALS IN THE BRINE AND ITS PRECIPITATES

Both ferrous and base metals concentrate in Recent marine chloride brines of high pH (Sonnenfeld *et al.*, 1977) but do not occur in neutral high-chloride brines (Truesdell and Jones, 1969) derived from the leaching of older salts. Hypersaline brines in a lake on Enderbury Island, Phoenix Island Group, Pacific Ocean, contain 3.6 mg/liter copper, over 1000-fold more than seawater (Brown and Gulbrandson, 1973). This is most certainly due to biogenic concentration. In particular, dissolved artinite [$Mg(OH)_2 \cdot MgCO_3 \cdot nH_2O$] is an effective sorbent of copper, zinc, and molybdenum in the early history of brine concentration (Barannik *et al.*, 1975). Alkaline sulfo- or chlorocarbonate brines of the western Great Basin in the United States carry significant trace metal contents as hydroxides or hydroxypolyions (Truesdell and Jones, 1969), but more concentrated alkaline sulfate brines in British Columbia do not (Crux, 1919; Hudec and Sonnenfeld, 1980).

All evaporites are large repositories of copper, lead, and zinc, and the incorporation of such extraneous cations is a function of the rate of precipitation (Thiede and Cameron, 1978; Dean and Anderson, 1978). Chalkophile and siderophile elements accumulate in gypsum. Iron and manganese thus appear to enter the crystal lattice of gypsum, whereas base metals are distributed in point form throughout the precipitate (Sonnenfeld *et al.*, 1976). Ultimately, the alteration of gypsum to anhydrite releases waters charged with the metals. The metal content of fossil anhydrites and halites is two to three orders of magnitude smaller than that of recent equivalents precipitated from ocean waters (Sonnenfeld *et al.*, 1977). If the gypsum releases its water to hygroscopic brine movements, the metals find a ready carrier in which to form complexes.

Some of the cations, such as manganese or lead in halite, act as catalyzers of crystallization. Crystallizing halite abstracts disproportionate amounts of cations from the brine (Yamamoto, 1938; Murata and Smith, 1946). Sea salt contains 0.8–1.3 mg lead per kilogram of salt, whereas seawater contains only 3.5–8.0 mg per cubic meter (Boury, 1938). Its concentration in fossil halites is often two to three orders of magnitude higher than in seawater (Moore, 1971). In the Stassfurt series of the Permian Zechstein evaporites, Herrmann (1961a) found a range of 60–560 ppm lead in halites and 200–1440 ppm in sylvites. Lead enters sylvite crystals in proportion to the velocity of crystallization and the lead concentration in the brine. Sylvite crystals containing lead grow to a larger size and tend to have octahedral rather than cubic faces (Kenat *et al.*, 1958). Later recrystallization of such halites or sylvites than allows the metals to be mobilized. Although halites and sylvites originally contain lead, carnallite and kainite are free of it (Born, 1934). In contrast, the zinc content of the Permian Upper Kama carnallites west of the Urals is more than five times as great as the bromide or calcium content (Morachevskii, 1940).

In saturated solutions of sodium or potassium chloride at 18°C, most chlorides dissolve only in small amounts (cf. Table 10-1), zinc chloride being an exception (Krasikov and Ivanov, 1927). However, very saline sodium–calcium chloride brines are potent solvents for base metals which favor the liquid phase of chloride complexing (White, 1968; Barnes, 1974). Most stable zinc complexes at ambient temperatures have the formula $2NaCl \cdot ZnCl_2$. Their complexing is accompanied by an increase in volume (Titov, 1939). Zinc complexes are apparently not resistant to sorption and precipitation processes. Galena is relatively insoluble in brines at atmospheric pressure, but its solubility increases substantially at 25°–90°C in sodium chloride brines with a reduced sulfur concentration and a pH of 8–13 (Hamann and Anderson, 1978).

Complexing aids the transport of metals. Lead carbonate is the dominant complex in normal seawater (Byrne, 1981), but sufficient quantities of carbonate ions are not available in more concentrated brines. Bisulfide complexes are not important near ambient temperatures (Herr and Helz, 1978). Hamann and Anderson (1978) questioned the possibility of lead transport as bisulfide or sulfide complexes even at temperatures around 100°C. Slightly acidic lead and zinc chloride complexes with a low sulfur content can lead to substantial accumula-

TABLE 10-1

Solubility of Selected Chlorides in Saturated NaCl and KCl Solutions[a]

Compound	NaCl solution (g/liter)[b]	KCl solution (g/liter)[b]
NaCl	—[c]	—
KCl	6.0	—
$MnCl_2$	4.0	4.0
$SrCl_2$	0.0[c]	5.6
$CaCl_2$	0.0	—
$MgCl_2$	0.0	—
$AlCl_3$	0.0	4.0
$FeCl_3$	0.0	—
$BaCl_2$	—	16.0
$ZnCl_2$	1164.0	1805.0

[a]After Krasikov and Ivanov (1927).

[b]Additional dissolution of individual salts leads to precipitation of halite or sylvite.

[c]Insoluble compounds are marked —. Soluble ones that immediately lead to NaCl or KCl precipitation are indicated by 0.0.

tions by a slight reduction in the pH, and thus a large reduction in the solubility (Anderson, 1973). The metals may then travel as simple chloride complexes or as compound organo-chloride complexes. The formation of complexes usually seems to increase the reactivity of compounds (Meerwein, 1930); for example, lead complexes with humic acids increase the solubility of galena (Leleu and Goni, 1974), and copper and zinc form soluble metallo-organic complexes that keep copper in solution, even in the presence of sulfide precipitation (Hallberg *et al.*, 1980). Dichloride and trichloride complexes of copper increase the solubility of this metal by up to two orders of magnitude (Rose, 1976). Since most of the base metals are bivalent, Meerwein (1930) arranged the chlorides according to increasing acidity, starting from practically neutral barium chloride: $BaCl_2 < SrCl_2 < CaCl_2 < MgCl_2 < MnCl_2 < ZnCl_2 < CuCl_2$. Ion pairing of sulfate ions involves cations of the same size, but chloride ion pairs involve cations 4–5 times larger (Johnson and Pytkowicz, 1979). The ion complexes form either ion pairs with electrostatic bonds between fully hydrated cations or coordination complexes. In these complexes, one or more hydration water molecules are replaced by covalent bonds to the anion (Truesdell and Jones, 1969), but metals can also travel as hydroxide complexes (Ganeyev, 1976). Lead and zinc migrating to the surface in the Cheleken Peninsula south of the Gulf of Kara Bogaz Gol do not move as large organometallic molecules. Instead, they appear to be complexed ions of very small radius, probably in the form of $PbCl_4^{2-}$. Zinc probably travels in a similar chlorine complex (Bugel'skiy *et al.*, 1969).

Desulfovibrio and other anaerobic sulfate reducers extract metals from the brine without necessarily requiring hydrogen sulfide to do so (Jones *et al.*, 1976). Some metals are also selectively removed by sessile algae and higher aquatic plants (Jennett and Wixson, 1975; Jackson, 1978), and are later released from decomposing organic materials (Truesdell and Jones, 1969). Organic matter may play a role, as amino acids greatly increase the solubility of heavy metals (W. E. Evans, 1964; Veitch and McLeroy, 1972; Rachid and Leonard, 1973). Biological activity produces tetramethyl lead. This compound competes with metal ions such as iron for the available hydrogen sulfide (Jarvie *et al.*, 1975) and prevents the precipitation of sulfides. Giordano and Barnes (1981) found that all inorganic complexes are insufficiently stable, and that their concentration in percolating brines is three to four orders of magnitude too low to be able to generate a lead deposit. Organometallic complexes, in contrast, can reach concentrations of up to 30,000 ppm, and concentrations of minute traces of organic acids are sufficient to transport quantities of lead adequate to form a deposit.

Evaporites even aid in the mobilization of iron and other metals disseminated in surrounding ordinary rocks (Sokolov and Pavlov, 1964). In addition to metals from inflowing seawater, a major contribution is made by cations adsorbed onto suspended clays of terrigenous provenance (Ongley *et al.*, 1981). Consequently, the metal content of shale and evaporitic dolomite is higher than that of halite and carnallite; specifically, copper and zinc are associated with the insoluble residue of evaporites (Thiede and Cameron, 1978). The same two metals are also concentrated by the biota in hypersaline brines.

DEPOSITION OF METAL SULFIDES

In dilute solutions (less than 1 molal), the pH above which metals precipitate increases in the following order: (Sn^{2+}, Zr, Fe^{3+}) $<$ Al $<$ (Zn, Cu, Cr^{3+}) $<$ (Fe^{2+}, Pb) $<$ (Co^{2+}, Ni^{2+}, Cd) $<$ Mn^{2+} $<$ Mg^{2+} (Pasztor and Snover, 1983). The mobilized metals precipitate where they encounter formation waters with a supply of sulfide ions, such as those associated with sour gas of carbonate or evaporite provenance, i.e., where the medium suddenly changes to a negative Eh and a high pH. Most formation waters in carbonates are enriched in hydrogen sulfide. A zonation of the metals occurs because the critical concentration of hydrogen sulfide is different for each metal and depends on the temperature of the solution (Ganeyev, 1976). Zechstein copper ores of southwestern Poland accumulated by ore-forming metasomatism along an oxidation–reduction front, which was controlled by the hydrogen sulfide concentration and the free enthalpy of sulfide formation (Niskiewicz, 1980). Fine laminations in sphalerite may be varves produced by annual mixing of ore-forming fluids (Roedder, 1968). Sulfides currently precipitate in the Greater Caucasus at the meeting point of metal-bearing brines and hydrogen sulfide-carrying surface waters. The hydrogen sulfide is thereby derived from biogenic gypsum reduction (Germanov *et al.*, 1979).

The distance of migration is inversely related to the solubilities of the metals (Sonnenfeld, 1964; Wedapohl, 1971). Leitz (1951) found strong replacement features among sulfides in the caprock of a salt dome: Dolomitization is here followed by precipitation of galena; sphalerite replaces in part precipitated galena but not calcite; and finally, either may be replaced first by pyrite and later by marcasite. The iron sulfides replace not only other sulfides but also calcite and dolomite. The metals are leached from the rock salt below, whereas the hydrogen sulfide is supplied by migrating sour petroleum.

Where adequate amounts of sulfur are not available in a brine of high ionic strength, the complexed metals may even be released in native form. Brines derived from the Jurassic Louann salt contain 100 mg/liter lead and 360 mg/liter zinc. Cooling of this brine to 25°C weakens the complex association, and native lead precipitates into pumps of brine disposal wells (Carpenter *et al.*, 1974). Metallic lead is currently also being deposited side by side with some sphalerite from hot brines reaching the surface, and downdip from the Kara Bogaz Gol brine basin east of the Caspian Sea (Lebedev, 1967a,b). These brines move along faults as chloride complexes. Although the amount of metallic lead at this locality is significant (7150 kg in 3 years), the amount of zinc is very small (Lebedev and Bugel'skiy, 1967). Lead is inert at the geochemical hydrogen sulfide barrier where zinc precipitates, because zinc chloride complexes are very unstable. Although $NaPbCl_4^-$ complexes do exist (Giordano and Barnes, 1981), $PbCl_2$ here does not form here a complex with sodium chloride (Gromov, 1940). Instead, lead appears to travel by itself as a chloride complex or in colloidal form with a small radius and negative charge. The brine temperature ranges from 30 to 60°C (Pavlov and Dvorov, 1977), which is in the range typical for solar heating of brines in density-stratified basins. The pH is 6.8; the brine contains ammonia, methane, and some hydrogen sulfide. It has a high magnesium and very low bicarbonate content, which again is normal for brines derived from evaporite basins but unusual for formation waters. The content of 24 mg/liter iodine and 450 mg/liter bromine (Tooms, 1970) indicates a high degree of preconcentration. Similarly, the Red Sea hot brines exit at the sea floor surface at a temperature range of 36–56°C (Tooms, 1970). Very significantly higher metal concentrations than reported from the Red Sea brines are found in pore waters of argillaceous rocks associated with Miocene potash deposits in Soviet Transcarpathia. They are derived from brines that seeped out of the evaporite basin and contained an average of 300,000 ppm dissolved solutes, especially 12 ppm lead,

38 ppm zinc, and 3 ppm copper. In that respect they are similar to brines from the Cheleken Peninsula or from the North Afghan-Tadzhik Basin (Bogasheva, 1983).

Secondary sylvite contains as much iron, manganese, nickel, and copper as carnallite, but much more than primary sylvite (Morachevskiy, 1940; Galakhovskaya, 1954). This suggests that conversion of carnallite into sylvite and leaching of magnesium chloride did not mobilize the accessory metals. These may actually remain adsorbed onto hematite needles that were derived from desiccation of goethite. Adsorption of copper, lead, or zinc onto goethite occurs as hydroxide and increases with the pH, particularly in the presence of chloride (Barrow *et al.*, 1981).

The metals show marked regional variation (Herrmann, 1958); the horizontal variation in the Stassfurt cyclothem of the Zechstein evaporites exceeds one order of magnitude. The base metal content varies by a factor of 3–5 in halites of different cyclothems within the same basin and differs from basin to basin (Leitz, 1951; Thiede and Cameron, 1978). Exchange reactions of the precipitated evaporites with brine seepage are probably responsible for these variations (Koch *et al.*, 1968).

A link between evaporite brines and commercial sulfide deposits is apparent worldwide (Davidson, 1965). It was postulated, e.g., for Elk Point evaporites with Middle Devonian lead–zinc deposits at Pine Point, Canada (Sonnenfeld, 1964; Schmalz, 1969) and for Carboniferous lead–zinc deposits of eastern Canada (van de Poll, 1978). The relation of lead–zinc deposits to evaporitic brines has been supported by their identical enrichment in heavy sulfur in the Devonian occurrences of northern Canada (Folinsbee *et al.*, 1966), in Germany (Buschendorff *et al.*, 1963), in Lower Permian deposits of the Donets Basin (Korenevskii *et al.*, 1968), and in Japan (Sasaki, 1970). Light sulfur incursions represent increased runoff from land (Gavrilov *et al.*, 1977; Pisarchik *et al.*, 1977); biochemical reduction concentrates heavy sulfur.

Base metal ores formed in sedimentary rocks throughout the Phanerozoic history of the earth, almost without interruption. Stratiform metalliferous deposits underlain by continental red beds or other oxidized strata and overlain by evaporites account for approximately 30% of world copper production (Renfro, 1974). An interesting observation is that sedimentary copper ores are unknown from Late Cretaceous strata (Strakhov, 1962), as are halite and potash deposits (Sonnenfeld, 1978). The same coincidence holds for a time interval of similar length within the Lower Paleozoic.

Pyrite

More readily soluble sulfides require a more negative Eh and a greater hydrogen sulfide supply to precipitate directly. Pyrite needs the most negative Eh, followed by sphalerite, galena, and chalcocite in decreasing order of redox potential and decreasing solubility. Iron hydroxides, in contrast, require a positive Eh. Pyrite and hauerite (MnS_2) are frequently associated with sapropelic streaks in evaporite sequences. In anhydrites, they are evidently secondary after expulsion of iron from the gypsum lattice. Pyrite occurs in gypsum, halite, and basal potash layers of the Zechstein salt; hematite follows in higher levels of the potash deposit (Borchert, 1959). In Permian potash deposits west of the Urals, however, pyrite occurs in primary precipitates, whereas hematite is restricted to secondary sylvite and recrystallized carnallite (Ivanov, 1963). In salt deposits of Morocco, some complex crystals of pyrrhotite have been reported in addition to pyrite (Kosakevitch, 1966).

Schneiderhoehn (1923, 1926) postulated the presence of sulfidized bacteria in the Kupferschiefer sulfide deposits. Love (1958) described two forms of fossil bacteria obtained as a residue from the dissolution of framboidal pyrite. Framboidal pyrite aggregates are formed when coacervate droplets of ferrous hydroxide come into contact with hydrogen sulfide (Kizil'shteyn and Minayeva, 1973). In an analogy to the formation of framboidal magnesioferrite, Taylor (1982) proposed that framboidal pyrite is generated by the ferromagnetic properties of a precursor FeS polymorph or van der Waal forces accentuated by the presence of charged ions in a strong electrolyte. The idea of a bacterial origin of framboidal pyrite has been both challenged and supported (Siegl, 1941; Jung and Knitzschke, 1976) and is still unresolved.

Lead and Zinc

Lead and zinc ores precipitate as sulfides, often in heavily dolomitized carbonates geographically not far from the evaporite pan. Frequently, they occur in a marginal facies of an evaporite basin separated geographically from the evaporites (Samama *et al.*, 1978). Dolomitization is thus only a specific form of metallization (Sonnenfeld, 1964) and precedes lead–zinc mineralization (Fritz, 1969). The dolomites are ferruginous, i.e., ankeritic, which is indicative of an anaerobic environment free of iron-oxidizing bacteria. In both reefal limestones and calcareous clay intercalations within halites, organic NH_4 salts play an important role as catalyzers in reactions of magnesium chloride on calcium carbonate (Linck, 1912). Experimentally, Sozinov *et al.* (1982) proved that lead complexes of humic acids and of other organic compounds of high molecular weight intensively promote the incorporation of magnesium into the calcite lattice and retard any calcium carbonate precipitation.

Lead and zinc favor an association with dolomites and calcareous dolomites, whereas copper, nickel, and uranium accumulate preferentially in terrigenous clastics. However, a 13-mm seam of zinc carbonate is associated with gypsum in Cumberland, England (Hollingworth, 1948). Stalactites internally filled with galena, or galena and cerussite, form at low temperatures with the aid of sulfate-reducing bacteria, provided that the lead concentration exceeds $10^{-4.9}$ (or

1.2589^{-5}). Experimentally, a 1-mm crust can be produced in 2 weeks (Leleu and Goni, 1974).

Metals that form complexes, notably lead, suppress undesirable nucleation centers of precipitating gypsum (Barta *et al.*, 1971). Some calcium sulfate is then carried to the sites of metal sulfide precipitation (Kinoshita, 1931). Galena precipitates in warm brines containing Pb–Cl–S–H_2O, whereas Zn–Cl–S–H_2O yields sphalerite. An admixture of Ca^{2+} produces anhydrite. The equilibrium partial pressure of oxygen was calculated at 10^{-27} Pa (Kiyosu and Nakai, 1971). Downward-percolating brines from evaporite accumulations precipitated anhydrite in conjunction with base metals in fissures of Hercynian massifs in France (Arnold and Guillou, 1980), where zinc sulfides, in particular, tend to be associated with calcium sulfates. Tiny sphalerite, galena, and pyrite aggregates, or secondary anhydrite infilling of vugs, are restricted in dolomitized Devonian reefs of Alberta, Canada, to voids below the oil–water or gas–water contact. This indicates a sequential relationship between sulfide and sulfate precipitation in the subsurface and the accumulation of hydrocarbons. Sphalerite in lead–zinc ores often contains significant quantities of organic matter, particularly alkanes (Kranz, 1968; Rickard *et al.*, 1975), as does fluorite (Garcia-Iglesias and Touray, 1976). A genetic relationship between lead–zinc–barium deposits and crude oil has, therefore, been suspected (Connan, 1979).

The organic matter in each Mississippi Valley-type ore body differs in composition and thermal maturity (Gize and Barnes, 1981), and the ores contain small amounts of petroleum as liquid inclusions or as seeps in the mine workings (Roedder, 1976). This gave rise to the idea that soluble metal sulfates, upon encountering organic matter, are reduced to insoluble sulfides (Siebenthal, 1915; Bastin, 1926; Germanov, 1965; Barton, 1967). It is not certain, however, that the metals travel as sulfates in the anaerobic medium. Evans *et al.* (1968) suggested that lead- and zinc-enriched brines travel behind hydrocarbons, precipitating after the latter have migrated into porous traps. Other subcropping metal occurrences may be lixiviated by the passing bitterns. Dozy (1970) produced a model of metallic brines and hydrocarbons generated together; when mobilized, the heavy metal-bearing brines then travel downward, and the hydrocarbons move updip.

The major ions found in liquid inclusions in minerals of zinc and lead ores in Silesia, Poland, are chloride, sodium, potassium, and calcium (Karwowski *et al.*, 1979), which are typical of an evaporitic bittern entering a carbonate bank. Nevertheless, Herrmann (1961a) thought that the base metals are not indigenous to an evaporitic brine, but enter with ascending oil field waters through Kupferschiefer marls and lower evaporite members to settle in the Stassfurt series, the second Zechstein evaporite cyclothem. Such an interpretation, of course, begs the question of the origin of oil field brines, which, no doubt, were enriched by hypersaline brines seeping into the substrate during evaporite deposition. The lead- and zinc-rich oil field brines underneath the Gulf Coast region of the United States offer Br/Cl and Br/K ratios that clearly associate them with hypersaline brines concentrated beyond halite saturation. Carpenter *et al.* (1974), therefore, thought that they are derived from the Jurassic Louann evaporite body and its numerous salt diapirs. Similarly, zinc, lead, and copper are sometimes enriched in the black muds separating lacustrine evaporites in Saskatchewan from underlying permeable sands (Last, 1984).

Copper

Copper concentration has been observed in halite and polyhalite (Gard, 1968) but also in other potash salts (Gloeckner, 1914). Biltz and Marcus (1909) found 0.5 ppm copper in prepotash halites and up to 4.5 ppm in anhydrite, whereas evaporitic clays concentrated by absorption up to 6.6 ppm. Copper precipitation in parallel planes is known from anhydrite caprock, producing a banding effect (Barnes, 1933). Nodules of copper pyrites the size of chicken eggs have been found in German Zechstein anhydrites (Gloeckner, 1914); they also occur in associated clay intercalations. Of course, the close association of copper and other base metals with some parts of the Kupferschiefer Formation in German Zechstein deposits is well known. The ore is very often tied to nodules of plant material. Permian red beds associated with the Upper Kama evaporites are likewise copper-bearing (Ivanov and Voronova, 1975). Six stratiform copper deposits are associated with the Permian Basin of Texas and New Mexico (Johnson, 1974), the product of shallow marine sedimentation. Anhydrite and gypsum are interbedded with three layers of tuff in Japan: The anhydrite is here associated with oolitic and laminated copper–iron–sulfide ores, and the phase transitions correspond to anhydrite zonations (Kajiwara, 1970). Evaporites in Uzbekistan are topped by red beds as a mark of regression; these red beds contain intercalations of halite, glauberite, mirabilite, bloedite, and cupriferous sandstones. An earlier transgressive phase contained anhydrite, halite, sylvite, and carnallite (Petrov and Chistyakov, 1972). Cupriferous sandstones have also been reported from the Ilga Formation and are related to Cambrian evaporites along the Upper Lena River, Siberia (Narkelyun and Berzodykh, 1966).

EVAPORITE-RELATED BASE METAL DEPOSITS

Hypersaline brines concentrate base metals proportionally faster than the major cations, probably due to a biogenic concentration. Evaporitic precipitates thus become major repositories of such metals, both in their crystal lattice and in point distribution. Such metals are remobilized upon recrystallization of the precipitate, probably travel as chloride complexes, and are redeposited in surrounding porous strata wherever they encounter hydrogen sulfide.

11

Organic Matter and Petroleum

The indraft of an evaporite basin is rich in biota, whereas the bottom brines are very inimical to life. The concentrating brine rapidly reduces the number of species whose bodies can maintain the osmotic pressure needed to counteract desiccation losses in hygroscopic brines. The concurrent decrease in oxygen solubility quickly eliminates bottom-dwelling faunas, leaving the field to anaerobic bacteria. Consequently, much of the swept-in organic material cannot oxidize or be recycled by scavengers, and must accumulate in the brine. Eventually, it will settle into impervious sediments, attach itself to precipitates, or be flushed through a permeable substrate.

DECOMPOSITION

Bottom waters, no longer in contact with the atmosphere, become anoxic and can then support only an anaerobic bacterial flora. The preservation of carbonized spores and pollen indicates a reducing environment during deposition of the evaporites. Such spores and pollen occur in most rock salt and potash deposits. They are not destroyed in later alteration of the host rock by epigenetic fluids. Planktonic or nektonic biota that accidentally cross the interface into bottom waters, or dead organic materials falling through into the anaerobic brine, are not oxidized but are desiccated by the hygroscopic brine and are only in part degraded. Suspended shreds of macerated organic material accumulate beneath the pycnocline of a density-stratified brine system, take up hydrogen from flocculating clay particles or from oxidation of hydrogen sulfide, and are stripped of COO radicals. They are preserved in the anaerobic environment, because there is much less scavenging by autotrophs and even heterotrophs than in oxygenated waters. Proteins decompose first, but macerated fatty acids are remarkably resistant to breakdown in this environment (Hecht, 1934). Decarboxylation is facilitated by the presence in the brine of calcium hydroxides, which are then precipitated as calcite (Yamasaki *et al.*, 1981). In salt marshes, lipids disappear downward. The fatty acids are lost faster than plant-produced hydrocarbons. Branched chains increase, whereas unsaturated compounds decrease (Johnson and Calder, 1973). Lipolytic anaerobes occur in recent marine sediments, in oil field brines, in tar sands, and in natural asphalt (Cox, 1946). In the absence of sulfate ions, the carbon dioxide and acetate are reduced to methane in antarctic lakes (Burton, 1981), and higher homologs of the n-alkane series are produced in subtropical lakes.

It was shown in Chapter 7 that many decomposition products of organic matter have a pronounced influence on the rate of precipitation and the crystal habit of both sulfates and chlorides. Because potassium chloride has positively charged surfaces, saturated and unsaturated straight chain amines attach to the crystal faces (Scroggin, 1978). Alcohols complex with magnesium chloride to produce a very hygroscopic brine. When magnesium chloride complexed with cyclohexanol is heated, hexanes, i.e., members of the alkane series, evolve (Kuchkarov and Shuykin, 1954).

The suspended organic matter settles out periodically. Impermeable beds adjacent to evaporites (shales or micritic limestones) are often bituminous; evaporite beds frequently contain streaks of organic matter or even faunal remains. Reuss (1867) reported an exceedingly rich fauna preserved in the Miocene salt in Poland and Romania. Black bituminous stringers occur in ancient anhydrites, polyhalites, halites, and kieserites (Rozsa, 1913b), as they do in contemporary gypsum and halite precipitates (Bradbury, 1971). Often the impurities in halite are connected by rods and branches of carbonaceous matter (Alling, 1928). Such markers can be followed within halite layers for scores of kilometers. The settling out is electrostatically induced during a seasonal change in electrolyte concentration.

STINKSTONES

In precipitating porous gypsum or halite, the accruing organic matter can be flushed into the subsurface. Gypsum from modern salinas is slower in setting than quarried gypsum, because it still contains unflushed organic matter (especially oil). Its setting time is quickened by simply washing in petroleum ether or by calcining to 200°C (Hiyama *et al.*, 1957). Gas has been reported in gypsum from Sicily (Sjoegren, 1893), and in anhydrite and underlying salt in the northern Caspian region. The latter is up to 94% burnable and emanates a faint hydrogen sulfide smell; anhydrite with halite veinlets often smells of gasoline or crude oil. These hydrocarbons are not strictly syngenetic, but are deemed to be derived from bituminous laminated anhydrites (Korobov and Tuchfatov, 1961). Calcite-lined fractures in Jurassic dolomites of the Russian Platform contain calcite-covered bitumen and, in turn, anhydrite; the anhydrite precipitated intermittently with the organic matter (Ripun, 1965).

In clays or micritic limestones, the organic matter is trapped. An anhydrite deposit in Bosnia, Yugoslavia, which is up to 100 m thick, is interrupted every 20–40 cm by a very thin, fetid black bituminous clay (Pamić, 1955). The clay laminae act here as blotters. A sapropelic shale member underlies the whole series of evaporitic cyclothems in the British Zechstein (Smith, 1970), but unlike the German one, it does not contain commercial quantities of base metals (Love, 1962). To preserve the sapropelic matter, the waters must have been oxygen deficient well before the first evaporite cyclothem started, since only the basal dolomites of each cyclothem are bituminous. Another example is the fine-grained, dark gray limestone intercalations in Frasnian salt deposits of the central part of the Dnieper–Donets syncline that have been subjected to extensive dolomitization and anhydritization. They are highly bituminous. The organic component increases with distance from the axial part of the syncline. However, there is a sharp distinction between these bitumina and the fracture fillings of younger oil and oil-resinous substances (Zhuze, 1972).

Organic matter forms pyrite-rich sapropels and is buried in stinkstones, bituminous nearshore limestones, and dolomites that originally lined the evaporite basin. The Cretaceous evaporites of Zipaquira, Colombia, are capped by black bituminous shales and are flanked by limestones containing alkane-base crude oil (Ujueta, 1969). The organic matter both adsorbs and incorporates uranium out of the brine (Amiel *et al.*, 1972). However, evaporites themselves are among the sediments that are poorest in uranium; they contain no more than a few parts per million.

In dolomites associated with the Mansfield Kupferschiefer Series, the bitumina are not leaked from adjacent shales but are primary components predating even the dolomitization event (Stutzer, 1933). There is a causal relationship between dolomite, dolomitization, and the organic matter present (Sonnenfeld, 1964; Davies *et al.*, 1975). To date, the bulk of natural gas production in Germany is derived from the dolomite intercalations in the evaporite cyclothems of the Zechstein evaporite (Richter-Bernburg, 1957a). The gas and oil are here related to a brine that has no connection to formation waters; otherwise, the gas would have leaked. The gas also occurs in closed caverns within the potash and halite horizons (Grimm, 1968).

HYDROCARBONS IN SALTS

The earliest record of a gas explosion and fire in a German salt mine dates back to 1664 (cf. Schauberger, 1960). Explosions and fires have also occurred in carnallitite mines near Solikamsk, west of the Urals, and in Miocene potash mines of Poland. The escaping gas is mainly a mixture of hydrogen, methane, and nitrogen, occasionally leaving behind a white, waxy substance (Chirvinskiy, 1946; Poborski, 1959). The earliest report of oil, in droplets and negative crystals, then called "mountain tar," was made by Fichtel in 1791 (cf. Pelikan, 1891). Inclusions of droplets and veinlets of oil and bitumen occur frequently in salt rocks at many localities in the northern Caspian region (Korobov and Tuchfatov, 1961). "Paraffin earth," a material of waxy consistency (Taylor, 1972), is apparently the by-product of bacterial metabolism, resulting in gypsum reduction to native sulfur, a process fueled by natural gas seepages into caprocks over salt.

Increasing salinity of brines and formation waters decreases the solubility of hydrocarbons (Faingersh, 1977), since nearly saturated levels of solute content alter the water structure to promote hydrophobic interactions of macromolecules (Borowitzka, 1981). Nonetheless, liquid and gaseous hydrocarbons are common in relatively undisturbed salt and potash sequences, and are evidently formed *in situ* (Peterson and Hite, 1969). Only in the upper parts of the evaporite sequence, in potash deposits, are streaks of organic matter often absent. Yet, interstitial and intercrystalline quantities of hydrocarbons are much greater than they are in underlying halite and anhydrite sequences. Both asphalt-free crude oil (Precht, 1907; Graefe, 1908) and H_2-enriched gaseous emanations (Erdmann, 1910a) are encountered in potash mines. In part they are derived from organic matter in the potash deposits and in part delivered by ascending solutions (Elert and Freund, 1969). Highly bituminous halite, containing burnable gases and emanating a crude oil smell, has been reported from Permian evaporites in Poland (Dette and Stock, 1957) and Germany (Hartwig, 1936). Crushed sylvite from the Pennsylvanian Paradox Basin, Utah, contains 0.30 weight % extractable oil, and many oil seeps are visible in a Utah potash mine (Peterson and Hite, 1969).

Among the hydrocarbons found in evaporites, there is a preference for even–C n-alkanes. A dominance of even–n C_{28}–C_{30} alkanes has been noted and could be due to metabolism of *Desulfovibrio* (Han and Calvin, 1969; Spiro, 1977; Sheng *et al.*, 1981). Isoprenoid hydrocarbons, such as phytane, dominate over pristane. Hence, a reducing environment can be inferred for their origin, because they are derived from phytol, an allyl alcohol that is particularly easily hydrated. Triterpanes of high sterane hopane series are subordinate or missing altogether, yet steranes occur in primitive algae; even–C n-alkanes are predominant in sulfate-reducing (*Desulfovibrio*) and sulfide-oxidizing bacteria.

The complete absence of norhopanes (C_{27}–C_{29}) points to a different maturation history from hydrocarbons in other source rocks (Hollerbach, 1980). Lower Cambrian rock salt in eastern Siberia contains gaseous alkanes in concentrations between 0.7 and 5.0 ppt, if there is no oil associated, but 53.3 ppt in samples with petroleum (Vdovykin and Shorokhov, 1980). Both members of the

C_nH_{2n+2} and the C_nH_{2n} series have been identified. Some of the gases are syngenetic, and some are epigenetic from underlying formations (Rabotnov *et al.*, 1980). Petroliferous bands in Devonian Elk Point anhydrites in western Canada often occur in vague veins, but bitumen in halite varves appears to spread into the adjoining salt (Klingspor, 1969). Permian halite west of the Ural Mountains contains 220–300 ppm bitumina (Ivanov, 1938). Similarly, the Silurian Salina salt of Michigan emanates a petroliferous odor from carbonaceous and pyritiferous laminar intercalations (Dellwig, 1955).

Schmalz (1969) assumed that a bituminous shale is characteristic of the onset of an evaporite cyclothem. However, within individual Zechstein cyclothems, dolomites are bituminous and the shale intercalations are not sapropelic, but indicative of a freshwater influx (Richter-Bernburg, 1953). In Oligocene evaporites of the Upper Rhine Valley the bituminous salt and bituminous marls occur only above the basal nonbituminous salt and nonbituminous dolomites (Gunzert, 1961). Even the clays are enriched in bitumina in the upper salts, where the halites and carnallites themselves contain 4 weight % bitumina (Wagner, 1953). The organic matter here consists of a yellow, waxy residue and smells of crude oil (Goergey, 1912).

Carnallite often contains a large amount of gases under high confining pressure in very fine, microscopic pore spaces that are not interconnected. When carnallite is dissolved in water, the pore walls tend to break under the gas pressure; thin lamellae form at first, and gas escapes along the laminations. When the salt is either strongly heated or quickly dissolved, it decrepitates. One can then hear a characteristic crackling (''crackle salt''). The tiny bubbles of gas that escape are primarily composed of nitrogen, hydrogen, and methane in the Upper Kama deposit west of the Urals. Both methane and higher homologs are trapped in significant quantities in carnallite crystals (Ivanov, 1963), but varicolored sylvite is at times much richer; striped sylvites have less gas, and red sylvites have the lowest gas content (Morachevskiy, 1938; Mueller and Heymel, 1956). ''Crackle salt'' is Miocene halite in Poland (Rose, 1839) has a weakly bituminous taste. In the German Zechstein deposit, the carnallite may contain a significant quantity of carbon dioxide. In the Zipaquira mine in Colombia, there is also a gassy halite that decrepitates with a crackling or popping sound when struck (McLaughlin, 1972).

Primary sylvite also contains nitrogen of exclusively biogenic origin (Morachevskiy *et al.*, 1937). This is related to the extensive decomposition of organic matter during sylvite precipitation, especially where it is secondary after carnallite. In other instances, carnallite contains the largest amount of gas (Aleksandrovich *et al.*, 1963), particularly if it occurs in fracture fillings (Polyakov *et al.*, 1971). Potash salts in the Permian basins west of the Urals contain 70 times as much gas as associated halite crystals (Morachevskiy, 1940; Ivanov, 1963). The gases contained here in crevices and cavities, or dispersed in crystals of the carnallites, contain 22–41% alkanes and 6–9% water. Hydrogen, nitrogen, and traces of noble gases (under 0.1%) constitute the remainder. Other analyses give higher percentages of methane and nitrogen, both as free gas and as inclusions in crystals (Sokolov, 1957). In sylvinites the gases contain no water, 0.15–0.22% noble gases, and 25–52% alkanes; the remainder is nitrogen. Alkanes make up only 21% in halites, but in addition to nitrogen, halites carry 31% H_2 (Andryukov, 1940; Nesmelova, 1961).

The gaseous microinclusions in salts of the Inder salt dome north of the Caspian Sea might be primary, since all of the surrounding strata are impervious. They are either composed of 94% methane and 6% heavier hydrocarbons, or contain up to 13.2% heavier hydrocarbons but are then coupled with up to 24% carbon dioxide (Kapchenko *et al.*, 1973). The N_2 and carbon dioxide contents are thereby mutually compensating in these Permian evaporites (Vil'denburg, 1975). However, primary Zechstein salts never contain carbon dioxide, as it is restricted to alteration zones (Grimm, 1968).

To argue that gases and heavier hydrocarbons in salt sequences are immature precursors of natural gas and crude oil, one must elucidate the fate of the nitrogen so dominant in many occurrences. If a major part of the hydrocarbons was consumed by bacteria in the early history of these deposits, the gases should have been enriched in nitrogen. Devonian evaporites in the Pripyat depression of Byelorussia, for instance, contain 81–93% nitrogen; sylvite contains more hydrocarbons, helium, and hydrogen than does halite (Dorogokupets *et al.*, 1972). In the Permian Upper Kama deposits, nitrogen constitutes more than 50% of the gaseous inclusions and up to 70% of mine gases; the remainder are hydrocarbons. Milky white sylvinites here are especially enriched in gaseous inclusions (Bol'shakov, 1972). No major sink for nitrogen, biogenic or not, occurs within the sedimentary column. Its eventual escape to the atmosphere would simultaneously allow all other volatiles to leave. Thus, these must represent residual gases unable to escape. Gases dissolved in highly concentrated brines associated with the evaporites of the eastern Russian Platform contain 70–98% methane and higher homologs (Zor'kin *et al.*, 1979), but do not appear to contain major amounts of nitrogen.

Immature, gasoline-range hydrocarbons, and gaseous hydrocarbons are seeping into overlying sediments from Upper Miocene evaporites beneath the Mediterranean Sea (McIver, 1973). Similarly, Burk *et al.* (1969) found immature, highly aromatic and asphaltic low-density (high-gravity) oils in the caprock of the Sigsby Knoll, associated with 19% native sulfur. The site is 320 km from shore, covered by 3600 m of waters of the Gulf of Mexico. Only because of its apparent immaturity was this oil interpreted to be very young, whereas underlying salt is Late Jurassic in age. The native sulfur is evidently the product of oxygenated seawater penetration into this caprock (Davis *et al.*, 1970).

Tectonically disturbed or squeezed evaporite sequences lose their hydrocarbons to adjacent aquifers. At least some of the hydrocarbons trapped in so many reservoirs abutting salt domes must have come out of the evaporite sequence itself. Several gas blowouts occurred during the drilling in salt in the western part of the Caspian Basin; the salt cores contained drops of oil. The salt was associated with oil-saturated anhydrites and limestones (Skrotskiy, 1971, 1974a). No immediate reservoir abuts the salt to serve as a provenance for the migration; any suspected faults would have their fault planes sealed by plastic halites. The hydrocarbon shows were, therefore, deemed to be evidence of a vertical migration through the salt.

Hydrocarbon Preservation

There is a distinct difference between the hydrocarbons accumulating in oxygenated surface waters and in anaerobic bottom brines. The former contain primarily undersaturated olefins, the latter saturated homologs (Linnenbom and Swinnerton, 1970). Sandy bottoms in well-oxygenated waters have a positive redox potential below the depositional interface (ZoBell, 1946) where organic material is oxidized and destroyed, but sediments beneath a density-stratified brine have a negative redox potential and preserve organic matter. Hydrocarbons can be preserved for any length of time only in an anaerobic environment. That precludes hydrocarbon accumulation in oxygenated waters or sediments of open marine bays or the world ocean, despite their massive production of organic debris. Anaerobic bottoms of oceanic depths that are created by overturning vertical water circulation generate sapropels, but such open areas are prone to seasonal flushing. An example is the offshore areas of Namibia, where excess hydrogen sulfide currently percolates to the water surface (Martin, 1963).

Deltas and evaporite basins are both marked by high subsidence rates and, consequently, high rates of sedimentation; both also offer high biomass production rates and density-stratified waters with anaerobic bottom brines. Here the grazers, scavengers, and other forms of higher benthonic life cannot survive, bioturbation does not take place, and the organic matter is subject to only limited bacterial decomposition. Evaporite basins also offer access to aquifers and reservoirs by vertical penetration or lateral migration through an initially very porous substrate. However, dissolved organic material does not concentrate in salt basins as rapidly as the salts themselves because part of it is oxidized. Krejčí-Graf (1962) pointed out that worldwide, one can find petroleum only extremely rarely in horizons that are unquestionably of freshwater origin. Even then its occurrence is related to migration patterns out of nearby marine evaporite sequences. Yet biota capable of producing crude oil components are equally present in both lakes and seas. The density-stratified, anoxic lacustrine environment, represented by the Green River shales of Wyoming, has not led to crude oil generation, but to other organic sediments. Evidently, a marine brine environment is essential to petroleum generation.

Even though deltas and areas around evaporite basins are prime areas of hydrocarbon generation and preservation, there are significant differences. After maturation, the typical hydrocarbon accumulation in a deltaic environment is a sweet crude oil or a sweet natural gas, i.e., one containing less than 5 ppt sulfur compounds. In contrast, hydrocarbon accumulations derived from evaporitic environments are normally sour crude oils or sour gases, i.e., ones having a substantial admixture of hydrogen sulfide. Before burial under a great load of overburden, the hydrocarbons in deltaic sediments tend to be rich in asphaltenes and higher homologs of various alkane and naphthene series, whereas immature crude oil from young evaporite beds tends to be a high-gravity gasoline-range petroleum. The organic matter in evaporites is not only derived from particular organisms but is also altered in a characteristic manner by the reducing environment. The vitrinite reflectivity, as a measure of maturation, sets oil genesis in clastic sediments at a value of 0.5–0.7. Equivalent maturation is reached in evaporites at a vitrinite reflectivity of less than 0.4 (Hollerbach, 1980).

Catalytic Reactions

Crude oil moving through rock salt is undergoing changes. About 0.11–0.37% is adsorbed. Resinous and asphaltene compounds increase by 3–4%, probably due to traces of sulfates present in the salt. Alkane structures increase their chain lengths (Mileshina and Komissarova, 1972). In potash beds within the Stassfurt Member of German Zechstein evaporites, a change is observed from heavy oils in one shaft to alkane-rich, light oil in an adjacent one; this is attributed to a filtration through the potash deposit (Graefe, 1910).

Although an anaerobic brine may be essential to the preservation of organic matter, the specific lithology of the brine floor is also of paramount importance in hydrocarbon generation. On a gypsum crust, any free fatty acids and hydrocarbons adsorb on calcium sites in the gypsum lattice; in natural environments they are preferentially branched chains (Barcelona and Atwood, 1979). Organic acids also adhere to calcite in a watery medium, but not to quartz (Ponahlo, 1979). Phosphates compete with fatty acids for adsorption onto calcium sites in the presence of magnesium ions (Lahann and Campbell, 1980). The finer the grain size of the carbonate sediment, the greater its specific surface area and the greater the adsorption of organic carbon, nitrogen, and phosphorus (Suess, 1970). Fatty acids attack calcium carbonate, and the reaction products disintegrate into members of the alkane series and into other heavy crude components at a temperature well below 180°C (Kuenkler and Schwedholm, 1908). Exothermic anaerobic degradation of algal matter by bacteria results in the formation of light hydrocarbons, specifically methane and higher homologs, that then can be stored in carbonate oolites (Ferguson and Ibe, 1981).

Biogenic reactions continue even after crude oil has migrated into a reservoir: Sparry calcite cement develops along the oil–water interface, and the amount seems to be proportional to the density of crude oil or inversely proportional to API gravity (Ashirov and Sazonova, 1962). In addition, acids produced by bacteria release adsorbed petroleum compounds in carbonates, increase voids, and increase the internal gas pressure (Cox, 1946). Aluminum chlorides, most certainly present in traces in brines capable of producing koenenite or loewigite, convert animal fats into a substance resembling crude oil (Ochsenius, 1898). There is an extensive literature on the cracking effects of aluminum chloride on a variety of organic compounds and the use of this knowledge in the refining process. Sodium tetrachloroaluminates solubilize at low-temperature kerogen components of Green River oil shales of Wyoming (Bugle *et al.*, 1978). Naphthalenes likewise are turned, under the influence of a small admixture of aluminum chloride, into saturated hydrocarbons at the expense of a residue of tar and carbon (Fischer, 1916).

Thus limestone, in reacting with fatty acids, fosters the generation of liquid bitumina that are mobilized in permeable carbonates, whereas clay polymerizes bitumina into solids (Stutzer, 1933). This leads to an initial immobilization of bitumina in shales and mudstones. Surface-active clays easily trap hydrocarbons: Palygorskite and sepiolite have active surface areas more than 100 times larger than the finest-grained carbonates (Suess, 1970; Serratosa, 1979). Polar organic molecules are bonded to interlayer sites. The potassium in muscovites can be replaced by NH_4 or by a variety of organic chains (Amphlett, 1964). Hydrocarbons are altered through proton-donor properties of montmorillonites. Catalytic cracking seems to occur where montmorillonite changes to mixed-layer varieties. Transformation of a long-chain molecule to smaller units by rupture of specific carbon–carbon bonds takes about 42 kJ/mole less energy in a montmorillonite system than in a uniquely hydrocarbon environment (Velde, 1977). Decarboxylation and hydrogenation are fostered in mixed-layer clays; illitization renders these clays inactive. The formation of mixed layer clays thus accompanies the transformation of dispersed organic matter into syngenetic hydrocarbon fluids. Even-C n-alkanes of high molecular weight are sorbed by mudstones, particularly montmorillonite, and odd-C n-alkanes are sorbed by dolomite, whereas odd-C n-alkanes of low molecular weight filter through both rock types (Safonova and Mileshina, 1972). The increase of mixed layer swelling clays is due to the flow of hydrocarbon fluids out of the beds. A montmorillonite cap forms later by oxidation of hydrocarbons issuing from the deposit (Petukhov *et al.*, 1981). Clay laminae may not only act as catalysts. Nesmelova (1959) thought that free gases in potash salts are related to biochemical processes in clay intercalations, as the amount and composition of the gases changes from layer to layer.

It has been noted that the red bed facies, which is closely related to evaporites, contains clays that have a pronounced catalytic effect on sedimentary organic matter and hydrocarbons (Veber and Gorbunova, 1969). They are the types of clays that a dust storm blows from surrounding flat lands into an evaporite basin. Hitherto, only the effect of high temperature on dehydration of montmorillonite to mixed-layer varieties has been investigated in terms of hydrocarbon maturation. The effects of strongly hygroscopic brines and bitterns on clays have not been studied in terms of hydrocarbon mobilization.

FLUSHING

Obviously, the organic materials settle out indiscriminately over the whole floor of the brine basin. When they settle out over clays or dense calcilutites, they are trapped, but not if they settle over a highly porous substrate, a skeletal or reefal limestone, or a gypsum crust. The quantity of bitumina found in black shales associated with evaporites can be high; there are up to 8% by weight of organic carbon in shales of the Pennsylvanian Paradox Basin (Peterson and Hite, 1969). However, this is not a measure of the total organic matter produced in an evaporite basin. It is only a measure of the fraction trapped in impervious brine floor sediments, and gives no indication of the amount flushed into aquifers and conduits through pervious brine floor strata.

It is most instructive that basement fractures in the Cretaceous Sergipe Basin of Brazil contain trapped oil wherever they are overlain by the now-impermeable evaporitic Ibura Member. The oil is certainly not derived from the basement, but from the evaporites above. Some of the commercial oil and gas is trapped in conglomeratic sandstones sandwiched between salts and crystalline basement rocks. However, wherever this sequence of sodium, potassium, and magnesium salts had been underlain by up to 50 m of clays, these were turned into bituminous, blackish-brown shales, locally containing numerous plant and animal remains (Meister and Aurich, 1972), but no free oil. The clay acted as a retaining sponge, but not as source rocks. Pockets of volcanic ash beds and clean sands overlying basalt and underlying Lower Cretaceous anhydrites and halites in Angola, correlative to the Sergipe evaporites, are producing oil, with no apparent source rocks. Black shale of Tertiary age above the salt has foraminiferal tests filled with oil and gas bubbles. Such bubbles also occur along bedding planes in that shale, and are judged to be *in situ* hydrocarbons (Brognon and Verrier, 1966). Chemical analysis of the hydrocarbons and other organic matter in a Zechstein carbonate reservoir in Bavaria, Germany, proved that not only do the hydrocarbons have no evident source rocks, but that the dolomite hostrock causes a petroleum maturation that is not shared by hydrocarbons trapped in impermeable anhydrite–dolomite interbeds (Leythaeuser *et al.*, 1981).

Bitumina are associated with metals concentrating in an evaporite pan. In a barite deposit in southern France, the bitumina are associated with metal sulfides, particularly galena. Most of the bitumina present do not resemble the chloroform extract of the presumed source rocks nearby, a bituminous shale (Connan and Orgeval, 1976). This may be due to biodegradation, which is directly related to the amount of sulfur present in the vicinity. Bitumina are similarly associated with galena and sphalerite at Les Malines, France (Connan, 1980), and in Sweden (Rickard *et al.*, 1975); with galena and sphalerite in fluorite in England (Pering, 1973); and with sulfate and halite in fluorite in Tunisia (Touray and Yajima, 1967). An association between organic matter and ore deposition (copper, silver, zinc) has also been noted repeatedly in the Zechstein Kupferschiefer horizon (Haranczyk, 1961; Tokarska, 1971; Wedepohl, 1971; Pering, 1973). This is certainly not accidental. Just as base metals travel as chloride complexes, so do some organic compounds. Complexed amino acids, probably attached to humic acids, increase their solubility with the rising salinity of the brine (Degens *et al.*, 1964). Furthermore, copper chlorides and bromides form complexes with long-chain aliphatic amines (Burkin, 1950).

High rates of salt precipition increase the geostatic pressure on the subsalt strata and the pore pressure; they increase the pressure gradient between hydrocarbon-bearing strata and reservoirs, and promote migration (Dzens-Litovskiy, 1967). The organic matter is subject to hydraulic pumping (Sonnenfeld and Hudec, 1977, 1978) and is thus capable of being flushed into the substrate. Filtration through the sediment at 20–40°C favors alkanes and naphthenes over aromatics (Kalinko *et al.*, 1978). The diffusion coefficients for potassium chloride, or sodium chloride, and amino acids are entirely related to their volume and shape of diffusing molecules and not to their molecular weight (Mehl and Schmidt, 1937). The temperature-dependent molal volumes and viscosities of brine components were determined by Fajans and Johnson (1942).

A relationship between gas pools and evaporites has been recognized (Goncharev and Kulibakina, 1971), since increasing salinity decreases the solubility of hydrocarbons in formation waters. They can then migrate in a separate phase of liquids (Polivanova, 1977) and gases first lightened isotopically by bacterial activity (Faingersh, 1977). Sodium chloride solutions concentrated to more than 58,500 ppm exert an electrostatic attraction for amino acids at 25°C. At the same time, repulsive forces are generated that increase with both the length of the hydrocarbon chains and the dielectric constant of the solute (Norman, 1935). Excess sodium ions in the percolating brines convert oil-wet host rocks to water-wet rocks. Excess sodium in formation waters can thus mobilize hydrocarbons and drive them into suitable conduits wherever an initial permeability is present. This is a feature utilized in driving crude oil out of oil-wet reservoirs by flooding pore spaces with water spiked with sodium hydroxide

(Leach *et al.*, 1962). Degens and Paluska (1979) thought that the interaction of hypersaline brines with organic detritus results in the generation of oil in oil shales. Fluid inclusions in recrystallized halite often contain a light petroleum as liquid phase (Kityk and Petrichenko, 1978). However, the sulfate cycle is also important. The bacterial reduction of sulfate to hydrogen sulfide prevents the formation of methane and stimulates the formation of bitumen by the epigenetic alteration of organic matter. Methane is common only in bottom waters of the Black, Azov, and Caspian seas, where hydrogen sulfide formation is absent (Yefremova, 1978).

Bromine concentrations have been used as an oil-finding tool in the region of the Kama River west of the Urals (Kapchenko and Soboleva, 1973), because hypersaline brines enriched in bromine take hydrocarbons with them when they move out of the evaporite basin. However, even a cursory examination of the differential solubilities of the individual components in any crude oil pool suggests that the oil has not traveled in simple solution, particularly if one keeps in mind that solubilities rapidly decrease with the rising salinity of a brine. If the hydrocarbons had been in solution, they would have required inordinately large volumes of brine as a carrier.

Methane solubility in brines appears to be independent of the anions in the solutes, but is substantially reduced in the presence of small quantities of dissolved nitrogen (Hanor, 1981). Its solubility increases with the increasing dominance of bivalent cations (Stoessell and Byrne, 1982; Byrne and Stoessell, 1982), according to the sequence $Na < K \ll Ca < Mg$. Magnesium solutions would thus have to dissociate from methane when they are stripped of their magnesium content by dolomitization or ion exchange with clays. This does not only apply to gaseous hydrocarbons. Calcium and magnesium salts of fatty acids yield petroleum as they are heated (Kuenkler and Schwedholm, 1909).

Most oil is emitted during peak hydrocarbon generation. The dikes of hydrocarbons in the Green River Basin of Wyoming and Colorado are not only proof of the existence of a separate phase migration but also indicate a migration early in the maturation history, other conditions being favorable (Jones, 1978). Postdepositional maturation of the hydrocarbons sets in, with progressively deeper burial of the fluids. If maturation were not sufficient, organic carbon would be present in substantial quantities, but hydrocarbon accumulations would not occur (Wardlaw and Reinson, 1971). The hydrocarbons cannot be flushed out from impervious, compacted claystones or from essentially impermeable halides. They must either move out before these rocks lose their permeability or wait until enhanced temperatures and pressures of a deeper burial convert some of these tight rocks into additional source rocks, and then move out before increasing compaction forestalls any further fluid movements. Experiments have shown that clays containing about equal amounts of crude oil and water keep almost 50% of the crude oil, even at a pressure of 5.2 MPa (Snarsky, 1962). However, hypersaline brine movements cause shales to become permeable by altering the clay structure. In some instances, they are thus able to flush hydrocarbons trapped in clays and utilize the catalytic cracking of clay particles.

Potonié (1928) had argued, a generation ago, that oil was not derived from oil shales. He pointed out that if crude oil really migrated out of bituminous shales under the influence of increasing pressures and temperatures, we should find individual fractionation products segregated geographically. They would migrate through aquifers separately as they were mobilized, one after another, according to the various threshold pressures and temperatures of fractionation. Instead, crude oil initially migrates continuously in very small traces.

All classes of sedimentary rocks, not only shale sequences, contain enough organic matter to be potential source rocks for oil and gas (Meinhold, 1980). The nonmobilized organic residue is converted to kerogen; the mobilized portion is exposed to catalysts in the pore water and in aquifer walls. Ultimately, it is converted to crude oil. Over three-quarters of all hydrocarbons known in Europe are stored under evaporites (Solowjow, 1978); the same is true of the prolific Middle East and the USSR (Kozlov, 1978). The paleo-latitudes of oil fields (Irving and Gaskell, 1962) cover a low-latitude belt almost identical to that of worldwide evaporite deposits (Irving and Briden, 1962; Gordon, 1975). The association of hydrocarbon accumulations in North America with evaporites in the Michigan, Paradox, Delaware, and Midland or Elk Point basins, to name only a few, is surely not accidental. This has repeatedly been pointed out (Woolnough, 1937; Moody, 1959; Hedberg, 1964). Weeks (1958) noted that evaporite basins are always very favorable for the formation and preservation of hydrocarbons on a grand scale, and that most crude oil reserves in carbonates are covered by evaporites. The correlation of salt sequences with crude oil occurrences is not new; it goes back to Harbort (1913) and was again revived by Szatmari (1980). A regional evaluation of paleo-migration patterns of percolating brines and bitterns from evaporite basins is a powerful exploration tool for stratigraphic traps containing crude oil or natural gas reservoirs.

SUMMARY OF ORGANIC MATTER IN EVAPORITES

Surface waters of hypersaline lagoons are extremely rich in plankton and nekton. In the photic zone occur photosynthesizing algae and bacteria. Below it the bottom brines, because of their lack of oxygen, are populated by anaerobic bacteria that largely decompose organic matter falling through the interface or discarded from the layer of photosynthetic biota. Resulting bituminous compounds are deposited in fine-grained sediments, such as stinkstones, or are trapped in halites and potash deposits. It is possible that bottom sediments, porous during evaporite deposition, allow bitumina to migrate with hot bitterns into aquifers after they have been catalytically altered by calcium and aluminum ions. These bituminous substances later mature into hydrocarbons at temperatures of deeper burial, without having been parked for a time in impervious strata.

Part IV

POSTDEPOSITIONAL ALTERATIONS

12

Syngenetic and Epigenetic Brine Movements

A precipitate is exposed to brines approaching from above, below, and laterally. These brines are of deeper formation origin, or represent residual bitterns from the precipitating brine, are younger seawater lying on top of covering strata, or are of meteoric origin (Fig. 12-1).

Syngenetic brines are derived from the same body of water that precipitated the original crystals. Over a period of time, these brines may have concentrated further and may even have lost some of their solute to ongoing precipitation. As very concentrated hypersaline brines, they become gravitationally unstable at the earth surface and penetrate down. As long as a permeability to brines persists,

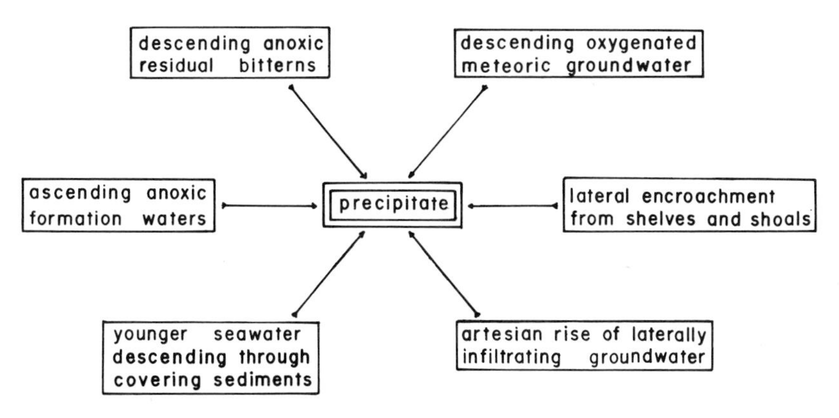

Fig. 12-1. Source of brines percolating through a precipitate.

these brines percolate past the precipitates and play a role in diagenetic or even epigenetic processes.

Epigenetic brines are derived from a body of water that did not crystallize the deposit. They enter a buried evaporite bed from above, below, or laterally along intercalations. They are slightly or substantially younger than the evaporite. Nonetheless, their effect at times is not distinguishable from alterations induced by syngenetic brines, particularly if they are derived from a new cycle of hypersaline brine formation. Therefore, the effects of the two types of brine are discussed together. *Epigenetic brines lead to the formation of pseudomorphs and crystal molds or casts. Syngenetic brines percolate before the precipitate has been consolidated and thus lead to recrystallization without the formation of pseudomorphs* (Stewart, 1956).

DIAGENESIS VERSUS METAMORPHISM

Borchert (1959) distinguished between diagenetic changes to freshly precipitated, metastable evaporite minerals and metamorphic changes to stable evaporite minerals that later equilibrate with unsaturated brines passing by. Ivanov (1953) defined metamorphic changes somewhat differently. He recognized three types:

1. Syngenetic metamorphism comprises changes that occur during or immediately following the crystallization process of salts, when the precipitate is still uncompacted, crumbly, and covered by brine. This approximates the diagenetic stage of Borchert (1959).

2. Diagenetic metamorphism consists of the reworking of the original salt deposits into evaporitic rocks. It includes changes due to percolating brines from the open evaporite basin.

3. Epigenetic metamorphism is produced by the action of external agents and forces that have no intrinsic genetic relationship with the evaporite deposit. It takes place after the formation of an evaporite deposit, and in some instances continues to the present. There are three basic types of epigenetic metamorphism:

a. Hydrometamorphism, induced by the action of meteoric waters or of migrating subsurface brines.

b. Tectonic metamorphism.

c. Thermometamorphism.

Evidence for the latter two types is usually not found in sedimentary evaporite rocks. Whereas limestones and dolomites find their equivalents in marbles of dynamic metamorphism, there are no such equivalents of chloride or sulfate rocks. Apparently, they are remobilized and squeezed out under the influence of incipient tectonic pressures and temperatures.

INITIAL POROSITY

A fresh precipitate of evaporite minerals is initially very porous; most of these minerals precipitate as a loose, watery slush that gradually firms up. *Freshly precipitated evaporites have a porosity of 10–50%.* This is true of Recent gypsum, but also of some ancient anhydrite. The gypsum in the sebkha El Melah, Tunisia, has a porosity of 25–54%, with permeabilities in the 10^{-3}–10^{-6} cm/s range (Perthuisot, 1975). Common along the Trucial coast, Persian Gulf, are up to 1-m-thick gypsum crystal mushes composed of flattened, discoid, or lensoid crystals (Kinsman, 1969). Large crystals intergrow during recrystallization and leave a considerable volume of connected pores and cavities filled with brine. Even ancient anhydrite deposits can have a volume of intercrystalline voids amounting to 34.15% of the total rock volume (Ul'masova, 1971). In Permian evaporites of northeastern England, the contacts between marl schlieren and halite are always sharp, but the boundaries between halite and anhydrite are irregular, as if the anhydrite had been porous when the halite was crystallizing. Moreover, the anhydrite is commonly more coarsely crystalline near halite contacts (Dunham, 1948), the mark of a decreasing rate of precipitation. Sapphire-blue halite crystals, up to 50 mm long and with intercrystalline void spaces up to 15 mm in diameter, on top of sylvites derived from leaching of carnallites on the Khorat Plateau, Thailand, represent a dissolution front that left a porosity of about 40% (Hite, 1982). The porosity has been retained only because of the near-surface position of the layer; deeper burial would have compacted the rock.

Recent halites are equally porous. Recent salt deposits in Tunisia possess 35–45% porosity that is filled up with brine to a depth of several meters (Busson and Perthuisot, 1977). Wafers of gypsum intercalations here seem to have no influence on the amount of available pore space. Permeability in this salt ranges from 1.22 to 24.5×10^6 nm^2, excluding solution cavities and geodes. The permeability of the salt is not significantly decreased even at a depth of 30 m, because the very dense residual brine prevents compaction (J. P. Perthuisot, personal communication, 1982). Quaternary halites in Eritrea and the Dead Sea area are as yet uncompacted and thus porous, vuggy, with euhedral crystals up to 10 cm along the edges, and distinctly soft (Holwerda and Hutchinson, 1968; Zak, 1974). The sparry crystal network of bedded halite in the Uyuni Basin of Bolivia leaves about 50% initial pore space. Considerable diagenetic changes are required to convert this loose aggregate into a massive, compact deposit (Eugster, 1980a). In Baja California, Shearman (1970) found layered halite, 1–8 cm thick, alternating with gypsum sand. The halite had recrystallized into a hard but cavernous rock with a vertical grain caused in part by small dissolution pipes. The surface salt layer in the Gulf of Kara Bogaz Gol, off the Caspian Sea, has a porosity of 10% and is saturated with interstitial brines. Underlying salt layers have high permeability and contain pressurized brines (Dzens-Litovskiy and

Vasil'ev, 1962). Within the top few tens of meters, the accumulated salts recrystallize into cemented aggregates, but maintain permeability and up to 20–30% porosity. Older (pre-Pliocene) salts are impermeable, and it is not known when the permeability disappears here (Dzens-Litovskiy, 1967). If crystallization does not involve removal of calcium sulfate or potassium chloride, some porosity is retained at specific concentrations of the admixture. From 4.5 to 5.0% porosity is retained at 0.4, 20.0, or 55.0 mole % of potassium chloride or at 0.5 or 3.2 mole % calcium sulfate. Porosity minima are encountered at 1.0 and 40.0 mole % of potassium chloride (Tollert, 1964). With rising anhydrite content, a porosity increase also occurs in synthetic rock salt (Price, 1982).

After burial and substantial compaction, the porosity of halite drops to less than 5%. The pore volume of Permian Zechstein halites and carnallites varies between 0.2 and 4.3% (Tollert, 1964). Permian halites and anhydrites in the Caspian depression have a porosity of 0.5–4.0% and variable permeabilities to oil (Yeventov *et al.*, 1971). Compacted rock salt contains 0.1–1.0% water, which is set free at temperatures above 100°C and then migrates along boundary surfaces, microfissures, and intergranular spaces (Jockwer, 1979). *Permeability is very good in a direction parallel to bedding in layered salt or along clay intercalations* (Heynke and Zaenker, 1970). Under low confining pressure, dry halite allows fluid flow along crystal boundaries and cleavage planes. Fluids, such as hydrocarbons, which do not interact with halite, flow more quickly and for longer periods of time than do saturated brines (Gevantman *et al.*, 1981). Salt solutions flow readily through compacted clays (McKelvey and Milne, 1962).

There is a difference in porosity and permeability between samples of ancient salt deposits tested in the laboratory and those evaluated on site. Salt is plastic, and its permeability decreases under pressure. However, this reduction is diminished as more anhydrite intercalations occur, because of the elastic reaction of the latter. Reynolds and Gloyna (1961) determined porosities of 1.71% for a salt dome at Grand Saline, Texas, and 0.59% in the salt of Hutchinson, Kansas, with no determinable permeability. Foerster (1974) tested salt samples in the laboratory and found a permeability to air varying from 3.75 to 26,450 nm^2, yet compressed-air tests on rock walls of solid salt showed no leakage. Salt apparently becomes tight under an overburden of about 100 m (Hofrichter, 1976). Pockets of brines under high pressure indicate that there cannot be any permeability at an overburden exceeding about 300 m (Baar, 1977). However, permeability can be generated by a temperature gradient or by uniaxial stress, as proven by the movement of fluid inclusions. In salinas, a precipitate of a few centimeters to decimeters of halite seems to produce a substrate that is impervious as long as an algal mat is alive. However, the gypsum and carbonate beds in the precondenser pans, an equivalent of the gypsum and carbonate shelves around evaporite basins, retain their permeability under a greater overburden.

When dry, the salt is still porous at shallow depths and remains permeable to nonaqueous fluids. This allows a potentially appreciable flow along crystal boundaries and cleavage planes, or through bands of impurities (Aufricht and Howard, 1961). Volcanic carbon dioxide gases entered Zechstein evaporites at pressures exceeding 10 MPa during Tertiary volcanism, using preexisting pathways. They have remained in interstitial spaces to this day, unable to leave. Thus, no *in situ* permeability to gas can be inferred in this case (Richter, 1953).

In contrast, Antonov *et al.* (1958) found in bench tests that pure methane diffuses through natural rock salt at the rate of 88.5 cm^3/m^2/year, and ethane twice as fast at 177 cm^3/m^2/year. Aufricht and Howard (1961) studied salts in several areas and found salt porosities ranging from 0.8 to 6.0%, the permeability being 0.099–775,000 nm^2 for dry gas and liquid hydrocarbons. This permeability decreased by two orders of magnitude under pressure. Without water, the permeability did not disappear even at pressures of 5.5 MPa. Where it disappeared at 34.5 MPa, it was reestablished when pressures fell back to 5.5 MPa.

Yeventov *et al.* (1973) found experimentally that of all the evaporitic salts, only anhydrite is impermeable to oil. By nature, it often contains higher amounts of bitumen than does halite. Dispersed bituminoids in halite amount to 3.1–400 ppm at a porosity of 0.5–4.0% and a permeability for oil ranging from 370 to 210 nm^2. Microfractures and admixtures loosen the mineral fabric and increase the permeability. That explains porosities of 11.3–19.1% in individual anhydrite samples. Salt beds above natural gas or crude oil occurrences are found to be saturated with hydrocarbon gases, but low in carbon dioxide or nitrogen content (Zoikin *et al.*, 1975).

In German Zechstein evaporites, there are several generations of brine stripes in sylvite (Siemens, 1961). Similarly, Richter (1962, 1964) has reported bleeding red color threads, i.e., hematite leaching by epigenetic brines. Submicroscopic lattice defects along the lines of brine invasion into crystals are due to material transport at the borderline between crystal and liquid (Gerlach and Heller, 1966).

A contact of salt with undersaturated water drastically reduces the permeability. Water increases the plasticity of the salt and closes the pores. Water also swells or hydrates shales or redeposits calcium sulfate. Even fractures with a permeability of 1.056×10^6 nm^2 healed in a few hours in the presence of water. Wetted salt has a high degree of plasticity and a noticeable increase in tensile strength. Most bedded salt contains moisture trapped along the grain faces and in cavities a few tens of microns to 1.5 mm in diameter located along grain boundaries. The moisture content of Silurian salt in the Detroit area is 1000–15,000 ppm; in Permian salts of central Kansas, 500–800 ppm; in Permian salts of New Mexico, 0–35,000 ppm; but in the Louisiana salt domes, as low as 30–80 ppm (Aufricht and Howard, 1961; Bays, 1963; Combs, 1975). However, Bradshaw *et al.* (1968) found as much as 1270–10,800 ppm water in Permian Hutchinson salt

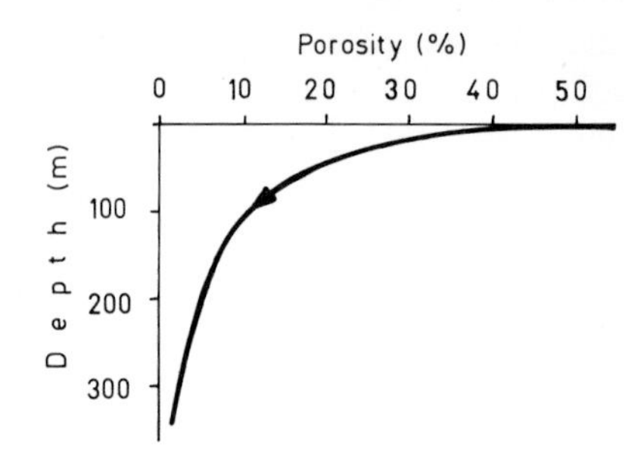

Fig. 12-2. Decrease in halite porosity with depth.

of Kansas, but only 400–1000 ppm in salt from a salt dome. In Permian evaporites of Texas, Powers *et al.* (1978) reported 1000–7000 ppm (average, 3600 ppm) as a minimum value, since most of the larger fluid inclusions were lost in the sampling process. Such amounts of trapped brine can be subject to significant diffusion effects in geologic time. The range in values can be explained by assuming that permeabilities in halite vary from place to place; percolating fluids would seek out a very complex routing.

Permeability is a function of the type and amount of impurities present, the crystalline structure and the presence of cleavage planes, the amounts of confining and overburden pressures, and, above all, the water content. Both porosity and permeability will vary from one evaporite deposit to another, and even within one salt bed, due to differential compaction, the original amount of brine in the slush, and the history of postdepositional stresses. *Permeability for water disappears soon after burial: large-scale vertical movements of syngenetic brines and consequent secondary alteration of evaporite minerals, end before burial to more than about 300 m* (Fig. 12-2). However, *evaporite deposits are not impermeable to nonaqueous fluids or saturated brines.* The movement of such fluids may thus continue even at burial deeper than 300 m.

CONNATE BRINES

Commonly encountered in potash mines are warm solutions rich in potassium and magnesium, with densities in excess of 1.28, that are of local origin. Their hydrostatic head diminishes quickly upon liberation, unlike that of intrusions of formation waters rich in sodium or magnesium (Dunlap, 1951; Litonski, 1967; Anisimov and Kisel'gof, 1972; Woods, 1979). A brine pocket of at least 4200 m^3 was opened in anhydrite after drilling through a Permian salt section in New Mexico (Peterson, 1982). Evaporites in the Caspian depression contain many lenses of very dense brine under pressures exceeding the hydrostatic pressure in surrounding beds by 13–15% (Kisel'gof, 1973). Some cavities in Middle Devonian evaporites of Saskatchewan harbor enough magnesium chloride brines or calcium–magnesium chloride brines to produce leakage flowing into mine workings unabated for several years (Baar, 1974). Because of their excess bromine content, they were interpreted to be residual brines, rather than brines derived from conversion of carnallite to sylvite. Such interstitial brines tend to have compositions similar to those of liquid inclusions in halite (Bogasheva *et al.*, 1971), which proves that crystals remain in chemical equilibrium with surrounding pore fluids. Interstitial moisture preferentially moves along crystal contacts, e.g., along anhydrite layers disseminated in halite (Dix and Jackson, 1982).

SYNGENETIC VERTICAL BRINE MOVEMENTS

Water volumes are lost not only by outflow over the entrance sill but also by hydraulic seepage into the subsurface or through the entrance sill. The ratio between the two modes of outflow will change with the seasons. A complete closure of the entrance barrier during the dry season would still allow considerable reflux into the basin from subsurface waters and from the sea through the barrier, because even a closed bay retains a hydrodynamic equilibrium with the sea akin to a set of communicating vessels (Sonnenfeld and Hudec, 1977). Lake Larnaca on the island of Cyprus is one example of such a closed lake that still maintains interstitial equilibrium with the sea.

In a closed system, a unit volume of seawater would yield halite at more than 22 times the volume of precipitated anhydrite, and about 3.66% potassium–magnesium salts. That leakage and outflow are essential to any model of evaporite deposition is demonstrated by the following observations:

1. In the British portion of the southern North Sea, the Werra anhydrite (first cyclothem of the Zechstein evaporites) comprises about one-sixth of the estimated basin volume. This amount of precipitation would require more than 400 water changes, and there would be insufficient space left in the basin to accept further seawater because of the volume occupied by the residual brine (Taylor, 1980). Consequently, continuous drainage of excess brine is indicated. King (1947) made the point that without such drainage, the gypsum in the Permian Basin of West Texas could be accounted for only by a seawater column 1000 km deep.

2. A deficiency of halite over gypsum–anhydrite volumes occurs in every evaporite deposit. If no inflow is altering the proportions in the sea salt, a marine lagoon should precipitate 1 weight % of the deposit as carbonate, 3% as gypsum, 69% as rock salt, 15% as sylvite and carnallite, and 12% as bischofite (Borchert, 1967). Expressed in terms of volume, one can expect to obtain from isochemical evaporation of 1 m^3 seawater 552 cm^3 calcium sulfate, 12,132 cm^3 halite, and 444 cm^3 sylvite and carnallite, giving a ratio of 1:22:0.8 between sulfates, halite, and potash salts (Valyashko, 1962). The last column of Table 1-2 translates into

21.5 dm^3 of precipitate per cubic meter of seawater, or 678 cm^3 gypsum, 12,880 cm^3 halite, 380 cm^3 sylvite, and 7492 cm^3 bischofite and epsomite, giving a ratio of gypsum:halite:sylvite:magnesium salts of 1:19:0.6:11, or a ratio of gypsum to all other salts of 1:30.6. This ratio is not attained in any ancient evaporite deposit. In a brine that developed a sulfate deficiency, the ratio of gypsum to halite by volume could surpass 1:175. The deficiency of halite over calcium sulfates is 6-fold in the Stassfurt horizon of the Zechstein evaporites (Krull, 1917) and 30-fold in the Permian Castile formation of Texas and New Mexico (King, 1947), with an average of about 4-fold throughout the seven Permian halite zones (Adams, 1963).

3. A deficiency of potash over halite volumes occurs, which is threefold in the same Zechstein horizon and up to twofold in Permian evaporites west of the Urals (Ivanov and Voronova, 1968). The Great Bittern Lake athwart the Suez Canal, cut off from the Red Sea 2600 years B.P. (or about 600 B.C.), contained far less bitter salts than would be expected from halite and gypsum thicknesses. There must have been a considerable leakage of bitterns during halite deposition (Grabau, 1920; Lotze, 1957). Theoretically, one would expect all potassium–magnesium chlorides in a basin with a sulfate deficiency to occupy about 27.6–27.9% of the chloride-bromide sequence.

4. The total absence of magnesium salts, or their deficiency over potassium salts, which is threefold in the mentioned Zechstein horizon, is common. Less than 2% of the magnesium entering the basin is precipitated in evaporite minerals. Yet, in any laboratory experiments, magnesium sulfates precipitate after halite and before potash salts. Despite the manifest marine origin of brines in Lake Eyre, Australia, there are no potassium or magnesium salts present. Subsurface seepage has been thought to be responsible for this deficiency (Bonython, 1956). The present assemblage of minerals in the Permian Basin of New Mexico not only contains minerals that are decidedly not primary, but also has a pronounced magnesium deficiency compared to other deposits. Magnesium-rich bitterns must have been able to escape, but the sink for magnesium cannot be identified (Powers *et al.*, 1978). In the Pripyat Basin of Byelorussia, in the artesian basin of Amu Darya east of the Sea of Aral, and at Angara-Lena in Siberia, the nonsulfatic potash salt beds contain no magnesium chloride in their formation waters (Korennov, 1974). All magnesium in solution, derived from the leaching of carnallite, has apparently been captured here by epigenetic dolomitization.

5. Secondary dolomitization may be related to leakage and outflow from an evaporite basin, and also secondary anhydrite void fillings in surrounding porous limestones, especially on the downdip side, eventually feathering out at some distance from the evaporite basin (Fig. 12-3) (Newell *et al.*, 1953; Sonnenfeld, 1964; Schlanger, 1965; Schenk, 1969). No carbonate lenses in halite have ever escaped such dolomitization. The highly concentrated bitterns are capable of dissolving calcium carbonate upon entering an aquifer. This increases the sol-

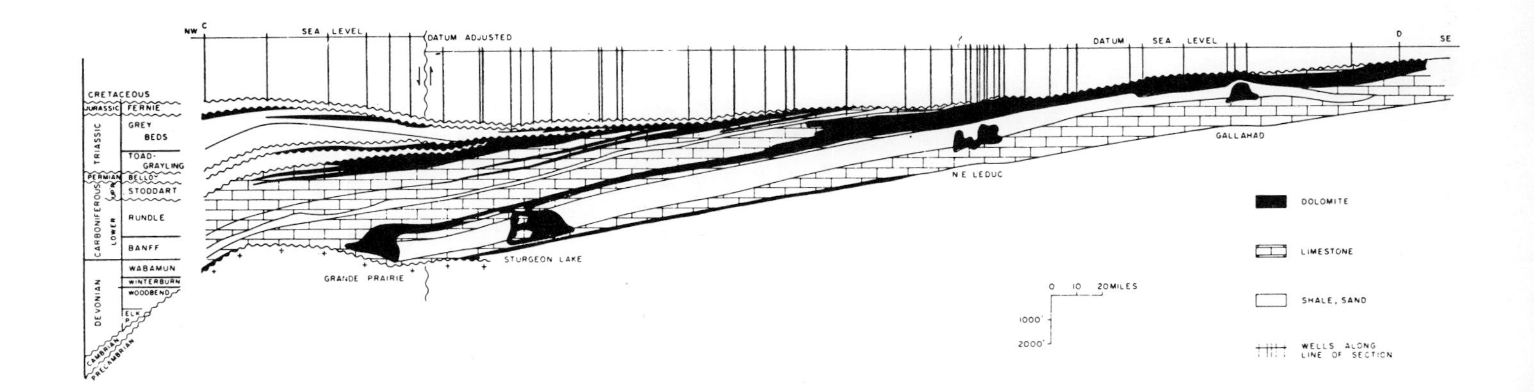

Fig. 12-3. Downdip dolomitization pattern (strike pattern) in the subsurface of Alberta, Canada (after Sonnenfeld, 1964).

ubility product of $[(Ca^{2+})(Mg^{2+})(CO_3^{2-})_2]$ and leads to supersaturation in respect to dolomite (Lippmann, 1973). It is this fresh supply of carbonate ions that induces dolomite formation at the interface between sebkha brines and fresh groundwater charged with atmospheric or biogenic carbon dioxide (Folk and Landes, 1975). However, groundwater charged with magnesium precipitates dolomite above and in advance of the seawater encroachment front; only high-magnesium calcite forms to the rear of that front (von der Borch *et al.*, 1975). This could be due to a reduction of magnesium solubilities in advancing brines caused by dilution by groundwater. Microcrystalline dolomite can be replaced by anhydrite, and the latter then turned into gypsum (Miropol'skiy, 1941). The descending magnesium chloride brines replace the calcium of limestone and thus generate a calcium chloride brine, with some of the calcium also used to fill voids with anhydrite (Poroshin, 1981). This suggests the descent of at least partially oxygenated brines.

Whereas virtually all recent dolomite is micro- to cryptocrystalline in grain size, most ancient secondary dolomites are fine to coarsely crystalline, of relatively uniform crystal size, giving rise to the descriptive terms "saccharoidal" and "sucrosic." Only in the presence of a brine saturated for dolomite, i.e., a brine rich in magnesium, could this sucrosic dolomite have originated from recrystallization of cryptocrystalline dolomite. Such a brine, however, is also capable of dolomitizing limestones, as evidenced by the ghost outlines of fossil fragments destroyed by recrystallization. In such cases, the precursor was likely comminuted skeletal material of aragonitic composition. The coarser crystallinity is a product of slow growth, i.e., an immersion of the original aragonite debris in slowly moving, magnesium-rich fluids. Since the subsurface brines of an evaporite basin are capable of producing brucite pillars in clay minerals, an excess of magnesium hydroxide must also exist in such brines when they enter the substrate. It is suggested that dolomitization will occur where such brines meet exposed calcium carbonate. Any available sulfate ions would result in gypsum precipitation. This then vindicates Friedman's (1980b) observation that dolomite always occurs in recent hypersaline ponds only in the substrate of gypsum precipitation. Thus, dolomite formation is not a sink for fresh magnesium sulfate solutions, as has been repeatedly intimated, but only for magnesium-bearing chloride solutions, preferably with a deficiency of chloride ions. It may be observed that no limestone intercalations in halites have escaped dolomitization.

DESCENDING BRINES

There are actually two types of descending waters that cause postdepositional changes:

1. Truly meteoric waters of rains or flash floods that enter a dried-out salt pan

at the end of an evaporitic cyclothem saturate by dissolution of exposed evaporitic minerals and then descend as brines. A current example is meteoric waters entering jointed anhydrite exposures, converting them to gypsum, and eventually percolating down to halite horizons. An ancient example is provided by Triassic salts of Germany, which consist of a lower salt with vertical banding of insolubles that contain inclusions of blocks of bedded salt, anhydrite, or clay. This partially leached salt, in turn, is covered by horizontally banded salt that often fills bowl-shaped depressions and is itself covered by an upper salt horizon. Banded salt is preserved only occasionally at the base of the lower salt. Interpretations of this sequence are varied (Dellwig, 1966). However, the boundaries between the three salt units can be interpreted as solution surfaces during substantial freshening of the brine, with some of the waters penetrating differentially downward. However, in the Stassfurt Member of the Zechstein evaporites, meteoric waters infiltrated along clay laminations, since the clayey streaks occurring at right angles to the bedding planes always start at the base of clay–anhydrite layers (Kroell and Nachsel, 1967).

Such downward movement is also found in lacustrine precipitates. In halite layers of Lake Baskunchak in the western part of the Caspian Lowlands, Valyashko (1962) recognized freshly deposited salt at the surface, 7–10 cm of salt from previous seasons underneath, and then a 10-m-thick layer of recrystallized salt. The recrystallized salt was composed of individual druses of halite stretched out vertically, separated by cylindrical vertical voids. Water undersaturated with respect to halite rises in winter and spring; the concentrated surface waters percolate down in the summer, precipitating halite on the way. Often the precipitation is caused by the encroachment of magnesium-rich bitterns from above onto brines saturated with sodium chloride, resulting in a salting out of halite.

Brines create voids by dissolving mineral species in their path. Anhydrites in the Polish Zechstein evaporites are often cut by numerous microfissures and contain open cavities suggestive of the migration of various brines. Preservation of dolomite crystals inside calcite grains in both the Werra and Stassfurt anhydrite members indicates a reactive solution that dedolomitized the interspersed dolomites or dissolved them. The resulting porosity and permeability then permitted the migration of oil and gas (Chlebowski, 1977).

Vertical fractures in evaporites filled with salts of greater solubility than the wall rock are quite common. In salt clays associated with Permian salts of Texas and New Mexico, there is a polygonal set of fissures filled with carnallite. The halite-bearing clay contains vertical carnallite stringers, and also blebs of carnallite and sylvite, together with clay pipes in the carnallitic salt clay. These carnallite fracture fillings must be nearly syngenetic, since the clay is overlain by solid halite (King, 1946). The sulfate minerals leonite, kainite, polyhalite, kieserite, and bloedite are exclusively gangue minerals in the Permian evaporites of New Mexico, occurring in association with carnallite, halite, and anhydrite (Cheeseman, 1978). Carnallite in the Devonian Prairie Evaporite Formation of

Saskatchewan, Canada, has been redistributed vertically, as Schwerdtner (1964) has shown. This occurred by solutions descending through cracks and fractures perpendicular to bedding planes. Only salt-filled vertical fractures are here hematite enriched and red; the host salt is clear or brown to orange (Wardlaw and Watson, 1966). Freshened inflow dumped clay interactions into these potash beds; the clay then infiltrated with the undersaturated solution into underlying halite. Here it formed irregular channels (Wardlaw, 1968). Similar crack fillings by halite occur in Triassic evaporites in Germany (Richter-Bernburg, 1977, 1979). In contrast, micropipes in Upper Triassic halites off the Newfoundland coast are filled with silty clay and rare dolomite crystals that even cut through halite crystals containing liquid inclusions (Jansa *et al.*, 1980).

2. Another type of descending brine is produced in an evaporite basin, covered by younger shales and carbonates, that is again inundated by the sea. The seawater column produces a hydrostatic head of oxygenated seawater which can carry new sulfate ions into the sediments below. Present analogs are the columns of Mediterranean Sea, Red Sea, or Gulf of Mexico waters moving about above older evaporites. The seawater is separated from the older evaporites by relatively thin layers of marine clays that are potentially subject to submarine syneresis. Most ancient evaporite basins have been exposed in time both to meteoric waters and to a hydraulic head of seawater encroaching in a subsequent geologic period. For example, the Devonian Elk Point Basin of Alberta, Canada, was covered by Mississippian seas, and the Permian Zéchstein by Triassic seas.

3. During the progressive concentration of the brine in the evaporite basin, a certain amount leaks out through the fresh precipitate and displaces formation fluids. These brines are generally less dense than the concentrated brines oozing in. Concentration fronts are set up in a cone shape open away from the basin floor. Some of the pressurized lighter formation waters are trapped by the brine, leading to gravitational instabilities (Polivanova, 1982). Moreover, since metals are more soluble in concentrated brines, some of them will precipitate along the dilution fronts set up by underlying formation waters.

Sometimes the seepage is not readily evident. The Ojo de Liebre lagoon of Baja California produces salt and gypsum in a series of interconnected brine pans. Halite accumulates in the far reaches of the system, where only episodic storm floods can deliver preconcentrated brines. No evidence has been discovered of any return flow of saline waters or of residual bitterns out of the lagoon (Phleger, 1969). Yet the solutes contained in such bitterns do not accumulate. In this case, only interstitial leakage through the basin floor can account for the flushing of the bitterns.

Precht (1898) and Everding (1907) were early proponents of resedimentation by descending waters that soaked in either during periods of temporary brine freshening, followed by clay deposition, or after completion of the evaporite cycle. Several examples can be cited: Baar and Kuehn (1962) found the lower part of the Tertiary potash deposits in the Upper Rhine Valley undisturbed; the upper part had changed through leaching and recrystallization. Herde (1955), and again Kuehn and Schwerdtner (1959), explained two zones of brecciation in the Leine (third) cyclothem of Zechstein evaporites by descending dissolution that allowed only the coarsest crystals to be preserved. The brecciated zones are underlain by undisturbed evaporites. Podemski (1972) used the same argument to explain very coarse crystallization in Stassfurt and younger salt horizons of Polish Zechstein evaporites. Moetzing (1968) revived the concept of resedimentation by descending brines, because he found that alterations became more extensive upward in his area of study of Zechstein evaporites. Additional magnesium chloride solutions, much later mobilized tectonically, affected only the sulfatized rim region. They produced fibrous carnallite, recarnallitized sylvite, or sylvitized carnallite. The distribution of secondary sylvite after carnallite in the Khorat evaporites of Thailand can be explained by vertical variations in the permeability of the original deposit, which facilitated later movements of altering solutions at some places, while hindering them at others (Hite and Japakasetr, 1979).

Concentration of brines leaking through the substrate continues due to the precipitation of hydrated minerals. Minerals with slow rates of precipitation can now form in the interstices or by reaction with the wall rock. The mineral species that grow from intercrystalline brines are thus distinctive and different from minerals precipitated on the floor of open brines (Strakhov, 1962). A mass transfer between evaporitic precipitates and salt-free rocks underneath produces secondary dolomite, anhydrite, and even halite in beds beneath and between evaporites of the Pripyat Basin in Byelorussia, and results in formation waters rich in iodine and bromine (Makhnach, 1981).

BRINE SEEPAGE

There are two types of fluids moving through the evaporite deposit after its precipitation: syngenetically, or early epigenetically descending and temporarily ascending fluids. Both types can be diverted into lateral flow by the anisotropy of the formations. After syngenetic and epigenetic brines have percolated through, the evaporite deposits experience varying degrees of diagenetic alteration. The present distribution of potash salts, then, does not coincide with their original distribution, although it approximates it (Stewart, 1956). Syngenetic brines generated during the precipitation of evaporitic minerals are similar in composition to those derived from soaking in seawater, except that sulfate ions are virtually absent. Heavy liquids produced during the evaporation process descend downdip into the subsurface, but liquids liberated by recrystallization of buried evaporites or formation waters expelled by compaction rise under the accumulating lithostatic pressure. Precipitation and metasomatism by encroaching bitterns are thus often stratigraphically separated from diagenetic alterations induced by crystal water migration.

Subsidence of a basin after it has been filled to an unconformity surface raises the piezometric surface and forces interstitial formation waters to the surface or subjects them to an artesian pressure related to the hydrodynamic gradient from a distant upland intake. The Ronnenberg seam of German Zechstein evaporites, originally a layered carnallitite, was altered at its base by rising saturated sodium chloride solutions squeezed out of underlying halite. The top of the Ronnenberg seam is sylvinite that was altered by a mother liquor infiltrating from halites that were precipitating above (Boehm, 1972). The saline springs in northeastern Alberta and adjacent Saskatchewan are produced by the slow updip movement of Devonian formation waters under the hydrostatic head of rainwater intake in the Rocky Mountains, over 1000 km away. This slow Cenozoic movement of formation waters has displaced the salinity maxima onto the eastern limb of the Alberta syncline (Sonnenfeld, 1964) (Fig. 12-4).

Quasi-syngenetic brine movements should be differentiated from truly epigenetic movements of waters entering from above, laterally, and from below (Fig. 12-5). The entering brine mixes with residual brines; its initial composition is determined by its mode of origin. There is a lot of evidence for epigenetic brine movements despite the apparent near-total loss of porosity and permeability in a halite bed subjected to overburden pressure. For instance, in the bottom evaporite cyclothem of the Permian Zechstein sequence in Germany, a change in facies occurs from massive anhydrite to some 200 m of halite in substantially less than 2000 m. This is interpreted to have been caused by secondary leaching of the anhydrite-covered sills that were raised by later tectonic movements (Weber, 1961). In the Triassic salts of northwestern England, halite is banded by paper-thin mudstone intercalations 2.5–10.0 cm apart, and more rarely by gypsum. Laterally, the banded salt passes into nonbanded salt where the mud intruded between halite crystals. The absence of banding is caused by recrystallization of banded salt (W. B. Evans, 1970) by epigenetic brine movements. Both the German and the English Zechstein potash salts, as well as the Permian deposits of New Mexico, seem to have been originally deposited as carnallite (Ingerson, 1968), and turned into sulfates later. The outlines of some original carnallite crystals are still preserved by hematite and clay flakes that coated their surfaces. Some of the present carnallite is secondary due to recarnallitization.

Circulating waters can produce a full cycle of alterations. It is thus often difficult to distinguish the mineralogical alterations caused by ascending brines from those caused by descending waters, unless one has done a thorough three-dimensional study of the whole evaporite basin. A complication is that subsurface brines are not homogeneous in composition, but vary laterally within the same aquifer. The combined activity of methane-producing bacteria creates a

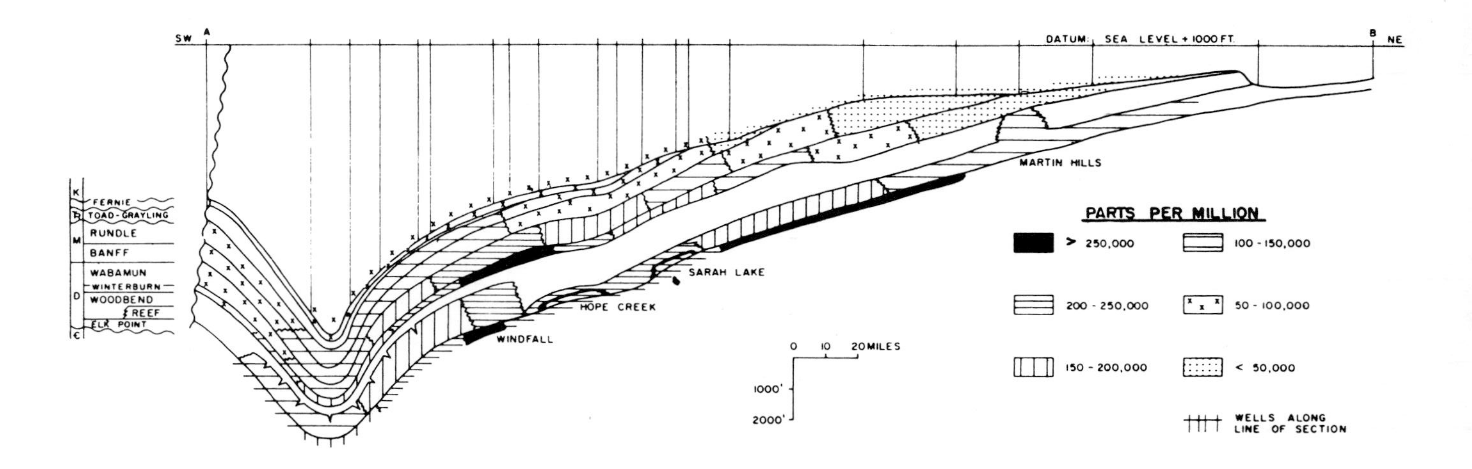

Fig. 12-4. Salinity of subsurface brines in Alberta, Canada (dip section, datum at sea level). The maximum salinity in each aquifer is displaced onto the northeastern limb of the Alberta syncline due to the post-Laramide hydrostatic head in the Rocky Mountains. The amount of displacement is proportional to the effective permeabilities in the individual aquifers (after Sonnenfeld, 1964).

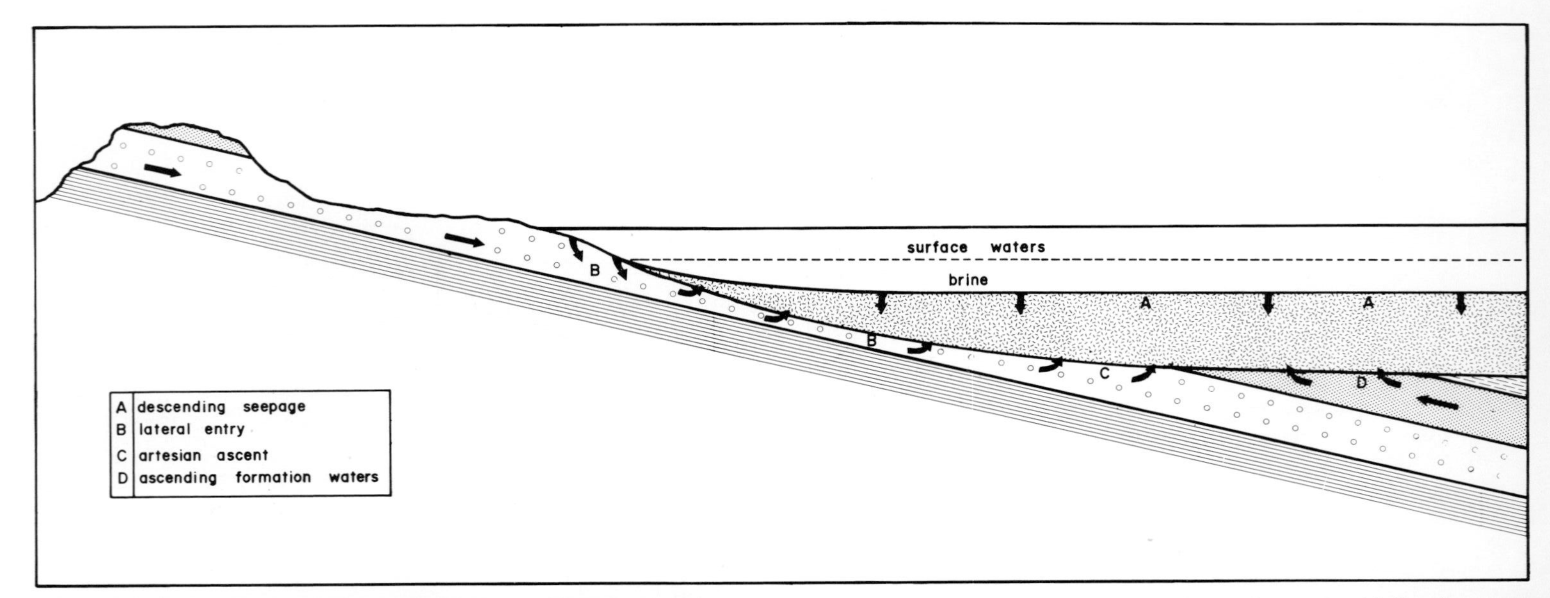

Fig. 12-5. Descending and ascending brines. A. Descending seepage of anoxic bitterns from the lagoon. B. Lateral entry of meteoric waters and infiltrating oxygenated surface waters. C. Artesian ascent of such waters. D. Ascending anoxic formation waters.

spotty pattern of calcium, magnesium, potassium, and sodium ions in subsurface waters. Oxidation of organic matter releases CH_3COOH and carbon dioxide, which are sources of hydronium ions and dissolve calcite (Germanov *et al.*, 1981). However, it appears that descending solutions can be recognized geochemically on the basis of their impoverishment in trace elements, their lower temperature regime, and their characteristic magnesium sulfate content (Koch and Vogel, 1980). Ascending solutions, whether warm formation waters or waters heated by geothermal sources, have a higher content of lead, boron, and other elements and are more prone to alter enclosed bitumina, but are initially deficient in magnesium. Finally, an artesian system can bring meteroic waters from elevated land into an evaporite deposit as ascending brines.

Multiple pseudomorphisms and mineral replacements at times complicate the task of unraveling the depositional and postdepositional history of an evaporite deposit. For instance, in the British Zechstein deposits, (see Zechstein map, Fig. 12-6) dolomite is replaced by anhydrite; gypsum by anhydrite, and that in turn by celestite. Anhydrite derived from gypsum is replaced by polyhalite. The polyhalite, in turn, is replaced by kieserite, which can then be replaced by aphthitalite or picromerite. Both anhydrite derived from gypsum and polyhalite derived from anhydrite can be replaced by glauberite. Glauberite can then be turned into mirabilite (Stewart, 1965). There is evident replacement of halite by anhydrite or gypsum in several stages, but halite also replaces gypsum, anhydrite, or even dolomite (Raymond, 1953; Dellwig, 1955). The Pegmatitic Anhydrite Member of the German Zechstein, an intergrowth of rock salt and anhydrite, contains anhydrite pseudomorphs after calcite, the interstices filled with halite, and halite pseudomorphs after gypsum (Zimmermann, 1907). Smith (1981) considered all halite in the Middle Marl of the British Zechstein evaporites as either displacive or replacive and thus secondary. In the same way, much of the anhydrite in the third cyclothem of the British Zechstein series could possibly be due to wholesale replacement of dolomite (Dunham, 1948). There is less pyrite in the anhydrite than in the dolomite. Either some pyrite has been removed in the anhydritization, or the pyrite-bearing dolomite is more resistant to replacement. Pyrite-bearing dolomite is also more resistant to dedolomitization (Perel'man, 1967). This sampling suffices; many other replacement reactions can be observed (Stewart, 1963b).

Hydraulic pumping of brines in and out of the subsurface of the bay occurs according to the seasonally changing hydrostatic head of the brine (Sonnenfeld and Hudec, 1977, 1978). The brine volume pumped into the substrate of a lagoon during the high water level period is much larger than the amount refluxed during the period of low water level; this outflow can take the form of seasonal dumping (Murray, 1969). Effectively, the brine becomes an early epigenetic brine capable of altering recent precipitates. In modern lagoons, such as at Los Roques, Venezuela, an outward movement can be documented by differ-

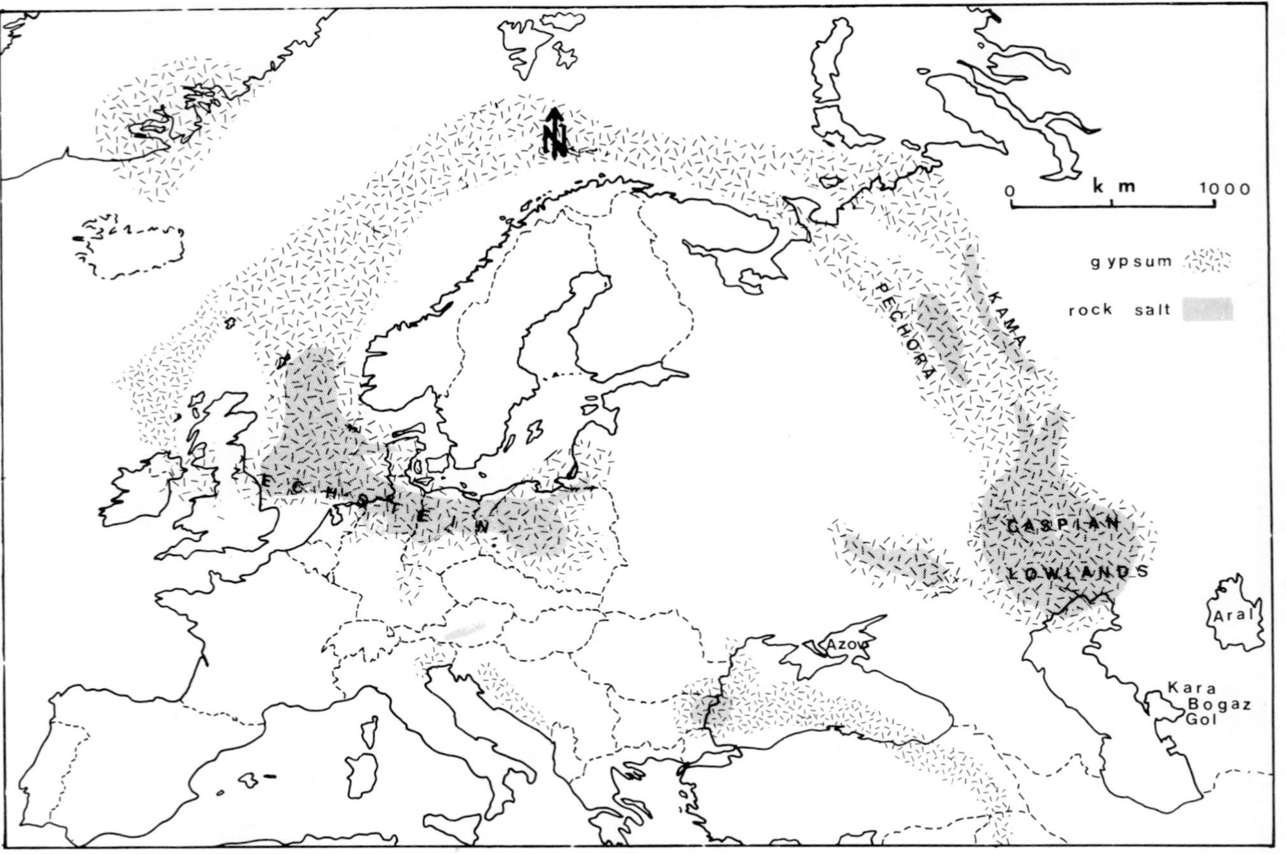

Fig. 12-6. Permian evaporites in Europe.

ent vertical density distributions in various partially isolated portions of the lagoonal waters and by the occurrence of tepee structures. The brine encroachment exceeds bottom outflow over the entrance sill of a marginal bay once the difference in density exceeds about 200–250 g/liter. Concentrated bitterns are too dense to be discharged above ground (W. A. White, 1961). The concept of subsurface leakage induced Adams and Rhodes (1960) to develop their seepage refluxion model of dolomitization to account for the evident deficiency of magnesium salts in the Permian basins of West Texas, and thus for the loss of magnesian bitterns. Although the inward or outward movement of brines is controlled in lagoons by the difference in hydrostatic head to a seawater column outside the lagoon, there is no such reference column in lakes. Thus, the brine seeps out primarily in the dry season, when it has become heavy and inflow cannot balance evaporation losses.

Many hypersaline lakes dry out seasonally because the inflow cannot replenish evaporation losses, although further evaporation is reduced about 170- to 200-fold, when the surface brine soaks into precipitating salts and loses an open water surface (Valyashko, 1958; Strakhov, 1962). The residual bittern becomes too heavy and percolates through a lowered groundwater table. By the time the bitterns are saturated for potash salts, they have lost up to two-thirds of their volume to subsurface leakage (Zabrodin, 1977). Around the Coorong lakes of South Australia, Muir *et al.* (1980) found that at the onset of the dry season, the evaporative loss of interstitial waters is much faster than the sluggish groundwater flow, and residual brine then percolates through the freshly precipitated dolomite. This leaves a dried-up lake surface until the onset of the next rainy season. Lakes in the Kyzylkum desert of Uzbekistan precipitate bloedite and halite, leaving a nearly pure magnesium chloride solution as residual brine. The distribution of the halite shows an outflow of the brine from the depression during the maximum decrease in lake level (Rubanov, 1964), i.e., when the brine is densest.

A rising brine level in a lagoon is due either to a seasonally enhanced influx of seawater or to increased runoff. In either case, the brine is diluted. Both periodic brine dilutions and continuous temperature changes act in concert to change the composition of the brine (Oka, 1944). Eventually, it bears little resemblance to a bittern produced in the laboratory under isothermal conditions. A subsurface brine from a halite-precipitating basin will be markedly enriched in magnesium, potassium, and bromine (Holser, 1966b), but also in iodine and various borates. Brine encroachment onto the groundwater is assisted only slightly by solute diffusion. In an aqueous solution, the diffusion of sodium chloride reaches a minimum at a concentration of 87,650 ppm and then rises steadily up to 350,000 ppm (Burrage, 1932). The observed outward pressure of a heavy hypersaline brine, acting in communication with the hydrostatic head of a lighter seawater outside a lagoon, is the inverse corollary of the inward pressure of seawater encroachment onto an evaporating freshwater lens, or onto a groundwater drawdown by coastal water wells.

LATERAL BRINE MOVEMENTS

Brines concentrated in an evaporite pan tend to be substantially heavier than most near-surface formation waters and, therefore, try to encroach on interstitial waters. In many areas, the latter have *in situ* densities exactly compensated for by increased salinity and temperature to yield a gravitationally stable column (Hanor, 1973). Seepage of brines and bitterns disturbs this equilibrium, and the bitterns spread into aquifers until they meet with formation waters of greater density and continue laterally. This was first suggested by Davidson (1966). Undersaturated waters advancing downward saturate quickly and then also spread laterally. The ion exchange with overlying or underlying waters is minimal because of very slow rates of diffusion (Storck, 1964).

Lateral brine movement is facilitated, because horizontal stresses in salt sequences average about 80% of vertical stresses. Both evaporite beds and clay seams cannot maintain high differential stresses for reasonably long periods. If pore fluid pressures exceed the rock stresses parallel to a bedding plane, then the permeability will be greatly increased. By the time the pore fluid pressures equal or exceed the confining pressure, the permeability has risen by several orders of magnitude. The time required either to dewater or to hydrate a bed will be shortened dramatically. If the fluid pressure exceeds the lateral stresses by an amount equal to the tensile strength of the rock, then the rock will fail in a series of vertical fractures, which then facilitate the upward migration of water (Price, 1975). Where anhydrite blebs have been precipitated in rocks adjacent to an anhydrite basin, although these rocks were previously devoid of evaporite minerals (Blatt *et al.*, 1980), lateral migration of syngenetic fluids into the country rock is proven. An impermeable barrier of even negligible thickness increases the lateral pressure tremendously, and an underground lake may form (Price, 1975). Such pockets of water have repeatedly been drilled into when penetrating basal anhydrites of an evaporite sequence.

Saturated brines and bitterns entering the adjacent strata can precipitate excess solute into open voids. Halite-filled vugs in reef trunks on the fringes of the Silurian salt of the Michigan Basin are one example. Another example is furnished by Tertiary coals in East Germany that were salted by percolating waters after deposition and burial (Leibinger *et al.*, 1964). Today the Dead Sea precipitates halite only in the deeper parts. Hot brines encroaching upon the subsurface from the Dead Sea are close to saturation for halite and undersaturated with regard to sylvite or carnallite (Lerman, 1967). However, lateral brine movement between two laminations can also lead to barren intervals (Siemens, 1961). Water saturated in calcium sulfate traveled through salt clay and bituminous shale intercalations of the Zechstein evaporites north and east of the Harz Mountains, but only along horizontal joints. Along the way, it hydrated anhydrites and precipitated coarse, sparry gypsum (Steinbrecher, 1959). Langbein and Seidel (1960a) cite an example of gypsification by lateral brine movement where both the base of the member and overlying units have remained as anhydrite. Likewise, the top of the Stassfurt potash member in the Permian Zechstein evaporites represents an intraformational residue area (Loeffler, 1962). Kieseritic–carnallitic lenses usually occur in anhydritic halite with low bromide content, covered by a halite with desiccation cracks. This suggests that the potash minerals were precipitated by seepage and that some of the potash solute migrated laterally toward a polyhalitization of anhydrites (Elert, 1968; Zaenker, 1979). Lateral replacement of anhydrite by polyhalite is also common in New Mexico (Jones, 1973).

EFFECTS OF CHANGING SEA LEVEL

Reducing the area of open water surface decreases total evaporation losses per unit of time. This happens when inflow no longer matches evaporation losses and leakage. A very slight drop in lagoon level will bare large areas of shoals and new beaches and allow subaerial exposure. The brine or bittern retreats toward the basin entrance, but the concentration of the brine rises only fractionally. In contrast, a slight increase in sea level will allow these concentrated brines to reflood previously exposed shoals with only minor amounts of redissolution. Local unconformities within the salt section near the basin margin are the result. A temporary reduction of the water level in the Silurian Michigan Basin exposed not only gypsum but also halite terrain on the basin margins. A rise in the halite-saturated brine level then renewed salt precipitation (Dellwig and Evans, 1969). Similarly, dissolution of carnallite along the basin rims and redeposition in the basin center could be deduced from rubidium enrichments in Tertiary Upper Rhine Valley carnallites (Kuehn, 1963; Braitsch, 1966).

If the longshore current continues to circulate in a somewhat shrunken bay, the waters also soak into the substrate and cause alteration of buried salts. Only in this way can one explain the early diagenetic alteration of Zechstein (Stassfurt horizon) carnallites to kieseritic and anhydritic sylvites in a fringe zone parallel to the ancient shores (Loeffler, 1963). The same is true of the Permian Salado Basin in New Mexico (Cunningham, 1934). This type of descending alteration front need not, however, penetrate the whole evaporite column.

ALTERATION OF CARNALLITE

Carnallite can either be stripped of its magnesium chloride content, reverting to secondary sylvite, or altered to any of the potassium–magnesium sulfates by circulating pore waters. Nearly all commercial potash deposits are the product of such secondary enrichment. This conversion is easily obtained in several ways: The presence of urea in a magnesium chloride solution is capable of dissolving practically all the magnesium chloride and leaving behind a sylvite (Olszewski, 1973), but carnallite will react with water to form sylvite, with or without concurrent evaporation of the bittern (Wardlaw, 1968). However, carnallite is capable of replacing both sylvite and halite (McIntosh and Wardlaw, 1968) in the presence of magnesium chloride bitterns (Braitsch, 1971); in the presence of precipitated bischofite, sylvite can turn into carnallite within a temperature range of −21°C to 152.5°C (Kuehn, 1952a). Carnallite dissolution and sylvite reprecipitation entail a volume loss of 78%. This volume loss is smaller if carnallitite is turned into sylvinite, as the halite fraction will be unaffected. A considerable salting out of halite further reduces the volume loss.

A layered distribution of sylvite in the Devonian Prairie Evaporite Formation in Saskatchewan, Canada, suggests that any alteration took place without considerable transfer of potassium chloride perpendicular to bedding planes (Schwerdtner, 1964). The erratic pattern of the bromide content is not explained by selective leaching of the more soluble bromides by percolating solutions that redistributed carnallite along vertical fractures. Instead, it is assumed that carnallite precipitated first, and was then dissolved by brines saturated for sodium chloride, which replaced it in succession by halite and sylvite with an irregular bromide content. Carnallite inclusions in both sylvinites and halites are cited in support of this model. An absence of primary sylvinites, albeit stripped of some of their bromide after deposition, would be surprising, since potassium chloride solubility (347 g/liter at 20°C) is much smaller than that of carnallite (645 g/liter at 20°C) and decreases with an increasing magnesium chloride content of the brine (Schwerdtner, 1964).

In the same evaporite deposit, a clear to yellowish-brown sylvite occurs encased in carnallite, with extensions of carnallite along halite and sylvite cleavage planes or grain boundaries. The penetration of carnallite along cleavage planes suggests that the carnallite is secondary (Wardlaw, 1968) or at least precipitated by percolating saturated bitterns. The presence of milky halite or of milky margins of clear halite at halite–sylvite contacts would suggest incorporation of tiny air bubbles into growing halite crystal faces, and thus temporary exposure above the brine–air interface. Rapid crystallization could then be the cause of the very coarse, milky halites that are often light blue due to excess sodium in the crystal lattice. Excess sodium, in turn, could signify a chloride ion deficiency.

Although primary hematite-rich sylvites have yet to be described, primary hematite-poor carnallites seem to be equally rare. Red, i.e., ferric iron-stained, sylvite almost always appears to be secondary. Descending oxygenated waters were able to leach magnesium chloride from precipitated carnallite. Red sylvite has an Fe_2O_3 content identical to that of adjacent preserved carnallite, but with disoriented hematite needles (Ivanov, 1963; Wardlaw, 1968). Sylvite with hematite prisms parallel to potassium chloride faces has been described from a Zechstein locality (Leonhardt and Tiemeyer, 1938). White carnallite in the Ronnenberg seam of the German Zechstein evaporites is derived from circulating brines. It thus has a low bromide content, is free of iron, and appears to be younger than red carnallite or rinneite. The purest carnallite contains no halite, but only anhydrite as needles that probably crystallized freely in a moving brine (Siemens, 1961). No sulfatic waters occur here above the Main Anhydrite Member, the basal unit of the Leine (third) cyclothem of evaporite precipitation (Siemeister, 1961) (Table 12-1).

TABLE 12-1

Permian Zechstein Horizons[a]

Northwestern basins:	Southeastern basins:
Zechstein-6 (Z-6, Friesland)	
	Friesland anhydrite
	Clay
Zechstein-5 (Z-5, Ohre)	
Saliferous marl	Boundary anhydrite Ohre salt
Top anhydrite	Layered anhydrite
	Clay-bank salt
Upper marl	Saltclod clay
Zechstein-4 (Z-4, Aller)	
Upper halite (unit E)	Upper Aller salt (rose salt)
Saliferous mudstone	Upper Aller anhydrite
Upper halite (units A + B)	Lower Aller salt (snow salt)
Upper anhydrite	Basal Aller salt
Upgang carbonate Carnallitic marl	Pegmatitic anhydrite Red salt clay
Zechstein-3 (Z-3, Leine)	
Upper Boulby salt (unit D)	Leine clay-flake salt
	Riedel group
	(Albert seam)
	(Riedel seam)
	(Streaky salt)
	(Salt with anhydrite partings)

(*continued*)

TABLE 12-1 *(Continued)*

Northwestern basins:	Southeastern basins:
Boulby potash horizon	Ronnenberg group
	(Varicolored salt)
	(Banded salt)
	(Seam Bergmannssegen)
	(bank salt)
	(Ronnenberg potash seam)
Lower Boulby salt (units A + B)	Leine banded salt
	(Orange eye salt)
	(Line salt)
	(Basal salt)
	(Anhydrite shell)
	(Black clay layer)
Billingham anhydrite	Main anhydrite
Seaham carbonate	Platy dolomite
Seaham clay	Gray salt clay
Zechstein-2 (Z-2, Stassfurt)	
Cover anhydrite	Banded cover anhydrite
Fordham salt { with anhydrite and polyhalite intercalations	Cover salt
	Stassfurt potash seam
	"Older" salt
	(Kieserite transition)
	(Hanging salt)
	(Main salt)
	(Basal salt)
Basal Fordham anhydrite	Basal anhydrite
Middle marl/Hartlepool dolomite	Main dolomite
Zechstein-1 (Z-1, Werra)	
	Brown clay wedges
Hartlepool/Hayton anhydrite	{ Upper Werra anhydrite
	Werra salt
	(Upper Werra salt)
	(Hessen seam)
	(Middle Werra salt)
	(Thueringen seam)
	(Lower Werra salt)
	Lower Werra anhydrite }
Magnesian limestone	Werra carbonate
Marl slate	Copper slate (Kupferschiefer)
	Zechstein conglomerate
Rotliegend	Cornberger sandstone

[a]After Smith (1981) and others.

Sylvite does not crystallize in contact with carnallite, but carnallite can dissolve in brines under concurrent salting out of sylvite, which then remains hematite-stained. Red sylvites are thus secondary after carnallite decomposition. They overlie the carnallitites of the Devonian Prairie Evaporite Formation in Saskatchewan, Canada, preferentially in structural lows, and decrease toward the ancient shores of the basin (Wardlaw, 1968). This could suggest reprecipitation of original carnallite that had been washed into the residual brine from its shores by solutions undersaturated with respect to this compound and the eventual salting out of sylvite. However, the sylvite–carnallite boundary here cuts across stratigraphic boundaries and is stratigraphically higher on the margins than in the center. The shape of this convex lens marks the original water table of infiltrating solutions (Holter, 1972); the red sylvite is then a residue after the leaching of magnesium chlorides out of the carnallite. The deposit thus contains both sylvite after carnallite and carnallite after sylvite. The lowest of four Elk Point potash horizons consistently contains a basal 2–3 m of unaltered sylvite (Fuzesy, 1982), indicating that any carnallitization of successive potash members must have occurred by magnesium chloride solutions percolating from above. There was no artesian drive in the area to raise such dense solutions from below. It is significant that bischofite has not been found in these evaporites.

SULFATIC BRINES

The entry of oxygenated, sulfatic waters indicates a late stage in the evaporite history. The roof salts of a potash deposit mark the gradual freshening of the brine after it was no longer saturated for potassium chloride, after it ceased to be saturated for halite, and eventually even ceased to precipitate gypsum. In this process of gradual freshening, the density stratification is overturned before the brine is reduced to about 1.5 times the concentration of seawater. The anaerobic brine then turns into an aerated, oxygenated one, and sulfate ions are no longer reduced and removed. At that time, the substrate has not yet been compacted enough to lose all permeability, and sulfatic brines can seep in.

The interface of density stratification usually does not extend to the limits of the water surface but hits bottom some distance out from the shore. There is then an overlap of oxygenated surface waters on the bay margins. As the salinity difference decreases, the position of the interface is gradually lowered. Seasonal changes in the hydrostatic head of the oscillating bay level can insert such oxygenated sulfate brines into the permeable substrate. *If the descending brines are derived from soaking in such overlying oxygenated seawater, they are charged with sulfate ions, are easily saturated with dissolving rock salt, carry along significant amounts of potassium, and have a high Mg/Ca ratio.* Sulfatization then proceeds from shoals, islands, and shorelines.

Berner (1970b) has strongly supported the continuous diffusion of sulfate into sediments from overlying seawater. In clays this leads to pyrite formation and, at times, even to minor gypsum and anhydrite precipitation. Where the diffusion front reaches a buried evaporite sequence, a sulfatic alteration of chlorides will result. This covers up the primary sulfate deficiency in the evaporite sequence. Such pyrite occurs in black sapropelic layers in anhydrite lenses within the potash seam of the Glueckauf mine in German Zechstein evaporites (Krueger, 1962). *In virtually all sedimentary basins, one can observe an increase in the percentage of sulfate ions in oil field waters toward the earth's surface, confirming that sulfatic brines have a near-surface origin and are an expression of the dilution of formation waters.*

As surface waters penetrate into underlying aquifers, they are eventually stripped of their oxygen and become anaerobic brines. Sulfatization thus appears to be possible only within a certain proximity of the source of the oxygenated brines. There is evidence that substantially younger waters have seeped through porous rocks at great depth. Much of the brine movements in Devonian strata of Alberta, Canada, appears to be traceable to the growing hydrostatic head of the rising Late Cretaceous and Tertiary Cordillera (Sonnenfeld, 1964). It is unlikely that no pre-Permian evaporites have ever been subject to percolation of younger waters. If they have not been sulfatized in any of the basins so far studied, the conclusion is by now warranted that sulfatization must be an early event in the diagenetic history of an evaporite basin, occurring well before compaction closes all pore spaces in the salts. This would preclude the assumption of very long time spans between deposition and sulfatization, e.g., the sulfatization of Permian strata by Late Mesozoic or younger percolating brines.

The formation of a native sulfur deposit requires large amounts of gypsum or anhydrite to be present and meteoric waters to have access to them. The reduction of sulfate to hydrogen sulfide in the presence of organic matter is an essential step. However, the occurrence of native sulfur in caprocks of salt stocks cannot be explained by oxidation involving sulfates and hydrogen sulfide alone. The reaction is unfavorable at a neutral-to-alkaline pH. The oxygen dissolved in groundwater is a more likely source (Davis *et al.*, 1970). This limits the formation of epigenetic sulfur deposits to a depth of about 300 m below the earth surface, but in each instance it is evidence of downward percolation of oxygenated waters. *Epigenetic sulfur deposits form from anhydrite if the circulating waters deliver the organic matter. Syngenetic sulfur deposits utilize beds rich in organic matter adjacent to the anhydrite beds* (Davis *et al.*, 1970).

GYPSUM BENEATH THE OCEAN FLOOR

Formation waters occasionally precipitate gypsum beneath the sea floor. In this case, the precipitation is induced by microbes and does not always lead to euhedral crystals, but at times to nodular aggregates. In the Kattegat, the entrance strait to the Baltic Sea, gypsum nodules biologically enriched in isotopically heavy sulfur are presently forming in sands and silts 1–3 m beneath the sediment surface. These gypsum nodules contain authigenic quartz and potash feldspar crystals of the adularia variety. The bacteria responsible for this precipitate derive their energy from a methane seep (Joergenson, 1980, 1981). Similarly, marine muds in lagoons and coastal areas along the Mediterranean coast of France contain gypsum crystals, although the overlying open waters are of variable salinity (Levy, 1972).

Gypsum is often precipitated after the redissolution of iron sulfides. Lignitic clays in northern France contain gypsum derived from pyrite and enclosing limestone (Visse, 1947). Similarly, pyrites in Lower Cretaceous clays in the Kuybyshev area, USSR, were oxidized by descending meteoric waters; the liberated sulfuric acid altered calcite to gypsum (Miropol'skiy and Kovyazin, 1950). Middle Oligocene marine clays in Belgium contain gypsum crystals 3 m below the surface; both marcasite and pyrite nodules occur and are at times overgrown by gypsum (Vochten and Stoops, 1978). In comparison to pyrite, marcasite needs a more acidic yet reducing environment. The walls of cracks penetrating this clay are covered with gel-like ferruginous oxyhydrates and small gypsum efflorescences. This suggests that iron hydrates are mobilized, whereas sulfide ions are in part converted to sulfates and trapped. For a gypsum–pyrite association beneath the ocean floor the reducing environment must be followed by a lowering of the pH to allow dissolution of skeletal material (Siesser, 1978).

Posthumous gypsification or anhydritization of skeletal remains of marine faunas has been widely reported (Chirvinskiy, 1928; Dabell, 1931; Adams, 1932; Aleksandrova, 1938; Clifton, 1944; Dunham, 1948; Walter, 1953). In most cases, it is probably due to sulfur-oxidizing bacterial activity utilizing the decomposing organic matter for food. A fossil example of such epigenetic gypsum formations is the gypsum crystals in Oligocene foraminifera in southern France (Rosset, 1965), which contain fluid inclusions of sodium chloride solutions. Okada and Shima (1973) explained the authigenic gypsum in cores from the Sea of Japan as reaction products of the oxidation of framboidal pyrite inside a variety of shell cavities in the presence of calcium from the shells. This begs the question, however, of where the oxidizing waters beneath the pyrite horizon come from.

Where anaerobic bottom waters deposit large amounts of sapropel, such as in depressions of the eastern Mediterranean Sea, the hydrogen sulfide may react with iron hydroxides to form hydrotroilites and ultimately pyrites. However, loose gypsum crystals may also be found in this oxygen-deficient medium (Zemmels and Cook, 1973; Sigl *et al.*, 1978). They represent up to 30% of the sand fraction, or 4% of the total sediment, occurring above a sequence of sapropels, but never below them (Robert and Chamley, 1974). Such gypsum crystals are due to penetration of oxygenated Mediterranean waters.

Gypsum crystals are commonly found in deep-sea cores a substantial distance below surficial pyrite horizons. Cores of sediments from the southwestern Indian Ocean (Criddle, 1974), the southern Arabian Sea (Guptha, 1980), and the central Pacific Ocean (Cook and Zemmels, 1972) contain gypsum crystals ($\leqslant 2 \times 10^{-3}$ mm), but only several hundred meters beneath the sediment–water interface. Authigenic gypsum was also found in Namibian continental slope sediments (Siesser and Rogers, 1976a,b), an area noted for its anoxic, hydrogen sulfide-producing bottom waters (Martin, 1963). Pyritized worm tubes are here encased in gypsum; the tubes may have formed during a Pleistocene regression. A fracture in basalts near the mid-Atlantic Ridge contains gypsum and even halite crystals some 625 m beneath the sea floor. The halite here may be derived from a local hypersalinity produced by water consumption in hydration of basalt (Drever *et al.*, 1979). Even anhydrite can form in void spaces in submarine basalts on the East Pacific Rise and in the Costa Rica rift (Kusakabe *et al.*, 1982; Alt *et al.*, 1983). Authigenic gypsum also forms in deep-sea manganese nodules of the central and eastern Pacific Ocean (Xavier and Klemm, 1979).

These occurrences may be derived from sulfate diffusion out of the overlying seawater, as suggested by Criddle (1974). However, where the gypsum does not occur near the interface between the sediment and sulfate-bearing seawater, but instead a significant distance below that level, it occurs below the level of activity of sulfate-reducing bacteria and is younger than pyrite precipitation. Kuprin and Potapova (1978) have shown that in the Black Sea, biochemical decomposition is overtaken by geochemical processes 45 m below the water–sediment interface. Below that level, the organic content of the sediment changes, easily hydrolyzed organic matter being gradually replaced by insoluble organic matter. Around the Falkland Plateau, southwest Atlantic Ocean, gypsum precipitated on pyrite crystals in Miocene host muds (Muza and Wise, 1983). Similarly, gypsum crystals of the selenite variety formed after pyrite are found beneath the sediment–water interface of an open shelf along the west coast of India, a shelf noted for very high organic activity (Hashimi and Ambre, 1979).

SALT HORSES

Postdepositional leaching by upward moving brines forms "salt horses," or barren zones in potash deposits (Keys and Wright, 1966; Linn and Adams, 1966), that are filled with secondary recrystallized halite. They can be produced by both descending and ascending waters. In the Salado Formation of New Mexico, such salt horses derived from ascending brines are called "floor horses," in contrast to the "back horses" produced by descending brines

(Adams, 1969). Back horses are here distinguished from floor horses by their lower bromide content.

Water dissolves 365.6 g of sodium chloride per liter at 25°C (Hodgman *et al.*, 1963); regardless of the NaCl/KCl ratio, a percolating brine will only dissolve 298 g of NaCl per liter and 163 g of KCl per liter at that temperature (Valyashko, 1968). The solution remains undersaturated if there is a deficiency of one of the two salts. Brines still saturated in sodium chloride but no longer saturated in potash salts will not only alter precipitated carnallite to sylvite (Wardlaw, 1968; Podemski, 1975) but will also selectively leach sylvite under accompanying crystallization of halite during the leaching process (McIntosh and Wardlaw, 1968) without complete destruction of primary layering. In that case, the bromide content of the salt horses exceeds that of adjacent sylvites. A certain salting out of halite occurs when sodium chloride brines partially or completely dissolve precipitated carnallite or sylvite. The sodium chloride brines can then recrystallize gypsum to anhydrite or allow authigenic quartz to form with liquid inclusions of a sodium chloride brine (Protopopov, 1969). Based on crystal habits, the composition of fluid inclusions and the low homogenization temperatures, Kovalevich (1977) surmised that in Miocene evaporites of the eastern Carpathians, such salting out of halite has been widespread during the alteration of potash seams.

Salt horses produced by epigenetic leaching must be distinguished from areas of impoverishment in potash seams that are due to syngenetic nondeposition. The latter are round or elliptical in plan view, and taper upward. Ferric iron predominates over ferrous iron, and there is less magnesium than in surrounding parts. Tret'yakov (1974) recognized these as ancient sills that extended upward into the zone of oxidized surface currents. Their subsidence below the interface of brine stratification may have been achieved only during or after the formation of a second potash seam in the deeper parts of the lagoon. Seidel (1966) observed that "hard salt," the epigenetic sulfatic alteration zone of a potash seam, occurs preferentially on sills, with carnallitite providing the basinal equivalent. The altered zone may give way to a totally depleted zone, a salt horse on the crest of the sill. On the gentle flank, the altered zone is mainly an anhydrite-rich mixture of potassium sulfates; on the steeper flank, it is a kieseritic mixture. Ivanov (1963) had earlier observed the same relationship of decomposed carnallite and secondary sylvite on sills, preserved carnallite, or primary sylvite in depressions of the Upper Kama deposit west of the Urals.

On the flanks of floor horses are lenses and pods of potash sulfates in juxtaposition to sylvite. Adams (1969) studied them in detail and found that a langbeinite core is surrounded by an outer mixed zone of leonite and kainite. Brines causing the alteration were oxygenated, contained magnesium and sulfate ions, and were warmer than 37°C, since langbeinite is unstable below that temperature. Leonite precipitates readily at that temperature, whereas kainite precipitates very slowly unless seeded (Lepeshkov and Bodaleva, 1949). The formation of salt horses appears to have occurred here before the alteration of the original carnallite to sylvite was completed. The sylvite contains four orders of magnitude more iron than it would obtain from concentrating seawater. The remaining carnallite is then probably secondary.

Borisenkov (1968a,b) differentiated two types of barren salt horses in German Zechstein evaporites: a layered and a massive type. The massive type often contains a large amount of kierserite; langbeinite and polyhalite are only sporadically present. This type is sometimes related to basalt veins. The layered type contains halite, bloedite and aphthitalite, and kieserite with kainite and sylvite at the base. Carnallite is separated from the barren zone by a sylvitic–kieseritic, a sylvitic–kieseritic–polyhalitic, and finally a langbeinitic–sylvitic–polyhalitic–kieseritic zone with some new anhydrite (Doehner and Elert, 1975). Secondary carnallite also occurs in the German Zechstein evaporites after kainite, which in turn formed after langbeinite. The langbeinite is probably a product of sulfatization of a primary carnallite (Moetzing, 1968).

Salt horses in the Stassfurt Member of the German Zechstein evaporites taper upward. If the waters were still undersaturated in sodium chloride when they reached halite beds, neoformations of secondary minerals in the barren zone would be difficult to notice. Fluids saturated for sodium chloride dissolved potassium chloride or magnesium chloride under concurrent but gradually decreasing salting out of halite and some anhydrite (Hentschel, 1961). Ascending brines seem to have leached carnallite locally in the Stassfurt Member, with a 50% volume reduction. They breached yellowish-red halite, so that the gray variety is closer to the barren area, which is marked by unclear layering and the absence of potash inclusions. Anhydritic halite clusters within the gray salt may represent a replacement of carnallite clusters (Heynke and Zaenker, 1970). The original thickness of the carnallite cannot be determined precisely because of brecciation (Baar, 1952) that produced a mushroom- or funnel-shaped barren zone, facing upward with its open end, and the overlying banded salt is sagging into the funnel. Kieseritic alterations preserved small rhythms as halite–kainite laminae wherever the carnallite is missing. Larger quantities of kainite under sylvinites are proof of more extensive soaking by ascending solutions. Eventually, a rim of secondary minerals formed around barren zones composed of kainite, leonite, aphthitalite, and some langbeinite.

For total replacement by polyhalite, the additional introduction of calcium sulfate must have taken place (Koch *et al.*, 1968). Such an additional epigenetic calcium sulfate supply has also been postulated for Carboniferous anhydrites of Spitsbergen because of the absence of compaction phenomena in both porous and nonporous anhydrite nodules. Initially, all nodules must have been porous (Kinsman, in Holliday, 1968).

ASCENDING FLUIDS

Ascending brines are under artesian pressure derived from hills and mountain ranges that may well be some distance away. Such waters produce mound springs in Lake Eyre, Australia (Madigan, 1930). Ascending brines try to reach their piezometric surface, and for short distances, may utilize the superior permeability parallel to bedding planes. They saturate quickly by dissolving salts, and produce brine chimneys or pipes (Anderson and Kirkland, 1980), or salt horses of exceptionally pure white halite rimmed by clays as selvage. Even if originally contaminated by meteoric waters, these brines have been sufficiently altered by ion exchange along the flow path to be indistinguishable from the normal range of formation waters. As such, they are charged with calcium and chloride ions and enriched in bromide and boron, but are grossly deficient in magnesium and potassium. Formation waters generally show Mg/Ca and K/Ca ratios < 1 (White *et al.*, 1963). Once the brines are saturated for sodium chloride at a given formation temperature, they supersaturate when rising and cooling. Thus, some of the rock salt can be precipitated as halite cement in sandstones and in carbonates.

Gravitational forces acting on solid phases of overlying rock increase pressures on interstitial waters during compaction. Thus, they will exceed hydrostatic pressures at comparable depths (de Sitter, 1947; Bredehoeft *et al.*, 1963). These waters flow selectively through lutitic rocks which act as semipermeable membranes holding back the salts in solution. An increasing salinity of formation waters is thus produced by ion filtration on charged-net clay membranes (Degens *et al.*, 1964). This was experimentally documented by Hanshaw and Coplen (1973) at a pressure of 33 MPa. The specific permeability of colloidal membranes increases with the concentration of sodium and potassium chloride, but decreases with the concentration of calcium and strontium chloride until a minimum is passed (Manegold and Hofmann, 1930). However, Manheim and Horn (1968) raised objections to a clay filtration model on two grounds:

1. A filtration system that can produce hypersaline brines would have to process enormous volumes of fluids through poorly permeable strata under virtually leakproof conditions, instead of moving them across permeable channels. This is a geologically unlikely scenario.

2. Fluid pressures that induce salt rejection by clay membranes must be sufficiently high to overcome the osmotic pressure differential between salt solutions on the input and output sides of the membranes, as well as the frictional resistance of the membrane to permeation. For a sea salt solution of 250,000 ppm, the pressure required to overcome osmotic forces created by the separation of salt and waters is 35 MPa at 25°C, rising to 40 MPa at 100°C. This is far in excess of the maximum pressure gradients observed in the normal sedimentary environment.

Sulfatic brines can enter laterally and preferentially affect deeper layers. In the Masli potassium salt deposit of Ethiopia (Ivanov, 1970), a basal potash deposit consists of kainite with a lesser amount of halite, evidently an alteration after carnallitite. The intermediate member contains halite and also anhydrite, kieserite, kainite, and small quantities of polyhalite and bischofite. Primary carnallite is preserved only in depressions. Finally, an upper member is composed entirely of sylvite that wedges out toward the edge of the basin. This sylvite has been altered from carnallite by atmospheric moisture and water flows, but has not been exposed to percolating sulfatic brines.

A mixture of mainly sulfatic potassium and magnesium salts in the German Zechstein evaporites contains sylvite with relic carnallite inclusions, halite corroded by sylvite, polyhalite after kieserite, langbeinite with relic kieserite, and sylvite grains (Meier, 1969a). This form of alteration is said to have been produced not by lateral introduction of brines rising through the precipitate, but by ascending brines saturated with calcium sulfate, according to Borchert (1940) and Baar (1944). Waters liberated in the anhydritization of gypsum were believed to be the source of kieserite precipitation in potash horizons (Borchert and Baier, 1953), because thick anhydrite occurs only where potash horizons are absent. However, thick anhydrites mark sills and shoals. It is likely that the bitterns precipitated potash salts on the slopes. Kieserite formation may well have occurred through brines, saturated with calcium sulfate, washed down from the gypsum shelves and migrating a short distance sideways, since permeabilities are generally greater along bedding planes than across them.

For a while, there was a considerable argument over the validity of this concept of sulfatization by ascending brines. Borchert (1940 and later works) allowed for no descending brines, but only for late (Cretaceous or Tertiary) epigenetic waters that were derived from the dehydration of basal gypsum layers. That would presuppose a late anhydritization of Permian gypsum, a mineral very unstable in the presence of hygroscopic brines. The presence of alkanes or other petroleum compounds admittedly reduces the hygroscopic nature of potassium chloride solutions (Ordonneau, 1950), but these compounds are soon removed by bacteria. With progressive burial of the gypsum under the growing rock salt thickness, the rock temperature is bound to increase according to the geothermal gradient. As the temperature of a descending brine rises, progressively smaller amounts of dissolved sodium or potassium chloride are able to retard and eventually stop the dehydration (Budnikoff and Schtschukareva, 1935). Consequently, it is much easier to dewater gypsum at an early stage of burial. Moreover, waters saturated with calcium sulfate percolating upward through dolomite intercalations would react with the host rock and induce large-scale dedolomitization, a feature not observed in evaporite basins.

The calcium sulfates are thinner underneath salt deposits than on their margins. A cubic meter of gypsum produces almost half a cubic meter of water upon dehydration. This water, saturated in respect to gypsum, carries less than 1 kg of sulfate ions in solution. In brines containing sodium chloride, the solubility of gypsum increases at first. However, by the time the waters saturate with respect to sodium chloride by percolating upward through halite sections, the solubility of gypsum has decreased by slightly more than 30%, thus raising the amount of water required to percolate through. Other ions present will slightly increase or decrease gypsum solubility, but will not change the order of magnitude. The highest strontium content is usually found in brines within an evaporite sequence or immediately below it (Korennov, 1974). That is evidently the consequence of diagenetic or epigenetic leaching of strontium, primarily from calcium sulfate and calcium carbonate sections.

The presence of a 1-cm-thick kainite layer requires all the sulfate ions mobilized in the dehydration of nearly 10 m of gypsum. The resulting brines, saturated with calcium sulfate but undersaturated with respect to sodium chloride, would dissolve about 60 cm of halite en route. The presence of a 1-cm layer of langbeinite requires all the sulfates mobilized in the dehydration of nearly 16 m of gypsum and the solution of almost 1 m of halite. In turn, 1 cm of kieserite requires the dehydration of 21 m of gypsum and the consequent solution of more than 1.25 m of salt (Fig. 12-7). The required thicknesses of basal gypsum beds

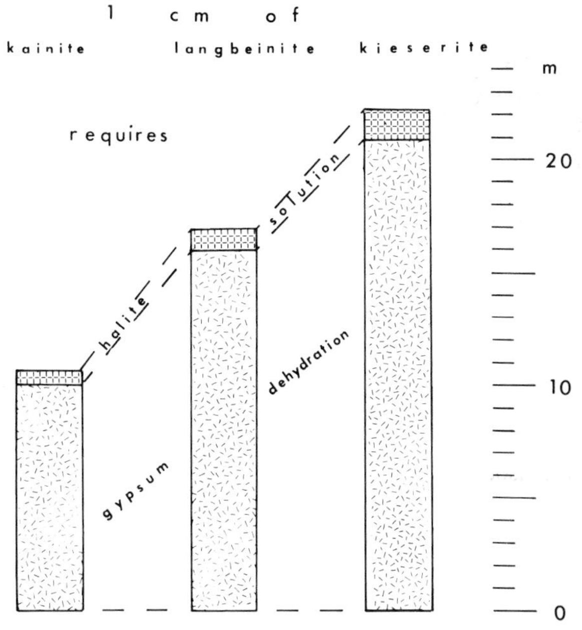

Fig. 12-7. Basal gypsum bank thicknesses required to dehydrate in order to sulfatize, by ascending saturated crystal waters, 1 cm of overlying potassium and magnesium minerals.

are not available, nor are the refilled or collapsed cavities found in intervening basal halites. Thus, it is impossible for ascending waters derived from gypsum dehydration to provide the sulfates required for sulfatization of a massive potash deposit. They can only dissolve potash deposits and replace them by the salting out of halite. It also has to be borne in mind that gypsum and calcium chloride solubilities decrease with temperature in potassium chloride solutions, particularly in the presence of sodium or magnesium, but at a constant temperature the solubility increases with an increasing potassium chloride content of the bitterns (Kruell, 1933). Similarly, the solubility of calcium sulfate in a solution of potassium sulfate decreases with dropping temperatures (Teraoka and Ito, 1958).

An assumed gypsum dehydration by ascending brines begs the question of their original composition. D'Ans and Kuehn (1960) envisaged small incursions of such brines into the evaporites from underneath—insufficient, though, to dissolve all the potash salts. Ascending fluids would have originated as calcium chloride brines from underneath the evaporites (Storck, 1953) and then dissolved some of the basal sulfates. Halite- and anhydrite-filled fractures indicate that some fluids saturated with halite and calcium sulfate were rising through the deposit (Storck, 1954). However, these brines could not have been very concentrated, since alkali chloride solubility is very low in calcium chloride solutions (d'Ans, 1933; Rza-Zade and Rustamov, 1961). Nonetheless, both ascending and lateral solutions entered the Zechstein potash seams.

BRINE EQUILIBRIUM

Evaporites remain in equilibrium with brines circulating through them, continuously reacting with even residual solutions (Niggli, 1952). This is evident from the difference in bromide content between brine and precipitated salts. This bromide content tends to equilibrate in carnallites and even halites (d'Ans, 1944; Schwerdtner, 1962), or the Pb/Na exchange in halite (Herrmann, 1958). There are four additional lines of geochemical evidence which show that precipitates remain in equilibrium with brines with which they are in contact: the study of isotopic ratios, the investigation of fluid inclusions, the multiple early diagenetic or epigenetic sulfatization cycles, and the determination of the bromide content of evaporite minerals.

ISOTOPIC RATIOS AND AGE DATING

Age dating of evaporite deposits on the basis of ratios of radioactive elements and their daughter isotopes has proven to be a failure, because evaporite minerals continue to act as open systems even after diagenesis. This applies to Rb/Sr ratios in whole-rock samples or individual minerals (Lippolt and Raczek, 1979a,b; Register and Brookins, 1980), and to K/Ar and K/Ca ratios (Ivanov, 1963; Wilhelm and Ackermann, 1972; Khrushchov, 1973; Shell Development Co., 1973). The rocks have acted as open systems either continuously or episodically: Different ages are determined from samples as little as 10 m apart due to heavy losses of radiogenic ^{87}Sr (Lippolt and Raczek, 1979a). Only occasionally do K/Ar ages correspond to stratigraphic ages, and then only at completely undisturbed sites. For instance, the Jurassic potash salts in Turkmenistan yield end-Jurassic/early Cretaceous ages in unchanged sections, but give Eocene ages in recrystallized portions (Levshin *et al.*, 1975). Undisturbed early Cambrian halite sections on the Siberian platform give correct K/Ar ages, whereas coeval halites in the Sayan Mountains to the south give ages that are one-sixth of the correct value (Tarasevich *et al.*, 1971).

Pink and red sylvites lose a significant amount of argon, whereas unrecrystallized salts lose less (Borshchevskiy and Borisova, 1963). There is thus a distinct lowering of K/Ar ages with increasing content of secondary sylvite (Schilling, 1973; Brookins, 1980a). Permian sylvites of the Upper Kama district west of the Urals contain only 1/30 of the argon to be expected on the basis of the geologic age of the formation (Gerling and Titov, 1949). The loss of argon may actually be a measure of the stress to which precipitated sylvites have been exposed (Brandt *et al.*, 1966). Moreover, the adsorption of argon onto (111) faces of octahedral sylvite differs considerably from that on the (100) plane of cubic sylvite (Young, 1952).

In German Zechstein evaporites, argon has leaked from sylvite, carnallite, and polyhalite (Pilot and Blank, 1967). However, langbeinite repeatedly gives Permian ages, because it does not lose its argon below 550°C, having a diffusion constant 15 orders of magnitude smaller than sylvite (Pilot and Roesler, 1967). The langbeinite-bearing alterations must have been formed shortly after deposition of the evaporites, i.e., sulfatic brines were able to enter before compaction eliminated permeabilities. Whereas some horizons yield Permian ages, others produce Jurassic dates (Oesterle and Lippolt, 1975a,b). Langbeinites of New Mexico likewise yield K/Ar dates close to the stratigraphic ages of the rock, but only if they are not mixed with halites (Brookins, 1980b). There may be a proportional relationship between the amount of lowering of the apparent age and the sodium content of the rock. Langbeinite alone does not give up more than 5% of its argon content if subjected to temperatures of up to 500°C (Lippolt and Oesterle, 1977).

Secondary polyhalites or rubble chimneys in southern New Mexico yield K/Ar dates that agree with those obtained as Rb/Sr ages, but are 30–70 million years younger than the stratigraphic age of the evaporite deposit. They record a dissolution–remobilization–final precipitation sequence in early to middle Triassic times. There was then no post-Triassic recrystallization except in very localized places (Brookins *et al.*, 1980). The clay layers act here as a sink for ^{87}Sr during early diagenesis. One pre-Permian age has been determined from such clay intercalations, whereas whole-rock data or data from the water-soluble fraction gave substantially younger ages (Powers *et al.*, 1978). Whereas Brookins (1980a) denied the possibility of either continuous or episodic leaching of ^{87}Sr or ^{40}Ar, Bodine (1978) maintained that the series had undergone extensive postdepositional reactions and recrystallizations. In addition to the wider range in isotopic age determinations, he cites as evidence the abundant replacement of anhydrite by polyhalite, the replacement of a primary bitter salt association by potash ores deficient in magnesium, and the great variability in bromide content, yielding abnormally low values.

A radiometric age of Florida anhydrites derived from a helium ratio gave a Late Cretaceous age corresponding to the stratigraphic position (Urry, 1937). Elsewhere the helium content bore little relation to age or radium content of the halite (Karlik and Kropf-Duschek, 1950). Often the He/Ar ratios give age determinations that are 100-fold too low because of ^{40}Ar enrichment (Nesmelova, 1961). The uranium and radium content of halite, sylvite, and carnallite shows no significant difference, averaging 6×10^{-7}/kg for uranium and 3×10^{-13}/kg for radium (Kemenyi, 1942). However, associated clays may contain 30 ppm (Bell, 1956). Although most of the radioactivity in evaporites is generated by ^{40}K, 4% are generated by radium and thorium in sylvites and 13% in carnallites. The radium and thorium contents are linear functions of the amount of clay present (Mishin, 1971).

FLUID INCLUSIONS

Inclusions in evaporites are voids usually filled with gaseous and liquid phases of fluids (Brewster, 1826); the voids are often in the form of negative crystals. A century ago, Hammerschmidt (1883) had noticed that negative crystals in anhydrite and gypsum always occur parallel to cleavage planes. The same holds true, however, for cavities and gas bubbles in rock salt (Privalova, 1971). The gas is frequently under pressure, yielding upon liberation about one and a half times the volume of the crystal it escaped from (Dumas, 1830). Fluid inclusions in carnallite and bischofite are commonly only a few microns across. They are slightly larger in sylvite, and progressively larger in polyhalite, langbeinite, and kainite (Petrichenko, 1973). Diagnetic solutions move along cleavage planes and either produce secondary inclusions or replace fluids in primary inclusions (Sabourand-Rosset, 1974). Gypsum crystals in Guardamar de Segura, Portugal, have solid centers. The inclusions increase toward the exterior, and clay particles outline planes within the crystals (Garrido, 1945). That fluid inclusions indicate rapid crystallization is corroborated by the occurrence of radial sets of anhydrite needles in these inclusions (Dubinina, 1951a). Whereas soluble phosphate reduces solid inclusions, heavy metals and certain organic compounds, e.g., carrageen Irish moss extract, reduce the size and frequency of fluid inclusions (Shuman, 1966).

Fluid inclusions tell us something about the nature of the last brines that have percolated through the evaporite deposit. A study of fluid inclusions in halites of the Pripyat Basin in Byelorussia and the Dnieper–Donets basins in the Ukraine showed that they had all been fed by identical chloride brines. As the concentration of the brine increased, so did the amount of dissolved potassium, but calcium and magnesium relatively declined (Kityk *et al.*, 1980). Fluid inclusions in Permian evaporites of West Texas are similarly very complex bitterns, and are not merely derived from groundwater containing sodium chloride (Powers *et al.*, 1978). The closer to the surface an inclusion occurs, the greater is its sulfate content (Petrichenko and Slivko, 1973). The K/Mg ratio decreases outward in halite inclusions because magnesium increases outward faster than potassium (Petrichenko, 1973). Both magnesium and the sulfate content of inclusions increase toward the periphery of the crystals, which proves that magnesium sulfates are a late addition to the system. They entered with postdepositional brine movements that lowered the pH and produced a positive redox potential. A study of fluid inclusions in evaporites of various ages produced pH values ranging from 4.3 to 8.92 and Eh values ranging from −260 to +408 mV in very local variations. Recrystallization of halite appears to have occurred under nearly neutral, slightly oxidizing conditions. The bromide content appears to be a good indicator of these alterations: Primary halite contains more than twice as much bromide as diagenetic halite (Petrichenko *et al.*, 1973a,b, 1974).

The vacuole disappearance temperature of fluid inclusions is not a measure of the temperature of the solution as a whole, but only of the temperature prevailing at the surface of the crystallizing compound (Dreyer *et al.*, 1949), or merely the measure of recrystallization (Peach, 1949). The homogenization temperature rises with increasing concentration of sodium or potassium chlorides (Shaposhnikov and Khetchikov, 1970). Factors affecting this local temperature have been critically discussed by Gerlach and Heller (1966). In primary inclusions, the temperature of disappearance is often less than 20°C, but it may be complicated by the presence of $FeCl_3$ (Touray, 1970). It also has to be corrected for overburden pressure (Powers *et al.*, 1978). In contrast to hydrothermal processes, in which the temperature of homogenization of fluid inclusions indicates the potential temperature of crystallization, in natural evaporites the temperature of homogenization is somewhat elevated (Sedletskiy *et al.*, 1971). Vacuole content is difficult to measure, as all available fluids are given up only at temperatures close to the melting point of the salts (Gevantman *et al.*, 1981) (Fig. 12-8).

The fluid inclusions are not necessarily in their original position. Small quan-

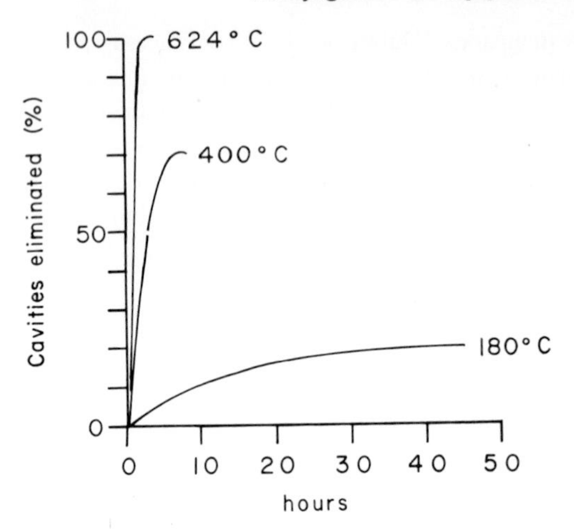

Fig. 12-8. Vacuole elimination at different temperatures (after Gevantman *et al.*, 1981).

tities of brine that are trapped in negative crystal voids can migrate. Migration toward a heat source was first observed by Nicol (1828). The driving force of this migration is the difference in solubility between the warmer and colder sides of the cavity. It is thus a function of temperature and is proportional to the temperature gradient. Solubility changes per 100 g of liquid water are only 0.03 g/liter/°C, and thus are very small. However, fluid inclusions move under uniaxial stress in minutes, changing shape and volume in the process (Powers *et al.*, 1978). In samples of Permian Wellington salt from Kansas, the migration rate for a gradient of 1°/cm varied from about zero at 20°C to 7.7 cm/year at 300°C (Bradshaw and Sanchez, 1969). It is still not understood why migration rates of different inclusions in the same salt sample vary by a factor of 3 (Roedder and Belkin, 1980). Droplets with more than 10 volume % of gas migrate down the thermal gradient toward cooler temperatures, whereas droplets with a smaller gaseous fraction migrate up the thermal gradient. The migration is hindered by the grain boundaries, where the salt can permanently entrap the droplet (Anthony and Cline, 1972).

SUMMARY OF BRINE MOVEMENTS

The difference between an isochemical distribution of seawater precipitates and the proportion of salts found in any evaporite basin is due to leakage and outflow. Much of the outflow occurs as brine seepage, and precipitated salts are subject to brine movements both during deposition and later. The brines may enter from above, laterally, or from below. Sulfatic brines entering the precipitates laterally or from above are oxygenated brines that alter the chloride precipitates into sulfates. They can also produce authigenic gypsum crystals in deep-sea sediments beneath the layer of anaerobic sulfate reduction. Formation waters entering from below are anoxic calcium chloride brines with a limited ability to alter the precipitates that continue to act as open systems for some time after burial. Fluid inclusions indicate the nature of the last brine with which the crystal remained in equilibrium, but they may have migrated along a temperature gradient.

13

Secondary Sulfates

DEHYDRATION OF GYPSUM (ANHYDRITIZATION)

Recent primary anhydrite is rare; most of it is produced by dehydration of precipitated gypsum. Subaerial desiccation of the environment, however, is not the primary factor in the formation of anhydrite, since there is evidence of subaqueous anhydritization (Amieux, 1980). Anhydrite forms at some depth below the surface, such as in the secondary anhydrite infilling in dolomitized reefs carried out by circulating epigenetic brines that are, no doubt, hygroscopic. The lower limit of anhydrite precipitation seems to lie in brines concentrated to 4.8 times seawater, but many brines of greater concentration still precipitate gypsum. Primary anhydrite is believed to precipitate directly, once the concentrating brine becomes sufficiently hygroscopic and the activity of the remaining water molecules has been sufficiently reduced. Bitterns available for seepage at the end of carnallite crystallization represent about 47% of the volume of precipitated sylvite and carnallite (Valyashko and Vlasova, 1975). Dehydration is then brought about by magnesium- and potassium-rich fluids that enter the subsurface and attack not only limestones but also gypsum and anhydrite. *A lengthy exposure to hygroscopic brines is necessary to alter gypsum to anhydrite* (Braitsch, 1971). Alkali earth metal chlorides in a slowly moving brine are the strongest inhibitors of hydration (Maslenitskiy, 1939) and thus act as the agents of dehydration; a rising magnesium chloride concentration increases the stability field of gypsum at the expense of anhydrite (Borchert, 1940).

A marine brine saturated with respect to sodium chloride will in time dehydrate gypsum at temperatures above 14°C (Braitsch, 1971), although the original precipitate in brines saturated for sodium or magnesium chlorides is gypsum rather than anhydrite (Spezia, 1888). Once some halite starts precipitating in the dry season, the residual brine remains saturated for sodium chloride until freshened. Such a descending hygroscopic brine has to play an important role in the penecontemporaneous diagenetic dehydration of gypsum to anhydrite underneath a halite bed. As precipitation progresses from halite to sylvinite or carnallitite, percolating bitterns become even more hygroscopic. They are responsible for the postdepositional dehydration of goethite and hydrogoethite to hematite needles in red halites, sylvites, and carnallites. The hygroscopic quality of some evaporite minerals is given in Table 13-1. It must be remembered, though, that the dehydration of hydrated minerals reduces the concentration of the brine (Langbein, 1979). The hydration of magnesium chloride, and for that matter magnesium sulfate, increases gradually with increasing halite content, but rather sharply once the halite content exceeds half the solute (Elizarov and Musanov, 1970). Because of the competition for available water molecules, the number of water molecules per magnesium chloride or sulfate molecule again decreases once most of the halite has precipitated. The brine has then become extremely hygroscopic.

Spezia (1888) was able to precipitate only gypsum even in the presence of saturated sodium chloride or magnesium chloride solutions. Vater (1900) showed that gypsum, rather than anhydrite, precipitates from calcium sulfate solutions saturated with sodium or magnesium chlorides. The greater the degree of supersaturation, the disproportionately greater is the growth rate of gypsum in a sodium chloride solution (Brandse *et al.*, 1977), but no anhydrite forms in this experiment. Along the Trucial coast in the Persian Gulf, gypsum remains stable up to chlorinities of 145 ppt (Butler, 1969). In order to achieve anhydrite precipitation in the laboratory (Mel'nikova *et al.*, 1977), the brine must contain at least 18% chlorides and be above 20°C. Suzuki (1943) confirmed the dehydration to be a single molecular reaction, on which the presence of sodium or potassium chloride has no effect. The rate of dehydration increases with rising vapor pressure. Dehydration takes 9 hr at a vapor pressure of 250 kPa, but only about 90 min at a vapor pressure of 810 kPa (Manzhurnet, 1939). Yet Šatava and Zbůžek (1971) found that the start of dehydration did not seem to depend on the

TABLE 13-1

Hygroscopic Coefficient of Some Evaporite Minerals[a]

Chlorides	%	Sulfates	%
Halite	74	Kainite	62.5
Sylvite	74	Picromerite	58
Carnallite	47	Langbeinite	57.5

[a]After Bezhenar *et al.* (1978).

water vapor pressure in the range of 2.6–165 kPa. Similarly, the dehydration of bassanite is also nearly independent of the partial pressure of water vapor at 69–80°C (Bobrov *et al.*, 1978).

Gypsum exposed to air on coastal flats at low water level tends to dehydrate to powdery bassanite, only to be reconstituted into gypsum when the brine level rises. It has been known for a century that any hydrate, such as gypsum, when heated in air, will lose water if the vapor pressure of the hydrate exceeds the partial pressure of the water vapor in the air in contact with it (Hannay, 1877). Prolonged outgassing over P_2O_5 turns bassanite into anhydrite even at room temperature (Powell and Way, 1962). Gypsum-precipitating lagoons in the Venezuelan and Dutch Antilles do not convert gypsum to anhydrite (Hudec and Sonnenfeld, 1974) as soon as the brine reaches and exceeds sodium chloride saturation. Brine temperatures of over 59°C have been recorded there. Only a bassanite veneer develops where polygonal ridges are exposed to air at low water level. Moiola and Glover (1965) observed that gypsum crystals in mud from Clayton Playa, Nevada, develop rinds of bassanite and anhydrite when exposed to air over a period of 11 months.

The occurrence of recent anhydrite appears to be related to the presence of organic decomposition products. In an inorganic environment, anhydrite cannot be produced even at salinities in excess of 200 ppt (Cody, 1976). Experimentally, it can be grown in the presence of several organic macromolecular compounds, and then it invariably aggregates into tiny nodules. Such surface-active compounds that strongly adsorb onto gypsum are disseminated throughout the gypsum crystal in concentrations insufficient to block crystal growth. Dissolution of the initial gypsum frees a large enough amount so that anhydrite might be able to nucleate without gypsum or bassanite precursors (Cody and Hull, 1980). However, any organic matter must first decompose or be flushed. Organic matter retards the dehydration and decreases the mechanical strength of gypsum (Logvinenko and Savinkina, 1967a,b); however, the presence of citric acid decreases the nucleation period of anhydrite (Šatava, 1977). Sills were likely sites of photosynthetic oxygen production and, for that reason, were conducive to gypsum precipitation. Later, the thicker gypsum on sills released more water upon anhydritization (Borchert and Muir, 1964).

In an arid climate, such as that prevailing along the Trucial coast, plant root molds are commonly cemented with euhedral gypsum arranged in a concentric pattern around former roots, irrespective of the type of cement found in the host rock. However, in a humid climate, such as that prevailing in Florida, the cement around plant root molds is calcite (Glennie, 1972). Supratidal anhydrite in Kuwait occurs around bushes of *Halocnemum strobilaceum* (Pall.) M.B. (Gunatilaka *et al.*, 1980). Its occurrence is restricted to mature plants, suggesting that the plant first had to shed a sufficient quantity of macromolecules into the soil. The anhydrite is rehydrated to gypsum upon the death of the plant, but the presence of these adsorbed organic macromolecules facilitates easy reconversion into cryptocrystalline anhydrite with a felty texture. The widespread presence of organic matter may facilitate the conversion of gypsum into a watery anhydrite slush below the surface of the Duqan sebkha in Qatar (Perthuisot, 1977), rather than at the prevailing high temperatures that exceed 30°C down to a depth of at least 50 cm beneath the surface. Conversion to anhydrite proceeds crystal by crystal, as witnessed by intact remnant gypsum crystals.

Gypsum is known to exist down to 1200 m (Murray, 1964; Mossop and Shearman, 1973), suggesting that neither the geothermal gradient nor the overburden pressure is the prime agent of anhydritization. Considerable alteration is evident in Miocene evaporites in Transylvania, Romania, in that polyhalite, kainite, leonite, and picromerite indicate intense sulfatization. However, only gypsum is found, not anhydrite (Pauca, 1968). Nonetheless, anhydritization is common in most ancient calcium sulfate deposits and is generally ascribed to the effects of overburden and geothermal temperature increase. Massive anhydrite deposits occur in Upper Miocene (Messinian) evaporites in the Mediterranean area, although many if not most of them have never been exposed to great lithostatic pressures or to temperatures elevated above the original brine temperature ranges. Despite high heat flow rates, bottom temperatures on the Mediterranean sea floor overlying these evaporites have not been much above 14°C since the early Pliocene. However, these anhydrites have been exposed to percolating, concentrated bitterns as halite precipitation in places gave way to Miocene potash deposition.

A localized pressure–temperature gradient is insufficient to dewater gypsum. Even at a pressure of 12.25 MPa, equivalent to burial at 5.43 km, no conversion by pressure alone takes place in laboratory experiments (Douglas and Goodman, 1957). Earlier, Spezia (1888) had found that a calcium sulfate solution still precipitates gypsum rather than anhydrite at pressures up to 51 MPa. Pressure dependence of gypsum/anhydrite has been discussed by Braitsch (1971), who noted that the transition temperature is raised by 1°C for every 4.1 MPa of pressure differential between lithostatic and hydrostatic pressures or for about every 160 m of overburden. Marsal (1952) undertook a theoretical study of the pressure dependence of the gypsum–anhydrite conversion. With increasing pressure, the solubility rises rapidly at low temperatures and then levels off. This is particularly noticeable at higher concentrations. The increasing permeability of overlying sediments decreases the required overburden, and the rising density of surrounding formation fluids decreases the transition temperature. Gypsum is thus no longer stable even at ambient temperatures if hypersaline brines can percolate through overlying permeable strata, such as through a porous halite mush. Anhydrite is consequently the favored form at great depths, even in the presence of only moderately hygroscopic brines.

Lengthy preheating at temperatures of 110–150°C produced partial dehydration of gypsum at 500 or even 200 MPa. Below 200 MPa, no transformation of gypsum occurred up to temperatures of 110°C (Heard and Rubey, 1966). These experiments are inconclusive, since it was not determined whether the dehydration product is bassanite or anhydrite proper. The reported instant partial rehydration to gypsum suggests the presence of only bassanite. Farnsworth (1924, 1925) had earlier found that bassanite forms at 110°C from gypsum, and anhydrite above 160°C, but that anhydrite does not hydrate even at pressures of up to 1.9 MPa and temperatures of up to 210°C. Shcherbina and Shirokikh (1971) found gypsum to be stable up to 110°C at pressures below 45.6 MPa. The hemihydrate bassanite was found at 110–155°C and anhydrite only above 155°C. However, at a very low partial pressure of water vapor, gypsum is altered directly to anhydrite (Mehta, 1976). The pronounced influence of gases on dehydration proves that the water of crystallization first passes through an adsorbed stage (Gardet *et al.*, 1976).

Mere stress relief cannot rehydrate anhydrite, as has been suggested by Schreiber *et al.* (1976). Yet gypsum derived from the hydration of anhydrite precipitated around plants, and thus soaked in organic decomposition products, reconverts to anhydrite under the hooves of a herd of camels (Gunatilaka *et al.*, 1980). This could be due to the grinding action of the hooves. Dundon and Mack (1923) observed that grinding removed three-quarters of the crystal water from gypsum, effectively producing bassanite. Upon standing for a while, only a small fraction of such ground material regains the crystal water (Steiger, 1910). Crushed gypsum has a lower dehydration temperature (Zolotov and Lavrov, 1960); so does finely ground gypsum as a result of kinetic factors (Powers *et al.*, 1978). Once the grain size reaches 130 μm, every slight decrease in diameter corresponds to an abrupt increase in the rate of dehydration (Khalil, 1982). Finely ground anhydrite, if left standing, will hydrate on its own. The speed of hydration, then, is inversely proportional to the grain size (Farnsworth, 1925). Hulett (1901) had noted that gypsum, because of its high porosity, is easily ground to powder, and that convex surfaces show greater solubility, and concave ones less solubility than planar surfaces. Hulett and Allen (1902), Tanaka *et al.* (1931), and Hara *et al.* (1934) then confirmed that poorly crystalline or very finely ground gypsum has a substantially increased solubility. Solubility of gypsum reaches a maximum around 0.2–0.5 μm (Hulett, 1901; Dundon and Mack, 1923); that of anhydrite reaches a maximum at 2.8 μm (Roller, 1931). Dehydration rates of monocrystals of selenite, polycrystalline gypsum, and precipitated gypsum differ. Dehydration in water proceeds topotactically. The nucleation rate apparently increases with the increasing specific area of gypsum (Šatava, 1977).

Hygroscopic chloride brines are not the only brines that can dehydrate gypsum. Native sulfur in gypsiferous caprocks can be reoxidized to sulfate during exposure to soil moisture in the oxidized zone. The resulting concentrated sulfatic brines not only have a very low pH, but are also hygroscopic and can cause dehydration of gypsum to a second generation of anhydrite. At a pH greater than 4, this form of dehydration yields only incomplete pseudomorphs; at a pH of 2 or less, all of the gypsum turns to anhydrite (Srebrodol'skiy, 1972, 1973). Some of these sulfatic waters of meteoric origin can penetrate deeper down and can alter precipitated potassium and magnesium chlorides.

A unique occurrence in the Kara-Kum desert east of the Caspian Sea is the free sulfuric acid dripping from a sandstone with a gypsiferous cement that also contains native sulfur (Klockmann, 1978). This is a bacterial decomposition product of gypsum preserved in the absence of drainage.

THE TRANSITION TEMPERATURE

The transition temperature from gypsum to anhydrite, or vice versa, has been under investigation for a long time; the results are still inconclusive. Some of the published determinations are given in Table 13-2. *One can observe two clusters of determinations, one around 38–42°C and the other around 50–58°C.*

Gibson and Holt (1933) found discontinuities at 73 ± 2°C and 92 ± 2°C, but Ostroff (1964) noted that gypsum does not change in pure calcium sulfate solutions until it reaches 97°C. However, other temperatures, e.g., 99°C, 103°C, and 105°C, have also been cited. This is the most widely quoted transition temperature range between gypsum and bassanite. *In vacuo* the transition temperature is lowered to 59°C (Krauss and Joerns, 1930), in saturated sodium chloride solutions to 76.1°C (d'Ans *et al.*, 1955). However, the temperature is above 110°C in rock gypsum, some 12–13°C higher than in precipitated gypsum (Buessen *et al.*, 1936). Other temperatures have been published by various authors.

In the dehydration of gypsum, an intimate mixture is formed of gypsum and bassanite, and not all of the water is expelled; less than 1% of it is still present even at 115°C (Parsons, 1927). However, if gypsum is slowly heated in currents of moist air, dehydration is practically complete at 97 ± 2°C (Balarew, 1926). The initial and final stages of dehydration show a slight shift due to an induction period and the presence of residual water retained in the anhydrite lattice. Consequently, the rate of dehydration rises sharply with the vaporization of this residual water and at 160°C attains almost twice the value it reaches at 100°C and nearly four times the value at 70°C (Khalil, 1982).

Furthermore, dehydration temperatures differ in dry air, humid air, and saturated steam (Murakami *et al.*, 1957). At a vapor pressure of less than 103 kPa, the transformation of gypsum proceeds directly to anhydrite without the intermediary step of bassanite (Lehmann *et al.*, 1969). Analogous variations exist in the gypsum–anhydrite transition temperatures in the presence of sodium chloride solutions of various concentrations (MacDonald, 1953; Bock, 1961; Zen, 1965;

TABLE 13-2

Some Gypsum–Anhydrite Transition Temperature Determinations

Temperature (°C)	Authors
140	Grengg (1914) (lowest temperature of anhydrite formation); also Tanaka and Sugimoto (1965)
90.5	Ostroff (1964)
90	Bel'yankin and Feodot'ev (1934)
73+	Tone (1934)
66	Budnikoff (1926)
63.5	van t'Hoff *et al.* (1903); d'Ans (1933)
58 ± 2	Hardie (1967)
56 ± 3	Blount and Dickson (1973)
55.5 ± 1.5	Knacke and Gans (1977)
>50	Cruft and Chao (1970)
49.5 ± 2.5	Innorta *et al.* (1980)
46 ± 25.0	Zen (1965)
45	Jauzein (1974) (dehydration starts at 44°C)
44	Shcherbina and Shirokikh (1971) (calcul.)
42.1–42.4 ± 1	Grigor'yev and Shamayev (1976)
42.01	d'Ans *et al.* (1955)
42.0	Hill (1937); Bock (1961); Marshall and Slusher (1966); d'Ans (1968)
42 ± 1	Posnjak (1938)
41 ± 1	Power *et al.* (1966)
40	Kelley *et al.* (1941)
38–39	Partridge and White (1929)
38	Toriumi and Hara (1938); Ramsdell and Partridge (1929)
37.5	Tanaka *et al.* (1931)
20	Nacken and Fill (1931) (*in vacuo* or in dry air)

Power *et al.*, 1966; Marshall and Slusher, 1966, 1968; Hardie, 1967; Braitsch, 1971; Blount and Dickson, 1973). Initially, the transition temperature was seen as a simple function of chlorinity (Fig. 13-1). However, a plot of a multiplicity of experimental data shows a belt of transition temperatures for any specific ionic strength of the brine (Fig. 13-2). Published thermodynamics constants of dissociation for almost all evaporite minerals similarly differ from author to author (Al-Droubi, 1976). However, the observation stands that in sodium chloride brines of rising temperature, the solubility of gypsum increases slightly, whereas that of anhydrite decreases slightly. Nakayama (1971) explained the enhanced

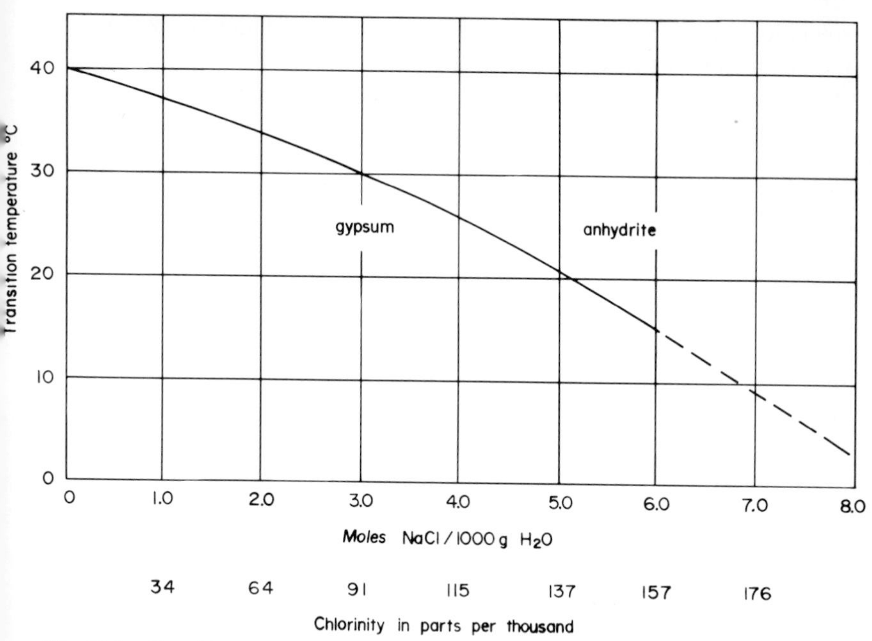

Fig. 13-1. Transition temperature of gypsum–anhydrite as a function of chlorinity and NaCl concentration (after MacDonald, 1953).

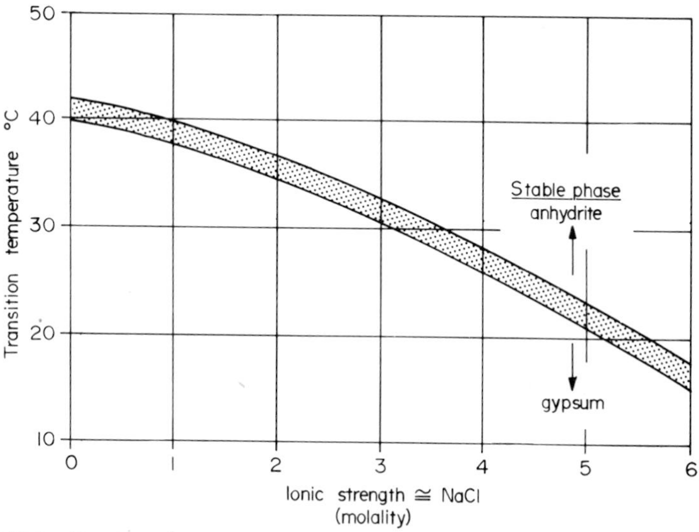

Fig. 13-2. Transition temperature of gypsum–anhydrite in aqueous NaCl solutions (after Marshall and Slusher, 1966) (Reprinted with permission from Chem. Engg. Data, Copyright 1966, Am. Chem. Soc.).

solubility of gypsum in concentrated sodium chloride solutions by assuming that the dissolution of gypsum results in both charged and uncharged calcium and sulfate species and in complexing of the uncharged, soluble ion pairs.

Gypsum solubility data for natural brines with an ionic strength in excess of 7.0 are still largely lacking (Lerman, 1967). The transition temperature in sodium chloride solutions at 18°C (d'Ans *et al.*, 1955; Braitsch, 1964) contrasts to the observation that bassanite can be prepared by dehydration only in an acid or salt solution above 45°C (Floerke, 1952). The transition point in calcium chloride solutions is given as 21% calcium chloride at 25°C, whereas at 50°C only anhydrite is stable at any concentration (Mel'nikova *et al.*, 1971). These experiments were not repeated with gypsum samples from different sources. Gypsum stability is maintained in magnesium sulfate solutions, but it is limited in magnesium, sodium, and calcium chloride and calcium sulfate brines up to 18–21.5 weight % of salt in solution (Mel'nikova and Moshkina, 1973).

The variability in data from one researcher to another cannot be explained by instrument errors or slipshod experimentation, e.g., as van t'Hoff had been accused by one author of purportedly using leaky vessels. There is such a wide range in determinations by reputable investigators that it is very likely that an important factor has been overlooked. Indeed, plasters obtained from selenite and by-products have different properties from plasters derived from rock gypsum. Gypsum deliveries from various sources have different calcination temperatures (Kelley *et al.*, 1941). Different samples of calcined gypsum show considerable individuality in response to retarders and accelerators, some being virtually unaffected and others being markedly affected (Ridge and Surkevicius, 1961). Water vapor initially accelerates and then retards dehydration (McAdie, 1964). There is a difference of up to fivefold between the heat of crystallization of natural gypsum and that of artificially precipitated calcium sulfate dihydrate (Schedling and Wein, 1955). Dehydration can be aided by the presence of glucose or potassium phosphates, which decrease the temperature of bassanite formation (Korol'kov and Krupnikova, 1953). The behavior of the calcium sulfate hydrates depends in part on their origin, previous history, and grain size and on the experimental conditions (Krauss and Joerns, 1930).

Gypsum is not a pure substance. It adsorbs a great variety of organic macromolecules and also takes into the crystal lattice a variety of cations that substitute for calcium and/or reside in interstitial positions among the structural water molecules. The liberation of these disseminated organic macromolecules will, of course, have a decisive effect on the transition temperature. A 1 ppt admixture of sodium oleate in a weakly alkaline solution collects up to 79% anhydrite and gypsum in flotation, and the absorption is irreversible (Glembotskiy and Uvarov, 1962). Intercrystalline layers in gypsum, about 2–10 μm thick, remain dark when the microscope stage is rotated (Zolotov, 1957). Crushing tests have proved that these were layers of minimum strength. Turbidity developed when such crystals were heated to 80–100°C. Shearman and Orti Cabo (1978) noted that various crystal segments of gypsum respond differently to ultraviolet light. Some do not fluoresce; others exhibit a strong yellow fluorescence. A yellow fluorescence is otherwise typical of crude oils, in which it is induced by the presence of minute traces of porphyrins, derivatives of various respiratory pigments.

During the growth process of gypsum crystals, some of the foreign cations may be trapped with part of their hydration spheres. These water molecules then become part of the structural water of the crystal (Kushnir, 1980a). Not only do strontium, magnesium, potassium, and sodium substitute regularly, but also iron, manganese, and some of the base metals (Sonnenfeld *et al.*, 1976). The relative amount of foreign cations depends on crystal growth rate, brine temperature, and brine composition and concentration (Kushnir, 1980b). Impurities change the velocity of dehydration; samples of finer crystallinity calcine more rapidly, and thus the dehydration properties distinguish the qualities of gypsum. Coarse- and fine-grained varieties of gypsum from the same quarry in the Ukraine differ in rate of dehydration, thermal, chemical and x-ray properties, and relative solubility (Zolotukhin, 1954).

Dehydration energies of different gypsum samples increase with increases in the bonding stabilities of the OH groups in the gypsum crystal lattice. Consequently, gypsum with less stable OH groups in the lattice is transformed into anhydrite at a much lower temperature than is gypsum with a more stable structure (Hrishikesh and Adhya, 1969). Since the water molecules are only weakly bonded to calcium sulfate, they leave gradually without forming well-defined phases (Vicq *et al.*, 1977). Gypsum with a more strained crystal lattice is more prone to shed its crystal water than is a purer variety. The dehydration of gypsum is also a function of crystal orientation, since a considerable anisotropy exists. Nucleation of bassanite is particularly intense along the separation lines between the inner and marginal parts of the crystals, leading to selective dehydration. The end products can thus be quite heterogeneous (Deicha and Deicha, 1955).

Overall, the behavior of the calcium sulfate hydrates depends in part on their previous history and grain size, and on the experimental conditions (Krauss and Joerns, 1930). Their transition temperature is not so much a function of a particular origin, but rather of particle size, textural differences, vapor pressure, heating rate, thermal conductivity, lattice imperfections coupled with the presence of impurities, and the effects of other compounds, such as fluorite, apatite, hematite, magnesite, dolomite, magnesia, silica, sodium fluoride, or phosphate ions (Hayashi and Sato, 1951; Murat and Comel, 1970; Murat, 1971; Bachiorrini *et al.*, 1976). The presence of minute amounts of fluorine is particularly important, since this stabilizes any soluble anhydrite.

Conversion of gypsum to anhydrite increases the pore pressure. Depending on the amount of primary void space in the gypsum, the resulting anhydrite and water together occupy about 10.5–11.5% more volume than the original gypsum. If this was a closed system and if the fluid pressure was not allowed to dissipate, the pressure buildup could approach lithostatic pressures high enough to induce flow structures (Blatt *et al.*, 1980) or intraformational breccias. The high permeability of gypsum is somewhat reduced in the postdepositional transformation to anhydrite, but it is not eliminated. Brines are able to enter the anhydrite bed, and they equalize pressures in all directions through the whole system. Excess pressures are, therefore, dissipated, and some of the dissolving sulfate is diffused. Thirty-eight percent of the gypsum volume is lost in the conversion to anhydrite once the water escapes. Observable settling and differential compaction effects in overlying sediments are negligible only if the accumulated overburden is still small at the time of the conversion, and if pore waters have the opportunity to escape. Varve-like marker beds in directly overlying halites are often traceable without disturbance for great distances. This can happen even in varved anhydrites (Richter-Bernburg, 1957a), negating any possibility that such differential settling was delayed. Occasionally observed enterolithic structures are caused by the dehydration of gypsum during an increase in the chloride content of superjacent brines. Overlying sediments then fill the developing fractures, which are widened by the removal of some calcium sulfate in departing waters (Levitskiy, 1961). Gypsum and calcium chloride solubilities decrease with temperature in potassium chloride solutions, particularly in the presence of sodium or magnesium, but at constant temperature the solubility increases with increasing potassium chloride content of the bitterns (Kruell, 1933). Similarly, the solubility of calcium sulfate in a solution of potassium sulfate decreases with dropping temperatures (Teraoka and Ito, 1958).

Anhydrite or halite pseudomorphs after gypsum (Stewart, 1953) are common in marine evaporite deposits. Anhydrite derived from the dehydration of gypsum retains the shape of the gypsum crystal. The dehydration is anisotropic, producing at first fibrous needles of anhydrite within the gypsum. No water is allowed to leave parallel to the long axes (Atoji, 1959). Fluid escape structures produced by fluidization have also been observed in ancient gypsum sequences (Ciarapica and Passeri, 1977). The pseudomorphs do not show any volume reduction. A topotactic arrangement of new anhydrite crystals in respect to the original gypsum crystal lattice can be deduced. The pseudomorphs must have formed, therefore, when the crystals were still in contact with productive sulfate- or halite-precipitating solutions (Langbein, 1979). Continued addition of further gypsum during anhydritization increases the whole-rock sulfate content and produces washed-out lithic boundaries and a cloudy sediment with no pseudomorphs (Langbein, 1978). The occurrence of anhydrite inclusions in euhedral quartz crystals in the Main Anhydrite Member of the German Zechstein evaporites (Kuehn, 1968) might prove that the quartz crystallized after the precipitated gypsum had turned into anhydrite, i.e., very early in the depositional history.

However, the grain size of quartz and other secondary silicates continues to increase during the sulfatization of carnallite (Hoffmann, 1961), and anhydrite inclusions can then be secondary.

Anhydritization must thus be a function of relatively shallow burial and occurs very early after burial through continued contact with hygroscopic sodium, magnesium, or calcium chloride *brines*. Dehydration in calcium chloride brines proceeds through the solution of gypsum. Nucleation and growth of crystals increase with increasing supersaturation, i.e., the rate of dehydration increases with temperature and calcium chloride concentration (Šatava and Prokop, 1963).

REHYDRATION OF ANHYDRITE (GYPSIFICATION)

Whereas anhydritization requires hygroscopic brines, secondary gypsification requires meteoric waters. Hydration necessitates the entry of waters either from below, or percolating from, the surface. Either way, these waters are initially undersaturated for sulfates. Interstitial waters left behind in an anhydrite supply insufficient volumes to rehydrate an anhydrite bed. The textures reveal the type of gypsification (Holliday, 1970; Mossop and Shearman, 1973): A porphyroblastic texture indicates volume-for-volume replacement under near-equilibrium conditions at an early stage; an alabastine texture is formed only under shallow overburden or at a late stage in outcrop. Remnants of single anhydrite crystals are still perfectly aligned and in the same orientation in Middle Carboniferous evaporites of Spitzbergen, although they are now separated by secondary gypsum (Holliday, 1967).

Descending meteoric waters, rather than laterally entering ones, are usually responsible for rehydration of anhydrite and resulting swelling phenomena. Gypsification of anhydrite produces a 61% volume increase, but a pressure of 60–150 kPa overcomes the hydration pressure of anhydrite and prevents expansion (M. N. Sahores, 1955; J. Sahores, 1962). That is the lithostatic pressure exerted by 60–75 m of overburden. Retarding agents, such as tartaric acid or calcium hydroxide, can decrease the expansion stresses by more than 90% (Kuhlmann and Ludwig, 1977). Gypsification is accelerated by the presence of alkali sulfates in the waters (Conley and Bundy, 1958). Similarly, the presence of barium, calcium, strontium, or trivalent metal ions in the water will accelerate the hydration of anhydrite (Budnikoff, 1928a,b). The rate of hydration is also controlled by the concentration of those salts in the solution, which increase the solubility of calcium sulfate while decreasing the potential rate of evaporation.

Many secondary gypsum occurrences show no increase in volume over the anhydrite precursor. This is true of the gypsum in a Louisiana salt dome (Goldman, 1952), southwestern Indiana (McGregor, 1954), and in the Blaine Formation of Oklahoma (Muir, 1934). The excess gypsum was probably removed mole for mole rather than volume for volume (Roth, 1937). Removal of some material from gypsum beds by solution is indicated by the observation of stylolites in gypsum sequences (Stockdale, 1936). If the excess volume had not been removed during rehydration, there should be evidence of deformation within the gypsified rock. Lack of such evidence indicates that the excess material has been removed in solution (Holliday, 1967). Indeed, the chloride content of secondary gypsum is always substantially lower than that of the original anhydrite; circulating brines must thus have removed the chlorine (Parfenov, 1967). Nonetheless, the removal of more soluble salts would only partially compensate for a lack in volume increase upon hydration of anhydrite (Withington, 1961). Finally, secondary gypsum is more porous than anhydrite (Murray, 1964).

Rehydration waters can enter the anhydrite only if some primary porosity and permeability were preserved. At a rate of penetration of 1 mm/year, the waters saturated with respect to gypsum will dissolve anhydrite at the rate of 0.4 mm/year. A 1% admixture of sodium chloride with the infiltrating waters will increase this rate by half. Consequently, the hydration of anhydrite does not generate great stresses; it merely acts to fill available voids (Farran and Orliac, 1953). Only when these voids are filled does further hydration also exert compaction pressures on overlying and underlying incompetent beds. Indeed, a corrugated, folded texture, which is especially noticeable along thin clay intercalations or selenite veinlets, occurs only in very shallow parts of a gypsified anhydrite in the Carpathian Foreland (Ripun, 1961).

Furthermore, the hydration of anhydrite to bassanite generates heat in the amount of 16.7–17.6 kJ/mole (Newman and Wells, 1938; Southard, 1940; Baouman, 1947) or 363–382 J/cm^3. Experimentally, Dubuisson (1950) and Michel (1955) found heats of 25.5–26.36 kJ/mole or 554–574 J/cm^3 for rehydration to gypsum. This heat must be dissipated rather rapidly, or it will raise the temperature of an equivalent volume of interstitial brine from its ambient temperature to well past its boiling point. The hydration of anhydrite also presupposes that overlying beds have become exposed subaerially and have become part of the dry land. Forced mixing of meteoric waters and brine is then greatest near the surface, where the brine is diluted and its solute dispersed (White, 1968). Such brine dispersion is then responsible for preventing swelling phenomena.

KIESERITE ($MgSO_4 \cdot H_2O$)

Kieserite, characterized by extreme kinetic values, cannot be produced directly out of solution and is a typical alteration mineral of salt beds. Magnesium sulfate is one of the early precipitates in halites during a laboratory evaporation of seawater, but an early sulfate deficiency in natural brines of marine origin prevents the direct precipitation of magnesium sulfates. The more hydrated magnesium sulfates, sanderite ($MgSO_4 \cdot 2H_2O$), leonhardtite ($MgSO_4 \cdot 4H_4O$), allenite ($MgSO_4 \cdot 5H_2O$), sakiite ($MgSO_4 \cdot 6H_2O$), and epsomite ($MgSO_4 \cdot 7H_2O$) are known in marine evaporite deposits only as weathering products on mine walls and other exposed surfaces, or by the action of Recent meteoric waters. At times, they have been considered to be precursors of kieserite, but no pseudomorphs have been found to substantiate that contention.

Sulfatic brines rich in potassium and magnesium are again able to form not too long after primary evaporite precipitation, possibly during the transgressive phase of a gradual return to normal brine salinities. These hygroscopic brines then soak into the precipitated sediment. Kieserite and polyhalite in the German Zechstein beds have often displaced anhydrite or gypsum while the sediment was still soft, and adjacent halite was forced to recrystallize (Schulze, 1960a). Very early polyhalitization has also been deduced for Permian evaporites in New Mexico from K/Ar data by Brookins (1982). Nevertheless, a primary coprecipitation of kieserite and polyhalite did not occur (Meier, 1969b). Since in sequential alteration kieserite can replace polyhalite, which in turn had earlier replaced an anhydrite (Armstrong *et al.*, 1951), Podemski (1972), following Meier *et al.* (1971), suggested that kieserite can also be secondary after carnallite because of the high bromide values of underlying halite. Heide (1968) concurred with this conclusion and observed that kieserite formation takes a very short time when carnallite is dissolving in a sulfatic influx, provided only that the number of water molecules is smaller than the concentration determined by the hydration numbers of the ions. Once the amount of water molecules exceeds a certain threshold, no more kieserite can form.

In the absence of magnesium chloride solutions, kieserite is stable in contact with air for years; in the presence of these solutions, however, kieserite decomposes at the surface by hydration (Heide, 1968). This cannot hold in the subsurface, as one finds in German Zechstein evaporites kieserite derived from migrating magnesium chloride brines alongside both secondary carnallite and a third generation of anhydrite (Kosmahl, 1969). In the Ciscarpathian Miocene trough, kieserite formed either simultaneously with langbeinite or by selective leaching of that mineral by magnesium chloride solutions:

$$K_2SO_4 \cdot 2MgSO_4 + MgCl_2(aq) \rightarrow 3\ MgSO_4 \cdot H_2O \downarrow + KCl\ (aq) \qquad (13\text{-}1)$$

Braitsch (1961) published a sequence of facies for the Stassfurt series of the Zechstein evaporites (Table 13-3). He tried to explain the sequence of mineral facies by assuming a periodic influx of seawater over a sill into a hypersaline lake with water temperatures above 35°C. The same sequence can also be explained by descending epigenetic sulfatic brines with a positive redox potential. They continued to replace carnallite with kieserite until the solution became saturated with potassium sulfate. Then they altered anhydrite intercalations to polyhalite

TABLE 13-3

Sequence of Facies in the Stassfurt Series, the Second Cyclothem of German Zechstein Evaporites[a]

Halite with carnallite and kieserite
Halite with kieserite
Halite with polyhalite
Halite with glauberite
Halite with anhydrite
Halite with basal anhydrite

[a]After Braitsch (1961).

and deposited some glauberite ($Na_2SO_4 \cdot CaSO_4$) nests into the upper part of the basal halite member. By then the sulfate content of the brine was spent, and it had become a chloride brine. The lower part of the anhydrite–halite member and the basal anhydrite remained unaltered by the exiting brines.

Evidently, the kieserite in the Ronnenberg seam (the Leine cyclothem of the Zechstein evaporites) is due to waters entering from the shelves, and is not derived from basinwide brine concentrations. The seam is composed mainly of sylvinite, and there is very little carnallite left; a secondary carnallite fills the cracks in anhydrite and halite. Polyhalite and anhydrite are present only in higher parts of the seam, whereas kierserite is restricted to the northern and northwestern peripheries (Sydow, 1959). The sulfatization seems to be most advanced in the Polish parts of the Zechstein Basin and around an island now represented by the Harz Mountains; it is minimal in the British portions of the Zechstein Basin. It is thus evidently related to the proximity to shores and shoals. Magnesium sulfates should be evenly distributed throughout the Zechstein Basin if they represent an early primary precipitate formed prior to potash deposition, as laboratory evaporation experiments with seawater suggest; nor should they be absent from other basins that contain only chloride salts.

POLYHALITE ($CaSO_4 \cdot K_2SO_4 \cdot MgSO_4 \cdot 2H_2O$).

Polyhalite has been precipitated from a brine containing potassium chloride, magnesium, and calcium sulfates (Yarzhemskiy, 1949b). Once formed, polyhalite is thereafter almost immune to any later overprinting (Elert, 1968). Polyhalite becomes more resistant to decomposition with increasing concentrations of potassium or magnesium chloride or potassium sulfate in a brine (Perova, 1961). Although primary polyhalite with inclusions of organic material is known in nature (d'Ans, 1967), most polyhalite is secondary. The first investigator to recognize that potassium and magnesium chloride solutions can be the cause of early diagenetic alteration of anhydrite to polyhalite or kieserite was Goergey (1915). The diagenetic formation of polyhalite from anhydrite exposed to solutions rich in potassium and magnesium is by now well documented (Weber, 1931; Schaller and Henderson, 1932; Holser, 1966b; Busson and Perthuisot, 1977). Secondary polyhalite is not oriented along bedding planes; at times it occurs near a clay intercalation or as a replacement of an anhydrite sheath of a sandstone intercalation (Yarzhemskiy, 1949b).

The formation of double and triple salts is induced in warm brines and inhibited in cool brines. These salts thus cannot form unless the brine is protected from nightly cooling. Such protection from nightly heat losses exists only where brines seep into the precipitates. Double salts, such as polyhalite, can replace anhydrite as the new stable phase in concentrated salt solutions above 42°C (Conley and Bundy, 1958). Polyhalitization commences above 25°C and reaches a maximum at 83°C (Kuehn, 1952a), i.e., well after burial. The main activators of the alteration of anhydrite to polyhalite are highly hydrated cations and anions (Heide, 1968). However, gypsum may remain indefinitely in its saturated solution below the hemihydrate transition temperature. This observation, then, turns into an argument in favor of having *all compound potassium–magnesium sulfate minerals in marine evaporite deposits derived from anhydrite as a precursor or from one of the chlorides, but not directly from an originally precipitated gypsum* without an intermediate transitional anhydrite phase. Dissolution of potash salts produces brines rich in potassium and magnesium that are no longer in equilibrium with anhydrite, and thus fosters the formation of polyhalite. Patches of anhydrite, polyhalite, and clay particles extending across grain boundaries are evidence of a moving brine. In the sebkha El Melah in Tunisia, however, polyhalite forms under a halite cover as the ultimate replacement of earlier gypsum. Brines rich in potassium and magnesium percolate through the unconsolidated gypsum layer during the precipitation of halite (Perthuisot, 1971). Whether they dehydrate it first, before altering it, is not known.

Precipitated calcium sulfates are exposed to more concentrated brines at high water level, when potassium-rich bitterns seep into the sediment. In many instances, the conversion of anhydrite to polyhalite is due to movement of bitterns during the precipitation of sylvinite or carnallitite in other parts of the basin. The waters then migrate mainly sideways (Elert, 1968) along permeable clay–anhydrite boundaries or through porous anhydrites. In the Stassfurt series of the Zechstein evaporites, the polyhalite distribution is antithetic to the quantity of anhydrite present (Braitsch, 1971) and is peripheral to the location of chlorides (Smith and Crosby, 1979). Often only the upper part of an anhydrite bed beneath gray halite is altered to polyhalite. The remaining anhydrite becomes iron-stained and red to yellow, whereas underlying anhydrites are dark gray (Weyrich, 1961). In the Permian Salado Formation of New Mexico, polyhalite forms bands and rims both above and below anhydrite. The anhydrite in the center is either massive or banded with magnesite (Adams, 1969). It is thus due to both ascending and descending brines that carried in magnesium and potassium ions. It would be interesting to find out whether the polyhalite veneer on top is of a different age than the one at the base of the anhydrite, or whether the same brine washed the anhydrite from all sides.

Polyhalitization is preferentially found on shelves and shoals, whereas anhydrite tongues, turbidites, and intercalations predominate in deeper parts of the basin. These anhydrite tongues are marks of temporary freshening of a highly concentrated brine. Any precipitate is stable only as long as percolating brines are saturated or supersaturated with respect to it. Polyhalitization is never found in basal anhydrites of an evaporite sequence. Consequently, *no potassium and magnesium brines ever penetrated down to basal anhydrites; instead, they moved away laterally.*

Elert (1979) distinguished polyhalitization of anhydrites in carnallite beds of the Zechstein evaporites by ascending solutions from anhydrites in the upper parts of the seams produced by descending solutions. The presence of anhydrite merely indicates that the percolating solutions were not yet saturated for potassium sulfates, since polyhalite can also form by the sudden introduction of calcium-rich runoff into a brine saturated for sylvite. Furthermore, polyhalite replaces all primary or secondary potash minerals during subaerial weathering (Strakhov, 1962) and its occurrence on evaporite shelves can thus signal a temporary drying out of the shelf area. Just as the original gypsum contained organic matter, the polyhalite is commonly associated with organic compounds (d'Ans, 1969).

The dissolution of kainite by infiltrating fresh water also fosters the precipitation of polyhalite. Secondary polyhalite can occur together with clay intercalations, anhydrite, kainite, kieserite, or magnesite; as a replacement after kaliborite and ascharite; after kainite at a clay contact; after aphthitalite ($Na_2SO_4 \cdot 3K_2SO_4$); or as a clay–polyhalite envelope of sylvite. However, it is very rarely in direct contact with langbeinite (Yarzhemskiy, 1954). Precipitated kaliborite or sylvite can be resorbed during polyhalite formation (Yarzhemskiy, 1949b). Polyhalite often occurs in traces in saline lakes. It may here represent an authigenic reaction or, on occasion, may be produced by direct precipitation (Irion, 1973). However, in saline lakes of Turkey, it occurs 60 cm below the surface, appears to be antithetic to gypsum, and is always found in horizons beneath the gypsum-bearing sediments (Irion and Mueller, 1968). This suggests a diagenetic origin. In Britain, the carnallite veins even cut through underlying anhydrite (Raymond, 1953). Solution cavities in potash horizons are filled with translucent white halite, which overlies anhydrite encrusted with hematite. More anhydrite is present than would accumulate as a residue from salt dissolution. The percolating brine, therefore, must have been a calcium sulfate brine that dropped some of its solute when picking up sylvite and carnallite (Woods, 1979).

Coulter and Reid (1980) advanced the argument that polyhalite beds in the second cyclothem of Zechstein evaporites in Britain are primary, because they can be correlated over some significant distance. However, evidence other than mere correlatability is needed as proof of primary origin. If polyhalitization affected an anhydrite bed of uniform thickness, the resulting polyhalite bed would be recognizable for significant distances without being of primary origin.

SYNGENITE ($CaSO_4 \cdot K_2SO_4 \cdot H_2O$)

Syngenite is a low-temperature alteration product of anhydrite or polyhalite in solutions low in magnesium chloride (Braitsch, 1971). It can form from polyhalite in a warm aqueous medium (Perthuisot, 1975) as

$$2CaSO_4 \cdot K_2SO_4 \cdot MgSO_4 \cdot 2H_2O + H_2O \rightarrow$$
$$CaSO_4 \cdot K_2SO_4 \cdot H_2O \downarrow + CaSO_4 \downarrow + Mg^{2+} + SO_4^{2-} + 2H_2O \quad (13\text{-}2)$$

It is also produced by a reaction of gypsum with potassium chloride (Kruell, 1933) whereby

$$2CaSO_4 \cdot 2H_2O + 2KCl = CaSO_4 \cdot K_2SO_4 \cdot H_2O \downarrow + CaCl_2 + 3H_2O \quad (13\text{-}3)$$

At a constant temperature, the amount of calcium chloride produced increases with the increasing potassium chloride content of the brine. However, as the temperature increases, the amount of calcium chloride decreases with increasing potassium chloride content, particularly if sodium or magnesium ions are present (Kruell, 1933). Temperature variations thus are more important than changes in potassium chloride concentration. The amount of alteration decreases with increasing temperature and is further decreased with rising quantities of sodium or magnesium chloride in the brine (Kruell, 1934). Nitrogen hydrides, specifically ammonia or hydrazine, are powerful catalysts in the conversion of gypsum or anhydrite in the presence of potassium chloride solutions to potassium sulfate minerals. Nitrogen hydrides can also yield a variety of compound sulfates, such as syngenite and aphthitalite (Fernandez Logano and Wint, 1982).

Syngenite often occurs in minor amounts in sulfatized potash beds. Small crystal druses of syngenite form in bloedite ($Na_2SO_4 \cdot MgSO_4 \cdot 5H_2O$) as a brine that is saturated for magnesium chloride and contains potassium chloride; it reacts with precipitated gypsum in a small lake near the Aral Sea. This lake precipitates halite, carnallite, and bischofite, as well as some bloedite, mirabilite ($Na_2SO_4 \cdot 10H_2O$), glauberite, epsomite, kieserite, and gypsum (Lepeshkov and Fradkina, 1958). Another lacustrine syngenite deposit has been reported from

Tibet (Liu and Jin, 1981). Nodular syngenite aggregates are found in the Polish Zechstein series in silt that is filling depressions in salt deposits (Fijal and Slanczyk, 1970). Evidently, the syngenite here is the residue of intense leaching.

At room temperature, syngenite is not stable, but decomposes readily in a watery environment:

$$K_2SO_4 \cdot CaSO_4 \cdot H_2O + H_2O \rightarrow CaSO_4 \cdot 2H_2O \downarrow + K_2SO_4 + H_2O \quad (13\text{-}4)$$

Goergeyite ($5CaSO_4 \cdot K_2SO_4 \cdot H_2O$) is a rare variant that is produced only in migrating brines with a very low magnesium chloride content, yet with some potassium sulfate present in solution.

PICROMERITE ($K_2SO_4 \cdot MgSO_4 \cdot 6H_2O$) AND LANGBEINITE ($K_2SO_4 \cdot 2MgSO_4$)

Picromerite increases its stability field as temperatures decrease toward the freezing point (Yanat'eva and Orlova, 1958). In contrast, langbeinite forms only at temperatures above 61°C (Kuehn, 1952a). To produce picromerite, langbeinite, or even kainite, the brine must be considerably altered: The solution of magnesium sulfate must be as concentrated as possible and relatively free of sodium chloride. Bloedite or aphthitalite would form otherwise (Koelichen and Przibylla, 1920). Solutions poor in sodium chloride can only travel laterally through anhydrite or clay beds, because solutions that have passed through halite beds are saturated with sodium chloride.

Ascending brines can produce langbeinitization followed by kieseritization during a stagnation of moving brines (Siemens, 1961). Descending sulfatic brines generate langbeinite concurrent with polyhalite formation (Kokorsch, 1960). Langbeinite and carnallite are thus mutually exclusive (Richter, 1964), and the former is never found below carnallite in the Leine (third) cyclothem of the Zechstein evaporites (Schulze and Seyfert, 1959). Picromerite, kainite, and epsomite then form after langbeinite and loewite ($6Na_2SO_4 \cdot 15H_2O$) as fracture fillings in langbeinite (Bobrov, 1973). In Ischl, Austria, langbeinite occurs with rims of kainite and bloedite (Mayrhofer, 1955), a case of alteration by descending, probably originally meteoric brines.

In order to calculate the volume required to generate langbeinite by sulfatization, a hypothetical incursion of seawater (with a salinity of 36.57 ppt) into a potash deposit may be visualized. The percolation of at least 303 m^3 of seawater would be required to convert 1 m^3 of carnallite to 0.424 m^3 of langbeinite, a 57.6% volume decrease. In contrast, only 213 m^3 of seawater can convert about 3 m^3 of carnallite to 1 m^3 of kieserite at a 67% reduction in volume. Much more brine is required for sulfatization of a unit volume before the brine saturates for potassium. Conversion of 1 m^3 of sylvite requires 717 m^3 of seawater and results

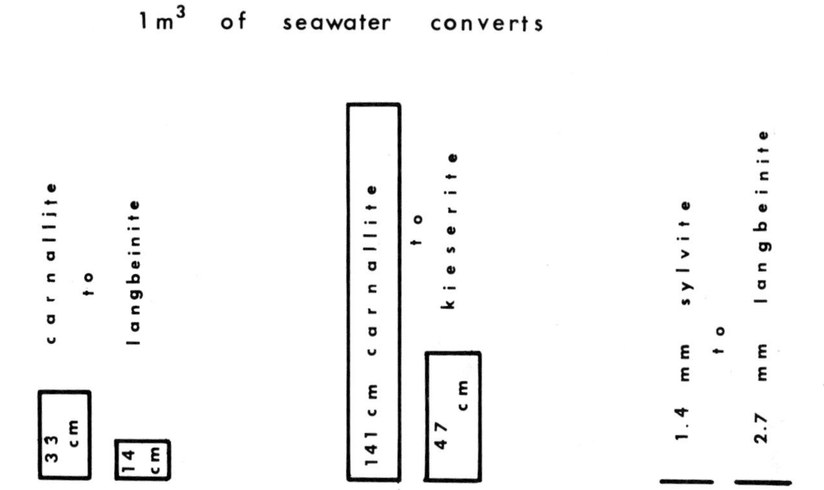

Fig. 13-3. Sulfatization of potash minerals by seawater and resulting thickness changes.

in 1.96 m^3 of langbeinite with a 96% volume increase. In both instances, the halite fraction in carnallitites and sylvinites is neglected (Fig. 13-3). Furthermore, one is here also neglecting for a moment the amount of dissolution produced by the entry of normal seawater. If the sulfate ions are delivered by meteoric waters, seven to eight orders of magnitude more water would be required. The presence of such quantities of meteoric waters would be reflected in salt solution phenomena. Thus, sulfatization can be produced only by the incursion of seawater, with concurrent moderate salt redissolution, or by a preconcentrated but well-oxygenated brine. *Sulfatization of carnallites generally leads to a volume reduction, whereas sulfatization of sylvites generates a volume increase.*

One can thus observe that sulfatization of sylvinites can, consequently, cause swellings and various intraformational deformations. This usually occurs in the upper part of the sequence (Korenevskiy and Donchenko, 1963). However, in New Mexico, the langbeinite deposit has the shape of a planoconvex lens, occurs on either side of a polyhalite-bearing orange halite marker, and is capped by sylvite (Dunlap, 1951). Lateral changes to sylvite are abrupt, just as elsewhere sylvite–carnallite contacts are often quite sharp. The pods of sulfate minerals (langbeinite, leonite, and kainite) near the rims convert anhydrite sheaths of clay interbeds to polyhalite, and discolor the gray clay with some of the hematite from dissolved potash minerals.

Langbeinite is altered by formation waters rich in calcium chloride as follows: If the calcium chloride/langbeinite ratio is smaller than unity, kieserite, sylvite, and polyhalite form according to the following reaction:

$$3(K_2SO_4{\cdot}2MgSO_4) + 2CaCl_2 + 7H_2O \rightarrow$$

$$5MgSO_4{\cdot}H_2O + 4KCl + 2CaSO_4{\cdot}K_2SO_4{\cdot}MgSO_4{\cdot}2H_2O \quad (13\text{-}5)$$

As the ratio rises, progressively more chlorides are produced until, at a ratio in excess of 3 and a corresponding reduction of the water/langbeinite ratio, only anhydrite and secondary carnallite are formed (Khodkov, 1971a,b) according to the following equation:

$$K_2SO_4{\cdot}2MgSO_4 + 3CaCl_2 + 12H_2O = 3CaSO_4\downarrow + KCl{\cdot}MgCl_2{\cdot}6H_2O\downarrow \quad (13\text{-}6)$$

Brecciated carnallite or carnallite rehealed by a soaking in passing brines is common in German Zechstein deposits as an alteration product of initially sulfatized potash horizons (Loeffler, 1960a). It also occurs as a fracture filling of such unaltered sulfatized horizons, proving its very late formation.

Leonite ($K_2SO_4{\cdot}MgSO_4{\cdot}4H_2O$) forms from langbeinite under concurrent crystallization of kieserite ($MgSO_4{\cdot}H_2O$):

$$K_2SO_4{\cdot}2MgSO_4 + 5H_2O \rightarrow K_2SO_4{\cdot}MgSO_4{\cdot}4H_2O + MgSO_4{\cdot}H_2O \quad (13\text{-}7)$$

KAINITE ($4KCl{\cdot}4MgSO_4{\cdot}11H_2O$)

Kainite can form from sylvite under the influence of magnesium sulfate solutions (Kuehn, 1952a). It is formed only if the ratio between magnesium sulfate and potassium chloride in the brine exceeds 1:9. Considerable leaching of potash salts thus precedes kainite precipitation. Only secondary carnallite is produced below that ratio. At a ratio of >1:9 but <1:3, kainite occurs mixed with carnallite; at a ratio <1:3, kainite is mixed with epsomite (Choudhari, 1968). In supergene zones of the Ciscarpathian Miocene trough, kainite and epsomite have formed after langbeinite by an early delivery of descendent waters and a redistribution of waters of hydration (Khodkova, 1968). Langbeinite itself is thus an early diagenetic alteration. Moetzing (1968) found that in Zechstein evaporites as well, kainite or kieserite replaced langbeinite. Here a carnallite with a secondary polyhalite coating was altered to red sylvite and eventually to halite with coarse-grained anhydrite. Surface waters commonly alter langbeinite either to kainite or to leonite (Lobanova, 1974), but only after deeper burial of the deposit; some of it later decomposes to kieserite (Khodkova, 1972). The opposite opinion is represented by Gemp and Korin (1980), who thought that langbeinite and kieserite are derived from an earlier generation of kainite by further input of sulfatic brines.

Precipitation of epsomite in sylvinite (Valyashko and Solov'eva, 1949) gives kainite according to the following reaction:

$$MgSO_4{\cdot}7H_2O + KCl = KCl{\cdot}MgSO_4{\cdot}3H_2O\downarrow + 4H_2O \quad (13\text{-}8)$$

At temperatures above 55°C, kainite can be formed from carnallite and sylvite under the influence of kieserite and water (Shlezinger *et al.*, 1940b):

$$KCl \cdot MgCl_2 \cdot 6H_2O + KCl + 2MgSO_4 \cdot H_2O + H_2O =$$
$$K_2SO_4 \cdot MgSO_4 \cdot MgCl_2 \cdot 6H_2O + MgCl_2 + 2H_2O \quad (13\text{-}9)$$

Kainite forms from carnallitite in the presence of sodium chloride and magnesium sulfate in solution only at temperatures below 72°C; sylvite and kieserite precipitate above that temperature (Herrmann *et al.*, 1980). Carnallite, experimentally equilibrated in a 1:4 ratio with bitterns of density 1.318 at 13°C yielded kainite within 16 days (Patel and Seshadri, 1975). Kainitization of carnallite in the roof of potash deposits and recarnallitization of partially leached zones are the product of later brine movements (Richter, 1962, 1964). Sulfatization of sylvite or halite is thus most effectively aided by the presence of hematite or other iron products (Vol'fkovich and Margolis, 1943).

SECONDARY ANHYDRITE

Intrusion of calcium chloride brines can cause the total removal of previously precipitated potassium and magnesium sulfates (Oka and Kaneko, 1942), replacing them by secondary sylvite and anhydrite or gypsum. In the case of kainite, the reaction proceeds (Bogasheva *et al.*, 1974) according to

$$2CaCl_2 + K_2SO_4 \cdot MgSO_4 \cdot MgCl_2 \cdot 6H_2O =$$
$$CaSO_4 \downarrow + 2KCl \downarrow + 2MgCl_2 + 6H_2O \quad (13\text{-}10)$$

or in the case of polyhalite in Triassic evaporites of the Alsace (Bonte and Celet, 1954) according to

$$2CaSO_4 \cdot K_2SO_4 \cdot MgSO_4 \cdot 2H_2O + 2CaCl_2 + 6H_2O =$$
$$4CaSO_4 \cdot 2H_2O \downarrow + 2KCl + MgCl_2 \quad (13\text{-}11)$$

or it can convert kieserite to anhydrite (Herrmann, 1961b):

$$MgSO_4 \cdot H_2O + CaCl_2\ (aq) = CaSO_4 \downarrow + MgSO_4\ (aq) \quad (13\text{-}12)$$

Dissolved magnesium sulfates do not precipitate in the presence of potassium or calcium chloride. Only gypsum is precipitated at ambient temperatures, but no other sulfates (Delecourt, 1946). The following reaction takes place:

$$MgSO_4 + CaCl_2 + 2H_2O = CaSO_4 \cdot 2H_2O \downarrow + MgCl_2 \quad (13\text{-}13)$$

$$K_2SO_4 + CaCl_2 + 2H_2O = CaSO_4 \cdot 2H_2O \downarrow + 2KCl \quad (13\text{-}14)$$

The dissolution of magnesium chloride on occasion can then lead to a secondary anhydrite generation when the magnesium chloride brine reacts with polyhalite (Zwanzig, 1928).

Replacement of halite by anhydrite or gypsum in several stages is common. Similarly, halite replaces gypsum, anhydrite, or even dolomite (Raymond, 1953; Dellwig, 1955). The Pegmatitic Anhydrite Member of the German Zechstein, an intergrowth of rock salt and anhydrite, contains anhydrite pseudomorphs after calcite, the interstices filled with halite, and halite pseudomorphs after gypsum (Zimmermann, 1907). Smith (1981) considered all of the halite in the Middle Marl of the British Zechstein evaporites as either displacive or replacive and thus secondary. In the same way, much of the anhydrite in the third cyclothem of the British Zechstein series could possibly be due to wholesale replacement of dolomite.

Secondary anhydritization of carbonates has been observed in isolated instances, but it is probably more common than that. The frequent vug filling of anhydrite in reefal dolomites below an oil–water contact is an indication of the late circulation of modestly sulfatic waters. Close to their source, they could have effected a much more intense anhydritization under concurrent liberation of carbon dioxide. *Such replacements are primarily observed along the margins of the bay, where oscillations in water level, salinity, or temperature would be most noticeable.*

SODIUM SULFATES IN MARINE EVAPORITES

The complete absence of sodium sulfates in marine evaporites could be due to the pumping action of solutions enriched in magnesium chloride, which would convert any precipitated amount to a magnesium sulfate as soon as it formed. Sodium sulfates are common, however, in paralic lakes (Valyashko, 1958). They survive here because a lake's water level is determined by the water supply and the presence of an impervious substrate. In a marine bay, the substrate need not be impervious, and very often it is not; the bay level is solely determined by the sea level.

Descending brines derived from meteoric waters are responsible for the nests of simple and compound sodium carbonates and sulfates occasionally found in the upper parts of an evaporite deposit of marine origin. Such brines are severely deficient in bromine and potassium, have a very low Mg/Ca ratio, are mainly charged with sodium ions, and are either carbonated, sulfatic, or both. They must contain no calcium to forestall gypsum precipitation. In sodium sulfate solutions of all concentrations, anhydrite is more soluble than gypsum. Above an 11% sodium sulfate concentration, the solubility exceeds that in water. Either gypsum or double salts then start precipitating (Ottemann, 1950). Glauberite occurring in polyhalite within the Permian Wellington salt of Kansas (Swineford and Runnels, 1953) is evidently derived from an epigenetic entry of sulfatic

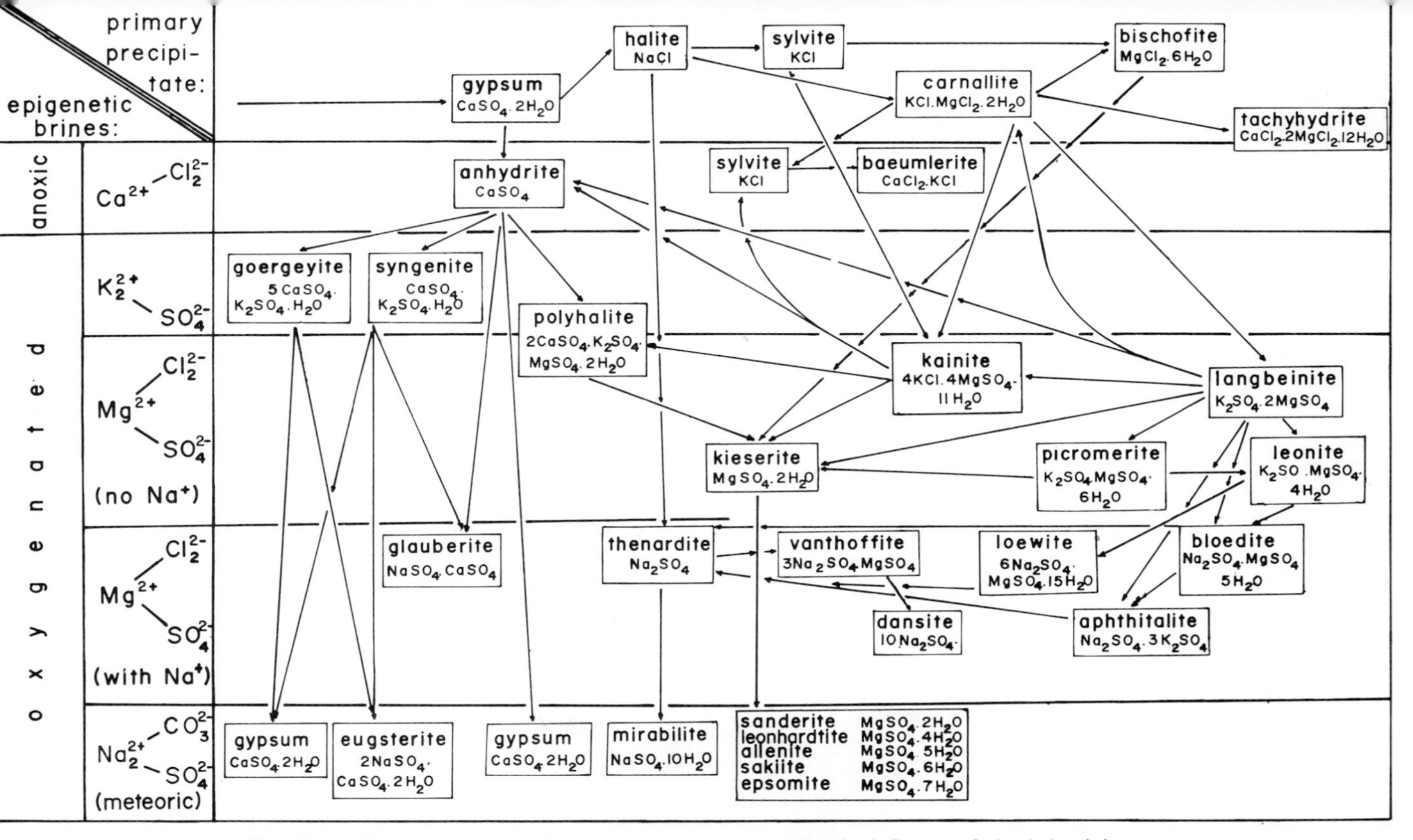

Fig. 13-4. Flow chart of alterations in marine evaporites under the influence of circulating brines.

brines. The nests of secondary sodium–magnesium sulfates in the Permian Salado Formation of New Mexico (Adams, 1970) or in German Zechstein evaporites (Braitsch, 1971) are likewise alterations due to sulfate ion influx.

In the Werra cyclothem of German Zechstein evaporites, Borisenkov (1968a,b) recognized two types of salt horses: One is a hard salt intercalation that gradually incorporated some kainite and langbeinite before turning laterally into kieserite and eventually halite. The other is composed of mounds of crumbly rock salt covered by kieserite, and this in turn by kainite. An irregular bed composed of a mixture of blöedite with aphthitalite caps the sequence. It is evidently derived from descending meteoric waters that supplied the sodium but were unable to supply chlorine. Here the sodium–potassium sulfates and potassium–magnesium sulfates overlie residual chlorides; the crumbly texture of the rock salt suggests removal of carnallite or sylvite intergrowths. Intergrowths with polyhalite, recrystallization, and partial redissolution of grains on their periphery, indicate a penetration by surface brines.

If large amounts of polyhalite, leonite, or aphthitalite show up in borehole samples, this may represent evidence of an exotic mineralogy, but it could also merely mean that certain cations were leached by the moving drilling mud, or that a sample of abnormal chemistry was generated by incongruent solubility of polyhalite (Powers *et al.*, 1978).

A flow chart (Fig. 13-4) can be constructed to show what happens to primary marine precipitates when they remain in equilibrium with circulating oxygenated sulfatic or anaerobic chloridic brines containing various cationic combinations. A variety of additional reverse reactions is also possible, but would clutter such a chart needlessly. *The greatest variety of secondary minerals is produced by circulating magnesium sulfate brines. Sodium sulfate brines are mostly meteoric infiltrations that produce only rare nests of minerals. Calcium chloride brines are derived from formation waters. They destabilize secondary minerals, and revert them to a second generation of chlorides and secondary anhydrite.*

SUMMARY OF SULFATIZATION

Since anhydritization requires a lengthy exposure to hygroscopic brines, most anhydrites are secondary. The transition temperature is dependent on the type and quantity of admixed organic compounds and thus varies from specimen to specimen. Rehydration requires meteoric waters of limited salinity. Both dehydration and rehydration usually occur under a very small overburden, judging by the lack of deformation due to volume changes. Anhydrite undergoes further changes in the path of hygroscopic circulating brines, producing polyhalite with brines containing both potassium and magnesium. Complex multiple changes are possible at slightly elevated temperatures, resulting in syngenite, langbeinite, or kieserite in intensely hygroscopic sulfatic brines, or of kainite or picromerite in less hygroscopic sulfatic brines. Secondary sodium sulfates occur in marine evaporites only where substantial quantities of meteoric waters have entered the deposit.

Part V

MORPHOLOGY OF THE EVAPORITE DEPOSITS

14

Major Synsedimentary (Paragenetic) Structural Features

Most textural and structural features of evaporite mineral assemblages are beyond the scope of this book. However, some structures are important in interpreting genetic relationships. Below the wave base or the interface with surface brines, the precipitate becomes laminated, indicating a cyclicity of precipitate accumulation. Thus, it has been used not only as a correlation tool over large distances, but also as a measure of elapsed time. Since not every precipitated lamina survives, such estimates of elapsed time have to be viewed with caution. Hopper crystals are formed at the air/water interface and signal the loss of density stratification, the inability of the inflow to cover the whole brine surface. Ripple marks likewise indicate a temporary loss of stratification. They occur in rock salt when the extremely rapid halite precipitation has temporarily ceased and waves can move eroded salt crystals. Nodular anhydrite can occur both as a diagenetic alteration in a submarine environment and as a subsoil concretionary formation. Consequently, it is useful as an interpretive tool only if its environment of formation can be ascertained in a detailed study. Tepee structures, in contrast, are invariably produced by changing hydrostatic heads of a salt slush overlain by a hardening crust, and thus are restricted to basin margins. Because a hypersaline brine has a greater density than seawater, the buoyancy of dropped-in debris is greater. Clays spread along density interfaces, but loosened gypsum or halite fragments will form turbidites. Gravity movements of loose precipitates can occur in much shallower water depths.

VARVITES, LAMINITES, OR RHYTHMITES

Ever since de Geer (1912) studied laminations in periglacial lakes, where seasonality is very pronounced, varve couplets have carried, for some investigators, the automatic connotation of annual winter and summer couplets. However, varved sediments also occur outside the winter–summer regime of temperate and polar climates. Consequently, Shearman (1970) suggested that laminated evaporites and carbonates should be referred to as "laminites," whereas Dean *et al.* (1975) suggested "rhythmites." However, the terms "varves" and "varved sediments" are used much too widely in the literature to be readily abandoned.

Polygons and tessellated soils indicate vertical movement of brine. Varves are either destroyed where such vertical water movement occurs or do not form in the first place. Varves are thus absent from the shallow shelf areas of rapidly oscillating hydrostatic heads, and are restricted to perennially water-covered regions with minimal changes in brine pressure. Nonetheless, physical conditions must fluctuate to produce interlaminations of monomineralic beds. These beds will change laterally into each other, as a laminated sediment is a natural product of rhythmic sediment accumulation by settling in any stratified water body subject to cyclic changes (Zen, 1960; Davies and Ludlam, 1973). True rhythmic precipitation due to chemical factors alone can be observed in the laboratory when a calcium chloride solution diffuses under a cover glass into an NaOH solution. Periodic variations in light intensity can cause rhythmic precipitation of calcium hydroxide (Fischer and Schmidt, 1926). A periodic precipitation can also be induced by the addition of concentrated hydrochloric acid, drop by drop, to a saturated sodium chloride solution. It leads to the formation of a precipitate of three sodium chloride layers (Kato, 1944). Ice cubes, partially melting in warm water, develop a laminated texture before dissolving completely. Large, compact blocks of ice likewise show a layered construction due to rhythmic accretion (Breitner, 1948).

Inorganic salts display a rhythmic crystallization in which the distance between bands is inversely proportional to the rate of formation and to the molecular weight (Schaaffs, 1954). Even the presence of starch or other colloids can produce an "annual" layering in halites; where the concentration of colloids becomes too great, the habit changes to herringbone crystals (Solé, 1954). For gypsum–calcite couplets to form, there must be a slight prevalence of sulfate over chloride ions. Anhydrite and calcite can coexist only if the SO_4/CO_3 ratio is smaller than 8650. If it exceeds this value, only gypsum forms (Babcan, 1980). A rapid oscillation from gypsum supersaturation to gypsum undersaturation is possible only in a small body of water. *Varves thus attest to very shallow water conditions* (Busson, 1978). In a lacustrine analogy, von der Borch *et al.* (1977) found bedded gypsum grading towards the basin margin into seasonally alternating gypsum–aragonite laminae under brines less than 2 m deep.

Varved sediments can be due to changes in salinity, variations in temperature, rates of evaporation, sea level or wind direction and strength (Scruton, 1953), or to long periods of repeated reworking of brine-pan halite and groundwater-brine diagenesis (Shearman, 1970). They can also be produced by the elastic streaming potential of outward-flowing interstitial brine during consolidation (Casagrande, 1947). Experimentally, Arthurton (1973) produced laminations by episodic agitation of brines, contrary to the generally held view (e.g., Dubinina, 1951a) that laminations represent a quiet water deposition. Fine laminations are possible only because of the absence of bioturbation. Well before the onset of gypsum precipitation, the waters have become toxic to most benthonic biota.

Varve thicknesses range from 0.01 to 10.0 cm and more, and grade into cycles of higher order. In the Permian Castile Formation of West Texas, the anhydrite laminae range from 0.02 to 0.7 cm in thickness and average 0.16 cm (Udden, 1924; King, 1947); in the Permian salt of Kansas, laminae range from 0.6 to 7.6 cm (Dellwig, 1963). Intercalations of anhydrite in the Silurian Salina Salt of Michigan are 1.5–8.0 cm apart, with the laminae less than 0.05 cm thick (Dellwig, 1955). The thicker intervals simply indicate that the brine remained saturated for the same solid phase over a longer period (Braitsch, 1962). The anhydrite couplets in the Stassfurt anticline average 10 cm in thickness, whereas the polyhalite couplets average only 7 cm (Lueck, 1913). This could be indirect proof that the polyhalite is a postdepositional alteration product after gypsum, since couplets in potash horizons should be thicker, not thinner, than couplets in halites. The presence of a transition layer with about 1% glauberite (Lueck, 1913) supports this contention, since primary glauberite normally does not form in purely marine evaporite sequences. Varves occur even in lacustrine evaporites, where they are to be expected if they are caused by periodic agitation of the brine. Halite–nahcolite couplets range from paper thin to 15 cm in thickness in the Green River Formation of Colorado; the nahcolite ($NaHCO_3$) laminae are only two-thirds as thick as their halite partners (Dyni *et al.*, 1970). They show a rhythmic stratification caused by brownish organic matter.

Laminated evaporite deposits show a regular change in brine concentration over wide areas. Braitsch (1971) observed that each anhydrite–halite lamination couplet runs through the complete concentration range, at least from the beginning of gypsum saturation to halite precipitation. Each couplet thus forms a small brine concentration cycle. The same can be said about dolomite–anhydrite laminae that represent temporary cessations of gypsum saturation and breaks in the continuity of deposition. The sequence of a clay film followed by a halite–sylvite couplet of varves in the Oligocene evaporites of the Upper Rhine Valley (Sturmfels, 1943) suggests that after a clay influx by rainwash, the clay spread along the brine interface. Its settling was concurrent with a downward advance of the mixing zone, which temporarily produced undersaturation to halite and sylvite. Kuehn (1954) observed that each anhydrite layer in an anhydrite–halite couplet shows corrosion features of the anhydrite. The overlying halite starts with a lower bromide content than the top portions of the underlying halite lamina.

Where thicknesses of couplets change, a progressive shallowing or deepening of the water column is indicated. Western Canadian varved dolomites are covered by varved anhydrites, and these in turn by varved halites. Each varve represents a petroliferous membrane rich in the remains of planktonic organisms (Klingspor, 1969). The laminates are restricted to the basinal areas between reef mounds (Dumestre, 1969). The plankton confirms the at-least seasonal existence of an oxygenated, low-salinity surface layer. These varves in the halite are at first 3–4 cm apart, and change upward to 10–20 cm. They consist of clay mineral, sulfate, and dolomite intercalations within the salt. The varves thus indicate a deepening of the brine column, as ephemeral freshening gradually occurs less often. Eventually, the varves become more frequent again, increasing to 3–4 cm apart, followed by varved anhydrite as caprock. Similarly, there is a repeated gradational upward decrease in calcium sulfate in varvites of the first cycle of Zechstein evaporites in the British Sector of the southern North Sea Basin (Taylor and Coulter, 1975). Again, this suggests progressively longer episodes of anaerobic bacterial decomposition of sulfates, or progressively longer persistence of a density stratification.

Within the basin, differential precipitation quickly tends to level out any irregularities in the basin floor, and a very orderly layering is produced. Varves can be correlated over distances of 25 km (Davies and Ludlam, 1973) or 160 km (Wardlaw and Reinson, 1971) in western Canada, 113 km in West Texas (Anderson *et al.*, 1972), and 300 km in Germany (Richter-Bernburg, 1957a,b, 1960). Even rhythmic alternations between halite and potash beds can be followed for scores of kilometers in Sicily (Decima and Wezel, 1971) and for 30–60 km in the Permian potash deposits west of the Urals (Ivanov, 1963; Vakhremeova, 1954). Similarly, rhythmic alternations of halite and sylvite beds in the Upper Rhine Valley can also be correlated over many kilometers (Sturmfels, 1943), with a planar basal contact of the halite and a comblike interfingering contact of halite with sylvite. The sylvite evidently precipitated here into a halite slush. Each halite–sylvite cycle starts with a thin clay film, which often contains minor dolomite, anhydrite, or halite.

Vertical repetitiveness of varve couplets indicates periodic oscillations in the environmental conditions, whereas horizontal correlatability of varved sediments over such great distances favors laterally homogeneous physical and chemical conditions prevailing in the brine environment. Basinwide environmental conditions are more uniform during gypsum saturation of the brine and concurrent precipitation than during any episodic freshening. The correlatability is, therefore, greater in anhydrite laminae than in associated calcite laminae, organic-rich layers, or pyrite crystals (Anderson and Kirkland, 1966). In contrast, halite stratification is often forcefully disturbed at the base of clay intercalations (Langler, 1959).

Lang (1950) assumed that organic layers in the Permian Castile and Salado Formations of West Texas represent winter mortality. He thought that anhydrite and halite precipitated during the following summer, and magnesite, sylvite, and gypsum in the winter. The lower Castile Formation consists of bituminous limestones alternating with anhydrite in fine varves, but the overlying Salado Formation displays cycles 50–900 cm thick, comprising a detrital layer followed by a thin anhydrite–polyhalite layer and a thick halite layer, capped by a mixed layer of halite and detrital material (Powers *et al.*, 1978). Both Borchert (1969) and Udden (1924) assumed that darker films in conjunction with a gypsum precipitate represent an annual freshening. The phytoplankton associated with the calcite in anhydrite–calcite couplets represents basinwide phytoplankton productivity or bloom (Dean and Anderson, 1978) and is thus indicative of a more or less normally saline surface layer of waters eventually increasing in salinity by diffusion, and anoxic bottom waters (Busson, 1978). A drop in the air temperature in the fall is by no means necessarily lethal to marine organisms. Much more damaging is an increase in salinity (Braitsch, 1962). Mass mortality ensues whenever surface waters have a chance to evaporate totally; in turn, gypsum is replaced by limestone whenever the water column is freshened.

Varved anhydrites in the German Zechstein evaporites are basinal, covering thin bituminous shales and underlying thick halite sequences, whereas basin margins are occupied by thick carbonate and anhydrite sequences (Richter-Bernburg, 1981). Taylor (1980) found laminations of displacive anhydrite and dark-colored, fetid, fibrous carbonate in the Werra cyclothem in the British portion of the southern North Sea. He suggested that these Zechstein anhydrite–calcite couplets were laid down "under water depths of up to at least 150 m and probably 300 m or more."

This depth figure can be evaluated. The presence of the couplets suggests that water salinities oscillated around the calcium sulfate supersaturation level. If we take them as annual couplets and assume a density range between 1.11 and 1.12, the increase in salinity requires the removal of 1.2 g/liter of water by evaporation in the dry season, or about 360 cm over a 300-m water column. Since evaporation also affects the freshening surface inflow in the wet season, the water loss applies only to the exposed brine in the dry season. Depending on the length of the wet season, removing 360 cm of water in the season of brine exposure translates into an improbable 6–7 m as an annual rate of evaporation. If, on the other hand, the calcite laminae represent decomposed gypsum in an anoxic, density-stratified environment, and the gypsum represents precipitation in an oxygenated brine, the mixing front between oxygenated inflow and anoxic brine must reach bottom semiannually. This translates into a rate of descent of 6.8

cm/hr for the interface to reach bottom, a value several orders of magnitude greater than the rates of interface propagation observed in recent hypersaline ponds. It is, therefore, more likely that the couplets were deposited in shallow waters below the wave base, i.e., in the range of no more than 50–100 m of water depth.

Tens of sinuous clay–carbonate laminae per millimeter of halite occur in the Devonian evaporites of the Byelorussian Pripyat depression on the southwestern margin of the Second Potash Horizon. Ripple marks suggest wave play in a coastal setting (Lupinovich and Kislik, 1965). The nature of the rhythmicity here separates a proximal from a distal facies, drier from wetter years, and irregularities in the hydrochemical regime (Shcherbina, 1962). In sandwiches of halite alternating with layers of either anhydrite, limestone, dolomite, or clastics, the halite rhythmically pinches out as a matter of paleotopography (Vysotskiy *et al.*, 1980). Halite precipitated in very shallow parts may have been washed into the depression prior to the next major freshening. Similarly, halite–clay varves in northeastern Spain are the result of repeated washings and precipitation of halite in a very shallow basin. The halite is coarsely crystalline, is low in bromine, potassium, and magnesium, and contains occasional anhydrite nodules (Orti Cabo and Pueyo Mur, 1977).

The early diagenetic anhydritization and compaction reduce porous gypsum beds to about 30% of their original thickness. Since dolomite and clay intercalations are not affected by such conversion and dehydration, there is a shift in thickness proportions in favor of intercalations, which sharpens the laminations (Langbein, 1978). Calcite and anhydrite laminations a few millimeters thick alternate for over 60 m in the lower part of the Permian Castile Formation in Texas (Udden, 1924; Adams, 1944). The anhydrite laminae thereby increase upward in thickness, representing initially only 20% and finally about 95% of the rock, whereas intercalated clacite thicknesses remain uniform. At the same time, there is an increase in $\delta^{18}O$ (Dean and Anderson, 1978). The precipitation of progressively larger quantities of calcium sulfate requires the concentration of progressively larger volumes of brine to go from calcite deposition to gypsum precipitation. Increasing volumes of water, however, indicate a gradual deepening of the water column, i.e., the rate of subsidence exceeds the rate of precipitation.

RATES OF PRECIPITATION

The rate of gypsum incrustation and precipitation in a herringbone (chevron) pattern in the Venezuelan and Dutch Antilles ranges from 0.5 to 1.5 mm/year. In southern Spain, a rate of 5–10 mm/year was observed over a period of 30 years, with growth rates reaching 50 mm/year in conduits of brine (Dronkert, 1977). In the Solar Lake along the Gulf of Aqaba, Krumbein and Cohen (1974) observed an optimal rate of about 10 mm/year. The occurrence of over 250,000 anhydrite–calcite couplets in 440 m of a stratigraphic section in the Permian Castile Formation in New Mexico and Texas (Dean and Anderson, 1978) is equivalent to about 1.75 mm per couplet. However, Hite and Buckner (1981) estimated anhydrite accumulation rates to be 0.8 mm/year. This would translate into an original gypsum accumulation of 1.3 mm/year. These rates must be average values, as the rate of gypsum precipitation varies between different textural patterns and crystal shapes, and is inverse to the amount of organic matter dissolved in the brine.

Halite precipitation is obviously linked to the rate of evaporation, i.e., the rate of removal of water from the brine. Theoretically, about 166.5 mm of halite can be precipitated for every 1000 mm of water evaporated. In the form of a halite–brine slush, this could translate into 333 mm of fresh sediment. Some sodium chloride remains in solution even after sylvite and carnallite precipitation; some of it coprecipitates with the latter two minerals. Consequently, the actual rates of halite precipitation are somewhat lower. Tasch (1970) interpreted Recent hopper crystal striations as diurnal and arrived at a rate of 36 mm of continuous halite precipitation per year. The Gulf of Kara Bogaz Gol off the Caspian Sea started precipitating halite only in August 1939. Since then it has been under continuous observation, and no period of temporary freshening and redissolution has occurred during this time interval. By 1953, some 1.37 m of halite had accumulated, with about 30% porosity (Yanshin, 1961). This represents less than 70 mm of net salt per year.

A wider range of estimates is found in evaluating ancient halite deposits. Based on a count of macroscopic laminations, Sozanskiy (1973) estimated the rate of halite precipitation to be 50–100 mm/year; Fiveg (1957) postulated 100–150 mm/year; and Wardlaw and Schwerdtner (1966) gave a range of 25–140 mm/year. At the other end of the scale, Jung and Lorenz (1968) carefully examined halite laminations subdivided further by microscopic anhydrite wafers and arrived at a rate of 5–10 mm of compacted rock salt precipitation per year. Carnallite or sylvite precipitation, then, is a multiple of these values. Band thicknesses of potash deposits are typically greater than those of halite, which, in turn, are thicker than gypsum laminae. The more concentrated the brine becomes, the greater is the amount of precipitate per unit volume of water extraction (Sonnenfeld, 1974).

Any freshening can easily dissolve more than one year's accumulated precipitation, and the preservation of varves becomes a rather fortuitous event. Episodic rainstorms, an influx of flood waters, increased seawater intake, or lowered rates of evaporation can lead to dilution and dissolution parallel to bedding planes, thereby reducing the accuracy of time estimates derived from salt thicknesses. A halite–bloedite–epsomite bed that formed in 15 years in an embayment of the Gulf of Kara Bogaz Gol off the Caspian Sea precipitated at a rate of 8–14 cm/year but produced only five to six annual layers (Strakhov,

1962). Similarly, 15–20 couplets in the Dead Sea represented 70 years of sedimentation (Neev and Emery, 1967). As a corollary, Eugster (1980b) found that 2- to 5-cm banding in trona deposits of Lake Magadi, Kenya, are periodically redissolved by flooding events.

Varve counts that assign each light–dark couplet to one season underestimate the age of the deposit by a factor of 3.0–4.5. In gypsum-precipitating lagoons off the coast of Venezuela, these laminations seem to represent time intervals punctuated by major rain or storm flood events. Not every rainy season is of equal duration, and some may even be negligible. The layering, then, represents 1 to several years.

Periodic variations in salinity need not always be equal in intensity or duration (Braitsch, 1962), and major dilutions, as occur decades or centuries apart, are thereby overlooked. From an analysis of micrometeorite material interspersed into Silurian halites of the Michigan Basin, Barnett and Straw (1983) arrived at a rate of halite accumulation (rate of halite preservation) of 0.1–4.0 mm/year. Salt thicknesses accumulating over brief periods of observation thus preserve only 22–33% of precipitated volumes; over longer intervals they do not exceed 10%. Consequently, overall rates of apparent accumulation represent considerably less than one-tenth of the time involved, and probably more than 90% of the salt is thus recycled. *Any rate of continuous accumulation of Recent salts is thus not suitable for direct extrapolation to ancient sequences of longer duration.*

Nonetheless, values of rock salt accumulation in the centimeters per year range represent tens of meters per 1000 years, or one to two orders of magnitude more rapid precipitation than subsidence, even in basins that are subject to tension and are sinking fast. Halite precipitation thus catches up with subsidence and the basin shallows. Thick halite sequences are commonly overlain by deposits laid down near either side of sea level.

Once the basin starts precipitating halite, its ratio of surface area to brine volume declines; consequently, the concentration of the brine will accelerate for a given rate of evaporation. That is why there are few halite basins that do not reach potash saturation. A major freshening is then required to reduce the salinity from potash to halite saturation.

VARVES AND ELAPSED TIME

All laminations represent cyclic breaks in sedimentation (diastems) of varying duration. The similarity of laminations in evaporites to silty varves in periglacial lakes is striking. Although the laminations have been interpreted as records of daily precipitation (Fauth, 1912), or as monthly couplets (Werner, 1938), Deicha (1942) wanted to recognize in these varves bimonthly, seasonal, and annual zones in gypsum deposits in the Paris Basin. Similarly, the darkest color of halite was related to increased clay content in the winter and spring of each year. This was deemed to be a period of decreased potassium chloride precipitation, but of increased halite crystallization preceded by increased gypsum generation (Borchert, 1969). Fiveg (1948) went a step further, correlating feathery halite crystals with the summer season, sparry halite layers with winter, and clayey anhydrite intercalations with spring. For anhydrite–halite couplets in the Devonian evaporites of western Canada, halite laminae have been interpreted as representing dilution, seasonal removal, and recrystallization when a new inflow reestablishes density stratification. The thickness of succeeeding convoluted anhydrite laminae is determined by seasonal reconcentration, and a black carbonaceous film between the laminae prevents total halite loss (Fuller and Potter, 1969).

It is certain that in many cases the layering represents the effects of episodic solution by freshening of the brine. The bromide content of the halite, a measure of brine concentration, takes a sharp dip immediately below each anhydrite lamina in the German Zechstein salt (Kuehn, 1954). However, in semiarid and arid regions, seasonal effects need not be regular annual effects. Contemporary formation of lamination is affected by slight changes in rainfall, runoff, and evaporation that prevent a seasonal salt crust from forming every year. During a wet year, a crust does not form; during a dry year, there are thin or no clastic interbeds. Solution and recrystallization at depth may also destroy evidence of seasonal sedimentation. Minor sea level changes due to lunar and solar tides or to seasonal increases in insolation will have a more marked effect on coastal areas than on offshore sites. An 11-year sunspot cycle was also suspected in these varves (Richter-Bernburg, 1953, 1964); this was disputed by Braitsch (1971). In support of that idea, Druzhinin *et al.* (1944) found that Lake Ebeity, which precipitates mirabilite, shows a 22-year cycle of periodic sodium chloride deposition.

Kislik *et al.* (1970) distinguished several orders of rhythmicity: simple varves, second-order varve groups, third-order secular alternations 0.6–2.0 m thick, and fourth-order ones of even longer periodicities typical of potash horizons, about 6–30 m in thickness. The time required to freshen a large basin and concentrate it again is two to three orders of magnitude larger than the time estimated for most varved evaporite sequences. Udden (1924) also visualized cycles of higher order (180, 250, or 340 couplets), as did Fiveg (1957) and Jung and Lorenz (1968). Holser (1979b) mentions 6-, 11-, 35-, 60-, 170-, and 400-unit cycles. Others have failed to corroborate any cyclicity in varve counts. In the Stassfurt Member of the Zechstein evaporites, the basal anhydrite–calcite couplets occur at a rate of 1500 per vertical meter of section, halite–anhydrite couplets at 150 per meter, and bituminous limestone at 20,000 per meter. A cyclicity of higher order is represented by a few halite intercalations in the potash horizon and by three carbonate–sulfate cycles in the bituminous limestone horizon. Cycles of even higher order can be distinguished, such as three cycles making up the Stassfurt salt itself. This is the second of up to seven cyclic evaporite events within the

Zechstein Group (Jung and Lorenz, 1968). Each chloride-precipitating event always represents several cycles of halite, with or without potash deposition, separated by anhydrites and dolomitized limestones that evidently formed uniformly thick, basinwide banks. Multiple cyclothems stacked on top of each other are common, but an even higher order of cyclicity is represented by the occurrence of multiple evaporite cyclothems in the same basin in more than one period of the earth's history.

Many estimates of time intervals based on a count of varve frequencies have been published. Wardlaw and Schwerdtner (1966) assigned approximately 4000 years to a lower portion of the Prairie Evaporite Formation, and Schmalz (1969) estimated 60,000 years for all of the Elk Point evaporites. Kuehn and Roth (1979) set a value of 11,000 years for 30 m of the Oligocene Alsatian evaporites. Ogniben (1957) estimated 90,000 years for the Upper Miocene evaporites of the Mediterranean area, and Richter-Bernburg (1950) assigned 4000 years to the Werra salt sequence and 6000–8000 years to the Stassfurt salt sequence of the Zechstein evaporites within a total Zechstein time interval of 500,000–1,000,000 years. Hite and Buckner (1981) calculated precipitation rates for halite at 40 mm/year, for anhydrite at 0.8 mm/year, for black shales at 0.69–2.08 mm/year, and for dolomite at 0.17 mm/year, arriving at a period of slightly over 100,000 years per individual evaporite cycle.

The assumption that laminations are semiannual or annual in cyclicity is not always warranted, as laminations are frequently not directly correlatable with given units of time. Questions then arise only when these varves are used to measure units of elapsed time. If laminations in gypsum or halite are not annual events, they cannot be used as dating mechanisms (Dellwig, 1955; Phleger, 1969; Shearman, 1970; Braitsch, 1971). Individual counts of couplets by different members of the East German Geological Survey were unsatisfactory, giving widely varying results (Schulze, 1960b). Moreover, the thickness of fine laminations in halite is always on the order of fractions of a millimeter, yet all observed rates of halite precipitation in Recent salinas and hypersaline lagoons are one to three orders of magnitude greater. Either much of the salt slush in natural lagoons is annually redissolved, leaving each time a residue of nearly uniform thickness, or the laminations represent events of ultimately meteorological origin that are not necessarily of annual regularity. Anhydrite couplets, 10–15 cm wide and containing gypsum pseudomorphs and limestone wafers in the Carboniferous Sverdrup Basin in arctic Canada (Davies and Nassichuk, 1975; Wardlaw and Christie, 1975), if studied in detail, suggest more than annual or seasonal variations in brine characteristics. Lambert (1967) found that halite–potash couplets varied in the Cretaceous Congo Basin from a few centimeters to several meters in thickness, hardly one season's precipitation.

Kaufmann and Slawson (1950) sprayed a varved salt face on a Detroit mine wall with water to dissolve some of the salt. In each centimeter of vertical distance, they found groups of four or five anhydrite lamellae separated by narrow zones of halite of variable thickness. Several lamellae often crowd each other, so that they become indistinguishable. One or two lamellae make up the dark portion of one varve couplet; several groups of four or five lamellae make up another one. The clear bands of halite between the dark layers contain the same thin anhydrite lamellae, but they become evident only upon leaching of the salt face. If each anhydrite lamella represents one rainy season, a varve couplet may represent a few or even a few dozen years. Each light–dark couplet alone represents a cyclicity covering several years. Similarly, Jung (1959) initially counted significantly more varves in Zechstein anhydrites than in adjacent correlative halites. Upon careful reexamination, however, it was found that the halite–anhydrite couplets contained 10–12 paper-thin anhydrite wafers in the halite portion (Jung and Lorenz, 1968), eliminating the difference. Every 10th to 12th anhydrite wafer is thick enough to be seen with the naked eye and thus is counted as a couplet. If we take as an analogy any locality along modern subtropical coasts, we find that it is exposed, on an average, about once in 10 years to hurricane damage and resulting storm floods. Storm-derived couplets have been observed in saline lakes (Hardie, 1968; Eugster and Hardie, 1973). The same idea was expressed by Clark (1980a), who suggested that many couplets are storm derived, settling below a storm wave base from a layer of turbid waters. The carbonate mud is thus produced by degradation of skeletal material in the marginal shallow marine zone.

If we call the couplets readily observable with the naked eye "macrovarves," we could consider the ones observable only under magnification (Kaufmann and Slawson, 1950; Jung and Lorenz, 1968) as "microvarves." If microvarves represent time intervals from one rainy season to the next one, or 1.0–1.5 years, then macrovarves represent time intervals ranging from at least 8 to 18 years. Estimates of time intervals of evaporite deposition based on a count of preserved macrovarves in halites are then too low by a factor of 10–20. Even then, the salt precipitation rates derived in this fashion are probably still too high, because no consideration is given to the lacunae represented by clayey anhydrite layers. These lacunae may cover longer time units than those that are represented by rocks (Zaenker, 1979). Furthermore, one must add to the time interval of actual evaporite precipitation the period of seawater concentration to saturation. Judging by the gradual increase in the chlorine content of sediments underlying the Barycz salt in Poland, Glogoczowski (1959) assumed that this time interval is substantially longer than the period of actual evaporite crystallization. Subsidence rates need not be exorbitant to keep up with such composite rates of precipitation.

For comparison, it might be mentioned in this context that Gize and Barnes (1981) studied the organic geochemistry of several Mississippi Valley–type ore deposits. Using zinc dispersion data, they came to the conclusion that the time needed for major ore body formation is 250,000 years. If metal-bearing fluids originate in evaporite basins, we have to look at this order of time interval for the active generation of hypersaline brines. Indeed, the presence of Br^-, Cl^-, and metallic aureoles around such deposits (Panno *et al.*, 1981) leaves little doubt that they are derived from hypersaline brines formed within an evaporite sequence. That is the same order of magnitude represented by 260,000 couplets in the Permian Castile varved evaporite sequence (Anderson *et al.*, 1972) if the couplets in this instance represent annual events.

HOPPER CRYSTALS AND CRUSTS

The evaporative water loss leaves behind a very concentrated surface layer of brine. When this surface layer reaches supersaturation for calcium sulfate, gypsum crystals may form; very occasionally, one can observe them floating. They can float only if they are attached to organic debris and use the buoyancy of the organic filament. By themselves, these gypsum crystals would normally be too heavy to float. At a density of 1.217, California salinas precipitate gypsum as a bottom layer and as a greenish to dirty brown crust on top of the brine (Elschner, 1923). This crust has a decided methylamine (CH_3NH_3) smell when dried. Upon disintegration of the crust, gypsum sands accumulate in depressions of the salina floor. Floating gypsum twins held merely by surface tension, without the aid of organic debris, have been produced in the laboratory (Borchert, 1959), admittedly in a brine, which was precipitating bischofite together with carnallite and sakiite (hexahydrite), i.e., a bittern well in excess of the densities normally encountered in brines saturated only for calcium sulfate.

Much more common than floating gypsum crystals are hopper crystals of halite growing down from the brine surface and eventually falling to the lagoon floor; sylvite and carnallite produce them in such quantities that they eventually form thick, floating crusts. Hopper crystals are graduated, cone-shaped, tetrahedral floats that form at the brine surface when crystallization is very rapid and wind mixing is at a minimum. As they fall to the brine floor, they produce tooth-shaped intergrowths with a zoned or pinnate internal structure. Such hopper crystals are best produced on wind-free days, or after a brief wind squall momentarily increased evaporation losses. Wave action can minimize or even prevent surface crystallization of hopper crystals. Wherever they are found in ancient rock salt seams, these hopper crystals are proof of primary crystallization from an open brine.

There are several shapes of hopper crystals representing varying physicochemical conditions in the surface brine (Gahm and Nacken, 1954). The length of time an air bubble and a halite particle remain in contact is shorter, the greater the concentration of amines in the brine, particularly of longer-chain species (Pavlyuchenko, 1967); this is further evidence that organic material speeds up halite precipitation. Hopper crystals form only with great difficulty, if the brine does not contain dissolved organic matter. However, even minor amounts of dissolved urea, sugars, gelatin, or some other organic compounds drastically reduce the surface tension of a saturated sodium chloride solution (Milone and Ferrero, 1947) and thus hinder hopper formation. Aluminum or iron chlorides also act in this manner. The adhesion of halite to air bubbles is retarded by the presence of barium, strontium, magnesium, cerium, or cadmium ions, whereas a concentration of 74×10^{-8} g-mol/liter of cadmium chloride lengthens the time interval 65 times (Pavlynchenko, 1967). Calcium and magnesium complexing with amines can slow down the contact of bubbles with halite particles and completely stop sylvite flotation. Rising temperatures break down these complexes. Hollow halite pyramids do not form in distilled water; they require the presence of calcium, magnesium, or sulfate ions. These pyramids are buoyed and sink unless river water brings in large amounts of bicarbonate of calcium, which causes an agglomeration of the halite crystals. A continuous crust then forms that prevents further evaporation (Basinski and Czerwinski, 1950). Anhydrite may be found inside these hollow halite crystals, aligned parallel to the horizontal halite faces, but not growing along vertical faces (Dubinina, 1951a).

In addition to hopper crystals, there are salt biscuits forming at the water surface. They continue to grow after dropping to the lagoon floor. Euhedral cubes are sprinkled over the upper surface of these biscuits; larger crystals grow on the underside (Mueller and Irion, 1969a). Both hopper crystals and salt biscuits occur only when the brine is not shielded from evaporation by a surface layer of lower salinity.

In the Silurian Salina Salt of the Michigan Basin, alterations of cloudy halite with hopper crystals and clear halite, which crystallized on the bottom, have been interpreted as depicting alternating cooler and warmer seasons (Dellwig, 1953, 1955). In the Devonian Prairie Evaporite of Saskatchewan, Canada, cloudy chevron halite grains grew competitively upward (Wardlaw and Schwerdtner, 1966); the same has been observed in the German Zechstein salt (von Gottesmann, 1963) and in Recent salt in Baja California (Shearman, 1970). The cooler seasons would thus represent the rainy seasons of maximum algal growth. In Dead Sea crystallization ponds, halite crystallizes as a hard white layer in shallow waters 20–30 cm deep, but as individual crystals in deeper (89+ cm) waters (Kenat, 1966). Floor-nucleated euhedral halite crystals, with zonations of terrigenous and chemical inclusions parallel to the (100) face, were also described from the Miocene Wieliczka salt in Poland (Pawlikowski and Ksiazek, 1975). Cloudy laminae arranged in chevrons that point upward were interpreted as diurnal growths in brine depths shallow enough to be exposed to a daily cycle (Holser, 1979b).

RIPPLE MARKS

Halite is generally precipitated at slightly greater water depths than gypsum, but interfingers laterally shoreward with gypsum and anhydrite. It normally precipitates on top of a gypsum layer that is later altered to anhydrite. The rare occurrences of ripple marks in halites may indicate a temporary halt in the precipitation prior to a transient freshening. Such ripple marks are known from the Silurian Salina salts of Michigan in a nearshore setting (Kaufmann and Slawson, 1950), and are preserved by anhydrite–dolomite laminae deposited on top of these undulations (Dellwig and Evans, 1969). Paper-thin anhydrite wafers also often cover, and preserve wave ripple marks where they do occur in halites in the Zechstein evaporites (Zimmermann, 1908; Simon, 1929). The anhydrite is then evidently derived from a fresh supply of calcium sulfates. Ripple marks were also found in halites in Triassic salts of Germany (Schachl, 1954) and in Byelorussian potash deposits. Here they occur in granular halite with a "sandy" texture, and are covered and preserved by a thin clay lamina (Lupinovich and Kislik, 1965). The wave play is an indication of a shallow water depth in the coastal parts along the southwestern margin of the second potash horizon in this evaporite sequence.

Not all wavy lines are ripple marks. A wind-driven brine diluted by rainwater dissolves furrows resembling wave ripples in marine salinas along the coast of Egypt (Walther, 1924). Ripple marks observed in Permian evaporites of west Texas (Porch, 1917; Lang, 1937) may be simulated, and may represent folded laminations probably caused by gravitational adjustment on the original slope of the basin (Anderson and Kirkland, 1966).

HASELGEBIRGE

The Permian red beds of northwestern Europe contain a special facies of claystones with a halite content that locally reaches 30% (Glennie, 1972), called *Haselgebirge* or "clay-flake salt" (*Tonflockensalz* of Richter-Bernburg, 1953), as distinguished from a "clay-clod salt" (*Tonbrockensalz*) and a "clay-matrix salt" (*Tonzwickelsalz*). Halite containing clay flakes occurs in German Zechstein evaporites both in the Leine cyclothem (Loeffler, 1960b) and in the Aller cyclothem (Langbein and Seidel, 1960b). The sequence is underlain by anhydrite-bearing bedded clays, in which the anhydrite component decreases downward. In addition to a tectonic interpretation (Mayrhofer, 1953; Schauberger, 1953, 1972), several others are available. It may represent a fossil case of a desert lake (Brunstrom and Walmsley, 1969), capillary groundwater efflorescences on playa-type desert mud flats, a fluidal redeposition of alternating clay/halite layers by river floods (Petraschek, 1947), or a mud flow

(Beyschlag, 1922). In the upper part (Leine cyclothem) of the Polish Zechstein (Dette and Stock, 1957), the *Haselgebirge* was equated the Polish term *zuber,* or "green salt clay." The *zuber* salt contains an admixture of mixed-layer magnesian clays that do not differ qualitatively from one horizon to another (Pawlikowski and Stasik, 1980). Coarsely crystalline halite with thin clay intercalations (*bedded zubers*) prevails in the lower part of the unit, whereas irregular intergrowths or pockets of clay are more common in the upper part. The thickness of zubers increases markedly toward potash beds.

The *Haselgebirge,* at the original sites in Austria, is a mixture of rock salt and clay, enclosing shale balls and some anhydrite in tectonic depressions (Medwenitsch, 1972). The *Haselgebirge* is here an Upper Permian pseudotectonic breccia composed of angular and subangular clay and sulfate fragments in a matrix of halite and finely ground clay (Medwenitsch, 1963). It is subdivided into gray, red, green, and variegated types, of which the last also contains tuffites. Layering sometimes persists for great distances. A black, silty, carbonaceous claystone with considerable salt residue is called *rute* in Colombia and is related to claystone clasts up to 1 m in diameter embedded in the laminated rock salt, with halos or fringes composed of halite wedges (McLaughlin, 1972).

It may well be that the term *Haselgebirge,* which in Austria applies to a tectonic rubble of green clays and red salt with a high bromide content, is misapplied elsewhere to red beds with halite crystal growth or halites with green clay contamination, for which different terms should be introduced. Other local miners' terms for specific types of dirty salt exist in Poland, Germany, Colombia, and elsewhere, but no uniform terminology has evolved.

NODULAR ANHYDRITE

There are at least three modes of origin of nodular anhydrite:

1. Formation in the aerated zone above a groundwater level. Nodular anhydrite as a syngenetic structure can be the product of gypsum precipitation in the capillary zone above the resting level of the groundwater table in the supratidal area (Shearman and Fuller, 1969), and thus indicates fossil groundwater levels (Shearman and Orti Cabo, 1978). Such supratidal anhydrite has a lower Sr/Ca ratio than laminated anhydrite (Butler, 1973). Triassic nodular anhydrites cover a belt nearly 350 km wide between continental claystones near the Baltic Sea and massive marine anhydrite banks in central Germany (Beutler and Schueler, 1979). Belts of Triassic and Jurassic supratidal nodular anhydrite are also known from Great Britain (Burgess and Holliday, 1974; West, 1975). In the Upper Keuper marls of Nottinghamshire, gypsum nodules are associated with veins of fibrous gypsum. The veins terminate at the gypsum nodules, always at the upper (never the lower) surfaces of gypsum masses (Richardson, 1920).

These veins are evidently produced by capillary brines from the nodule horizon toward the ancient sediment surface.

2. Formation in a subaqueous environment as an early diagenetic phenomenon. A variety of contorted structures, some resembling nodules, can also be produced in several ways after the sediment has been formed, as truly epigenetic changes. Riley and Byrne (1961) proved in laboratory experiments that compaction and the weight of overburden can deform a still soft evaporite layer. Flowage of the semifluid sediment then produces a mosaic pattern. Various textures associated with such features have been classified by Maiklem *et al.* (1969). Prolonged lateral flux of calcium sulfate-bearing fluids first causes anhydrite nucleation in organic-rich compounds of the carbonate sediment. Porphyroblasts of sparry anhydrite with initially square, rectangular, or blocky outlines and lath-crystal nodules expand and coalesce.

There are many fossil examples. Nodular anhydrites in beds underlying Lower Permian halites of Kansas occur in conjunction with acicular crystals (Holdoway, 1978) that may be the remnant of more extensive gypsum, which decomposed while overlying salts were repeatedly dissolved by rainwash. A Carboniferous sequence of alternating limestones and anhydrites in the Arctic Archipelago of Canada has been laid down entirely underwater. No desiccation features common in supratidal environments are to be found. However, both chickenwire and nodular anhydrites are present, and grade upward into laminated anhydrite with swallow-tail pseudomorphs (Wardlaw and Christie, 1975). The nodules are here the product of early diagenetic growth below the sediment–water interface in a mixed sulfate–carbonate host rock. A mosaic fabric is then caused by the coalescence of sulfate nodules overlain by bedded subaqueous gypsum (Davies and Nassichuk, 1975). Middle Carboniferous nodular anhydrites of Spitsbergen replace dolomite as an early diagenetic phenomenon. Even younger lath-shaped anhydrites project into, displace, or replace dolomite and cut across granular anhydrites (Holliday, 1968). The granular anhydrite is obviously diagenetic, since no recent granular anhydrite is known (Kinsman, in Holliday, 1968). Nodular anhydrite as a replacement for dolomite was also found in Permian evaporites of Cumberland, and is commonly interbedded with varved anhydrite in west Cumberland, England (Arthurton, 1971; Arthurton and Hemingway, 1971).

Nodular anhydrite in blackish-brown, finely bedded bituminous carbonates characterizes the rim and the sills of the lowermost cyclothem of the Permian Zechstein Basin in Germany. In each instance, it thickens toward the basin (Weber, 1961). The rims may have been exposed occasionally to supratidal drying out. If these nodular anhydrites of the sills did not form in a subaqueous environment, their presence is proof of an extremely shallow Zechstein Sea, at least during the first cyclothem of evaporite deposition. Massive anhydrite is missing precisely where the salt thickness exceeds 300 m, and marls with nodular anhydrite occur in its place (Weber, 1961). Nodular anhydrite, immediately overlying an oolite zone in Middle Devonian beds of the Michigan Basin (Gardner, 1974), juxtaposes a high-energy subaqueous environment to a supratidal environment unless the nodules were also formed underwater. A similar problem is raised by five nodular anhydrite–gypsum horizons in an Upper Jurassic evaporite sequence in Sussex, England. They are intercalated with calcite laminites that had been produced by bacterial gypsum replacement under stagnant reducing conditions (Holliday and Shephard-Thorn, 1974).

3. Formation by later recrystallization. Anhydrite nodules in the lower part of the Permian Salado Formation of New Mexico were formed by recrystallization of preexisting laminae below the onset of halite precipitation. They are not related to subaerial exposure (Dean and Anderson, 1978), but probably indicate basinwide salinity changes. Basinwide nodular zones occur here 1000–3000 laminar couplets apart; not all of them underlie halite beds, but no halite beds occur without subjacent nodular anhydrite beds (Anderson *et al.*, 1972).

Mosaic gypsum crusts 1–7 m in diameter in the underlying Seven River formation may represent a very late gypsification of anhydrite nodules, but are also interpreted as an early diagenetic growth below the sediment–water interface (Sarg, 1981). Nodular alabastine gypsum is a recrystallization due to gypsification of anhydrite along tectonic zones (Schreiber and Decima, 1976). Nodules that are possibly diagenetic have also been described from Upper Miocene evaporites in Spain (Orti Cabo and Shearman, 1977). Jurassic nodules in fine-grained gypsum in Morocco are believed to be a postdepositional recrystallization of saccharoidal gypsum in the presence of salt; the rounded form is due to a slow rate of diffusion relative to the more rapid rate of crystallization (Gaudefroy, 1956). Dense nodular mosaics mimicking those found in sebkhas have been found by replacement at shallow depths of burial of subtidal high-energy carbonates over 13 m thick (Jacka, 1981). These secondary anhydrite beds seem to be correlatable for long distances, because they formed in highly permeable carbonates capable of vigorous water circulation. West of the Urals are Permian gypsum nodules up to 2.5 cm^3 in size, with thin veinlets of gypsum between them. They are interpreted as resulting from a reaction between local calcite and circulating magnesium sulfate brines; in this case, iron and silica diffused in the same direction (Kuznetsov and Ignat'ev, 1948). The recrystallization of rows of pallisade gypsum crystals into anhydrite nodules, and the subsequent recrystallization of anhydrite into gypsum nodules, can represent a multiple stage of nodule formation.

Porphyroblasts sometimes replace nodules (Arthurton, 1971). They are often caused by recrystallization. Initially, there is an electrostatic disequilibrium in the force fields between the individual, differently oriented lattices of gypsum crystals in fine-grained gypsum rocks. This disequilibrium tries to change into an equilibrium by parallel reorientation of crystals. A watery solvent is required to facilitate both recrystallization and the reactions leading to the establishment of that equilibrium. Finally, the formation and site of the prophyroblasts are a function of the unequal distribution of factors causing the recrystallization (Noll, 1934).

TEPEE STRUCTURES

In all evaporite environments one finds polygonal welts with a tessellated pattern. Figure 14-1 shows an example from Gran Roque, Venezuelan Antilles, where gypsum tessellated into polygons a little over 1 m in diameter and about 20 cm high, often fractured on top. This area is the preferred breeding ground of marine flies; in the dry season, the polygons are formed by halite that later redissolves. The polygons are antiform structures that have an overthrust slab appearance and circumscribe saucer-shaped depressions. These structures were first described in detail by Adams and Frenzel (1950) in the Capitan barrier reef of Texas, and were then named "tepee structures." Today this term is occasion-

Fig. 14-1. Polygonal pattern of gypsum crusts in a gypsum- and halite-precipitating lagoon in Gran Roque, Venezuelan Antilles. The diameter of the polygons is about 1 m.

ally spelled "teepee." The authors interpreted these features as those caused by intrastratal flowage of beds of unconsolidated sediment displaced by differential loading. Assereto and Kendall (1977) also described them from peritidal carbonates. The polygonal pattern, as such, occurs in lithographic limestones and clays alike, as a product of subaqueous shrinkage (Twenhofel, 1923; Rich, 1951) in contact with a hypersaline brine. Tepee structures are also found in halite or potash precipitates. They occur today in episodically drying-out hypersaline lakes, but also in shallow hypersaline lagoons that are perennially underwater. These structures appear to exist in effectively two size ranges: the 0.5–3.0-m range, with very shallow depressions and with cracks several decimeters deep; and the 25–300-m-diameter range, with depressions about 1 m deep and cracks several meters deep. Their large depth/diameter ratios in evaporites make it unlikely that the polygons could be derived from desiccation alone (Lang, 1943; Christiansen, 1963; Neal and Motts, 1967; Watson, 1979). The coefficient of thermal expansion of rock salt is at 40°C, equal to 40.4×10^{-6} per 1°C (Hunt *et al.*, 1966).

That desiccation cracks may have, at best, only a subordinate supportive role is corroborated by an example from melting snow. Duecker (1966) found polygons generated by windblown sand of a nonuniform grain size distribution that settled onto a layer of snow. Melting of the snow rendered the layer unstable. The excess mass at the surface (the dust) caused the meltwaters to move up, while the heavier sand was sinking with a convective motion. Horizontal extension of the cells produced an irregular honeycomb pattern due to the pressure exerted by the various cells against each other. No desiccation cracks were available to dictate the location of the polygons. They were solely a product of the upward movement of a less competent layer.

Any loading on a light medium (such as a precipitate-brine mush) by a heavier medium (such as a more consolidated rock) induces a gravitational instability (Artyushkov, 1963a,b; Dżułyński, 1966), which can lead to the development of polygonal structures in the upper medium (Fig. 14-2). "Quicksand" conditions occur in the Umm-as-Samim inland sebkha in Arabia, where a polygonal halite crust, thin near the recharge entrance, is weakened after a flooding and can be broken through by an unwary man or beast into the underlying waterlogged, silty clay (Beydoun, 1980). Bustillo *et al.* (1978) observed the formation of tepee structures in lakes of the Toledo region, Spain: the subjacent material was annually injected into the surface crust during the summer. In most cases, the upward movement of moisture can be documented. Soriano *et al.* (1977) observed the similar genesis of tepees southeast of Madrid.

Alternating clay and halite layers of a playa lake in the Kavir Desert, Iran, form polygonal structures in the dry season. A brine–mud slurry remains beneath 32 cm of surface sediments (Grabau, 1920) and can rise by capillary action or ooze out because of thermal expansion and pressure of the overlying crust. It

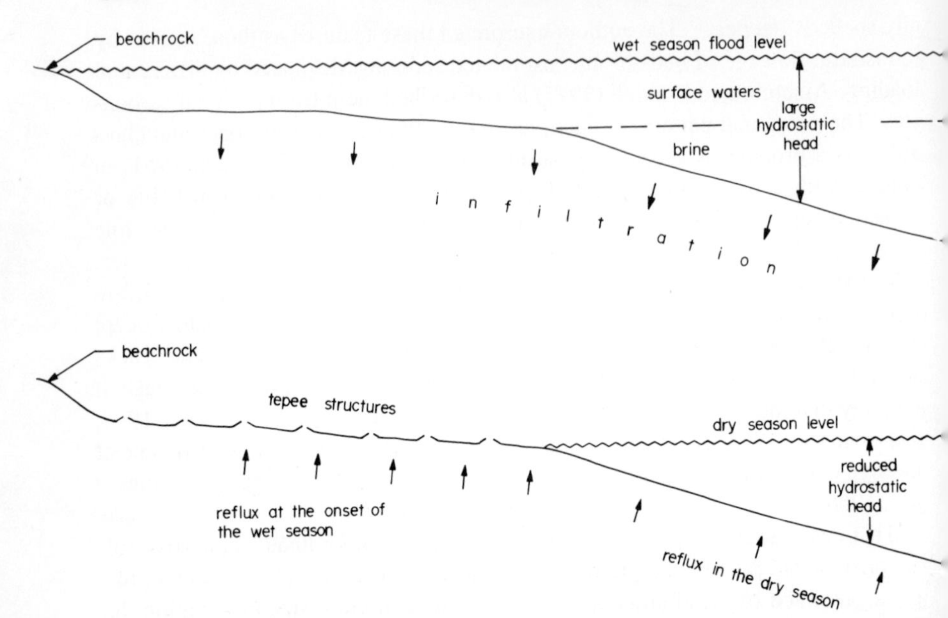

Fig. 14-2. Growth of tepee structures. A reduction of the hydrostatic head that had been created at high water level allows brines pushed into the subsurface to ooze out in response to capillary suction by evaporation, or in trying to maintain an isostatic equilibrium with an open brine that continues to get denser.

eventually forms black dykes. Groundwater may initially rise to the surface through desiccation cracks. However, the polygonal cracks persist between inundations because salt accumulates here more rapidly than it is removed. Ascending fresh brines are less saline than the indigenous residual brines and evaporate more easily. During the heat of the day, the water bubbles out along the cracks and produces salt blisters or blossoms. Hollow salt tubes or cones that taper upward are frequently seen; they are about 20–40 cm high and usually somewhat less across. Since the halite slabs between the cracks grow more rapidly on the windward side, the windward edge eventually overlaps the leeward edge. This gives the illusion of overthrusting. Uneven escape of underlying mud slurry tilts the slabs even further (Krinsley, 1970). When hypersaline solar (heliothermal) lakes dry out in the Irtysh River area of Siberia, a salt crust forms and fractures. Small volcanoes arise that discharge under pressure a gelatin-like salt porridge (Brecht-Bergen, 1908). Such pressurized, trapped brines also occur in salt flats along the Trucial coast (J. L. Wilson, personal communication).

Where the substrate has greater permeability, either because of jointing or because of differential grain packing, the reflux is faster, and gypsum crystals grow more quickly. Since the ridges receive more sunlight, they become preferential sites for photosynthesizers and thus for more vigorous calcium carbonate precipitation and conversion to gypsum. The crusts are thicker than those in interridge areas, despite the seasonal fracturing and bassanitic erosion. Moreover, since the underlying soft material can be rich in dissolved organic matter, the tepee fractures are often the sites of stromatolitic overgrowths. Even in summer there is increased soil moisture along the cracks in polygons, as witnessed by the preferential placement of halophytes; brine-covered polygons are filled with precipitate.

Filled collapse fissures marking the outline of polygonal structures have been observed in Recent clay plains (Vinogradov, 1955) and ancient halite sequences (Richter-Bernburg, 1979). Fissure depths may extend 15 m below the surface (Neal, 1975). They are evidently produced by water oozing up from the water table. In Tunisia they are 1–2 m deep, and the polygons have a columnar structure in profile, whereby the sides of the columns are tightly packed microcrystalline gray gypsum. The cores are composed of coarser material that is soft and damp, but hardens upon exposure to the dry atmosphere. The waters themselves have an alkaline pH of 8.0–8.5. They may exude on slopes of up to 20° (Watson, 1979) or even 37° (Hunt *et al.*, 1966). Perthuisot (1975) also observed another format, i.e., artesian solution cavities surrounded by cracked edges.

Tessellated ground, i.e., ground with polygonal welts, is restricted in northern Iraq to a narrow belt along the pediment of outcrops of alternating gypsum and limestone (Tucker, 1978). It disappears in the alluvial plain, evidently marking the zone of groundwater seepage. The polygons, 50–200 cm across, separated by 30-cm-deep cracks, are composed of a surficial layer of compact gypsum 1–3 cm thick, covering about 10 cm of powdery gypsum, possibly a bassanite alteration in the dry season. Gypsum replaces rootlets or occurs as euhedral discoid crystals. In recent dolomites, von der Borch and Lock (1979) interpreted the tepees as groundwater resurgence structures. During lake flood cycles, the indurated dolomite crusts in the Coorong lakes of southern Australia come under considerable pressure with the rising water table. Fractures are then loci for upwelling groundwater, and are usually associated with a localized buildup of a halite welt during the summer, only to be dissolved again later. Polygons do not develop wherever the water table is so deep that capillarity cannot moisten the upper reaches of the ground (Neal and Motts, 1967).

On the island of Gran Roque, Venezuelan Antilles, the shelf area of an enclosed pond is covered with a gypsified, laminated algal crust. This crust produces nearly circular mounds at the inland edge of the pond, where the substrate is particularly coarse and porous. The mounds gradually coalesce into garlands and ridges as one moves down the slope of the shelf. The garlands then combine into a polygonal pattern, 30–70 cm in diameter, of ridges that are about 10 cm high and 20 cm across. Open joints and connected pores in the sediment in deeper parts of the pond are perennially brine covered and allow the waters seasonally to reverse their direction of flow without hindrance. At low water level, the ridge crests become exposed to air, and the gypsum dehydrates to the powdery hemihydrate bassanite under a 26% volume reduction. Lesser reductions can also occur (Cooke and Smalley, 1968). This is a much greater contraction than that of the desiccation cracks in a dolomite or aragonite shelf.

At the same time, the ridges become unsupported by underlying brine and crack open parallel to the crests. Fissures develop due to the lowering of the piezometric surface. Gypsum crystals continue to grow for a while on the still-submerged flanks of these ridges, until they also dry out. The broken flanks slide over each other and produce antiform tepee structures consisting of unsupported, upended slabs of gypsum crusts leaning against each other. Rising water levels in Gran Roque, Venezuelan Antilles, remove the bassanite at the onset of the wet season, leaving a very smooth, polished surface. Bassanite is about 30% more soluble than gypsum, and most of it goes into solution. The smooth, polished surface is probably caused by the interaction of wave play with a natural plaster-of-paris veneer of a bassanite–water mixture that covers and mends the cracks. A theoretical volume reduction notwithstanding, bassanite expands slightly upon hardening into gypsum (Chatterji and Jeffery, 1963), and thus seals any cracks. Only the presence of substantial amounts of calcium chloride, aluminum, or potassium sulfates, or other acids, would markedly prevent this expansion (Hiyama and Fukui, 1951), thus eliminating volume changes in the presence of a hypersaline residual bittern.

Complete closing of the cracks prevents the rising waters from escaping. The pressurized liquid induces tensile stresses along the margins of the injected liquid. Fractures are created that spread like cracks propagating in asphalt, and will continue to grow as long as the mass flow is greater than the leakage into the fractured crust (Glass, 1976). A new layer of centipetal arrowhead-shaped gypsum crystals forms only after the high water level of the lagoon is passed. This preserves the morphology of the upended ridges, and possibly even accentuates it.

In the Atacama Desert in northern Chile, the crystallization from below raises slabs steeply to a man's height. Bassanite replacing gypsum, which itself is a pseudomorph after glauberite, cements the slabs. The nightly fog has a pronounced gluing effect (Wetzel, 1928; Stoertz and Ericksen, 1974). Similar slabs of halite, propped together in pairs, have been observed on Lake Eyre, Australia (Madigan, 1930). Such polygonal structures also occur in Death Valley, California (Hunt *et al.*, 1966), between sand dunes of Rajasthan, India (Srivastava, 1970), in Tunisian playas (Bellair, 1957; Plet-Lajoux *et al.*, 1971), in sodium carbonates along the shores of Lake Magadi in Kenya (Eugster, 1980b), and elsewhere. They occur in salt pans of the Nile delta, but only after the pans dry out (Grabau, 1920). Irregular 30–40-cm polygonal patterns with edges a few millimeters high have been observed in the Ojo de Liebre lagoon (Phleger, 1969). They are apparently caused both by the upward migration of brine along cracks and by the deposition of halite at the surface. Similar polygons also occur in abandoned salinas in Curaçao in the dry season, and in many other salinas. Two endorheic lakes on the Chilean altiplano, four times deeper than any other lakes in the vicinity, produce damlike ramparts of gypsum in the form of 5- to 15-m-high polygonal plates that buckle up into pressure ridges and overthrust slabs. They accumulate iceberglike masses of gypsum on their leading edges. Because of the ramparts, both lake surfaces lie 1.2 m above the salar (Hulbert *et al.*, 1976). Since evaporation rates are the same for all the lakes, these two lakes must be able to obtain a larger groundwater supply.

Groundwater oozes through marginal sediments of many lakes in the Kulunda steppe of Kazakhstan (Strakhov, 1962). Here the groundwater seepage is charged with sodium carbonate, but the major anion in the lakes is the sulfate. Calcite, dolomite, or magnesite concretions form in the conduits of rising groundwater. They create complex polygonal patterns on the surface of that part of the lake floor which is exposed at low water level (Fig. 14-3). There is evidence that the hydrostatic head of the brine in the lake prevents the formation of polygons. Numerous pores and cavities in the concretions form a network of channels along which the groundwater clearly had risen. Sodium sulfate waters passing upward through sodium chloride brines form aggregates of thenardite. They are elongated in the vertical direction, and are composed of well-formed, strongly bonded crystals with vertical tubes filled with mud. These tubes had been used as a conduit for the groundwater.

In ancient halites, the polygonal structures are always underlain by 40–50 cm of coarse, recrystallized salt (Dellwig, 1972) or by cauliflower structures, previously discussed in the context of groundwater seepage. Varves are then absent underneath; insolubles are interspersed above for 2.5–10.0 cm, indicating a basin margin position. The banding in Permian Zechstein halites becomes irregular, or the halites become mottled toward the episodically exposed shelf areas of calcium sulfate deposition (Marr, 1959). In the Stassfurt Member, the polygons are wedge-shaped, pointing downward, and are filled either with a clay–anhydrite material or a coarse halite–carnallite mixture (Zaenker, 1970, 1979). At least 15–20 desiccation horizons can be recognized in the uppermost parts of the sequence. Such polygons can have roots up to 15 m long (Richter-Bernburg, 1980). In the Devonian Prairie Evaporite Formation of Saskatchewan, the polygons are restricted to halite horizons immediately underlying potash seams (Baar *et al.*, 1971). The same is true of the Boulby halite in the third cyclothem of the British Zechstein (Woods, 1973). Anhydrite polygons also occur in the roof of the Permian salt of Kansas. They are 1–3 m in diameter and are outlined by welts of anhydrite 18 cm across and 5 cm thick (Dellwig, 1963). The underlying halite must still have been very permeable at the time of their formation. Underneath the polygons are depressions in the shale intercalations caused by salt solution, and filled with anhydrite considerably mixed with clay. Clay is also dispersed into the salt overlying the polygons (Dellwig, 1968), and is then not layered.

Fig. 14-3. Polygonal pattern induced by carbonate concretions in a sulfate lake (after Strakhov, 1962).

Selenitic anhydrites, composed of giant crystals of pseudomorphs after gypsum that are several tens of centimeters long, are the equivalent of similar giant gypsum crystals in modern lagoons (cf., e.g., Vonder Haar, 1976). Their exact mode of origin has not been studied, but they could be subaqueous sites of slowly oozing out saturated interstitial waters.

TURBIDITES AND OTHER GRAVITY FLOWS

Gypsum, and for that matter all other evaporites deposited on the shelves or sills of evaporite basins, are subject to gravity flow, grain flow, debris flow, and turbidity currents (Catalano *et al.*, 1976). The sedimentation of gypsum is so rapid that relatively steep slopes form toward the basin (Herrmann and Richter-Bernburg, 1955). Current strength is evidenced by the occurrence of cross-bedding, occasionally even in laminated sediments. Gravity flows can start over very modest submarine slopes. Turbidites occur in much shallower basins filled with hypersaline brine rather than in the open sea because of the greater buoyancy provided by the dense brine. However, the higher viscosity of the medium slows down the the relative velocity and decreases the kinetic energy of the particles. Such redeposited gypsum breccias are known from many evaporite basins. Only the coarser fraction of torn-off carbonate and gypsum fragments will cascade down the coastal slope. The finer fraction remains suspended at the density interface and then sinks slowly, a good part of it even redissolving due to its higher solubility in concentrated brines and bitterns.

Basinwide gypsum intercalations signal a freshening of the brine. However, tongues or thin banks of gypsum or anhydrite wedging out basinward originate as turbidity currents from marginal shelves and concurrent wider dispersion of suspended fines. Such gypsum–anhydrite turbidites extend 25–30 km basinward in the Zechstein evaporites (Meier, 1977), but are also known elsewhere (Richter-Bernburg, 1953; Herrmann and Richter-Bernburg, 1955; Bernardini, 1969; Ricci-Lucchi, 1969, 1973; Parea and Ricci-Lucchi, 1972; Catalano *et al.*, 1976; Schlager and Bolz, 1977). A halite turbidite has been observed at Caltanissetta, Sicily (F. C. Wezel, personal communication, 1980). Synsedimentary flowage marked by rounded anhydrite and dolomite fragments is evident in Silurian salt of New York State (Treesh and Friedman, 1974) and in Miocene evaporites in the Carpathian Foreland (Garlicki, 1980). In addition to turbidites, gypsum often develops slump structures (Meier, 1975). An occurrence of anhydritic conglomerates and fragmented fossils, as well as other indications of high current velocities, are found in Triassic carbonates and evaporites in Germany (Wilfarth, 1934). Most halite initially precipitates as a wet slush that eventually hardens into rock salt. A slush is liable to creep or slide on inclined shelf slopes even of gentle dips, resulting in a variety of slump structures. Halite turbidites must not be confused with tectonically produced, rounded brecciation fragments in diapiric salt stocks.

Contemporary halite precipitation produces a salt sand (haloarenite) in turbulent waters (Bradbury, 1971). Halolites (halite oolites) and rippled structures form near the shores of the Dead Sea by wave agitation (Weiler *et al.*, 1974) due to shoreward-blowing winds. Cross-bedding in Silurian salt of New York or embedded shale balls up to 1 m in diameter (Dellwig and Evans, 1969) suggest that even halite precipitation can be accompanied by currents of significant velocities. The foreset beds of cross-bedding dip away from the basin margin (Dellwig, 1972) as the brine slides off the shelf into the deeper parts of the basin. The current energy during gypsum or halite precipitation can be quite substantial; one has to keep in mind that, for equal current velocities, concentrated brines contain more energy than seawater or freshwater currents. Fast-flowing currents can carry major amounts of gypsum crystals as coarse sand or even erode a gypsum breccia. Slope-controlled flow, i.e., either fluvial currents or debris flow, move and resediment autochthonous gypsum and selenite (Vai and Ricci-Lucchi, 1978). Miocene examples include current-deposited selenite crystals in evaporite basins of northern Sicily (Lo Cicero and Catalano, 1976), a coarse gypsarenite covering the intra-Messinian unconformity between two evaporite cyclothems (Decima and Wezel, 1971), gypsum conglomerates (Catalano *et al.*, 1976), and gypsum fanglomerates in the northern Appennines (Vai and Ricci-Lucchi, 1977). In contrast, because of the greater buoyancy potential of concentrated brines, lutitic aragonite can be spread out along stratification interfaces to a uniform distribution across the basin.

Enterolithic layers due to subaqueous slides, current-derived inclined bedding in channels, and corrosion surfaces due to dilution events penetrating the whole water column down to the sea floor are common in Zechstein anhydrites (Kosmahl, 1969). Enterolithic structures may also be caused in other ways, e.g., by corroding fluids meandering through a sulfate bed (Lomonosov, 1971), by circulating epigenetic fluids that produce coarsely crystalline secondary anhydrites with relic unreplaced grains (Lomonosov, 1972; Clark, 1980b), or by volume changes due to dehydration or rehydration. Finally, enterolithic structures can also be caused by creep due to tectonic stresses. Gypsum is strongly anisotropic: It has only 1.8 times the plasticity index of calcite on the (010) face, but 4.6 times the value for calcite on the (100) face (Koyfman *et al.*, 1969). This overall 2.5-fold difference from face to face is responsible for some early diagenetic deformation. Dana (1951) referred to halite as brittle and gypsum as very brittle. However, halite has a plasticity index nearly 2.7 times greater than that of calcite. A harder, more brittle surface layer is formed in halite by nitrogen that penetrates rock salt very easily (Aerts and Dekeyser, 1956). No deformation of dry salt under a load takes place; deformation is easily achieved only if the salt is underwater or at least wet on the side of load application (Ewald and Polanyi, 1924).

Local unconformities can be produced by scouring waves or internal seiches. Dellwig (1972) photographed an oscillating unconformity within a rock salt section with a wavelength of 12–14 m and an amplitude of a few tens of centimeters. Synsedimentary folds are known from many potash deposits. Fluidization structures in Permian evaporites of the Upper Kama region west of the Urals now occur at depths of 200–300 m, and thus cannot have been induced by excessive overburden or a great increase in geostatic temperature (Kopnin, 1962). The Cretaceous tachyhydrites of Brazil are folded above a prominent anhydrite marker and unfolded underneath it (Wardlaw, 1972b). If the gypsum, precursor to the anhydrite, was deposited in the photic zone, its growth rates would accentuate any preexisting differences in water depth. The resulting very gentle slopes would suffice to induce creep, particularly if lithification of tachyhydrite is delayed.

A variety of potash crystals is precipitated in the high supratidal zone (Butler, 1969). Such salts of higher solubility that precipitate on occasion in small depressions on temporarily exposed shelves, or in the soil horizons of the supratidal zone, are washed down into deeper waters if rain or new inflow does not completely redissolve them, and they are remobilized as turbidity or density flows, or at least as slumping (Richter-Bernburg, 1979). Hematite-enriched halite nodules in the uppermost layers of potash-bearing beds in several German mines may represent reworked and transported broken salt crusts (Schauberger and Kuehn, 1959; Loeffler, 1962; Schwerdtner, 1962). Nonetheless, augen-like structures can also be primary (Gunatilaka *et al.*, 1980).

15

The Evaporitic Environment

RECAPITULATION OF REQUIREMENTS

Any model of marine evaporite deposition must meet all of the observational data on atmospheric, oceanographic, geochemical, and morphological conditions; the pertinent parameters of changing basin morphology, brine volume, and alimentation; sequence and cyclicity of precipitates; and facies distribution. Environmental conditions to be fulfilled for large-scale evaporite precipitation to occur are very specific, yet have often been met, judging by the appearance of large evaporite deposits in almost all epochs of Phanerozoic history. The basic prerequisites that must be met for evaporite precipitation in a marginal bay fed by ocean waters have been detailed in previous chapters. They can be recapitulated as follows:

Original Depression

To initiate brine concentration, the presence of a water-filled basin with access to the world ocean is necessary, i.e., *a gentle depression must preexist.* Subsidence of depressions proceeds at a slightly faster rate than subsidence of surrounding sills and bay margins (Busson, 1980). Similarly, deltas also start only at the lowest point of the coastline and then continue to depress it under their own weight at a fairly rapid rate. The mere loading of reservoir lakes with fresh water depresses the land noticeably and in short order. If we regard coal seams as peat compacted at a ratio of 10:1, we obtain some idea of the magnitude of subsidence induced by the loading of paralic swamps. Rapid precipitation of evaporites similarly exerts pressure on the crust underneath, a pressure certainly greater than that represented by fresh water or by a water–wood swampy mixture. Just as coal seams often show a rhythmic alternation between subsidence and temporary emergence, evaporitic sequences do likewise.

The sea that precipitated the Kupferschiefer series underlying the Permian Zechstein evaporites probably had a positive water balance, akin to a modern fjord in its pattern of water circulation. The marls constituting the metalliferous series average 25–45 cm in thickness and reach 80–200 cm only in a narrow belt along the southern margin. The overall constancy of thickness suggests that the sea floor had an even or a very slightly undulating relief; the unit thins over sills. Early Upper Permian epeirogenetic movements caused an inversion of the relief. The barren zone is restricted to shoals where the redox potential remained positive. Hematite occurs on the flanks of shallows followed in the waters of negative redox potential by a horizontal zonality of copper to lead–zinc from shoals toward the deep part of the basin (Jung and Knitzschke, 1976). The base of the "Rote Faeule" barren zone thus indicates the original interface between oxygenated surface waters abounding with plankton (Pompeckij, 1914) and periodically oxidized mixed brines. The boundary between hematite and base metals indicates the top of the perennially anoxic bottom brines. A corresponding vertical zonality signals the gradual rise of the interfaces.

Deficit in the Water Budget

At least a moderate deficit in the water budget of this basin must prevail, coupled with very low atmospheric humidity during part of the year: *Evaporation rates must exceed rates of atmospheric precipitation and runoff.* Return flow and seepage must be smaller than inflow and surface drainage (Baker, 1929). By the time the concentration of the brine reaches saturation for halite, its rate of evaporation has decreased by about 30%. With rising concentration, its specific heat is reduced and the brine warms up. As a warm brine increases its concentration further, the decrease in its rate of evaporation accelerates. Figure 2-9 shows the rapid steepening of vapor pressure decline curves for warm seawater concentrates. If the atmospheric conditions and entrance strait configuration remain unaltered, the demand for water replenishment decreases and the inflow slows down and allows a greater outflow. Consequently, the salinity in the basin decreases slightly and the precipitation moves out of the potash saturation field back to halite precipitation. The halite covers and preserves any potassium–magnesium salts (Fig. 15-1). Similarly, primary tachyhydrite is always encased in carnallite, which in turn is covered by halite. Accordingly, one may find more than one potash horizon sandwiched into halite. For thick halite or potash beds to form, the bottom brine must remain in the saturation field of one or the other mineral assemblages for a reasonable length of time. It cannot reduce its concentration without producing corrosion planes; it may, however, increase its

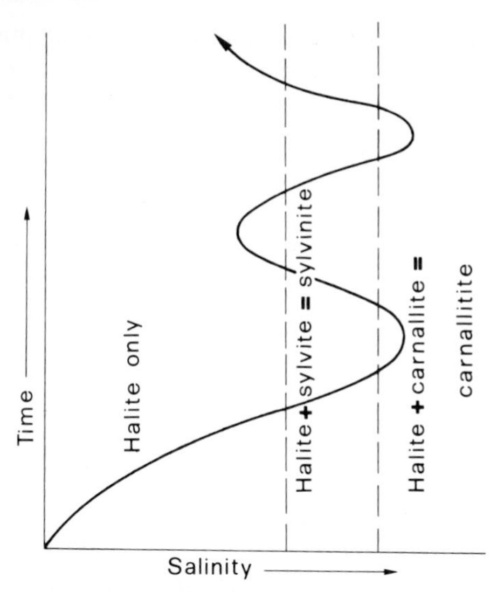

Fig. 15-1. Curve of oscillating saturation. Saturation for halite is maintained for some time after sylvinite or carnallitite has ceased to form. Either of these occurs on top of a sequence only in playa lakes where the brine receded and the surface dried out.

concentration temporarily. Potash minerals precipitated on halite and redissolved again when the brine concentration slightly drops leave no trace. The bromide curve has a sufficiently wide scatter to accommodate such variations in concentration, even where the bromide curve is not altered by later leaching. The Caspian Lowlands are underlain by a Permian evaporite sequence that contains 10 potash horizons, 3 above a main anhydrite sequence and 7 beneath it. Each potash horizon is separated from the next one by up to 100 m of halite (Diarov, 1971). In other words, although the brines oscillated episodically within a density range under 1.22 and over 1.33, they circulated for a significant time within these limits without further dilution or further concentration.

Starved Basin

A scarcity of runoff promotes a scarcity of continental clastic intercalations. This is easier to obtain in low-lying areas than in territory abutting hills or mountain chains. Many evaporite basins are underlain by clastics that are fining upward until they are replaced by limestone or dolomite banks that show the prevalence of clean, transparent waters. Evaporite basins are actually starved basins, lacking a clastic sediment supply. Their true nature is masked by the rapid chemical precipitation. No continuous runoff supplies major quantities of terrigenous materials; whatever siliciclastic matter enters originates in episodic flash floods that drain surrounding land areas. During the time when the basin has open waters, any rainstorm produces a freshwater layer on the brine that eventually evaporates. A rainwash of the surrounding countryside dumps terrestrial clays that spread along the density interface in the form of a submarine delta and then slowly descend and intermingle with the precipitate. No beds of clastics grading upward from rock salt to sand to silt to clay have ever been found inside an evaporite sequence. Nor does a lateral gradation occur from salts to clays. Either the sands all dissolved, or they were never delivered while there was an open water surface. Each successive potash bed in the Permian Castile Formation of New Mexico contains more clay (Kroenlein, 1939). This represents either the increasing frequency of dust storms or more numerous periods of rainwash. The latter is improbable, because it would have led to a dilution of the brine and cessation of potash precipitation. Increasing aridity reduces any meager vegetation cover and bares more soil for the wind to pick up, but also produces an increasing frequency of dust storms.

During the time when the basin is full of precipitate and has no open water surface, any rainstorm will dissolve some salt. Rainwash sweeps in a layer of freshwater clays distributed as tongues offshore. Such rainwash is responsible for the occurrence of tree trunks in halites and carnallites, and for detrital sands and silts associated with the clays. Terrigenous deposits are usually encased in anhydrite, suggesting either the delivery of new supplies of sulfate ions or the oxidation of hydrogen sulfide.

Shape of Basin

The shape of the evaporite basin (Fig. 15-2), its orientation to prevailing winds, and its inflow are of great importance for its water budget. The shore irregularity is defined (Reeves, 1968; Håkanson, 1981) as a shoreline factor F by the formula

$$F = L/2\sqrt{A\pi} \tag{15-1}$$

where L is the normalized (scale-independent) shoreline length (in kilometers) and A is the lagoon area (in square kilometers). The shoreline factor illustrates the relationship between the actual length of the shoreline and the circumference of a circle with an area equal to that of the lake or lagoon (A). The F value is closely related to lake or lagoon bottom roughness (R), which is quite understandable for morphological reasons: A lake or lagoon with an irregular bottom will also have an irregular shoreline (Håkanson, 1981). In an evaporite basin, however, the roughness of the lagoon floor is quickly evened out by continued precipitation, whereas the irregularity of the shoreline is maintained.

A perfectly circular basin has an shoreline factor of 1. If inflow is subparallel to wind direction, density stratification will be maintained more easily over the

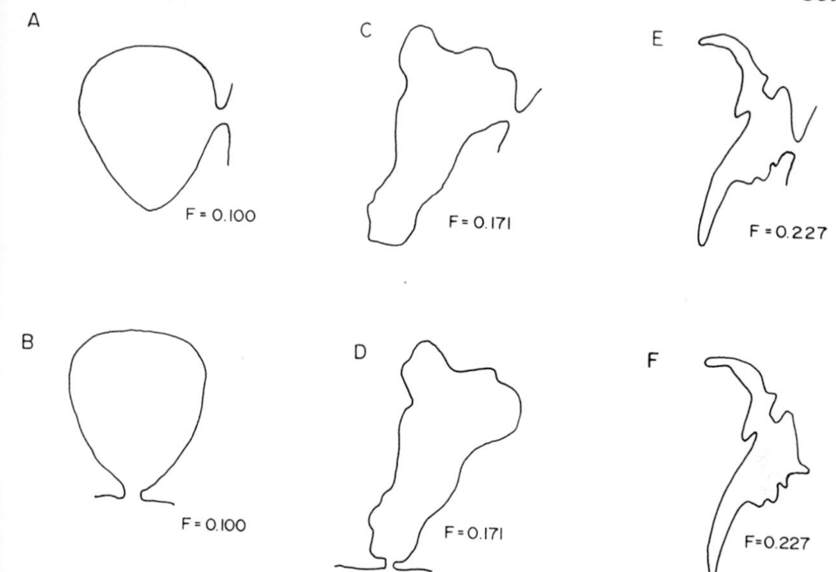

Fig. 15-2. Basin shapes. The F value is the same for inflow from one side or from one end. The degree of mixing is greater in the former (modified after Lotze, 1957).

whole brine surface than if inflow is at right angles or at 180° to the wind direction. Shoreline factors larger than 10 are rare. The higher the F value, the lower usually is the fetch of the wind and the shallower is the wave base. An example of a basin with a low F value is the Michigan Basin (Fig. 15-3). In contrast, lagoons with high F values do not allow passing winds to saturate and thus maintain high rates of evaporation.

Another measure is the shape factor (Håkanson, 1981), which is the ratio between the maximum length of the water surface over which winds and waves may act without interruption from land or islands and a line perpendicular to this length, also connecting the most distant points of the shoreline without touching land. The meaning of shape factor differs, depending on the direction of the prevailing winds.

The Entrance Strait

The entrance strait controls the water exchange. For larger evaporite basins, an open entrance channel is required. Periodic connections to the sea or seepage through porous soils can supply only small depressions with sufficient amounts of brine to keep an open water surface. However, *the inflow channel must possess an effective restriction* in order to obstruct the inflow sufficiently, and by

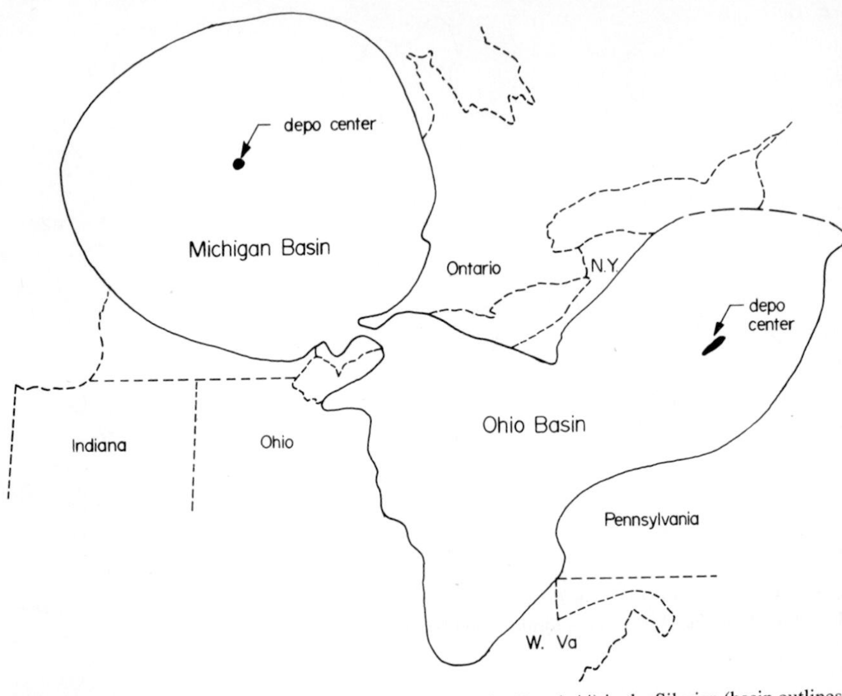

Fig. 15-3. Michigan Basin ($F = 1.05$) and Ohio Basin ($F = 1.44$) in the Silurian (basin outlines from Alling and Briggs, 1961).

so doing, to create an increased inflow velocity that will severely impede the outflow. Reefal organisms have a tendency to settle in these entrance straits of fast-flowing, nourishing waters, and they restrict the water exchange even further. The bumpier, the longer, and the more convoluted the entrance strait, the less water is allowed into the basin. If the entrance strait is located on a shallow shelf, a variety of reef builders will have settled on either side of the entrance, and reef debris will gradually accumulate. This detritus extends the entrance strait outward and thus inhibits the water exchange through it. The Permian evaporite basin of West Texas developed broad, shallow shelves of back reef limestone as the reefs prograded at a rate of 5 m/1000 years (Adams, 1963). The cross-sectional area of the inlet will have to become about eight orders of magnitude smaller than the surface area of the basin before saturation for evaporite precipitation can be reached. That is, the connecting channel to the ocean would generally be so small as to be geographically insignificant (Lucia, 1972).

The amount of inflow is controlled by the need to maintain a common sea level; inflow and outflow then have a tendency to find an equilibrium. Any increase in the water deficit or in the volume of the basin, or any further restriction of the inflow channel, reduces the outflow. For the precipitation of more soluble salts, *the inflow/outflow ratio must not reach a constant value,* as it is this ratio that controls the degree of concentration. Where this ratio stabilizes below gypsum saturation, no evaporite precipitation will occur. This is the case in all major gulfs and seas today. Where the ratio stabilizes short of halite saturation, only gypsum and gypsiferous clastics will occur. The inflow/outflow ratio can stabilize for a time even at higher concentrations, giving rise to thick monomineralic precipitates.

If, during evaporite precipitation, the basin manages to strike a balance for a while between inflowing and outflowing salt volumes without reducing the salinity of the dead water, and without significantly reducing the area covered by brine, then a corrosion plane forms basinwide. Multiple surfaces of this type within each salt basin are very common (cf., e.g., Jones and Madsen, 1968). Consequently, a way of increasing the inflow/outflow ratio is required. There are four ways of accomplishing this:

1. The rates of evaporation rise continuously. This is an unlikely scenario.
2. The entrance strait is affected by tectonic movements that further constrict the flow and reduce the cross-sectional area, and with it the hydraulic radius. This is happening in this century in the Gulf of Kara Bogaz Gol. A progressive reduction of the cross-sectional area is critical as one method of raising the basin salinity through increased drag on the outflow.
3. Continued subsidence of the basin reduces the outflow and thus raises the inflow/outflow ratio. A rapid rate of subsidence along the basin axis promotes an increase in the ratio between water-covered shelf areas and brine-filled deeps.
4. A slight eustatic lowering of sea level has little effect inside the basin, but substantially affects the water exchange through the shallow entrance strait. Such eustatic changes have many causes, especially the overall cooling of the ocean surface, and the capture of large water masses in mountain glaciations or in a lifted groundwater table in rising mountain chains. An example of the order of magnitude of sea level changes involved is a calculation of putative ocean temperature reduction. World air temperatures decline 6–8°C during glacial periods. A concomitant reduction in ocean temperatures that affects mainly surface waters is to be expected. An overall reduction in the temperature of seawater in the world oceans by only 1°C results in a lowering of sea level in excess of 125 m, based on a seawater volume of 1303.51×10^6 km^3, a surface area of 323.78×10^6 km^2, and a volume change of 3.1524^{-5} m^3/m^3/°C. This could be achieved in 10^5 years by a reduction in solar radiation of only 6–7%.

If the cross-sectional area of the entrance strait is not altered, and if rates of evaporation do not significantly decrease, the inflow/outflow ratio increases with increasing rates of subsidence inside the basin. Thus, the Erevan Basin in Soviet Armenia contains a 30-m unit of gypsum and gypsiferous clays, and a separate evaporite cyclothem with 933 m of halite (Voronova, 1968). Similarly, two cyclothems in the Pennsylvanian Paradox Basin of Utah, which contain halite sandwiched between anhydrite, are two to three times as thick as a cyclothem containing only anhydrite (Herman and Barkell, 1957). However, once halite starts precipitating, the rate of sedimentation usually exceeds the rate of subsidence and the basin volume becomes smaller, altering the inflow/outflow ratio.

Water Supply

A perennial or nearly perennial supply of water is required, preferably but not necessarily of marine origin, in order to prevent desiccation. For every cubic meter of deposited halite, some 70–90 m^3 of seawater have to enter, evaporate, precipitate gypsum and halite, and flush excess solutes. A temporary lowering of the water level or the shutting out of additional supplies of water increases desiccation; marginal deposits then become subject to erosion (Baker, 1929). As long as the evaporite basin maintains a water level identical to that of sea level in the ocean outside the bay, a two-way water exchange will take place. The interface between inflow and outflow will dip both poleward and to the west under the influence of the vertical Coriolis parameter.

Increments in evaporitic water loss are compensated for by increasing rates of inflow. Increasing inflow velocities depress the outflow until it ceases. Further brine losses are then restricted to seepage losses. That happens just prior to potash saturation and can lead to a lowering of the brine level in the lagoon. Due to frictional effects, the inflow can no longer sustain a velocity sufficient to maintain a common sea level. This happened to the Gulf of Kara Bogaz Gol in the early part of this century despite a concurrent drop in the level of the Caspian Sea. Shortly after the cessation of any outflow through the entrance neck, the progressive appearance of sundry potash minerals around the gulf margins was recorded.

A continuous outflow or seepage is required to dispose of highly soluble components; the proportion of seepage increases with the rising density of the brine. Worldwide, all evaporite basins contain a proportion of evaporite minerals substantially different from the proportion of such compounds dissolved in seawater. Only outflow or seepage can continuously adjust the proportions in the resident brine to the proportionality of precipitates.

Surface Area

The surface area of the exposed brine must be several times the area eventually covered by precipitated chlorides. Water surfaces over reefal carbonates, other limestone banks, and gypsum shelves serve as marginal concentrators. This is what Richter-Bernburg (1957a) meant by "saturation shelves'' (Fig. 15-4). The ratio of shelf area to halite surface depends on the amount of annual water deficit of the basin. This ratio can be expressed as

$$(h + s)/h: cp/e \qquad (15\text{-}2)$$

where h is the surface area of halite deposition, s is the area of shelves and shoals, c is a constant (< 1.087) derived from water extraction in halite saturation, p is the rate of halite precipitation, and e is the rate of evaporation. A reasonable estimate of the shelf and shoal area required is about 2.5–4.0 times the halite surface. Indeed, van der Zwaan (1982) observed a repeated onlap or gradual transgression of the early Messinian sea, i.e., an enlargement of the water surface area, at the onset of evaporite deposition in the Mediterranean region.

An example of such a calculation is offered as an illustration: If we assume a brine surface temperature of 30°C and a rate of evaporation from a freshwater surface of 1 m/year, the rate of evaporation drops by slightly over 50% by the time seawater has concentrated to reach saturation for potash salts (cf. Fig. 2-9). The averaged rate of evaporation has then removed about 750 mm/year from the inflowing seawater. Since 96% of the water has to be removed (cf. Fig. 4-2) and 12.7 mm halite is precipitated from each meter of seawater reaching saturation for potash salts (cf. Fig. 4-5), 9.9 mm of halite are produced annually from 750 mm of evaporative water loss. It follows that 3.6–4.0 times the brine surface is needed for a precipitation of 36–40 mm halite per year. In other words, at a rated evaporation loss of 1 m/year, shelves and brine-covered shoals must represent at least 2.6–3.0 times the area occupied by depressions with ongoing halite precipitation. If we assume higher rates of precipitation than those determined by Tasch (1970) or Hite and Buckner (1981), the shelf must have been correspondingly larger. This is the case in solar ponds, where the aggregate water surface of several condenser pans, together with the surface of the precipitating pans, is at least an order of magnitude more extensive than the area of halite precipitation. Commonly this arrangement then gives rates of precipitation in the order of 8–15 cm/year. A change in brine surface temperature causes only a very small correction. The presence of broad, shallow shelves suggests the general absence of cliffy coasts. These shelves also increase the impact of episodic flash floods on the isotopic composition of nearshore carbonates and sulfates.

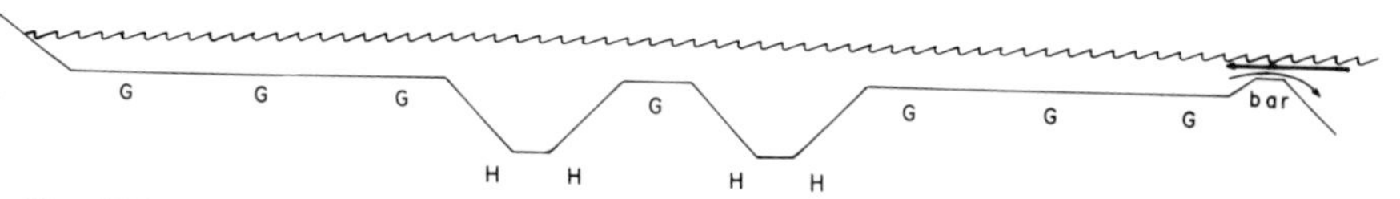

Fig. 15-4. Evaporite basin with extensive saturation shelves (G) precipitating calcium carbonates and sulfates, and smaller depressions (H) for the accumulation of chlorides.

In a series of interconnected basins, the bottom reflux or seepage attains the greatest concentration in the depression closest to the inlet to the whole set of basins. The surface inflow here has the least concentration producing the greatest density difference across the interface. Anaerobic conditions in the bottom brine, induced both by bacterial consumption and by low oxygen solubility, contrast to the positive redox potential in the surface brine. As the bacteria convert all sulfate ions to hydrogen sulfide, the electrochemical gradient aids in expelling this gas. An anion deficiency thus remains. Cations move from the positively charged surface waters into the negatively charged bottom brines, increasing the anion deficiency there (cf. Chapter 4). The total solute in the surface waters (but not the chlorinity) is reduced, causing an increase in the vapor pressure and with it a slightly increased rate of evaporation.

Brine Volume

The volume of brine to be concentrated is in reciprocal relationship to the time available during which, within certain limits of oscillation, environmental conditions remain reasonably stable. Initially very deep, i.e., voluminous basins have a very low probability of ever reaching saturation for gypsum, much less for halite or salts of even higher solubility. Otherwise, we could expect halites to form in open deep waters; this has never occurred. There are no open marine halites (Delecourt, 1946). Only because they could not visualize a rapid, synsedimentary subsidence, did Fiveg (1964) and Pannekoek (1965) advocate a deep-water model of halite precipitation.

That a deep, very voluminous basin takes more time to saturate than environmental conditions allow has been documented. During the Würm-Wisconsin glaciation (11,000–13,000 B.P.), the Red Sea was a hot, highly saline body of water; the water exchange over the sill to the Arabian Sea was severely limited by the eustatic lowering of sea level. Yet these conditions did not produce saturation for any mineral other than aragonite in the time available (Friedman, 1972). Aragonite saturation, however, is common even today in tropical ocean waters (Dietrich *et al.*, 1975). Whatever gypsum was precipitated along the shallow shores of the Late Pleistocene Red Sea was then scavenged by sulfur bacteria and turned into calcite. Gypsum from former sea margin flats or drowned shoals met with the same fate. Another example is provided by the eastern Mediterranean Sea, which in the same period suffered a significant reduction in water exchange across the Sicilo-Tunisian sill. This led to stagnant waters and sapropel production (Sigl *et al.*, 1978), and to lithified hard layers comprising low-magnesium calcilutites probably drived from bacterial decomposition of gypsum (Friedman, 1972). Rates of evaporation must have been significantly lower in the Levantine or Ionian seas than in the Red Sea farther south, yet the same effects of a dry climate occurred.

An Adequate Time Interval

A sufficient time interval is required in which tectonic forces will be active to control entrance strait configuration, but during which the tectonic and climatic rates of change remain relatively stable. An increasingly arid climate speeds up evaporite precipitation; a climate rapidly alternating between very dry and very humid periods either does not develop brine saturation or destroys evaporites that have been precipitated episodically.

Continued Subsidence

The rate of subsidence controls the type of precipitation. Epeirogenetic movements of three different orders of magnitude act on an evaporite basin (Fiege, 1940): Large- and small-scale regional movements, as well as local vertical movements, determine the morphology of the basin. When the rates of subsidence increase, inflow velocities will increase correspondingly to maintain a common sea level, provided that inflow channel dimensions and rates of evaporation remain constant. Thus, it is not important whether the inflow is continuous or tied to seasonal floods. An enlarged inflow increases drag and friction on the outflow, and as the basin continues to subside, the inflow/outflow ratio rises and the brine salinity increases even more rapidly. The more rapid the rate of subsidence, the greater is the imbalance between inflowing and outflowing solute.

Precipitation rates usually exceed subsidence rates after halite saturation; the floor of the basin evens out, and the shrinking basin volume diminishes inflow and with it friction on outflow. It counteracts somewhat the impact of the progressive restriction of the inlet. Since the coastline also changes, the ratio of the basin volume to its area undergoes a complex development. Further precipitation would cease when the basin fills up. To prevent that from happening, high rates of subsidence are apparently essential during halite precipitation to counteract the high rates of precipitation. For instance, the Oligocene salts of southeastern France are found only in strongly subsiding grabens. On platforms separating these grabens, halite is usually absent, and gypsum predominates (Truc, 1980). Derumaux (1980) showed convincingly that in the thick Permian salt sections under the North Sea, the subsidence is centered in those sections. Accumulation of salts in the Permian Zechstein (Figs. 15-5 and 15-6) or the Tertiary north Carpathian basins is also restricted to the parts of the basin with the most intensely subsiding floor. Ivanov and Levitskiy (1960) evaluated all of the then-known potash deposits in the USSR and observed that they are found only where evaporite basins have been subjected to intensive subsidence during salt deposition. Busson (1978) also included the Upper Miocene (Messinian) evaporites of the Mediterranean region in this scenario. With few exceptions, one now finds

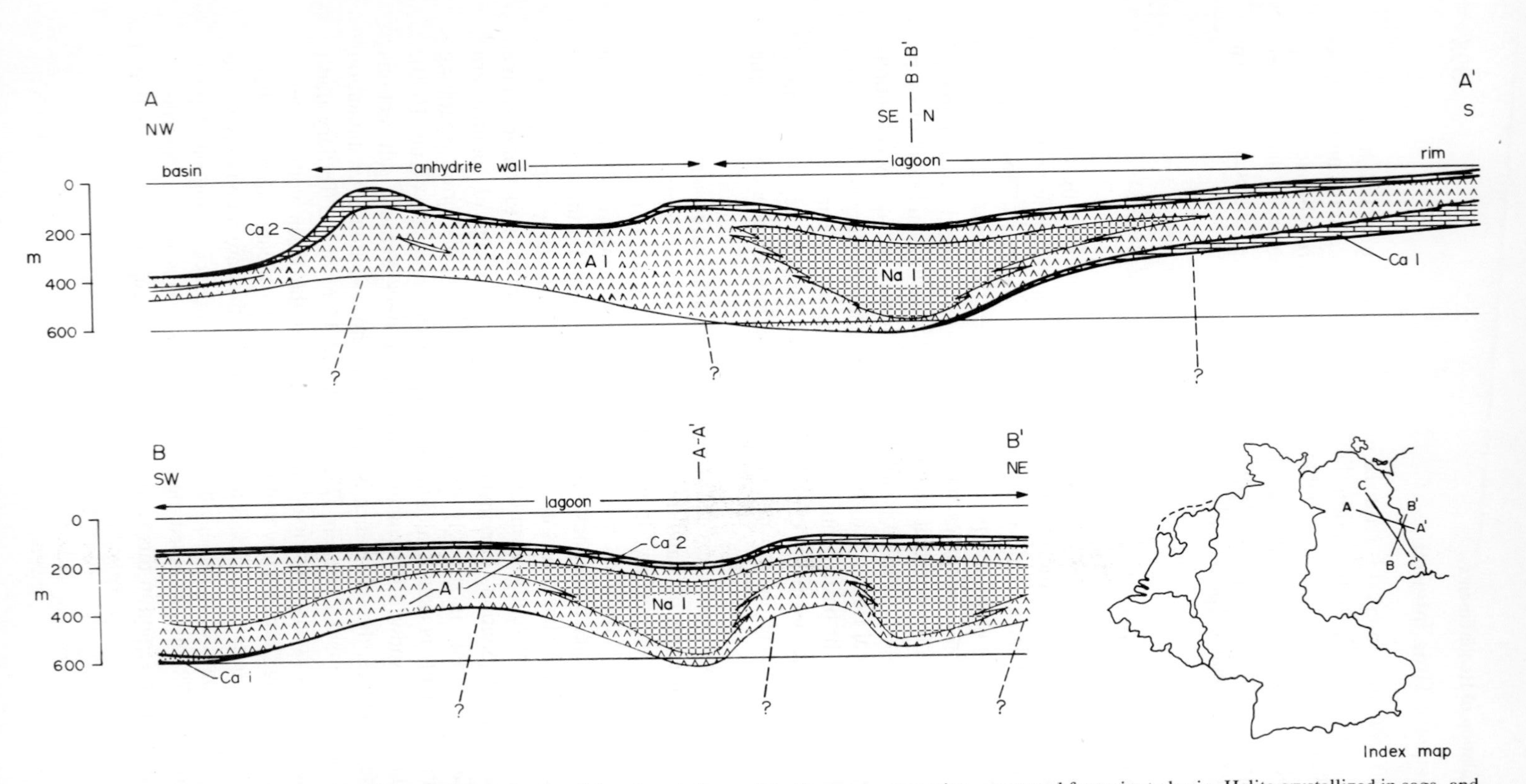

Fig. 15-5. Cross-sections A-A′ and B-B′ through the first (Werra) cyclothem of the Zechstein evaporites, westward from rim to basin. Halite crystallized in sags, and anhydrite precipitated around sills that may have been fault controlled. Anhydrite thins wherever halite thickens. Section C-C′ is shown in Fig. 15-6 (after Rockel and Ziegenhardt, 1979).

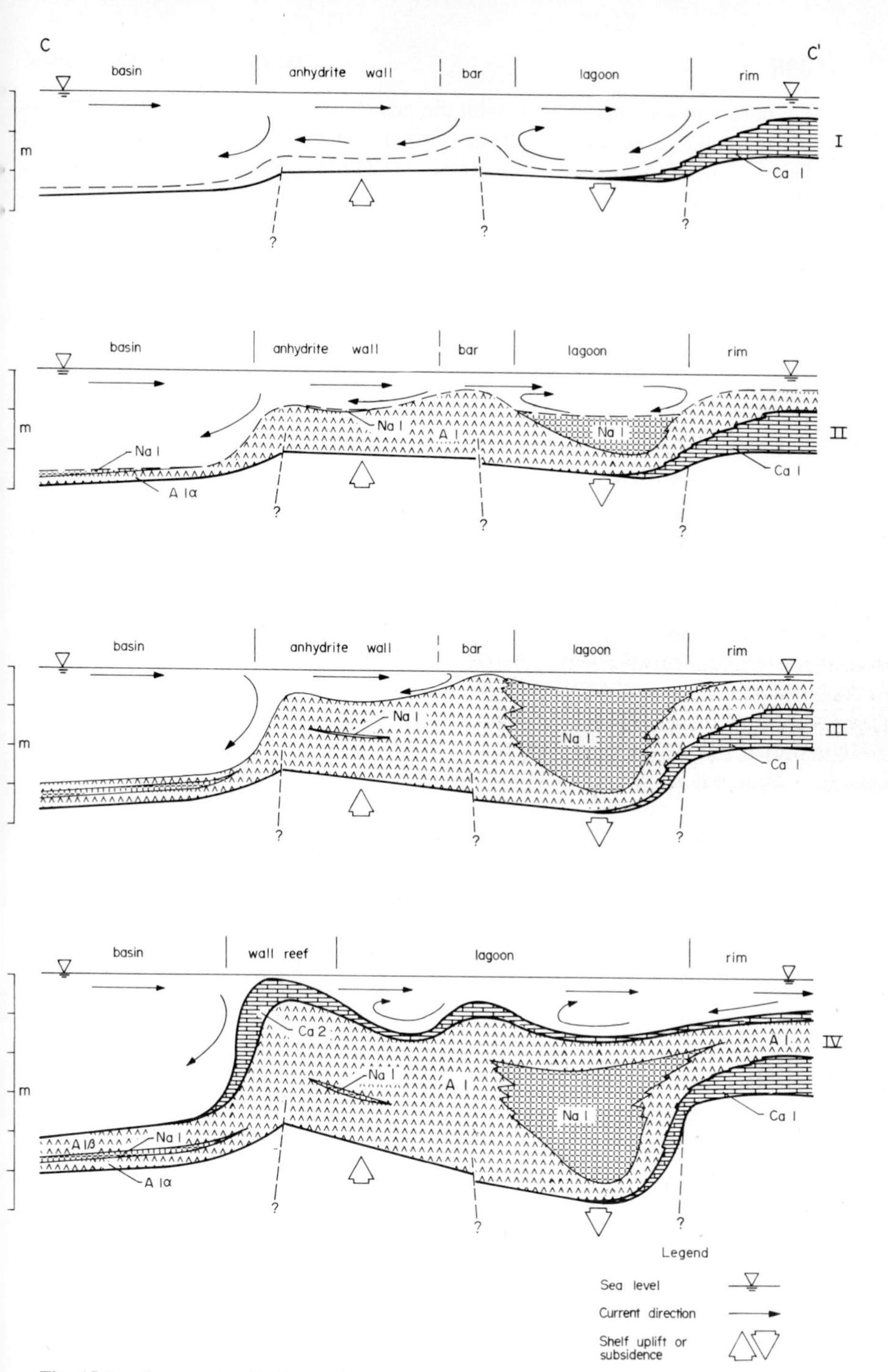

Fig. 15-6. Cross section C-C′ through the first (Werra) cyclothem of the Zechstein evaporites, westward from rim to basin (location indicated in Fig. 15-5). Slower subsidence of the lagoon entrance allows the building of an anhydrite wall capped eventually by an entrance reef. Halite precipitated in downthrown fault blocks (after Rockel and Ziegenhardt, 1979).

the thick salt sections in areas with the greatest rates of subsidence, in grabens and depressions of abyssal plains. Tripoli and gypsum occur in intermediate basins, whereas marine algal limestones are found on the very margins in Spain, Tunisia, and the western Sinai.

The basal anhydrite of the Khorat evaporites in Thailand covered a surface of low relief, but further subsidence during halite deposition created a 50-m relief on the marginal anhydrite (Hite and Japakasetr, 1979). The distribution of ephemeral halite in Lake Eyre, South Australia, suggests such an instability of the lake bed. During several years of observation, a marked uphill transfer of salt occurred (Dulhunty, 1974). Clark (1980a) made the point that isostatic adjustments during the deposition of gypsum and carbonate rims of an evaporite basin alter the thickness ratio between shelf and basin. If placed on a datum beneath the evaporites, the shelf slope appears to be steepened into a wall that may have been only a gentle slope during deposition. The difference is on the order of 750 kPa/100 m in favor of the shelf deposits. The sag may be caused by an isostatic depression under the load of prograding carbonates and anhydrites (Smith, 1981). Continued subsidence as a gentle downwarp would affect surrounding low-lying shores and allow the sea to transgress onto the tidal flats. This increases the area subject to evaporation without changing the configuration of the entrance strait, and thus increases the rates of inflow. Many halite occurrences show such a transgressive nature. One example is the expansion of the Permian Castile Sea in Texas into the much larger Salado Sea (Fig. 15-7).

Salt pans cannot accumulate substantial thicknesses of salt unless local tectonic movements involve the salt pan as distinct from the supply basin (Grabau, 1920). Layered halite in depressions of supratidal flats off the mouth of the Colorado River, California, occur only where strong linear trends suggest fault control and where downward displacement of various facies by several meters can be traced into the subsurface (Shearman, 1970). A synsedimentary epeirogenic subsidence is thus required (Zaenker, 1979), and the frequently observed linear outline of evaporite basins is presumed to be due to normal faults. If that is so, it is evidence of relatively rapid bursts of vertical movement. The subsidence occurs almost entirely along vertical block movements. Synsedimentary basinward tilting of the beds then provides for thickening of the precipitates in that direction. Salt basins are thus restricted to areas of crustal tension. Valyashko (1951) had earlier observed that potash salts, in particular, are preferentially located in such faulted basins. The tectonic instability is frequently controlled by normal faults that can and do act as conduits for the basaltic lavas so often associated with thick evaporite sequences. The continuity of facies over wide areas without any significant deformation corroborates such vertical subsidence of blocks of great areal extent.

Many evaporite basins, indeed, seem to be fault controlled, such as the Tertiary evaporites of the Upper Rhine Valley (Rozsa, 1920) and the Cretaceous

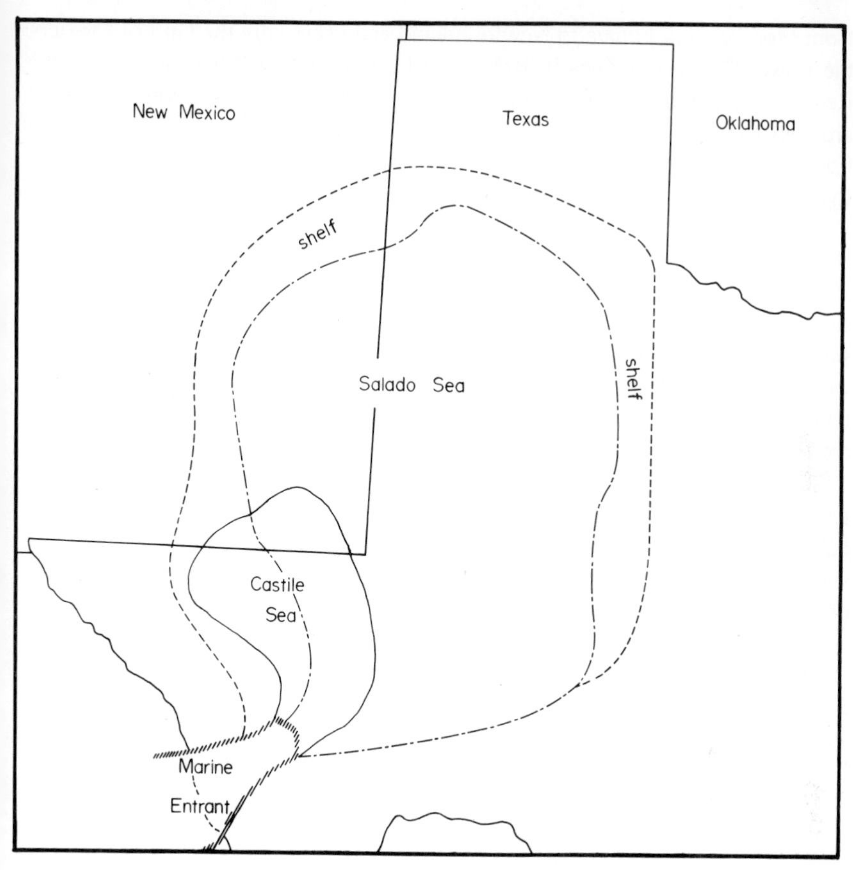

Fig. 15-7. Areal extent of Permian Castile and Salado seas (after Hills, 1942).

evaporites of Gabon and Brazil (Borchert, 1977). The Bukhara fault in western Uzbekistan revived at the beginning of Upper Jurassic evaporite precipitation with multiple cycles (Ismatullayev *et al.*, 1974). The Upper Devonian evaporites of the Dnieper-Donets depression were deposited in a large, highly differentiated, shallow basin (Galabuda, 1978). Abandoned arms of aulacogens and grabens due to major wrench (strike–slip) fault systems are prime localities for the generation of a salt basin, provided they are able to produce an entrance gateway, a narrow neck preferentially populated by a reef chain. Such fault control would indicate rapid subsidence. It is likely that the basins in West Texas belong into this category.

On a larger scale, the incipient separation of two continental slabs leads to rapid subsidence of a central trough: The Juro-Triassic evaporites that extend from Morocco and France to Newfoundland and eventually the Gulf of Mexico; the Lower Cretaceous ones in Brazil and Gabon; and the Miocene ones in the Red Sea are examples of that type. In the same vein, foredeeps to orogenic belts are also zones of rapid subsidence. The evaporite basins on either side of the Carpathian arc on the west side of the Ural Mountains and on the north side of the Pyrenees are excellent examples.

The limit of subsidence seems to lie at a depth of about 4–5 km, close to the chord of the curvature of the earth, as measured across the basin. In the example shown in Fig. 15-21, the depth of the basin—over 800 m—could not have subsided more than about another 180 m without first widening the area of subsidence. Such widening would appear as a transgression of the shelf area and would lead to further marginal faulting. Rouchy (1981) found such distension faults in the late stages of evaporite basin development in backreef areas around the rim of Upper Miocene (Messinian) evaporites (Fig. 15-8).

Short-term rates of subsidence of 5–6 m/1000 years are not unusual in collapsing basins (Sonnenfeld, 1974, 1975; Angelier, 1979). Flemming (1978) found that isostatic or tectonic causes can maintain a rate of 2–10 m/1000 years, whereas a rate of 2–4 mm/year is typical only for stable areas. Higher rates prevail in mobile areas, reaching as high as 15–30 mm/year in the Alps and Caucasus. Along the east coast of North America, Passamaquoddy Bay, Maine, is subsiding at a rate of 9 m/1000 years (Simon, 1981), whereas in Chesapeake Bay the mean sedimentation rates are currently 2–4 m/1000 years (Schubel, 1968) and in several basins off California are 2–4 m/1000 years (Lasaga and Holland, 1976).

These are examples of rapidly subsiding basins with a matching accumulation of clastic sediments. Without such massive input of clastics, they would have become starved basins of progressively greater water depths, unless chemical precipitation covered the shortfall in sedimentation. The early Pleistocene salts in the Dead Sea accumulated in a depression that subsided 4–5 km during the Pleistocene alone (Zak and Bentor, 1972), or 2.2–2.8 m/1000 years. The gypsum-bearing salt marsh of Sečovlje in Yugoslavia displays a rate of sediment accumulation and a rate of subsidence of 3 m/1000 years (Ogorelec *et al.*, 1981). A modern extreme example has been provided by Dzens-Litovskiy and Vasil'ev (1962): The Gulf of Kara Bogaz Gol off the Caspian Sea had a mean water depth of 10 m in 1939, which was reduced to less than 3.5 m by 1956. In the same period of time, up to 38 m of salts were precipitated. That would require a rate of subsidence of about 2.6 m/year, or 2.6 km/1000 years, during episodes of salt precipitation. Rates of subsidence of such magnitude have not been documented in any other basin. Although isostatic loading might be responsible for some of the subsidence, other tectonic causes have to be invoked in this instance. It is obvious that such extreme rates of subsidence or sedimentation can be main-

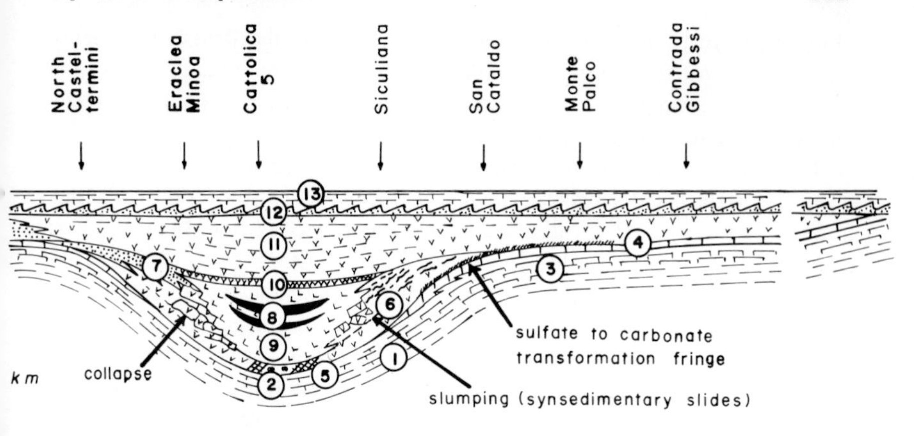

1. Tortonian to Messinian marls
2. White pre-evaporitic marls
3. Diatomites and diatomaceous laminites
4. Basal limestone
5. Anhydrite and anhydrite-marl breccias
6. Lower gypsum-anhydrite member
7. Gypsarenites
8. K- and Mg-salt intercalations
9. Halite
10. Roof anhydrite and gypsum
11. Upper gypsum-anhydrite member
12. Post-evaporitic Arrenarola Formation
13. Pliocene Trubi marls

Fig. 15-8. Onlap of the Upper Anhydrite Member onto an earlier cyclothem in the Upper Miocene Messinian Caltanissetta Basin, Sicily. Bacterial transformation of gypsum into carbonate is typical of basin margins. Distension faults and slumping mark ancient hinge lines (after Rouchy, 1982).

tained for only the briefest geologic time intervals, although they can occur repeatedly. Such extremes are, however, not really necessary. Moderately rapidly subsiding basins are much more favored if their water depth does not exceed a few tens to a few hundreds of meters at a time. The interfingering facies at basin margins suggests that even the thickest evaporite sequences did not form in extremely deep waters. As rapid as such rates of subsidence may seem to be, they are one order of magnitude slower than halite precipitation.

Intervals of rapid subsidence are evidently followed by intervals of relative stillstands. The evaporite-bearing Sverdrup Basin in arctic Canada reached long-term average rates of 8–11 cm/1000 years several times in its history (Sweeney, 1976, 1977). In reality, these are episodes of rapid subsidence averaged with periods of very slow subsidence or even temporary slight uplift. The estimate of

8 cm/1000 years of subsidence for the marl slate in the British Zechstein sequence (Magaritz and Turner, 1982) is of the same order of magnitude. Although it is a typical estimate for carbonate sedimentation rates, it is obviously too low for the rapid subsidence that occurs during halite precipitation.

The uniform thickness of carbonate banks sandwiched between lens-shaped evaporite sequences provides evidence in many basins for periods of reduced rates of subsidence. Thin carbonate banks can be traced from Michigan through Ohio and Pennsylvania into New York State, a distance of 1000 km, with minimum thickness variations (Rickard, 1970). That indicates not only a lack of differential subsidence during carbonate sedimentation but also a contemporaneity of underlying evaporites in the Michigan, Ohio, and Alleghany basins. The inflow/outflow ratio increased and then found a new temporary equilibrium. Since carbonate intercalations extend over the whole Zechstein Basin and the whole Elk Point Basin (Klingspor, 1969), the precipitation of carbonates must have started everywhere in the basin at the same time. There was thus no significant horizontal salinity gradient in operation.

Carbonate banks of this type are frequently reefal. There would be no reason for a biostrome to restrict its vertical growth in a basin deepening toward the center, unless the limiting factor had really been sea level. Thus, the reefs prograded into the basin, not a difficult task during a stillstand in subsidence. The near parallelity between top and bottom surfaces suggests very shallow conditions. Photosynthesizers associated with these carbonate banks would have found it difficult to survive in the deeper parts of the basin.

Instead of depressing only the flanks of one basin, continued subsidence can also affect the shelf areas of that basin. These shelf areas then turn into subsidiary lagoons and create a series of preconcentrator basins for moving brines. An example of such gradual spread of subsidence is provided by the Devonian Elk Point Basin in Alberta, Canada. Initially only a very small basin, it gained areal extent by a collapse of the shelves along putative fault planes (Figs. 15-9 and 15-11). From Figs. 15-5 through 15-11, a pattern seems to be emerging: An initial small basin with broad shelves precipitates the first cycle of chlorides. Then the former shelf areas collapse and join the initial basin as chloride-precipitating depressions separated by subsidiary sills (Fig. 15-10). This pattern seems to apply to the Devonian Elk Point Basin, to the Permian basins of West Texas and northern Europe, and to the Pechora–Upper Kama–Caspian Lowland series of Permian basins. It is likewise indicated in the Caltanissetta subbasin of the Upper Miocene Messinian Mediterranean evaporites. On a larger scale, only one salt phase is known in Balearic, Tyrrhenian, Ionian, or Aegean occurrences, but two distinct salt phases separated by a thick carbonate–sulfate sequence appear to be present in the Levantine Sea (Finetti and Morelli, 1973). It is uncertain whether the lower salt cycle is Upper or Middle Miocene in age; in the Red Sea and Suez area, thick Middle Miocene, pre-Messinian evaporites occur.

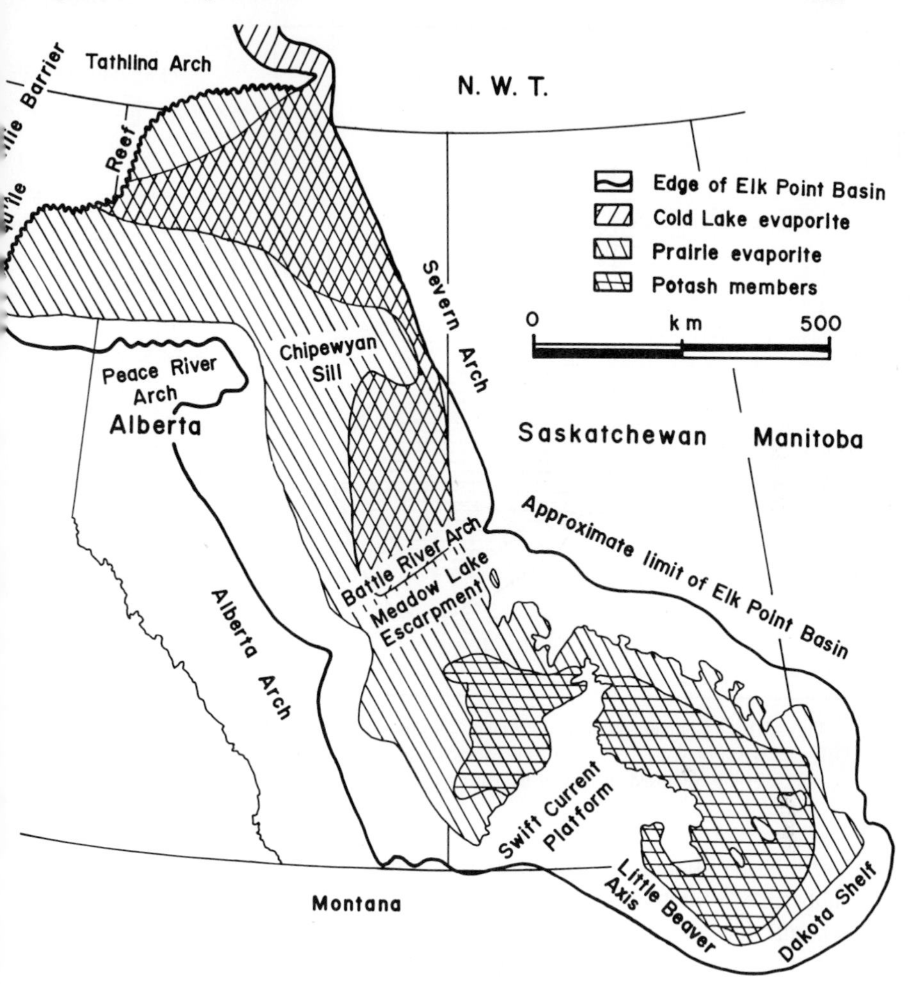

Fig. 15-9. The Middle Devonian Elk Point Basin of western Canada. Outline of Cold Lake (Lower Elk Point) evaporites and more extensive Prairie (Upper Elk Point) evaporites. Fringing carbonates and basin margin clastics are not shown. The Cold Lake evaporites abut in the northwest against the Rainbow reef chain proximal to the inflow from open seas. Reefs were also scattered throughout the distal parts in the southeastern portion of the basin. Exposure of the Swift Current Platform and formation of the irregular northeastern rim of the evaporites are due to later salt solution by percolating waters of meteoric origin. Potash members of the Prairie Evaporite occur only in the distal part of the basin against northeastern shoaling.

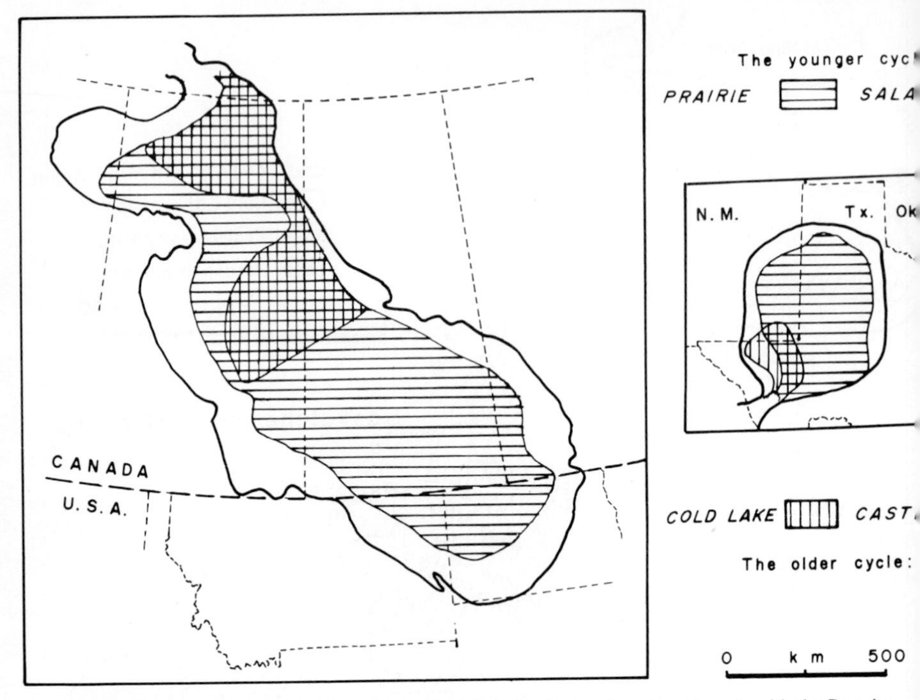

Fig. 15-10. Comparison of the Devonian Elk Point Basin of western Canada with the Permian Delaware Basin of West Texas and New Mexico using the same scale. In both cases, a later cycle extends over the collapsed shelves of the earlier cycle.

The Silurian Michigan Basin must be seen in this regard as merely the proximal subbasin of a Michigan–Ohio–Alleghany Basin system, since the first halite (the A1 salt) does not extend beyond it.

Subsidence takes place both inside a basin and along its margins, and depresses gradually wider areas. Compensation by precipitation leads to ostensibly transgressive salt sequences (Fig. 15-12)—in other words, an onlap sequence. In many salt basins, halite thus seems to mark a transgressive phase. Nowhere is this better documented than in the Cretaceous Cuanza Basin of Angola, where the halite precipitation onlaps step by step onto an ancient alluvial plain, slowly erasing the existing relief (Brognon and Verrier, 1966). Likewise, the relief on the Zechstein floor that allowed thick evaporite sequences to be deposited was

---

Fig. 15-11. Middle Devonian formations in the Elk Point Basin, Alberta, Canada. Early subsidence involved only a small depression in central Alberta. Further subsidence created a series of younger evaporite basins on former shelf areas flanking the basin on either side (north is to the left of the section) (after Pearson, 1963).

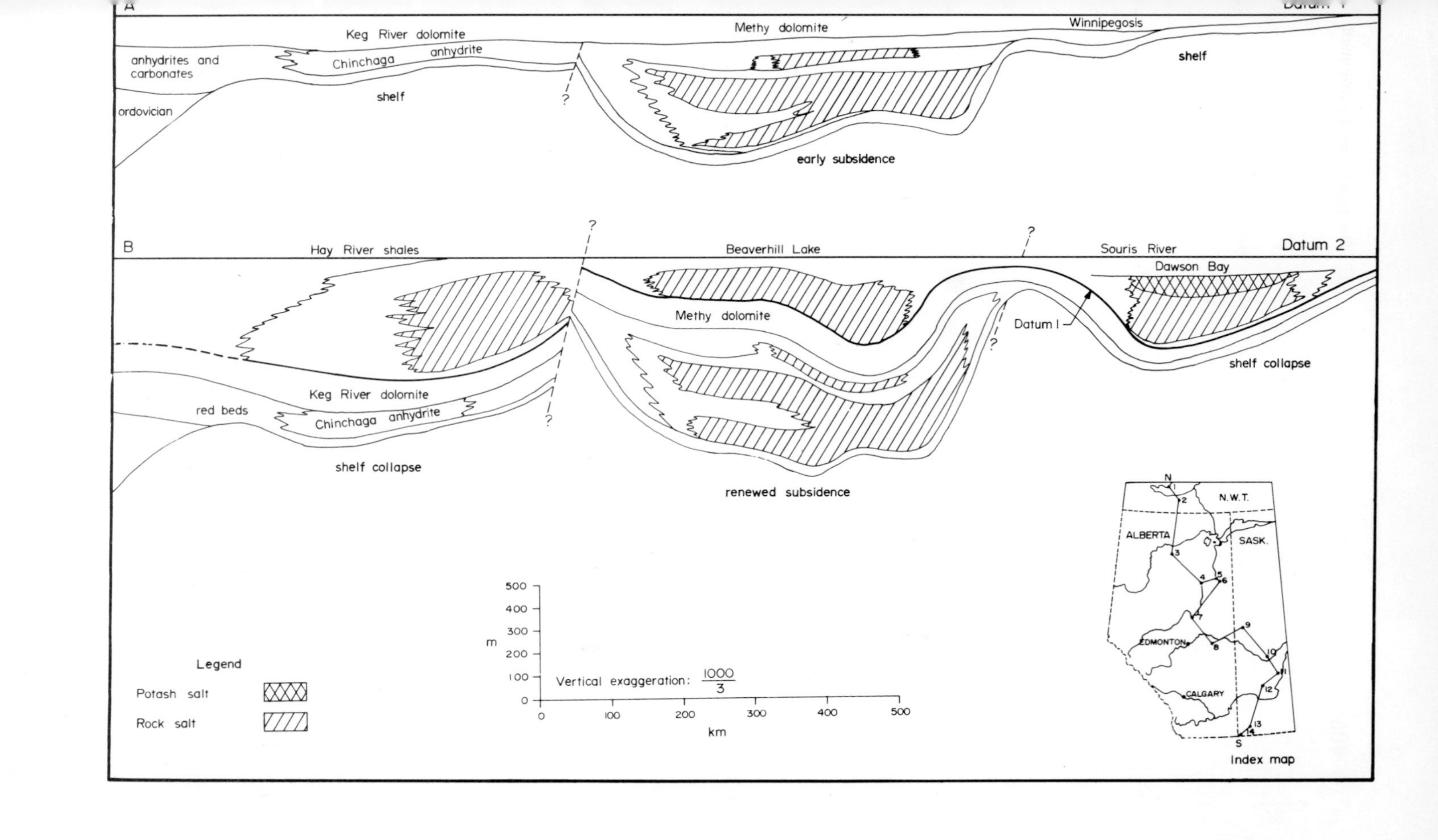
A
Keg River dolomite
Methy dolomite
Winnipegosis
anhydrites and carbonates
Chinchaga anhydrite
shelf
ordovician
shelf
early subsidence
B
Hay River shales
Beaverhill Lake
Souris River
Datum 2
Dawson Bay
Methy dolomite
Datum I
shelf collapse
Keg River dolomite
red beds
Chinchaga anhydrite
shelf collapse
renewed subsidence
N
N.W.T.
ALBERTA
SASK.
EDMONTON
CALGARY
S
Index map
m
500
400
300
200
100
0
Vertical exaggeration: 1000/3
0
100
200
300
400
500
km
Legend
Potash salt
Rock salt

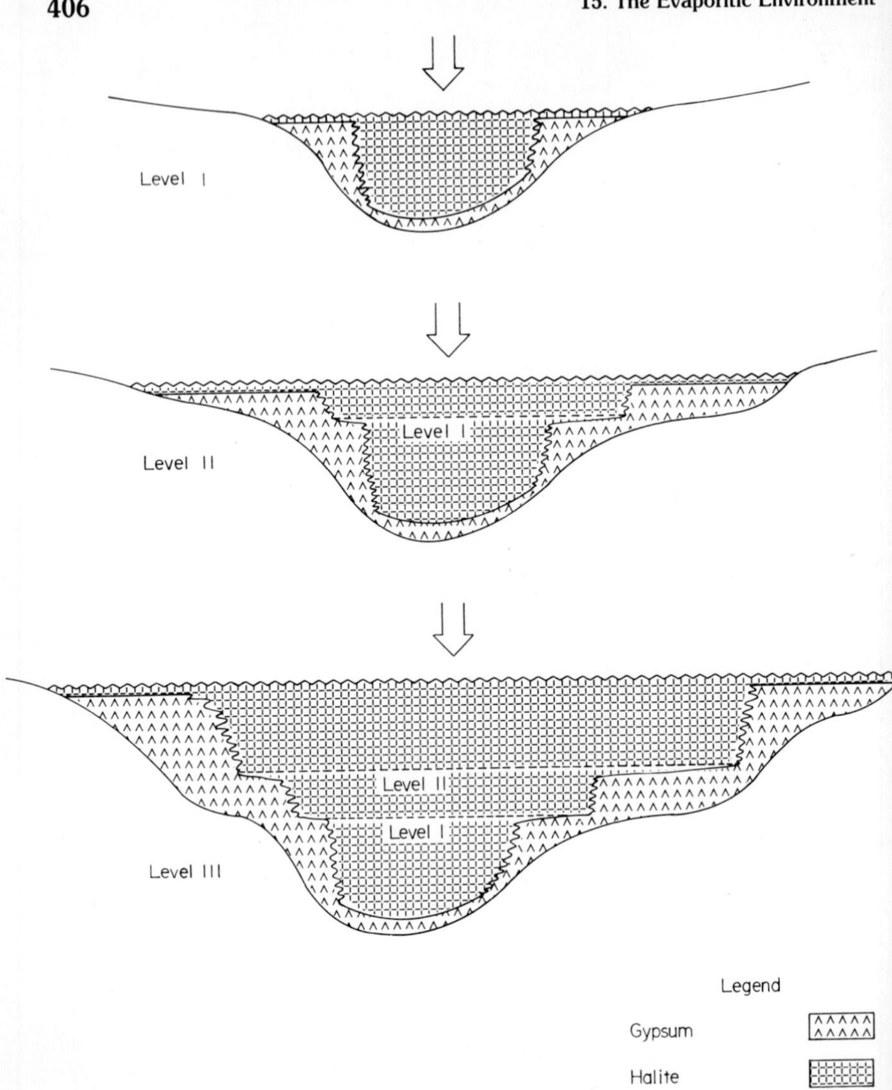

Fig. 15-12. Transgression simulated by subsidence.

the result of synsedimentary sagging. It was not a preexisting deep depression. In other words, it is a normal transgressive sequence (Jung and Knitzschke, 1976). Regressive tendencies (offlaps) are locally apparent only near the "Rote Faeule" margins, where the sills were subject to intermittent and slight uplift. By the time the Stassfurt salt of the Zechstein evaporite sequence started precipitating, the sea floor had developed dips of 2–3°, allowing salt thicknesses to increase about

10 times over a distance of 15–20 km (Simon, 1967). The transgression refers to the onlap of the sediment, but it need not represent a phase of transgression of the open sea outside the entrance strait. Prior to halite precipitation, rates of subsidence often exceed rates of sedimentation. During halite precipitation, the rates of subsidence may continue at the same rate, yet will lag behind the rates of sedimentation; the basin shallows and eventually may fill up.

We are here concerned not so much with absolute rates of subsidence as with the differential settling between the entrance sill and the center of the basin. The greater the difference in favor of the basin, the less outflow is permitted, and thus the faster the concentration rises. Intermediate sills and basin rims subside at a lesser rate than depressions. The difference in the rate of subsidence between gypsum-precipitating marginal shelves and halite-precipitating basin centers has frequently been on the order of 1:5 to 1:7. Eventually, however, the rate of precipitation catches up with the rate of subsidence or exceeds it, the water body shrinks, parts of the basin fill up, and the surface dries out. The evaporite sequence is terminated by an unconformity, a dissolution surface, or at least a desiccation horizon (Zaenker, 1979). An apparently regressive series results even in the face of a gradual increase in the rates of subsidence.

Dissolution and desiccation surfaces are telltale signs of slowing rates of precipitate accumulation. In the Pennsylvanian Paradox Basin of Utah, each evaporite cyclothem ends with a potash bed, which has a sharp contact with overlying anhydrite. Peterson and Hite (1969) interpreted these contacts as dissolution surfaces or disconformities along which some halite and probably also some potash had been removed. Such dissolution surfaces with alternations of carnallite to sylvite or reprecipitation of carnallite in low-lying pods, washed in by rainstorms, are known from Devonian potash horizons in Saskatchewan (Baar, 1974) and from the Permian Fore-Ural (Strakhov, 1962). Corrosion surfaces in the Permian Salado Formation of New Mexico are laterally persistent, converge northward, and truncate underlying beds as they approach the distal shores (Jones, 1973). Hopper crystals truncated by dolomite–anhydrite laminae in the Silurian Salina salt of the Michigan Basin (Dellwig, 1955) or chevron grains truncated by clay layers in the Devonian Elk Point Basin (Wardlaw and Schwerdtner, 1966) record such dissolution surfaces. The potash beds in the first megacycle of the Michigan Basin evaporites are truncated, and a corrosion surface evidently extends over most of the surface of the first evaporite cyclothem (Matthews and Egleson, 1974).

As subsidence continues, the brine will reenter, oozing out at first from the porous substrate in response to a rising piezometric surface. Flow over the entrance sill follows in the form of a transgression. A small body of water near the entrance, which may never have dried out and may not have lost contact with the open sea, now merely expands its areal extent. The result is a small unconformity between older and younger evaporites over most of the outlying parts of the basin. The mixing in of incoming seawater dilutes rising interstitial brines below potash saturation and causes further halite precipitation. Flooding by subsurface or oceanic waters delivers clean waters and produces an additional evaporite sequence without significant terrigenous admixture.

If we take the Devonian Elk Point Basin of western Canada as an example, the subsidence seems to migrate towards the distal part of the evaporite basin from one cyclothem to the other. When evaporite precipitation ceased either because of climatic changes or because of an alteration in the configuration of the en-

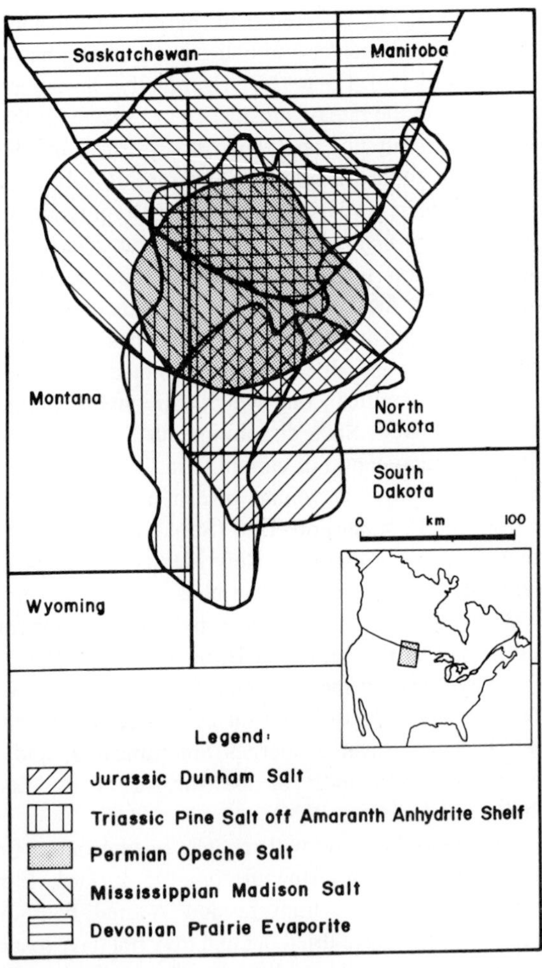

Fig. 15-13. The migration and onlap of younger salt basins onto the distal margin of the Devonian Prairie Evaporite (data from Johnson and Gonzales, 1978).

trance strait, subsidence was not arrested. The center of subsidence continued to migrate and eventually crossed over the distal margin of the Elk Point Basin. Renewed evaporite precipitation followed as soon as climatic conditions or entrance restrictions again became auspicious. Younger evaporite basins cluster around the distal rim of the Elk Point Basin (Fig. 15-13) with only partial overlap over each other. All of these basins seem to have been marginal bays of a sea to the west, rather than being fed from an entrance in the very distant northwest.

DEPTH ESTIMATES

Individual estimates of water depth vary widely for each evaporite basin, with estimates derived from potash deposits generally much lower than those from halite occurrences. Many fruitless arguments have been started between proponents of the same model because of a failure to first define the limits of the terms "shallow" and "deep," with "deep" meaning below the wave base of a very viscous bittern (less than 10 m) or below more substantial brine columns of lesser density. Therefore, the meaning of these two terms will first have to be evaluated.

Cognizant of the vertical instability of evaporite basins and the resulting variable depths, Fabricius *et al.* (1978) coined the term "dynamic model of shallow basins." The best definition of such a basin has been offered by Davies and Ludlam (1973), who called it, "a basin with water depth great enough relative to water transparency, wave fetch, and other hydrographic parameters, to permit development of stable, chemically controlled density stratification where photosynthesis below the chemocline was slight or non-existent." The basin has to be barred (Woolnough, 1937) in order to maintain anaerobic conditions in bottom brines. Purely on sedimentological grounds, Vai and Ricci-Lucchi (1978) found it convenient to place the boundary between "shallow" and "deep" water at the wave base, since the terms "deep water" and "quiet water" are being confused. Such a basin is subject to high rates of subsidence that are compensated for in part by high rates of sediment accumulation. During preevaporitic carbonate or clastic sedimentation, and during the early evaporitic phase of gypsum accumulation, the rate of subsidence would have exceeded the rate of sediment accumulation. The rapid precipitation of chlorides, however, would be able to catch up and cause a shallowing of the basin.

Based on a study of fluid inclusions, Petrichenko *et al.* (1973a,b) suggested a water depth of 3–4 mm during sylvinite accumulation in the Solotvinsk depression in the Carpathian Mountains. Similarly, Jung and Lorenz (1968) argued for a very shallow basin from an annual rate of 5–10 mm of compacted rock salt accumulation in the Stassfurt Member of the German Zechstein evaporites. Only a small volume of saturated brine, with a depth of 50–150 cm, would yield such a small annual precipitate. Valyashko (1962) evaluated evaporation experiments at Lake Inder in the Caspian depression, which contains 61.5 kg potassium chloride per cubic meter of brine. If that brine is cooled on average to 0°C, it yields 1 cm sylvite per meter of brine depth. If it is not cooled to such a low temperature, the sylvite yield is considerably reduced.

The Salado evaporites of the Ochoan series in the Permian Basin of West Texas were estimated to have precipitated in a sea less than 50 m deep, probably only 5–10 m (Adams, 1969), shallowing to 2–10 m in many parts, since influx repeatedly could dissolve evaporite rock (Adams, 1970). Each flooding of fresh water diluted the total water column, leaving multiple clay seams as corrosion surfaces. Consequently, the volume of water to be diluted could not have been very large (Adams, 1969). A frequent dissolution of potash beds has also been observed in Devonian Prairie Evaporite beds in Saskatchewan (Baar, 1974).

In the Pennsylvanian Paradox Basin of Utah, Hite (1970) correlated each evaporite cyclothem across the salt basin into an equivalent cycle within the carbonate facies on the southwestern shelf of that basin. This gave him a depth of no more than 20+ m from the onset of halite precipitation to subaerial salt exposure. Low-energy oscillation ripples in resedimented Upper Miocene gypsum crusts in the northern Appennines led Vai and Ricci-Lucchi (1978) to estimate a water depth of only a few meters and no more than a few tens of meters. Arkhangel'skaya and Grigor'yev (1960) postulated a depth of no more than 30–50 m for the Lower Cambrian salt basin along the Lena River of eastern Siberia. Eroshina and Kislik (1980) likewise estimated a moderate depth of less than 30 m up to 40 m for the Upper Frasnian evaporite basin in the Pripyat area of Byelorussia. On the basis of algal structures, interbedded oolites, traces of raindrops, tracks of terrestrial animals, desiccation features, and intraformational water erosion, Ivanov (1967) argued for very shallow deposition of the Permian potash deposits west of the Ural Mountains. The same argument was offered by Amieux (1980) for Lower Oligocene evaporites in southeastern France. He estimated water depths to have been there in the centimeter to meter range on the basis of gypsarenite levels, desiccation cracks, and erosion of white gypsiferous dolomite.

In contrast to the very shallow Salado Sea, Adams (1969) estimated the depth of the Permian Castile Sea that predated the Salado Sea but covered a much smaller area to be 150–190 m. Previous depth estimates ranged from 150 to 700 m, calculated from the present-day relief between the top of the Capitan reef and the base of the Castile Formation (Anderson *et al.*, 1972). On the basis of variations in the bromide content, Holser (1966a) calculated the depth of the Permian Castile Sea to have been about 80 m; the Permian Wellington Sea in Kansas, 110 m; the Silurian Salina Sea in Michigan, 50 m; the Middle Devonian Elk Point Sea in Saskatchewan, 100–250 m during precipitation of Prairie Evaporites; and the Zechstein Sea in Germany, 20–210 m while precipitating the Stassfurt Member.

Concurrent with precipitation was continued subsidence. Pompeckij (1914) and Schuchert (1915) estimated a water depth of less than 200 m but more than 100 m for the German Kupferschiefer marls at the base of the Zechstein evaporites, because they thought that high waves would replenish oxygen in shallower waters and terminate reducing conditions. Fiege (1939) noted that Kupferschiefer sediments indicate a very shallow Zechstein basin. He believed that several cycles of subsidence then occurred, akin to the evaporite cycles in the Alps, which are also centered in rapidly subsiding areas. Rentsch (1964) postulated a depth of 50–100 m as the maximum depth of the sea during deposition of this unit. Desiccation cracks in the basal Kupferschiefer series, and the presence of oxidized barren zones separating ore zones, were used by Oelsner (1959) to deduce a water depth less than 100 m. Swayed by arguments in favor of water depth estimates in excess of 500 m (Borchert and Muir, 1964), Wedepohl (1964) raised his original maximum depth estimate to 300 m. Smith (1981) argued in favor of the Zechstein Sea flooding a depression initially 200–250 m deep. However, Kupferschiefer thicknesses average 22 cm and vary from only 2 to 50 cm, and the overlying Zechstein limestone Ca-1 averages 3.3–10.5 m (Rockel and Ziegenhardt, 1979). These figures do not suggest any significant relief on the floor of the Zechstein sea prior to the onset of evaporite precipitation.

Reefs grew along the margins of the Silurian Sea in the Michigan Basin, which contain faunas precluding a water depth greater than 10–17 m (Jodry, 1969). Several times these reefs were exposed and experienced karstification of their tops. Matthews and Egleson (1974) espoused a "deep" water origin of the same evaporites, assigning a 400-m depth to the lowermost of the evaporite units, whereas stratigraphically higher evaporite horizons were deposited in a progressively shallower sea. However, the aggregate maximum thickness of the basal halite, potash layer, and cover salt does not exceed 200 m. Briggs *et al.* (1980) assume that the reefs did not grow during evaporite precipitation, but in their full present height predate evaporative drawdown of the brine in a basin 180 m deep at its margins and 300 m in its center. After a drop in sea level exposed the sill level and initiated evaporite deposition, depths ranged from 46 m at the margins to 240 m in the center, decreasing there to about 30 m during potash deposition. Water flowed in over a sill 90 km wide and 72–107 m deep. At comparable flow velocities, 2.5 times more water would have flowed annually into the Silurian Michigan Basin than is flowing presently into the much larger Mediterranean Sea, a sea of roughly 10 times the surface area and 70 times the volume. Evaporation losses per unit area would have had to be 25 times greater in the Silurian Michigan Basin than in the present Mediterranean Sea, and that is unrealistic.

In contrast, a minimum water depth of up to 400 m in Carboniferous evaporites of the Canadian Arctic Archipelago was postulated on the basis of depositional relief (Davies, 1977a,b). Limestone beds toe out here as interbeds between some anhydrite units at the base of an apparently depositional slope. Neev (1979) likewise spoke of "deep-water" deposition of Miocene gypsum in the coastal plain of Israel, meaning a depth of about 350 m. He assumed that well-aerated, oxygen-saturated waters overturned and thereby prevented the activity of sulfate-reducing anaerobic bacteria. Nesteroff (1973a) thought of a water depth on the order of 100–500 m for Upper Miocene (Messinian) evaporites of the Mediterranean region. Fabricius *et al.* (1978) concurred by assigning a water depth of up to 500 m to the same sequence. Based on isostatic calculations but neglecting density differences in precipitating brines and the effects of concurrent subsidence of a basin, Fabbri and Curzi (1979) arrived at similar figures. However, Ogniben (1957) came to the conclusion that Messinian carbonates of Italy were deposited in some tens of meters of water, basal limestone and gypsum in much less.

The speculative nature of some of these figures is best illustrated by the estimate by Schreiber and Friedman (1976) of 600 m for the deep waters of the Sicilian portion of the Messinian Mediterranean basins. In the same year, essentially the same authors gave a minimum of 175 m for the same basin (Schreiber *et al.*, 1976). Consequently, the warning is justified (Schreiber *et al.*, 1982) that evaporite facies are components of a regional depositional fabric, and that models based on preconceived notions of water depths obtained on the basis of pre- or postevaporitic conditions must stand the test of observation, lest the models be wrongly applied. The two depth estimates given above contain more than a threefold difference in the hydrostatic head that would be generated to initiate syngenetic hygroscopic leakage into subjacent beds. These depth estimates also contain more than a threefold difference in the brine volume to be kept at a given concentration, and thus the time required to reach that concentration. In contrast, based on an evaluation of selenitic gypsum crystals, Pagnier (1978) concluded that Messinian evaporites in southeastern Spain were deposited in waters at least 10 but no more than 70 m deep. The same facies can also be found in Sicily, the Ionian Islands, and the islands of Crete and Cyprus.

Direct criteria have not yet been developed to determine the precise depths of ancient evaporite basins; thus, any estimates are arbitrary. On the basis of the proportionality of precipitate to mother liquor developed earlier by M. G. Valyashko, Strakhov (1962) assigned a brine depth of 2–6 m to potash deposits west of the Urals, in the Alsace, and elsewhere. For halite deposition, one could calculate the amount of isostatic adjustment produced by halite rock compared to a brine of halite saturation. Fabbri and Curzi (1979) made an ingenious attempt to calculate ancient water depths. Following their suggestions, one can calculate that

$$H_h\rho_h = H_b\rho_b + (H_h - H_b)\rho_r \tag{15-3}$$

where H_h is the thickness and ρ_h the density of halite, H_b and ρ_b are the thickness and density of the original brine that was saturated with sodium chloride, and ρ_r is the density of the underlying rock column of the crust to the asthenosphere. With ρ_r varying between 2.6 and 3.4, the depth of a brine ranging in density from 1.212 to 1.3235 would have to be 30–60 m to generate 100 m of halite. An error in such a calculation lies in the fact that precipitated halite has to be continuously averaged into the values for total crustal density, thus reducing the depth of the isostatically balanced water column required to about 25–30 m per 100 m of halite. Average depth has little meaning in a basin with wide shelves all around and one or more central depressions. A present-day average depth of 1450 m is of little value in assessing the diverse sedimentary facies of the Mediterranean Sea in depths ranging from the nearshore environment to 5121 m in the Vavilov deep off the Greek coast. Moreover, the equation is valid for only an isostatic equilibrium caused by crustal loading with halite.

However, halite accretes in depressions that subside independently of loading, because of mass movements in the mantle underneath. The precipitates then fill the basin only if the rate of precipitation is able to catch up with and surpass the rate of subsidence. Otherwise, a starved basin results, as the supply of terrigenous material decreases with increasing aridity. A good example is the depressions in the present Mediterranean Sea that continued to subside rapidly in the Plio-Quaternary, despite a very small sedimentary loading. The rapid subsidence is aided by the drastically increased thermal conductivity of evaporite sequences compared to other rocks, which allows an increased heat flow and thus eventual cooling and contraction of the substrate.

Even if the rate of halite precipitation is estimated to be a minimum of 35 mm/year (following Tasch, 1970), it is much more rapid than any reasonable estimate of rates of subsidence. If we were to assume continuous precipitation of a halite deposit, we would be forced to conclude that a very deep depression must have preexisted. However, we should not forget that each clay lamina, gypsum wafer, or dolomite stringer, often spaced only millimeters apart, represents a period of brine freshening, nondeposition, or even corrosion, parallel to bedding planes and thus unrecognizable. The cumulative time interval represented by these many diastems and lacunae is a multiple of the time interval represented by preserved precipitation. A ratio of 1:8 or 1:10 is a very conservative time estimate for deposition versus nondeposition, whether seasonal, storm derived, or secular. This brings the rate of continuous subsidence within the time range of evaporite precipitation. It is not necessary to invoke the prior existence of deep depressions that are uncompensated for isostatically.

The Shallowing Basin

It was thought that continued subsidence would not normally be tuned finely enough to the rate of deposition, so that several hundred meters of uninterrupted basinwide laminations would be preserved (Dean and Anderson, 1978). However, a slow epeirogenetic regional subsidence slightly in excess of anhydrite precipitation rates could achieve just that. After all, the area covered by the evidently shallow Salado Sea is a multiple of that covered by the Castile Sea (Fig. 15-7). This indicates that subsidence prior to Salado halite precipitation led to a transgression onto surrounding coastal flats. Cys (1978) found flow rolls and pillows in limestones overlying the Castile Formation, suggesting somewhat deeper waters outside the area of Castile evaporites. In the Salado Formation above the Castile anhydrites, one finds halite cubes in clays that originated as mud flats (Powers *et al.*, 1978). Consequently, the water depth must have been extremely shallow, although Anderson *et al.* (1972) advocated a water depth of 650 m for the Castile anhydrites. A sebkha model for the varved Castile anhydrites would require a sudden drop of some 600 m, which is not evident. No unconformity is present either at the base, at the top, or within the formation. A model of gradual deepening eliminates this difficulty and accounts for the rapid shallowing of the Salado Sea when the rate of halite precipitation overtook the rate of subsidence.

Varvites overlap onto the bases of reef mounds, and are thus younger than the onset of reef growth in the Devonian Elk Point Basin. Since the varvites contain planktonic debris preserved in a euxinic environment (G. Busson, personal communication, 1982), they formed after a density stratification was estalished. The reef tops were subject to erosion. If they had completed their growth before the onset of evaporite precipitation (Fuller and Potter, 1969a,b), the water depth would have been 50 m at the onset of varvite formation, but much less if the mounds were still growing during the precipitation of gypsum or anhydrite laminations (Davies and Ludlam, 1973). Within the Permian Castile Formation of West Texas, the rate of sedimentation appears to have been remarkably constant at 1.8–1.9 mm for anhydrite–calcite couplets for over 260,000 couplets (Dean and Anderson, 1978), by inference annual laminations. A progressive change is noted from a single-bed halite intercalation within the anhydrite, to halites with five minor anhydrite bands interbedded into an anhydrite member, to halites interrupted by six major anhydrite beds. At the same time, the anhydrite couplets thicken very gradually upward (Anderson *et al.*, 1972). Possible explanations include the gradually increasing impact of climatic oscillations, brine freshening, or a progressively smaller body of water, i.e., a gradually shallowing basin. Alternatively, one would have to invoke prolonged trends in climatic changes or sea level positions.

Sedimentation rates in excess of subsidence rates will decrease water depths. Heavy brines sliding into the deeper parts of the basin precipitate there proportionately greater quantities of salts per unit of time. The gradient toward the deeper part thus decreases. Wherever a close enough grid of radioactive logs has been used (West Texas, western Canada), they show a thickening of even the tiniest traceable interval between markers, which usually are dolomite or anhydrite bands (Stewart, 1963a; Klingspor, 1969). The cumulative thickness increment of a dolomite, anhydrite, and halite sequence, a measure of concurrent basin subsidence, leads researchers to assume a relatively deep water level along the axis of maximum subsidence.

Virtually all depth estimates derived from layers of potassium, magnesium, or calcium chlorides differ by three to six orders of magnitude from those derived from the estimation of anhydrite or halite layers. To bring the two sets of estimates into accord, the conclusion is inescapable that halite precipitation rates are in excess of basin subsidence rates, i.e., a basin is rapidly shallowing during halite precipitation. A bed of sylvites or carnallites is preferentially preserved on the flanks of original sills, because potassium chlorides become lighter upon cooling in a sodium chloride solution. Consequently, potash horizons do not indicate the deepest parts of the ancient depression, but occupy an asymmetric position in a basin against ancient shoals and shelf slopes. When water temperatures drop, solutions containing both potassium and sodium chlorides decrease in density, and the solubility of potassium chloride declines. As they become heavier with rising temperatures, the bitterns tend to descend toward the deeps during the daytime and rise along the slopes at night. Such bitterns will thus tend to rise toward shallow parts of the basin prior to precipitation. Potash precipitation into a halite slush along the basin slope invites creep. Frequently, the potash layers are severely distorted, while overlying and underlying halites show uniform lamination.

Once the brine reaches saturation for potash minerals, it is concentrated enough to form a continuous crust of floating crystals, which reduces evaporation losses to a minimum. That crust is disturbed only by strong winds or storms, which push it toward the leeward shoreline, aided by currents, tides, or wave action. Potash beds often offer evidence of substantial brine agitation. The shallower the brine pool under the crust, the less it is that a windstorm will be able to disturb the crust, but the more likely is the development of surges of interstitial waters from below. A typical example of the asymmetric distribution of sylvinites is furnished by the potash beds of the Michigan Basin, which comprise several 1- to 4-m-thick bands in the center of the basin, but which coalesce into one bed 15–20 m thick over the shoals around a reef chain along the distal northwestern basin margin (Fig. 15-14). Both the marginal carbonate bank and the onlapping anhydrite thicken at the point where the potash salts terminate (Elowski, 1980).

The occurrence of 3–4 km of Middle Miocene salts, postulated from seismic data underneath Upper Miocene evaporites under the Red Sea (Stoffers and Kuehn, 1974a,b), indicates that the Middle Miocene evaporite had filled any

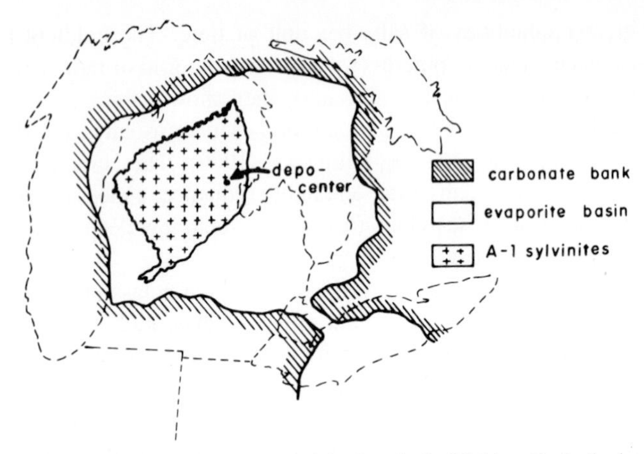

Fig. 15-14. Location of the Salina A-1 potash horizon in the Michigan Basin (basin outline after Alling and Briggs, 1961; potash after Elowski, 1980).

depression completely, even if we assumed a preexisting water depth on the order of 1–4 km. Consequently, it is not surprising that Upper Miocene deposits display features commensurate with extremely shallow water depths.

Most of the time, the evaporite basin may have turned into a series of emergent brine-logged salt flats (Valyashko, 1951; Friedman, 1972) with corrosion planes (Baar, 1974) by the time the brine reached saturation for potassium chlorides. A seasonally drying out sebkha (such as envisaged by Holwerda and Hutchinson, 1968) would be covered by normal seawater, which first precipitates gypsum and halite before resuming potash precipitation. For every meter of water depth created by incoming seawater, over 80 cm of carnallite would be dissolved. Seasonal drowning by seawater, typical of a coastal sebkha, therefore could not produce potash sequences tens of meters thick. An occurrence of individual pods of bischofite in a continuous carnallite that is locally leached in part to red sylvite (Korenevskiy, 1963) would, instead, support the concept of residual brine pools on a sebkha-like terrain. This bischofite may have been precipitated as tachyhydrite that decomposed upon cooling or exposure to air. Tachyhydrite cannot be preserved in seasonally drying out lagoons. Occurrences of primary tachyhydrite, therefore, presuppose a permanent brine cover until gradual dilution of the bittern covers the tachyhydrite with carnallite, together with halite. Bischofite likewise disintegrates eventually upon drying out, as it is also hygroscopic, adsorbs atmospheric humidity, and dissolves in its own water.

A small reduction in sea level or inflow leads to a large reduction in brine surface area. Vast rims will be exposed in sebkha fashion. A small rise in sea level or a reduction in evaporation at constant inflow lets the bittern spread out again and transgress over the coastal corrosion surface. The Great Salt Lake of

Utah covered 4500 km^2 in 1850 at an average depth of 4 m. A rise in water level of only 1.5 m flooded an additional 1100 km^2 (Lotze, 1957). That required a 3.9-fold increase in water volume in the lake. Conversely, the entrance strait to the Gulf of Kara Bogaz Gol off the Caspian Sea underwent in 18 years (1938–1956) tectonic movements that reduced the inflow by 69% (Dzens-Litovskiy and Vasil'ev, 1962), with most of the reduction in inflow occurring in 1937–1941 (Fedin, 1979). The surface area of the gulf then shrank by 45%, its volume by 73%, and its maximum depth from over 10 to under 3.5 m.

LATERAL AND VERTICAL ZONATION

Hardly any halite sequence is less than 20 m thick (Richter-Bernburg, 1977); most of them are several times thicker. Only where individual halite seams onlap onto basin margins may the thickness of a halite bed pinch out to 1–2 m in a borehole (Ivanov and Voronova, 1975). However, there also seems to be an upper limit below 1000 m of continuous monomineralic halite precipitation within an individual evaporite cyclothem (Yanshin, 1961). In contrast, the aggregate thickness of accumulated salt in modern salt flat environments is normally less than 20 m, consisting of calcium sulfate and halite (Schmalz, 1969). The same is true in paleolacustrine environments (Reeves, 1968).

The thickness of a halite bed corresponds to a given amount of seawater inflow. Twenty meters of halite precipitation require a through flow of at least 1400–1800 m^3 of seawater per square meter of halite surface area. Inflow of such quantities of seawater not only maintains a density stratification but also supplies new quantities of calcium. Under the influence of the Coriolis force, the inflow follows the shores and thus the shelf areas, and there reaches saturation for calcium sulfate. Such shelves were aptly named "saturation shelves" (Richter-Bernburg, 1953). Most fossil halites grade laterally (shoreward) into anhydrite and then into dolomite (Hollingworth, 1942). *Shoreward, the total halite thickness is represented by an equivalent anhydrite sequence* (Richter-Bernburg, 1957a). Marr (1959) saw in this lateral facies change the reaction rim between the quiet waters of the deeper bay and the high-energy environment of the shelf area. The basal Zechstein sequence was deposited on an essentially leveled-out relief. The relief-forming anhydrite walls and transverse sills were thus formed, at the earliest, during sulfate precipitation, as even the basal anhydrite shows no relief differentiation. The congruence of axes of sills with the strike of older fracture zones is thus noteworthy (Rockel and Ziegenhardt, 1979).

In a strict analogy to the lateral and vertical stacking of evaporite minerals in salt pans (Hunt, 1960) and groundwater efflorescences, lagoonal evaporites show a similar distribution. A basal anhydrite is overlain by rock salt that grades laterally into anhydrite on shelves and shoals, and is capped by anhydrite. Potash horizons are tucked into the rock salt sequence; magnesium or calcium chlorides, if present, are tucked into the potash sequence. Van Werveke (1924), who studied Upper Triassic evaporites in Lorraine, France, was the first to observe the lateral equivalence of halite beds to anhydrite banks and the vertical succession of a basal anhydrite into a halite-bearing anhydrite followed by anhydrite. The Permian Zechstein beds of northern England similarly pass laterally from red beds on land to lenticular limestone into contemporary anhydrite, and the latter into halite in the deeper part of the basin. Carbonate intercalations then separate the individual cyclothems. The same is true in the Devonian Elk Point Basin and elsewhere. However, direct lateral gradations from rock salt into claystones or other siliciclastics are not known; a transition zone of occasionally marly limestones, dolomites, or anhydrites is always placed in between.

The carbonate banks can grade into thin shales if the supply of calcium carbonate is limited. The thin shales then represent time equivalents in a sediment-starved environment. An example is supplied by a very thin layer of varved bituminous shale and anhydrite couplets that represent a lateral equivalent of the highly bituminous and productive Platy Dolomite Member in the German Zechstein beds (Zimmermann, 1914). A stunted fauna in these bituminous shales indicates continuing high salinities, but somewhat reduced from those existing during the preceding evaporite cyclothem. Such lateral zonation is also caused by runoff from land, notably around a river mouth. A belt of carbonates, mainly reefal, frequently develops offshore (Hollingworth, 1942).

Anhydrite thicknesses are always much smaller under thick halite sequences and increase laterally as the halite grows thinner. To a small extent, the thinning under halite may be due to redissolution. Gypsum solubility increases with rising sodium chloride concentration in the brine (Hem, 1959). However, for the most part, gypsum continues to precipitate in nearshore areas, whereas halite starts accumulating basinward, often steepening a shelf-to-basin slope. This relationship is explained by the rapid submergence of the offshore part of the gypsum layer below the photic zone into an anaerobic medium. The greater the rate of subsidence, the shorter is the relative time interval when the lagoon floor remains within the habitat of photosynthetic oxygen suppliers. If one calculates the number of years represented by the basal anhydrite, a subsidence rate of twice the rate of deposition would require a halite sequence three times as thick as the basal anhydrite to fill the depression (Fig. 15-15). Isostatic adjustment during deposition is invoked to explain the thickness differences in the second cyclothem of the British Zechstein sequence, reducing the calculated depth of water to 60–80 m at the base of the shelf slope and to slightly over 100 m in the basin center (Clark, 1980a). In part of the density-stratified basin that is beneath the photic zone when concentration exceeds gypsum saturation, gypsum will not precipitate because of lack of oxygen. Eventually, halite precipitates directly into

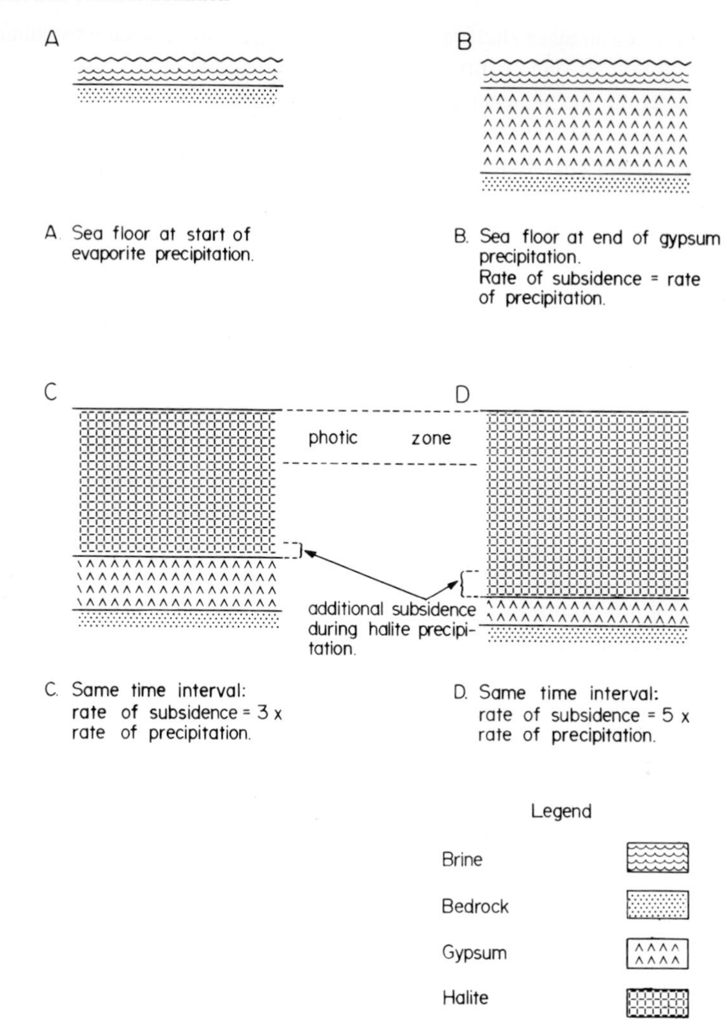

Fig. 15-15. Proportional gypsum and halite precipitation at different rates of subsidence.

the substrate without an intervening gypsum layer. This appears to have happened in parts of the Triassic evaporite basin in North Africa (Busson, 1968, 1980).

Lateral facies changes can be seemingly abrupt. Jung (1966) cites the example of an anhydrite that laterally replaces halite within a few meters on shoals. At basin margins where a shallow shelf and a deeper brine pool meet, a complete facies change occurs in less than 2000 m and may appear as an unconformity as contacts between shoal or shelf facies and basin precipitate replace more than one stratigraphic unit. If one postulates that an oxygen supply is necessary for gypsum precipitation, the presence of a thin basal gypsum or anhydrite unit indicates that basin subsidence commenced within the zone of photosynthesis. However, a preexisting depression extending beneath the photic zone could commence evaporite deposition with halites directly overlying pelagic sediments. Anhydrites then form a basal tongue wedging out from basin margins. Well-documented examples of this absence of basal calcium sulfates are hard to find.

SUBDIVIDED BASINS

The basin may contain sills that effectively subdivide it into several smaller depressions. This was the case in the Cretaceous Gabon-Congo Basin (Belmonte *et al.*, 1965), the Devonian Elk Point Basin of western Canada (Pearson, 1963; Wardlaw and Reinson, 1971), and the Silurian Michigan–Ohio–Alleghany Basin system (Alling and Briggs, 1961). This led Branson (1915) to suggest that a *sequential arrangement of interconnected basins fascilitates the preconcentration of precipitating brines.* Branson's concept is based on the example of Permian evaporites in West Texas, where a decided southwestward migration of salt in time can be documented (Lang, 1937; King, 1942). The same idea was later proposed for north Carpathian salt basins (Fiveg, 1956), and for the transport of concentrated brines in Permian seas on the flanks of the Ural Mountains. A northeastern basin is filled with polyhalite, halite, and sylvite, whereas a southwestern basin contains carnallite, kieserite, kainite, and sundry borates (Lepeshkov, 1946). The Upper Jurassic evaporites of central Asia also formed in a series of connected tectonic depressions that had but a single outlet (Popov, 1968). Interconnected basins are often marked by lacunae in the sequence of evaporite minerals, such as potassium–magnesium salts directly overlying anhydrites without intervening halites. Borchert (1967) defended the concept of preconcentrating basins by pointing out that basins 500–1000 m deep cannot move in short-term cycles from limestone saturation to potash saturation and back to mere carbonate deposition.

Interconnected basins can be classified (Skrotskiy, 1974b) into the following:

1. Those having a perennial connection to the outside water supply. They have progressively denser bottom brines.

2. Those connected seasonally, i.e., at the time of high water level. They experience seasonal aeration and oxidation of the total brine column.

3. Those supplied with water sporadically, i.e., during catastrophic storms. They have a tendency to dry out periodically.

Shallow sills between depressions are also frequent sites of continuous gypsum precipitation, whereas halite precipitates on either flank, thinning toward each sill. Where the structural highs were the site of gypsum precipitation, rock salt is the concurrent precipitate in the lower parts of the basin (Hollingworth, 1948). The margins of such shelves or sills are the site of coarse halite crystals (Hofrichter, 1960), indicating very slow growth. As long as the chemocline between inflowing and outflowing waters remains above the sill crests, no reefal organisms can take hold. If the crests do rise above the chemocline, reefs may start growing, even if their dead trunks are later submerged below the chemocline into the concentrated anaerobic brine. Where the sills were the sites of limestone deposition, the flanks are occupied by contemporary anhydrite. Such separation occurs even in continental deserts. The hills in the Chilean desert are composed of gypsum and secondary anhydrite, and the valleys of rock salt; gypsum also occupies the rim of the desert (Wetzel, 1928).

Precipitation of salts tends to level out any preexisting topography. Nonetheless, Goldsmith (1969) found elevational differences of up to 60 m on the floor of the Permian Basin in New Mexico during advanced stages of brine concentration. Potash beds are restricted to depressions, yet no gravitational movement of halite slush from the local highs is observable. This suggests that the elevation differences formed, at the earliest, only after potash deposition had ceased and the lagoon had returned to basinwide halite precipitation. A reduction in the inflow/outflow ratio is responsible for this change in sedimentation; the reduction itself may also have had tectonic causes.

EVAPORITE FACIES DISTRIBUTION

Shoal Facies and Coastal Flats

If the bottom counterflow reaches a finite density short of gypsum saturation, then no massive precipitates will accrue, although subsidence would continue. The bay floor, as a starved basin, very slowly accumulates some clays. A flora of sulfur bacteria takes hold beneath the water–sediment interface. Iron leached from the clays and iron hydroxides dissolved in interstitial waters react within the sediment to form various hydrotroilites. The result is a very black, pyritiferous shale, rich in organic material, which may also be calcareous due to the incorporation of planktonic biota. Black euxinic shales in Lower Oligocene beds of southeastern France (Amieux, 1980) are laminites of calcareous shales and organic matter (lignite?). They pass into gypsiferous black shales and eventually into gypsum. The lower part of this sequence represents sedimentation before the waters surpassed the threshold of gypsum saturation; episodic fluviatile freshwater incursions continued during the deposition of gypsiferous black shales.

If, however, the concentration of the bottom brine continues to increase until precipitation of some solutes commences, then the solubility of oxygen in bottom waters becomes negligible. This allows the anaerobic sulfur bacteria to migrate into the brine, and the sulfate–sulfide boundary rises above the sediment interface. The bacteria feed on sulfate ions brought in by the inflow and do not have to first solubilize already precipitated gypsum. Most bacterial hydrogen sulfide is not trapped in pyrites, but either escapes or is converted back to SO_4^{2-} in the photic zone. The migration of the sulfide boundary into the brine explains why the accreting precipitates are very low in pyrite content and at the same time devoid of hematite needles.

On the shallow shelf, the sulfate ions attack precipitated algal carbonates; in deeper water, they are converted back to hydrogen sulfide beneath the shallow photic zone. Thus, there is a rapid buildup of gypsum on the shelf, which accentuates any depth differential and creates a steeper shelf-to-basin slope for slumping and turbidity currents to descend. As long as the sulfate ions can reach bottom at least seasonally, gypsum laminae will precipitate even in deeper water. Once the continued subsidence has removed the deeper parts from seasonal exposure to SO_4^{2-}, gypsum precipitation retreats to the flanks of the basin. Jung (1966) maintained that in the Stassfurt cyclothem of Permian Zechstein deposits, the gypsum precipitation continued as shoal facies at some distance from the coastline, even during halite and potash deposition in the deeper parts of the basin.

The basins are usually surrounded by extensive flats that are inundated for great distances by a very small rise in water level. A rising sea level inundates a wide strip (Fig. 15-16). Rising waters in Lake Eyre, South Australia, produce a saturated brine layer extending 1 km beyond the limits of bottom halites, a high-salinity layer 5 km beyond the edge of the salt, and a 5-m-thick surface layer of even greater areal extent, reaching several tens of kilometers farther (Dulhunty, 1976). If the inflow constrained by friction and by an upper limit to inflow velocity, is unable to cover all of the marginal basin surface, the rims will start drying out. This reduces evaporative water losses and thus demands on the inflow. No matter how deep the basin is, the shoreline will retreat, and wide mud flats or salt flats will be exposed, resembling a vast coastal sebkha, even around

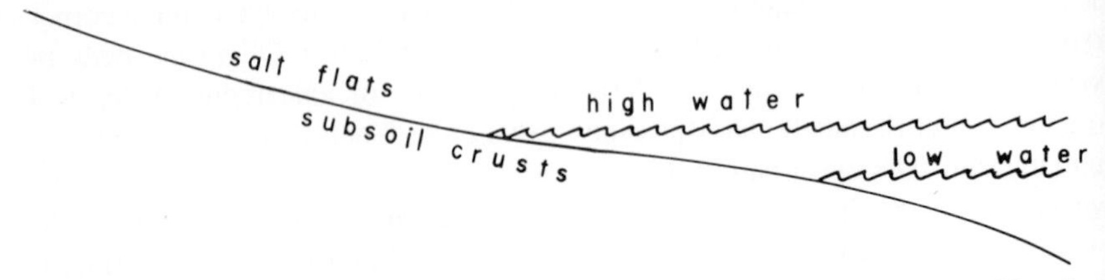

Fig. 15-16. Margin of an evaporite basin. A slight change in sea level inundates a wide strip. Salt flats and subsoil crusts form on the landward side.

basins with a substantial maximum depth in the center of subsidence. At high water level, much of this shore area may become inundated again, later giving the impression, in cross section or along a mine wall, of a wedge of mini-disconformities. High water levels, due to a local rise in ocean level, are coincident with a greater influx of less soluble salts and induce the formation of intercalations, of carbonate in gypsum, of carbonate and gypsum in halite, and so forth. If, however, the high water level is caused by increasing freshwater influx or rainwash, the basin is flooded by suspended clays. Thus, the mineralogy of the intercalation reveals whether the waters came primarily from the sea or from land.

Some Permian Zechstein deposits have been taken as such a growth of extended salt flats: Halite precipitated on extensive salt flats in Yorkshire after a temporary inundation caused redissolution and redeposition. Residual concentrates gravitated to depressions, eventually sinking into the sediment and precipitating potash minerals (Smith, 1973). Similarly, the sea withdrew temporarily from the top of the gypsum shelf in the first (Werra) cyclothem in Germany. Subaerial exposure resulted in karstification while deposition continued in the center of the basin (Sannemann *et al.*, 1978). This would explain the occurrence of potash and other salts in a distended and distorted slush of fine clays, carbonate, and carbonaceous sediment. Clay films protect the precipitate and represent either a residue from dissolved salt or a deposit due to a freshening that cut off the potassium and magnesium precipitation. These films are always followed by halite. As halite precipitation exceeded the rate of subsidence, the salt flats prograded and ultimately merged to isolate large lagoons, until these, too, were eventually eliminated. Rainfall and runoff then continued to redistribute the salts. A brownish-red clay, previously considered to be a basal unit of the Stassfurt cyclothem, is really a shoreline facies of the Werra cycle, wedging into the anhydrites (cf. Table 12-1); a thin Stassfurt series transgressed over this unit (Jungwirth and Seifert, 1966).

The greatest concentration of bittern salts coincides with the zone of greatest pore fluid concentration, which occurs a short distance below the sediment surface (Strakhov, 1962). It is overlain by a thin zone of halite that is relatively poor in potassium and magnesium salts. Similarly, the Permian evaporite basin in the Caspian Lowlands shrank considerably as the brine turned into a bittern. The area of halite crystallization differs very little from the area of anhydrite occurrence. However, the area of sylvinite precipitation is significantly smaller, and the carnallite deposition is restricted to less than half the original areal extent of the basin. Extensive coastal salt flats mark the former outline of the basin, a playa lake or dried-up lake, as envisaged by Valyashko (1972), that probably flooded seasonally to provide a large evaporating surface for bittern concentration.

Potash minerals are probably not precipitated from open waters on the floor of primary depressions, but rather from bitterns draining a larger area into the halite slush at the low point of each depression. This is indicated by the occurrence of channel fill or remnants removed from crests. In New Mexico, the potash salts are concentrated where stratigraphic clay marker beds are farthest apart. They are absent, or present only in traces, where the marker beds come closer together (Dunlap, 1951). Many halite layers here form wedges pinching out northward toward the shelf area. The overall thinning is interrupted by halite lenses filling troughlike depressions aligned parallel to a hinge line between the Delaware Basin and the northwestern shelf (Jones and Madsen, 1968). Red varicolored halitic or anhydritic clays define top and bottom of each of three potash horizons in the Devonian Prairie Evaporite Formation of Saskatchewan, Canada. These clays mark dissolution surfaces and coat all potash crystals in the uppermost potash bed. Crystal size increases from the lowest to the highest potash bed (Klingspor, 1966), indicating a gradual slowing down of growth rates and of rates of brine concentration (Fig. 15-17).

A rise in the water level re-forms an interface to residual brines. On the shoreward side, the surface waters touch shallow bottom and can alter the precipitated salts (cf. Fig. 4-5). Reducing the bromide content is one example. Kieseritic or anhydritic alterations of former potash horizons are seen as early diagenetic rims produced by longshore currents (Loeffler, 1963) or shore-derived waters (Langbein, 1964). Slumping and turbidity currents can move much of the early precipitate, even over a very gently inclined slope. The Verotyshchensk series of potash deposits in the Carpathian region is an example, which comprises five cyclothems, each with 28 individual lithostratigraphic units, lying only on a thin layer of halite-bearing breccias (Donchenko, 1964). These breccias are not produced by solution, since the potash beds are unaffected. Thus, they must represent the proximal part of a subaqueous slide.

As the brine surface shrinks, the belt of periodic flooding widens and potash minerals from dried-out coastal ponds are washed into the residual basin. Whereas surface brines display a small lateral concentration gradient, bottom brines are nearly constant in lateral variation, and are only vertically stratified. This is equally true today of both the Red and Mediterranean seas. The precipitates on the bay floor remain in continuous equilibrium with the brines washing them. A residual bittern concentrated to potash precipitation will protect potash minerals already precipitated, but will attack gypsum or anhydrite. If the bittern is diluted below that concentration, it will protect anhydrite and dissolve potash minerals. The rate of redissolution is a function of brine composition. It decreases for sodium chloride with increasing magnesium chloride concentration. The rate of potassium chloride dissolution in concentrated magnesium chloride solutions becomes 3.5 times as fast as that of sodium chloride. Carnallite dissolves even faster than sylvite and rock salt. The sulfates as a group dissolve more slowly than any of the chlorides. Potassium sulfates dissolve faster than epsomite,

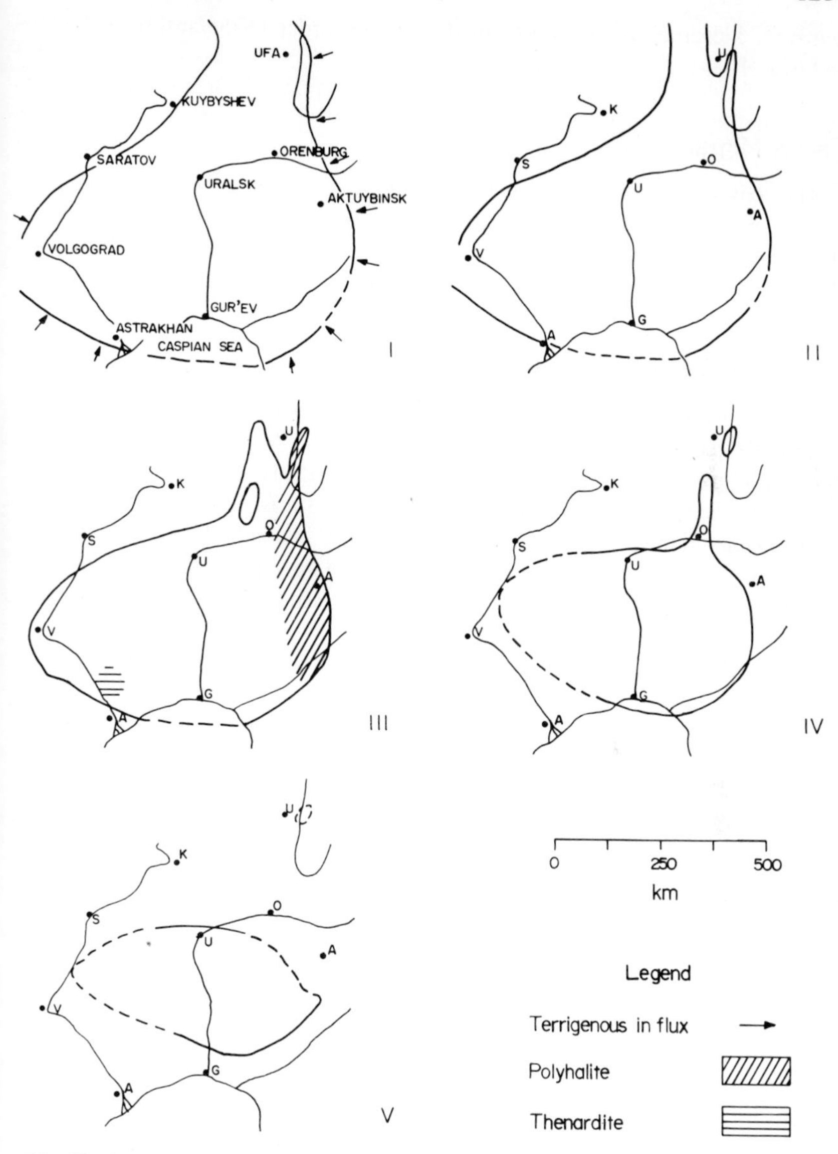

Fig. 15-17. The evolution of the Permian evaporite basin in the Caspian Lowlands. Shown are the outlines of the basin at the onset of I. clastic and carbonate deposition (arrows indicate the direction of inflow); II. gypsum-anhydrite precipitation; III. halite crystallization; IV. sylvinite precipitation; and V. carnallite deposition. Bischofite occurs in small pods along the rim of this shrunken lagoon (after Korenevskiy, 1963).

gypsum, kieserite, or anhydrite in descending order (Viluyanskiy and Menshikova, 1933).

Facies Migration

The axis of subsidence frequently migrates during the precipitation of the various evaporite cyclothems. If the regression (or transgression) is diachronous, the nature of regressive (transgressive) facies of sedimentation requires early diagenetic alterations also to be diachronous (Kinsman, in Holliday, 1968).

Symmetric Facies Distribution

Only an evaporite basin completely bordered by wide shelves will develop a symmetric pattern of deposition. For example, the Michigan Basin shows a concentric pattern of deposition of evaporite minerals. Framed by carbonate banks and barrier reefs on almost all sides, the basin is rimmed by anhydrites. Halite occurs in the center of the basin, sandwiched into anhydrite beds. Residual brines can drain into this depression from a much larger area. Potash deposits are then found off center, on the rim of the area of maximum subsidence. However,

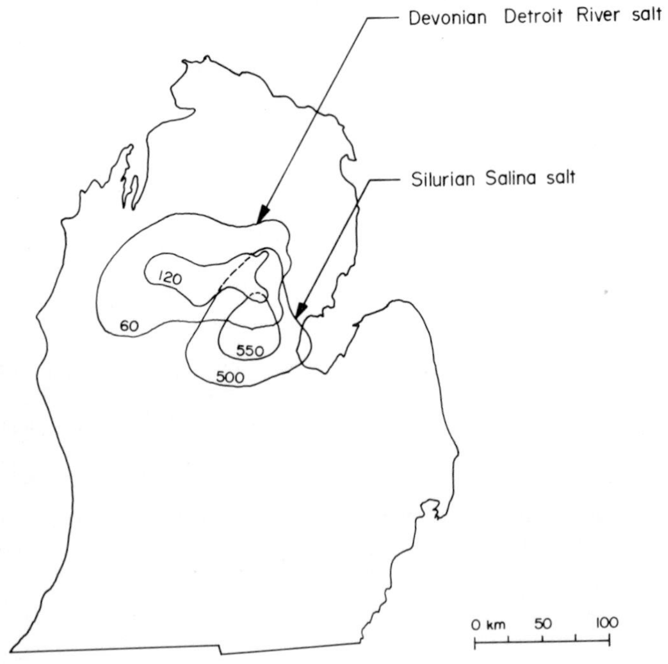

Fig. 15-18. Shift in Silurian to Devonian time of salt deposition centers in the Michigan Basin.

the maximum subsidence in the Devonian does not occur in the same locality as in the Silurian; this means that the lowest point in the basin shifted slightly in time (Fig. 15-18). The Permian halite sequences in the Amazon Basin are similarly embedded in a carbonate platform (Szatmari *et al.*, 1979) and show a similar shift in subsidence. The opposite occurs in the Upper Kama Basin, where Ivanov and Voronova (1975) noted that the greatest subsidence did not occur along the axis in a bull's-eye pattern, but along the periphery.

If the basin has a round shape (its *F* value approaches unity), then a bull's-eye pattern of facies is possible. If the basin is elongate (such as the Elk Point Basin), the facies will align themselves parallel to the long axis of the basin. In a basin that has its chief source of detrital sediments at one end, at the bayhead, and its opening to the ocean at the other end, one might expect, at first sight, to see transverse contours of grain size, facies distribution, and water salinity. In the Persian Gulf, a good example of such a configuration, the isofacies contours are not transverse, but instead run parallel to the axis of the basin (Emery, 1956; Lange, 1970). In the narrow Gulf of Suez, the trend of Miocene evaporites is likewise parallel to the gulf axis (Zaghloul *et al.*, 1977). *Bottom currents redistribute sediments into an alighment with the basin axis.*

Asymmetric Facies Distribution

A basin possesses an asymmetrical facies distribution if the marginal gypsum banks are missing along part of the rim. Frequently, one finds pure clays, clays mixed with carbonates, or clays with minor anhydrite. The Triassic salt basin of northwest Africa is such an example (Salvan, 1968; Busson, 1969). South of Gibraltar, the carbonate shelf is bordered by a thick anhydrite sequence that was originally precipitated as gypsum. It must have been deposited in the sunlit, oxygen-rich surface waters and thus represents the shallow shelf area. Southward the anhydrite grades into a halite province, but south and southwest of the halite province there is no anhydrite, only an anhydrite-bearing clay province. If that was the steeper coast, rainwash and runoff would provide clays from time to time, but a greater water depth would prevent the formation of gypsum crusts, and would instead foster the bacterial decomposition of any accumulating gypsum grains.

In clays episodically washed by freshened waters, the sulfate salts that are weakly adsorbed to clays (Swoboda and Thomas 1965; Aylmore *et al.*, 1967) are subject to bacterial breakdown to hydrogen sulfide. Potassium concentrations in the water facilitate the desorption process (Bornemisze and Llanos, 1967). As the brine concentrates to halite saturation, it produces euhedral halite crystals in the clay. This would occur in clays onto which brines have transgressed, which were saturated with sodium chloride.

Triassic basins of North Africa, western Europe, and offshore Newfoundland are marked by halite, gypsum, and dolomite in the center; the margins are the sites of sand and clay, and only small amounts of carbonates (Busson, 1980; Jansa *et al.*, 1980). A similar example is furnished by the Triassic evaporites in the Paris Basin. The chlorides and sulfates here appear to have been precipitated into a mud, and not into open waters (Ricour *et al.*, 1958), whereas the subsidence was evidently faster toward the Newfoundland Grand Banks. An analogous situation also prevailed in salt in Cheshire and Shropshire, England. The deepest part of the basin does not coincide with its widest part. Northward the basin shallows and spreads out into a wide plain. Here, too, there are two halite horizons separated and covered by marl. However, gypsum forms as lenses in overlying marls; this gypsiferous brown marl is consequently referred to locally as "horse beans" (Sherlock, 1921). It is very likely that the originally precipitated gypsum was bacterially decomposed and is the source of the carbonate fraction of the marls.

Warrington (1974) suggested that all the sulfates and carbonates that should have been associated with Keuper halites in Britain are now represented by marls. He took issue with the interpretation of Bonney (1906), repeated by Brunstrom and Kent (1967), that the salt is of continental origin. Although conceding the predominance of continental influences, he thought that the bulk of the Keuper salt was nonetheless derived from marine sources, namely, an incursion of the Tethys Sea through France. The same had been suggested earlier by Wills (1970) and W. B. Evans (1970). Cross-stratification, small-scale ripple marks, and mud cracks indicated extremely shallow conditions. Holser and Wilgus (1981) later corroborated this contention on the basis of bromide content. The amount of bromides rises from Poland, near the portal to the Tethys (Tokarski, 1965), toward the Netherlands, i.e., toward the distal parts of the basin. The British Keuper must then have constituted a shoreline facies subject to ephemeral drying out. The low bromide content of Triassic salts is due to later recrystallization (Schachl, 1954).

The Devonian Pripyat Basin in Byelorussia became asymmetric due to nonuniform sinking of separate parts. The south limb is the shallower one; the north limb dips steeply. Each subsequent cyclothem, therefore, covers a smaller area (Kirikov, 1963). The maximum thickness of carnallite is found in the axis of the Devonian Prairie Evaporite Basin within the Williston Basin area, but the basin axis migrates westward (Holter, 1972). All potash members thin toward the southwest part of the basin (Anderson and Swinehart, 1979; Worsley and Fuzesy, 1979). The relative rate of subsidence in any one locality thus varied in time during deposition. The four potash horizons in this deposit extend basinward from the ancient shelf slope (Fig. 15-19). They are overlain by barren halite even where truncated, suggesting leaching from above. The barren halite cover (called "salt back") increases in thickness southeastward (Fuzesy, 1982). Another example is offered by the Oligocene evaporite basin in the Upper Rhine

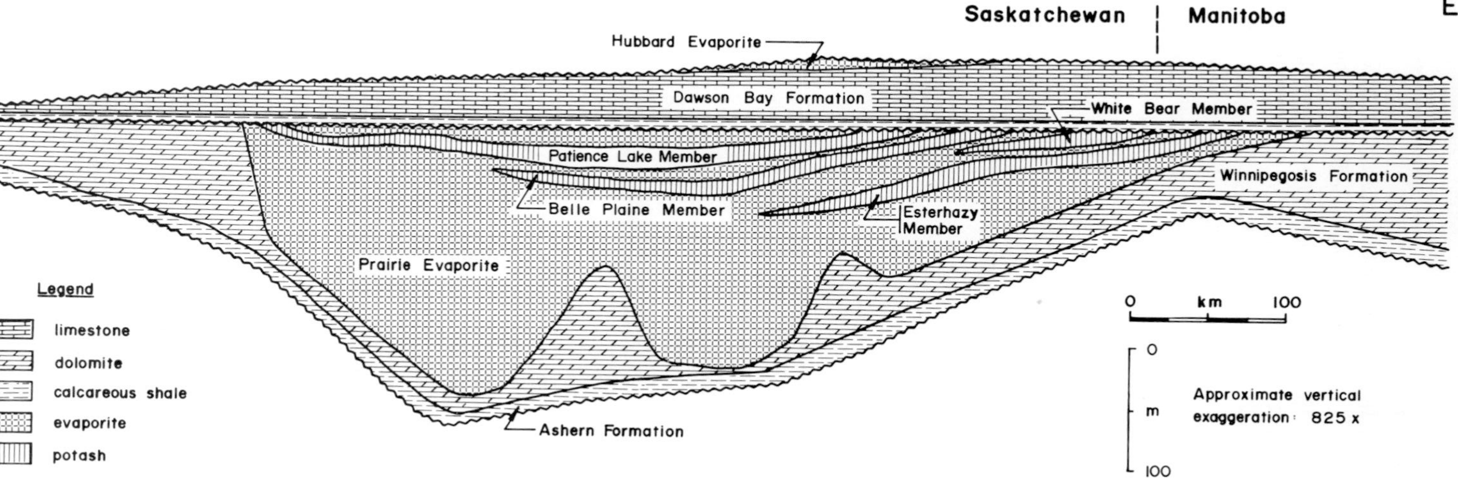

Fig. 15-19. Diagrammatic cross-section through the Devonian Elk Point Basin in southern Saskatchewan, Canada (expanded from Fuzesy, 1982).

Valley, where the lower potash horizon onlaps in places onto underlying marls (Harbort, 1913), i.e., migrated to shoals. The potash beds in the Permian Basin of West Texas and New Mexico likewise are concentrated toward the shallower margin (Moore, 1960). In the Caspian depression, Skrotskiy (1974b) found K–Mg salts widely deposited on top of the anhydrite shelf marginal to the basin.

In the Castile Formation, each major anhydrite member is thickening to the east, whereas each halite intercalation either thickens to the north or finds a series of deeper depressions to fill immediately to the west of the eastern margin (Anderson *et al.*, 1972). That very strongly suggests the occurrence of differential subsidence during deposition in an asymmetric basin. This is also the case in the Zechstein sequence along the Dutch border, where the potash minerals were precipitated onto the Werra shelf shoreward from a shoal (Fig. 15-20). Both the basal carbonate of the overlying Stassfurt Member and the copper shale underneath contain a fauna typical of shallow waters, and were thus not deposited in waters of a very deep depression (Teichmueller, 1958). The northern shelf of the

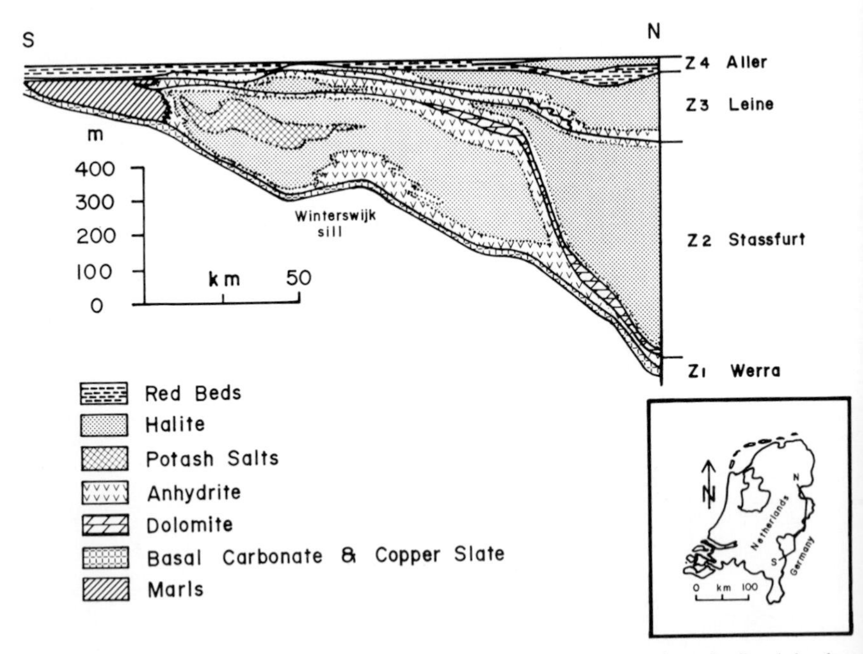

Fig. 15-20. Diagrammatic cross-section through the Zechstein Basin along the Dutch border. Note the position of the potash salts shoreward from the shoals of an anhydrite mount called the "Winterswijk sill." The basal limestone of the Stassfurt cyclothem shows no deep-water features and was probably laid down originally in a nearly horizontal shallow-water position. The northern shelf of the Werra cyclothem then subsided most rapidly during Stassfurt salt deposition (after Teichmueller, 1958).

Werra Member thus seems to have collapsed during the Strassfurt cycle after the basal carbonate member was deposited.

Busson (1978) thought that marginal limestones grow into surface waters that differ only slightly from ocean waters. As the brine concentrates or the water level drops during a temporary separation from the open ocean, the limestones tend to prograde toward the interior of the basin and dip lower and lower. Busson (1978) cited as an example the Upper Miocene (Messinian) limestones of Almeria, Spain. The Upper Carboniferous carbonates of the Sverdrup Basin of arctic Canada, described by Davies (1977a), appear to follow the same pattern. They occupy an axial position in the basin. The depositional relief gradually increased from a few meters to over 400 m. Steeply dipping tongues of shelf foreslope carbonates (mainly limestones) were interbedded with anhydrites (Davies, 1977a). The residual brines seeped only downward into the substrate. The updip limestones remained unaltered.

Renewed subsidence may tilt precipitated layers to a new axis of subsidence before deposition commences again. The apparent angle of an unconformity or the amount of truncation can then become significant. Bonython (1956) described such an onlap of younger salts onto barely older ones from Lake Eyre, Australia. Hite (1970) recognized several such horizons in the Pennsylvanian

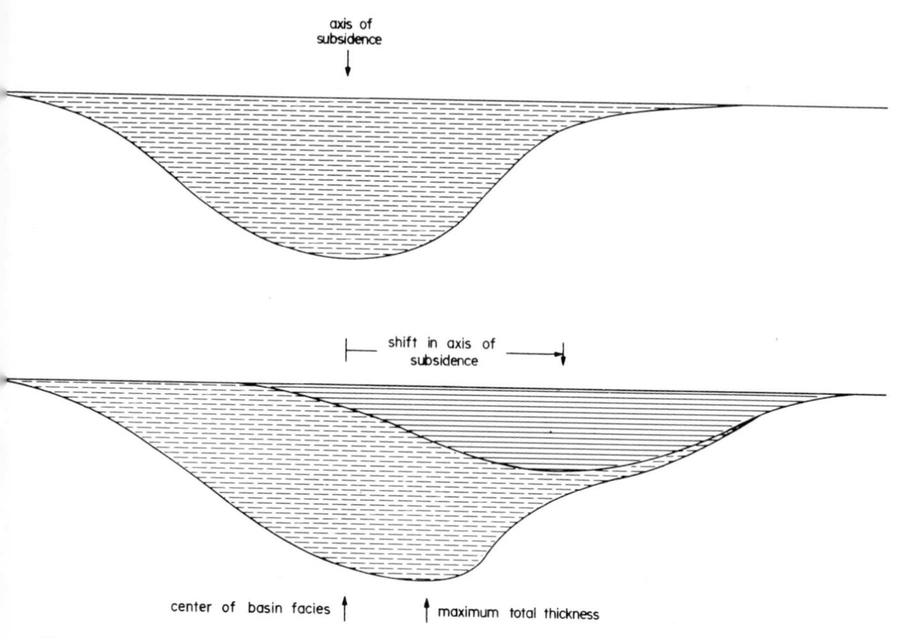

Fig. 15-21. Effect of a shifting axis of subsidence. Isopach maxima are offset against centers of basinal facies.

Paradox Basin of Utah, where as much as several meters of halite are probably missing under each anhydrite cap. Decima and Wezel (1971) described a similar unconformity in the Upper Miocene (Messinian) evaporites of Sicily. However, Rouchy (1981) showed that this unconformity is restricted to the collapsing fringe of a rapidly deepening basin. Toward the shelf, this unconformity is replaced by a solution breccia. If the axis of maximum subsidence shifts in time, the earlier cyclothems or portions thereof need not correspond to maxima in overall thickness (Fig. 15-21). In the Triassic evaporite sequence in the Sahara, the facies do not correspond to thickness maxima; the greatest expanse of halite is offset against the axis of greatest thickness of the formation (Busson, 1968, 1980). In an analogous way, the axis of greatest subsidence shifted northwest during sedimentation of the Stassfurt horizon of Zechstein evaporites in Germany (Schulze, 1960a).

REVERSAL OF PRECIPITATION

It is typical to find potash layers covered by halite, and halite in turn covered by gypsum (later altered to anhydrite), signaling a progressive freshening of the brine. There are many examples: The Upper Miocene (Messinian) potash layers of Sicily are covered by halite and anhydrite. Eventually, the Mediterranean area generally dried out in pre-Pliocene times, or terminated in Tunisia, Italy, Crete, and Cyprus with brackish or freshwater materials. The same symmetry of potash deposits underlain and overlain by halite, and the latter in turn underlain and overlain by anhydrite, can be observed in the Silurian Salina salt of Michigan (Matthews and Egleson, 1974), the Permian Basin of West Texas (Adams and Frenzel, 1950), the Permian evaporites of the Urals (Korenevskiy, 1963), and the German Zechstein deposits (Richter-Bernburg, 1953). The precipitation of gypsum follows when halite ceases to precipitate, and it covers the halite (Fig. 15-22). Being much less soluble than halite, it does not dissolve easily in inflowing normal seawater, despite its increasing solubility in concentrated sodium chloride solutions. Clay intercalations in the Oligocene salts of Alsace, France, are always separated from halite by a layer of anhydrite (Goergey, 1912). *Continued precipitation of the evaporite mineral with the next lower solubility preserves salts of higher solubility and shields them from redissolution.* Freshening and brine concentration are very gradual processes. It is nearly impossible to produce in nature rapid (short-term) sequences of gypsum–anhydrite to potash and back to gypsum. The larger the volume of brine in the bay, the longer it takes for a readjustment of precipitation from one mineral to another.

A reversal of the precipitation is induced by a dilution of the brine (Fulda, 1923, 1924) and can have climatic or geologic causes. An increase in the rate of subsidence would merely cause further concentration of the brine by reducing the

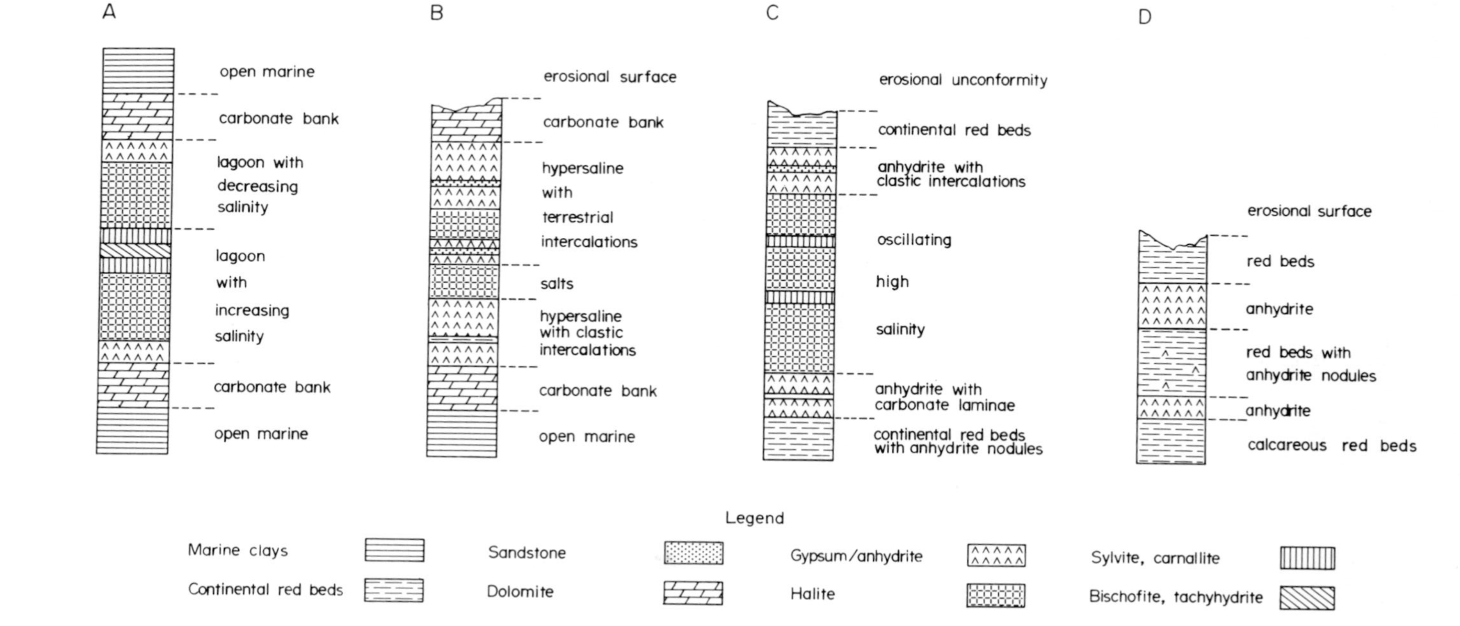

Fig. 15-22. Sequence of marine precipitation. A. An open marine environment becomes hypersaline and then gradually reverts to an open marine environment. B. An open marine environment becomes hypersaline and then freshens and becomes land. C. Continental red beds are inundated, turn into a hypersaline lagoon, and then revert to a continental environment. D. Sebkha environment (*sensu stricto*): Red beds are inundated by the sea, which leads to gypsum precipitation. The region eventually turns again into a clastic province.

outflow. If the cross-sectional area of the entrance strait is not progressively reduced by reef growth or tectonic events, an increase in the surface area of the brine (Fig. 15-12) by basin subsidence alters the ratio between the two values. As the ratio between the cross-sectional area of the entrance strait and the brine surface is reduced, the rate of brine concentration increases. The onset of magnesium salt deposition marks the minimum in that ratio. Gradual reduction in the brine surface, possibly caused by a slowdown in the rate of subsidence, leads to a reversal in the sequence of precipitates, since a reduction in the surface area lowers the evaporation losses. Other causes of reversals in brine concentration include an opening of the inflow channel, a rise in sea level, and flash floods from surrounding lands in response to greater rainfall. Duff *et al.* (1967) have pointed out that there is usually a pronounced asymmetry between the thickness of evaporitic rocks underlying a potash horizon and those covering it. Apparently, freshening of the brine usually takes much less time than the preceding concentration phase.

If flash floods are capable of substantially altering the salinity of a basin, the ratio between incoming flood volume and resident brine volume must be large, i.e., only a very shallow water column can be present. In every instance, however, the salinity drops only very slowly in a large volume of water. There is a significant inertia in the response of marginal bays to oscillations in environmental factors. The precipitation of halite continues when the waters are no longer saturated for potash salts. At the same time, inflowing waters redissolve halites near the entrance shelves and reprecipitate this rock salt upon reaching saturation further into the bay. It should be noted that the waters causing dilution of the hypersaline brine have the same source as the waters feeding into the basin during the brine concentration phase.

Climatic causes reduce rates of evaporation, and thus the water deficit, in both wetter and cooler years. The inflow continues and slows down, but has the opportunity to sweeten the brine. At the same time that the rate of evaporation drops, the input of continental waters rises. A deepening or broadening of the entrance narrows, similarly leading to a reduction of inflow velocities and of drag on any undercurrent. Inflow volumes increase and bottom brines are flushed more extensively. Again, a reverse order of precipitation is the result.

Vertical stacking of salts according to their solubilities occurs in basins of progressive brine concentration until the basin is filled and precipitation ceases. Vertical stacking in reverse order signals either a reduction in evaporation rates or a decrease in subsidence rates. Multiple thin anhydrite–halite–anhydrite interbeds are related to oscillations in rates of subsidence. They are frequent in the substrate of the thickest evaporite accumulations. Alternating sequences such as gypsum–halite–gypsum–(bituminous) dolomite–gypsum–halite–(potash–halite)–gypsum–dolomite sequences are very common, and signal climatic variability, tectonic instability, or both.

Transgressive and Regressive Sequences

The reversal of precipitation represents the "phase of regressive salinity" of Richter-Bernburg (1977, 1979), but is the "transgressive phase" of Hite (1970). The time span represented by this phase is usually much shorter than that represented by progressive concentration of the brine (the "regressive" phase of Hite, 1970). Evaporite sequences overall can thus be either regressive or transgressive. A sequence starting with open marine shales or carbonates and ending with continental sediments, red beds, or deltaic freshwater sediments is evidently regressive. Conversely, a series starting with red beds and ending with open marine sediments is evidently transgressive. A third case is represented by sequences starting with open marine sediments and ending with similar ones, or starting and ending with continental red beds. Many examples could be cited for all three of these cases. However, this subdivision is valid only as a first-order approximation (cf. Krumbein, 1951). It is the changing rate of subsidence which controls the sedimentation. Despite a high rate of subsidence, the filling up of a basin is the regressive phase, and the reentry of seawater is the transgressive phase, yet it produces a reversed sequence of precipitates. The gradual dilution of the brine produces a regressive salinity (Richter-Bernburg, 1977). The sequence can then repeat itself, giving rise to several potash horizons within a halite sequence.

Eventually, subsidence stops or slows down. If the basin is filled at the time, it will be overrun by terrigenous clastics that are being delivered to the coastal

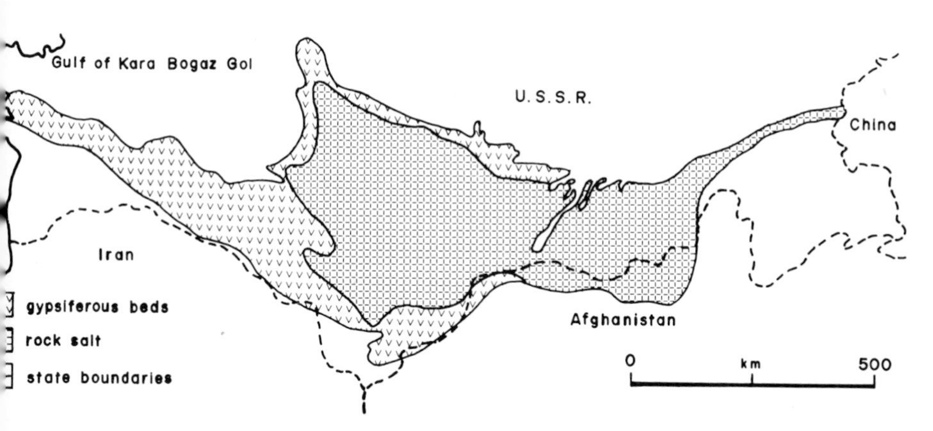

Fig. 15-23. The Upper Jurassic basins of central Asia. The total water surface expanded in time over a much larger area than the depression receiving salt precipitation. Potash deposits formed on the eastern flanks (simplified after Popov, 1968b).

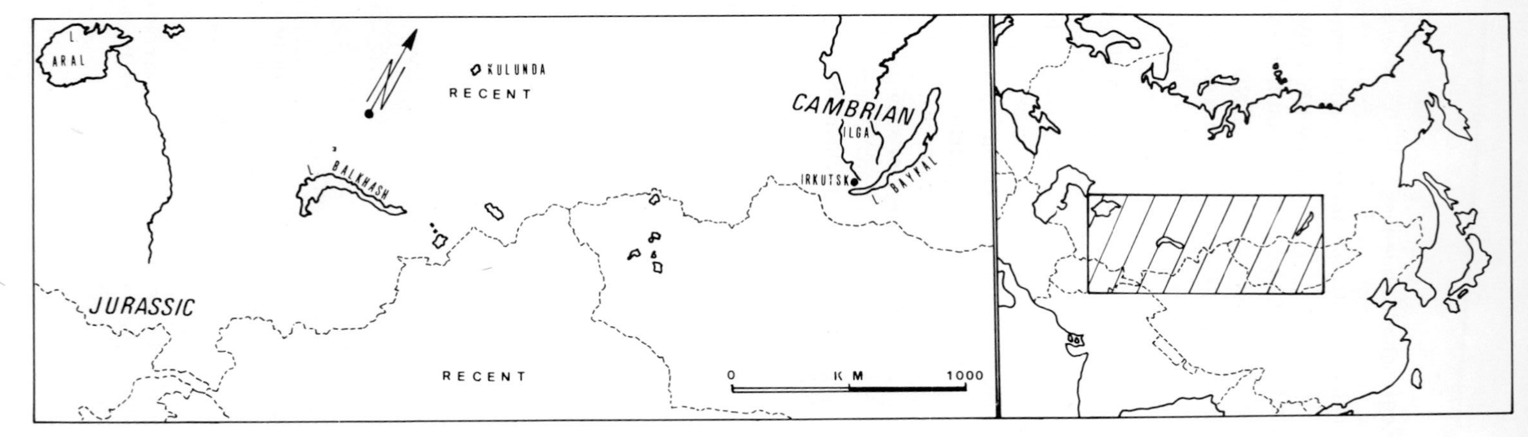

Fig. 15-24. Evaporite basins in southern Siberia.

plain, either as red beds or as deltaic deposits. If, on the other hand, the sill collapses during subsidence, the basin gradually reverts to an open marine environment, i.e., the evaporite sequence is capped by open marine limestones and shales. If subsidence resumes and the entrance strait has opened, or the climate has changed, the basin will not fill with precipitates. A starved basin results whenever the supply of terrigenous materials is not stepped up. For example, the Louann salt in the Gulf of Mexico is underlain by red beds and capped by normal marine sediments (Evans, 1978). The Permian Welligton salt of Kansas is underlain by shales and red beds (Jones, 1965). An example of green shales and silts, and a prograding delta covering an evaporite basin, is offered by the Carboniferous Sverdrup Basin in arctic Canada (Davies and Nassichuk, 1975). Both the Michigan and Elk Point basins can serve as examples of basins that continued to precipitate carbonates after the evaporite phase had terminated. Both basins, however, show final shale and sand intercalations prior to resumption of a carbonate–shale sequence. Upper Jurassic evaporite basins of central Asia (Fig. 15-23) formed a series of interconnected, small basins (Popov, 1968b) that eventually coalesced. The marginal anhydrite beds transgressed onto surrounding continental deposits, while salt and potash precipitation was restricted to deeper depressions. The approximate location of Mesozoic and Paleozoic evaporite basins in Siberia is shown in Fig. 15-24.

Repeated Cyclothems

Wherever chloride salts were precipitated, there is always more than one cycle of such chloride deposition. The Zechstein sequence is customarily subdivided into cyclothems starting with clay intercalations or carbonates, followed by basal anhydrites and rock salt with intercalations of potash salts, and ending with halite or anhydrite. Thus, the horizons of maximum dilution are placed each time at the beginning of an evaporite cycle.

Four intercalations of gypsiferous carbonate oozes with a Caspian fauna are found in the halite-bearing evaporite sequence of the Gulf of Kara Bogaz Gol off the Caspian Sea (Dzens-Litovskiy and Vasil'ev, 1962). Triassic evaporites in the Ebro Valley, Spain, represent four cyclothems (Castillo Herrador, 1974). Three such major cyclothems are recognized in the Cuanza Basin of Angola (Brognon and Verrier, 1966) and six in the German Zechstein (Richter-Bernburg, 1953; Reichenbach, 1970; Kaeding, 1978), four of which extend into Poland; but as many as seven occur in the British sequence (Raymond, 1953). It is to be expected that the number of cyclothems is highest near the entrance portal and lowest at the distal end of a basin system. Eight such halite cyclothems can be recognized in the Devonian Elk Point Basin of western Canada (Zharkov, 1978), and at least 10 more evaporite cyclothems precipitating only anhydrite continued in the same region throughout Mississippian and even Jurassic times (Reed, 1963).

The Permian evaporites of the Anadarko Basin in western Oklahoma contain three cyclothems (Jordan and Vosburg, 1963). For the Pennsylvanian Paradox Basin between New Mexico and Utah, three megacyclothems (Herman and Barkell, 1957) comprising as many as 29 individual cyclothems have been postulated (Hite, 1961, 1970). In the Pripyat Sag of southern Byelorussia, 20 cyclothems of Upper Devonian halite deposition, including four discrete potash horizons, are restricted to the most depressed parts of the sag, but still show regular transition phases from shore to shore in all directions (Kirikov, 1963). In the Devonian salts of the Dnieper-Donets depression, 6 large cyclothems with smaller rhythms are recognized (Galabuda, 1978). Five or possibly 7 cyclothems comprise the Cambrian evaporites in the southern Siberian Platform. The southern limit of each overlying evaporite sequence shifts gradually northward, away from the marginal foothills towards the platform interiors (Zharkov, 1978). Several cyclothems of Ordovician and Silurian anhydrites overlie these salt sequences. Very thick intercalations of salt-free rocks are marks of the periodic reappearance of open communication with the sea (Voronova, 1960).

In short, *multiple cyclothems of halite,* with or without potash deposition, *separated from each other by anhydrite, dolomite, and shales, are the rule, rather than the exception.* Downwarping of such a salt basin is thus a periodic event (Dellwig, 1955). The similarity of this scenario to the cyclicity of paralic coal measures caused by an analogous nervosity of vertical movements was not lost on Ochsenius (1892). However, he felt compelled to deny the existence of any autochthonous incoaling, and thus his views fell into oblivion.

Figure 15-25 shows a cross section through the Ilga depression west of Lake Baykal in southern Siberia, a subsidiary basin to the large Cambro-Ordovician Angara-Lena Basin. There are over 13 halite cyclothems, each of which thickens from shelf to basin and thins again toward the entrance sill. Each dolomite intercalation is uniform in thickness throughout and marks a stillstand eposide in basin subsidence. Eventually, the sill was covered by a thick unit consisting of mainly dolomites. Rates of subsidence were greatest in the halite-precipitating parts of the basin; they were less in the anhydritic shelf rocks and least in the carbonate beds. Halite precipitation always coincided with maximized rates of subsidence; evidently there was an interrelationship between subsidence and the degree of constriction of brine exchange with connected basins. The rate of subsidence of the entrance sill lagged not only behind the rate in the basin but even behind the rate prevailing on the distal shelf, eventually inviting minor transgressions of halite precipitation onto the shelf. Individual carbonate banks within unit A do not show any thickening toward the basin center and seem to mark stillstand periods in subsidence. Unit B was deposited during a shallowing phase and marks a period of erosion, first over the sill, and finally over the whole area, as indicated by a sandy facies. Unit C was deposited while the basin topography was flat. A rejuvenation of subsidence in the basin during deposition

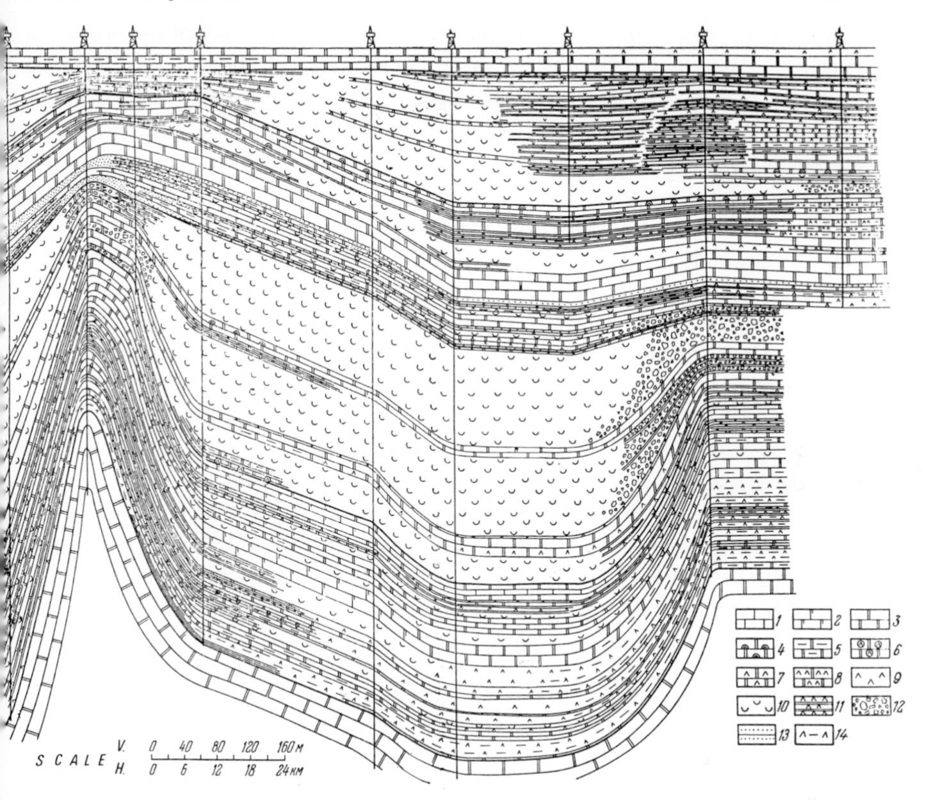

Fig. 15-25. Cross-section through the Ilga depression west of Lake Baykal in southern Siberia. Legend: 1: limestone, 2: limy dolomite, 3: dolomite, 4: reefal dolomite, 5: argillaceous dolomite, 6: dolomite with anhydrite inclusions, 7: anhydritic dolomite, 8: dolomitic anhydrite, 9: anhydrite, 10: rock salt, 11: varved beds, 12: carbonate breccias, 13: sandstones, 14: argillaceous anhydrite. The entrance sill, prominent during the deposition of unit A, collects beach sands during unit B deposition and experiences some uplift. It disappears again during the deposition of the uniformly thick unit C. Unit D shows rejuvenation of the sill as a topographic feature. Unit E again covers it with beds of uniform thickness over both basin and sill (after Adamov *et al.*, 1970).

of unit D coincides with small bursts of uplift in the entrance sill area; these bursts tilted the surrounding halite horizons. Unit E was again deposited over a filled basin.

The Ilga depression could produce a multiplicity of evaporite cycles only because the rates of subsidence of both the entrance sill and the basin center oscillated within narrow limits. If after the second or third evaporite cycle the entrance channel had been widened, the balance between inflow and outflow would have been altered. The same effect would result if tectonic events opened up a second strait into the basin, and thus increased the amount of inflow but decreased its velocity. Further evaporite deposition would cease once the ratio between inflow and outflow equilibrated at some brine density value short of chemical precipitation. Clastic sedimentation would have to take over; the rate of sedimentation would then be only a fraction of that achieved during evaporite precipitation. If a high rate of subsidence remained constant, the basin center would be covered by a progressively greater depth of seawater. We must ask ourselves whether this did not happen in the Gulf of Mexico, where the sea floor covers a salt sequence, or in the Mediterranean Sea, where an opening of the Strait of Gibraltar in the early to middle Pliocene supplied new quantities of water. A rate of subsidence on the order of 1 m/1000 years would be more than sufficient to create the present water depths.

If we utilize the example provided by the Ilga depression (Fig. 15-25), we arrive at a model of an evaporite basin with oscillating rates of subsidence without significant migration of the axis of subsidence. This model could apply to the Michigan Basin with its Silurian, Devonian, and Mississippian evaporites. If, however, we take as a model the Elk Point Basin (Fig. 15-11), we notice an initial basin that fills up and is covered by a carbonate bank of reasonably uniform thickness (the Keg River-Methy dolomite). Renewed subsidence involves not only the original basin but also its former shelf areas, producing a series of basins separated by intermediate sills. The same situation seems to have prevailed in the Zechstein evaporites, where the first cyclothem yielded Werra salts in the German and Polish subbasins but only anhydrites in the British basin margins. Further collapse of former shelves occurred prior to deposition of salts in subsequent cyclothems. Figure 3-9 indicates basins that continue to subside, without the subsidence spreading farther afield. Figure 15-26 shows the model of a basin where subsidence spreads in time from a single axis of subsidence (Fig. 15-26, stage I) to a multiple set of axes of subsidence with onlap onto distal shores. The Triassic evaporites appear to onlap progressively on a shallow shelf and marginal redbeds in Morocco, creating subsidiary axes of subsidence (Salvan, 1974).

The Cambrian evaporite basin in the Angara area of Siberia experienced several cycles of evaporite precipitation, spanning several stages of the Lower and Middle Cambrian. Set into a carbonate platform, it received its water supply through a neck in the northeast, and reefal buildups in that area witness the continuous indraft of marine waters. Repeatedly, the outline of the evaporite deposits shrank and then expanded, suggesting the distinction between deeper basin and shallower shelf areas (Fig. 15-27). The final stage of the evaporite

Fig. 15-26. Evaporite basin with spreading subsidence. After the original basin is filled, the waters transgress over surrounding lowlands. New sites of rapid submergence and chloride precipitation are developed. Two-way flow is maintained through the entrance strait.

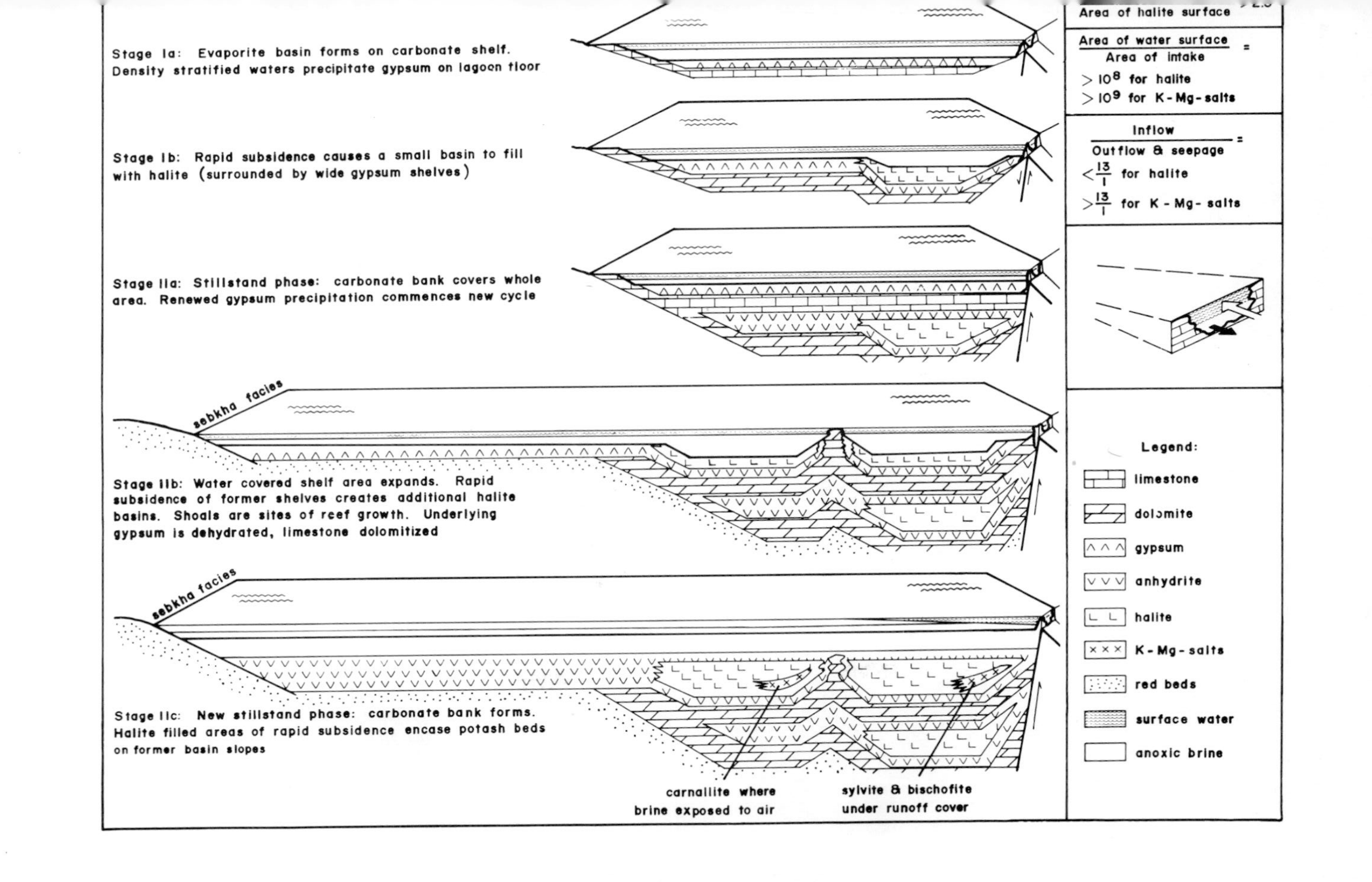

Area of halite surface
Area of water surface / Area of intake =
> 10^8 for halite
> 10^9 for K-Mg-salts
Inflow / Outflow & seepage =
< 13/1 for halite
> 13/1 for K-Mg-salts
Legend:
limestone
dolomite
gypsum
anhydrite
halite
K-Mg-salts
red beds
surface water
anoxic brine
Stage Ia: Evaporite basin forms on carbonate shelf. Density stratified waters precipitate gypsum on lagoon floor
Stage Ib: Rapid subsidence causes a small basin to fill with halite (surrounded by wide gypsum shelves)
Stage IIa: Stillstand phase: carbonate bank covers whole area. Renewed gypsum precipitation commences new cycle
sebkha facies
Stage IIb: Water covered shelf area expands. Rapid subsidence of former shelves creates additional halite basins. Shoals are sites of reef growth. Underlying gypsum is dehydrated, limestone dolomitized
sebkha facies
Stage IIc: New stillstand phase: carbonate bank forms. Halite filled areas of rapid subsidence encase potash beds on former basin slopes
carnallite where brine exposed to air
sylvite & bischofite under runoff cover

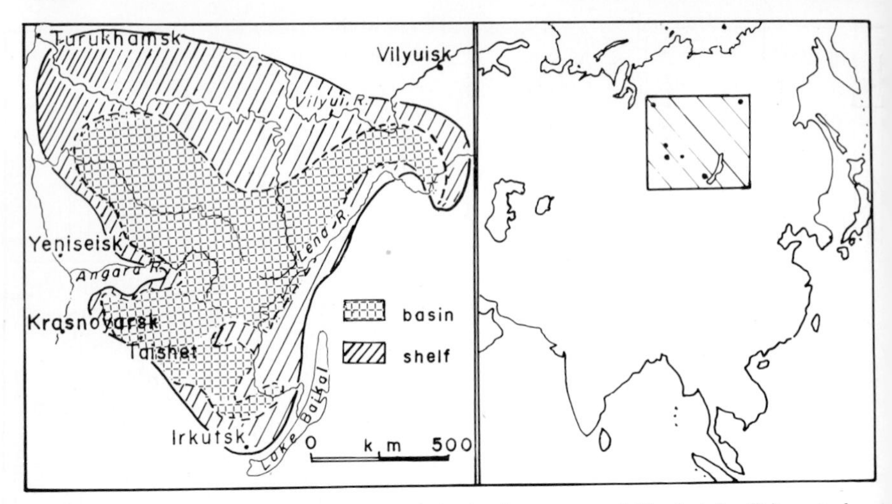

Fig. 15-27. The Cambrian evaporite basin in the Angara area of Siberia (after Britan *et al.*, 1977).

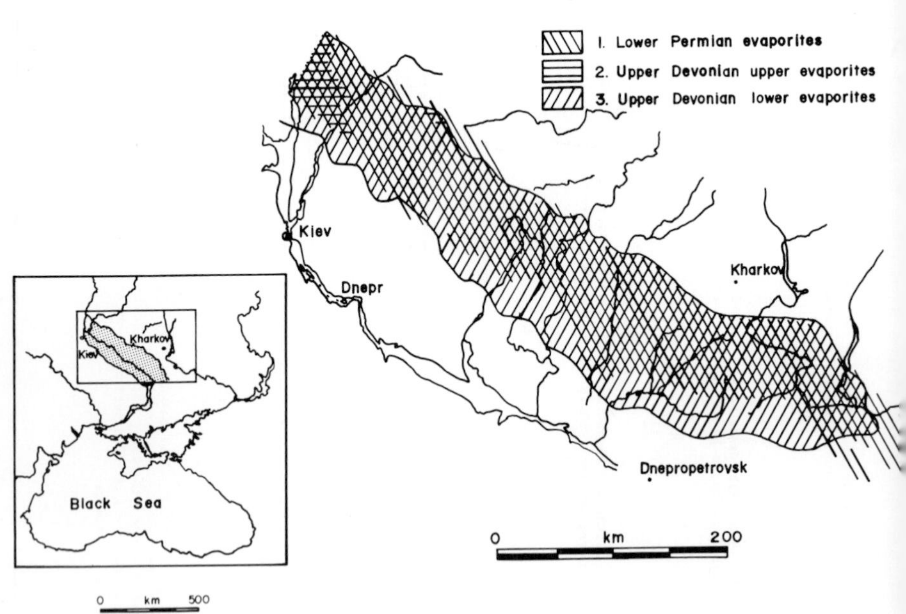

Fig. 15-28. The evaporite basin in the Dnieper-Donets depression: an example of an elongate graben type ($F = 1.82$) that is possibly an abandoned aulacogen (after Kityk and Petrichenko, 1974).

sequence is represented by a very small lagoon in the entrance area and a drying-out lake in the center of the basin. In a graben structure, as developed in the Donets Basin in the Ukraine (Fig. 15-28) and in the Rhine River Valley, an expansion of the evaporite basin is not possible.

FABRIC ANALYSIS

In evaporites there are three types of rock fabric to be considered: fabrics produced during mineral precipitation, fabrics produced by compaction and drainage of interstitial fluids, and fabrics produced by recrystallization in circulating or intercrystalline fluids, by hygroscopic brines or meteoric waters initiating solution and recrystallization. Selenitic gypsum crystals belong in this category; halite develops enlarged crystals only in the absence of anhydrite. Immobilization of halite grain boundaries by anhydrite crystals could inhibit grain enlargement by a process called ''pinning'' (Dix and Jackson, 1982). Until the minute variations in the fabric of precipitates in different types of salinas have been studied in detail, the study of the other two types of fabric continues to have only a theoretical base.

Deeper Burial

Burial under the increasing weight of overburden plasticizes rock salt and to a lesser extent anhydrite. Gypsum does not flow even under pressures of 4.15 GPa. Halite becomes ductile under 12 km of overburden, sylvite under 10 km, and carnallite under 3 km (Fulda, 1928). Although limestones and dolomites have marbles as their corollary in metamorphic rocks, no such high-pressure corollary exists for anhydrite and other evaporites. Well before pressures and temperatures rise to the point of recrystallizing silicate minerals, the halides and sulfates become mobilized and are squeezed out, often as diapirs. Only fluid inclusions in metamorphic rocks bear witness to the former presence of evaporites in their vicinity. There is some doubt whether evaporite sequences have ever been covered by more than about 5 km of rocky overburden. Unless some bending is applied to that overburden, the rock salt must stay in place. Neither the Elk Point Basin of western Canada nor the Michigan and Ohio basins in the Great Lakes area in the United States have produced salt domes or even significant salt pillows. However, wherever the overburden has been exposed to either orogenic or epeirogenic stresses, the salt has been mobilized. It has been suspected of being a gliding or lubricating agent in several fold systems. Salt domes are ubiquitous where the coastal plain is gradually being bent down toward the Gulf of Mexico, where the Arctic Archipelago faces the deepening Arctic Ocean,

Fig. 15-29. Salt mining at Wieliczka in Poland. Lateral pressure release due to mining operations allows halite walls to curve outward (from an old Austro-Hungarian engraving, author unknown).

or where the North German Lowlands subside toward the North Sea. In each case, the strike of salt dome trends roughly parallel to the coastline, parallel to the strike of the downwarp.

Even during mining operations, the salt can creep. An excellent example is provided by the now convex walls of a salt mine near Wieliczka in Poland (Fig. 15-29). Busch (1907) undertook detailed observations and found that rock salt expands into voids only under at least 300 m of overburden, whereas carnallite and sylvite begin to expand sooner. The absence of significant porosity values in many salt mines is thus not primary, but is a function of deep enough burial. Moreover, most percolating fluids that induced sulfatization and other oxidation phenomena would have passed through well before the overburden became 300 m thick.

Salt tectonics is largely a postdepositional phenomenon, and as such is a major subject by itself. A thorough understanding of its causes can be attained only if the formation of the evaporite basin and its precipitates is completely understood.

Part VI

MINERALOGICAL DATA

16

Evaporitic Minerals

Brief notes are in order on minerals encountered in an evaporite sequence, as well as on their main diagenetic alterations, to understand some of the relationships involved in precipitation. A very small number of cations and anions participate in the precipitation of evaporite minerals. Very few evaporite mineral species are composed of more than three to four elements and water. The major cations involved are sodium, potassium, calcium, and magnesium; the major anions are chloride, sulfate, and to some extent carbonate. The variety of sulfate minerals in marine sediments is much greater than that of chlorides, but their aggregate volume is smaller. The composition of some of the minerals in marine evaporites is given in Table 16-1. Iron, rubidium, cesium, strontium, thallium, and to a very minor degree ammonium sometimes substitute for other cations. Borates, bromides, and fluorides are very subordinate anions. Aluminum as a cation, and silicates and hydroxides as anions, occur mainly in conjunction with clay intercalations. Manganese, copper, zinc, and lead are concentrated in the brine, but are for the most part flushed into formation waters and precipitated elsewhere. Lithium, boron, and iodine also remain largely in solution. With the exception of radiogenic helium and argon and biogenic nitrogen all other elements occur only in the parts per billion range in lattice imperfections. Phosphates are for the most part almost antithetic to evaporite precipitation.

Excluded from this list are simple and compound iron and aluminum sulfates and carbonates, which have so far been recognized only in weathering environments and not as evaporitic precipitates. For details of crystal habit, indices of refraction, hardness, thermal curves, and x-ray diffraction patterns, the reader is referred to appropriate textbooks on mineralogy.

TABLE 16-1

Composition of Some Minerals in Marine Evaporites (in kg/m^3)

Mineral	Formula	Na	K	Ca	Mg	Sr	Cl	SO_4	CO_3
Anhydrite	$CaSO_4$			877				2103	
Aphthitalite	$Na_2K_6(SO_4)_4$	184	937					1534	
Aragonite	$CaCO_3$			1173					1757
Baeumlerite	$KCaCl_3$		485	497			1318		
Bassanite	$Ca_2(SO_4)_2 \cdot H_2O$			746				1786	
Bischofite	$MgCl_2 \cdot 6H_2O$				192			559	
Bloedite	$Na_2Mg(SO_4)_2 \cdot 5H_2O$	292			155			1222	
Calcite	$CaCO_3$			1086					1625
Carnallite	$KMgCl_3 \cdot 6H_2O$		226		140		613		
Celestite	$SrSO_4$					864		2070	
Dolomite	$CaMg(CO_3)_2$			312	189			2371	
Epsomite	$MgSO_4 \cdot 7H_2O$				165			653	
Glauberite	$Na_2Ca(SO_4)_2$	471		411				1968	
Goergeyite	$K_2Ca_5(SO_4)_6 \cdot H_2O$		248	636				1828	
Gypsum	$CaSO_4 \cdot 2H_2O$			539				1294	
Halite	$NaCl$	852					1313		
Kainite	$K_4Mg_4Cl_4(SO_4)_4 \cdot 11H_2O$		340		212		309	836	
Kieserite	$MgSO_4 \cdot H_2O$				451			1783	
Langbeinite	$K_2Mg_2(SO_4)_3$		533		332			1965	
Leonite	$K_2Mg(SO_4)_2 \cdot 4H_2O$		469		146			1154	
Loewite	$Na_{12}Mg_7(SO_4)_{13} \cdot 15H_2O$	340			210			1540	
Mirabilite	$Na_2SO_4 \cdot 10H_2O$	213						443	
Picromerite	$K_2Mg(SO_4)_2 \cdot 6H_2O$		418		130			1025	
Polyhalite	$K_2Ca_2Mg(SO_4)_4 \cdot 2H_2O$		360	369	112			1678	
Strontianite	$SrCO_3$					2,204			1576
Sylvite	KCl		1041				943		
Syngenite	$K_2Ca(SO_4)_2 \cdot H_2O$		618	317				1519	
Tachyhydrite	$CaMg_2Cl_4 \cdot 12H_2O$			129	157		685		
Thenardite	Na_2SO_4	873						1825	
Vanthoffite	$Na_6Mg(SO_4)_4$	680		119				1894	

MARINE CALCIUM OR STRONTIUM CARBONATES

Aragonite $CaCO_3$ $\rho = 2.93$

Solubility: at 25°C, 0.0153 g/liter
at 75°C, 0.0190 g/liter

Aragonite is the prevalent form of calcium carbonate precipitated in marine brines. In continental lakes, where magnesium interference is low, calcite is the common form. Kinsman and Holland (1969) showed that aragonite is the preferred form of crystallization in waters above 30°C.

Calcite $CaCO_3$ $\rho = 2.711$

Solubility: at 25°C, 0.014 g/liter
at 75°C, 0.018 g/liter

Most recent calcites are high-magnesium calcites (with magnesium present in solid solution), ancient ones are not. Magnesian calcites are known only from Cenozoic localities. A siderite ($FeCO_3$) content of 1–15% produces ferrocalcite. Scalenohedral, tabular, prismatic, or rhombohedral crystals are common. The mineral often forms fine-grained, distinctly granular masses. According to Kinsman and Holland (1969), a mixture of aragonite and magnesian calcite forms at 16°C or above. At 3°C it is a mixture of low-magnesium calcite, vaterite, and monohydrocalcite ($CaCO_3 \cdot H_2O$). Calcite crystals grow rapidly sideways in freshwater environments and produce stubby crystals. As the salinity increases, the lateral growth is curtailed and the calcite crystals become more elongated.

Vaterite $CaCO_3$ $\rho = 2.56$

Vaterite forms fibrous hexagonal tabular or lenticular crystals. It occurs naturally as shell material of several gastropods. In time, it alters generally to aragonite. In an aqueous medium, vaterite can change to calcite at 40°C (Bayer and Wiedemann, 1979). In lake waters, vaterite forms when the pH rises from around 7.2 to 8.5–9.2, due to the high level of algal photosynthesis (Rowlands and Webster, 1971). Cleavage plates of gypsum immersed at room temperature into a 7.5% Na_2CO_3 solution completely converted within a week to pseudomorphs of vaterite, calcite, and a little gaylussite. Powdered gypsum yielded only calcite in a similar experiment (Floerke and Floerke, 1961).

Monohydrocalcite $CaCO_3 \cdot H_2O$
Hexahydrocalcite $CaCO_3 \cdot 6H_2O$

So far, these minerals have been identified in antarctic lakes. Since the presence of sodium chloride favors the transformation of hydrocalcite to aragonite (Marschner, 1969), hydrocalcite is not stable in marine brines.

Strontianite $SrCO_3$ $\rho = 3.780$

Solubility: at 18°C, 0.011 g/liter
at 100°C, 0.65 g/liter

Pale green, white, or yellowish brown in color, strontianite forms orthorhombic crystals in geodes and veinlets in dolomite. Yellowish-green strontium aragonite [$CaSr(CO_3)_2$] has been discovered in association with galena and pyrite in China (Kao *et al.*, 1981).

Dolomite $CaCO_3 \cdot MgCO_3$ $\rho = 2.872$

Solubility: at 18°C, 0.32 g/liter

Dolomite forms cryptocrystalline masses, saccharoidal coarser crystal aggregates, or euhedral rhombohedra in calcite masses. Almost all dolomites contain some iron that substitutes for magnesium in the crystal lattice. When the Mg/Fe ratio approaches 1.5, the mineral is called "ankerite." Ankerite is particularly common in hydrocarbon-bearing beds (Teodorovich, 1958).

SIMPLE CALCIUM OR STRONTIUM SULFATES

Gypsum (formerly karstenite) $CaSO_4 \cdot 2H_2O$ $\rho = 2.317–2.33$

Solubility: at 0°C, 2.41 g/liter
at 100°C, 2.22 g/liter

Gypsum occurs as monoclinic-prismatic crystals or as massive, fine-grained bodies. The crystals have various habits: tabular, columnar, angular, and lenticular. Thick platy or bladed crystals are often very large. Dense gypsum is fine to coarse-grained. The variety called "selenite" occurs in veinlets, has a fibrous habit, and is yellow or pink, with a silky luster. In the crystallized state, gypsum is colorless and transparent. In dense masses it is white, gray, pink, red, or brown, sometimes spotty or marbled. Gypsum is poorly soluble in fresh water; its solubility increases slightly in concentrated brines. When roasted or exposed to hot, dry air, it gradually loses its water, turns into the polyhydrate (bassanite), and may even change into anhydrite.

Bassanite (formerly vibertite, miltonite) $2CaSO_4 \cdot H_2O$ $\rho = 2.753–2.761$

Solubility: at 20°C, 3.0 g/liter; decreases with rising temperature

Bassanite crystallizes in the ditrigonal scalenohedral system and occurs as a white powder wherever gypsum is exposed to very dry air. It may lose enough water to acquire the composition $CaSO_4 \cdot 0.02H_2O$ (Kendall and Skipwith, 1969). There is a great difference in the hydration rate between powdered material and the compact mineral hydrated only superficially. Two forms of the hemihydrate are known in the plaster industry. The α-hemihydrate cannot be produced at atmospheric pressure (Kuntze, 1965), and thus does not occur in nature. Bassanite has been found in continental desert deposits of central Asia and also as thin layers in oil sands (V. I. Popov and Vorobiev, 1947), 100 m below two desert basins in southern California (Allen and Kramer, 1953), in fracture fillings of deeply buried Hungarian bauxites (Bardossi *et al.*, 1979), or even in coal mines (Koritnig, 1967). Porphyroblasts and nodules of bassanite occur in massive Miocene gypsum of Bulgaria in a fine-grained ground mass of anhydrite (Trashliev, 1971), here as a product of incomplete rehydration. Frik and Kuzel (1982) recognized a variety with the formula $CaSO_4 \cdot 0.52H_2O$, crystallizing in the hexagonal system and a variety with the formula $CaSO_4 \cdot 0.48H_2O$, crystallizing in the orthorhombic system. Na-polyhalite $[(Na,K)_{(2x)}Ca_{(1-x)}SO_4 \cdot 5H_2O]$, $\rho = 2.75$, may be a rare variety (Gudowius and von Hodenberg, 1979).

Anhydrite (formerly muriazite) $CaSO_4$ $\rho = 2.963–2.98$

Solubility: at 20°C, 2.98 g/liter
at 30°C, 2.09 g/liter
at 100°C, 1.619 g/liter

Anhydrite usually forms continuous, very dense and fine-grained, almost monomineralic rocks of various textures. Sometimes it forms lamelliform or fibrous aggregates. There are cleavage planes at right angles to each other, simulating a cubic form of fragments, although anhydrite crystallizes in the rhombic system. Its color is white, gray, bluish-gray, yellow, tan, pink, red, or purple, and sometimes spotty or marbled. Anhydrite dissolves poorly in water, but under the prolonged influence of moisture it alters to gypsum. Anhydrite is widespread in evaporitic sequences and occurs disseminated in various saline rocks, as interbeds, laminations, nests, blocks, and as independent deposits of large areal extent and great thickness. Most ancient anhydrites formed through diagenetic dehydration of precipitated gypsum. Anhydrite also occurs in contemporary lakes as small grains and crystals that are undoubtedly of primary, authigenic origin. Some of the individual anhydrite crystals found in sebkha environments are also probably primary.

Some anhydrites show a small calcium deficiency. In addition to $CaO \cdot SO_3$, $CaO \cdot (SO_3)_2$ and $CaO \cdot (SO_3)_3$ have also been identified by x-ray diffraction (Abaulina *et al.*, 1976).

Celestite $SrSO_4$ $\rho = 3.971$

Solubility: at 0°C, 0.113 g/liter
at 30°C, 0.114 g/liter

Celestite frequently contains admixtures of calcium and barium. There is a complete isomorphous series ranging from celestite through barytocelestite to barite ($BaSO_4$). Calcium may substitute up to 12%, but more than 2–3% substitution is rare. Orthorhombic crystals are tabular or prismatic. The color is deep blue, pale bluish-gray, or colorless. If the color fades in sunlight, it is restored by x-ray irradiation (Zaritskiy, 1960). Celestite is insoluble in hydrochloric acid but dissolves in strong sulfuric acid. It occurs in dolomite, gypsiferous shales and marls, and anhydrites, particularly those forming intercalations in halites.

A variety is *kalistrontite* [$K_2Sr(SO_4)_2$, $\rho = 3.2$–3.32], a transparent, colorless mineral with a glassy luster. Hexagonal, prismatic crystals grow up to 15 mm. It occurs as a replacement after anhydrite and is isostructural with *palmierite* [$(K,Na)Pb(SO_4)_2$], but is distinctly different from aphthitalite [$K_3Na(SO_4)_2$] (Voronova, 1962).

ROCK-FORMING CHLORIDE MINERALS

Halite NaCl $\rho = 2.135$–2.16366–2.1647

Solubility: at 0°C, 357.0 g/liter
at 100°C, 391.2 g/liter
Specific heat: at 0°C, 105.5 J/m^3
at 100°C, 112.2 J/m^3

Halite crystals are cubic if they can grow freely. Halite possesses perfect cubic cleavage and a conchoidal fracture. It is a brittle mineral, but easily becomes plastic at elevated pressures and temperatures. Additions of very small amounts of sodium ferrocyanide [$Na_4Fe(CN)_6 \cdot 10H_2O$] produce dendritic crystals; addition of carboxymethyl cellulose produces euhedral octahedra (Shearman, 1966). The addition of more than 12 g of glycerine to 100 g of a sodium chloride solution produces rhombic dodecahedra of halite, because the glycerine is adsorbed onto the (100) face (Seifert, 1952). Halite that fills fractures or small voids in rocks and forms veinlets has a fibrous habit. There is 50–60% pore space in scaly halite.

Hydrohalite (formerly maakite) $NaCl \cdot 2H_2O$ $\rho = 1.54$

Hydrohalite crystallizes from concentrated saltwater at temperatures below −0.1°C and from seawater below −23°C. In the presence of sulfate ions, while mirabilite is precipitating, hydrohalite will form only below −3.5°C (Pel'sh, 1940). Hydrohalite can take on a selenitic appearance (Deicha, 1947). Only when it crystallizes slowly does hydrohalite produce whole crystals. Rapid crystallization produces a dendritic pattern (Deicha, 1976). When warming up, hydrohalite recrystallizes to halite, losing 53.4% of its volume. It reverts to halite and water at temperatures above +0.2°C (d'Ans, 1933).

Sylvite (formerly leopoldite, schaetzelite, or hoevelite) KCl $\rho = 1.984–1.9897$

Solubility: at 20°C, 347.0 g/liter
at 100°C, 567.0 g/liter

Primary sylvite usually forms intimate intergrowths with halite in massive, fine to coarse crystalline accumulations. The resulting rock is called "sylvinite." Euhedral crystals of sylvite are much rarer than those of halite. They form either simple cubes or cubes combined with octahedra. Very rarely, one may find hopper crystals, indicating direct precipitation from bitterns (Dubinina, 1951b). Sylvite assumes a fibrous habit in fracture fillings. Sylvite has a very burning, salty taste and a cubic cleavage, and is less brittle and more plastic than halite.

Pure sylvite is colorless and transparent, but most of the time it is discolored pink, brick red, flesh-colored, brownish-red, deep or pale yellow, or milky white, sometimes with a red rim around the crystals. The discoloration is caused by finely disseminated hematite needles and platelets; the exact hue is determined by the amount of hematite present. Unlike halite, sylvite almost never shows blue colors. Radioactive breakdown of the K—Cl bond produces a purplish-violet color (Przibram, 1927a). The milky white color is due to very fine interstices filled with nitrogen and methane, with a small admixture of bitumina (Ivanov, 1953). A bituminous odor is exuded whenever a milky white crystal is scratched.

Oriented growths of cryptomelane ($K_2O \cdot MnO \cdot 15MnO_2$) have been reported in sylvites of Carlsbad, New Mexico (Sun, 1962).

Anhydrite with sylvite and halite is called ''anhydritic hard salt'' in Germany. It is distinguished from kieseritic and langbeinitic hard salt on the basis of the sulfate mineral, which in each case is secondary.

Carnallite $KCl \cdot MgCl_2 \cdot 6H_2O$ $\rho = 1.602$

Solubility: at 19°C, 645.0 g/cm^3
at 25°C, disintegrates in fresh water

Carnallite usually forms granular masses, with particle sizes ranging from a few millimeters to more than 5–6 cm. The carnallite grains are rounded or subangular. Carnallite crystallizes in the rhombic (pseudohexagonal) system; its taste is unpleasantly bitter, burning, and salty. Often the crystals are intimately intergrown with halite, but they also occur together with sylvite and other precipitated minerals. Primary carnallite intergrown with halite is called ''carnallitite.'' In veinlets, carnallite has a fibrous habit. Euhedral crystals are barrel-shaped and very rare. Colorless and transparent carnallite can occasionally be found, but most of the time the mineral is stained various shades of red, orange, and yellow. This staining is caused by inclusions of very fine needles and hexagonal platelets of hematite. The hematite needles cause a peculiar sheen when dark-colored grains are rotated, and a silky luster in the yellow varieties. Carnallite is very hygroscopic. It does not possess any cleavage and has an irregular or conchoidal fracture. A feature peculiar to carnallite alone is a squeaky noise, resembling that of a grinding rocksaw, which is heard whenever the mineral is scratched with any metal object (''crackle salt'').

Rubidium, cesium, thallium, or ammonium may substitute in traces for potassium, and iron for magnesium. Some substitution of iron for magnesium occurs as $KCl \cdot FeCl_2 \cdot 6H_2O$, which dehydrates to $FeCl_2 \cdot KCl \cdot 6H_2O$ at 38.3°C (Boeke, 1911).

Bischofite $MgCl_2 \cdot 6H_2O$ $\rho = 1.56–1.604$

Solubility: 1670 g/liter in cold water
3670 g/liter in hot water
at 116°C, decomposes

Bischofite forms by alteration of carnallite in water. However, it can also form as a primary mineral after the potassium supply has been exhausted in a concentrating brine. It is a monoclinic (pseudohexagonal) mineral, white or colorless, with a vitreous luster. It occurs as granular, leafy, or sometimes fibrous masses. Its taste is stingingly bitter, and it is deliquescent.

Antarcticite $CaCl_2 \cdot 6H_2O$ ρ = 1.700–1.715

Solubility: at 0°C, 2790 g/liter
at 20°C, 5360 g/liter
at 29.92°C, decomposes

Colorless, prismatic crystals have been found in Victoria Land, Antarctica, and in Bristol Dry Lake, California (Dunning and Cooper, 1969).

Tachyhydrite (or tachhydrite, or tachydrite) $CaCl_2 \cdot 2MgCl_2 \cdot 12H_2O$ ρ = 1.667–1.669

Solubility: 8000 g/liter in warm water

The crystals are rhombohedral, with a waxy to honey yellow color and a glassy luster. The yellow color is derived from a partial replacement of $MgCl_2$ by $FeCl_2$ at the rate of 1 part of $FeCl_2$ for every 10 parts in solution (Kling, 1915). It occurs in antarctic lakes and occasionally as a secondary mineral in marine evaporites. Major deposits occur in Brazil, the Congo, and Thailand; minor nests occur in Permian Zechstein evaporites of Germany.

In the quinary system of $CaCl_2$–$MgCl_2$–KCl–NaCl–H_2O, Meyer *et al.* (1949) coprecipitated $CaCl_2 \cdot 4H_2O$ with tachyhydrite at 35°C, but such a compound has not yet been found in naturally occurring tachyhydrites.

ACCESSORY HALIDE MINERALS

Baeumlerite (or hydrophilite or chlorocalcite) $CaCl_2 \cdot KCl$ ρ = 2.3

Solubility: extremely soluble

This is a white, transparent mineral, cubic/pseudocubic, and deliquescent. It is found in association with halite and sylvite in fumaroles, and is also intergrown with tachyhydrite and halite in German Zechstein evaporites.

Sal Ammoniac NH_4Cl ρ = 1.527

Solubility: at 0°C, 107.4 g/liter
at 85°C, 425.4 g/liter

Sal ammoniac has been found in halite and potash sequences (Ochsenius, 1898). White sylvites contain about 70–1400 ppm sal amoniac, whereas pink carnallite contains about 6 ppt (Apollonov, 1976). *Kremersite* $[(NH_4,K)_2Cl_2 \cdot FeCl_3 \cdot H_2O]$ is a compound variety.

Chloraluminite $AlCl_3 \cdot 3H_2O$

Chloraluminite is found in potash salts (Ochsenius, 1898) and has been described by Reventos *et al.* (1974).

Scacchiite $MnCl_2$

Scacchiite is found as a deliquescent mineral associated with halite and sylvite in fumaroles. Reported with it are *chloromagnesite* ($MgCl_2$) and *chloro-manganokalite* ($4KCl \cdot MnCl_2$). It is uncertain whether any one of them occur in traces in evaporites.

Rinneite $FeCl_2 \cdot 3KCl \cdot NaCl$ $\rho = 2.3–2.347$

Rinneite, which is colorless, rose, yellow, or violet, crystallizes in the hexagonal system. It has a vitreous luster, is brittle, and shows a conchoidal fracture. It often contains substitutions from the iron group of elements, such as manganese, titanium, or copper (Pustyl'nikov, 1975). The minimum temperature of formation is 26.4°C (Boeke, 1911) in the system $NaCl–KCl–FeCl_2–H_2O$. It is dehydrated to $FeCl_2 \cdot KCl \cdot 2H_2O$ at 38.3°C (Kolosov and Pustyl'nikov, 1970b). Rinneite occurrences are thus usable as indicators of paleotemperature limits.

Although it substitutes for magnesium in carnallite, iron may also substitute for potassium. At 22.8°C, $FeCl_2 \cdot MgCl_2 \cdot 8H_2O$ can form as a metastable mineral.

Douglasite $KCl \cdot FeCl_2 \cdot 2H_2O$ or $FeCl_2 \cdot 2KCl \cdot 2H_2O$

Douglasite is green and coarsely granular. It quickly turns brown, and becomes covered with a rust-brown crust.

Rokuehnite $FeCl_2 \cdot 2H_2O$ $\rho = 2.358$

Rokuehnite has recently been described (cf. Herrmann *et al.*, 1980).

Molysite $FeCl_3$ $\rho = 2.90–3.04$

Molysite is yellowish- to brownish-red, but green in reflected light.

Erythrosiderite $FeCl_3 \cdot 2KCl \cdot H_2O$ $\rho = 2.32–2.372$

Erythrosiderite, which crystallizes in the rhombic system, can be red due to the presence of ferric iron. It is an alteration product of rinneite, which is often covered by it. Otherwise, erythrosiderite occurs as yellow flakes on halite grains (Kolosov and Pustyl'nikov, 1967). Ammonia can substitute for part of the potassium, leading to kremersite, with $NH_4 > K$.

Cotunnite $PbCl_2$ $\rho = 5.81$

Cotunnite is colorless, white, yellowish to greenish, or transparent. Found in saline lakes, it also forms when lead utensils are left in seawater. Occasionally, it occurs with *pseudocotunnite* ($2KCl \cdot PbCl_2$).

Eriochalcite ($CuCl_2 \cdot 2H_2O$) $\rho = 2.55$

Eriochalcite forms bluish-green, lichenlike aggregates around efflorescences on volcanic soil.

Fluorite CaF_2 $\rho = 3.180$

Fluorite is colorless, and occurs in occasional nests or crusts in the upper layers of evaporite deposits. In German Zechstein dolomites, fluorites occur in layers within dark dolomites that are rich in organic matter. The fluorite here is a late diagenetic addition (Ziehr *et al.*, 1980).

Sellaite MgF_2 $\rho = 3.08–3.15$

Solubility: at 18°C, 0.076 g/liter

Sellaite, colored faintly violet, has been found in carnallite (Kuehn, 1952b), but also in paragenesis with bitumina in the Main Anhydrite Member of the third cyclothem of Zechstein evaporites (Heidorn, 1931), or dependent on remnant bitumina in dolomitic stinkstones of the Zechstein. It is absent in clean oolitic dolomites (Heinrich, 1966).

Villiaumite NaF
Neighborite $NaF \cdot MgF_2$

These minerals have hitherto not been identified in evaporites.

Carobbite KF

Carobbite has been described by Reventos *et al.* (1974).

Sulfohalite $2Na_2SO_4 \cdot NaCl \cdot NaF$ $\rho = 2.5$

Sulfohalite occurs in evaporitic sediments of Kara Bogaz Gol, off the Caspian Sea (Andryasova and Lepeshkov, 1971).

Isokite $CaMg(PO_4)F$

Isokite is monoclinic and white, with some substitution of arsenic for phosphorus.

Wagnerite $Mg_2(PO_4)F$ $\rho = 3.15$

Wagnerite is monoclinic, yellow, red, or green. Iron or calcium may substitute for magnesium.

Fluorapatite $Ca_5(PO_4)_3F$ $\rho = 3.1–3.2$

Fluorapatite crystals are hexagonal. The three phosphatic minerals, fluorapatite, isokite, and wagnerite, have been reported on very rare occasions, in addition to lueneburgite, a phosphatic borate mineral.

COMPOUND POTASSIUM SULFATES

With the exception of gypsum, all other sulfate minerals in marine sequences commonly form nests and layers of secondary alteration, and are rarely primary. In paralic and continental lakes, however, they are often the prime evaporite minerals. No anhydrous or hydrated, simple potassium sulfate occurs in the evaporitic environment. Potassium sulfates are always compound salts of potassium–calcium, potassium–magnesium, or potassium–calcium–magnesium combinations. Precipitation of such compound potassium sulfates is fostered in the laboratory by the presence of tartrates in a brine containing sulfate, calcium, and potassium ions (Combe and Smith, 1965). The following minerals are of major significance.

Polyhalite (formerly mamanite) $K_2SO_4 \cdot 2CaSO_4 \cdot MgSO_4 \cdot 2H_2O$ $\rho = 2.775–3.0$

Solubility: incongruent

Polyhalite usually forms massive, very fine-grained aggregates in which individual crystals cannot be distinguished with the naked eye. It is also found in coarsely fibrous, scaly, or columnar varieties, as small nests or nodules, or as seams composed entirely of polyhalite; these seams are a few centimeters to a few meters thick. No euhedral triclinic (pseudorhombic) crystals have been found so far. The fracture is uneven, conchoidal, or hackly; the mineral is transparent on the edge and sometimes has a silky luster. It is poorly soluble in water and has no taste. The color of polyhalite varies over very short distances. It includes different hues of red, brownish-red, pink, pinkish-violet, orange-red, yellowish-orange, yellow, gray, grayish-white, and white. Gray and white polyhalites outwardly resemble anhydrite and even have the same hardness. Polyhalite occurs in all potassium sulfate deposits, and is often the only potash salt in rock salt sequences. Polyhalitization is widespread, and undoubtedly most of the polyhalite is secondary in origin. However, in thin seams and laminae extending for great distances, it is of primary (syngenetic) origin and often has organic material included in the crystal lattice. Krugite ($K_2SO_4 \cdot 4CaSO_4 \cdot MgSO_4 \cdot 2H_2O$) is not a mineral but a mixture of polyhalite and anhydrite.

Syngenite (formerly kaluszite) $K_2SO_4 \cdot CaSO_4 \cdot H_2O$ $\rho = 2.597$

Solubility: 2.5 g/liter in cold water
decomposes in hot water

Syngenite is a secondary monoclinic mineral derived from the reaction of potassium chloride brines with gypsum. In Antarctica, syngenite replaces gypsum in encrustations on vesicular basalt (Lindholm *et al.*, 1969).

Goergeyite $K_2SO_4 \cdot 5CaSO_4 \cdot H_2O$ $\rho = 2.77$

Solubility: incongruent

Goergeyite is a rare secondary mineral after gypsum.

Kainite $4KCl \cdot 4MgSO_4 \cdot 11H_2O$ $\rho = 2.131–2.190$

Solubility: at 18°C, 795.6 g/liter

Kainite usually occurs in massive, fine-grained form, closely intergrown with halite and other minerals, mainly sulfates. The rare euhedral monoclinic crystals have a prismatic habit. Fracture fillings with a fibrous habit are very rare. Kainite is typically light yellow or honey yellow, and rarely yellowish-green, yellowish-gray, or brownish-gray. Kainite is very soluble in water, but is less hygroscopic than other potash minerals. It is the only cryophile sulfate among the evaporite minerals, migrating toward the cool end of the solution. The taste is weakly salty and bitter, resembling that of cheap soap. Kainite is seldom present in commercial quantities. It often forms the caprock of secondary alteration after kieserite and carnallite or sylvinite, bypassing halites (Lotze, 1957). It is impossible to precipitate kainite at 25°C; in the laboratory, it is produced only by rapid cooling of a very hot brine (Braitsch and Herrmann, 1963). Above 83°C kainite is unstable (Fulda, 1928), but Dubil *et al.* (1981) succeeded in crystallizing it from aqueous solutions below 95°C.

Anhydrokainite $K_2SO_4 \cdot MgSO_4 \cdot MgCl_2$ $\rho = 2.2$

Solubility: incongruent

Anhydrokainite is the anhydrous equivalent of kainite which has been stripped of its crystal water. Anhydrokainite is, therefore, much less common than kainite. There is some question of whether anhydrokainite, synthesized in the laboratory, actually occurs in nature (Linstedt, 1952).

Langbeinite $K_2SO_4 \cdot 2MgSO_4$ $\rho = 2.829$

Solubility: incongruent above 61°C

Langbeinite usually occurs in coarse- to very coarse-grained masses, in which individual irregular-shaped cubic crystals grow up to several centimeters long. Rare euhedral crystals are tetrahedra up to 2 cm in diameter. Cleavage is absent, and the fracture is irregular or conchoidal. The color is light gray, grayish-white, colorless, pale pink, or pinkish-violet, with a glassy luster. Some varieties resemble quartz in appearance. They are easily distinguished from halite by their lack of cleavage, greater hardness, and lower solubility. In cold water langbeinite dissolves more slowly than halite, sylvite, or carnallite, but it takes up atmospheric moisture very easily. Even at a relatively dry site, langbeinite is quickly covered with a fine film of white or light yellow, powdery picromerite and epsomite (Ivanov, 1953). The taste is slightly salty. Langbeinite is thermoluminescent. If struck with a pick or hammer, it produces a bluish-green luminescence that is distinctly visible for an instant, even under poor light conditions. If a powder is placed on a heated metal plate, one can observe the same distinct green luminescence. Langbeinite is stable above 37°C (Fulda, 1928). Thus, it cannot occur as primary mineral in brines cooling even temporarily below that temperature. It is probably derived from the interaction of carnallite with oxygenated brines carrying sulfate ions. Langbeinite rarely occurs in large quantities; the largest commercial deposit is in New Mexico. Another widespread occurrence is in the Carpathian foothills (Lobanova, 1953). Langbeinite with sylvite and halite is called "langbeinitic hard salt" in Germany.

Leonite $K_2SO_4 \cdot MgSO_4 \cdot 4H_2O$ $\rho = 2.201$

Solubility: very slight in cold water

Leonite is a secondary monoclinic mineral after carnallite or langbeinite. It is colorless, white, or yellowish, with a vitreous luster. It occurs in small nests (Weber, 1931) in the Stassfurt Member of German Zechstein evaporites. The stability limits of leonite are 18°C and 61.5°C (Fulda, 1928). Therefore, it occurs only in relatively shallow horizons down to about 1200 m.

Picromerite (formerly schoenite) $K_2SO_4 \cdot MgSO_4 \cdot 6H_2O$ $\rho = 2.028–2.15$

Solubility: at 0°C, 192.6 g/liter
at 20°C, 250.0 g/liter
at 75°C, 598.0 g/liter
above 72°C, decomposes

In macroscopically discernible occurrences, picromerite is very fine-grained or fibrous, with a glassy luster and a perfect cleavage. The color is usually yellow, somewhat brighter than kainite, colorless, or white. An argillaceous admixture produces a dirty yellow color. Where it weathers out, it is quickly covered by a white film of epsomite. Although there is a stability field for picromerite in solar evaporation of open bitterns at temperatures up to 26°C (Eberhardt, 1971), in nature it is usually a secondary mineral derived from the hydration of langbeinite in the presence of water vapor (Yarzhemskiy, 1950) or the leaching of kainite (Aksenova and Khodakovskiy, 1973). It can also form from sakiite in the presence of potassium sulfate solutions at temperatures below 47.2°C (Kuehn, 1952a).

MAGNESIUM SULFATES

Kieserite (formerly wathlingenite) $MgSO_4 \cdot H_2O$ $\rho = 2.57$

Solubility: 419 g/liter in hot water
slowly soluble in cold water

Kieserite is found in white to gray massive bodies as blocks, nodules, impregnations, and veinlets. It can also be colorless, yellowish, greenish, or reddish, with a glassy luster. More often, kieserite forms microscopic particles inside other evaporite minerals. Euhedral monoclinic crystals are rare. Kieserite is hygroscopic. It becomes cloudy in humid air and substantially increases its volume, gradually disintegrating into a white powder of epsomite. Similar kieserite–epsomite grains and small nodules form on the weathered walls of mine workings in sulfatic potash ores. Kieserite is typical of individual zones within a potash deposit, such as the kieseritic sylvinite in the Zechstein of Germany or in the Ozin deposit of Saratov, USSR. In both instances, it is a secondary alteration product created by percolating solutions. However, kieserite is also found in very small quantities in contemporary lakes. The temperature here must exceed 18°C at all times, as this is the lower limit for the stability of this mineral (Fulda, 1928).

Kieserite with halite and sylvite is called "kieseritic hard salt" in Germany. It is stable only at temperatures above 72°C (Fulda, 1928).

Sanderite $MgSO_4 \cdot 2H_2O$
Leonhardtite (formerly starkeyite) $MgSO_4 \cdot 4H_2O$
Allenite (formerly pentahydrate) $MgSO_4 \cdot 5H_2O$
Sakiite (formerly hexahydrite) $MgSO_4 \cdot 6H_2O$

These four minerals occur as an efflorescence on kieserite in mines. Sakiite is stable between 13 and 31.5°C. (Leonhardtite was misidentified as hydrated iron sulfate and as such was given the name "starkeyite.")

Epsomite (formerly reichhardtite, pickrite) $MgSO_4 \cdot 7H_2O$

ρ= **1.636–1.677**

Solubility: at 20°C, 710 g/liter
at 40°C, 910 g/liter
at 150°C, disintegrates

Epsomite occurs in contemporary lake deposits as prismatic, angular, acicular, and fibrous rhombic (pseudotetragonal) crystals or as massive encrusted or earthy aggregates. Sometimes it forms beds several meters thick, or it occurs as an admixture to mirabilite or bloedite. Epsomite is white, greenish or gray, yellowish or reddish, but can also be colorless and transparent. The taste is bitter and salty. In dry air, epsomite gradually loses water, becomes cloudy, and disintegrates into a powder. It is stable only below 27°C (Fulda, 1928) or 28°C (Braitsch, 1964). The presence of borax in the brine reduces the number of epsomite crystals that precipitate, but the otherwise acicular crystals become thicker with the increasing presence of borax in solution. Aluminum sulfate doubles the yield, whereas iron, copper, manganese, ammonium, or sodium sulfates decrease it. The effect on the yield is proportional to the difference between the ionic potential of magnesium and that of the added cation (Stroitelev, 1960). In ancient deposits, epsomite occurs as an epigenetic mineral only after minerals containing magnesium sulfate, such as kainite, langbeinite, kieserite, and picromerite, have weathered out and have become hydrated.

CALCIUM–MAGNESIUM SULFATES

No calcium–magnesium sulfates are known as rock-forming evaporite minerals.

Tatarskite

Tatarskite, a transparent, odorless or pale yellowish mineral forming orthorhombic aggregates has been described. It occurs together with anhydrite, halite, bischofite, magnesite, and hilgardite in anhydritic potassium-magnesium salts. Its density is 2.341. Its composition is given as $(K,Na)_{0.1} \cdot Ca_{3.1} \cdot Mg_{0.8} \cdot CO_3 \cdot SO_4 \cdot Cl_2 \cdot (OH)_2 \cdot 3.5H_2O$ or $CaCO_3 \cdot CaCl_2 \cdot CaSO_4 \cdot$-$Mg(OH)_2 \cdot 3.5H_2O$, making it akin to the hanksite of Searles Lake, California (Lobanova, 1963).

SIMPLE AND COMPOUND SODIUM SULFATES

Mirabilite (formerly Glauber's salt, exanthalite) $Na_2SO_4 \cdot 10H_2O$ ρ = 1.464–1.490

Solubility: at 0°C, 110 g/liter
at 32.384°C, decomposes

Mirabilite occurs in granular or massive bodies, crusts, and films. Monoclinic crystals are mostly columnar or prismatic. They are often euhedral, with a smooth surface, and are up to 3–4 cm long. In fracture fillings, mirabilite has a fibrous habit. The luster is glassy. Mirabilite is colorless and transparent. Impurities (mainly clay) give the crystals a gray, greenish, yellowish, milky white, or black color. Mirabilite is very soluble in water. Its taste is weakly bitter, salty, and cooling. Halite and epsomite alter in the presence of water to mirabilite and aqueous magnesium chloride (Klockmann, 1978). After brief exposure to dry air, mirabilite clouds over, gradually loses all of its crystal water, and disintegrates into a fine white powder of anhydrous sodium sulfate (thenardite). Mirabilite occurs mainly in contemporary lakes. In polar regions, it can crystallize from concentrating seawater. Occasionally, it occurs together with thenardite and halite. Epigenetic mirabilite is found in the upper parts of sulfatic potash salts affected by circulating waters. It is uncertain whether pustynite ($Na_2SO_4 \cdot 7H_2O$) is a partially dehydrated mirabilite.

Thenardite Na_2SO_4 $\rho = 2.664$–2.698

Solubility: at 0°C, 47.6 g/liter
at 100°C, 427 g/liter
Specific heat: at 0°C, 48.27 kJ/t
at 100°C, 52.57 kJ/t

Thernardite forms druses and granular aggregates, as well as massive crystalline bodies, seams, lenses, and crusts. Rhombic crystals have a bipyramidal or lamellar habit. Thenardite is colorless and transparent. Clayey admixtures discolor it gray, dark gray, or charcoal gray; iron gives it a yellowish or reddish hue. Thenardite is very soluble in water; its taste is salty and cooling. In moist air, thenardite becomes hydrated and is covered by a white film of hydrous sodium sulfate (= mirabilite) that in time disintegrates into a powder. Thenardite occurs in contemporary lake deposits, often as a product of mirabilite dehydration. Very commonly, it occurs together with mirabilite and halite. It precipitates directly in the presence of concentrated sodium chloride brines. However, it is stable only at temperatures above 13.5°C (Fulda, 1928). All sodium halides (NaCl, NaBr, NaI) and glucose are equally effective in lowering the transition temperature of mirabilite to thenardite (Norton and Johnston, 1926).

Glauberite (formerly brogniartine) $Na_2SO_4 \cdot CaSO_4$ $\rho = 2.85$

Solubility: incongruent

Glauberite is monoclinic and occurs mostly as lamelliform, earthy, or dense accumulations. Glauberite is colorless, white, reddish-brown, yellowish, gray, or blackish. The taste is that typical of bitter salts. If hydrated, it turns into gypsum and mirabilite. In contemporary lakes, it forms a slushy, crumbly sediment. It is also found as an epigenetic mineral in ancient evaporites, but rarely in large accumulations. Noteworthy are a large Neogene deposit near the southern tip of the Ural Mountains and the contemporary precipitation in the Gulf of Kara Bogaz Gol, off the Caspian Sea.

An alteration product of glauberite is hydroglauberite ($5Na_2SO_4 \cdot 3CaSO_4 \cdot 6H_2O$), with a density of 1.510. It is snow white, with a silky luster and a fibrous habit (Shynsareva, 1969; Fleischer, 1970). It occurs together with halite, bloedite, mirabilite, and polyhalite, and decomposes in the presence of water to gypsum.

Eugsterite $2Na_2SO_4 \cdot CaSO_4 \cdot 2H_2O$

Eugsterite is a monoclinic mineral, found in lacustrine environments in Kenya and central Turkey (Vergouwen, 1981), which has long been known as a synthetic compound under the name ''Fritzsche's salt'' (cf. Fritzsche, 1857). It was first found in limestone caves of Kentucky, and was there named ''bruckerite'' (Kuehn, 1972).

Aphthitalite (formerly glaserite) $Na_2SO_4 \cdot 3K_2SO_4$
$\rho = 2.656$

Solubility: slightly soluble

Aphthitalite is hexagonal–trigonal, and usually forms granular accumulations or vitreous clusters in which individual crystals are hard to recognize. Sometimes it forms euhedral prismatic crystals up to 1 cm long. Aphthitalite has a white or pale blue to gray or greenish color, a glassy luster, and an uneven or conchoidal fracture. Outwardly, it sometimes resembles langbeinite. The taste is weakly bitter-salt-like. Aphthitalite always forms very small accumulations of epigenetic origin. It is an alteration product of thenardite under the influence of potassium sulfate solutions at temperatures above 1.8°C (Kuehn, 1952a). It can similarly be produced by heating mirabilite in potassium sulfate solutions (Rassonskaya *et al.*, 1968). There are parageneses of aphthitalite with hanksite, or with langbeinite, leonite, and picromerite (Batalin and Stankevich, 1975). The mineral has been found in the German Zechstein potash salts and in the Miocene potash deposits in the Soviet Carpathians.

Bloedite (or simonyite, warthite, astrakhanite) $Na_2SO_4 \cdot MgSO_4 \cdot 5H_2O$
$\rho = 2.23–2.28$

Bloedite is monoclinic, with a vitreous luster. It is transparent, colorless, reddish, or greenish in color, with a conchoidal fracture. It forms white crusts in humid air. Bloedite is stable only in the range of 4–59.5°C (Braitsch, 1964). It occurs in small nests in carnallite in the Stassfurt horizon of German Zechstein salts (Weber, 1931), or in langbeinite that encloses sylvite (Elert, 1977) in the center of the Stassfurt Member toward its roof. It has also been identified in several desert playas, e.g., in the Chad Basin (Maglione, 1974b). An unstable variety, first described by Friedel (1976), is called "konyaite" (van Doesburg *et al.*, 1982).

Loewite $6Na_2SO_4 \cdot 7MgSO_4 \cdot 15H_2O$ $\rho = 2.423$

Loewite is colorless, red, or yellow, with a glassy luster and forms granular masses. It occurs in small nests in halite enclosing polyhalite in the roof of the Stassfurt Member toward the center of the Zechstein Basin (Elert, 1977). Loewite is stable only at temperatures above 43°C (Fulda, 1928).

Vanthoffite $3Na_2SO_4 \cdot MgSO_4$ $\rho = 2.694$

Vanthoffite, a monoclinic mineral, has been synthesized in the laboratory, but has hitherto been found in nature only in small, evidently secondary crystal aggregates that formed at a temperature of 60–90°C (Kuehn, 1958; Elert, 1977) in the roof of the Stassfurt Member or together with a variety of secondary sulfates in the Miocene evaporites north of the Carpathian Mountains (Nguyen *et al.*, 1973). Vanthoffite is an alteration product of thenardite under the influence of magnesium sulfate solutions (Kuehn, 1952a). It can also be produced by heating mirabilite with epsomite (Rassonskaya *et al.*, 1968). Vanthoffite is stable only above 46°C (Fulda, 1928).

D'ansite $9Na_2SO_4 \cdot MgSO_4 \cdot 3NaCl$ or $10Na_2SO_4 \cdot MgCl_2 \cdot NaCl$ $\rho = 2.65$

This cubic mineral, first described by Autenrieth and Braune (1958), has hitherto been found only as an alteration product of vanthoffite (Moetzing, 1978) in an environment deficient in magnesium chloride. It is also known to occur in salt dust from sea air (Radczewski, 1968), from evaporites in southern Austria (Goergey, 1909), and in China (Qu *et al.*, 1975). D'ansite is stable only in the range of 28–46°C (Braitsch, 1964).

ACCESSORY SULFATE MINERALS

Boussingaultite $(NH_4)_2SO_4 \cdot MgSO_4 \cdot 6H_2O$ $\rho = 1.723$

Solubility: at 0°C, 176.8 g/liter
at 100°C, 1305.8 g/liter
at 120°C, melts

Boussingaultite is colorless to yellowish-pink and transparent.

Krausite $Fe_2(SO_4)_3 \cdot 2H_2O$ $\rho = 2.840$

Krausite is monoclinic and vitreous, pale lemon yellow, or transparent. It often contains considerable amounts of potassium.

The following minerals are extremely rare and found mainly in lacustrine environments:

Hanksite (makite)	$KCl \cdot 9Na_2SO_4 \cdot 2Na_2CO_3$	$\rho = 2.57$
Uklonskovite	$NaOH \cdot MgSO_4 \cdot 2H_2O$	$\rho = 2.45$
Burkeite	$2Na_2SO_4 \cdot Na_2CO_3$	$\rho = 2.57$
Kogarkoite	$2Na_2SO_4 \cdot 2NaF$	$\rho = 2.78$
Sulfohalite	$2Na_2SO_4 \cdot NaCl \cdot NaF$	$\rho = 2.50$
Galeite	$5Na_2SO_4 \cdot NaCl \cdot 4NaF$	$\rho = 2.60$
Schairerite	$7Na_2SO_4 \cdot NaCl \cdot 6NaF$	$\rho = 2.67$

Oldhamite CaS $\rho = 2.18$

Solubility: at 15°C, 1.21 g/liter; decomposes
at 100°C, 4.614 g/liter; decomposes

Oldhamite is listed here, although it is not an evaporitic mineral in the true sense. It is a putative intermediary in the anaerobic bacterial decomposition of gypsum. It does not occur in any terrestrial rocks because of the extremely low partial pressure of oxygen required for its stability. In nature it is found only in meteorites (Larimer, 1968). However, it does occur in vacuoles of lunar soil, together with halite and sylvite (Ashikhmina *et al.*, 1978).

BORATES

Table 16-2 gives a list of borates that have been reported from evaporite deposits. No attempt has been made to separate minerals with a constellation of $2B_2O_3 \cdot 3H_2O$ from those of $B_2O_3 \cdot 2B(OH)_3$. The acids H_3BO_3, H_4BO_5, $H_4B_4O_9$, $H_3B_3O_6$, $H_4B_3O_7$, and $H_5B_2O_6$ can all be written as $(B_2O_3)_x(H_2O)_y(OH)_{2z}$, where $x = 1\text{–}3$, $y = 1\text{–}4$, and $z = 0\text{–}1$.

Anhydrous borates generally do not occur in the sedimentary environment. The exceptions are the silica-bearing danburite ($CaO \cdot B_2O_3 \cdot 2SiO_2$) in carnallite breccias (Kuehn and Baar, 1955; Klockmann, 1978) and the chloride-bearing boracites ($Me_3B_7O_{13}Cl$ with Me = Mg, Fe, Mn) (Wendling *et al.*, 1972). Boracite *sensu stricto* contains only magnesium, congolite iron, and ericaite, a mixture of magnesium, iron, and manganese. Tennyson (1963) produced a classification on the basis of crystal chemistry.

Kaliborite (hintzeite) ($K_2O \cdot 4MgO \cdot 12B_2O_3 \cdot 18H_2O$) is known from German

TABLE 16-2
Borate Minerals in Evaporite Deposits

Hydrous borates	
Tectoborates	
Metaborite	$B_2O_3 \cdot H_2O$
Phylloborates[a]	
Heidornite	$2CaSO_4 \cdot 2NaCl \cdot 3CaO \cdot 5B_2O_3 \cdot 2H_2O$
Strontiohilgardite	$SrCl_2 \cdot 3CaO \cdot 5B_2O_3 \cdot 2H_2O$
Ivanovite	$CaCl_2 \cdot CaO \cdot B_2O_3 \cdot 6H_2O$
Hilgardite	$3CaCl_2 \cdot 3CaO \cdot 5B_2O_3 \cdot 2H_2O$
Parahilgardite	$3CaCl_2 \cdot 3CaO \cdot 5B_2O_3 \cdot 2H_2O$
Hydrochlorborite	$3CaCl_2 \cdot CaO \cdot 4B_2O_3 \cdot 22H_2O$
Volkovite	$SrO \cdot CaO \cdot 2B(OH)_3 \cdot 6B_2O_3 \cdot 5H_2O$
Ginorite (cryptomorphite)	$2CaO \cdot 2B(OH)_3 \cdot 6B_2O_3 \cdot 5H_2O$
Strontioginorite	$2(Sr,Ca)O \cdot 7B_2O_3 \cdot 8H_2O$
Veatchite	$SrO \cdot 3B_2O_3 \cdot 2H_2O$
Tunellite	$SrO \cdot 3B_2O_3 \cdot 6H_2O$
Korshinskite	$CaO \cdot B_2O_3\ H_2O$
Tyretskite	$4CaO \cdot 5B_2O_3 \cdot 3H_2O$
Fabianite	$2CaO \cdot 3B_2O_3 \cdot H_2O$
Gowerite	$CaO \cdot 3B_2O_3 \cdot 5H_2O$
Nobleite	$CaO \cdot 3B_2O_3 \cdot 10H_2O$
Soroborates	
Lueneburgite	$MgO \cdot (P_2O_5) \cdot B_2O_3 \cdot 5H_2O$
Satimolite	$2Al_2O_3 \cdot KCl \cdot 2NaCl \cdot 3B_2O_3 \cdot 13H_2O$
Borcaite	$2CaO \cdot MgO \cdot 2(CO_2) \cdot B_2O_3 \cdot 3H_2O$
Carboborite	$CaO \cdot MgO \cdot (CO_2) \cdot B_2O_3 \cdot 7H_2O$
Solongite	$3CaO \cdot CaCl_2 \cdot 3B_2O_3 \cdot 4H_2O$
Fedorovskite	$MgO \cdot CaO \cdot 2B_2O_3 \cdot 3H_2O$
Inderborite	$MgO \cdot CaO \cdot 3B_2O_3 \cdot 11H_2O$
Wardsmithite	$MgO \cdot 5CaO \cdot 6B_2O_3 \cdot 15H_2O$
Sibirskite	$2CaO \cdot B_2O_3 \cdot H_2O$
Uralborite	$CaO \cdot B_2O_3 \cdot 2H_2O$
Pandermite	$5CaO \cdot 6B_2O_3 \cdot 9H_2O$
Tertschite	$4CaO \cdot 5B_2O_3 \cdot 20H_2O$
Priceite	$4CaO \cdot 5B_2O_3 \cdot 7H_2O$
Inyoite	$2CaO \cdot 3B_2O_3 \cdot 13H_2O$
Mayerhoffite	$2CaO \cdot 3B_2O_3 \cdot 5H_2O$
Ascharite (szaibelyite, camsellite) (probable alteration product after boracite)	$2MgO \cdot B_2O_3 \cdot H_2O$
Pinnoite	$MgO \cdot B_2O_3 \cdot 3H_2O$
Kurnakovite (inderite)	$2MgO \cdot 3B_2O_3 \cdot 15H_2O$
Lesserite	$2MgO \cdot 3B_2O_3 \cdot 15H_2O$
Hungchaoite	$MgO \cdot 2B_2O_3 \cdot 9H_2O$
Halurgite	$2MgO \cdot 4B_2O_3 \cdot 5H_2O$
Macallisterite	$2MgO \cdot 6B_2O_3 \cdot 15H_2O$
Aksaite	$MgO \cdot 3B_2O_3 \cdot 5H_2O$

(continued)

TABLE 16-2 *(Continued)*

Hydrous borates	
Paternoite	$MgO \cdot 4B_2O_3 \cdot 4H_2O$
Santite	$K_2O \cdot 5B_2O_3 \cdot 7H_2O$
Ulexite	$Na_2O \cdot 2CaO \cdot 5B_2O_3 \cdot 16H_2O$
Rivadavite	$3Na_2O \cdot MgO \cdot 12B_2O_3 \cdot 22H_2O$
Ezcurrite	$2Na_2O \cdot 5B_2O_3 \cdot 7H_2O$
Nasinite	$2Na_2O \cdot 5B_2O_3 \cdot 7H_2O$
Borax	$Na_2O \cdot 2B_2O_3 \cdot 10H_2O$
Tincalconite	$Na_2O \cdot 2B_2O_3 \cdot 3H_2O$
Ameghinite	$Na_2O \cdot 3B_2O_3 \cdot 4H_2O$
Sborgite	$Na_2O \cdot B_2O_3 \cdot 10H_2O$
Inoborates	
Chelkarite	$MgCl_2 \cdot CaO \cdot B_2O_3 \cdot 6H_2O$
Braitschite	$(Ca,Na_2)O \cdot REE_2O_3 \cdot 12B_2O_3 \cdot 6H_2O$
Kurtangaite	$SrO \cdot CaO \cdot 2B_2O_3 \cdot H_2O$
Hydroboracite	$MgO \cdot CaO \cdot 3B_2O_3 \cdot 6H_2O$
Pentahydroborite	$CaO \cdot B_2O_3 \cdot 5H_2O$
Frolovite	$2CaO \cdot 2B_2O_3 \cdot 7H_2O$
Vimsite (nifontovite)	$CaO \cdot B_2O_3 \cdot 2H_2O$
Colemanite	$2CaO \cdot B_2O_3 \cdot 5H_2O$
Volkovskite	$CaO \cdot 3B_2O_3 \cdot 3H_2O$
Preobrazhenskite	$6MgO \cdot 10B_2O_3 \cdot 9H_2O$
Kaliborite	$K_2O \cdot 4MgO \cdot 12B_2O_3 \cdot 18H_2O$
Probertite	$Na_2O \cdot 2CaO \cdot 5B_2O_3 \cdot 10H_2O$
Aristavainite	$Na_2O \cdot MgO \cdot 3B_2O_3 \cdot 5H_2O$
Kernite	$Na_2O \cdot 2B_2O_3 \cdot 6H_2O$
Biringuccite	$2Na_2O \cdot 5B_2O_3 \cdot 4H_2O$
Sassolite	$B_2O_3 \cdot 3H_2O$
Nesoborates	
Searlesite	$Na_2O \cdot 4SiO_2 \cdot B_2O_3 \cdot 2H_2O$
Garrelsite	$(Ba,Ca,Mg)O \cdot SiO_2 \cdot 3B_2O_3 \cdot 3H_2O$
Datolite	$2CaO \cdot 2SiO_2 \cdot B_2O_3 \cdot H_2O$
Bakerite	$8CaO \cdot 6SiO_2 \cdot 5B_2O_3 \cdot 5H_2O$
Howlite	$4CaO \cdot SiO_2 \cdot 5B_2O_3 \cdot 5H_2O$
Sulfoborite ($MgSO_4$ is removed in water)	$2MgO \cdot 2MgSO_4 \cdot 2B_2O_3 \cdot 9H_2O$
Teepleite	$Na_2O \cdot 2NaCl \cdot B_2O_3 \cdot 4H_2O$
Sakhaite	$7MgCO_3 \cdot 9CaO \cdot CaCl_2 \cdot 3B_2O_3 \cdot 9H_2O$
Olshanskyite	$3CaO \cdot 2B_2O_3 \cdot 9H_2O$
Wightmanite	$9MgO \cdot B_2O_3 \cdot 8H_2O$
Anhydrous borates	
Boracite (Stassfurtite)	$MgCl_2 \cdot 5MgO \cdot B_2O_3$
Ericaite (Mn-Boracite)	$MgCl_2 \cdot 5(Mg,Fe,Mn)O \cdot B_2O_3$
Danburite	$CaO \cdot 2SiO \cdot B_2O_3$

[a] Several phylloborates replace Ca with Sr in ratios of 35 : 65 or 25 : 75.

Zechstein evaporites and as a massive deposit in Inder Lake, USSR, and several lakes in California. It is a gray to colorless mineral with irridescent cleavage planes and a glassy luster. Isometric or prismatic crystals occur cemented with a fine-grained matrix of anhydrite, halite, or clay particles. Kaliborite is very often the source of all other epigenetic borate minerals, such as borax ($Na_2B_4O_7 \cdot 10H_2O$), ulexite ($Na_2Ca_2B_{10}O_{18} \cdot 16H_2O$), and many others. Kaliborite is very soluble in water and forms a white crust if exposed to humid air. Hydrated borates are found as massive commercial accumulations only in intracontinental lakes fed by thermal springs or by groundwater in contact with volcanic emanations. Boron is enriched in both lacustrine and marine evaporite rocks wherever these precipitated during a period of active volcanism. In the German Zechstein evaporites, boron is distributed rather irregularly (Biltz and Marcus, 1911), possibly because of the circulation of hydrothermal solutions and gases of a Tertiary volcanism (cf. the discussion in Kuehn, 1968). In ancient evaporites, borates also occur at times, due to the influence of percolating boron-enriched oilfield brines (Herrmann and Hoffmann, 1961). In marine evaporites, borates occur only as small nodules or smears, and often go undetected because of their superficial similarity to other evaporite minerals, i.e., to their host rock. For the most part, they are recognized only in the insoluble residue of the respective salt layer.

LACUSTRINE CARBONATES

Table 16-3 gives a list of simple and complex carbonates that have been reported from lacustrine environments, but that also form along mine openings or in caves.

TABLE 16-3

Lacustrine Carbonate Minerals

Mineral	Formula	Mineral	Formula
Kalicinite	$KHCO_3$	Siderite	$FeCO_3$ with Mg:Fe under .05
Fairchildite	$K_2CO_3 \cdot CaCO_3$	Sideroplesite	$(Mg,Fe)CO_3$ with Mg:Fe = .05 – .30
Buetschliite	$K_2CO_3 \cdot CaCO_3 \cdot 2H_2O$	Pistomesite	$(Mg,Fe)CO_3$ with Mg:Fe = .30 – 1.0
Thermonatrite	Na_2CO_3	Mesitite	$(Mg,Fe)CO_3$ with Mg:Fe = 1 – 3
Soda (natron)	$Na_2CO_3 \cdot 10H_2O$	Breunerite	$(Mg,Fe)CO_3$ with Mg:Fe = 3 – 20
Nahcolite	$NaHCO_3$	Magnesite	$MgCO_3$ with Mg:Fe over 20
Wegscheiderite	$Na_2CO_3 \cdot 3NaHCO_3$	Barringtonite	$MgCO_3 \cdot 2H_2O$
Trona	$Na_2CO_3 \cdot NaHCO_3 \cdot 2H_2O$	Nesquehonite	$MgCO_3 \cdot 3H_2O$
Nyerereite	$Na_2CO_3 \cdot CaCO_3$	Lansfordite	$MgCO_3 \cdot 5H_2O$
Shortite	$Na_2CO_3 \cdot 2CaCO_3$	Artinite	$2MgCO_3 \cdot Mg(OH)_2 \cdot 3H_2O$
Pirssonite	$Na_2CO_3 \cdot CaCO_3 \cdot H_2O$	Dypingite	$4MgCO_3 \cdot Mg(OH)_2 \cdot 3H_2O$
Gaylussite	$Na_2CO_3 \cdot CaCO_3 \cdot 5H_2O$	Hydromagnesite	$3MgCO_3 \cdot Mg(OH)_2 \cdot 3H_2O$ or
Burbankite	$Na_2CO_3 \cdot (Ca,Ba,Sr,La,Ce)_4(CO_3)_4$		$4MgCO_3 \cdot Mg(OH)_2 \cdot 4H_2O$
Eitelite	$Na_2CO_3 \cdot MgCO_3$	Huntite	$3MgCO_3 \cdot CaCO_3$
Northupite	$Na_2CO_3 \cdot MgCO_3 \cdot NaCl$	Dolomite	$MgCO_3 \cdot CaCO_3$
Tychite	$Na_4Mg_2(CO_3)_2(SO_4)_3$	Calcite, aragonite	$CaCO_3$
Bradleyite	$Na_3PO_4 \cdot MgCO_3$	Monohydrocalcite	$CaCO_3 \cdot H_2O$
Dawsonite	$NaAl(OH)_2CO_3$	Trihydrocalcite	$CaCO_3 \cdot 3H_2O$
Burkeite	$Na_2CO_3 \cdot 2Na_2SO_4$	Pentahydrocalcite	$CaCO_3 \cdot 5H_2O$
Hanksite	$Na_{22}K(CO_3)_2(SO_4)_9Cl$	Ikaite	$CaCO_3 \cdot 6H_2O$

Bibliography

Abaulina, L. I., Leonova, V. V., and Andrianov, V. F. (1976). Anhydrite from the Krylatovskoye and the Levikha deposits of the Urals. *Dokl. Akad. Nauk SSSR* **231**(1), 199–201; *Dokl. Acad. Sci. USSR, Earth Sci. Sect. (Engl. Transl. Am. Geol. Inst.)* **231**(1–6), 217–219.

Abd-El-Malek, and Rizk, K. (1963). Bacterial sulfate reduction and the development of alkalinity. *J. Appl. Bacteriol.* **26**(1), 7–26.

Abramova, S. A., and Marchenko, O. K. (1964). Results of a palynological study of the Upper Kama salt deposits in the Kungurian stage (in Russian). *Tr. Vses. Nauchno-Issled. Proektn. Inst. Galurgii* **45,** 75–121.

Acheson, D. T. (1963). Vapor pressures of saturated aqueous solutions. *Proc. Int. Symp. Humid. Moisture,* vol. 3, pp. 521–530.

Adamov, E. A., Tsobin, V. A., and Chechel', E. I. (1970). Some features of the geologic structure and development of the Ilga depression with reference to potash prospects (in Russian). *Tr. Inst. Geol. Geofiz., Akad. Nauk SSSR, Sib. Otd.* **116,** 100–110.

Adams, J. E. (1932). Anhydrite and associated inclusions in the Permian limestone of West Texas. *J. Geol.* **40**(1), 30–45.

Adams, J. E. (1936). Oil pool of open reservoir type. *Bull. Am. Assoc. Pet. Geol.* **20**(6), 780–796.

Adams, J. E. (1944). Upper Permian Ochoa series of Delaware Basin, West Texas and southeastern New Mexico. *Bull. Am. Assoc. Pet. Geol.* **28**(11), 1596–1625.

Adams, J. E. (1963). Permian salt deposits of West Texas and eastern New Mexico. *In* "Symp. on Salt" (A. C. Bersticker, K. E. Hoekstra, and J. F. Hall, eds.), pp. 124–130. N. Ohio Geol. Soc., Cleveland, Ohio.

Adams, J. E., and Frenzel, H. N. (1950). Capitan barrier reef, Texas and New Mexico. *J. Geol.* **58**(4), 289–312.

Adams, J. E., and Rhodes, M. L. (1960). Dolomitization and seepage refluxion. *Bull. Am. Assoc. Pet. Geol.* **44**(12), 1912–1920.

Adams, S. S. (1969). Bromine in the Salado Formation, Carlsbad, New Mexico. *Bull. N. M. Bur. Mines Miner. Resour.* **93,** 1–122.

Adams, S. S. (1970). Ore controls, Carlsbad potash district, Southeast New Mexico. *In* "Symp. on Salt, 3rd." (J. L. Rau and L. F. Dellwig, eds.). vol. 1, pp. 246–257. N. Ohio Geol. Soc., Cleveland, Ohio.

Adams, T. C. (1964). Salt migration in the northwest body of Great Salt Lake, Utah. *Science* **143**(3610), 1027–1029.

Aerts, E., and Dekeyser, W. (1956). Gases in rock salt and the Joffé effect. *Acta Metall.* **4,** 557–558; *Chem. Abstr.* **53,** 21,001.

Aharon, P., Kolodny, Y., and Sass, E. (1977). Recent hot brine dolomitization in the "Solar Lake," Gulf of Elat; isotopical, chemical and mineralogical study. *J. Geol.* **85**(1), 27–48.

Åkerlof, G. (1934). The calculation of the composition of an aqueous solution saturated with an arbitrary number of highly soluble strong electrolytes. *J. Am. Chem. Soc.* **56**(7), 1439–1443.

Aksenova, T. D., and Khodakovskiy, I. L. (1973). Thermodynamic analysis of potash salt bed deposition and metamorphism; the kainite–schoenite equilibrium. *Geokhimiya* No. 12, pp. 1864–1870; *Engl. Transl. of Abstr.: Geochem. Int.* **10**(6), 1393.

Albrecht, F. (1949). Über die Wärme und Wasserbilanz der Erde. *Ann. Meteorol.* **2,** 129–143.

Albrecht, F. (1951). Monatskarten des Wasserhaushaltes und der Verdunstung des Indischen und Stillen Ozeans. *Ber. Dtsch. Wetterdienst U.S. Zone* **29,** 1–46.

Alderman, A. R. (1965). Dolomitic sediments and their environment in the South-East of South Australia. *Geochim. Cosmochim. Acta* **29**(12), 1355–1365.

Alderman, S. S., Jr. (1984). Geology of the Owens Lake evaporite deposit. *In* "Symp. on Salt, 6th." in press. N. Ohio Geol. Soc., Cleveland, Ohio.

Al-Droubi, A. (1976). Géochimie des sels et des solutions concentrées par évaporation. Modèle thermodynamique de simulation. Application aux sols salées du Tchad. *Univ. Strasbourg, Mém. Sci. Géol.* **46,** 177 pp.

Aleksandrova, E. P. (1938). Dolomitization and anhydritization of limestone in the Ishimbaev petroleum province (in Russian). *Tr. Vses. Neft. Nauchno-Issled. Geologorazved. Inst., Sect. A* **101,** 13–26.

Aleksandrovich, Kh.M., and Pavlyuchenko, M. M. (1966). "Potash Salts of Byelorussia, Their Processing and Utilization" (in Russian). Izd. Nauka–Tekhnika, Minsk, 304 pp.

Aleksandrovich, Kh.M., Pavlyuchenko, M. M., Markin, A. D., Povoroznyuk, L. I., and Labetskiy, V. A. (1963). Trace inclusions of gases and moisture in naturally occurring potassium salts (in Russian). *In* "Kaliinye Soli i Metody ikh Pererabotky," pp. 82–94. Akad. Navuk BSSR, Inst. Obshch. Neorg. Khim., Minsk.

Alexander, K. F. (1951). Über die Bestimmung von Soretkoeffizienten mit Hilfe von Glasfrillen. *Z. phys. Chem.* **197,** 233–238.

Alexandersson, T. (1972). Intergranular growth of marine aragonite and Mg–calcite: Evidence of precipitation from supersaturated sea water. *J. Sediment. Petrol.* **42**(2), 441–460.

Alha, A. R. (1946). Studies on the relative resistance of germs to the salts of group I of the periodic table. *Acta Pathol. Microbiol. Scand., Suppl.* **65,** 1–74.

Al-Hashimi, W. S. (1972). A study of dolomitization by scanning electron microscopy. *Proc. Yorkshire Geol. Soc.* **38**(25), 593–606.

Allen, P., and Kramer, H. (1953). Occurrence of bassanite in two desert basins in SE California. *Am. Mineral.* **38**(11–12), 1266–1268.

Alling, H. L. (1928). The geology and origin of the Silurian salt of New York State. *Bull. N. Y. State Mus.* No. 275, pp. 1–139.

Alling, H. L., and Briggs, L. I., Jr. (1961). Stratigraphy of Upper Silurian Cayugan evaporites. *Bull. Am. Assoc. Pet. Geol.* **45**(4), 515–547.

Allmann, R., Lohse, H. H., and Hellner, E. (1963). Die Kristallstruktur des Könenits, eine Doppelschichtstruktur mit zwei inkommensurablen Teilgittern. *Z. Kristallogr. Mineral.* **126**(1–3), 7–22.

Alt, J. C., Honnorez, J., Hubberten, H. W., and Saltzman, E. (1983). Occurrence and origin of anhydrite from Deep–Sea Drilling Project Leg 70, Hole 504B, Costa Rica Rift. *In* Intl. Rep.

Deep–Sea Drlg. Proj., vol. 69 (J. R. Cann, L. N. Stout *et al.*, eds.), pp. 547–550. U.S. Govt. Print. Off., Washington, D.C.

Amiel, A. J., Miller, D. S., and Friedman, G. M. (1972). Uranium distribution in carbonate sediments of a hypersaline pool, Gulf of Elat, Red Sea. *Isr. J. Earth Sci.* **21,** 187–191.

Amieux, P. (1980). Exemple d'une passage des "black shales" aux évaporites dans le Ludien (Oligocène inferieur) du bassin de Mormoiron (Vaucluse, Sud-Est de France). *Bull. Cent. Rech. Explor.–Prod. Elf–Aquitaine* **4**(1), 281–307.

Amphlett, C. B. (1964). "Inorganic Ion Exchangers." Elsevier, Amsterdam, 141 pp.

Anati, D. A., Grad, A., and Thompson, R. O. R. Y., (1977). Laboratory models of sea straits. *J. Fluid Mech.* **81**(2), 341–351.

Anderle, J. P., Crosby, K. S., and Waugh, D. C. E. (1979). Potash at Salt Springs, New Brunswick. *Econ. Geol.* **74**(2), 389–396.

Anderson, G. M. (1973). The hydrothermal transport and deposition of galena and sphalerite near 100°C. *Econ. Geol.* **68**(4), 480–492.

Anderson, H. (1945). Adaptation of various bacteria to growth in the presence of sodium chloride. *J. Int. Soc. Leather Trades' Chem.* **29,** 215–217.

Anderson, R. Y., and Kirkland, D. W. (1966). Intrabasin varve correlation. *Bull. Geol. Soc. Am.* **77**(3), 241–255.

Anderson, R. Y., and Kirkland, D. W. (1980). Dissolution of halite by brine density flow. *Geology* **8**(2), 66–69.

Anderson, R. Y., Dean, W. E., Jr., Kirkland, D. W., and Snider, H. I. (1972). Permian Castile varved evaporite sequence, West Texas and New Mexico. *Bull. Geol. Soc. Am.* **83**(1), 59–86.

Anderson, S. B., and Swinehart, R. P. (1979). Potash salts in the Williston Basin, U.S.A. *Econ. Geol.* **74**(2), 356–376.

Andrée, K. (1912). Über ein blaues Steinsalz. *Kali* (Halle) **6**(20), 497–501.

Andryasova, G. M., and Lepeshkov, I. N. (1971). Distribution of fluorine, bromine, and iodine in salt deposits and silts of the Kara Bogaz Bay (in Russian). *Izv. Akad. Nauk Turkm. SSR, Ser. Fiz.–Tekh. Khim. Geol. Nauk* No. **1,** pp. 47–55.

Andryukov, N. A. (1940). An investigation of the gas content in Upper Kama potash deposits (in Russian). *Tr. Mater. Sverdlov. Gorn. Inst.* No. 5, pp. 11–15.

Angelier, J. (1979). Recent Quaternary tectonics in the Hellenic arc: Examples of geologic observations on land. *Tectonophysics* **52**(1–4), 267–275.

Anghelescu, D. (1953). Empirical formulae for the calculation of specific heat of liquids (in Romanian). *Riv. Univ. Politeh. "C.I. Parhon" (Bucarest), Ser. Stiinti Nat.* **3,** 132–136.

Anisimov, L. A., and Kisel'gof, S. M. (1972). Mode of occurrence and chemical composition of sedimentary brine in evaporites of the Caspian basin. *Dokl. Akad. Nauk SSSR* **202**(4), 932–934; *Dokl. Acad. Sci. USSR, Earth Sci. Sect. (Engl. Transl. Am. Geol. Inst.)* **202**(1–6), 222–224.

Anthony, T. R., and Cline, H. E. (1972). Thermomigration of biphase vapor–liquid droplets in solids. *Acta Metall.* **20**(2), 247–255.

Antonov, P. L., Gladysheva, G. A., and Kozlov, V. P. (1958). The diffusion of hydrocarbon gases through rock salt (in Russian). *Geol. Nefti* **20**(2), 47–49.

Antonyuk, E. S. (1959). Gypsum and opal as products of recent weathering of amphibolites in the Greater Lapot region (in Russian). *Mater. Mineral. Kol'sk. Poluostrova (Kirovsk)* No. 1, pp. 127–129. *Referat. Zhur. Geol.* **10** V 213 (1961).

Antsiferov, A. S. (1979). Formation of Jurassic basins of the Sayan region. *Dokl. Akad. Nauk SSSR* **244**(2), 421–424; *Dokl. Acad. Sci. USSR, Earth Sci. Sect. (Engl. Transl. Am. Geol. Inst.)* **244**(1–6), 76–78.

Apollonov, V. N. (1976) Ammonium ions in sylvite of the Upper Kama. *Dokl. Akad. Nauk SSSR*

231(3), 709–710; *Dokl. Acad. Sci. USSR, Earth Sci. Sect. (Engl. Transl. Am. Geol. Inst.)* **231**(1–6), 101.

Apollonov, V. N. (1980). Rubidium and ammonium in sylvinite deposits of the Upper Kama district (in Russian). *Dokl. Akad. Nauk SSSR* **255**(4), 961–963.

Apollonov, V. N., Kudryavtseva, G. P., and Borisenkov, V. I. (1977). A study of the distribution of bromine in sylvite by x-ray spectrophotometry (in Russian). *Dokl. Akad. Nauk SSSR* **237**(4), 945–946.

Aprodov, V. A. (1945). Blue halite from the Solikamsk potash deposit (in Russian). *Dokl. Akad. Nauk SSSR* **48**(6), 346–348.

Arbey, F. (1980). Silica forms and evaporite identification in cherts. *Bull. Cent. Rech. Explor.–Prod. Elf–Aquitaine* **4**(1), 309–365.

Arkhangel'skaya, N. A., and Grigor'yev, V. N. (1960). Conditions of formation of halogenic zones in marine basins illustrated by the example of the Lower Cambrian evaporite basin of the Siberian Platform. *Izv. Akad. Nauk SSSR, Ser. Geol.* **4,** 41–54; transl. in: "Marine Evaporites: Origin, Diagenesis, and Geochemistry." (D. W. Kirkland and R. Evans, ed.), Dowden, Hutchinson & Ross, Inc., Stroudsburg, Pennsylvania, (1973), pp. 130–144.

Armstrong, G., Dunham, K. C., Harvey, C. O., Sabine, P. A., and Waters, W. F. (1951). The paragenesis of sylvite, carnallite, polyhalite and kieserite in Eksdale borings Nos. 3, 4 and 6, northeast Yorkshire. *Mineral. Mag.* **29**(214), 667–689.

Arnold, A. (1981). Salzmineralien in Mauerwerken. *Schweiz. Mineral. Petrogr. Mitt.* **61**(1), 147–166.

Arnold, M., and Guillou, J. J. (1980). Dépôt d'anhydrite dans 17 filons métallifères des massifs hercyniens français, conséquences métallogèniques. *C.R. Hebd. Séances Acad. Sci. Paris,* [D] **290**(3), 155–157.

Arnold, M., and Guillou, J. J. (1981). Les filons métallifères hercyniens. Origine de l'anhydrite et mécanisme de la pseudomorphose subséquente: Proposition d'un modèle. *Sci. Terre* **24**(2), 173–195.

Arrhenius, G. (1963). Pelagic sediments. *In* "The Sea" (M. N. Hill, ed.), vol. 3, pp. 655–718. Wiley (Interscience), New York.

Arth, G., and Chrétien, C. (1906). La solubilité du sulfat de chaux en solution du chloride de sodium. *Bull. Soc. Chim. Fr.* **35,** 778–779.

Arthurton, R. S. (1971). The Permian evaporites of the Langwathby borehole, Cumberland. *Rep. Inst. Geol. Sci. (U.K.)* No. 71/17, pp. 1–18.

Arthurton, R. S. (1973). Experimentally produced halite compared with Triassic layered halite rock from Cheshire, England. *Sedimentology* **20**(1), 145–160.

Arthurton, R. S., and Hemingway, J. E. (1971). The St. Bees evaporites: A carbonate–evaporite formation of Upper Permian age in West Cumberland, England (with discussion). *Proc. Yorkshire Geol. Soc.* **38**(4), 565–591.

Artyushkov, E. V. (1963a). The possible genetics and general laws governing the growth of a convection instability in sedimentary rock (in Russian). *Dokl. Akad. Nauk SSSR* **153**(1), 162–165.

Artyushkov, E. V. (1963b). Basic shapes of convectional structures in sedimentary rock (in Russian). *Dokl. Akad. Nauk SSSR* **153**(3), 412–415.

Ashikhmina, N. A., Gorshkov, A. I., Mokhov, I. A., Obronov, V. G. (1978). Sylvite and halite in lunar soil (in Russian). *Dokl. Akad. Nauk SSSR* **243**(5), 1258–1260.

Ashirov, K. B., and Sazonova, I. V. (1962). Biogenic sealing of oil deposits in carbonate reservoirs. *Microbiology (Engl. Transl.)* **31,** 555–557.

Assaf, G., and Hecht, A. (1974). Sea straits: A dynamic model. *Deep-Sea Res.* **21**(11), 947–958.

Assarsson, G. O. (1950). Equilibria in aqueous systems containing K^+, Na^+, Ca^{++}, Mg^{++}, and Cl^-. *J. Am. Chem. Soc.* **72,** 1433–1444.

Assarsson, G. O. (1957). Crystallization phenomena and paragenesis in the systems alkali chlorides–alkaline earth chlorides–water. *Årsb., Sver. Geol. Unders.* **51**(7), 556–572.

Assarsson, G. O., and Balder, A. (1955). The polycomponent aqueous systems containing the chlorides of Ca^{++}, Mg^{++}, Sr^{++}, K^{+}, and Na^{+} between 18°C and 93°C. *J. Phys. Chem.* **59,** 631–633.

Assereto, R. L. A. M., and Kendall, C. G. St. C. (1977). Nature, origin and classification of peritidal teepee structures and related breccias. *Sedimentology* **24**(2), 153–210.

Atoji, M. (1959). Anhydrite obtained by the dehydration of gypsum. *J. Chem. Phys.* **30**(1), 341–342.

Atwood, D. K., and Bubb, J. N. (1970). Distribution of dolomite in a tidal flat environment. *J. Geol.* **78**(4), 499–505.

Auden, J. B. (1952). Some geological and chemical aspects of the Rajasthan salt problem. *Bull. Natl. Inst. Sci. India* **1,** 53–67.

Aufricht, W. R., and Howard, K. C. (1961). Salt characteristics as they affect storage of hydrocarbons. *J. Pet. Technol.* **13**(8), 733–738.

Augusthitis, S. S. (1980). On the textures and treatment of the sylvinite ore from the Danakili Depression, Salt Plane (Piano del Sale), Tigre, Ethiopia. *Chem. Erde* **39**(1), 91–95.

Autenrieth, H., and Braune, G. (1958). Ein neues Salzmineral, seine Eigenschaften, sein Auftreten und seine Existenzbedingungen im System der Salze ozeanischer Ablagerungen. *Naturwissenschaften* **45,** 362–363.

Avias, J. (1949). Note préliminaire sur quelques phénomènes actuels ou subactuels de pétrogénèse et autres, dans les marais côtiers de Moindou et de Canala (Nouvelle Calédonie). *C. R. Somm. Séances Soc. Géol. Fr.* No. 13, 277–280.

Avias, J. (1958). Note préliminaire sur l'existence de vases bariolées gypsifères actuelles dans les marais côtiers de la Nouvelle Calédonie. *C. R. Somm. Séances Soc. Géol. Fr.*, No. 15–16, pp. 396–397.

Aylmore, L. A. G., Karin, M., and Quirk, J. P. (1967). Adsorption and desorption of sulfate ions by soil constituents. *Soil Sci.* **103**(1), 10–15.

Ayrton, S. (1974). Rifts, evaporites, and the origin of certain alkaline rocks. *Geol. Rundsch.* **63**(2), 430–450.

Azizov, A. I. (1979). Distribution and formation of bischofite deposits (in Russian). *Sov. Geol.* No. 8, pp. 115–118.

Azzaroli, A., and Guazzone, G. (1979). Terrestrial mammals and land connections in the Mediterranean before and during the Messinian. *Palaeogeogr., Palaeoclimatol., Palaeoecol.* **29**(3–4), 155–167.

Baar, C. A. (1944). Entstehung und Gesetzmässigkeiten der Fazieswechsel im älteren Kalilager am westlichen Südharz unter besonderer Berücksichtigung des Kaliwerkes Bismarckshall. *Kali Verw. Salze Erdöl* (Halle) **38,** 175, 189, 207; and **39,** 3.

Baar, C. A. (1952). Entstehung und Gesetzmässigkeiten der Fazieswechsel im Kalisalzlager am Südharz. *Bergakademie* **4**(4), 128–150.

Baar, C. A. (1963). Der Bromgehalt als stratigraphischer Indikator in Steinsalzlagerstätten. *Neues Jahrb. Mineral., Monatsh.* No. 7, pp. 145–153.

Baar, C. A. (1974). Geological problems in Saskatchewan potash mining due to peculiar conditions during deposition of potash beds. *In* "Symp. on Salt, 4th." (A. H. Coogan, ed.), vol. 1, pp. 101–118. N. Ohio Geol. Soc., Cleveland, Ohio.

Baar, C. A. (1977). "Applied Salt Rock Mechanics I. The In Situ Behavior of Salt Rocks." Elsevier, Amsterdam, 283 pp.

Baar, C. A., and Kuehn, R. (1962). Der Werdegang der Kalisalzlagerstätten am Oberrhein. *Neues Jahrb. Mineral., Abh.* **97,** 289–336.

Baar, C. A., von Hodenberg, R., and Kuehn, R. (1971). Gelbes lichtempfindliches Steinsalz von

Esterhazy, Saskatchewan, und gelber lichtempfindlicher Borazit von Lehrte/Niedersachsen. *Kali Steinsalz* **5**(13), 460–472.

Baas-Becking, L. G. M., and Kaplan, I. R. (1956). The microbiological origin of the sulphur nodules of Lake Eyre. *Trans. R. Soc. South Aust.* **79,** 52–65.

Babčan, J. (1980). Die theoretische und experimentelle Modellierung der Entstehung von Magnesit in Evaporiten der Westkarpaten. *Geol. Zborn.* (Bratislava) **31**(1–2), 65–74.

Bachiorrini, A., Negro, A., and Murat, M. (1976). Transition zone between anhydrite III and anhydrite II. *Cim. Betons, Plâtres, Chaux* **703,** 347–353; *Chem. Abstr.* **87,** 10436.

Badiozamani, K. (1973). The dorag dolomitization model. Application to the Middle Ordovician of Wisconsin. *J. Sediment. Petrol.* **43**(4), 965–984.

Baker, B. H. (1958). Geology of the Magadi area. *Rep. Geol. Surv. Kenya* **42,** 1–81.

Baker, C. L. (1929). Depositional history of the red beds and saline residues of the Texas Permian. *Bull. Univ. Tex.* No. 2901, pp. 9–72.

Baker, P. A., and Kastner, M. (1981). Constraints on the formation of sedimentary dolomite. *Science* **213**(4504), 214–216.

Bakker, J. P. (1968). Die Flussterrassen Surinams als Hinweise auf etwas trockenere Klimate während der quartären Eiszeiten. *Acta Geogr. Debrecina* **7,** 9–17.

Balarew, D. (1926). Über die Entwässerung von Gips. *Z. Angew. Chem.* **156,** 258–260.

Bancroft, M. E. (1957). Salt deposits at Malagash and Pugwash, Nova Scotia. *Proc. Commonw. Min. Metall. Congr. 6th,* "Geology of Canadian Industrial Mineral Deposits," pp. 215–218. Can. Inst. Min. Metall., Montreal.

Banin, A., Rishpou, J., and Margulies, L. (1981). Composition and properties of the Martian soil as inferred from Viking biology data and simulation experiments with smectite clays. *Lunar Planet. Inst., Contrib.* **441,** 16–18.

Bannister, F. A., and Hey, M. H. (1936). Report on some crystalline components of the Weddell Sea deposits. *'Discovery' Rep.* **13,** 60–69.

Baouman, A. (1947). Thermodynamic study of plaster. *Rev. Mater. Constr. Trav. Publics,* [C], pp. 327–331. *Chem. Abstr.* **42,** 4726.

Barannik, V. P., Zhorov, V. A., Sovga, E. E., Bezborodov, A. A., Boguslavskaya, A. S., and Solov'eva, L. V. (1975). Hydrochemical studies of the Sivash (in Russian). *Morsk. Gidrofiz. Issled.* No. 2, pp. 210–216.

Barber, D. M., Malone, P. G., and Larson, R. J. (1975). Effect of cobalt ion on nucleation of calcium carbonate polymorphs. *Chem. Geol.* **16**(3), 239–241.

Barbier, J. (1976). Sur la signification paléogéographique de certaines minéralisations filonniennes à fluorine–barytine. *Mém. (Hors Sér) Soc. Géol. Fr.* **7,** 85–94.

Barcelona, M. J., and Atwood, D. K. (1978). Gypsum–organic interactions in natural seawater: Effect of organics on precipitation kinetics and crystal morphology. *Mar. Chem.* **6**(2), 99–115.

Barcelona, M. J., and Atwood, D. K. (1979). Gypsum–organic interactions in the marine environment: Sorption of fatty acids and hydrocarbons. *Geochim. Cosmochim. Acta* **43**(1), 47–53.

Bardossi, G., Dosza. L., Gecse, E., Kenyeres, J., and Siklosi L. (1979). Bassanite and metabasaluminite in Hungarian bauxite (in Hungarian). Földt. Közl. **109**(1), 111–119.

Barley, M. E., Dunlap, J. S. R., Glover, J. E., and Groves, D. I. (1979). Sedimentary evidence for an Archean shallow water volcanic–sedimentary facies, eastern Pilbara block, western Australia. *Earth Planet. Sci. Lett.* **43**(1), 74–84.

Barnes, H. (1974). Processes of hydrothermal ore deposition. *Fortschr. Mineral.* **52**(Beih. 2), 88–89.

Barnes, R. O., and Goldberg, E. D. (1976). Methane production and consumption in anoxic marine sediments. *Geology* **4**(5), 297–300.

Barnes, V. E. (1933).Metallic minerals in anhydrite caprock, Winnfield salt dome, Louisiana. *Am. Mineral.* **18**(8), 335–340.

Barnett, J. M., and Straw, W. T. (1983). Sedimentation rate of salt determined by micrometeorite analysis. *Geol. Soc. Am., Abstr. Prog.* **15**(6), 521.

Barrett, E. C. (1974). "Climatology from Satellites." Methuen, London, 418 pp.

Barrett, W. T., and Wallace, W. E. (1954). Studies of NaCl–KCl solid solutions I/II. *J. Am. Chem. Soc.* **76**(1), 366–373.

Barrow, N. J., Bowden, J. W., Posner, A. M., and Quirk, J. P. (1981). Describing the adsorption of copper, zinc and lead on a variable charge mineral surface. *Aust. J. Soil Res.* **19**(4), 309–321.

Barta, C., Zemlicka, J., and Rene, V. (1971). Growth of $CaCO_3$ and $CaSO_4 \cdot 2H_2O$ crystals in gels. *J. Cryst. Growth* **10**(2), 158–162.

Barton, P. B., Jr. (1967). Possible role of organic matter in the precipitation of Mississippi Valley ores. *In* Genesis of stratiform lead–zinc–barite–fluorite deposits (J. S. Brown, ed.), *Econ. Geol. Mem.* **3,** 371–378.

Baseggio, G. (1974). The composition of sea water and its concentrates. *In* "Symp. on Salt, 4th." (A. H. Coogan ed.), vol. 2, pp. 351–358. N. Ohio Geol. Soc., Cleveland, Ohio.

Basinski, A., and Czerwinski, Z. (1950). Rate of vaporization of brines (in Polish). *Przegl. Gorn.* **6,** 269–276.

Bastin, E. S. (1926). A hypothesis of bacterial influence on the genesis of certain sulphide ores. *J. Geol.* **34**(8), 773–792.

Batalin, Yu. V., and Stankevich, Ye. F. (1975). Parageneses of saline minerals and the geochemical types of salt formation (in Russian). *Izv. Akad. Nauk SSSR, Ser. Geol.* No. 8, pp. 88–94.

Bates, R. G., Staples, B. R., Robinson, R. A. (1970). Ionic hydration and single ion activities in unassociated chlorides at high ionic strengths. *Anal. Chem.* **42**(8), 867–871.

Bauld, J., Bubela, B., Plumb, L. A., et al. (1978). Simulated evaporative sedimentary environment. *Annu. Rep. Baas–Becking Geobiol. Lab.*, pp. 20–23.

Baumann, E. W. (1973). Determination of pH in concentrated salt solutions. *Anal. Chim. Acta,* **64**(2), 284–288.

Bayer, G., and Wiedemann, H. G. (1979). Stabilität und Übergangsverhalten von Vaterit ($CaCO_3$). *Experientia, Suppl.* **37,** 9–22.

Bays, C. A. (1963). Use of salt solution cavities for underground storage. *In* "Symp. on Salt, 5th." (A. C. Bersticker, K. E. Hoekstra, and J. F. Hall, eds.), vol. 1, pp. 564–578. N. Ohio Geol. Soc., Cleveland, Ohio.

Baysal, O., and Ataman, G. (1980). Sedimentology, mineralogy, and geochemistry of a sulphate series (Sivas, Turkey). *Sediment. Geol.* **25**(1–2), 67–81.

Belkin, N. I. (1940). The ascent of salt solutions in the soil (in Russian). *Sb. Nauchn. Rabot Dnepropetr. S-kh. Inst.* No. 2, pp. 3–10. *Chem. Abstr.* **37,** 2861.

Bell, K. G. (1956). Uranium in precipitates and evaporites. *U.S. Geol. Surv. Prof. Paper* **300,** 381–386.

Bellair, P. (1954). Sur l'origin des dépôts de sulfate de calcium actuels et anciens. *C. R. Hebd. Séances Acad. Sci. Paris* **239**(17), 1059–1061.

Bellair, P. (1957). Sur les sols polygonaux du Chott Djerid (Tunisie). *C. R. Hebd. Séances Acad. Sci. Paris* **244**(1), 101–103.

Belmonte, Y., Hirtz, P., Wenger, R. (1965). The salt basins of the Gabon and the Congo (Brazzaville). A tentative paleogeographic interpretation. *In* "Salt Basins around Africa" pp. 55–74. Institute of Petroleum, London.

Belyankin, D. S., and Feodot'ev, K. M. (1934). Dehydration of gypsum (in Russian). *Tr. Petrogr. Inst. Akad. Nauk SSSR* **6,** 453–460.

Bénard, H. (1901). Les tourbillons cellulaires dans une nappe liquide transportant de chaleur par convection en régime permanent. *Ann. Chim. Phys.* **23**(7), 62–144.

Benavides, V. (1968). Saline deposits of South America. *In* "Saline Deosits" (R. B. Mattox *et al.*, eds.). *Spec. Paper Geol. Soc. Am.* **88,** 249–290.

Bennett, A. C., and Adams, F. (1972). Solubility and solubility product of gypsum in soil solutions and other aqueous solutions. *Proc. Soil Sci. Soc. Am.* **36**(2), 288–291.

Berdesinski, W. (1952). Über die Synthese aluminium–reicher Könenite. *Neues Jahrb. Mineral., Abh.* **84,** 147–188.

Bernardini, F. (1969). Studio sedimentologico della serie alto–ascolana. *Atti Accad. Gioenia Sci. Nat. Catania,* [7] **1,** 353–394.

Berner, R. A. (1970a). Low temperature geochemistry of iron. *In* ''Handbook of Geochemistry,'' vol. II-1, Sect. 26 (K. H. Wedepohl, ed.). Springer-Verlag, New York.

Berner, R. A. (1970b). Sedimentary pyrite formation. *Am. J. Sci.* **268**(1), 1–23.

Berner, R. A. (1971). ''Principles of Chemical Sedimentology.'' McGraw-Hill, New York, 240 pp.

Berner, R. A. (1975). The role of magnesium in the crystal growth of calcite and aragonite from seawater. *Geochim. Cosmochim. Acta* **39**(4), 489–504.

Bernstein, F. (1960). Distribution of water and electrolyte between homoionic clays and saturating NaCl solutions. *Clays Clay Mineral.* **8,** 122–149.

Bertrand, G. (1937). Présence et répartition du bore dans les sels potassiques d'Alsace. *C. R. Hebd. Séances Acad. Sci. Paris* **205**(11), 473–476.

Bertrand, J. P., and Jelisejeff, A. (1974). Formation d'évaporites par des processes d'évaporation capillaire. *Rev. Géogr. Phys. Géol. Dyn.,* [2] **16**(2), 161–170.

Bestougeff, A., and Combaz, A. (1974). Action d'H_2S et de S sur quelques substances organiques actuelles et fossiles. *Proc. Int. Meet. Adv. Org. Geochem, 6th. (1973),* pp. 747–759.

Bettenay, E. (1962). The salt lake systems and their associated aeolian features in the semi-arid regions of Western Australia. *J. Soil Sci.* **13**(1), 11–17.

Beutler, G., and Schueler, F. (1979). Über Vorkommen salinarer Bildungen in der Trias im Norden der D.D.R. *Z. Geol. Wiss.* **7**(7), 903–912.

Beydoun, Z. R. (1980). Some Holocene geomorphological and sedimentological observations from Oman and their palaeogeological implications. *J. Pet. Geol.* **2**(4), 427–437.

Beyschlag, F. (1922). Der Salzstock von Berchtesgaden als Typus alpiner Salzlagerstätten verglichen mit norddeutschen Salzhorsten. *Z. Prakt. Geol.* No. 1, pp. 1–6.

Bezhenar, V. P., Alekseyevich, S. N., Marusyak, R. A. (1978). Hygroscopicity of water soluble potassium minerals (in Russian). *Zh. Prikl. Khim.* **51**(10), 2342–2343.

Bhat, G. D., and Oza, M. R. (1977). Storage of bitterns. *Salt Res. Ind. (India)* **13**(1–2), 1–5.

Bhat, G. D., Oza, M. R., and Sanghavi, J. R. (1979). Production of magnesium chloride hexahydrate from mixed salt end-liquor by solar evaporation. *Salt Res. Ind. (India)* **15**(2), 10–18.

Bianco, Y. (1951). Formation des chlorures basiques de magnésium de 50° à 175°C par voie aqueuse. *C.R. Hebd. Séances Acad. Sci. Paris* **232**(11), 1108–1110.

Bien, E., and Schwartz, W. (1965). Über das Vorkommen von wohlerhaltenen lebendigen und toten Bakterienzellen im Steinsalz. *Z. Allg. Mikrobiol.* **5**(3), 185–205.

Billy, C., Blanc, P., and Rouvillois, A. (1976). Aragonite et association bactérienne en milieu marin. *Trav. Lab. Micropaléont. (Univ. Pierre-et-Marie-Curie)* **6,** 91–109.

Biltz, W., and Marcus, E. (1908). Über das Vorkommen von Ammoniak und Nitrat in den Kalisalzlagerstätten. *Z. Anorg. Allg. Chem.* **62**(3), 183–202; and **64**(3), 215–216 (1910); reprinted in *Kali* (Halle) **3**(9), 189–194; and **5**(22), 497–501 (1910).

Biltz, W., and Marcus, E. (1909). Über ein Vorkommen von Kupfer in Stassfurter Salzlagerstätten. *Z. Anorg. Allg. Chem.* **64**(3), 236–244.

Biltz, W., and Marcus, E. (1911). Über die Verbreitung von borsauren Salzen in den Kalisalzlagerstätten. *Z. Anorg. Allg. Chem.* **74**(4), 302–312.

Bird, C. W., and Lynch, J. M. (1974). Formation of hydrocarbons by microorganisms. *Rev. Chem. Soc.* **3**(3), 309–328.

Bird, J. R., Rose, A., and Wilkins, R. W. T. (1981). Decoration of dislocations by proton irradiation of halite. *Nucl. Instrum. Methods Phys. Res.* **191**(1–3), 19–22.

Birina, L. M. (1974). A defence of a hypothesis on the plutonic origin of salts (in Russian). *Geol. Zh.* **34**(1), 85–91.

Birnbaum, S. J., and Coleman, M. (1979). Source of sulphur in the Ebro Basin (northern Spain). Tertiary non-marine evaporite deposits as evidenced by sulphur isotopes. *Chem. Geol.* **25**(1–2), 163–168.

Bischof, F. (1864). "Die Steinsalzwerke bei Stassfurt." Pfeffer-Verlag, Halle. 2nd ed. (1875), 70 pp.

Bischoff, J. L. (1968). Catalysis, inhibition and the calcite–aragonite problem II. The vaterite–aragonite transformation. *Am. J. Sci.* **266**(2), 80–90.

Blank, F., and Smekal, A. (1930). Über den Einfluss kleiner Verunreinigungen auf die Kohäsionsgrenzen der Steinsalzkristalle. *Naturwissenschaften* **18,** 306–307.

Blank, F., and Urbach, F. (1927). Colloidal gold in alkali halite crystals. *Naturwissenschaften* **15,** 700.

Blasdale, W. C. (1918). Equilibria in solutions containing mixtures of salts. I. The system water and the sulfates and chlorides of sodium and potassium. *J. Ind. Eng. Chem.* **10**(5), 344–347.

Blatt, H., Middleton, G., Murray, R. (1980). "Origin of Sedimentary Rocks." 2nd ed. Prentice-Hall, Englewood Cliffs, New Jersey, 782 pp.

Bliefert, C. (1978). "pH–Wert–Berechnungen." Verlag Chemie, Weinheim, 255 pp.

Bloch, M. R., and Schnerb, J. (1953). On the Cl^-/Br^- ratio and the distribution of Br-ions in liquids and solids during evaporation of bromide-containing chloride solutions. *J. Res. Counc. Isr.* **3,** 151–158.

Bloch, R., Kertes, V., and Lowy, D. (1952). The decomposition of carnallite. *Bull. Res. Counc. Isr.* **2**(2), 115–117.

Block, J., Waters, O. B., Hunter, J. A., Gillam, W. S., and Leiserson, L. (1968). Precipitation of sulfate salts from saline solutions. *U.S. Off. Saline Water, Res. Dev., Progr. Rep.*, 59 pp.

Blount, C. W., and Dickson, F. W. (1973). Gypsum–anhydrite equilibria in system $CaSO_4$–H_2O and $CaSO_4$–NaCl–H_2O. *Am. Mineral.* **58**(3–4), 323–331.

Blumenthal, M. M. (1955). Geologie des Hohen Bolkardag, seiner nördlichen Randgebiete und westlichen Ausläufer (Südanatolischer Taurus). *Publ. Miner. Res. Explor. Inst. (Ankara),* [D], No. 7, pp. 1–169.

Bobek, H. (1937). Die Rolle der Eiszeit in Nordwestiran. *Z. Gletscherk. Eiszeitforsch. Gesch. Klimas* **25,** 130–183.

Bobrov, B. S., Zhigun, I. G., and Kiseleva, L. V. (1978). Kinetics of the dehydration of calcium sulfate dihydrate (in Russian). *Izv. Neorg. Mater. Akad. Nauk SSSR* **14**(7), 1333–1337.

Bobrov, V. P. (1973). The mineralogy of potash salts in the salt-bearing formation of the Donbass (in Russian). *Mineral. Sb. (L'viv)* **27**(3), 275–281.

Bobrov, V. P., Korenevskiy, S. M., and Ryabichenko, O. P. (1968). Lithology and stratigraphy of the Kramatorsk suite in the Donbass and the mineralogic–petrographic characteristics of rocks in its potash horizons (in Russian). *In* "Geology of Salt and Potash Deposits" (S. M. Korenevskiy, ed.). *Tr. Vses. Nauchno–Issled. Geol. Inst.* **161,** 80–116.

Bock, E. (1961). On the solubility of anhydrous calcium sulfate and of gypsum in concentrated solutions of sodium chloride at 25°C, 30°C, 40°C, and 50°C. *Can. J. Chem.* **39**(9), 1746–1751.

Bodine, M. W., Jr. (1976). Magnesium hydroxychloride; a possible pH buffer in marine evaporite brines? *Geology* **4**(2), 76–80.

Bodine, M. W., Jr. (1978). Clay–mineral assemblages from drill cores of Ochoan evaporites, Eddy county, New Mexico. *Circ. N. M. Bur. Mines Miner. Resour.* No. 159, pp. 21–30.

Bodine, M. W., Jr., and Laskin, J. P. (1979). Aluminum-deficient silicate residues from polyhalite beds in the Salado Formation, Carlsbad potash district, New Mexico. *Geol. Soc. Am., Abstr. Prog.* **11**(7), 390.

Bodine, M. W., Jr., and Standaert, R. R. (1977). Chlorite and illite compositions from Upper Silurian rock salts, Retsof, New York. *Clays Clay Miner.* **25**(1), 57–71.

Boecher, T. W. (1949). Climate, soil and lakes in continental West Greenland in relation to plant life. *Medd. om Grønl.* **147**(2), 1–63.

Boehm, G. (1972). Zur petrographischen Ausbildung und Genese des Kalisalzlagers ''Flöz Ronnenberg'' auf der Scholle von Calvoerde. *Chem. Erde* **31**(1), 113–115.

Boeke, H. (1908). Über das Kristallisationsschema der Chloride, Bromide, Iodine von Natrium, Kalium, and Magnesium sowie über das Vorkommen von Brom und das Fehlen von Jod in den Kalisalzlagerstätten. *Z. Kristallogr. Mineral.* **45,** 346–391.

Boeke, H. (1911). Die Eisensalze der Kalilagerstätten. *Neues Jahrb. Mineral. Geol. Paläont.* **1,** 48–76.

Bogasheva, L. G. (1983). Metal-bearing pore solutions of evaporite formations. *Dokl. Akad. Nauk SSSR* **269**(4), 932–933.

Bogasheva, L. G., Valyashko, M. G., and Sadykov, L. Z. (1971). Pore solutions in argillaceous rocks of the Stebnik potash salt deposit (in Russian). *In* ''Materials on Hydrogeology and the Role of Groundwater'' (A. E. Khodkov, ed.), Leningrad Univ. Press, pp. 165–173.

Bogasheva, L. G., Borisenkov, V. I., Sadykov, L. Z., Valyashko, M. G., and Volkova, N. N. (1974). Besonderheiten der Metamorphose von Porenlösungen der Salztone im Bereich der sulfatischen Kali–Magnesium–Salzlagerstätten. *Z. Angew. Geol.* **20**(4), 152–156.

Bogomolov, G. V., Kudel'skiy, A. V., and Kozlov, M. F. (1970). The ammonium ion as an indicator of oil and gas. *Dokl. Akad. Nauk SSSR* **195**(4), 938–940; *Dokl. Acad. Sci. USSR Earth Sci. Sect. (Engl. Transl. Am. Geol. Inst.)* **195**(1–6), 202–204.

Bogorodskiy, A. Ya., and Dezideriev, G. P. (1935). Specific heat of concentrated aqueous lithium, sodium and potassium chlorides (in Russian). *Tr. Akad. Nauk Kaz. SSR, Kirov Inst. Khim. Tekhnol.* No. 4–5, pp. 29–40.

Bokiy, G. B., and Kachalov, A. I. (1963). Crystal form of bloedite (in Russian). *Vestn. Mosk. Univ., Ser. 4: Geol.* **18**(21), 58–66.

Bol'shakov, Ya. Ya. (1972). On the characteristics of gas accumulations in evaporite formations of the Solikamsk Basin. *Dokl. Akad. Nauk SSSR* **204**(5), 1222–1224; *Dokl. Acad. Sci. USSR Earth Sci. Sect. (Engl. Transl. Am. Geol. Inst.)* **204,** 213–214.

Bonatti, E. (1966). North Mediterranean climate during the last Wurm glaciation. *Nature (London)* **289**(5027), 984–985.

Bonney, T. G. (1906). On the origin of the British Trias. *Proc. Yorkshire Geol. Soc.* **16**(1), 1–14.

Bonte, A., and Celet, P. (1954). Sur une transformation de la polyhalite en gypse. *Ann. Soc. Geol. Nord* **74,** 53–67.

Bonython, C. W. (1956). The salt of Lake Eyre—its occurrence in Madigan Gulf and its possible origin. *Trans. R. Soc. South Aust.* **79,** 66–92.

Bonython, C. W., and King, D. (1956). The occurrence of native sulphur at Lake Eyre. *Trans. R. Soc. South Aust.* **79,** 121–130.

Borchert, H. (1933). Die Vertaubungen der Salzlagerstätten und ihre Ursachen. *Kali Verw. Salze, Erdöl (Halle)* **27,** 97–100, 105–111, 124–127, 139–141, 148–150; **28,** 290–296; **29,** 1–5.

Borchert, H. (1940). Die Salzlagerstätten des deutschen Zechsteins. *Reichsst. f. Bodenforsch., Archiv Lagerstättenforsch.* **67,** 196 pp.

Borchert, H. (1959). Grundzüge der Entstehung und der Metamorphose ozeaner Salzlagerstätten. *Freiberger Forschungsh.,* [A] **123,** 11–40.

Borchert, H. (1967). Geologische, strukturelle und physikochemische Grundfragen bei der primären Ausscheidung und der sekundären Faziesentwicklung der ozeanischen Kalilager. *Geologie* **16**(9), 1031–1044.

Borchert, H. (1969). Principles of oceanic salt deposition and metamorphism. *Bull. Geol. Soc. Am.* **80**(5), 821–864.

Borchert, H. (1977). On the formation of Lower Cretaceous potassium salts and tachhydrite in the Sergipe Basin (Brazil) with some remarks on similar occurrences in West Africa (Gabon, Angola, etc.). *In* "Time- and Strata-Bound Ore Deposits" (D. D. Klemm and H.-J. Schneider, eds.), pp. 94–111. Springer-Verlag, New York.

Borchert, H., and Baier, E. (1953). Zur Metamorphose ozeaner Gipsablagerungen. *Neues Jahrb. Mineral., Abh.* **86**(1), 103–152.

Borchert, H., and Muir, R. O. (1964). "Salt Deposits; the Origin, Metamorphism and Deformation of Evaporites." Van Nostrand-Reinhold, Princeton, New Jersey, 338 pp.

Borchert, W. (1948). Über die Verfärbung von Steinsalz durch Röntgenstrahlen. *Beitr. Mineral. Petrogr.* **1,** 203–212.

Bordovskiy, O. K. (1964). "The Accumulation and Transformation of Organic Matter in Marine Sediments" (in Russian). Izd. Nedra, Moscow, 128 pp.

Borisenkov, V. I. (1968a). Geologischer Aufbau und Mineralogie einiger Vertaubungszonen des Kaliflözes Hessen (Werra–Serie) der Schachtanlage Meikus (Rhoen). *Z. Angew. Geol.* **14**(1), 7–10.

Borisenkov, V. I. (1968b). Some data on the mineralogy and petrography of rocks in the depleted zone of the Hesse potash formation, Werra series, German Democratic Republic (in Russian). *Vestn. Mosk. Univ., Ser. 4: Geol.* **4**(23), 114–119.

Born, H. J. (1934). Der Bleigehalt der norddeutschen Salzlager und seine Beziehung zu radioaktiven Fragen. *Chem. Erde* **9**(1), 66–87.

Bornemisze, E., and Llanos, R. (1967). Sulfate movement, adsorption and desorption in three Costa Rica soils. *Proc. Soil Sci. Soc. Am.* **31**(3), 356–360.

Borowitzka, L. J. (1981). The microflora. Adaptations to life in extremely saline lakes. *Hydrobiologia* **81,** 33–46.

Borschchevskiy, Yu. A., and Borisova, S. L. (1963). Retention of radiogenic argon by sylvite. *Geokhimiya* No. 11, pp. 1055–1057; *Geochemistry (USSR) (Engl. Transl.)* **11,** 1099–1102.

Boss, G. (1941). Niederschlagsmenge und Salzgehalt des Nebelwassers an der Küste Deutsch Südwest Afrikas. *Meteorol. Z., Bioklimatol. Beibl.* **8,** 1–15.

Bottini, O. (1933). The capillary rise of solutions of electrolytes in sodium soils. *Ann. Tec. Agrar. (Rome)* **6**(5–6), 1–11. *Chem. Abstr.* **29,** 6683.

Boury, M. (1938). Le plomb dans un milieu marin. *Rev. Trav. Off. Pêches Marit. (Paris)* **11,** 157–165.

Bouwer, E. J., and McCarty, P. L. (1983). Transformation of halogenated organic compounds under denitrifying conditions. *Appl. Environ. Microbiol.* **45**(4), 1295–1299.

Bowler, J. M. (1970). Late Quaternary Environments: A Study of Lakes and Associated Sediments in Southeast Australia. Ph.D. Dissertation. Aust. Natl. Univ., Canberra, A.C.T.

Bowser, C. J., and Dickson, F. W. (1966). Chemical zonation of the borates at Kramer, California. *In* "Sympos. on Salt, 2nd." (J. L. Rau, ed.), vol. 1, pp. 122–132. N. Ohio Geol. Soc., Cleveland, Ohio.

Bowser, C. J., Rafter, T. A., and Black, R. F. (1970). Geochemical evidence for the origin of mirabilite deposits near Hobbs Glacier, Victoria Land, Antarctica. *Mineral. Soc. Am., Spec. Paper* **3,** 261–272.

Boyé, M., Marmier, F., Nesson, C., and Trecolle, G. (1978). Les dépôts de la sebkha Mellala. *Rev. Géomorph. Dyn.* **27**(2–3), 50–62.

Boyko, T. F. (1966). Distribution of rare elements in evaporite deposits. *Dokl. Akad. Nauk SSSR* **171**(2), 457–460; *Dokl. Acad. Sci. USSR Earth Sci. Sect. (Engl. Transl. Am. Geol. Inst.)* **171**(1–6), 212–215.

Bradbury, J. P. (1971). Limnology of Zuni Salt Lake, New Mexico. *Bull. Geol. Soc. Am.* **82**(2), 379–398.

Braddock, W. A., and Bowles, C. G. (1963). Calcitization of dolomite by calcium sulfate solutions in the Minnelusa Formation, Black Hills, South Dakota and Wyoming. *U.S. Geol. Surv., Prof. Paper* No. **475C,** pp. 96–99.

Bradshaw, R. L., and Sanchez, F. (1969). Migration of brine cavities in rock salt. *J. Geophys. Res.* **74**(17), 4209–4212.

Bradshaw, R. L., Empson, F. M., Boegly, W. J., Jr., Kubota, H., Parker, F. L., and Struxness, E. G. (1968). Properties of salt important in radioactive waste disposal. *In* ''Saline Deposits'' (R. B. Mattox *et al.*, eds.). *Spec. Paper Geol. Soc. Am.* **88,** 643–658.

Brady, H. T., Leckie, R. M., and White, R. (1979). Cape Spirit mirabilite beds. *Antarct. J. U.S.* **14**(5), 50–52.

Braitsch, O. (1958). Über den Mineralbestand der wasserunlöslichen Rückstande von Salzen der Stassfurt–Serie im südlichen Leinetal. *Freiberger Forschungsh.,* [A] **123,** 160–165.

Braitsch, O. (1960). Die Borate und Phosphate im Zechsteinsalz Südhannovers. *Fortschr. Mineral.* **38,** 190–191.

Braitsch, O. (1961). Zur Entstehung der Stassfurter Salzfolge. *Naturwissenschaften* **48**(10), 402.

Braitsch, O. (1962). Die Entstehung der Schichtung in rhythmisch geschichteten Evaporiten. *Geol. Rundsch.* **52**(1), 405–417.

Braitsch, O. (1964). The temperature of evaporite formation. *In* ''Problems in Paleoclimatology'' (A. E. M. Nairn, ed.), pp. 479–490. Wiley (Interscience), New York.

Braitsch, O. (1966). Bromine and rubidium as indicators of environment during sylvite and carnallite deposition of the Upper Rhine Valley evaporite. *In* ''Symp. on Salt, 2nd.'' (J. L. Rau, ed.). vol. 1, pp. 293–301. N. Ohio Geol. Soc., Cleveland, Ohio.

Braitsch, O. (1967). Rubidium und Brom als Indikatoren der Bildungsbedingungen der Oberrhein–Kalisalze. *Proc. Int. Kalisymposium, 3rd* (1965), vol. 2, pp. 67–85. VEB Dtsch. Verlag Grundstoffind., Leipzig.

Braitsch, O. (1971). ''Salt Deposits, Their Origin and Composition.'' Springer-Verlag, Berlin, 297 pp.

Braitsch, O., and Herrmann, A. G. (1963). Zur Geochemie des Broms in salinaren Sedimenten. Teil I. Experimentelle Bestimmung der Bromverteilung in verschiedenen natürlichen Salzsystemen. *Geochim. Cosmochim. Acta,* **27**(4), 361–391.

Braitsch, O., and Herrmann, A. G. (1964a). Zur Geochemie des Broms in salinaren Sedimenten. Teil II. Die Bildungstemperaturen primärer Sylvin–Carnallit–Gesteine. *Geochim. Cosmochim. Acta* **28**(7), 1081–1109.

Braitsch, O., and Herrmann, G. (1964b). Konzentrations–, Dichte–, und Temperaturverteilung in der unteroligozänen Salzlagune des Oberrheins. *Geol. Rundsch.* **54**(1), 344–356.

Brandse, W. P., van Rosmalen, G. M., and Brouwer, G. (1977). The influence of sodium chloride on the crystallization rate of gypsum. *J. Inorg. Nucl. Chem.* **39**(11), 2007–2010.

Brandt, S. B., Petrov, B. V., and Kriventsov, P. P. (1966). Radiogenic argon migration from sylvite under the effect of stress. *Geokhimiya,* No. 11, p. 1365; *Engl. Transl. Geochem. Int.* **3**(6), 1089–1090.

Branson, E. B. (1915). Origin of thick gypsum and salt deposits. *Bull. Geol. Soc. Am.* **26**(2), 231–242.

Brasack, F. (1867). Notiz über mikroskopische Quartzkristalle im Stassfurter Steinsalz. *Naturwissenschaften,* **29,** 91.

Brecht-Bergen, R. (1908). Das Salz—und Bitternseengebiet zwischen Irtysch und Ob. *Globus (Braunschweig)* **93**(9), 133–139.

Bredehoeft, J. D., Blyth, C. R., White, W. A., and Maxey, G. B. (1963). A possible mechanism for the concentration of brines in subsurface formations. *Bull. Am. Assoc. Pet. Geol.* **47**(2), 211–223.

Breger, I. A. (1970). What you don't know can hurt you: Organic colloids and natural waters. *In*

"Symp. on Organic Matter in Natural Waters" (D. W. Hood, ed.). *Occas. Publ. Univ. Alaska, Inst. Mar. Sci.* **1,** 563–574.

Breitner, H. J. (1948). Rhythmische Kristallization durch Frieren von Wasser. *Kolloid-Z.* **111,** 80–82.

Brewster, Sir D. (1826). On the existence of two new fluids in the cavities of minerals which are immiscible and possess remarkable physical properties. *Trans. R. Soc. Edinburgh* **10,** 1–42.

Briggs, L. I., Jr. (1957). Quantitative aspects of evaporite deposition. *Pap. Mich. Acad. Sci., Arts Lett.* **42,** 115–123.

Briggs, L. I., and Pollack, H. N. (1967). Digital model of evaporite sedimentation. *Science* **155**(3761), 453–456.

Briggs, L. I., Gill, D., Briggs, D. Z., and Elmore, R. D. (1980). Transition from open marine to evaporite deposition in the Silurian Michigan Basin. *In* "Hypersaline Brines and Evaporitic Environments" (A. Nissenbaum, ed.), pp. 253–270. Elsevier, Amsterdam.

Britan, I. V., Zharkov, M. A., Kovitskiy, M. L., Kolosov, A. C., Mashovich, Ya. G., and Chechel', E. I. (1977). The structure and the conditions of formation of Cambrian evaporite deposits in the USSR (in Russian). *In* "Problems of Salt Accumulation" (A. L. Yanshin and M. A. Zharkov, eds.), vol. 2, pp. 203–227. Izd. Nauka, Sib. Otd., Novosibirsk.

Brodtkorb, J. (1980). Some sodium chloride deposits from Patagonia, Argentina. *In* "Symp. on Salt, 5th." (A. H. Coogan and L. Hauber, eds.), vol. 1, pp. 31–39. N. Ohio Geol. Soc., Cleveland, Ohio.

Brognon, G. P., and Verrier, G. R. (1966). Oil and geology in Cuanza basin of Angola. *Bull. Am. Assoc. Pet. Geol.* **50**(1), 108–158.

Brongersma-Sanders, M. (1971). Origin of major cyclicity of evaporites and bituminous rocks; an actualistic model. *Mar. Geol.* **11**(2), 123–144.

Brookins, D. G. (1980a). Polyhalite K–Ar radiometric ages from southeastern New Mexico. *N. M. Bur. Mines Miner. Resour., Isochron/West* No. 29, pp. 29–30.

Brookins, D. G. (1980b). Use of evaporite minerals for K–Ar and Rb–Sr geochronology. Evidence from bedded evaporites, southeastern New Mexico, U.S.A. *Naturwissenschaften* **67**(12), 604–605.

Brookins, D. G. (1982). Study of polyhalite from the WIPP site, New Mexico. *Proc. Mater. Res. Soc. Symp.* **6,** 257–264.

Brookins, D. G., Krueger, H. W., and Register, J. K. (1980). Potassium–argon dating of polyhalite in southeastern New Mexico. *Geochim. Cosmochim. Acta* **44**(5), 635–637.

Broughton, P. L. (1972). Monohydrocalcite in speleothems; an alternative interpretation. *Contrib. Mineral. Petrol.* **36**(2), 171–174.

Brown, B. J., and Farrow, G. E. (1978). Recent dolomitic concretions of crustacean burrow origin from Loch Sunart, west coast of Scotland. *J. Sediment. Petrol.* **48**(3), 825–833.

Brown, D. W., and Gulbrandson, R. A. (1973). Chemical composition of a saline lake on Enderbury Island, Phoenix Island Group, Pacific Ocean. *J. Res., U.S. Geol. Surv.* **1**(1), 105–111.

Brownell, G. M. (1942). Quartz concretions in gypsum and anhydrite. *Univ. Toronto Stud. Geol.* No. 47, pp. 7–18.

Brunskill, G. J., and Vallentyne, J. R. (1966). Amino acids in the Silurian evaporite. *Verh. Int. Ver. Theor. Angew. Limnol.* **16,** 490–491.

Brunstrom, R. G. W., and Kent, P. E. (1967). Origin of the Keuper salt in Britain. *Nature (London)* **215**(5109), 1474.

Brunstrom, R. G. W., and Walsmley, P. J. (1969). Permian evaporites in North Sea Basin. *Bull. Am. Assoc. Pet. Geol.* **53**(4), 870–883.

Budnikoff, P. P. (1926). Die Bildungsgeschwindigkeit des unlöslichen Anhydrits. *Z. Anorg. Allg. Chem.* **159,** 87–95.

Budnikoff, P. P. (1928a). Die Beschleuniger und Verzögerer der Abbindungsgeschwindigkeit des Stuckgipses. *Kolloid-Z.* **44,** 242–249.

Budnikoff, P. P. (1928b). Some interesting experiments with gypsum. *Pit Quarry* **16**(1), 72–76.

Budnikoff, P. P., and Schtschukareva, L. A. (1935). Kinetik der Gipsentwässerung. *Kolloid-Z.* **73,** 334–339.

Buecking, E. (1911). Magnesit und Pyrit in Steinsalz und Carnallit. *Kali (Halle)* **5**(11), 221.

Buessen, W., Cosmann, O., and Schuster, C. (1936). Gipsentwässerung. *Sprechsaal* **69,** 405–406, 421–422, 433–434, 443–445.

Bugel'skiy, Yu. Yu., Lebedev, L. M., Nikitina, I. B., and Stepashkina, V. M. (1969). Some information about the forms of migration of lead and zinc in hot Cheleken brines. *Dokl. Akad. Nauk SSSR* **184**(5), 1189–1190; *Dokl. Acad. Sci. USSR Earth Sci. Sect. (Engl. Transl. Am. Geol. Inst.)* **184**(1–6), 185–186.

Bugle, A. C., Wilson, K., Olsen, G., Wade, L. G., Jr., and Osteryoung, R. A. (1978). Oil shale kerogen: Low temperature degradation in molten salts. *Nature (London)* **274**(5671), 578–580.

Burcik, E. J. (1954). The inhibition of gypsum precipitation by sodium polyphosphates. *Gypsum Prod. Mon.* **19**(1), 42–44.

Burgess, I. C., and Holliday, D. W. (1974). The Permo–Triassic rocks of the Hilton borehole, Westmoreland. *Bull. Geol. Surv. G. B.* **46,** 1–34.

Burk, C. A., Ewing, M., Worzel, J. L., Beall, A. O., Jr., Berggren, W. A., Bukry, D., Fisher, A. G., and Pessagno, E. A., Jr. (1969). Deep-sea drilling into the Challenger Knoll, central Gulf of Mexico. *Bull. Am. Assoc. Pet. Geol.* **53**(7), 1338–1347.

Burke, W. H., Denison, R. E., Hetherington, E. A., Koepnick, R. B., Nelson, H. F., Otto, J. B. (1982). Variations of seawater strontium-87/strontium-86 throughout Phanerozoic time. *Geology* **10**(10), 516–519.

Burkin, A. R. (1950). Complexes between metal salts and long-chain aliphatic amines. *J. Chem. Soc.* pp. 122–136.

Burksev, E. S. (1954). Salinity of Quaternary deposits in the southern part of the Ukraine (in Russian). *Gidrokhim. Mater.* **22,** 71–74.

Burollet, P. F. (1971). Remarques géodynamiques sur le Nord-Est de la Tunisie. *C. R. Somm. Séances Soc. Géol. Fr.* No. 8, pp. 411–413.

Burrage, L. J. (1932). The diffusion of sodium chloride in aqueous solution. *J. Phys. Chem.* **36,** 2166–2174.

Burton, H. R. (1981). Chemistry, physics, and evolution of Antarctic saline lakes. *Hydrobiologia* **82,** 339–362.

Burton, H. R., and Barker, R. J. (1979). Sulfur chemistry and microbiological fractionation of sulfur isotopes in a saline Antarctic lake. *Geomicrobiol. J.* **1**(4), 329–340.

Busch, B. (1907). Etwas über die Expansionskraft des Salzes. *Z. Prakt. Geol.* **15**(11), 369–371.

Busch, F. (1959). Über Magnesiumchlorid - Spaltung. *Freiberger Forschungsh.,* [C] **123,** 448–456.

Buschendorff, F., Nielsen, H., Puchelt, H., Ricke, W. (1963). Schwefel–Isotopen–Untersuchungen am Pyrit–Sphalerit–Baryt-Lager Meggen/Leune (Deutschland) und an verschiedenen Devon–Evaporiten. *Geochim. Cosmochim. Acta* **27**(5), 501–523.

Bush, P. (1973). Some aspects of the diagenetic history of the Sabkha in Abu Dhabi, Persian Gulf. *In* "The Persian Gulf" (B. H. Purser, ed.), pp. 395–422. Springer-Verlag, New York.

Busson, G. (1968). La sédimentation des évaporites. Comparaison des données sahariennes à quelques théories, hypothèses et observations classiques ou nouvelles. *Mém. Mus. Natl. Hist. Nat.,* [C] **19**(3), 125–169.

Busson, G. (1969). Le salifère principal (trias superieur et lias p.p.) du Sahara algéro-tunésien. *C. R. Hebd. Séances Acad. Sci. Paris,* [D] **268**(2), 251–254.

Busson, G. (1978). L'unité des faciès confinés en milieu de plate–forme carbonatée. *In* "Livre

Jubilaire Jacques Flandrin, Les Sédiments, Leur Histoire, Leur Environment et Leur Devenir.'' *Fac. Sci. Lyon, Lab. Géol., Doc. (Hors Sér.)* **4,** 87–112.

Busson, G. (1980). Large evaporitic pans in aclastic environment: how they are hollowed out, how they are filled. *Bull. Cent. Rech. Explor.-Prod. Elf-Aquitaine* **4**(1), 557–588.

Busson, G., and Perthuisot, J. P. (1977). Interêt de la Sebkha El Melah (Sud-Tunesien) pour l'interprétation des séries évaporitiques anciennes. *Sediment. Geol.* **19**(2), 139–164.

Bustillo, M. A., Garcia, M. A., Marfil, R., Ordonez, S., and de la Pena, J. A. (1978). Estudio sedimentologico de algunas lagunas de la region manchega, sector Lillo–Villacanas–Quero (provincia de Toledo). *Estud. Geol. (Madrid)* **34**(2), 187–191.

Bustillo Revuelta, M. A. (1976). Estudio petrologico de las rocas siliceas miocenas de la Cuenca del Tajo. *Estud. Geol. (Madrid)* **32**(5), 451–497.

Butler, G. P. (1969). Modern evaporite deposition and geochemistry of coexisting brines, the Sabkha, Trucial Coast, Arabian Gulf. *J. Sediment. Petrol.* **39**(1), 70–89.

Butler, G. P. (1973). Strontium geochemistry of modern and ancient calcium sulphate minerals. *In* ''The Persian Gulf'' (B. H. Purser, ed.), pp. 423–452. Springer-Verlag, New York.

Butler, G. P., Krouse, R. H., and Mitchell, R. (1973). Sulphur isotope geochemistry of an arid supratidal evaporite environment, Trucial Coast. *In* ''The Persian Gulf'' (B. H. Purser, ed.), pp. 453–462. Springer-Verlag, New York.

Butlin, K. R., and Adams, M. E. (1947). Autotrophic growth of sulfate-reducing bacteria. *Nature (London)* **160**(4057), 154–155.

Butlin, K. R., and Postgate, J. R. (1955). Microbial formation of sulfide and sulfur. *Proc. Int. Congr. Microbiol., 6th, Suppl.*, pp. 126–143.

Byrne, P. A., and Stoessell, R. K. (1982). Methane solubilities in multisalt solutions. *Geochim. Cosmochim. Acta* **46**(11), 2395–2397.

Byrne, R. H. (1981). Inorganic lead complexation in natural seawater determined by UV-spectroscopy. *Nature (London)* **290**(5806), 487–489.

Cagle, F. R., Jr., and Cruft, E. F. (1970). Gypsum deposits of the coast of Southwest Africa. *In* ''Symp. on Salt, 3rd'' (J. L. Rau and L. F. Dellwig, eds.), vol. 1, pp. 156–180. N. Ohio Geol. Soc., Cleveland, Ohio.

Cairns, D. H. (1948). Western Australia's potash lakes. *J. Fertil. Feed. Stuffs Farm Supplies* **34,** 671–674. *Chem. Abstr.* **43,** 1920.

Caldwell, D. R. (1973a). Measurement of negative thermal diffusion coefficients by observing the onset of thermohaline convection. *J. Phys. Chem.* **77**(16), 2004–2008.

Caldwell, D. R. (1973b). Thermal and Fickian diffusion of sodium chloride in a solution of oceanic concentration. *Deep-Sea Res.* **20,** 1029–1039.

Cameron, E. M. (1982). Sulphate and sulphate reduction in early Precambrian oceans. *Nature (London)* **296**(5853), 145–148.

Cameron, F. K. (1901). Solubility of gypsum in aqueous solutions of sodium chloride. *J. Phys. Chem.* **5**(8), 556–570.

Cameron, F. K. (1907). Calcium sulfate in aqueous solution of sodium chloride. *J. Phys. Chem.* **11**(6), 495–496.

Cane, R. F. (1977). Coorongite, balkhashite, and related substances; an annotated bibliography. *Trans. R. Soc. South Aust.* **101**(5–6), 153–164.

Carames Lorite, M., Lopez Aguayo, F., and Martin Vivaldi, J. L. (1973). Nota sobre la mineralogia del sondeo de Tielmes en el Terciario de la cuenca del Tajo. *Estud. Geol. (Madrid)* **29,** 307–313.

Carenas, B., and Marfil, R. (1979). Petrografia y geoquimica de gasos actuales continentales. *Estud. Geol. (Madrid)* **35,** 77–91.

Caroll, D., and Starkey, H. C. (1960). Effects of sea water on clay minerals. *Clays Clay Miner.* **7,** 80–101.

Carpelan, L. H. (1957). Hydrobiology of the Alviso salt ponds. *Ecology* **38**(3), 375–390.

Carpenter, A. B., Trout, M. L., and Picket, E. E. (1974). Preliminary report on the origin and chemical evolution of lead- and zinc-rich oilfield brines in central Mississippi. *Econ. Geol.* **69**(8), 1191–1206.

Casagrande, D. J., Idowu, G., Friedman, A., Rickert, P., Siefert, K., and Schlenz, O. (1979). H_2S incorporation in coal precursors: Origin of organic sulphur in coal. *Nature (London)* **282**(5739), 599–601.

Casagrande, L. (1947). Structures produced in clays by electric potentials and their relation to natural structures. *Nature (London)* **160**(4066), 470–471.

Castillo Herrador, F. (1974). Le Trias évaporitique des bassins de la Vallée de l'Ebre et de Cuenza. *Bull. Soc. Géol. Fr.*, [7] **16**(6), 666–676.

Catalano, R., Renda, P., and Slaczka, A. (1976). Redeposited gypsum in the evaporite sequence of the Ciminna Basin (Sicily). *Mem. Soc. Geol. Ital.* **16,** 83–93.

Cavayé Hazen, E., and Hoyoz Ruiz, A. (1953). Iodide content of samples of Spanish salt. *An. Bromatol.* **5,** 277–286.

Chamard, P. C. (1973). Monographie d'une sebkha continentale du sud-ouest saharien: la sebkha de Chemchane (Adrar de Mauritanie). *Bull. Inst. Fodam. Afr. Noire,* [A] **35**(2), 207–243.

Chanu, J. (1958). L'effet Soret en solutions ioniques. *J. Chim. Phys.* **55,** 733–742, 743–753.

Chanu, J. (1967). Thermal diffusion of halides in aqueous solution. *Adv. Chem. Phys.* **13,** 349–367.

Chanu, J., and Lenoble, J. (1955). Mésure de l'effet Soret dans les solutions de chlorure de potassium. *C. R. Hebd. Séances Acad. Sci. Paris* **241,** 1115–1117.

Chanu, J., and Lenoble, J. (1956). Mésure de l'effet Soret dans les solutions de chlorure de potassium. *J. Chim. Phys.* **53,** 309–315.

Chanu, J., and Mousselin, L. (1961). Coefficient Soret des solutions aqueuses de bromure de potassium. *C. R. Hebd. Séances Acad. Sci. Paris* **252**(6), 855–857.

Chapman, V. J. (1974). "Salt Marshes of the World." 2nd ed. J. Cramer Publ., Lehre, 494 pp.

Charmandarian, M. O., and Andronikova, N. N. (1952). Le comportement du colloïde en présence d'electrolyte pendant le processus de sa formation. *Bull. Soc. Chim. Fr.* **19,** 97–99.

Chatterji, S., and Jeffery, J. W. (1963). Crystal growth during the hydration of $CaSO_4 \cdot 2H_2O$. *Nature (London)* **200**(4905), 463–464.

Chave, K. E., and Suess, E. (1970). Calcium carbonate saturation in sea water, effects of dissolved organic matter. *Limnol. Oceanogr.* **15**(4), 633–637.

Cheeseman, R. J. (1978). Geology and oil/potash resources of Delaware Basin, Eddy and Lea Counties, New Mexico. *Circ. N. M. Bur. Mines Miner. Resour.* **159,** 7–14.

Chen, J., Zhao, R., Huo, W., Yao, Y., Pan, S., Shao, M., and Hai, C. (1981). Sulfur isotopes of some marine gypsum (in Chinese). *Sci. Geol. Sin.* No. 3, pp. 273–278.

Chen, Y., Wang, X., Sha, Q., and Zhang, N. (1979). Experimental studies on the system of Ca^{2+}–Mg^{2+}–HCO_3^{-1}–H_2O at room temperature and pressure (in Chinese). *Sci. Geol. Sin.* No. 1, pp. 22–36.

Chepelevetskiy, M. L., and Evzlina, B. B. (1937). The dynamics of crystal growth and change in form with gypsum and calcium carbonate as examples (in Russian). *Tr. Nauchno-Issled. Inst. Udobren. Insekt. Akad. Nauk SSSR (Sci. Inst. Fert. Insectofungicides)* **137,** 127–144.

Chesnokov, N. A. (1962). The viscosity of aqueous solutions of inorganic salts (in Russian). *Tr. Vses. Nauchno-Issled. Inst. Metrol.* **62,** 44–51.

Chilingar, G. V., Zenger, D. H., Bissell, H. J., and Wolf, K. H. (1979). Dolomites and dolomitization. *In* "Diagenesis in Sediments and Sedimentary Rocks" (G. Larsen and G. V. Chilingar, eds.), pp. 425–536. Elsevier, Amsterdam.

Chipman, J. (1926). The Soret effect. *J. Am. Chem. Soc.* **48**(10), 2577–2589.

Chirkov, S. K. (1935). Distribution of bromine in the water–sylvinite system (in Russian). *Kalii* **4**(10), 19–24.

Chirkov, S. K., and Shnee, M. S. (1936). The effect of admixtures during crystallization on the distribution of isomorphic substances. The distribution of bromide in the crystallization of potassium chloride, sodium chloride, and magnesium chloride from aqueous solutions (in Russian). *Kalii* **5**(9), 25–28.

Chirkov, S. K., and Shnee, M. S. (1937). Distribution of iodine and bromine in crystallization of potassium and sodium chlorides from aqueous solutions (in Russian). *Kalii* **6**(4), 15–26.

Chirkov, S. K., and Shnee, M. S. (1939). The distribution of bromide ion in the system water–carnallite (in Russian). *Zh. Prikl. Khim.* **12,** 209–219.

Chirvinskiy, P. N. (1943). The blue rock salt of the Solikamsk district (in Russian). *Zap. Vseross. Mineral. O-va.* **72**(1), 51–55.

Chirvinskiy, P. N. (1945). Blue rock salt as a possible indicator of the presence of rock salt deposits (in Russian). *Zap. Vseross. Mineral. O–va.* **74**(4), 313–314.

Chirvinskiy, P. N. (1946). An instance of change by fire of carnallite in the Solikamsk mine (in Russian). *Zap. Vseross. Mineral. O–va.* **75**(2), 149–150.

Chirvinskiy, V. N. (1928). Einige Worte über Gipspseudomorphosen nach Rostren der Belemniten aus der Umgebung von Traktemirov, District Kanev. *Zap. Kiiv. Tov. Prirodozn.* **27**(3), 119–122.

Chlebowski, R. (1977). Dedolomitization in Zechstein anhydrites of the pre-Sudetic monocline (in Polish). *Przegl. Geol.* **25**(8–9), 445–450.

Choppin, G. (1965). Water—H_2O or $H_{180}O_{90}$? *Chemistry* **38**(3), 7–11.

Chorower, C. (1941). Evaporation of saltwater in relation to meteorological factors. II. Coefficient of slow evaporation of salt pans of the Union Salinera de España. *An. R. Soc. Esp. Fis. Quim.* **37,** 69–113.

Chou, I. M., and Haas, Jr., J. L. (1981). Strontium chloride solubility in complex brines. *Sci. Basis Nucl. Waste Manage.* **3,** 499–506.

Choudhari, B. P. (1968). Magnesium sulfate/potassium chloride ratio in brine as a guide for predicting composition of potassic mixed salt. *Salt Res. Ind. (India)* **5**(1), 19–21.

Choudhari, B. P. (1969). Potassic mixed salts and crystalline bischofite from inland (Kutch) bitterns, case study of Kharaghoda and Udoo bitterns. *Indian J. Appl. Chem.* **32**(6), 355–359.

Christiansen, F. W. (1963). Polygonal fracture and fold systems in the salt crust, Great Salt Lake Desert, Utah. *Science* **139**(3555), 607–609.

Churinov, M. V. (1945). The presence of gypsum in the Lower Cretaceous deposits of the Volga region. *Dokl. Akad. Nauk SSSR* **47**(2), 214–216; *Dokl. Acad. Sci. USSR Earth Sci. Sect. (Engl. Transl. Am. Geol. Inst.)* **47**(1–6), 208–210.

Ciarapica, G., and Passeri, L. (1977). Deformazioni de fluidificazione ed evoluzione diagenetica della formazione evaporitica di Burano. *Boll. Soc. Geol. Ital.* **95**(5), 1175–1195.

Cifrulak, S., and Cohen, A. J. (1969). Infrared evidence of crystalline ammonium chloride in fluid inclusions contained in sylvite (abstr.). *EOS, Trans. Am. Geoph. Union* **50**(4), 357.

Cita, M. B. (1973). Mediterranean evaporite; paleontological arguments for a deep-sea desiccation model. *In* ''Messinian Events in the Mediterranean'' (C. W. Drooger *et al.*, eds.), pp. 206–228. North-Holland, Amsterdam.

Cita, M. B., Wright, R. C., Ryan, W. B. F., and Longinelli, A. (1978). Messinian paleo-environments. *In* Intl. Rep. Deep-Sea Drlg. Proj., vol. **42A** (K. J. Hsu, L. Montadert, *et al.*, eds.), pp. 1003–1035. U.S. Govt. Print. Off., Washington, D.C.

Clark, D. N. (1980a). The sedimentology of the Zechstein–2 carbonate formation of eastern Drenthe, The Netherlands. *Contrib. Sedimentol.* **9,** 131–165.

Clark, D. N. (1980b). The diagenesis of Zechstein carbonates. *Contrib. Sedimentol.* **9,** 167–203.

Clarke, F. W. (1924). Data of geochemistry. *Bull. U.S. Geol. Surv.* **770,** 841 pp.

Claypool, G. E., Holser, W. T., Kaplan, I. R., Sakai, H., and Zak, I. (1980). The age curves of sulfur and oxygen isotopes in marine sulfate and their mutual interpretation. *Chem. Geol.* **28**(3–4), 199–260.

Clayton, R. N., Skinner, M. C. W., Berner, R. A., and Robinson, M. (1968). Isotopic composition of recent South Australian lagoonal carbonates. *Geochim. Cosmochim. Acta* **32**(9), 983–988.

Clifton, R. L. (1944). Paleoecology and environment inferred for some marginal Middle Permian marine strata (Kansas, Oklahoma, Texas). *Bull. Am. Assoc. Pet. Geol.* **28**(7), 1012–1031.

Cody, R. D. (1976). Growth and early diagenetic changes in artificial gypsum crystals grown with bentonite muds and gels. *Bull. Geol. Soc. Am.* **87**(8), 1163–1168.

Cody, R. D. (1979). Lenticular gypsum occurrences in nature, and experimental determinations of effects of soluble green plant material on its formation. *J. Sediment. Petrol.* **49**(3), 1015–1028.

Cody, R. D., and Hull, A. B. (1980). Experimental growth of primary anhydrite at low temperatures and water salinities. *Geology* **8**(10), 505–509.

Cohen, Y., Krumbein, W. E., Goldberg, M., and Shilo, M. (1977). Solar Lake (Sinai). *Limnol. Oceanogr.* **22**(4), 597–608, 609–620, 621–634, 635–656.

Cole, G. A., Whiteside, M. C., and Brown, R. J. (1967). Unusual meromixis in two saline Arizona ponds. *Limnol. Oceanogr.* **12**(4), 584–591.

Collot, L. (1880). Description géologique des environs d'Aix-en-Provence. PhD Dissertation, Univ. Paris.

Combaz, A. (1974). La matière algaire et l'origine du pétrole. *Proc. Int. Meet. Adv. Org. Geochem. 6th. (1973),* 423–438.

Combe, E. C., and Smith, D. C. (1964). The effects of some organic acids and salts on the setting of gypsum plaster. I. Acetates. *J. Appl. Chem.* **14**(12), 544–553.

Combe, E. C., and Smith, D. C. (1965). The effects of some organic acids and salts on the setting of gypsum plaster. II. Tartrates. *J. Appl. Chem.* **15**(8), 367–372.

Combe, E. C., and Smith, D. C. (1966). The effects of some organic acids and salts on the setting of gypsum plaster. III. Citrates. *J. Appl. Chem.* **16**(3), 73–77.

Combs, D. W. (1975). Mineralogy, petrology, and bromine chemistry of selected samples of the Salado Salt, Lea and Eddy Counties, New Mexico; a potential horizon for the disposal of radioactive waste. *Rep. Oak Ridge Natl. Lab. SUB* (U.S.) **3670-5,** 87 pp.

Conley, R. F., and Bundy, W. M. (1958). Mechanism of gypsification. *Geochim. Cosmochim. Acta* **15**(1–2), 57–72.

Connan, J. (1979). Genetic relation between oil and ore in some lead–zinc–barium deposits. *Spec. Publ. Geol. Soc. S. Afr.* **5,** 263–274.

Connan, J., and Orgeval, J. J. (1976). Relationship between hydrocarbons and mineralizations: The Saint-Irvat barite deposit (Lodève Basin), France. *Bull. Cent Rech. Pau* **10**(1), 359–374.

Cook, H. E., and Zemmels, I. (1972). X-ray mineralogy data from the central Pacific, Leg 33, Deep Sea Drilling Project. *In* Intl. Rep. Deep-Sea Drlg. Proj., vol. 33 (Schlanger, S. O., Jackson, E. D. *et al.,* eds.), pp. 539–556. U.S. Govt. Print. Off., Washington, D.C.

Cook, P. J. (1973). Supratidal environments and geochemistry of some Recent dolomite concretions, Queensland, Australia. *J. Sediment. Petrol.* **43**(4), 998–1012.

Cooke, R. U., and Smalley, I. J. (1968). Salt weathering in deserts. *Nature (London)* **220**(5173), 1226–1227.

Cooke, R. U., and Warren, A. (1973). "Geomorphology in Deserts." Univ. of California Press, Berkeley, 394 pp.

Cooper, H. H., Jr., Kohout, F. A., Henry, H. R., and Glover, R. E. (1964). Seawater in coastal aquifers. *U.S. Geol. Surv. Water-Supply Paper* **1613–C,** 84 pp.

Coradossi, N., and Corazza, E. (1976). Geochemistry of Messinian clay sediments from Sicily: A preliminary investigation. *Mem. Soc. Geol. Ital.* **16,** 45–54.

Corbel, J., Murat, M., and Gallo, G. (1970). Étude physico-chimique et hypothèses de formation d'une gypse cristalline découvert sur les neiges de Spitsberg. *C. R. Hebd. Séances Acad. Sci. Paris,* [D] **270**(24), 2887–2890.

Cornec, E., and Krombach, H. (1932). Le équilibre de l'eau, chlorure de potassium et chlorure de sodium entre −23°C et 90°C. *Ann. Chim. (Paris)* **18,** 5–31.

Cornu, F. (1907). Über den färbenden Bestandteil des grünen Salzes von Hallstadt. *Österr. Z. Berg-Hüttenw.* **55**(47), 571–572.

Cornu, F. (1908a). Mineralogische und minerogenetische Beobachtungen. *Neues Jahrb. Mineral. Geol. Paläont.* No. 1, pp. 22–58.

Cornu, F. (1908b). Über das Färbungsmittel des grünen Salzes von Hallstadt. *Österr. Z. Berg-Hüttenw.* **55,** 571–572.

Cornu, F. (1910). Zur Frage der Färbung des blauen Steinsalzes. *Centralbl. Mineral.,* pp. 324–330.

Coulter, V. S., and Reed, G. E. (1980). Zechstein-2 Fordon evaporites of the Atwick No. 1 borehole, surrounding areas of N.E. England and the adjacent southern North Sea. *Contrib. Sedimentol.* **9,** 115–129.

Courel, L. (1962). Les faciès à évaporites rapportées au Trias sur la bordure nord-est du Massif Central. *C. R. Somm. Séances Soc. Géol. Fr.* No. 1, pp. 20–21.

Coutinho, M. G. N., and Fernandez, G. (1973). Analise geologica e petrografica comparativa da silvinita de Carmopolis e Matarandiba (abstr.). *In Resumo das Communicaçaos, Simposio de evaporitos do Brasil, Congr.* **27**(2), 59–61.

Cox, B. B. (1946). Transformation of organic matter into petroleum under geologic conditions ("the geologic fence"). *Bull. Am. Assoc. Pet. Geol.* **30**(5), 645–659.

Craig, J. R., Fortner, R. D., and Weand, B. L. (1974). Halite and hydrohalite from Lake Bonney, Taylor Valley, Antarctica. *Geology* **2**(8), 389–390.

Criddle, A. J. (1974). A preliminary description of microcrystalline pyrite from the nannoplankton ooze at site 251, Southwestern Indian Ocean. *In* Intl. Rep. Deep-Sea Drlg. Proj., vol. 26 (B. P. Luyendyk, T. A. Davies, *et al.,* eds.), pp. 603–607. U.S. Govt. Print. Off., Washington, D.C.

Cruft, E. F., and Chao, P. C. (1970). Nucleation kinetics of the gypsum–anhydrite system. *In* "Sympos. on Salt, 3rd" (J. L. Rau and L. F. Dellwig, eds.), vol. 1, pp. 109–118. N. Ohio Geol. Soc., Cleveland, Ohio.

Crutzen, P., Taylor, J. E., and Smith, R. C. (1970). "Measurement of Spectral Irradiance under Water." Gordon & Breach, New York, 103 pp.

Crux, G. C. (1919). Magnesium sulfate deposits at Basque, British Columbia. *Can. Chem. J.* **3,** 179–183.

Cuff, C. (1969). Lattice disorder in recent anhydrite and its geologic implications. *Proc. Geol. Soc. London* No. 1659, pp. 326–332.

Cunningham, W. A. (1934). The potassium sulfate mineral polyhalite in Texas. *Univ. Tex. Bull.* **3401,** 833–867.

Curtis, M. T. (1974). Minerals from the upper evaporite horizon of the Keuper marl at Yate. *Proc. Bristol Nat. Soc.* **33,** 73–78.

Curtis, R., Evans, G., Kinsman, D. J. J., Shearman, D. J. (1963). Association of dolomite and anhydrite in the recent sediments of the Persian Gulf. *Nature (London)* **197**(4868), 679–680.

Cvijić, J. (1908). Grundlinien der Geographie und Geologie von Mazedonien und Altserbien. *Petermann's Mitt., Erg. H.* **162,** 392 pp.

Cys, J. M. (1978). Transitional nature and significance of the Castile–Bell Canyon contact. *Circ. N. M. State Bur. Mines Mineral. Resour.* **159,** 53–56.

Dabell, H. (1931). Salt occurrences in Egypt. *J. Inst. Pet. Technol.* **17,** 346–348.

Daber, R. (1960). Paläobotanische Bemerkungen zur Kupferschieferpaläogeographie. *Geologie* **9,** 930–936.

Damuth, J. E., and Fairbridge, R. W. (1970). Equatorial Atlantic deep-sea arkosic sands and ice age aridity in tropical South America. *Bull. Geol. Soc. Am.* **81**(1), 189–206.

Dana, J. D. (1951). "The System of Mineralogy," 7th ed., 2 vols. Wiley, New York.

Danil'chenko, P. T., and Ponizovskiy, A. M. (1958). Complex utilization of the salt reserves of the Sivash (Gniloe More) and the lakes of the Perekop group (in Ukrainian). *Kompleksn. Ispol'z. Solyanykh Resur. Sivasha Perekop. Ozer. Akad. Nauk Ukr. RSR,* Kiev, pp. 36–48.

Danil'chenko, P. T., Ponizovskiy, A. M., and Globin, N. I. (1953). Surface tension of seawater and brines of salt lakes. *Tr. Krym. Fil. Akad. Nauk SSSR* **4**(1), 69–73.

d'Ans, J. (1933). "Die Lösungsgleichgewichte der Systeme der Salze ozeanischer Salzablagerungen." Ackerbau Verlag, Berlin, 254 pp.

d'Ans, J. (1944). Die metastabilen Löslichkeiten in Systemen der Salze ozeanischer Salzablagerungen. *Kali Verw. Salze Erdöl (Halle)* **38,** 42–49, 69–73, 86–92, 181–815.

d'Ans, J. (1947). Über die Bildung und Umbildung der Kalisalzlagerstätten. *Naturwissenschaften* **34**(10), 295–301.

d'Ans, J. (1961). Über die Bildungsmöglichkeiten des Tachhydrits in Kalisalzlagerstätten. *Kali Steinsalz,* **3**(4), 119–125.

d'Ans, J. (1965). Die Löslichkeit der Calziumsulfate in Kochsalzlösungen. *Kali Steinsalz,* **4**(4), 109–111.

d'Ans, J. (1967–69). Bemerkungen zu Problemen der Kalisalzlagerstätten. *Kali Steinsalz,* **4**(11), 369–386 and **5**(5), 152–157.

d'Ans, J. (1968). Der Übergangspunkt Gips–Anhydrit. *Kali Steinsalz,* **5**(3), 109–111.

d'Ans, J., and Freund, H. E. (1954). Versuche zur geochemischen Rinneitbildung. *Kali Steinsalz* **1**(6), 3–9.

d'Ans, J., and Katz, W. (1941). Magnesiumhydroxid Löslichkeiten, pH–Zahlen und Pufferung im System H_2O–$MgCl_2$–$Mg(OH)_2$. *Kali Verw. Salze Erdöl (Halle)* **35**(1), 37–41.

d'Ans, J., and Kuehn, R. (1938). Über wolkigen Sylvin. *Kali Verw. Salze Erdöl* **32,** 152–155.

d'Ans, J., and Kuehn, R. (1940). Über den Bromgehalt des Steinsalzes in den Kalilagerstätten. *Kali Verw. Salze Erdöl* **34,** 42–46, 59–64, 77–83, also **38,** 167–169 (1944).

d'Ans, J., and Kuehn, R. (1960). Bemerkungen zur Bildung und zu Umbildungen ozeaner Salzlagerstätten. *Kali Steinsalz* **3**(3), 69–84.

d'Ans, J., Bredtschneider, D., Eick, H., and Freund, H. E. (1955). Untersuchungen über die Calciumsulfate. *Kali Steinsalz* **1**(9), 17–38.

d'Ans, J., Busse, W., and Freund, H. E. (1966). Über basische Magnesiumchloride. *Kali Steinsalz,* **4**(8), 3–7.

Darmois, G. (1957). Sur une possibilité de calcul de la quantité d'eau fixée par les ions d'un sel. *C. R. Hebd. Séances Acad. Sci. Paris* **244,** 601–604.

Dauzère, C. (1912). Sur la stabilité des vortices cellulaires. *C. R. Hebd. Séances Acad. Sci. Paris* **154,** 974–976.

Davidson, C. F. (1965). A possible mode of origin of strata-bound copper ores. *Econ. Geol.* **60**(5), 942–954.

Davidson, C. F. (1966). Some genetic relationships between ore deposits and evaporites. *Trans. Inst. Metall., Sec. B,* **75**(717), 216–225; (720), 300–305; **76**(729), 175–177; **76**(730), 7.

Davies, G. R. (1977a). Carbonate–anhydrite facies relationships, Otto Fiord Formation (Mississippian–Pennsylvanian), Canadian Arctic Archipelago. *In* "Reefs and Evaporites" (J. H. Fisher, ed.), *Am. Assoc. Pet. Geol. Stud. Geol.* No. 5, pp. 145–167.

Davies, G. R. (1977b). Carbonate–anhydrite facies relations in Otto Fiord Formation (Mississippian–Pennsylvanian), Canadian Arctic Archipelago. *Bull. Am. Assoc. Pet. Geol.* **61**(11), 1929–1949.

Davies, G. R., and Ludlam, S. D. (1973). Origin of laminated and graded sediments, Middle Devonian of western Canada. *Bull. Geol. Soc. Am.* **84**(11), 3527–3546.

Davies, G. R., and Nassichuk, W. W. (1975). Subaqueous evaporites of the Carboniferous Otto Fiord Formation, Canadian Arctic Archipelago: A summary. *Geology,* **3**(5), 272–278.

Davies, P. J., Ferguson, J., and Bubela, A. (1975). Dolomite and organic material. *Nature (London)* **255**(5508), 472–474.

Davis, J. B. (1968). Paraffinic hydrocarbons in the sulfate reducing bacteria *Desulfovibrio desulfuricans. Chem. Geol.* **3**(2), 155–160.

Davis, J. B., Stanley, J. P., and Custard, H. C. (1970). Evidence against oxidation of hydrogen sulfide by sulfate ions to produce elemental sulfur. *Bull. Am. Assoc. Pet. Geol.* **54**(12), 2444–2447.

Davis, J. S. (1974). Importance of microorganisms in solar production. *In* "Symp. on Salt, 4th." (A. H. Coogan, ed.), vol. 2, pp. 369–372. N. Ohio Geol. Soc., Cleveland, Ohio.

Dean, R. S., and Ross, G. J. (1976). Anomalous gypsum in clays and shales. *Clays Clay Miner.* **24**(2), 103–104.

Dean, W. E., and Anderson, R. Y. (1978). Salinity cycles; evidence for subaqueous deposition of Castile Formation and lower part of Salado Formation, Delaware Basin, Texas and New Mexico. *Circ. N. M. State Bur. Mines Miner. Resour.* **159,** 15–20.

Dean, W. E., Davies, G. R., and Anderson, R. Y. (1975). Sedimentological significance of nodular and laminated anhydrite. *Geology,* **3**(7), 367–372.

Deardorff, D. L. (1963). Eocene salt in the Green River Basin, Wyoming. *In* "Symp. on Salt" (A. C. Bersticker, K. E. Hoekstra, and J. F. Hall, eds.), pp. 176–195. N. Ohio Geol. Soc., Cleveland, Ohio.

Debenedetti, A. (1976). Messinian salt deposits in the Mediterranean: Evaporites or precipitates? *Bull. Soc. Geol. Ital.* **95**(5), 941–950.

Debenedetti, A. (1982). The problem of the origin of the salt deposits in the Mediterranean and of their relations to the other salt occurrences in the Neogene formations of the contiguous regions. *Mar. Geol.* **49**(1), 91–114.

de Bruijn, H. (1973). Analysis of the data bearing upon the correlation of the Messinian with the succession of land mammals. *In* "Messinian Events in the Mediterranean Sea" (C. W. Drooger *et al.*, ed.), pp. 260–262. North-Holland, Amsterdam.

Decima, A. (1976). Initial data on the bromine content distribution in the Miocene salt formation of southern Sicily. *Mem. Soc. Geol. Ital.* **16,** 39–43.

Decima, A., and Wezel, F. C. (1971). Osservazioni sulle evaporiti messiniane della Sicilia centromeridionale. *Riv. Mineral. Sicil.* **12**(130–132), 172–187.

de Geer, G. (1912). A geochronology of the last 12,000 years. *Proc. Int. Geol. Congr., 11th, Sweden 1912, Sect. S,* pp. 241–253.

Degens, E., and Paluska, A. (1979). Hypersaline solutions interact with organic detritus to produce oil. *Nature (London)* **281**(5733), 666–668.

Degens, E., Hunt, J. M., Reuter, J. H., and Reed, W. E. (1964). Data on the distribution of amino acids and oxygen isotopes in petroleum brine waters of various geologic ages. *Sedimentology* **3**(3), 199–225.

de Groot, K. (1967). Experimental dedolomitization. *J. Sediment. Pet.* **37**(4), 1216–1220.

Deicha, B., and Deicha, G. A. (1955). Comportement à la cuisson de cristaux de sélénite laminaire des plâtrières de la région parisienne. *Bull. Soc. Fr. Minéral. Cristallogr.* **78**(4–6), 249–256.

Deicha, G. (1942a). Zones bimensuelles, saisonnières et annuelles dans le gypse parisien. *C. R. Somm. Séances Soc. Géol. Fr.* **9,** 83.

Deicha, G. (1942b). Sur les conditions de dépôt dans le golfe du gypse parisien. *C. R. Hebd. Séances Acad. Sci. Paris* **214**(21), 863–864.

Deicha, G. (1947). Hydrohalite et halite anhydre. *Bull. Soc. Fr. Minéral. Cristallogr.* **70,** 172–176.

Deicha, G. (1976). Inclusions et metastabilité des édifices cristallins; le cas de l'hydrohalite artificielle étudiée par microcinématographie. *Bull. Soc. Fr. Minéral. Cristallogr.* **99**(2–3), 95–97.

Delecourt, J. (1946). Géochimie des oceans, des bassins clos et des gîtes salifères, mers et lacs contemporains. *Mém. Soc. Belge Géol. Paléontol. Hydrol.* **1,** 1–177.

Dellwig, L. F. (1953). Hopper crystals of halite in the Salina of Michigan. *Am. Mineral.* **38**(7–8), 730–731.

Dellwig, L. F. (1955). Origin of the Salina salt of Michigan. *J. Sediment. Petrol.* **25**(2), 83–110.

Dellwig, L. F. (1963). Environment and mechanics of deposition of the Permian Hutchinson salt member of the Wellington shale. *In* "Symp. on Salt" (A. C. Bersticker, K. E. Hoekstra, and J. F. Hall, eds.), pp. 74–85. N. Ohio Geol. Soc., Cleveland, Ohio.

Dellwig, L. F. (1966). The Muschelkalk salt at Heilbronn, Germany. *In* "Symp. on Salt, 2nd." (J. L. Rau, ed.), vol. 1, pp. 328–334. N. Ohio Geol. Soc., Cleveland, Ohio.

Dellwig, L. F. (1968). Significant features of deposition in the Hutchinson salt, Kansas, and their interpretation. *In* "Saline Deposits" (R. B. Mattox *et al.*, eds.). *Spec. Paper Geol. Soc. Am.* **88,** 421–427.

Dellwig, L. F. (1972). Primary sedimentary structures of evaporites. *In* "Geology of Saline Deposits" (G. Richter-Bernburg, ed.), U.N.E.S.C.O., Paris, Earth Sci. Ser. **7,** 53–60.

Dellwig, L. F., and Evans, R. (1969). Depositional processes of Salina salt in Michigan, Ohio, and New York. *Bull. Am. Assoc. Pet. Geol.* **53**(4), 949–956.

Dellwig, L. F., and Kuehn, R. (1980). Depositional mechanism for the Muschelkalk salt. *In* "Symp. on Salt, 5th." (A. H. Coogan and L. Hauber, eds.), vol. 1, pp. 41–48. N. Ohio Geol. Soc. Cleveland, Ohio.

Del Monte, M., and Sabbioni, C. (1980). Authigenic dolomite on marble surface. *Nature (London)* **288**(5789), 350–351.

Demangeon, P. (1966). À propos des quartz authigènes des terrains salifères. *Bull. Soc. Fr. Minéral. Cristallogr.* **89**(4), 484–487.

Denman, W. L. (1961). Maximum reuse of cooling water. *J. Ind. Eng. Chem.* **53**(10), 817–822.

de Quervain, F. (1945). Verhalten der Bausteine gegen Witterungseinflüsse der Schweiz. *Beitr. Geol. Schweiz, Geotech. Ser.*, No. 23, pp. 1–56.

Derumaux, F. (1980). The evaporitic Permian of the North Sea; relations between tectonics and sedimentology. *Bull. Cent. Rech. Explor.–Prod. Elf–Aquitaine* **4**(1), 495–510.

de Ruyter, P. A. C. (1979). The Gabon and Congo Basins salt deposits. *Econ. Geol.* **74**(2), 419–431.

de Sitter, L. U. (1947). Diagenesis of oilfield brines. *Bull. Am. Assoc. Pet. Geol.* **31**(11), 2030–2040.

Dette, K., and Stock, F. (1957). Der Zechstein in Polen. *Geologie* **6**(2), 170–179.

Diarov, M. (1971). Sedimentation features of potassium salts in the Caspian Sea basin (in Russian). *In* "Problems of Western Kazakhstan Geology" (Sh. E. Yesenov, ed.), pp. 205–212. Izd. Nauka, Alma Ata.

Dietrich, G., Kalle, K., Krauss, W., and Siedler, G. (1975). "Allgemeine Meereskunde." 3rd ed. Gebr. Borntraeger, Stuttgart, 593 pp.

di Franco, S. (1942). Mineralogia di Monte Etna. *Mem. Accad. Gioenia Sci. Nat. Catania,* [5] **9,** 1–175.

Dix, O. R., and Jackson, M. P. A. (1982). Lithology, microstructures, fluid inclusions, and geochemistry of rock salt and of the cap-rock contact in Oakwood dome, East Texas: Significance for nuclear waste storage. *Rep. Invest. (Univ. Tex. Bur. Econ. Geol.)* No. 120, 60 pp.

Doehner, C., and Elert, K. H. (1975). Genetische Probleme im Stassfurt Salinar. *Z. Geol. Wiss.* **3**(2), 121–141.

Doelter, C. (1910). Blaues Steinsalz. *Tschermak's Mineral. Petrogr. Mitt.* **28,** 559–560.

Doelter, C. (1920). Kolloidpigmente in Mineralien. *Kolloid-Z.* **26,** 23–27.

Doelter, C. (1925). Die Färbung der Minerale durch Strahlung. *Tschermak's Mineral. Petrogr. Mitt.* **38,** 456–463.

Doelter, C. (1929). Blaues Steinsalz. *Neues Jahrb. Mineral., Monatsh.* No. 12, pp. 304–310.

Dombrowski, H. J. (1966). Geological problems in the question of living bacteria in Paleozoic salt deposits. *In* "Symp. on Salt, 2nd." (J. L. Rau, ed.), vol. 1, pp. 215–220. N. Ohio Geol. Soc., Cleveland, Ohio.

Donchenko, K. B. (1964). Geologic structure and genesis of potash deposits in the Verotyshchensk series in the Carpathian region (in Russian). *Litol. Polezn. Iskop.* No. 3, pp. 5–19.

Dorogokupets, T. I., Zor'kin, L. M., and Vortov, G. I. (1972). Gases of evaporite beds of the Dankov–Lebedyan horizon in the Pripyat depression (in Russian). *Dokl. Akad. Navuk BSSR* **16**(10), 922–925.

Dort, W., Jr., and Dort, D. S. (1970). Low temperature origin of sodium sulfate deposits, particularly in Antarctica. *In* "Symp. on Salt, 3rd." (J. L. Rau and L. F. Dellwig, eds.), vol. 1, pp. 181–203. N. Ohio Geol. Soc., Cleveland, Ohio.

Douglas, G. V., and Goodman, N. R. (1957). The deposition of gypsum and anhydrite. *Econ. Geol.* **52**(7), 831–837.

Dozy, J. J. (1970). A geological model for the genesis of the lead–zinc ores of the Mississippi Valley, U.S.A. *Trans. Inst. Min. Metall., Sect. B* **79,** 163–170.

Dravert', P. (1908). Expedition to the Suntara salt-bearing region (in Russian). *Tr. Yakutsk. Oblast. Statist. Komit.* **1,** 1–80.

Dregne, H. E. (1968). Appraisal research on surface materials of desert environments. *In* "Deserts of the World: An Appraisal of Research into their Physical and Biological Environments" (W. G. McGinnies, B. J. Goldman, and P. Taylor, eds.), pp. 290–377. Univ. Arizona Press, Tucson.

Dreizler, I. (1962). Mineralogische Untersuchungen an zwei Gipsvorkommen der Werraserie (Zechstein). *Heidelb. Beitr. Mineral. Petrogr.* **8,** 323–338.

Drever, J. L., and Smith, C. L. (1978). Cyclic wetting and drying of the soil zone as an influence on the chemistry of ground water in arid terrains. *Am. J. Sci.* **278**(10), 1448–1454.

Drever, J. L., Lawrence, J. R., and Armstrong, R. C. (1979). Gypsum and halite from the Mid-Atlantic Ridge, DSDP Site 395. *Earth Planet. Sci. Lett.* **42**(1), 98–102.

Dreyer, R. M., Garrels, R. M., and Howland, A. L. (1949). Liquid inclusions in halite as a guide to geologic thermometry. *Am. Mineral.* **34**(1), 26–34.

Driessen, P. M., and Schoorl, R. (1973). Mineralogy and morphology of salt efflorescence on saline soils in the Great Konya Basin, Turkey. *J. Soil Sci.* **24**(4), 436–443.

Dronkert, H. (1976). Late Miocene evaporites in the Sorbas Basin and adjoining areas. *Mem. Soc. Geol. Ital.* **16,** 341–361.

Dronkert, H. (1977). A preliminary note on a recent sabkha deposit in southern Spain. *Publ. Barcelona Inst. Invest. Geol.* **32,** 153–166.

Dropsy, U. (1938). A hopper-shaped cast after halite from Beaume de Venise (Vaucluse). *Bull. Soc. Fr. Mineral.* **61,** 205–208.

Droste, J. B. (1961). Clay minerals in sediments of Owens, China, Searles, Paramint, Bristol, Cadiz, and Danby lake basins, California. *Bull. Geol. Soc. Am.* **72**(11), 1713–1722.

Droste, J. B. (1963). Clay mineral composition of evaporite sequences. *In* "Symp. on Salt" (A. C. Bersticker, K. E. Hoekstra, and J. F. Hall, eds.), pp. 47–54. N. Ohio Geol. Soc., Cleveland, Ohio.

Droste, J. B., and Shaver, R. H. (1982). The Salina Group (Middle and Upper Silurian) of Indiana. *Spec. Rep. Indiana Geol. Surv.* **24,** 1–41.

Druzhinin, I. G., Druzhinin, I. E., and Godzev, E. V. (1944). Glauber's salt and halite of Lake Ebeity (in Russian). *Zh. Priklad. Khim.* **17,** 144–150.

Dubil, E. S., Martynets, I. P., and Saakyan, L. S. (1981). Study of kainite crystallization from solutions of potassium and magnesium sulfates and chlorides (in Ukrainian). *Visn. L'viv. Politekh. Inst.* **149,** 92–93.

Dubinina, V. N. (1951a). Halite from the Upper Kama deposit (in Russian). *Dokl. Akad. Nauk SSSR* **79**(5), 859–862.

Dubinina, V. N. (1951b). The question of sylvite genesis (in Russian). *Dokl. Akad. Nauk SSSR* **80**(2), 233–236.

Dubuisson, A. (1950). Étude de plâtre. *Rev. Mater. Constr. Trav. Publics, Ed. C,* **418,** 228–232; **419,** 259–264; **420,** 282–287.

Duecker, A. (1966). Polygonal gemusterte rezente Flugsandabsätze in Schleswig–Holstein. *Bundesanst. Bodenforsch., Geol. Jahrb.* **84,** 193–202.

Duff, E. J. (1972). Transformation of gypsum into fluorite in aqueous fluoride solutions. *J. Appl. Chem. Biotechnol.* **22**(4), 487–489.

Duff, P. M. D., Hallam, A., and Walton, E. K. (1967). "Cyclic Sedimentation." Elsevier, Amsterdam, 280 pp.

Dulhunty, J. A. (1974). Salt crust distribution and lake bed conditions in southern areas of Lake Eyre North. *Trans. R. Soc. South Aust.* **98**(3), 125–134.

Dulhunty, J. A. (1976). Salt crust solution during fillings of Lake Eyre. *Trans. R. Soc. South Aust.* **101**(5–6), 147–151.

Dumas, J. P. (1830). Note sur une variété de sel gemme qui décrepite au contact de l'eau. *Ann. Chim. Phys.,* [2] **43,** 316–320.

Dumestre, A. (1969). Relations entre hydrocarbures et environment évaporitique à Rainbow, Alberta (Canada). *Rev. Assoc. Fr. Technol. Pet.* **194,** 29–46.

Dumont, H. J. (1981). Kratergol, a deep, hypersaline crater lake in the steppic zone of western Anatolia (Turkey), subject to occasional limno–meteorological perturbations. *Hydrobiologia* **82,** 271–279.

Dundas, I. D., and Larsen, H. (1962). The physiological role of the carotenoid pigments of *Halo – bacterium salinarium. Arch. Mikrobiol.* **44,** 233–239.

Dundon, M. L., and Mack, E., Jr. (1923). The solubility and surface energy of calcium sulfate. *J. Am. Chem. Soc.* **45**(11), 2479–2485.

Dunham, K. C. (1948). A contribution to the petrology of the Permian evaporite deposits of northeastern England. *Proc. Yorkshire Geol. Soc.* **27,** 217–227.

Dunlap, J. C. (1951). Geological studies in a New Mexio salt mine. *Econ. Geol.* **46**(8), 909–923.

Dunning, G. E., and Cooper, J. F., Jr. (1969). Second occurrence of antarcticite from Bristol Dry Lake, California. *Am. Mineral.* **54**(7–8), 1018–1025.

Dutt, G. R., and Low, P. F. (1962). Diffusion of alkali chlorides in clay–water systems. *Soil Sci.* **93**(4), 233–240.

Dyni, J. R., Hite, R. J., and Raup, O. B. (1970). Lacustrine deposits of bromine-bearing halite, Green River Formation, northwestern Colorado. *In* "Symp. on Salt, 3rd." (J. L. Rau and L. F. Dellwig, eds.), vol. 1, pp. 166–180. N. Ohio Geol. Soc., Cleveland, Ohio.

Dzens-Litovskiy, A. I. (1962). Underground waters of the gypsum–mirabilite weathering crust of mountain valleys in Tien Shan (in Russian). *Sb. Statei po Vopr. Gidrogeol. i Inzhin. Geol., Moscow Univ.,* pp. 190–210.

Dzens-Litovskiy, A. I. (1966). The problem of Kara Bogaz Gol. *Litol. Polezn. Iskop.* No. 1, pp. 88–96; *Lithol. Mineral. Resour. (Engl. Transl.)* No. 1, pp. 70–76.

Dzens-Litovskiy, A. I. (1967). "Kara Bogaz Gol" (in Russian). Nedra, Leningrad, 96 pp.

Dzens-Litovskiy, A. I. (1968). "Saline Lakes of the USSR and Their Mineral Resources" (in Russian). Nedra, Leningrad, 119 pp.

Dzens-Litovskiy, A. I., and Vasil'ev, G. V. (1962). Geologic conditions of formation of bottom sediments in Kara Bogaz Gol in connection with fluctuations of the Caspian sea level. *Izv. Akad. Nauk SSSR, Ser. Geol.* **3,** 79–86; transl. in "Marine Evaporites: Origin, Diagenesis and

Geochemistry'' (D. W. Kirkland and R. Evans, eds.), pp. 9–16. Dowden, Hutchinson & Ross, Inc., Stroudsburg, Pennsylvania, 1973.

Dżułyński, S. (1966). Sedimentary structures resulting from convection-like pattern of motion. *Rocz. Pol. Towar. Geol.* **31**(1), 3–21.

Eardley, A. J. (1938). Sediments of Great Salt Lake, Utah. *Bull. Am. Assoc. Pet. Geol.* **22**(10), 1385–1411.

Eardley, A. J. (1962). Gypsum dunes and evaporite history of the Great Salt Lake desert. *Spec. Stud. Utah Geol. Mineral. Surv.* **2,** 1–27.

Eardley, A. J., and Stringham, B. (1952). Selenite crystals in clays of the Great Salt Lake. *J. Sediment. Petrol.* **22**(4), 234–238.

Eaton, G. P., Petersen, D. L., and Schumann, H. H. (1972). Geophysical, geohydrological and geochemical reconnaissance of the Luke salt body, central Arizona. *U.S. Geol. Surv. Prof. Paper* **753,** 1–28.

Eberhardt, A. (1971). Kristallisation von Salzen aus den Laugen des Grossen Salzsees in Theorie und Praxis. *Kali Steinsalz* **5**(10), 327–334.

Eckstein, Y. (1970). Physicochemical limnology and geology of a meromictic pond on the Red Sea shore. *Limnol. Oceanogr.* **15**(3), 363–372.

Edinger, S. E. (1973). The growth of gypsum. An investigation of the factors which affect the size and growth rates of the habit faces of gypsum. *J. Crystal Growth* **18**(3), 217–224.

Ehrlich, H. L. (1978). Inorganic energy sources for chemolithotrophic and mixotrophic bacteria. *Geomicrobiol. J.* **1**(1), 65–84.

Eigen, M., and Wicke, W. (1951). Hydration of ions and specific heats of aqueous solutions of electrolytes. *Z. Elektrochem.* **55,** 354–363.

Eilert, A. (1914). Das Ludwig-Soretsche Phänomen. *Z. Anorg. Allg. Chem.* **88**(1), 1–37.

Elderfield, H. (1976). Hydrogenous material in marine sediments excluding manganese nodules. *In* ''Chemical Oceanography'' (J. P. Riley and G. Skirrow, eds.), 2nd. ed., vol. 5, pp. 137–215. Academic Press, New York.

Elert, K. H. (1963). Mineralogische und geochemische Untersuchung der drei Faziesbezirke des Stassfurt–Kalilagers auf der Grube Neusollstedt des Kaliwerkes ''Karl Marx''. *Freiberger Forschungsh.,* [C] **145,** 1–93.

Elert, K. H. (1968). Erscheinungsformen sekundärer Sulfatminerale im Kaliflöz ''Stassfurt''. *Bergakademie* **20**(3), 447–454.

Elert, K. H. (1977). Natrium–magnesium–doppelsulfatminerale im oberen Zechstein–2. *Z. Angew. Geol.* **23**(11), 582–589.

Elert, K. H. (1979). Zum Kalziumhaushalt bei der Umbildung des Kaliflözes. *Z. Angew. Geol.* **25**(12), 576–581.

Elert, K. H., and Freund, W. (1969). Untersuchungen zum Auftreten von feindispers verteilten Kohlenwasserstoffgasen im Kaliflöz ''Stassfurt.'' *Bergakademie* **21**(10), 584–589.

Elizarov, M. V., and Musenov, A. M. (1970). Effect of halite content on the hydration of magnesium chloride and sulfate (in Russian). *Tr. Kaz. Politekh. Inst.* **31,** 106–108.

Elowski, R. C. (1980). Potassium salts (potash) of the Salina A–1 evaporite in the Michigan Basin. *Rep. Invest. Mich. Geol. Surv.* No. 25, 15 pp.

Elschner, C. (1923). Beiträge zur Kenntnis von natürlichen und künstlichen Seewasserlagunen. *Geol. Rundsch.* **14**(4), 351–354.

Emden, R. (1940). Zum Temperaturproblem der Seen. *Helv. Phys. Acta* **13**(5), 396–434.

Emery, K. O. (1956). Sediments and water of the Persian Gulf. *Bull. Am. Assoc. Pet. Geol.* **40**(10), 2354–2383.

Emons, H. H. (1967). Untersuchungen an Salzsystemen in gemischten Lösungsmitteln. *Proc. Int. Kalisymposium, 3rd* (1965), vol. 1, pp. 257–268. VEB Dtsch. Verlag Grundstoffind., Leipzig.

Erdey-Gruz, T. (1974). ''Transport Phenomena in Aqueous Solutions.'' Halsted Press, New York, 512 pp.

Erdmann, E. (1907). ''Die Chemie und Industrie der Kalisalze.'' *Kgl. Preuss. Geol. Landesanst., Abh.*, No. 52, 123 pp.

Erdmann, E. (1910a). Zwei neue Gasausströmungen in deutschen Kalisalzlagerstätten. *Kali (Halle)* **4**(7), 137–142.

Erdmann, E. (1910b). Über das Vorkommen von Iod in Salzmineralien. *Z. Angew. Chem.* **123,** 343–347.

Ericksen, G. E. (1961). Rhyolite tuff, a source of the salts of northern Chile. *U.S. Geol. Surv. Prof. Paper* **424C,** 224–226.

Ericksen, G. E., and Mrose, M. E. (1972). High purity veins of soda–niter, $NaNO_3$, and associated saline minerals in the Chilean nitrate deposits. *U.S. Geol. Surv. Prof. Paper* **800B,** 43–49.

Ermolenko, N. F., and Levitman, S. Ya. (1950). High molecular compounds in solutions of salt mixtures according to refractometric data (in Russian). *Zh. Obshch. Khim.* **20,** 31–37.

Ernst, W. G., and Calvert, S. E. (1969). An experimental study of the recrystallization of porcelanite and its bearing on the origin of some bedded cherts. *Am. J. Sci.* **267A,** 114–133.

Eroshina, D. M., and Kislik, V. Z. (1980). The conditions for the formation of potash salts in an Upper Frasnian salt-bearing formation of the Pripyat Basin (in Russian). *Sov. Geol.* No. 10, pp. 43–50.

Estefan, S. F., Awadalla, F. T., and Yousef, A. A. (1980). Solar evaporation of Qarun Lake brine. *Salt Res. Ind. (India)* **16**(1), 1–11.

Eucken, A., and Hertzberg, G. (1950). Der Aussalzungseffekt und Ionenhydration. *Z. Phys. Chem.* **195**(1), 1–23.

Eugster, H. P. (1980a). Geochemistry of evaporitic lacustrine deposits. *Annu. Rev. Earth Planet. Sci.* **8,** 35–63.

Eugster, H. P. (1980b). Lake Magadi, Kenya, and its precursors. *In* ''Hypersaline Brines and Evaporitic Environments'' (A. Nissenbaum, ed.), pp. 195–232. Elsevier, Amsterdam.

Eugster, H. P., and Hardie, L. A. (1973). Saline Lakes. In ''Lakes, Chemistry, Geology, Physics'' (A. Lerman, ed.), pp. 237–293. Springer-Verlag, New York.

Eugster, H. P., and Jones, B. F. (1979). Behavior of major solutes during closed-basin brine evolution. *Am. J. Sci.* **279**(6), 609–631.

Eugster, H. P., Harvie, C. E., and Weare, J. H. (1980). Mineral equilibrium in a six-component seawater system Na–K–Mg–Ca–SO_4–Cl–H_2O at 25°C. *Geochim. Cosmochim. Acta* **44**(9), 1335–1347.

Evans, G., and Shearman, D. J. (1964). Recent celestine from the sediments of the Trucial Coast of the Persian Gulf. *Nature (London)* **202**(4930), 385–386.

Evans, G., Schmidt, V., Bush, P., and Nelson, H. (1969). Stratigraphy and geologic history of the sabkha, Abu Dhabi, Persian Gulf. *Sedimentology* **12**(1–2), 145–159.

Evans, R. (1970). Genesis of sylvite- and carnallite-bearing rocks from Wallace, Nova Scotia. *In* ''Symp. on Salt, 3rd.'' (J. L. Rau and L. F. Dellwig, eds.), vol. 1, 239–245. N. Ohio Geol. Soc., Cleveland, Ohio.

Evans, R. (1978). Origin and significance of evaporites in basins around the Atlantic margin. *Bull. Am. Assoc. Pet. Geol.* **62**(2), 223–234.

Evans, T. L., Campbell, F. A., and Krouse, H. R. (1968). A reconnaissance study of some western Canadian lead–zinc deposits. *Econ. Geol.* **63**(4), 349–359.

Evans, W. B. (1970). The Triassic salt deposits of northwestern England. *Qu. J. Geol. Soc. London* **126,** 103–123.

Evans, W. E. (1964). The organic solubilization of minerals in sediments. *In* ''Advances in Organic Geochemistry'' (U. Colombo and G. D. Hobson, eds.), pp. 263–270. Pergamon Press, New York.

Everding, H. (1907). Zur Geologie der deutschen Zechsteinsalze. *In* ''Deutschlands Kalibergbau, Festschrift, X. Allgem. Bergmannstag Eisenach.'' *Abh. Kgl. Preuss. Geol. Landesanst.* No. 52, pp. 25–133.

Ewald, W., and Polanyi, M. (1924). Plasticity and strength of rock salt under water. *Z. Phys.* **28,** 29–50.

Ezrokhi, L. L. (1949). Rate of solution of rock salt and of sylvinite in mixed solutions (in Russian). *Zh. Priklad. Khim.* **22,** 24–32.

Ezrokhi, L. L. (1952). Viscosity of aqueous solutions of the individual salts of seawater systems. *Zh. Prikl. Khim.* **25,** 838–849; *J. Appl. Chem. USSR (Engl. Transl.)* **25,** 917–926.

Fabbri, A., and Curzi, P. (1979). The Messinian of the Tyrrhenian Sea: Seismic evidence and dynamic implications. *Giorn. Geol.,* [2] **43**(1), 215–248.

Fabbri, A., and Selli, R. (1973). The structure and stratigraphy of the Tyrrhenian Sea. *In* ''The Mediterranean Sea: A Natural Sedimentation Laboratory'' (D. J. Stanley, ed.), pp. 75–81. Dowden, Hutchinson & Ross, Inc., Stroudsburg, Pennsylvania.

Fabricius, F. H. (1977). Origin of marine ooids and grapestones. *In* ''Contributions to Sedimentology'' vol. 7, (H. Fuechtbauer *et al.* eds.), 113 pp. Schweizerbart, Stuttgart.

Fabricius, F. H., and Hieke, W. (1977). Neogene to Quaternary development of the Ionian Basin (Mediterranean): Considerations based on a ''dynamic shallow basin model'' of the Messinian salinity event. *In* ''Structural History of the Mediterranean Basins'' (B. Biju-Duval and L. Montadert, eds.), pp. 391–400. Ed. Technip, Paris.

Fabricius, F. H., Heimann, K. O., and Braune, K. (1978). Comparison of site 374 with circum-Ionian land sections: Implications for the Messinian ''salinity crisis'' on the basis of a ''dynamic model.'' *In* Int. Repts. Deep-Sea Drlg. Proj., vol. 42A (K. J. Hsu, L. Montadert *et al.,* eds.), pp. 927–942. U.S. Govt. Print. Off., Washington, D.C.

Fahey, J., and Mrose, M. E. (1962). Saline minerals of the Green River Formation. *U.S. Geol. Surv. Prof. Paper* **405,** 50 pp.

Faingersh, L. A. (1977). Particularities of gas accumulations in Paleozoic subsalt deposits of the Arctic syneclise (in Russian). *In* ''Problems of Salt Accumulation'' (L. A. Yanshin and M. A. Zharkov, eds.), vol. 2, pp. 308–311. Izd. Nauka, Akad. Nauk SSSR, Sib. Otd., Novosibirsk.

Fajans, K., and Johnson, O. (1942). Apparent volumes of individual ions in aqueous solution. *J. Am. Chem. Soc.* **64,** 668–678.

Farnsworth, M. (1924). Effects of temperature and pressure on gypsum and anhydrite. *Rep. Invest. U.S. Bur. Mines* No. 2654, pp. 1–3.

Farnsworth, M. (1925). The hydration of anhydrite. *J. Ind. Eng. Chem.* **17,** 967–970.

Farran, J., and Orliac, M. (1953). Sur la transformation d'anhydrite au gypse. *C. R. Congr. Soc. Savantes, Sér. Sci.* **78,** 189–192. *Chem. Abstr.* **49,** 8049.

Fauth, P. (1912). ''Horbiger's Glazialkosmogonie, eine neue Entwicklungsgeschichte des Weltalls und des Sonnensystems.'' H. Kaiser, Kaiserslautern, 790 pp., reprinted 1925.

Fedin, V. P. (1979). Salzführende Ablagerungen im Kara–Bogaz–Gol und ihre Bildungsgeschichte. *Z. Geol. Wiss.* **7**(4), 843–846.

Feigelson, I. B. (1940). Evaporation of the Elton Lake brines under natural conditions, Part 2 (in Russian). *Tr. Vses. Nauchno-Issled. Proektn. Inst. Galurgii* No. 16, pp. 41–51.

Fenchel, T. M., and Joergenson, B. (1977). Detritus food chains of aquatic ecosystems: The role of bacteria. *Adv. Microb. Ecol.* **1,** 1–58.

Ferguson, J., and Ibe, A. C. (1981). Origin of light hydrocarbons in carbonate oolites. *J. Pet. Geol.* **4**(1), 103–107.

Fernandez Lozano, J. A., and Wint, A. (1982). Double decomposition of gypsum and potassium chloride catalyzed by aqueous ammonia. *Chem. Eng. J.* **23**(1), 53–61.

Ferreiro, E. A., and Helmy, A. K. (1974). Flocculation of Na–montmorillonite by electrolytes. *Clay Miner.* **10**(3), 203–213.

Ferroniere, G. (1901). Études biologiques sur la faune supralittorale de la Loire-Inferieure. *Bull. Soc. Sci. Nat. Ouest Fr.*, [2] **1,** 1–451.

Feth, J. H. (1981). Chloride in natural continental water. *U.S. Geol. Surv. Water–Supply Paper* No. 2176, 30 pp.

Fiege, K. (1939). Die zyklische Sedimentation in der Salzfazies des deutschen Zechsteins und die Grossflutenhypothese. *Zentralbl. Mineral. Geol. Paläont., Abh.*, [A], No. 9, pp. 353–390.

Fiege, K. (1940). Bildung der Salzfazies im deutschen Zechstein. *Z. Prakt. Geol.* **48**(4), 37–47.

Fiestas y Contreras, B. (1966). El origen del salpitre de Chile. *Bol. Soc. R. Esp. Hist. Nat., Secc. Geol.* **64**(1), 47–56.

Fijal, J., and Stanczyk, I. (1970). Syngenite from Inowroclaw (in Polish). *Pr. Mineral., Pol. Akad. Nauk, Kom. Nauk Mineral. Oddzial Krakowie,* No. 24, pp. 22–28.

Filippova, G. R., Vlasov, N. A., and Zorina, T. A. (1969). Fluorine in salt formations of the Borzinsk lake (in Russian). *Izv. Irkutsk. Univ., Nauchno–Issled. Inst. Nefti–Uglekhim. Sin.* **11**(2), 155.

Finaton, C. (1934). Les dépôts lagunaires et le gypse du bassin de Paris. *Rev. Géogr. Phys. Géol. Dynam.* **7**(4), 357–378.

Finetti, I., and Morelli, C. (1973). Geophysical exploration of the Mediterranean Sea. *Boll. Geofis. Teor. Applic.* **15**(60), 263–341.

Fischbeck, R., and Mueller, German (1971). Monohydrocalcite, hydromagnesite, nesquehonite, dolomite, aragonite, and calcite in speleothems of the Fränkische Schweiz, western Germany. *Contrib. Mineral. Petrol.* **33**(2), 87–92.

Fischer, F. (1916). Die Gewinnung von flüssigem Hydrokarbon durch Einwirkung von Aluminiumchlorid auf Naphthalen unter Druck. *Ber. Kais. Wilh. Inst. Kohlenforsch.* **49,** 252–259.

Fischer, W. M., and Schmidt, A. (1926). Rhythmische Fällung von Kalziumhydroxid. *Roczn. Chem.* **6,** 404–414.

Fisk, H. N. (1959). Padre Island and the Laguna Madre Flats, coastal Texas. *In* "Coastal Geography Conference, 2nd." (R. J. Russel, ed.), pp. 103–151. U.S. Natl. Acad. Sci.—Natl. Res. Coun., Washington, D.C.

Fiveg, M. P. (1948). Seasonal cycle of rock salt sedimentation in the Upper Kama deposits (in Russian). *Dokl. Akad. Nauk SSSR* **61,** 1087–1090.

Fiveg, M. P. (1956). The geological conditions of sedimentation in the generation of saline formations (in Russian). *Vopr. Mineral. Osad. Obraz.* No. 3–4, 155–161.

Fiveg, M. P. (1957). Geologic conditions of sedimentation of salt-bearing formations and their potash horizons. *Proc. Int. Geol. Congr., 20th., India* **5**(1), 17–37.

Fiveg, M. P. (1964). The significance of brine level oscillations in evaporite basins during the deposition of a salt-bearing series. *Tr. Vses. Nauchno–Issled. Proektn. Inst. Galurgii* **45,** 61–69.

Fiveg, M. P., and Rayevskiy, V. I. (1971). Potash rocks and ores from the Upper Kama and cis–Carpathian basins (in Russian). *In* "Flotation of Soluble Salts" (V. A. Glembotskiy, ed.), pp. 15–19. Izd. Nauka i Tekhnika, Minsk.

Flannery, W. L., Doetsch, R. N., and Hansen, P. A. (1952). Salt desideratum of *Vibrio costicolus,* an obligate halophilic bacterium.—Ionic replacement of sodium chloride requirements. *J. Bacteriol.* **64,** 713–717.

Fleet, M. E. L. (1965). Preliminary investigations into the sorption of boron by clay minerals. *Clay Miner.* **6,** 3–16.

Fleischer, M. (1970). New mineral names. *Am. Mineral.* **55**(1–2), 317–323.

Flemming, N. C. (1978). Holocene eustatic changes and coastal tectonics in the northeastern Mediterranean: Implications for models of coastal consumption. *Phil. Trans. R. Soc. London,* [A] **289** (1362), 405–458.

Flint, R. F., and Gale, W. A. (1958). Stratigraphy and radiocarbon dating at Searles Lake, California. *Am. J. Sci.* **256**(10), 689–714.

Floerke, O. W. (1952). Kristallographische und röntgenometrische Untersuchungen im System $CaSO_4$–$CaSO_4 \cdot 2H_2O$. *Neues Jahrb. Mineral., Abh.* **84**(2), 189–240.

Floerke, W., and Floerke, O. W. (1961). Über Bildung von Vaterit aus Gips in Natriumkarbonatlösung. *Neues Jahrb. Mineral., Monatsh.* No. 3, pp. 179–181.

Flohn, H. (1964). Grundfragen der Paläoklimatologie im Lichte einer theoretischen Klimatologie. *Geol. Rundsch.* **54**(2), 504–515.

Flohn, H. (1978). Abrupt events in climatic history. *In* ''Climatic Change and Variability'' (A. B. Pittock *et al.*, eds.), pp. 124–134. Cambridge Univ. Press, London.

Florschutz, F., Menendez-Amor, J., and Wijmstra, T. A. (1971). Palynology of a thick Quaternary succession in southern Spain. *Palaeogeogr. Palaeoclimat. Palaeoecol.* **10**(4), 233–264.

Foerster, S. (1974). Durchlässigkeits- und Rissbildungsuntersuchungen zum Nachweis der Dichtheit von Salzkavernen. *Neue Bergbautechnik* **4**(4), 278–283.

Folinsbee, R. E., Krouse, R., and Sasaki, A. (1966). Sulphur isotopes and the Pine Point lead-zinc deposits, N. W. T., Canada (abstr.). *Econ. Geol.* **61**(7), 1307–1308.

Folk, R. L. (1974). The natural history of crystalline calcium carbonate: Effect of magnesium content and salinity. *J. Sediment. Petrol.* **44**(1), 40–53.

Folk, R. L., and Landes, L. S. (1975). Mg/Ca ratio and salinity: Two controls over crystallization of dolomite. *Bull. Am. Assoc. Pet. Geol.* **59**(1), 60–68.

Fontes, J. C. (1966). Interêt en géologie d'une étude isotopique de l'evaporation cas de l'eau de mer. *C. R. Hebd. Séances Acad. Sci. Paris,* [D] **263**(25), 1950–1953.

Fontes, J. C. (1968). Le gypse du bassin de Paris; historique et données récentes. *Mém. Bur. Rech. Géol. Min.* **58,** 359–386.

Fontes, J. C., and Gonfiantini, R. (1967a). Fractionnement isotopique de l'hydrogene dans l'eau de crystallisation du gypse. *C. R. Hebd. Séances Acad. Sci. Paris,* [D] **265**(1), 4–6.

Fontes, J. C., and Gonfiantini, R. (1967b). Comportement isotopique au cours de l'evaporation de deux bassins sahariens. *Earth Planet. Sci. Lett.* **3**(3), 258–266.

Fontes, J. C., and Letolle, R. (1976). ^{18}O and ^{34}S in the Upper Bartonian gypsum deposits of the Paris Basin. *Chem. Geol.* **18**(4), 285–295.

Fontes, J. C., Gonfiantini, R., and Tongirogi, S. (1963). Composition isotopique du Bassin de Paris. *C. R. Somm. Séances Soc. Géol. Fr.* No. 3, pp. 92–96.

Fontes, J. C., Gauthier, J., and Kulbicki, G. (1966). Paramètres isotopiques et géochimiques du gypse parisien (étude préliminaire). *C. R. Somm. Séances Soc. Géol. Fr.* No. 3, pp. 119–121.

Fontes, J. C., Fritz, P., Gauthier, J., and Kulbicki, G. (1967). Minéraux argilleux, éléments-traces et compositions isotopiques ($^{18}O/^{16}O$ et $^{13}C/^{12}C$) dans les formations gypsifères de l'Éocène supérieur et de Oligocène de Coreilles-Parisis. *Bull. Cent. Rech. Pau* **1**(2), 315–366.

Forestier, H., and Kremer, G. (1952). Influence des cations étrangères en solution sur le faciès cristalline du gypse précipité. *C. R. Hebd. Séances Acad. Sci. Paris* **234,** 941–943.

Foret, J. (1943). Inexistence d'un prétendu paradoxe observé à propos de la solubilité du gypse dans les solutions de chlorure. *C. R. Hebd. Séances Acad. Sci. Paris* **217**(11), 264–266.

Foronoff, N. P. (1962). Physical properties of sea water. *In* ''The Sea'' (M. N. Hill, ed.), vol. 1, pp. 3–30. Wiley (Interscience), New York.

Fournier, R. O., Rosenbauer, R. J., and Bischoff, J. L. (1982). The solubility of quartz in aqueous sodium chloride solutions at 350°C and 180 to 500 bars. *Geochim. Cosmochim. Acta* **46**(10), 1975–1978.

Franz, E. (1967). Zur Frage der Genese authigener Quarze im Salinar. *Z. Angew. Geol.* **13**(3), 157–159.

Fraser, A. R., and Tilbury, L. A. (1979). Structure and stratigraphy of the Ceduna Terrace region, Great Australian Bight Basin. *J. Aust. Pet. Explor. Assoc.* **19**(1), 53–65.

Friedel, B. (1976). $Na_2Mg(SO_4)_2 \cdot 5H_2O$, ein metastabiles Salz der Krustenbildung auf Böden. *Neues Jahrb. Mineral., Abh.* **126**(2), 187–198.

Friedman, G. M. (1972). Significance of Red Sea in problem of evaporites and basinal limestones. *Bull. Am. Assoc. Pet. Geol.* **56**(6), 1072–1086.

Friedman, G. M. (1978). Importance of microorganisms in sedimentation. *In* "Environmental Biogeochemistry and Geomicrobiology" (W. E. Krumbein, ed.), vol. 1, pp. 323–326. Ann Arbor Sci. Publ., Ann Arbor, Michigan.

Friedman, G. M. (1980a). Review of depositional environments in evaporite deposits and the role of evaporites in hydrocarbon accumulation. *Bull. Cent. Rech. Explor.–Prod. Elf–Aquitaine* **4**(1), 589–608.

Friedman, G. M. (1980b). Dolomite is an evaporite mineral: Evidence from the rock record and from sea–marginal ponds of the Red Sea. *In* "Concepts and Models of Dolomitization" (D. H. Zenger, J. B. Dunham, and R. L. Etherington, eds.), *Spec. Publ. Soc. Econ. Paleontol. Mineral.* **28,** 69–80.

Friedman, G. M., and Brenner, I. B. (1977). Progressive diagenetic elimination of strontium in Quaternary to Late Tertiary coral reefs of Red Sea; sequence and time scale. *In* "Reefs and Related Carbonates; Ecology and Sedimentology" (S. H. Frost, M. P. Weiss, and J. B. Saunders, eds.), *Am. Assoc. Pet. Geol. Stud. Geol.* **4,** 353–355.

Friend, J. N., and Allchin, J. P. (1939). Colour of celestine. *Nature (London)* **144**(3649), 633.

Friend, J. N., and Allchin, J. P. (1940a). Colloidal gold, a coloring principle in minerals. *Mineral. Mag.* **25**(170), 584–596.

Friend, J. N., and Allchin, J. P. (1940b). Blue rock salt. *Nature (London)* **145**(3668), 266–267.

Frik, M., and Kuzel, H. J. (1982). Röntgenographische und thermoanalytische Untersuchungen an Calciumsulfat–Halbhydrat. *Fortschr. Mineral.* **60,** Suppl. 1, 79–80.

Frisch, R. (1927). Der Einfluss langsamer Kathodenstrahlen auf Steinsalz. *Sitzungsber. Akad. Wiss. Wien, Naturwiss. Kl.,* [2a] **136,** 57–64.

Fristrup, B. (1953). High Arctic Deserts. *Proc. Int. Geol. Congr., 19th, Algiers 1952, Sect. 7,* No. 7, 91–99.

Fritz, P. (1969). The oxygen and carbon isotope composition of carbonate from the Pine Point lead–zinc ore deposits. *Econ. Geol.* **64**(7), 733–742.

Fritz, P., and Frape, S. K. (1982a). Comments on the ^{18}O, ^{2}H, and chemical composition of saline groundwaters on the Canadian Shield. *In* "Isotope Studies of Hydrologic Processes" (E. C. Perry, Jr., and C. W. Montgomery, eds.), pp. 57–63. Northern Ill. Univ. Press, DeKalb.

Fritz, P., and Frape, S. K. (1982b). Saline groundwaters in the Canadian Shield—a first overview. *Chem. Geol.* **36**(1–2), 179–190.

Fritzsche, J. (1857). Über die Bildung von Glauberit auf nassem Wege und ein zweites Doppelsalz aus schwefelsaurem Natron und schwefelsaurem Kalke. *J. Pract. Chem. (St. Petersburg)* **72,** 291–297.

Fuchs, T. (1874). Intorno alla existenza presso Siracusa di strati miocenici che presentano i caratteri del piano Sarmatico. *Boll. R. Comit. Geol. Ital.* **5,** 373–377.

Fuchs, Y. (1978). Sur un exemple de rélation entre une minéralisation barytique et un milieu à évaporites. *Sci. Terre* **22**(2), 127–146.

Fuechtbauer, H. (1958). Die petrographische Unterscheidung der Zechsteindolomite im Emsland durch ihren Säurerückstand. *Erdöl Kohle* **11,** 689–693.

Fuechtbauer, H., and Goldschmidt, H. (1959). Die Tonminerale der Zechsteinformation. *Beitr. Mineral. Petrogr.* **6**(5), 320–345.

Fuechtbauer, H., and Mueller, G. (1977). "Sedimente und Sedimentgesteine" 3rd. ed., Part II. Schweizerbart, Stuttgart, 784 pp.

Fujiwara, S. (1979). The basis of concentration of salt in sea water (in Japanese). *Geochem. J. (Jpn.)* **13**(5), 225–226.

Fulda, E. (1923). Zur Entstehung des deutschen Zechsteins. *Z. Dtsch. Geol. Ges.* **75**(1), 1–13.

Fulda, E. (1924). Studie über die Entstehung der Kalisalzlagerstätten des deutschen Zechsteins. *Z. Dtsch. Geol. Ges.* **76**(1–4), 7–30.

Fulda, E. (1928). "Das Kali." Enke, Stuttgart, 400 pp.

Fulda, E. (1930). Die Barrentheorie von K. Ochsenius und ihre Bedeutung für die Geologie des Zechsteins. *Kali Verw. Salze Erdöl (Halle)* **24**(5), 71–74.

Fulda, E. (1937). Die Entstehung der Zechsteinsalze nach der Grossflutenhypothese von Martin Wilfarth. *Kali Verw. Salze Erdöl* **31**(1), 1–3 and (2), 17–19.

Fulda, E. (1938). Ein Koniferenstamm aus dem Älteren Kalilager (Oberen Zechstein) des Kaliwerks Bismarckhall bei Bischofferode. *Kali Verw. Salze Erdöl (Halle)* **32**(1), 1–2.

Fuller, J. G. C. M., and Potter, J. W. (1969a). Evaporite formations with petroleum reservoirs in Devonian and Mississippian of Alberta, Saskatchewan and North Dakota. *Bull. Am. Assoc. Pet. Geol.* **53**(4), 909–926.

Fuller, J. G. C. M., and Potter, J. W. (1969b). Evaporites and carbonates: Two Devonian basins of western Canada. *Bull. Can. Pet. Geol.* **17**(2), 182–193.

Fuzesy, A. (1982). Potash in Saskatchewan. *Sask. Geol. Surv. Rep.* No. 181, 44 pp.

Gac, J. Y., Al-Droubi, A., Paquet, H., Fritz, P., and Tardy, Y. (1979). Chemical model for origin and distribution of elements in salts and brines during evaporation of waters. Application to some saline lakes of Tibesti, Chad. *In* "Origin and Distribution of the Elements" (L. H. Ahrens, ed.). *Phys. Chem. Earth* **11,** 149–160.

Gahm, J., and Nacken, R. (1954). Skelettkristallbildung bei den Alkalihalogeniden, besonders beim Steinsalz. *Neues Jahrb. Mineral., Abh.* **86,** 309–366.

Gaines, A. M. (1974). Protodolomite synthesis at 100°C and atmospheric pressure. *Science* **183**(4124), 518–520.

Gaines, A. M. (1980). Dolomitization kinetics: Recent experimental studies. *In* "Concepts and Models of Dolomitization" (D. H. Zenger, J. B. Dunham, and R. L. Etherington, eds.), *Soc. Econ. Paleontol. Mineral. Spec. Publ.* **28,** 81–86.

Galabuda, N. I. (1978). Geology of the Frasnian salt formation of the Dnieper–Donets depression. *Pet. Geol. (Engl. Transl.)* **15**(12), 539–544.

Galakhovskaya, T. V. (1954). Spectroscopic determination of small quantities of copper, lead, manganese, iron, and nickel in the soluble portion of halite, sylvite, and sylvinite (in Russian). *Tr. Vses. Nauchno-Issled. Proektn. Inst. Galurgii* **27,** 230–238.

Gallup, W. B., and Hamilton, G. J. (1953). The orogenic history of the Williston Basin, Saskatchewan. *Billings Geol. Soc., Annu. Field Conf. Guidebook* **4,** 123–136.

Ganeyev, I. G. (1976). Properties of hydrothermal solutions and the form in which mineral components are transported in them. *Dokl. Akad. Nauk SSSR,* **227**(2), 458–460; *Dokl. Acad. Sci. USSR, Earth Sci. Sect. (Engl. Transl. Am. Geol. Inst.)* **227**(1–6), 205–207.

Garavelli, C. L. (1958). Presence of ferric chloride hexahydrate among newly formed minerals in the ore deposits of Rio Marina, Elba. *Period. Mineral. (Roma)* **27,** 211–214. *Chem. Abstr.* **52,** 16,127.

Garcia-Iglesias, J., and Touray, J. C. (1976). Hydrocarbures liquides en inclusion dans le fluorite du gîsement de "La Cabaña" (Berbes, Asturias, Espagne). *Bull. Soc. Fr. Minéral. Cristallogr.* **99**(2–3), 117–118.

Gard, L. M., Jr. (1968). Geologic studies, project Gnome, Eddy County, New Mexico. *U.S. Geol. Surv. Prof. Paper* **589,** 33 pp.

Gardet, J. J., Guilhot, B., and Soustelle, M. (1976). The dehydration kinetics of calcium sulfate dihydrate. Influence of the gaseous atmosphere and the temperature. *Cem. Concr. Res.* **6**(5), 697–705.

Gardner, L. S., Haworth, H. F., and Chiangmai, P. N. (1967). Salt Resources of Thailand. *Rep. Investig. Thailand Dep. Mineral Resour.* No. 11, 100 pp.

Gardner, W. C. (1974). Middle Devonian stratigraphy and depositional environments in the Michigan Basin. *Mich. Basin Geol. Soc. Spec. Paper* **1,** 1–46.

Garlicki, A. (1979). Sedimentation of Miocene salt in Poland (in Polish). *Pr. Geol., Pol. Akad. Nauk, Kom. Nauk Geol., Oddzial Krakowie,* No. 119, pp. 1–67.

Garlicki, A. (1980). On some sedimentary structures of anhydrite within Miocene evaporites in the Carpathian Foreland area, Poland. *In* ''Symp. on Salt, 5th.'' (A. H. Coogan and L. Hauber, eds.), vol. 1, 49–53. N. Ohio Geol. Soc., Cleveland, Ohio.

Garlicki, A., and Wiewiorka, J. (1981). The distribution of bromine in some halite rock salts of Wieliczka salt deposit (Poland). *Rocz. Pol. Towar. Geol.* **51**(3–4), 353–359.

Garrels, R. M., and Lerman, A. (1981). Phanerozoic cycles of sedimentary carbon and sulfur. *Proc. Natl. Acad. Sci. U.S.A.* **78**(8), 4652–4656.

Garrido, J. (1945). Sur des cristaux de gypse présentant des ''fantomes'' de croissance. *Bol. Soc. Geol. Port.* **4**(3), 213–214.

Gaudefroy, C. (1956). Questions de cristallogénie phénocristaux de gypse inclus dans le gypse saccharaïde de Safi. *Notes Mém. Serv. Géol. Morocco* No. 135, 139–145.

Gaudefroy, C. (1960). Hypothèse sur la formation des phénocristaux de dolomite au voisinage des cours d'eau salés du Maroc. *Neues Jahrb. Mineral., Abh.* **94**(2), 1191–1199.

Gavish, E. (1975). Geochemistry and mineralogy of a recent sabkha along the coast of Sinai, Gulf of Suez. *Sedimentology* **21**(3), 397–414.

Gavish, E. (1980). Recent sabkhas marginal to the southern coast of Sinai, Red Sea. *In* ''Hypersaline Brines and Evaporitic Environments'' (A. Nissenbaum, ed.), pp. 233–252. Elsevier, Amsterdam.

Gavrilov, Ye. Ya., Pankina, R. G., Snakhtina, A. M., and Osipova, M. G. (1977). Sulfur isotopic composition of the Amu–Darya saline basin. *Geokhimiya* No. 5, pp. 782–788; *Geochem. Int. (Engl. Transl.)* **14**(3), 74–81.

Gawel, A. (1947). Geological conditions for the origin of blue salt, amethyst and violet fluorite (in Polish). *Rocz. Pol. Towar. Geol.* **17,** 39–60.

Geilmann, W., and Barttlingek, H. (1942). Die Bestimmung von Jodspuren in Salzen. *Mikrochem. Verein. Microchim. Acta* **30,** 217–225.

Geisler, D. (1981). Génèse et évolution des gypses des marais salants de Saline de Giraud (Camarge). *Bull. Minéral.* **104**(5), 625–629.

Gelpi, E., Schneider, H., Mann, J., and Oro, J. (1970). Lipids of geochemical significance in microscopic algae. *Phytochemistry* **9**(3), 603–612 and 613–617.

Gemp, S. D., and Korin, S. S. (1980). Role of metamorphism in the formation of the mineral composition of potassium salts of the Ciscarpathians (in Russian). *Sov. Geol.* No. 3, 104–110.

Genovese, S. (1962). Sulle condizione fizico–chimiche dello stagno di Faro in seguito dell'operatura di un nuovo canale. *Atti Soc. Peloritana Sci. Fis. Mat. Nat.* **8,** 67–72.

Gerlach, H., and Heller, S. (1966). Über künstliche Flüssigkeitseinschlüsse in Steinsalzkristallen. *Ber. Dtsch. Ges. Geol. Wiss.,* [B] **11,** 195–214.

Gerling, E. K., and Titov, N. E. (1949). The decay of potassium by K-capture (in Russian). *Izv. Akad. Nauk SSSR, Otd. Khim. Nauk,* pp. 128–133.

Germanov, A. I. (1965). Geologic significance of organic matter in the hydrothermal process (in Russian). *Geokhimiya* **7,** 834–843.

Germanov, A. I., Yasupova, I. F., and Borzenkov, I. A. (1979). Contemporary sulfide mineralization in paragensis with hydrogen sulfide waters (in Russian). *Sov. Geol.* No. 10, pp. 118–120.

Germanov, A. I., Borzenkov, I. A., and Yasupova, I. F. (1981). Biogenic sulfate reduction and methane formation in a subsurface water–rock system (in Russian). *Sov. Geol.* No. 8, pp. 116–121.

Gevantman, L. H., Lorenz, J., Haas, J. L., Jr., Clynne, M. A., Potter, R. W., II, Schafer, C. M., Tomkins, R. P. T., Shakoor, A., Hume, H. R., Yang, J. M., Li, H. H., and Matula, R. A.

(1981). "Physical Properties Data for Rock Salt." *Natl. Bur. Stand. (U.S.), Monogr.* **167,** 282 pp.

Gevers, T. W., and van der Westhuyzen, J. P. (1932). The occurrences of salt in the Swakopmund area, Southwest Africa. *Trans. Geol. Soc. S. Afr.* **34,** 61–80.

Geyh, M. A., and Rohde, P. (1972). Weichselian chronostratigraphy, ^{14}C dating and statistics. *Proc. Int. Geol. Congr., 25th, Montreal, Sect 12* **24,** 27–36.

Gibson, C. S., and Holt, S. (1933). Hydrates of calcium sulfate. *J. Chem. Soc.*, pp. 638–640.

Gibson, G. W. (1962). Geological investigations in southern Victoria Land, Antarctica. Pt. 8. Evaporite salts in the Victoria Valley region. *N. Z. J. Geol. Geophys.* **5**(3), 361–374.

Gill, D., and Briggs, L. I. (1970). Silurian reefs in Michigan Basin, stratigraphic, facial and reservoir property analysis (abstr.). *Bull. Am. Assoc. Pet. Geol.* **54**(5), 848–849.

Gillespie, J., and Breck, S. (1941). Thermal diffusion in ternary liquid mixtures, particularly aqueous solutions containing ferrous chloride. *J. Chem. Phys.* **9,** 370–374.

Giordano, T. H., and Barnes, H. L. (1981). Lead transport in Mississippi Valley–type ore solutions. *Econ. Geol.* **76**(8), 2200–2211.

Gistl, R. (1940). Tintenstriche und Gipsbildung an Kalksteinen. *Zentralbl. Bakteriol. Parasitenkd. Infektionskr. Hyg.*, [2] **102,** 486–492.

Gize, A. P., and Barnes, H. L. (1981). The organic geochemistry of three Mississippi Valley–type deposits. *Geol. Soc. Am. Abstr. Prog.* **13**(7), 459.

Glaccum, R. A., and Prospero, J. M. (1980). Saharan aerosols over the tropical North Atlantic—mineralogy. *Mar. Geol.* **37**(3), 295–321.

Glass, I. I. (1976). Utilization of Geothermal Energy. *Rep. Univ. Toronto Inst. Aerospace Stud.* No. 40, 84 pp.

Glembotskiy, V. A., and Uvarov, V. S. (1962). Flotation properties of anhydrite and gypsum (in Russian). *Tr. Akad. Nauk Tadzh. SSR, Otd. Geol.–Khim. Tekhnol. Nauk* No. 1, 50–55.

Glennie, K. W. (1972). Permian Rotliegendes of Northwest Europe interpreted in the light of modern desert sedimentation studies. *Bull. Am. Assoc. Pet. Geol.* **56**(6), 1048–1071.

Glennie, K. W., and Evans, G. (1976). A reconnaissance of the recent sediments of the Ranns of Kutch, India. *Sedimentology* **23**(5), 625–647.

Glew, D. N., and Hames, D. A. (1970). Gypsum, disodium pentacalcium sulfate and anhydrite solubilities in concentrated sodium chloride solutions. *Can. J. Chem.* **48**(23), 3733–3738.

Gloeckner, F. (1914). Ein Vorkommen von Kupferkies in Kalisalzen. *Kali (Halle)* **8**(13), 307–308.

Glogoczowski, J. J. (1959). Geochemical observations in the vicinity of the Barycz salt (in Polish). *Bull. Pol. Akad. Nauk, Ser. Chem. Geol. Geogr.* **7,** 833–836 and 837–843.

Glueckauf, E. (1955). Influence of ionic hydration on activity coefficients in concentrated electrolyte solutions. *Trans. Faraday Soc.* **51,** 1235–1244.

Goergey, R. (1909). Salzvorkommen aus Hall im Tirol. *Tschermak's Mineral. Petrogr. Mitt.* **28,** 334–346.

Goergey, R. (1912). Zur Kenntnis der Kalisalzlager von Wittelsheim im Oberelsass. *Tschermak's Mineral. Petrogr. Mitt.* **31,** 339–468.

Goergey, R. (1915). Über die Kristallform des Polyhalites. *Tschermak's Mineral. Petrogr. Mitt.* **33,** 48–102.

Golding, L. Y., and Walter, M. R. (1979). Evidence of evaporite minerals in the Archean Black Flag beds, Kalgoorlie, western Australia. *J. Aust. Geol. Geophys.* **4**(1), 67–71.

Goldman, M. I. (1952). Deformation, metamorphism, and mineralization of gypsum–anhydrite caprocks of sulfur salt dome in Louisiana. *Mem. Geol. Soc. Am.* **50,** 1–169.

Goldsmith, L. H. (1969). Concentration of potash salts in saline basins. *Bull. Am. Assoc. Pet. Geol.* **53**(4), 790–797.

Golikova, S. M. (1930). Über eine Gruppe von obligat halophiler Bakterien, die in einem Natriumchlorid–reichen Nährboden wachsen. *Zentralbl. Bakteriol. Parasitenkd. Infektionskr. Hyg.*, [2] **80,** 35–41.

Goncharev, E. S., and Kulibakina, I. V. (1971). The relationship between gas pools and evaporites. *Dokl. Akad. Nauk SSSR,* **197**(5), 1150–1152; *Dokl. Acad. Sci. USSR, Earth Sci. Sect. (Engl. Transl. Am. Geol. Inst.)* **197**(1–6), 240–241.

Gonfiantini, R. (1965). Effetti isotopici nell'evaporazione di acque salate. *Atti Soc. Toscana Sci. Nat., Pisa, Mem.,* [A] **72**(2), 550–569.

Gonfiantini, R., and Fontes, J. C. (1963). Oxygen isotope fractionation in the water of crystallization of gypsum. *Nature (London)* **200**(4907), 644–646.

Gordon, W. A. (1975). Distribution by latitude of Phanerozoic evaporite deposits. *J. Geol.* **83**(6), 671–684.

Goudie, A. (1971). Climate, weathering, crust formation, dunes and features of the Central Namib Desert near Gobabeb, South-West Africa. *Madogua (Windhoek)* **2,** 15–31.

Grabau, A. W. (1920). ''Principles of Salt Deposition.'' Vol. 1 of ''Geology of the Non-Metallic Mineral Deposits Other Than Silicates.'' McGraw-Hill, New York, 435 pp.

Graefe, E. (1908). Das Erdöl aus dem Kalisalzbergwerk Desdemona bei Alfeld an der Leine. *Kali (Halle)* **2**(21), 468–470.

Graefe, E. (1910). Über Erdölvorkommen in deutschen Kalisalzlagerstätten. *Kali* (Halle) **4**(10), 261.

Grebe, H. (1957). Zur Mikroflora des niederrheinischen Zechsteins. *Bundesanst. Bodenforsch., Geol. Jahrb.* **73,** 51–54.

Grebennikov, N. P., and Yermakov, V. A. (1980). Some characteristics of the distribution of K- and Mg-salts in the western section of the Caspian Sea basin (in Russian). *Tr. Akad. Nauk SSSR, Sib. Otd.* **439,** 44–47.

Gregg, M. C., and Cox, C. S. (1972). The vertical microstructure of temperature and salinity. *Deep-Sea Res.* **18,** 355–376.

Grengg, R. (1914). Dehydration products of gypsum. *Z. Anorg. Allg. Chem.* **90,** 327–336.

Greschuchna, R. (1976). Crystals of cement paste in sulfate acid solutions. *Gidratatsiya Tverd. Tsem.* **2**(1), 337–339. *Chem. Abstr.* **87,** 10367.

Grigor'yev, A. P., and Shamayev, P. P. (1976). Determination of gypsum–anhydrite equilibrium temperature (in Russian). *Izv. Akad. Nauk SSSR, Sib. Otd., Ser. Khim.* **104**(5), 104–107.

Grimm, W. D. (1962a). Ausfällung von Kieselsäure in salinar beeinflussten Sedimenten. *Z. Dtsch. Geol. Ges.* **114**(3), 590–619.

Grimm, W. D. (1962b). Idiomorphe Quarze als Leitmineralien für salinare Fazies. *Erdöl Kohle Erdgas Petrochem.* **15**(11), 880–887.

Grimm, W. (1968). ''Kali und Steinsalzbergbau.'' 2 vols. VEB Dtsch. Verlag Grundstoffind., Leipzig.

Grokhovskiy, L. M., and Grokhovskaya, M. A. (1980). ''Search and Exploration for Deposits of Mineral Salts'' (in Russian). Izd. Nedra, Moscow, 163 pp.

Gromov, B. V. (1940). The formation of complexes in the system $PbCl_2$–NaCl–H_2O (in Russian). *Zh. Prikl. Khim.* **13,** 337–344.

Grube, G., and Brauning, W. (1938). Über die Entwässerung von Magnesiumchloridhydrat und Carnallit. *Z. Elektrochem.* **44,** 134–143.

Gubin, V., and Tsekhomskaya, V. (1930). Über die biochemische Sodabildung in den Sodaseen. *Zentralbl. Bakteriol. Parasitenkd. Infektionskr. Hyg.,* [2] **81,** 396–401.

Gudowius, E., and von Hodenberg, R. (1979). ''Natrium–polyhalit,'' eine dem Bassanit und dem ?–$CaSO_4$ verwandte Phase. *Kali Steinsalz* **7**(12), 501–504.

Guillou, J. J. (1972). La série carbonatée magnésienne et l'évolution de l'hydrosphère. *C. R. Hebd. Séances Acad. Sci. Paris,* [D] **274**(22), 2952–2955.

Gunatilaka, A., Saleh, A., and Al-Temeemi, A. (1980). Plant-controlled supratidal anhydrite from Al Khiran, Kuwait. *Nature (London)* **288** (5788), 257–260.

Gunzert, G. (1961). Die ''obere bituminöse Zone'' im Bereich der Kalisalzlagerstätte von Buggingen (Baden). *Kali Steinsalz,* **3**(4), 111–118.

Guptha, M. V. S. (1980). Authigenic gypsum in a deep-sea core from southeastern Arabian Sea. *J. Geol. Soc. India* **21**(11), 568–571.

Gusseva, A. N., and Faingersh, L. A. (1974). La formation des pétroles soufrés dans les zones d'hypergenèse ancienne. *Proc. Int. Meet. Adv. Org. Geochem, 6th (1973),* pp. 741–746.

Guthrie, F. C. (1929). Blue rock salt. *Nature (London)* **123,** 130.

Gvirtzman, G. (1969). The Saqiye Group (Late Eocene to Early Pleistocene) in the coastal plain and Hashephela regions, Israel. *Bull. Geol. Surv. Isr.* **51,** (2 vols.).

Gvirtzman, G., and Buchbinder, B. (1978). The late Tertiary of the coastal plain and continental shelf of Israel and its bearing on the history of the eastern Mediterranean. *In* Int. Repts. Deep-Sea Drlg. Proj., vol. 42A (D. A. Ross, Y. P. Neprochnov *et al.* eds.), pp. 1181–1183. U.S. Govt. Print. Off., Washington, D.C.

Haberfeld, M. (1933). Über das Auftreten und Verschwinden von Färbung in Steinsalzkristallen unter Druck. *Sitzber. Akad. Wiss. Wien,* [2a] **142,** 135–154.

Haegele, G. (1939). Gipskristalle aus basaltischem Andesit von Stromboli. *Zentralbl. Mineral., Paläontol., Geol.,* [A] **8,** 254–256.

Hahn, H. H., and Stumm, W.(1970). The role of coagulation in natural waters. *Am. J. Sci.* **268**(4), 354–368.

Hahn-Weinheimer, P., Fabricius, F., Mueller, J., and Sigl, W. (1978). Stable isotopes of oxygen and carbon in carbonates and organic material from Pleistocene to Upper Miocene sediments at site 374 (DSDP Leg 42A). *In* Intl. Rep. Deep-Sea Drlg. Proj., vol. 42A (K. J. Hsu, L. Montadert *et al.,* eds.), pp. 483–488. Govt. Print. Off., Washington, D.C.

Håkanson, L. (1981). On lake bottom dynamics—the energy–topography factor. *Can. J. Earth Sci.* **18**(5), 899–909.

Hallberg, R. O., Bubela, B., and Ferguson, J. (1980). Metal chelation in sedimentary systems. *Geomicrobiol. J.* **2**(2), 99–114.

Haltendorf, M., and Hofrichter, E. (1972). Feinstratigraphische Fazies und Bromgehalte isochroner Schichten des Leinesalzes (Zechstein 3) im zentralen Teil des Zechsteinbeckens (Raum Hannover). *Bundesanst. Geowiss., Geol. Jahrb.* **90,** 1–61.

Ham, W. E., Mankin, C. J., and Schleicher, J. A. (1961). Borate minerals in Permian gypsum of west-central Oklahoma. *Bull. Okla. Geol. Surv.* **92,** 1–77.

Hamann, R. J., and Anderson, G. M. (1978). Solubility of galena in sulfur-rich sodium chloride solutions. *Econ. Geol.* **73**(1), 96–100.

Hammer, U. T. (1981). Primary production in saline lakes: A review. *Hydrobiologia* **81,** 47–58.

Hammerschmidt, F. (1883). Beiträge zur Kenntniss [sic] des Gyps- und Anhydritgesteines. *Tschermak's Mineral. Petrogr. Mitt.* **5**(3), 245–284.

Hammond, A. L. (1976a). Paleoclimate: Ice age earth was cool and dry. *Science* **191**(4226), 455.

Hammond, A. L. (1976b). Solar variability, is the sun an inconstant star? *Science* **191**(4232), 1159–1160.

Hamner, W. M. (1981). Strange world of Palau's salt lakes. *Natl. Geogr. Mag.* **161**(2), 264–282.

Han, J., and Calvin, M. (1969). Hydrocarbon distribution of algae and bacteria, and microbiological activity in sediments. *Proc. Natl. Acad. Sci. U.S.A.* **64**(2), 436–443.

Hannay, J. B. (1877). Examination of substances by the time method. *J. Chem. Soc.* **32,** 381–395.

Hanor, J. S. (1973). The role of *in situ* densities in the migration of subsurface brines. *Geol. Soc. Am., Abstr. Prog.* **5**(7), 651–652.

Hanor, J. S. (1981). The solubility of methane in sedimentary pore waters: Effect of other dissolved gas species. *Geol. Soc. Am., Abstr. Prog.* **13**(7), 467.

Hanshaw, B. B., Back, W., and Deike, R. G. (1971). A geochemical hypothesis for dolomitization by groundwater. *Econ. Geol.* **66**(5), 710–724.

Hanshaw, B. B., and Coplen, T. B. (1973). Ultrafiltration by a compacted clay membrane. *Geochim. Cosmochim. Acta* **37**(10), 2311–2327.

Hara, R., Nakamura, K., and Higashi, K. (1932). Specific gravity and water pressure of concentrated sea water at 0°C–175°C. *Technol. Rep. Tohoku Imp. Univ.* **10,** 433–452.

Hara, R., Tanaka, Y., and Nakamura, K. (1934). On the calcium sulphate in sea water. I. Solubilities of dihydrate and anhydrite in sea waters of various concentrations at 0°C–200°C. *Technol. Rep. Tohoku Imp. Univ.* **11,** 199–221.

Haranczyk, C. (1961). Correlation between organic carbon, copper and silver content in Zechstein copper-bearing shales from the Lubon–Sieroszowice region (Lower Silesia) (in Polish). *Bull. Pol. Akad. Nauk, Ser. Geol. Geogr.* **9**(4), 183–189.

Harbeck, G. E., Jr. (1955). The effect of salinity on evaporation. *U.S. Geol. Surv. Prof. Paper* **272A,** 1–6.

Harbort, E. (1913). Zur Frage der Genesis der Steinsalz-und Kalisalzlagerstätten im Tertiär vom Ober–Elsass und von Baden. *Z. Prakt. Geol.* **21**(2), 189–198.

Harder, H. (1959). Beitrag zur Geochemie des Bors. II. Bor in Sedimenten. *Nachr. Akad. Wiss. Göttingen, Math. Phys. Kl.* **6,** 123–183.

Harder, H. (1961). Einbau von Bor in detritische Tonminerale. *Geochim. Cosmochim. Acta* **21**(3–4), 284–294.

Hardie, L. A. (1967). The gypsum–anhydrite equilibrium at one atmosphere pressure. *Am. Mineral.* **52**(1–2), 171–200.

Hardie, L. A. (1968). The origin of the Recent non-marine evaporite deposit of Saline Valley, California. *Geochim. Cosmochim. Acta* **32**(12), 1279–1301.

Hardie, L. A., and Eugster, H. P. (1970). The evolution of closed-basin brines. *Mineral. Soc. Am., Spec. Paper* **3,** 273–298.

Harding, S. T. (1949). Evaporation from free water surfaces. *In* ''Physics of the Earth'' (O. E. Meinzer, ed.), vol. 9, pp. 56–82. McGraw-Hill, New York.

Harkins, W. D., and McLaughlin, H. M. (1925). Surface tension and adsorption for aqueous solution of chloride. *J. Am. Chem. Soc.* **47**(8), 2083–2089.

Harris, H. J. H., Cartwright, K., and Torii, T. (1979). Dynamic chemical equilibrium in a polar desert pond; a sensitive index of meteorological cycles. *Science* **204**(4390), 301–303.

Hartwig, G. (1936). Über das Sylvinitgebiet von ''Einigkeit I.'' *Kali Verw. Salze Erdöl (Halle)* **30,** 111–114, 121–125, 131–133, 141–144, 151–154.

Hartwig, G. (1954). Zur Kontrolle von Kohlendioxid in den Werra– und Fulda–Kalisalzlagerstätten. *Kali Steinsalz* **1**(5), 3–26.

Hartwig, G. (1955). Über die Petrographie und Kreuzschieferung der tieferen Lagen des Zechstein 2 unter Solling–Elfas und Duenn–Hainleite–Eck mit einem Ausblick auf die Situation unter der Ostflanke des Göttinger Leinetales. *Kali Steinsalz* **1**(8), 8–29.

Harvie, C. E., Eugster, H. P., and Weare, J. H. (1982). Mineral equilibriums in the six-component seawater system sodium–potassium–magnesium–calcium–sulfate–chloride–water at 25°C. II. Composition of the saturated solutions. *Geochim. Cosmochim. Acta* **46**(9), 1603–1618.

Hashimi, N. H., and Ambre, N. V. (1979). Gypsum crystals in the inner shelf sediments off Maharashtra, India. *J. Geol. Soc. India* **20**(4), 190–192.

Haslam, J., Allberry, E. C., and Moses, G. (1950). Bromine content of the Cheshire salt deposit and of some borehole and other brines. *Analyst (London)* **75,** 352–356.

Haude, R. (1970). Die Entstehung von Steinsalz–Metamorphosen. *Neues Jahrb. Geol. Paläontol., Monatsh.,* No. 1, pp. 1–10.

Hausmann, W., and Krumpel, O. (1930). Durchlässigkeit von Gips und Glimmer im Ultraviolet. *Strahlentherapie* **35,** 387–390.

Hayashi, T., and Sato, S. (1951). Dehydration properties of gypsum. *Gypsum (Jpn)* **1,** 137–141.

Heard, H. C., and Rubey, W. W. (1966). Tectonic implications of gypsum dehydration. *Bull. Geol. Soc. Am.* **77**(7), 741–760.

Hecht, F. (1934). Die chemische Zersetzung der tierischen Substanz während der Einbettung in marine Sedimente (ein Beitrag zur Urbitumenbildung). *Kali Verw. Salze Erdöl (Halle)* **28**(17), 209–215.

Hedberg, H. D. (1964). Geologic aspects of petroleum. *Bull. Am. Assoc. Pet. Geol.* **48**(11), 1755–1803.

Heide, K. (1968). Zum Mechanismus der Umbildungsvorgänge in Salzgesteinen. *Chem. Erde* **27**(4), 353–369.

Heide, K., Franke, H., and Brueckner, H. P. (1980). Vorkommen und Eigenschaften von Borazit in Zechsteinsalzlagerstätten der D.D.R. *Chem. Erde* **39**(3), 201–232.

Heidorn, F. (1931). Über ein Vorkommen von Sellait (MgF_2) in Paragenese mit Bitumen aus dem Hauptdolomit des mittleren Zechsteins bei Bleicherode. *Zentralbl. Mineral. Geol. Paläontol.,* [A] pp. 356–364.

Heinrich, J. (1966). Ein Vorkommen von Sellait (MgF_2) im Stassfurtkarbonat von Süd–Rügen. *Ber. Dtsch. Ges. Geol. Wiss.,* [B] **11**(2), 225–228.

Heinrichs, T. K., and Reimer, T. O. (1977). A sedimentary barite deposit from the Archean Fig Tree Group of the Barberton Mountain Land (South Africa). *Econ. Geol.* **72**(8), 1426–1441.

Helgeson, H. C. (1964). "Complexing and Hydrothermal Ore Deposition." Pergamon Press, New York, 128 pp.

Helgeson, H. C. (1968). Evaluation of irreversible reactions in geochemical processes involving minerals and aqueous solutions, I. *Geochim. Cosmochim. Acta* **32**(8), 853–877.

Helgeson, H. C., and MacKenzie, F. T. (1970). Silicate—sea water equilibria in the ocean system. *Deep-Sea Res.* **17**(5), 877–892.

Helgeson, H. C., Garrels, R. M., and MacKenzie, F. T. (1969). Evaluation of irreversible reactions in geochemical processes involving minerals and aqueous solutions, II. *Geochim. Cosmochim. Acta* **33**(4), 455–481.

Heller, F. (1976). Ein Riesen–Steinsalzcrystalloid aus dem Coburger Bausandstein von Zeil bei Hassfurt/Main. *Geol. Bl. Nordost–Bayern* **26**(1), 49–55.

Heller-Kallai, L., Yariv, S., and Riemer, M. (1973). The formation of hydroxy interlayers in smectites under the influence of organic bases. *Clay Miner.* **10**(1), 35–40.

Helmy, A. K. (1963). Calculation of negative and positive adsorption in some clay electrolyte systems. *J. Soil Sci.* **14**(2), 217–224.

Helwig, H. (1972). Stratigraphy, sedimentation, paleogeography, and paleoclimates of Carboniferous ("Gondwana") and Permian of Bolivia. *Bull. Am. Assoc. Pet. Geol.* **56**(6), 1008–1033.

Hem, J. D. (1959). Study and interpretation of the chemical characteristics of natural water. *U.S. Geol. Surv. Water-Supply Paper* No. 1473, 269 pp.

Hemmann, M. (1972). Zur feinstratigraphischen Gliederung des Leinesteinsalzes im Ostteil des Subherzynen Beckens. *Zentr. Geol. Inst. (D.D.R.), Jahrb. Geol.* **4**(1968), 291–300.

Henderson, R. A., and Southgate, P. N. (1980). Cambrian evaporite sequences from the Georgina Basin. *Search (Sydney)* **11**(7–8), 247–249.

Hentschel, J. (1961). Die Faziesunterschiede im Flöz Stassfurt des Kalisalzbergwerks Königshall–Hindenburg. *Kali Steinsalz* **3**(5), 137–157.

Herde, W. (1955). Progressive und deszendente Vorgänge bei der Sedimentation der Riedel–Gruppe (Zechstein-3). *Z. Dtsch. Geol. Ges.* **105**(4), 680–686.

Herman, G., and Barkell, C. A. (1957). Pennsylvanian stratigraphy and productive zones, Paradox salt basin. *Bull. Am. Assoc. Pet. Geol.* **41**(5), 861–881.

Herman, P., Vanderstappen, R., and Hubaux, A. (1961). Sublimates of Nyiragongo (Kivu). *Bull. Séances Acad. R. Belge Sci. Outre–Mer* **6,** 961–971.

Herr, F. L., and Helz, G. R. (1978). Possibility of bisulfide ion pairs in natural brines and hydrothermal solutions. *Econ. Geol.* **73**(1), 73–81.

Herrick, C. L. (1900). The geology of the White Sands of New Mexico. *J. Geol.* **8**(2), 112–126.

Herrmann, A. G. (1958). Geochemische Untersuchungen an Kalisalzlagerstätten im Südharz. *Freiberger Forschungsh.*, [C] **43,** 1–111.

Hermann, A. G. (1961a). Über die Einwirkung Cu–, Sn–, Pb–, und Mn–haltiger Erdölwässer auf die Stassfurtserie des Südharzbezirkes. *Neues Jahrb. Mineral., Monatsh.* No. 1, pp. 60–67.

Herrmann, A. G. (1961b). Geochemie des Strontiums in der Stassfurt–Serie des Zechstein–Salinars im Südharz. *Chem. Erde* **21,** 137–194.

Herrmann, A. G. (1976). Modelle fur die Bestimmung der Bildungstemperaturen primärer Sylvinite und Carnallitite im $MgSO_4$–freien Meerwassersystem mittels der Br–Verteilung. *Fortschr. Mineral. Beih.* **54**(1), 31–32.

Herrmann, A. G. (1977). Modelle für die Bestimmung der Bildungstemperaturen primärer Sylvinite und Carnallitite im $MgSO_4$–freien Meerwassersystem mittels der Br–Verteilung. *Kali Steinsalz* **7**(4), 134–146.

Herrmann, A. G., and Hoffmann, R. O. (1961). Zur Genese einiger Borate in den Salzablagerungen der Stassfurt–Serie des Südharzbezirkes einschliesslich der Grube Königshall–Hindenburg. *Neues Jahrb. Mineral., Monatsh.* No. 1, pp. 52–60.

Herrmann, A. G., and Richter-Bernburg, G. (1955). Frühdiagenetische Störung der Schichtung und Lagerung im Werra–Anhydrit (Zechstein 1) am Südwestharz. *Z. Dtsch. Geol. Ges.* **105**(4), 659–702.

Herrmann, A. G., Knake, D., Schneider, J., and Peters, H. (1973). Geochemistry of modern seawater and brines from salt pans: Main components and bromide distribution. *Contrib. Mineral. Petrol.* **40**(1), 1–24.

Herrmann, A. G., Kaeding, K. C., and Struenser, G. V. (1980). Kalisalzlagerstätten des Sulfat–Typs: Hattorf (Werra–Fulda Bezirk) und Salzdetfurth (Bezirk Südhannover). Entstehung, Umbildung, Bergbau, und wirtschaftliche Nutzung von Salzlagerstätten. *Fortschr. Mineral.* **58**(2), 33–53.

Heykes, K. (1924). Einige Bemerkungen zur "Studie über die Entstehung der Kalisalzlagerstätten des deutschen Zechsteins" von Ernst Fulda. *Kali (Halle)* **18**(19), 281–283.

Heyne, G. (1913). Die Eisenchlorürdoppelsalze des Rubidiums und Cäsiums und Untersuchungen über Vorkommen und Verteilung des Rubidiums in deutschen Kalisalzlagerstätten. Ph.D. Dissertation, A. Lax Verlag, Hildesheim, 89 pp.; also in *Neues Jahrb. Mineral., Monatsh.* (1913), No. 1, p. 365.

Heynke, A., and Zaenker, G. (1970). Zur Ausbildung und Leitbankgliederung des Stassfurtsteinsalzes im Südharzrevier. *Z. Angew. Geol.* **16**(7–8), 344–356.

Hidalgo, A. F., and Orr, C., Jr. (1968). Method for predicting the properties of supersaturated solutions of the alkali chlorides. *J. Chem. Eng. Data* **13**(1), 49–53.

Hill, A. E. (1937). The transition temperature of gypsum to anhydrite. *J. Am. Chem. Soc.* **59**(11), 2242–2244.

Hill, A. E., and Wills, J. H. (1938). Ternary systems: Calcium sulfate, sodium sulfate and water. *J. Am. Chem. Soc.* **60**(6), 1647–1655.

Hill, C. A. (1979). Recent anhydrite and bassanite from caves in Big Bend National Park, Texas. *Bull. Natl. Speleol. Soc.*, 41(4), 126–127.

Hiller, J. E., and Keller, P. (1965). Untersuchungen an den Lösern der Kalisalzlagerstätte Buggingen. *Kali Steinsalz* **4**(6), 190–203.

Hills, J. M. (1942). Rhythm of Permian seas—a paleogeographic study. *Bull. Am. Assoc. Pet. Geol.* **26**(2), 217–255.

Hingston, F. J. (1964). Reactions between boron and clays. *Aust. J. Soil Res.* **2,** 83–95.

Hintze, G. (1915). "Mineralogie." Veit & Co., Leipzig, (with addenda to 1969).

Hirano, K. (1954). The rate of nucleation in supersaturated solution. *Sci. Rep. Tohoku Univ.*, [1] **38,** 97–116.

Hirano, T., and Oki, Y. (1978). Importance of Mg^{2+} and SiO_2 (aq) reaction for pH control in seawater–rock interaction. *EOS, Am. Trans. Geophys. Union* **59**(12), 1220.

Hirota, K. (1941). Thermal diffusion of binary salt solution. A note on the paper by Gillespie and Breck. *Bull. Chem. Soc. Jpn* **16,** 232–234.

Hirota, K. (1942). The thermal diffusion of solutions (in Japanese). *Bull. Chem. Soc. Jpn* **63,** 999–1006, 1061–1067; **64,** 16–22, 112–119 (1943).

Hirsch, P. (1980). Distribution and pure culture studies of morphologically distinct solar lake microorganisms. *In* "Hypersaline Brines and Evaporitic Environments" (A. Nissenbaum, ed.), pp. 41–60. Elsevier, Amsterdam.

Hite, R. J. (1961). Potash-bearing evaporite cycles in the salt anticlines of the Paradox Basin, Colorado and Utah. *U.S. Geol. Surv. Prof. Paper* **424D,** 135–138.

Hite, R. J. (1970). Shelf carbonate sedimentation, controlled by salinity in the Paradox Basin, Southeast Utah. *In* "Symp. on Salt, 3rd." (J. L. Rau and L. F. Dellwig, eds.), vol. 1, pp. 48–66. N. Ohio Geol. Soc., Cleveland, Ohio.

Hite, R. J. (1978). Possible genetic relationships between evaporites, phosphorites, and iron-rich sediments. *Mt. Geol.* **15**(3), 97–107.

Hite, R. J. (1982). Progress Report on the potash deposits of the Khorat Plateau, Thailand. *U.S. Geol. Surv. Open-File Rep.* No. 82–1096, 70 pp.

Hite, R. J., and Buckner, D. H. (1981). Stratigraphic correlations, facies concepts, and cyclicity in Pennsylvanian rocks of the Paradox Basin. *In* "Geology of the Paradox Basin" (Del L. Wiegand, ed.), *Field Conf. Guidebook, Rocky Mt. Assoc. Geol.*, pp. 147–159.

Hite, R. J., and Japakasetr, T. (1979). Potash deposits of the Khorat Plateau, Thailand and Laos. *Econ. Geol.* **74**(2), 448–458.

Hiyama, S., and Fukui, H. (1951). Expansion of gypsum by adding materials (in Japanese). *Gypsum (Jpn)* **1,** 73–83, 142–145.

Hiyama, S., and Fukui, H. (1952). A consideration of the hydration mechanism of calcined gypsum (in Japanese). *Gypsum (Jpn)* **1,** 233–235.

Hiyama, S., Kobayashi, T., Takatsu, M., and Nagai, S. (1957). Setting of gypsum from salterns (in Japanese). *Gypsum Lime (Jpn)* **1,** 1517–1521.

Hodgman, C. D., Weat, R. C., Shankland, R. S., and Selby, S. M. (eds.) (1963). "Handbook of Chemistry and Physics," 44th ed., Chem. Rubber Publ. Co., Cleveland, Ohio, 3604 pp.

Hoering, P. C., and Parker, P. L. (1961). The geochemistry of the stable isotopes of chlorine. *Geochim. Cosmochim. Acta* **23**(3–4), 186–199.

Hoffmann, J. (1934). Über Mineralfärbungen. *Z. Angew. Allgem. Chem.* **219**(2), 197–202.

Hoffmann, R. O. (1961). Die Mineralzusammensetzung der in Wasser schwer löslichen Rückstande von Filterschlämmen und Rohsalzen einiger mitteldeutscher Kaliwerke. *Bergakademie* **13,** 237–248.

Hofrichter, E. (1960). Zur Stratigraphie, Fazies und Genese der Ronneberg-Gruppe und Anhydritmittelzone (Zechstein 3) in Nordwestdeutschland. Ph.D. Dissertation, Univ. Kiel.

Hofrichter, E. (1976). Zur Frage der Porosität und Permeabilität von Salzgesteinen. *Erdöl Erdgas* **92**(3), 77–80.

Holdoway, K. A. (1974). Behavior of fluid inclusions in salt during heating and irradiation. *In* "Symp. on Salt, 4th." (A. H. Coogan ed.), vol. 1, pp. 303–312. N. Ohio Geol. Soc., Cleveland, Ohio.

Holdoway, K. A. (1978). Deposition of evaporites and red beds of the Nippewalla Group, Permian, western Kansas. *Bull. Kans. State Geol. Surv.* **215,** 1–43.

Holland, H. D. (1974). Marine evaporites and the composition of sea water during the Phanerozoic. *Soc. Econ. Paleontol. Mineral. Spec. Publ.* **20,** 187–192.

Holland, T. H. (1912). The origin of desert salt deposits. *Proc. Liverpool Geol. Soc.* **11**(53), 227–250.

Holland, T. H., and Christie, W. A. K. (1909). The origin of the salt deposits of Rajputana. *Rec. Geol. Surv. India* **38**(2), 154–186.
Hollerbach, A. (1980). Charakteristische Kohlenwasserstoffe in Evaporaten. *Erdöl, Kohle, Erdgas, Petrochem.* **33**(7), 327.
Holliday, D. W. (1967). Secondary gypsum in Middle Carboniferous rocks of Spitsbergen. *Geol. Mag.* **104**(2), 171–176.
Holliday, D. W. (1968). Early diagenesis in Middle Carboniferous nodular anhydrite of Spitsbergen. *Proc. Yorkshire Geol. Soc. Pt. 3,* **36**(16), 277–292.
Holliday, D. W. (1970). Petrology of secondary gypsum rocks: A review. *J. Sediment. Petrol.* **40**(2), 734–744.
Holliday, D. W., and Shepherd-Thorn, E. R. (1974). Basal Purbeck evaporites of the Fairlight borehole, Sussex. *Rep. Inst. Geol. Sci. (U.K.),* No. 74–4, 14 pp.
Hollingworth, S. E. (1942). The correlation of gypsum-anhydrite deposits and the associated strata in the north of England. *Proc. Geol. Assoc. London* **53,** 141–151.
Hollingworth, S. E. (1948). Evaporites. *Proc. Yorkshire Geol. Soc.* **27**(3), 192–198.
Holloway, H. L. (1945). Salt deposits in Eritrea. *Min. Mag.* **73,** 211–216.
Holser, W. T. (1966a). Bromide geochemistry of salt rocks. *In* "Symp. on Salt, 2nd." (J. L. Rau, ed.), vol. 1, pp. 248–275. N. Ohio Geol. Soc., Cleveland, Ohio.
Holser, W. T. (1966b). Diagenetic polyhalite in recent salt from Baja California. *Am. Mineral.* **51**(1–2), 99–109.
Holser, W. T. (1977). Catastrophic chemical events in the history of the ocean. *Nature (London)* **267**(5610), 403–408.
Holser, W. T. (1979a). Rotliegend evaporites. Lower Permian of northwestern Europe. *Erdöl, Kohle, Erdgas, Petrochem.* **32**(4), 159–162.
Holser, W. T. (1979b). Mineralogy of evaporites. *In* "Marine Minerals" (R. G. Burns, ed.), *Rev. Mineral.,* vol. 6, pp. 211–294. Miner. Soc. Am., Washington, D.C.
Holser, W. T. (1979c). Trace elements and isotopes in evaporites. *In* "Marine Minerals" (R. G. Burns, ed.), *Rev. Mineral.,* vol. 6, pp. 295–346. Mineral. Soc. Am., Washington, D.C.
Holser, W. T., and Wilgus, C. K. (1981). Bromide profiles of the Roet salt, Triassic of northern Europe, as evidence of its marine origin. *Neues Jahrb. Mineral., Monatsh.,* No. 6, pp. 267–276.
Holser, W. T., Wardlaw, N. C., and Watson, D. W. (1972). Bromide in salt rocks: Extraordinarily low content in the Elk Point Salt, Canada. *In* "Geology of Saline Deposits" (G. Richter-Bernburg, ed.), U.N.E.S.C.O. (Paris) Earth Sci. Ser. **7,** 69–75, cf. discussion.
Holter, M. (1972). Geology of the Prairie Evaporite Formation. *In* "Geology of Saline Deposits" (G. Richter-Bernburg, ed.), U.N.E.S.C.O. (Paris) Earth Sci. Ser. **7,** 183–190.
Holwerda, J. G., and Hutchinson, R. W. (1968). Potash-bearing evaporites in the Danakil area, Ethiopia. *Econ. Geol.* **63**(2), 124–150; (5), 572–573; (8), 978–979.
Horne, R. A. (1969). "Marine Chemistry." Wiley (Interscience), New York, 568 pp.
Horovitz-Vlassowa, L. M. (1931). Über die Rotfärbung gesalzener Därme (der rote Hund). *Zentralbl. Bakteriol. Parasitenkd. Infektionskr. Hyg.,* [2] **85,** 12.
Hovi, V., and Hyvoenen, L. (1951). The free energy of formation and the gap of solubility of alkali halide solid solutions. *Ann. Acad. Sci. Fenn., Ser. AI: Math.–Phys.* **106,** 1–16.
Hrishikesh, R., and Adhya, S. K. (1969). Dehydration of gypsum and ints influence on the transformation of anhydrite. *Technology, (Coimbatore, India)* **6**(2–3), 87–90.
Hsu, K. J. (1972). Origin of saline giants: A critical review after the discovery of the Mediterranean evaporite. *Earth Sci. Rev.* **8,** 371–396.
Hsu, K. J. (1973). The desiccated deep-basin model for the Messinian events. *In* "Messinian Events in the Mediterranean" (C. W. Drooger *et al.,* eds.), pp. 60–67. North Holland, Amsterdam.

Hus, K. J., and Siegenthaler, C. (1969). Preliminary experiments on hydrodynamic movement induced by evaporation and their bearing on the dolomite problem. *Sedimentology* **12**(1–2), 11–26.

Hsu, K. J., Cita, M. B., and Ryan, W. B. F. (1973a). The origin of the Mediterranean evaporites. *In* Intl. Rep. Deep-Sea Drlg. Proj., vol. 13 (W. B. F. Ryan, K. J. Hsu *et al.*, eds.), Part 2, pp. 695–708. U.S. Govt. Print. Off., Washington, D.C.

Hsu, K. J., Ryan, W. B. F., and Cita, M. B. (1973b). Late Miocene desiccation of the Mediterranean. *Nature (London)* **242**(5395), 240–244.

Hsu, K. J., Montadert, L., Bernoulli, D., Cita, M. B., Erickson, A., Garrison, R. E., Kidd, R. B., Melieres, F., Muller, C., and Wright, R. (1978). History of the Mediterranean Salinity Crisis. *In* Intl. Rep. Deep-Sea Drlg. Proj., vol. 42A (K. J. Hsu, L. Montadert *et al.*, eds.), pp. 1053–1078. U.S. Govt. Print. Off., Washington, D.C.

Huang, T. C., and Stanley, D. J. (1973). Western Alboran Sea: sediment dispersal, ponding and reversal of currents. *In* "The Mediterranean Sea: A Natural Sedimentation Laboratory" (D. J. Stanley, ed.), pp. 521–560. Dowden, Hutchinson & Ross Inc., Stroudsburg, Pennsylvania.

Huang, T. C., Stanley, D. J., and Stuckenrath, R. (1972). Sedimentological evidence for current reversal at the Straits of Gibraltar. *J. Mar. Technol. Soc.* **6,** 25–33.

Huang, W. H., and Keller, W. D. (1972). Organic acids as agents of chemical weathering of silicate minerals. *Nature (London) Phys. Sci.* **239**(96), 149–151.

Hudec, P. P., and Sonnenfeld, P. (1974). Hot brines on Los Roques, Venezuela. *Science* **185**(4149), 440–442 and **186**(4169), 1074–1075.

Hudec, P. P., and Sonnenfeld, P. (1980). Comparison of Caribbean solar ponds and inland solar lakes of British Columbia. *In* "Hypersaline Brines and Evaporitic Environments" (A. Nissenbaum, ed.), pp. 101–114. Elsevier, Amsterdam.

Hulburt, E. O. (1928). Penetration of ultraviolet light into pure water and sea-water. *J. Opt. Soc. Am.* **17**(1), 5–22.

Hulett, G. A. (1901). Beziehungen zwischen Oberflächenspannung und Löslichkeit. *Z. Phys. Chem.* **37,** 385–406.

Hulett, G. A., and Allen, L. E. (1902). The solubility of gypsum. *J. Am. Chem. Soc.* **24**(7), 667–669.

Hunt, C. B. (1960). The Death Valley salt pan, study of evaporites. *U.S. Geol. Surv. Prof. Paper* **400B,** 456–458.

Hunt, C. B., Robinson, T. W., Bowles, W. A., and Washburn, A. L. (1966). Hydrologic basins, Death Valley, California. *U.S. Geol. Surv. Prof. Paper* **494B,** 138 pp.

Huppert, H. E., and Linden, P. F. (1979). On heating a stable salinity gradient from below. *J. Fluid Mech.* **95,** 431–464.

Hurlbert, S. H., Berry, R. W., Lopez, M., and Pezzani, S. (1976). Lago Verde and Lago Flaco: Gypsum-bound lakes of the Chilean Altiplano. *Limnol. Oceanogr.* **21**(5), 637–645.

Hurle, D. T. J., and Jakeman, E. (1969). The significance of the Soret effect on the Rayleigh–Jeffreys' problem. *Phys. Fluids* **12**(12), 2704–2705.

Hurle, D. T. J., and Jakeman, E. (1971). Soret-driven thermosolutal convection. *J. Fluid Mech.* **47**(4), 667–687.

Hvid-Hansen, N. (1951a). Sulfate reducing hydrocarbon-producing bacteria in groundwater. *Acta Pathol. Microbiol. Scand.* **29,** 314–334.

Hvid-Hansen, N. (1951b). Anaerobic actinomyces (*Actinomyces israeli*) in groundwater. *Acta Pathol. Microbiol. Scand.* **29,** 335–338.

Igelsurd, I., and Thompson, T. G. (1936). Equilibria in the saturated solutions of salt occurring in sea water. II. Quaternary system $MgCl_2$–$CaCl_2$–KCl–H_2O at 0°C. *J. Am. Chem. Soc.* **58,** 2003–2009.

Ikeda, T. (1959). New method of determining the Soret effect. *J. Chem. Phys.* **30**(1), 345–346.

Ikeda, T., Ohe, H., and Kawamura, S. (1968). Relative determination of Soret coefficients of alkaline earth chlorides (in Japanese). *Bull. Chem. Soc. Jpn* **41**(12), 2858–2860.

Imelik, B. (1948). Altération bactériologique des pétroles. *C. R. Hebd. Séances Acad. Sci. Paris* **226**(11), 922–933.

Imhoff, J. F., Hashwa, F., and Trueper, H. G. (1978a). Isolation of extremely halophilic phototrophic bacteria from the alkaline Wadi Natrun, Egypt. *Arch. Hydrobiol.* **84**(3), 381–388.

Imhoff, J. F., Sahl, H. G., Soliman, G. S. H., and Trueper, H. G. (1978b). Wadi Natrun—chemical composition and microbial mass developments in alkaline brines of eutrophic desert lakes. *Geomicrobiol. J.* **1**(3), 219–234.

Ingerson, E. (1968). Deposition and geochemistry work sessions. *In* "Saline Deposits" (R. B. Mattox *et al.*, eds.). *Spec. Paper Geol. Soc. Am.* **88,** 671–681.

Innorta, G., Rabbi, E., and Tomadin, L. (1980). The gypsum–anhydrite equilibrium by solubility measurements. *Geochim. Cosmochim. Acta* **44**(12), 1931–1936.

Irion, G. (1973). Dei anatolische Salzseen, ihr Chemismus und die Entstehung ihrer chemischen Sedimente. *Arch. Hydrobiol.* **71,** 517–557.

Irion, G., and Mueller, G. (1968). Huntite, magnesite and polyhalite of Recent age from Tuz Gölü, Turkey. *Nature (London)* **220**(5174), 1309–1310.

Irving, E., and Briden, J. C. (1962). Paleolatitude of evaporite deposits. *Nature (London)* **196**(4853), 425–428.

Irving, E., and Gaskell, T. F. (1962). The paleogeographic latitude of oilfields. *Geophys. J., R. Astron. Soc.* **7**(1), 54–64.

Ismatullayev, Kh. K., Salyamova, S. K., and Kuychibayev, T. K. (1974). Some features of internal structure of Upper Jurassic evaporite formation in western Uzbekistan (in Russian). *Uzbek. Geol. Zh.* No. 5, pp. 54–57.

Ivanov, A. A. (1933). Results of exploratory work on the Upper Kama potash deposit in 1931–32 (in Russian). *Tr. Vses. Geol.-Razved. Obedin.* **345,** 86 pp.

Ivanov, A. A. (1938). The Shumki deposit of rock salt in the Ural Mountains (in Russian). *Mem. Soc. Russ. Mineral.* **67**(2), 367–384.

Ivanov, A. A. (1953). "Fundamentals of Geology, Methods of Exploration, Prospecting and Evaluation of Mineral Salt Deposits" (in Russian). Gosgeoltekhizdat, Moscow, 204 pp.

Ivanov, A. A. (1963). The varicolored sylvinites of the Upper Kama deposits of potash salts (in Russian). *In* "Geology of Potash Salt Deposits" (A. A. Ivanov, ed.). *Tr. Vses. Nauchno–Issled. Geol. Inst.,* [*N.S.*] **99,** 153–180.

Ivanov, A. A. (1967). The depth of salt-bearing basins in the geologic past. *Litol. Polezn. Iskop.,* No. 2, pp. 16–19; *Lithol. Mineral. Resour. (Engl. Transl.),* No. 2, pp. 161–171.

Ivanov, A. A. (1970). The Masli potassium salt deposit in Ethiopia (in Russian). *Sov. Geol.* **13**(7), 118–123.

Ivanov, A. A., and Levitskiy, Yu. F. (1960). Geology of the evaporite formations of the USSR (in Russian). *Tr. Vses. Nauchno–Issled. Geol. Inst.,* [*N.S.*] **35,** 1–424.

Ivanov, A. A., and Voronova, M. L. (1963). The sylvinite caprock of the Upper Kama River potash salt deposit (in Russian). *In* "Geology of Potash Salt Deposits" (A. A. Ivanov, ed.). *Tr. Vses. Nauchno–Issled. Geol. Inst.,* [*N.S.*] **99,** 181–190.

Ivanov, A. A., and Voronova, M. L. (1968). Geology of the Upper Pechora saliferous basin (in Russian). *Tr. Vses. Nauchno–Issled. Geol. Inst.,* [*N.S.*] **161,** 3–79; transl.: *Dep. State (Can.),* 131 pp.

Ivanov, A. A., and Voronova, M. L. (1972). "Evaporite Formations" (in Russian). Nedra, Moscow, 328 pp.

Ivanov, A. A., and Voronova, M. L. (1975). The Upper Kama deposits of potash salts (stratigraphy, mineralogy and petrography, tectonics, genesis) (in Russian). *Tr. Vses. Nauchno–Issled. Geol. Inst.*, [*N.S.*] **232,** 1–219.

Ivanov, A. A., Sheshukov, N. G., and Salrykin, F. L. (1963). Woody remains in fossil salt deposits. *Sov. Geol.* **6**(8), 107–111; *Int. Geol. Rev. (Engl. Transl.)* **7**(8), 1358–1359 (1965).

Jacka, A. D. (1981). Observations on the replacement of carbonates by sulfates. *Geol. Soc. Am., Abstr. Prog.* **13**(7), 479.

Jacka, A. D., and Franco, L. A. (1974). Deposition and diagenesis of Permian evaporites and assorted carbonates and clastics on shelf areas of the Permian basin. *In* "Symp. on Salt, 4th." (A. H. Coogan, ed.), vol. 1, pp. 67–89. N. Ohio Geol. Soc., Cleveland, Ohio.

Jackson, T. A. (1978). The biochemistry of heavy metals in polluted lakes and streams at Flin Flon, Canada, and a proposed method for limiting heavy metal pollution of natural waters. *Environ. Geol.* **2**(3), 173–189.

Jackson, T. A., and Bischoff, J. L. (1971). The influence of amino acids on the kinetics of the recrystallization of aragonite to calcite. *J. Geol.* **79**(4), 493–497.

Jacobs, W. C. (1951). The energy exchange between sea and atmosphere and some of its consequences. *Bull. Scripps Inst. Oceanogr.* **6,** 27–122.

Jaenecke, E. (1915). "Die Entstehung der deutschen Kalisalzlagerstätten." 2nd ed. Vieweg, Braunschweig, 109 pp.

Jansa, L. F., Bujak, J. P., and Williams, G. L. (1980). Upper Triassic salt deposits of the western North Atlantic. *Can. J. Earth Sci.* **17**(5), 547–559.

Jarvie, A. W. P., Marshall, R. N., and Potter, H. R. (1975). Chemical alkylation of lead. *Nature (London)* **255**(5505), 217–218.

Jauzein, A. (1974). Les données sur le système $CaSO_4$–H_2O et leurs implications géologiques. *Rev. Géogr. Phys. Géol. Dynam.* **16**(2), 151–159.

Jauzein, A. (1982). Deuterium et oxygène-18 dans les saumures: Modèlisation et implications sédimentologiques. *C. R. Hebd. Séances Acad. Sci. Paris,* [2] **294**(11), 663–668.

Jayaraman, N. (1940). Origin of celestin in the Cretaceous beds of Trichy. *Qu. J. Indian Inst. Sci.* **3,** 25–59 (*Mineral Abstr.* 9–34).

Jennett, J. C., and Wixson, B. G. (1977). The new lead belt: Aquatic pathways. *Proc. Int. Conf. Heavy Met. Environ. 1st* (1975) **2,** 247–255.

Jensen, I. A. D. (1889). Undersøgelse af Grønlands vestkyst fra 64° til 67° N.B. *Medd. Grønl.* **8,** 33–122.

Jockwer, N. (1979). Investigations on the migration and the release of water within rock salt. *Proc. "Workshop on the Migration of Long-Lived Radionucleides in the Geosphere,"* pp. 249–254. Org. Econ. Coop. Dev., Nucl. Energy Agency, Comm. Eur. Commun., Paris.

Jodry, R. L. (1969). Growth and dolomitization of Silurian reefs, St. Clair county, Michigan. *Bull. Am. Assoc. Pet. Geol.* **53**(4), 957–981.

Joergenson, B. B. (1978). A comparison of methods for the quantification of bacterial sulfate reduction in coastal marine sediments. *Geomicrobiol. J.* **1**(1), 29–47, 49–64.

Joergenson, N. O. (1980). Gypsum formation in recent sediments from Kattegat, Denmark. *Chem. Geol.* **28**(3–4), 349–353.

Joergenson, N. O. (1981). Authigenic K–feldspar in recent submarine gypsum concretions from Denmark. *Mar. Geology* **39**(1–2), M21–25.

Johannes, W. (1966). Experimentelle Magnesitbildung aus Dolomit und $MgCl_2$. *Contrib. Mineral. Petrol.* **13,** 51–58.

Johannesson, J. K., and Gibson, G. W. (1962). Nitrate and iodate in Antarctic salt deposits. *Nature (London)* **194**(4828), 567–568.

Johns, W. D. (1963). Die Verteilung von Chlor in rezenten marinen und nicht-marinen Sedimenten. *Fortschr. Geol. Rheinl.-Westfalen* **10,** 215–230.

Johnsen, A. (1909). Regelmässige Verwachsung von Carnallit und Eisenglanz. *Zentralbl. Mineral., Geol., Paläontol.* pp. 168–173.

Johnson, D. (1939). "The Origin of Submarine Canyons—A Critical Review of Hypotheses." Columbia Univ. Press, New York; reprinted Hafner, New York, 1967, 126 pp.

Johnson, K. S. (1974). Permian copper shales of southwestern United States. *In* "Gîsements Stratiformes et Provinces Cuprifères" (P. Bartholomé, I. de Magnée, P. Evrard, and J. Moreau, eds.), pp. 383–393. Soc. Belge Géol. Paléont. Hydrol., Liège.

Johnson, K. S., and Gonzales, S. (1978). Salt deposits in the United States and regional geological characteristics important for storage and radioactive waste. *Rep. Office Waste Isolation, U.S. Dept. Energy,* No. Y/OWI/SUB-7414/1//W-7405-eng-26. Washington, D.C.

Johnson, K. S., and Pytkowicz, R. M. (1979). Ion association of chloride and sulphate with sodium, potassium, magnesium and calcium in sea water at 25°C. *Mar. Chem.* **8**(1), 87–93.

Johnson, R. W., and Calder, J. A. (1973). Early diagenesis of fatty acids and hydrocarbons in a salt marsh environment. *Geochim. Cosmochim. Acta* **37**(8), 1943–1955.

Johnston, R. M. (1983). The conversion of smectite to illite in hydrothermal systems: A literature review. *Rep. Atomic Energy Can.* No. 7792, 27 pp.

Jones, B. F. (1966). Geochemical evolution of closed basin water in the western Great Basin. *In* "Symp. on Salt, 2nd." (J. L. Rau, ed.), vol. 1, pp. 181–200. N. Ohio Geol. Soc., Cleveland, Ohio.

Jones, C. L. (1965). Petrography of evaporites from the Wellington Formation near Hutchinson, Kansas. *Bull. U.S. Geol. Surv.* **1201,** 1–67.

Jones, C. L. (1973). Salt deposits of Los Medanos area, Eddy and Lea counties, New Mexico. *U.S. Geol. Surv. Open-File Rep.* No. 1862, 67 pp.

Jones, C. L. (1974). Salt deposits of the Mescalero Plains area, Chaves County, New Mexico. *U.S. Geol. Surv. Open-File Rep.* No. 74-190, 21 pp.

Jones, C. L., and Madsen, B. M. (1968). Evaporite geology of Fifth Ore Zone, Carlsbad District, southeastern New Mexico. *Bull. U.S. Geol. Surv.* **1252-B,** 21 pp.

Jones, H. E., Trudinger, P. A., Chambers, L. A., and Pyliotis, N. A. (1976). Metal accumulation by bacteria with particular reference to disseminatory sulfate-reducing bacteria. *Z. Allg. Mikrobiol.* **16**(6), 425–435.

Jones, L. M., and Faure, G. (1967). Origin of the salts in Lake Vanda, Wright Valley, southern Victoria Land, Antarctica. *Earth Planet. Sci. Lett.* **3**(2), 101–106.

Jones, R. W. (1978). Some mass balance and geological restraints on migration mechanisms. *Contin. Educ. Course Notes, Am. Assoc. Pet. Geol.* **8,** A1–A43.

Jordan, L., and Vosburg, D. L. (1963). Permian salt and associated evaporites in the Anadarko Basin of the western Oklahoma–Texas Panhandle region. *Bull. Okla. Geol. Surv.* **102,** 1–76.

Juengst, H. (1934). Zur geologischen Bedeutung der Synärese. Ein Beitrag zur Entwässerung der Kolloide im werdenden Gestein. *Geol. Rundsch.* **25**(5), 312–325.

Jung, W. (1959). Das Steinsalzequivalent des Zechsteins 1 in der Sangerhäuser und Mansfelder Mulde und daraus resultierende Bemerkungen zum Problem der "Jahresringe." *Ber. Geol. Ges. D.D.R.* **4**(4), 313–325.

Jung, W. (1966). Nochmals zum Sangerhäuser Anhydrit (Zechstein 2). *Geologie* **15**(4–5), 443–460.

Jung, W., and Knitzschke, G. (1976). Kupferschiefer in the German Democratic Republic with special reference to the Kupferschiefer deposit in the southeastern Harz Foreland. *In* "Handbook of Strata-Bound and Stratiform Ore Deposits" (K. H. Wolf, ed.), vol. 6, pp. 352–406. Elsevier, Amsterdam.

Jung, W., and Lorenz, S. (1968). Die Ausbildung des Stassfurtsalzes (Na2) im südostlichen Harzvorland. *Bergakademie* **20**(9), 509–515.

Jungwirth, J., and Seifert, J. (1966). Zur Stratigraphie und Fazies des Zechsteins in Südwest Thüringen. *Geologie* **15**(4–5), 421–433.

Jurgan, H. (1969). Sedimentologie des Lias der Berchtesgadener Kalkalpen. *Geol. Rundsch.* **58**(2), 464–501.

Kachhara, R. C. (1977). Potash-rich subsoil brine in the Great Rann of Kutch. *Miner. Wealth* **13**(1), 4–6; *Chem. Abstr.* **88,** 123943.

Kaeding, K. C. (1978). Stratigraphische Gliederung des Zechsteins im Werra–Fulda Becken. *Landesamt Hessen, Geol. Jahrb.* **106,** 123–130.

Kahanowicz, M. (1932). Über die seitliche Strahlung und die Natur der färbenden Substanz im natürlichen blauen Steinsalz. *Z. Phys.* **76,** 283–292.

Kaiser, E., and Neumaier, F. (1932). Sand-Steinsalz-Kristallskelette aus der Namib Südwestafrikas. *Zentralbl. Mineral., Geol., Paläontol.,* [A], No. 6, pp. 177–188.

Kajiwara, Y. (1970). Gypsum–anhydrite ores and associated minerals from the Motoyama deposits of the Hanawa mine (in Japanese). *In* "Volcanism and Ore Genesis" (T. Tatsumi, ed.), pp. 207–213. Univ. Tokyo Press, Tokyo.

Kalinko, M. K. (1973). Sources of sodium chloride for the accumulation of thick salt-bearing sequences. *Sov. Geol.* No. 2, pp. 93–104; *Int. Geol. Rev. (Engl. Transl.)* **16**(1), 93–102 (1974).

Kalinko, M. K., Mileshina, A. G., and Safonova, G. I. (1978). Migration rates of petroleum components (experimental data) (in Russian). *Tr. Vses. Neft. Nauchno-Issled. Geologorazved. Inst.* **205,** 107–113.

Kamienski, M., and Glinska, S. (1965). Tuffites with halite from the salt mine of Bochnia (in Polish). *Arch. Mineral.* **26**(1–2), 77–79.

Kane, G. P., and Kulkarni, G. R. (1949). Acceleration of solar evaporation by dyes. *Trans. Indian Inst. Chem. Eng.* **3,** 105–108.

Kao, T. S., Chang, Y. M., and Kung, K. H. (1981). Discovery of strontium aragonite [$CaSr(CO_3)_2$] in China (in Chinese). *Bull. Sci. (China)* **26**(15), 932–935.

Kapchenko, L. N., and Soboleva, N. S. (1973). Genesis and prospecting for petroleum by bromine in subsurface deep-lying Permian brines in the Kama River region (in Russian). *Tr. Vses. Neft. Nauchno–Issled. Geologorazved. Inst.* **338,** 47–62.

Kapchenko, L. N., Rogozina, E. A., and Sokolova, N. Ya. (1973). Gaseous microinclusions in salts of the Inder salt dome (Ciscaspian depression) (in Russian). *Geol. Nefti Gaza* No. 5, pp. 71–75.

Kaplan, I. R., and Friedman, A. (1970). Biological productivity in the Dead Sea. Pt. I. Microorganisms in the water column. *Isr. J. Chem.* **8,** 513–528.

Karavaiko, G. I., Ivanov, M. V., and Srebrodol'skiy, B. I. (1962). Oxidation of stored sulfur ore (in Russian). *Sov. Geol.* No. 5, pp. 133–139.

Karlik, B., and Kropf-Duschek, F. (1950). Determination of helium content of rock salt specimens. *Acta Phys. Austriaca* **3,** 448–451.

Karsten, O. (1954). Lösungsgeschwindigkeit von Natriumchlorid, Kaliumchlorid und Kieserit in Wasser und in wässrigen Lösungen. *Z. Anorg. Allg. Chem.* **27**(5–6), 247–266, 267–274.

Karwowski, L., Kozlowski, A., and Roedder, E. (1979). Gas–liquid inclusions in minerals of zinc and lead ores from the Silesia–Cracow region (in Polish). *In* "Research on the Genesis of Zinc–Lead Deposits of Upper Silesia, Poland" (E. Chabelska, ed.). *Pr. Pol. Inst. Geol.* No. 59, pp. 87–96.

Kato, N. (1944). Periodic precipitation of sodium chloride. *Bull. Chem. Soc. Jpn* **65,** 678–680.

Katz, A. (1971). Zoned dolomite crystals. *J. Geol.* **79**(1), 38–51.

Katz, A., and Matthews, A. (1977). The dolomitization of calcium carbonate: An experimental study at 252–295°C. *Geochim. Cosmochim. Acta* **41**(2), 297–308.

Kaufmann, D. W., and Slawson, C. B. (1950). Ripple marks in rock salt of the Salina Formation. *J. Geol.* **58**(1), 24–29.

Kawano, Y. (1948). A new occurrence of anhydrite phenocrysts in the glassy rocks from Himeshima, Oita Pref., Japan. *Rep. Geol. Surv. Jpn* **126,** 1–6.

Kazakov, A. V., and Sokolova, E. I. (1950). Conditions of fluorite formation in sedimentary rocks (fluorite system) (in Russian). *Tr. Akad. Nauk SSSR Inst. Geol. Nauk, Ser. Geol.* **114**(40), 22–64.

Kazantsev, O. D., Yermakov, V. A., and Grebennikov, N. B. (1974). Discovery of a bischofite deposit in the Lower Volga region (in Russian). *Sov. Geol.* No. 7, pp. 124–132.

Keller, W. D. (1963). Diagenesis in clay minerals—a review. *Proc. 11th Natl. Conf. Clays, Clay Miner. (Ottawa, 1962),* **11,** 136–157 (Int. Ser. Mons. Earth Sci., v. 13, MacMillan, New York).

Kelley, K. K., Southard, J. C., and Anderson, C. F. (1941). Thermodynamic properties of gypsum and its dehydration products. *U.S. Bur. Mines Tech. Paper* No. 625, 73 pp.

Kemenyi, E. (1942). Über Uranium- und Radiumgehalt im Steinsalz und Sylvin. *Sitzungsber. Akad. Wiss. Wien, Math.–Naturw. Kl.,* [2a] **150,** 193–207.

Kemp, E. M. (1978). Tertiary climatic evolution and vegetation history in the Southeast Indian Ocean region. Palaeogeogr. Palaeoclimat. Palaeoecol. **24**(2), 169–208.

Kenat, J. (1966). The production of potassium chloride from the Dead Sea by crystallization. *In* "Symp. on Salt, 2nd." (J. L. Rau, ed.), vol. 2, pp. 195–203,. N. Ohio Geol. Soc., Cleveland, Ohio.

Kenat, J., Glasner, A., and Bloch, M. R. (1958). The absorption of lead by potassium chloride crystallizing from a solution containing small amounts of lead chloride. *Bull. Res. Counc. Isr.,* [A] **7,** 224–225.

Kendall, C. G. St. C., and Skipwith, Sir P. A. d'E. (1969). Holocene shallow-water carbonate and evaporite sediments of Khor-al-Bazam, Abu Dhabi, Southwest Persian Gulf. *Bull. Am. Assoc. Pet. Geol.* **53**(4), 841–869.

Kennard, T. G., Howell, D. H., Yaeckel, M. P. (1931). Spectrographic examination of colorless and blue halite. *Am. Mineral.* **22,** 65–67.

Kern, R. (1952a). Influence de la vitesse d'évaporation des solutions aqueuses d'halogènures alcalins sur le faciès des cristaux précipités. *C. R. Hebd. Séances Acad. Sci. Paris* **234**(9), 970–971.

Kern, R. (1952b). Influence de la vitesse d'évaporation des solutions non-aqueuses d'halogènures alcalins sur le faciès des cristaux obtenus. *C. R. Hebd. Séances Acad. Sci. Paris* **234**(13), 1379–1380.

Kerr, S. D., Jr., and Thompson, A. F. (1968). Origin of nodular and bedded anhydrite in Permian shelf sediments, Texas and New Mexico. *Bull. Am. Assoc. Pet. Geol.* **47**(9), 1726–1732.

Kettner, R. (1948). On bat guano and guano corrosion in the Domica cave (in Czech). *Sb. Ústř. Úst. Geol.* **15,** 41–67.

Keys, D. A., and Wright, J. Y. (1966). Geology of the I.M.C. potash deposit, Esterhazy, Saskatchewan. *In* "Symp. on Salt, 2nd." (J. L. Rau, ed.), vol. 1, pp. 95–101. N. Ohio Geol. Soc., Cleveland, Ohio.

Keys, J. R., and Williams, K. (1981). Origin of crystalline, cold desert salts in the McMurdo region, Antarctica. *Geochim. Cosmochim. Acta* **45**(12), 2299–2309.

Khalil, A. A. (1982). Kinetics of gypsum dehydration. *Thermochim. Acta* **55**(2), 201–208.

Khismatullin, S. D. (1961). Saline soils of the river valleys of the Upper Angara area (in Russian). *Proc. Sib. Conf. Soil Sci.,* 1st., pp. 298–314.

Khodkov, A. Ye. (1971a). Role of the discharge of post-sedimentary waters in the formation of carnallite rocks in the Fore-Carpathians (in Russian). *Vestn. Geol. Geogr. Leningr. Univ.* No. 18, 74–78.

Khodkov, A. Ye. (1971b). The processes of epigenetic alteration of rocks in the langbeinite–kainite series (in Russian). *In* "Materials on the Hydrogeology and the Geological Role of Groundwater" (A. Ye. Khodkov, ed.), pp. 91–96. Leningr. Univ. Press, Leningrad.

Khodkova, S. V. (1968). Langbeinite of Ciscarpathia and its parageneses, on the example of the Stebnik deposit. *Litol. Polez. Iskop,.* No. 6, pp. 73–85; *Lithol. Miner. Resour.* **3**(6), 710–718.

Khodkova, S. V. (1972). Kieserite, kainite, and halite in the Stebinsk deposit (in Russian). *Tr. Vses. Nauchno-Issled. Proektn. Inst. Galurgii* **60,** 51–69.

Khrushchov, D. P. (1973). Absolute age of potassium salt from the Dnieper-Donets Basin and Donbass (in Ukrainian). *Dopov. Akad. Nauk Ukr. RSR, [B]* **35**(12), 1082–1084.

Khrushchov, D. P., and Stroev, V. M. (1973). A find of sylvite-bearing sodium chloride in the Novosenzhansk structure (in Russian). *Geol. Zh.* **33**(1), 145–146.

King, P. B. (1942). Permian of West Texas and southeastern New Mexico. *Bull. Am. Assoc. Pet. Geol.* **26**(4), 535–763.

King, P. B. (1948). Geology of the southern Guadeloupe Mountains, Texas. *U.S. Geol. Surv. Prof. Paper* No. 215, 183 pp.

King, R. H. (1946). Carnallite-filled mud cracks in salt clay. *J. Sediment. Petrol.* **16**(1), 14.

King, R. H. (1947). Sedimentation in Permian Castile Sea. *Bull. Am. Assoc. Pet. Geol.* **31**(3), 470–477.

Kinoshita, K. (1931). The ''kuroko'' (black ore deposits) (in Japanese). *Jpn J. Geol.* **8,** 281–352.

Kinsman, D. J. J. (1966). Gypsum and anhydrite of recent age, Trucial Coast, Persian Gulf. *In* ''Symp. on Salt, 2nd.'' (J. L. Rau, ed.), vol. 1, pp. 302–326. N. Ohio Geol. Soc., Cleveland, Ohio.

Kinsman, D. J. J. (1967). Huntite from a carbonate–evaporite environment. *Am. Mineral.* **52**(9–10), 1332–1340.

Kinsman, D. J. J. (1969). Modes of formation, sedimentary associations, and diagnostic features of shallow-water and supratidal evaporites. *Bull. Am. Assoc. Pet. Geol.* **53**(4), 830–840.

Kinsman, D. J. J. (1976). Evaporites: Relative humidity control of primary mineral facies. *J. Sediment. Petrol.* **46**(2), 273–279.

Kinsman, D. J. J., and Holland, H. D. (1969). The coprecipitation of cations with $CaCO_3$. 4. The coprecipitation of Sr^{2+} with aragonite between 16°C and 96°C. *Geochim. Cosmochim. Acta* **33**(1), 1–17.

Kinsman, D. J. J., Boardman, M., and Borcsik, M. (1974). An experimental determination of the solubility of oxygen in marine brines. *In* ''Symp. on Salt, 4th.'' (A. H. Coogan, ed.), vol. 1, pp. 325–327. N. Ohio Geol. Soc., Cleveland, Ohio.

Kirchheimer, F. (1976). Über Steinsalz und sein Vorkommen im Neckar- und Oberrheingebiet. *Bundesanst. Geowiss., Geol. Jahrb.,* [D] **18,** 1–139.

Kirchheimer, F. (1978). Das blaue Steinsalz. *Kali Steinsalz* **17**(7), 271–287.

Kirikov, V. P. (1963). The mode of occurrence of potash horizons in the Starobinsk deposit (in Russian). *In* ''Geology of Potash Salt Deposits'' (A. A. Ivanov, ed.). *Tr. Vses. Nauchno–Issled. Geol. Inst.,* [*N.S.*] **99,** 233–245.

Kirkland, D. W., Bradbury, J. P., and Dean, J. E., Jr. (1966). Origin of Carmen Island salt deposit, Baja California, Mexico. *J. Geol.* **74**(6), 932–938.

Kisel'gof, S. M. (1973). Effect of saline beds on the hydrochemical conditions of the profile (in Russian). *Tr. Volgogr. Gos. Nauchno–Issled. Proektn. Inst. Neft. Prom–sti.* **19,** 3–9.

Kislik, V. Z., Lupinovich, Yu. I., and Yeroshima, D. M. (1970). On the regularity of structure of evaporite formation in the Pripyat depression and the lithological–geochemical features of its potash horizons (in Russian). *Tr. Inst. Geol. Geofiz. Akad. Nauk SSSR, Sib. Otd.* **116,** 240–251.

Kiss, J. (1981). Dolomitization, dedolomitization, recalcitization under hydrothermal conditions. *Acta Geol. Acad. Sci. Hung.* **24**(2–4), 161–216.

Kitano, Y., and Hood, D. W. (1965). The influence of organic material on the polymorphic crystallization of calcium carbonates. *Geochim. Cosmochim. Acta* **29**(1), 29–41.

Kitano, Y., Okumura, M., and Idogaki, M. (1975). Incorporation of sodium, chloride and sulfate with calcium carbonate. *Geochem. J. (Jpn)* **9**(2), 74–84.

Kitano, Y., Kanamori, N., and Yoshioka, S. (1976). Adsorption of zinc and copper ions on calcite and aragonite, and its influence on the transformation of aragonite to calcite (in Japanese). *Geochem. J. (Jpn)* **10**(4), 175–179.

Kitano, Y., Tokuyama, A., and Arakaki, T. (1979a). Magnesian calcite synthesis from calcium bicarbonate solution containing magnesium and barium ions. *Geochem. J. (Jpn.)* **13**(4), 181–186.

Kitano, Y., Okumura, M., and Idogaki, M. (1979b). Influence of borate boron on crystal form of calcium carbonate. *Geochem. J. (Jpn.)* **13**(5), 223–224.

Kityk, V. I. and Petrichenko, O. I. (1974). The salt bearing formations of the Ukraine, the mineral resources related to them, and the prospects of their exploitation in the national economy (in Russian). *In* "Geology and Mineral Resources of Salt Sequences" (V. I. Kityk, ed.), pp. 5–14. Izd. Nauka, Moscow.

Kityk, V. I., and Petrichenko, O. I. (1978). Use of inclusions in minerals to explain the conditions of formation of petroleum and gas deposits (in Ukrainian). *Visn. Akad. Nauk Ukr. RSR* No. 1, pp. 55–60.

Kityk, V. I., Galabuda, N. I., Petrichenko, O. I., and Shaidetskaya, V. S. (1980). A study of Upper Devonian salt accumulations in the Pripyat depression and the Dnieper–Donets basin (in Russian). *In* "Lithology and Geochemistry of Salt Sequences" (V. I. Kityk, ed.), pp. 77–95. Izd. Naukova Dumka, Kiev.

Kiyosu, Y., and Nakai, N. (1971). Synthesis of lead and zinc sulfides in hydrothermal systems. *J. Earth Sci. Nagoya Univ.* **19,** 67–80.

Kizil'shteyn, L. Ya., and Minayeva, L. G. (1973). Origin of pyrite framboids. *Dokl. Akad. Nauk SSSR,* **206**(5), 1187–1189; *Dokl. Acad. Sci. USSR, Earth Sci. Sect. (Engl. Transl. Am. Geol. Inst.)* **206**(1–6), 144–145.

Klaus, W. (1953). Alpine Salzmikropaläontologie (Sporendiagnose). *Paläontol. Z.* **27**(1), 52–56.

Kling, P. (1915). Tachyhydritvorkommen in den Kalisalzlagerstätten der Mansfelder Mulde. *Centralbl. Mineral. Geol. Paläont.,* pp. 11–17, 44–50.

Klingspor, A. M. (1966). Cyclic deposition of potash in Saskatchewan. *Bull. Can. Petrol. Geol.* **14**(2), 193–207.

Klingspor, A. M. (1969). Middle Devonian Muskeg evaporites of western Canada. *Bull. Am. Assoc. Pet. Geol.* **53**(4), 927–948.

Klockmann, F. F. H. (1978). "Lehrbuch der Mineralogie," 16th ed. 1978, (R. Ramdohr and H. Strunz, eds.), Enke, Stuttgart, 876 pp.

Klopotovskiy, B. A. (1949). Relic gypsum solonchak in southern Georgia, Caucasus (in Russian). *Pochvovedenie* No. 2, pp. 110–114.

Knacke, O., and Gans, W. (1977). The thermodynamics of the system $CaSO_4$–H_2O. *Z. Phys. Chem.* **104**(1–3), 41–48.

Koch, K., and Vogel, J. (1980). Zu den Beziehungen von Tektonik, Sylvinitbildung und Basaltintrusion im Werra–Kaligebiet. *Freiberger Forschungsh.,* [C] **347,** 1–104.

Koch, K., Kockert, W., and Grunewald, V. (1968). Geochemische Untersuchungen an Salzen und Salzlösungen (Laugen) von Salzlagerstätten in der D.D.R. *Geologie* **17**(6–7), 792–803.

Koelichen, K., and Przibylla, C. (1970). Über den Kalimagnesiumsulfatprozess. *Mitt. Kali–Forsch. Anst.,* pp. 113–131.

Koeppen, W. (1931). "Grundriss der Klimakunde." de Gruyter, Berlin, 388 pp.

Koester, E. A. (1972). Thick halite beds surprise many Arizona watchers. *Oil Gas J.* **70**(31), 152–155.

Kohout, F. A. (1967). Ground water flow and the geothermal regime of the Floridian Plateau. *Trans. Gulf Coast Assoc. Geol. Soc.* **7,** 339–354.

Koide, H., and Nakamura, T. (1943). Growth of crystals in the presence of colloids (in Japanese). *Proc. Imp. Acad. (Tokyo)* **19,** 202–204.

Kokorsch, R. (1960). Zur Kenntnis von Genesis, Metamorphose und Faziesverhältnissen des Stassfurtlagers im Grubenfeld Hildesia–Mathildenhall, Diekholzen bei Hildesheim. *Bundesanst. Bodenforsch., Geol. Jahrb., Beih.* **41,** 140 pp.

Kolosov, A. S., and Pustyl'nikov, A. M. (1967). Occurrence of composite iron halides in Devonian salt deposits of Tuva. *Dokl. Akad. Nauk SSSR,* **172**(4), 946–948; *Dokl. Acad. Sci. USSR, Earth Sci. Sect. (Engl. Transl. Am. Geol. Inst.)* **172**(1–6), 196–198.

Kolosov, A. S., and Pustyl'nikov, A. M. (1970a). Regional extent of rinneite. *Dokl. Akad. Nauk SSSR,* **195**(2), 451–453; *Dokl. Acad. Sci. USSR, Earth Sci. Sect. (Engl. Transl. Am. Geol. Inst.)* **195**(1–6), 190–191.

Kolosov, A. S., and Pustyl'nikov, A. M. (1970b). Distribution of bromide between sylvite and halite in Cambrian salts of the Kan–Taseyeva Basin. *Dokl. Akad. Nauk SSSR,* **195**(4), 944–947; *Dokl. Acad. Sci. USSR, Earth Sci. Sect. (Engl. Transl. Am. Geol. Inst.)* **195**(1–6), 206–209.

Kolosov, A. S., and Pustyl'nikov, A. M. (1972). Principal results of investigations on the geochemistry of evaporite deposits in the southwestern part of the Siberian Platform (in Russian). *In* "Prospects for Potassium Ores in Siberia" (M. A. Zharkov and A. L. Yanshin, eds.), pp. 46–57. Izd. Nauka, Moscow.

Kolosov, A. S., Pustyl'nikov, A. M., and Zharkova, T. M. (1968). Complex iron manganese chlorides in Cambrian evaporites of the Kansk–Taseyevka depression. *Dokl. Akad. Nauk SSSR,* **181**(6), 1472–1475; *Dokl. Acad. Sci. USSR, Earth Sci. Sect. (Engl. Transl. Am. Geol. Inst.)* **181**(1–6), 213–216.

Kolosov, A. S., Pustyl'nikov, A. M., and Fedin, V. P. (1974). Precipitation of glauberite and conditions of glauberite-bearing sediments in the Kara Bogaz Gol. *Dokl. Akad. Nauk SSSR,* **219**(6), 1456–1460; *Dokl. Acad. Sci. USSR, Earth Sci. Sect. (Engl. Transl. Am. Geol. Inst.)* **219**(1–6), 184–186.

Kopnin, V. I. (1962). The salt content of Upper Kungurian rocks in the area of the Upper Kama salt deposit (in Russian). *Dokl. Akad. Nauk SSSR* **144**(5), 1123–1125.

Kopnin, V. I. (1966). Distribution of constant components in sediments of the Upper Kama potassium deposit (in Russian). *Geokhimiya,* No. 6, pp. 715–725.

Kopnin, V. I., and Maloshtanova, N. E. (1980). The problem of the mineral composition of sylvinite ores of the Upper Kama deposits (in Russian). *Tr. Inst. Geol. Geofiz. Akad. Nauk SSSR, Sib. Otd.* **439,** 44–47.

Koppenol, W. H., Arnoldus, R., Kreulen, R., and Schuiling, R. D. (1977). Recent dolomitization on Naxos. *Colloq. Geol. Aegean Reg., 6th 1977,* vol. 3, pp. 985–994.

Korenevskiy, S. M. (1960). Composition of salt, inter-salt, and country rocks of the Solotvin halite deposit (in Russian). *Tr. Vses. Nauchno–Issled. Proektn. Inst. Galurgii* **40,** 216–244.

Korenevskiy, S. M. (1963). The potash content of the evaporitic Kungurian of the southern Fore-Urals (in Russian). *In* "Geology of Potash Salt Deposits" (A. A. Ivanov, ed.). *Tr. Vses. Nauchno–Issled. Geol. Inst.,* [*N.S.*] **99,** 191–214.

Korenevskiy, S. M. (1970). Manganese in halide formations and the beds containing them. *Litol. Polezn. Iskop.* **5**(4), 71–77; *Lithol. Miner. Resour. (Engl. Transl.)* **5**(4), 456–461.

Korenevskiy, S. M., and Donchenko, K. B. (1963). Geology and the conditions of formation of the potash deposits of the Soviet Carpathian foreland (in Russian). *In* "Geology of Potash Salt Deposits" (A. A. Ivanov, ed.). *Tr. Vses. Nauchno–Issled. Geol. Inst.,* [*N.S.*] **99,** 3–152 (contains also 340 references on the history of Carpathian Foreland evaporite investigations).

Korenevskiy, S. M., Bobrov, V. P., Supronyuk, K. S., and Khrushchov, D. P. (1968). "The Salt Formations of the Northwestern Donbass and the Dnieper–Donets Depression and Their Potash Content" (in Russian). Izd. Nauka, Moscow, 240 pp.

Korennov, Yu. F. (1974). Genesis of strontium in highly mineralized subsurface brines (in Russian). *Tr. Vses. Nauchno–Issled. Inst. Gidrogeol. Inzh. Geol.* **84,** 37–41.

Koritnig, S. (1951). Ein Beitrag zur Geochemie des Fluors. *Geochim. Cosmochim. Acta* **1**(1), 89–116.

Koritnig, S. (1967). Bassanit, $CaSO_4 \cdot 0.5H_2O$ von Poettsching, Burgenland. *Karinthin* **56,** 249–252.

Korobov, S. S., and Tuchfatov, K. O. (1961). On the natural gas content of the salt series of the northern Caspian region (in Russian). *Nov. Neft. Gazov. Tekhnol., Ser. Geol.,* No. 11, pp. 21–23.

Korol'kov, I. I., and Krupnikova, A. V. (1953). Transition of calcium sulfate hydrates. *Zh. Prikl. Khim.* **26,** 907–911; *J. Appl. Chem. USSR (Engl. Transl.)* **26,** 837–840.

Korol'kov, I. I., and Tyagunova, Z. A. (1956). The effects of kolloids on the crystallization of gypsum (in Russian). *Gidroliz. Lesokhim. Prom–sti.* **9**(8), 8–9.

Kosakevitch, A. (1966). Authigenic pyrrhotite from Souk–el-Arba du Rharb. *Notes Mém. Serv. Géol. Morocco* **198,** 126–127.

Koshurnikov, G. S., and Mokievskiy, V. A. (1948). Effect of addition of organic matter on the crystallization of alkali halides (in Russian). *Zh. Obshch. Khim.* **18,** 569–571.

Kosmahl, W. (1969). Zur Stratigraphie, Petrographie, Paläogeographie, Genese, und Sedimentation des gebänderten Anhydrits (Zechstein-2), grauen Salztones und Hauptanhydrits (Zechstein-3) in Nordwestdeutschland. *Bundesanst. Geowiss. Geol. Jahrb., Beih.* **71,** 1–129.

Kostenko, I. F. (1973). A possible reason for the decrease in the bromine–chlorine ratio. *Dokl. Akad. Nauk SSSR* **208**(4), 951–953; *Dokl. Acad. Sci. USSR, Earth Sci. Sect. (Engl. Transl. Am. Geol. Inst.)* **208**(1–6), 210–211.

Kostenko, I. F. (1974). Distribution of bromide between brine and halite. *Geokhimiya,* No. 7, pp. 1099–1100; *Geochem. Int. (Engl. Transl.)* **11**(4), 765–766.

Kostenko, I. F. (1982). Hydration ions as an index of halogenesis (in Russian). *Geol. Zh.* **42**(1), 21–28.

Kotel'nikova, Ye. N., Kotov, N. V., and Frank–Kamenetskiy, V. A. (1979). Conditions of formation of tosudite, chlorite, and accompanying phases of hydrothermal transformation of kaolinite in magnesium chloride solutions (in Russian). *In* Crystal Chemistry and Structural Mineralogy'' (V. A. Frank–Kamenetskiy, ed.), pp. 81–92. Izd. Nauka, Leningrad.

Koulumies, R. (1946). The action of the metal chlorides of the second group of the periodic system on various microorganisms. *Acta Pathol. Microbiol. Scand., Suppl.* **64,** 1–89.

Kovalevich, V. M. (1977). The salting out of halite in Miocene evaporite deposits of the eastern Ciscarpathian region (in Russian). *In* ''Geology and Geochemistry of Salt Formations in the Ukraine'' (V. I. Kityk, ed.), pp. 48–53. Izd. Naukova Dumka, Kiev.

Koyama, T., and Sugawara, K. (1953). Separation of the components of atmospheric salt and their distribution (in Japanese). *Bull. Chem. Soc. Jpn.* **26,** 123–126.

Koyfman, M. I., Senatskaya, G. S., Solomina, I. A., and Kvashnina, O. I. (1969). New information on the plasticity of halite, gypsum and calcite crystals. *Dokl. Akad. Nauk SSSR* **184**(5), 1174–1176; *Dokl. Acad. Sci. USSR, Earth Sci. Sect. (Engl. Transl. Am. Geol. Inst.)* **184**(1–6), 123–125.

Kozary, M. T., Dunlap, J. C., and Humphrey, W. E. (1968). Incidence of saline deposits in geologic time. *In* ''Saline Deposits'' (R. B. Mattox *et al.,* eds.). *Spec. Paper Geol. Soc. Am.* **88,** 43–57.

Kozlov, A. L. (1978). The characteristics of salt basins that cause local wedging out of evaporites and control the distribution of petroleum and gas deposits (in Russian). *Sov. Geol.* No. 8, pp. 15–21.

Kranz, R. I. (1968). Participation of organic compounds in the transport of ore metals in hydrothermal solution. *Trans. Inst. Min. Metall., Sect. B* **77,** 26–36.

Krasikov, I. I., and Ivanov, I. T. (1927). The solubility of salts in saturated solutions of other salts of different composition (in Russian). *Zap. Byeloruss. Gos. Akad. Sel'skogo Khozyaistva* **4,** 238–260.

Krauss, F., and Joerns, G. (1930). Über die Hydrate des Kalziumsulfats. *Tonind.-Ztg.* **54,** 1467–1468, 1483–1484.

Krayushkin, V. A., and Osadchiy, V. G. (1965). Halite in the Carpathian flysch (in Ukrainian). *Geol. Zh.* **25**(3), 89–95.

Krejčí-Graf, K. (1962). Über Bituminierung und Erdölentstehung. *Freiberger Forschungsh.*, [C] **123,** 5–34.

Krinsley, D. B. (1970). A geomorphological and paleoclimatological study of the playas of Iran. *Final Sci. Rep. Contr. CP70-800,* 2 vols., 486 pp. U.S. Geol. Surv.

Kroell, D., and Nachsel, G. (1967). Zur Ausbildung des Stassfurt-Steinsalzes im Südharz–Kalirevier. *Geologie* **16**(3), 269–279.

Kroenlein, G. A. (1939). Salt, potash and anhydrite in Castile Formation of Southeast New Mexico. *Bull. Am. Assoc. Pet. Geol.* **23**(11), 1682–1693.

Kropachev, A. M., Kropacheva, T. S., and Rayevskiy, V. I. (1972). Distribution characteristics of accessory lithium in the Upper Kama evaporite formations (in Russian). *Tr. Vses. Nauchno-Issled. Proektn. Inst. Galurgii* **56,** 12–18.

Krotov, P. B. (1935). Genesis of the Kuybyshev sulphur deposit (in Russian). *In* "Sb. Statei. Akad. Nauk SSSR" (A. E. Fersman and D. I. Shcherbakova, eds.), pp. 1–15. Izd. Nauka, Moscow.

Krueger, P. (1962). Ein Vorkommen von Pyrit in der Kalisalzlagerstätte der Grube Glückauf, Sondershausen, Südharz–Bezirk. *Bergakademie* **14,** 425–430.

Kruell, F. (1933). Über die Bildung von Chlorcalzium in den Laugen der Salzlagerstätten. *Kali Verw. Salze Erdöl (Halle)* **27**(6), 67–69 and (7), 84–88.

Kruell, F. (1934). Über die Bildung von Chlorkalzium in den Laugen der Salzlagerstätten. II. Die Wirkung von Chloridlösungen auf Anhydrit. *Kali Verw. Salze Erdöl (Halle)* **28,** 161–165, 173–176.

Kruemmel, O. (1911). "Handbuch der Ozeanographie." 3rd ed., 2 vols., J. Englehorn Nachf., Stuttgart.

Krull, O. (1917). Beiträge zur Geologie der Kalisalzlager. *Kali (Halle)* **11**(14), 227–231.

Krull, O. (1919). Die Geologie der deutschen Kalisalzlager nach dem heutigen Stand der Forschung. *Kali (Halle)* **13**(18), 277–281, 296–304; (19), 317–322.

Krumbein, W. C. (1951). Occurrence and lithographic association of evaporites in the U.S. *J. Sediment. Pet.* **21**(2), 63–81.

Krumbein, W. C., and Garrels, R. M. (1952). Origin and classification of chemical sediments in terms of pH and oxidation–reduction potentials. *J. Geol.* **60**(1), 1–33.

Krumbein, W. E., and Cohen, Y. (1964). Biogene, klastische und evaporitische Sedimentation in einem mesothermen monomiktischen ufernahen See (Golf von Aqaba). *Geol. Rundsch.* **63**(2), 1035–1065.

Krumgalz, B. (1980). Salt effect on the pH of hypersaline solutions. *In* "Hypersaline Brines and Evaporitic Environments" (A. Nissenbaum, ed.), pp. 23–40. Elsevier, Amsterdam.

Krupkin, P. I. (1963). Movement of salt solutions in soils and soil materials (in Russian). *Pochvovedenie,* No. 6, pp. 70–79; *Sov. Soil Sci. (Engl. Transl.)* **1,** 567–574.

Kuchkarov, A. B., and Shuykin, N. I. (1954). Complex compounds of metal halides and alcohols (in Russian). *Izv. Akad. Nauk SSSR, Otd. Khim. Nauk,* pp. 470–477.

Kudryavtsev, N. A. (1971). Evaporites and petroleum: Discussion. *Bull. Am. Assoc. Pet. Geol.* **55**(11), 2033–2061.

Kuehn, R. (1951). Zur Kenntnis des Könenits. *Neues Jahrb. Mineral., Monatsh.*, No. 1, pp. 1–16.

Kuehn, R. (1952a). Reaktionen zwischen festen, insbesondere ozeanischen Salzen. *Heidelb. Beitr. Mineral. Petrogr.* **3**(3), 147–168.

Kuehn, R. (1952b). Sellait als salinares Mineral. *Fortschr. Mineral.* **29–30,** 390.

Kuehn, R. (1953). Tiefenberechnung des Zechsteinmeeres nach dem Bromgehalt der Salze. *Z. Dtsch. Geol. Ges.* **105**(4), 645–663.

Kuehn, R. (1954). Petrographische Studien an Jahresringen im Steinsalz. *Fortschr. Mineral.* **32**(1), 90–92.

Kuehn, R. (1955). Mineralogische Fragen der in den Kalisalzlagerstätten vorkommenden Salze. *Kalisymp. Bern 1955,* pp. 51–105.

Kuehn, R. (1958). Über die Mineralogie des Vant'Hoffits. *Kali Steinsalz* **2**(7), 217–222.

Kuehn, R. (1961). Die chemische Zusammensetzung des Könenits nebst Bemerkungen über sein Vorkommen und über Faserkönenit. *Neues Jahrb. Mineral., Abh.* **97,** 112–141.

Kuehn, R. (1963). Rubidium als geochemisches Leitelement bei der lagerstättenkundlichen Charakterisierung von Carnalliten und natürlichen Salzlösungen. *Neues Jahrb. Mineral., Monatsh.,* No. 5, pp. 107–115.

Kuehn, R. (1966). Mineralogisch-petrographische Beiträge zu drei Kalibergwerkexkursionen. *Fortschr. Mineral.* **43**(2), 122–144, 153–187, and 194–210.

Kuehn, R. (1968). Geochemistry of the German potash deposits. *In* "Saline Deposits" (R. B. Mattox *et al.,* eds.). *Spec. Paper Geol. Soc. Am.* **88,** 427–504.

Kuehn, R. (1969). Zum Tachhydritvorkommen im Flöz Stassfurt. *Kali Steinsalz* **5**(5), 166–170.

Kuehn, R. (1972). Zur Kenntnis der Rubidiumgehalte von Kalisalzen ozeaner Salzlagerstätten nebst einigen lagerstättenkundlichen Ausdeutungen. *Bundesanst. Geowiss., Geol. Jahrb.* **90,** 127–158.

Kuehn, R., and Baar, A. (1955). Ein ungewöhnliches Vorkommen von Danburit. *Kali Steinsalz* **1**(19), 17–21.

Kuehn, R., and Hsu, K. J. (1974). Bromine content of Mediterranean halite. *Geology* **2**(5), 213–216; discussion by A. C. Kendall *ibid.* **2**(11), 524–526.

Kuehn, R., and Hsu, K. J. (1978). Chemistry of halite and potash salt cores, DSDP sites 374 and 376, leg 42A, Mediterranean Sea. *In* Intl. Rep. Deep-Sea Drlg. Proj., vol. 42A (K. J. Hsu and L. Montadert, eds.), pp. 613–619. U.S. Govt. Print. Off., Washington, D.C.

Kuehn, R., and Roth, R. H. (1979). Beitrag zur Kenntnis der Salzlagerstätte am Oberrhein. *Z. Geol. Wiss.* **7**(8), 953–966.

Kuehn, R., and Schwerdtner, W. (1959). Nachweis deszendenter Vorgänge während der Entstehung der Leine–Serie des deutschen Zechsteinsalzes. *Kali Steinsalz* **2**(11), 380–383.

Kuenkler, A., and Schwedholm, H. (1908–9). Die Bildung von Erdöl auf fettsauren Erdalkalien. Seifensieder–Ztg., **34,** 165–196 (1908); **35,** 1285–1288, 1341–1342, 1365–1366, 1393–1394 (1909).

Kuhlmann, J., and Ludwig, U. (1977). Volume stability of gypsum building plasters. *Zem.–Kalk–Gips* **30**(5), 214–218. *Chem. Abstr.* **87,** 72,487.

Kukla, J. (1970). Correlations between loesses and deep-sea sediments. *Geol. Foren. Stockholm Forh.,* **92,** 148–180.

Kuleliev, K. I., and Kostov, B. T. (1960). Geochemical studies in the Radnovo gypsum-bearing region. I. The origin of the gypsum (in Bulgarian). *God. Vissh. Minno-Geol. Inst., Sofia* **6,** 101–126.

Kulke, H. (1974). Zur Geologie und Mineralogie der Kalk- und Gipskrusten Algeriens. *Geol. Rundsch.* **63**(3), 970–998.

Kulke, H. (1978). Tektonik und Petrographie einer Salinarformation am Beispiel der Trias des Atlassystems (NW–Afrika). *Geotekton. Forsch.* **55,** 158 pp.

Kulke, H. (1979). Sédimentation, diagénèse et métamorphisme léger dans un milieu sursalé, exemple du Trias Magrebin. *Sci. Terre* **23**(2), 39–74.

Kullenberg, B. (1952). On the salinity of the water contained in marine sediments. *Goteborgs K. Vetensk. Vitterhets–Samh. Handl.,* [B] **6**(6), 3–38.

Kunasz, I. (1970). Significance of laminations in the Upper Silurian evaporite deposits of the Michigan Basin. *In* "Symp. on Salt, 3rd." (J. L. Rau and L. F. Dellwig, eds.), vol. 1, pp. 67–77. N. Ohio Geol. Soc., Cleveland, Ohio.

Kuntze, R. A. (1965). Effect of water vapor on the formation of $CaSO_4 \cdot 0.5H_2O$ modifications. *Can. J. Chem.* **43**(9), 2522–2529.

Kuntze, R. A. (1966). Retardation of the crystallization of calcium sulfate dihydrate. *Nature (London)* **211**(5047), 406–407.

Kuprin, P. N., and Potapova, L. I. (1978). Early diagenetic conversion of organic matter in sediments (in Russian). *In* "Fundamental Conditions of Oil and Gas Generation and Accumulation" (V. D. Nalivkin *et al.*, eds.), pp. 17–35. Izd. Nauka, Moscow.

Kurilenko, V. V., and Frolovskii, E. E. (1982). Hydrogeochemical and hydrodynamic regime of intercrystalline subsurface brines of the Kurguzul Inlet (in Russian). *In* "New Data on the Geology and Geochemistry of Subsurface Waters and of Economic Deposits in Evaporite Basins" (A. L. Yanshin and M. A. Zharkov, eds.), vol. 2, pp. 67–72. Izd. Nauka, Novosibirsk.

Kusakabe, M., Hitoshi, C., Hiroshi, O. (1982). Stable isotopes and a fluid inclusion study of anhydrite from the East Pacific Rise at 21°N. *Geochemical J. (Jpn)* **16**(2), 89–95.

Kushnir, J. (1980a). The coprecipitation of strontium, magnesium, sodium, potassium and chloride ions with gypsum; an experimental study. *Geochim. Cosmochim. Acta* **44**(10), 1471–1482.

Kushnir, J. (1980b). The coprecipitation of seawater-cations with gypsum in a Recent sea–marginal sabkha. *In* Proc. Int. Symp. on Water–Rock Interact., 3rd. 1980, (A. R. Campbell, ed.), pp. 27–28. Alberta Res. Council, Edmonton.

Kushnir, J. (1982). The composition and origin of brines during the Messinian desiccation event in the Mediterranean basin as deduced from concentration of ions coprecipitated with gypsum and anhydrite. *Chem. Geol.* **35**(3–4), 333–350.

Kuznetsov, A. M., and Ignat'ev, N. A. (1948). Nodular formation of gypsum (in Russian). *Dokl. Akad. Nauk SSSR* **63,** 433–436.

Kuznetsov, V. P., and Kutovaya, A. A. (1977). pH values of condensation and formation waters in gas condensate wells (in Russian). *Korroz. Zashch. Neftegazov. Prom–sti.* No. 1, 7–9; *Chem. Abstr.* **87,** 8313.

Lacombe, H. (1961). Contribution a l'etude du régime du détroit de Gibraltar. *Cah. Océanogr.* **13**(2), 73–107.

Lacombe, H., and Tchernia, P. (1960). Quelques traits généraux de l'hydrologie méditerranéenne d'après diverses campagnes hydrologiques récentes en Méditerranée, dans la proche Atlantique et dans le détroit de Gibraltar. *Cah. Océanogr.* **12**(8), 527–548.

Lagny, P. (1980). The stratiform ore deposits associated with evaporites: Location with respect to time and space in the sedimentary basin. *Bull. Cent. Rech. Explor.–Prod. Elf–Aquitaine* **4**(1), 445–478.

Lahann, R. W., and Campbell, R. C. (1980). Adsorption of palmitic acid onto calcite. *Geochim. Cosmochim. Acta* **44**(5), 629–634.

Lamanon de Paul, R. (1782). Description de divers fossiles trouvés dans les carrières de Montmartre. *J. Phys.* **19**(1), 173–194.

Lamb, H. H. (1977). "Climate, Present, Past and Future" pp. 353–389. Methuen, London.

Lambert, A. (1967). Esquisses géologiques du bassin potassique Congolais. *Ann. Mines* **11,** 709–723.

Lambert, I. B., Donnelly, T. H., Dunlop, J. S. R., and Groves, D. I. (1978). Stable isotope composition of early Archaean sulphate deposits of probable evaporitic and volcanogenic origins. *Nature (London)* **276**(5690), 808–810.

Landes, K. K. (1963). Origin of salt deposits. *In* "Symp. on Salt" (A. C. Bersticker, K. E. Hoekstra, and J. F. Hall, eds.), pp. 3–10. N. Ohio Geol. Soc., Cleveland, Ohio.

Lang, W. B. (1937). The Permian formations of the Pecos valley of New Mexico and Texas. *Bull. Am. Assoc. Pet. Geol.* **21**(7), 833–898.

Lang, W. B. (1943). Gigantic drying cracks in Animas Valley, New Mexico. *Science* **98**(2557), 583–584.

Lang, W. B. (1950). Comparison of the cyclic deposits of the Castile and the Salado Formations of the Southwest. *Bull. Geol. Soc. Am.* **61**(12), 1479–1480.

Langauer, D. (1932). Influence of magnesium salts on the solubilities of potassium and sodium chlorides (in Polish). *Rocz. Chem.* **12,** 258–269.

Langbein, R. (1961). Zur Petrographie des Hauptanhydrits (Z3) im Südharz. *Chem. Erde* **21**(2), 248–264.

Langbein, R. (1964). Geochemische Untersuchungen an Salztonen des Zechsteins im Südharz–Revier. *Chem. Erde* **23**(1), 1–70.

Langbein, R. (1968). Zur Petrologie des Anhydrits. *Chem. Erde* **27**(1), 1–38.

Langbein, R. (1978). Petrologisch bedingte Veränderungen der Zusammensetzung von Anhydritgesteinen. *Z. Geol. Wiss.* **6**(7), 889–896.

Langbein, R. (1979). Petrologische Aspekte der Anhydritbildung. *Z. Geol. Wiss.* **7**(7), 913–926.

Langbein, R., and Seidel, G. (1960a). Zur Frage des "Sangerhauser Anhydrits." *Geologie* **9**(7), 778–787.

Langbein, R., and Seidel, G. (1960b). Zur Geologie im Gebiete des Holunger Grabens (Ohmgebirgsgrabenzone). *Geologie* **9**(1), 36–57.

Langbein, W. B. (1961). Salinity and hydrology of closed lakes. *U.S. Geol. Surv. Prof. Paper* **412,** 1–20.

Lange, J. (1970). Geochemische Untersuchungen an Sedimenten des Persischen Golfes. *Contrib. Mineral. Petrol.* **28**(4), 288–305.

Langier, R. (1959). Observations pétrographiques nouvelles sur les niveaux salifères du Trias moyen de Lorraine. *Bull. Soc. Géol. Fr.*, [7] **1**(1), 31–39.

Lapinski, S. (1906). Über Gipskristalle im menschlichen Harn. *Wien. Klin. Wochenschr.* **19**(45), 1348–1349.

Larimer, J. W. (1968). An experimental investigation of oldhamite, CaS, and the petrologic significance of oldhamite in meteorites. *Geochim. Cosmochim. Acta* **32**(9), 965–982.

LaRock, P. A., Lauer, R. D., Schwarz, J. R., Watanabe, K. K., and Wiesenburg, D. A. (1979). Microbial biomass and activity distribution in an anoxic hypersaline basin. *Appl. Environ. Microbiol.* **37**(3), 466–470.

Larsen, H. (1980). Ecology of hypersaline environments. *In* "Hypersaline Brines and Evaporitic Environments" (A. Nissenbaum, ed.), pp. 23–40. Elsevier, Amsterdam.

Lasaga, A. C., and Holland, H. D. (1976). Mathematical aspects of non-steady-state diagenesis. *Geochim. Cosmochim. Acta* **40**(3), 257–266.

Last, W. M. (1984). Sedimentology of playa lakes of the northern Great Plains. *Can. J. Earth Sci.* **21**(1), 107–125.

Lattman, L. H., and Lauffenburger, S. K. (1974). Proposed role of gypsum in the formation of caliche. *Z. Geomorphol., N. S., Suppl.* **20,** 140–149.

Leach, R. O., Wagner, O. R., Wood, H. W., and Harpke, C. F. (1962). A laboratory and field study of wettability adjustment in water flooding. *J. Pet. Technol.* **14**(2), 206–212.

Lebedev, L. M. (1967a). Contemporary deposition of native lead from hot Cheleken thermal brine. *Dokl. Akad. Nauk SSSR,* **174**(1), 197–200; *Dokl. Acad. Sci. USSR, Earth Sci. Sect. (Engl. Transl. Am. Geol. Inst.)* **174**(1–6), 173–176.

Lebedev, L. M. (1967b). Modern growth of sphalerite. *Dokl. Akad. Nauk SSSR,* **175**(4), 920–923; *Dokl. Acad. Sci. USSR, Earth Sci. Sect. (Engl. Transl. Am. Geol. Inst.)* **175**(1–6), 196–199.

Lebedev, L. M., and Bugel'skiy, Yu. Ya. (1967). Metal–bearing properties of highly mineralized thermal waters of Cheleken (in Russian). *Geol. Rudn. Mestorozhd.* **9**(3), 82–87.

Lebedev, V. I. (1952). Correct and incorrect use of the crystal lattice energy concept in geochemistry (in Russian). *Vestn. Leningr. Univ., Ser. Biol., Geogr.,Geol.* **7**(10), 107–119.

Lecointre, G. (1960). Observation au sujet de la note de M. Ricour sur la génèse des niveaux salifères. *C. R. Somm. Séances Soc. Géol. Fr.* No. 6, p. 137.

LeDred, R., and Wey, R. (1965). Formation et application de complexes mica–vermiculite–chlorure de sodium. *Clays Clay Miner.* **13**(2), 177–186.

Lee, R. F., and Loeblich, A. R., III (1971). Distribution of 21:6 hydrocarbon and its relation to 22:6 fatty acid in algae. *Phytochemistry* **10**(3), 593–602.

Lee, W. B., and Egerton, A. C. (1923). Heterogeneous equilibria between the chlorides of calcium, magnesium, potassium, and their aqueous solutions. *J. Chem. Soc.* **123,** 706–716.

Lefond, S. J. (1969). "Handbook of World Salt Resources." Plenum Press, New York, 384 pp.

Legrand, R. (1947). Composition chimique de stalactites récentes. *Bull. Soc. Belge Géol. Paléont. Hydrol.,* [B] **55**(2–3), 210–211.

Legros, J. C., Platten, J. K., and Poty, P. G. (1972). Stability of a two-component fluid layer heated from below. *Phys. Fluids,* **15**(8), 1383–1390.

Legun, A. S., and Rust, B. R. (1981). Coal accumulation on a semi-arid alluvial plain: The Carboniferous Clifton Formation of northern New Brunswick. *Geol. Assoc. Can., Abstr.* **6,** A65.

Lehmann, H., Mathiak, H., and Kurpiers, P. (1969). Die Umwandlung von Kalziumsulfatdihydrat zu Hemihydrat und die Entwässerung zu Anhydrit III während des Wasserverlustes von Gips. *Tonind.-Ztg. Keram. Rundsch.* **93**(9), 318–327.

Leibinger, H., Pester, L., and Pommerenke, J. (1964). Stand und Perspektive der Erkundung der Salzkohlefelder der D.D.R. *Z. Angew. Geol.* **10**(9), 474–478.

Leitmeier, H. (1953). Die Entstehung der Spatmagnesite in den Ostalpen. *Tschermak's Mineral. Petrogr. Mitt.* **3,** 305–331.

Leitmeier, H., and Siegel, W. (1954). Untersuchungen an Magnesiten am Nordrande der Grauwackenzone Salzburg und ihre Bedeutung fur die Entstehung der Spatmagnesite der Ostalpen. *Berg- Hüttenmänn. Monatsh.* **9,** 201–208.

Leitz, J. (1951). Sulfidische Klufterze im Deckgebirge des Salzstockes Reitbrook. *Mitt. Hamb. Geol. Staatsinst.* **20,** 110–118.

Leleu, M., and Goni, J. (1974). Sur la formation biogéochimique de stalactites de galène. *Mineral. Deposita* **9**(1), 27–32.

Leonhardt, J., and Berdesinski, W. (1953). Semisalinare (assimilierte) Mineralkomponenten der Salzlagerstätten. *Chem. Erde* **16**(1), 22–26.

Leonhardt, J., and Tiemeyer, R. (1938). Über Sylvin mit Hämatiteinschlüssen. *Naturwissenschaften* **26,** 410–411.

Lepeshkov, I. N. (1946). Theory of salt basin formation and salt crystallization in the southern zone of the former Permian sea (in Russian). *Byull. Otd. Khim. Nauk Akad. Nauk SSSR,* pp. 601–606.

Lepeshkov, I. N., and Bodaleva, N. V. (1949). Solubility isotherm of the aqueous reciprocal system $K_2Cl_2 + MgSO_4 \rightleftharpoons K_2SO_4 + MgCl_2$ at 25°C (in Russian). *Izv. Akad. Nauk SSSR, Inst. Obshch. Neorg. Khim., Sekt. Fiz.-Khim. Anal.* **17,** 338–344.

Lepeshkov, I. N., and Bodaleva, N. V. (1952). The order of crystallization of salts on evaporation of Aral Sea water (in Russian). *Dokl. Akad. Nauk SSSR* **83,** 583–584.

Lepeshkov, I. N., and Fradkina, K. B. (1958). Carnallite and syngenite in sediments of the salt lake Dzhaksa–Klych (Aral Sea area) (in Russian). *Dokl. Akad. Nauk SSSR* **120,** 83–85.

Lepeshkov, I. N., Shaposhnikova, A. N., and Zaitseva, I. S. (1970). Lithium distribution in natural salts (in Russian). *Geokhimiya* No. 11, pp. 1322–1328.

Lerman, A. (1967). Model of chemical evolution of a chloride lake, the Dead Sea. *Geochim. Cosmochim. Acta* **31**(12), 2309–2330.

Lerman, A. (1970). Chemical equilibria and evolution of chloride brines. *Mineral. Soc. Am. Spec. Paper* **3,** 291–306.

Lerman, A. (1979). "Geochemical Processes: Water Sediment Environments." Wiley, New York, 481 pp.

Leroux, M. (1976). La circulation atmosphérique générale et les oscillations climatiques tropicales. *In* "La Désertification au Sud de Sahara." *Mém. Inst. Fondam. Afr. Noire* **90,** 82–88.

Leung, I. S. (1981). Intercellular mineral deposition in bluegreen algae. *Bull. Geol. Soc. Am., Abstr. Prog.* **13**(7), 496.

Levey, M. (1958). Gypsum salt and soda in ancient Mesopotamian technology. *Isis* **49,** 336–341.

Levitskiy, Yu. F. (1961). Anhydrite with enterolithic texture (in Russian). *Tr. Vses. Nauchno–Issled. Geol. Inst.* **68,** 87–91.

Levshin, B. A., Fartukov, M. M., and Lymarov, V. A. (1975). Results of determinations of the absolute age of potash salts of Gourdak (Turkmenian SSR). *Litol. Polezn. Iskop.*, No. 2, pp. 108–109; *Lithol. Mineral. Resour. (Engl. Transl.)* **10**(2), 229–230.

Levy, A. (1972). Données préliminaires sur la formation du gypse dans les milieux marginolittoraux polyhalins et euhalins de la bordure du Golfe du Lyon. *C. R. Hebd. Séances Acad. Sci. Paris,* [D] **275**(23), 2579–2582.

Levy, Y. (1977). Description and mode of formation of the supratidal evaporite facies in northern Sinai coastal plain. *J. Sediment. Petrol.* **47**(1), 463–474.

Levy, Y. (1980). Evaporitic environments in northern Sinai. *In* "Hypersaline Brines and Evaporitic Environments" (A. Nissenbaum, ed.), pp. 131–144. Elsevier, Amsterdam.

Lewis, W. T., Incroper, F. P., and Viskanta, R. (1982). Interferometric study of stable salinity gradients heated from below and cooled from above. *J. Fluid Mech.* **116,** 411–413.

Leythaeuser, D., Schaeffer, R. G., and Radke, M. (1981). Genese und Migration von Kohlenwasserstoffen in der Tiefbohrung Vorderriss 1. *Geol. Bavarica* **81,** 159–192.

Liebermann, O. (1967). Synthesis of dolomite. *Nature (London)* **213**(5073), 241–245.

Liebezeit, G., Boelter, M., Brown, I. F., and Dawson, R. (1980). Dissolved free amino acids and carbohydrates at pycnocline boundaries in the Sargasso Sea and related microbial activity. *Oceanol. Acta* **3**(3), 357–362.

Liermann, H., and Rexer, E. (1932). Über das Wesen des blauen Steinsalzes. *Naturwissenschaften* **20,** 561.

Liesegang, R. E. (1916). Zur Theorie der heissen ungarischen Salzseen. *Int. Rev. Gesamt. Hydrobiol.* **7**(6), 469–471.

Lightfoot, W. J., and Prutton, C. F. (1946). Equilibria in saturated solutions. I. The ternary systems $CaCl_2$–$MgCl_2$–H_2O, $CaCl_2$–KCl–H_2O and $MgCl_2$–KCl–H_2O at 35°C. *J. Am. Chem. Soc.* **68**(6), 1001–1008.

Lightfoot, W. J., and Prutton, C. F. (1947). Equilibria in saturated solutions. II. The ternary systems $CaCl_2$–$MgCl_2$–H_2O, $CaCl_2$–KCl–H_2O, and $MgCl_2$–KCl–H_2O at 75°C. *J. Am. Chem. Soc.* **69**(9), 2098–2100.

Lightfoot, W. J., and Prutton, C. F. (1948). Equilibria in saturated solutions. III. The quaternary system $CaCl_2$–$MgCl_2$–KCl–H_2O at 35°C. *J. Am. Chem. Soc.* **70**(12), 4112–4115.

Lightfoot, W. J., and Prutton, C. F. (1949). Equilibria in saturated solutions. IV. The quaternary system $CaCl_2$–$MgCl_2$–KCl–H_2O at 75°C. *J. Am. Chem. Soc.* **71**(4), 1233–1235.

Lin, A., and Lee, S. M. (1974). Study of exchange of brines through dykes with viscous flow analogy. *In* "Symp. on Salt, 4th." (A. H. Coogan, ed.), vol. 2, pp. 399–402. N. Ohio Geol. Soc., Cleveland, Ohio.

Linck, G. (1912). Arbeiten über die Synthese des Dolomites. As quoted in "Handbuch der Mineralchemie" (C. A. Doelter y Cisterich, ed.), Steinkopff, Dresden (1912) v.1, p. 395, and in H. E. Boeke and W. Eitel (1923), "Grundlagen der physikalisch–chemischen Petrographie." Gebr. Borntraeger, Berlin, p. 451.

Linck, G. (1937). Bildung des Dolomits und Dolomitisierung. *Chem. Erde* **11**(2), 278–286.

Linck, G. (1942). Beobachtungen und ihre Ergebnisse an Gesteinen des mittleren Zechsteins (Hauptdolomit und Grauer Salzton) in Thüringen. *Chem. Erde* **14,** 312–357.

Linck, O. (1946). Die sogenannten Steinsalzmetamorphosen als Kristall–Relikte. *Abh. Senckenb. Naturforsch. Ges.* **470,** 1–50.

Linck, O. (1948). Wie werden die sogenannten Steinsalzpseudomorphosen gebildet? *Natur und Volk* **78,** 103–109.

Lindberg, M. L. (1946). Occurrence of bromine in carnallite and sylvite from Utah and New Mexico. *Am. Mineral.* **31,** 486–494.

Lindholm, R. C., Siegel, F. R., Dort, W., Jr. (1969). Diagenetic syngenite from Victoria Land, Antarctica. *Antarctic J. U.S.* **4**(4), 130–131.

Lindsay, D. A., Glanzman, R. K., Naeser, C. W., and Nichols, D. J. (1981). Upper Oligocene evaporites in basin fill of Sevier Desert region, western Utah. *Bull. Am. Assoc. Pet. Geol.* **65**(2), 251–260.

Link, G. B., and Ottemann, J. (1968). Zur Frage des Einbaues von Strontium in Gips. *Landesamt Baden-Württemberg Jahrb.* **10,** 175–178.

Linn, K. O., and Adams, S. S. (1966). Barren halite zones in potash deposits, Carlsbad, New Mexico. *In* "Symp on Salt, 2nd." (J. L. Rau, ed.), vol. 1, 59–68. N. Ohio Geol. Soc., Cleveland, Ohio.

Linnenbom, V. J., and Swinnerton, J. W. (1970). Low molecular weight hydrocarbons and carbon monoxide in sea water. *In* "Symp. Organic Matter in Nat. Waters" (D. W. Hood, ed.). *Occas. Publ. Univ. Alaska Inst. Marine Sci.* **1,** 455–467.

Linstedt, H. (1952). Anwendung der gravimetrischen Methode auf das Anhydrokainit-Problem. *Z. Phys. Chem.* **199**(5–6), 345–349.

Lippmann, F. (1973). "Sedimentary Carbonate Minerals." Springer-Verlag, Berlin, 228 pp.

Lippmann, F., and Savascin, M. Y. (1969). Mineralogische Untersuchungen an Lösungsrückständen eines württembergischen Keupergipsvorkommens. *Tschermak's Mineral. Petrogr. Mitt.* **13**(2), 165–190.

Lippolt, H. J., and Oesterle, F. P. (1977). Argon retentivity of the mineral langbeinite. *Naturwissenschaften* **64**(2), 90–91.

Lippolt, H. J., and Raczek, I. (1979a). Cretaceous Rb–Sr total rock ages of Permian salt rocks. *Naturwissenschaften* **66**(8), 422–423.

Lippolt, H. J., and Raczek, I. (1979b). Rinneite dating of episodic events in potash salt deposits. *J. Geophys.* **46**(2), 225–228.

Lisitsin, E. (1974). "Sea-Level Changes." Elsevier, New York, 286 pp.

Litonski, A. (1967). Laugengefahr und Sicherungsvorkehrungen in Salzbergwerken. *In* "Int. Kalisymp., 3rd 1965," Pt. 2, pp. 330–351. Deutscher Verlag Grundstoffindustrie, Leipzig.

Little-Gadow, S. (1974). Influence of NaCl on sedimentation. *Senckenb. Marit.* **6**(2), 129–133.

Liu, C., and Lindsay, W. T., Jr. (1972). Thermodynamics of sodium chloride at high temperatures. *J. Solution Chem.* **1**(1), 45–69.

Liu, S. T., and Griffith, D. W. (1979). Adsorption of alpha-aminomethyl-phosphonic acids on the calcium sulfate dihydrate. *Proc. Int Symp. Oilfield Geotherm. Chem.*, pp. 15–22. Am. Inst. Mech. Eng., Houston, Texas.

Liu, S. T., and Nancollas, G. H. (1973). Crystal growth of calcium sulfate dihydrate in the presence of additions. *J. Colloid Interface Sci.* **44**(3), 422–429.

Liu, S. T., and Nancollas, G. H. (1975). Kinetic and morphological study of the seeded growth of calcium sulfate dihydrate in the presence of additives. *J. Colloid Interface Sci.* **52**(3), 593–601.

Liu, W., and Jin, W. (1981). A study of the Xizang (Tibet) salt minerals (in Chinese). *J. Chin. Acad. Geol. Sci.* **2**(1), 75–101.

Livingstone, D. A. (1963). The chemical composition of rivers and lakes. *U.S. Geol. Surv. Prof. Paper* **440,** 1–6.

Lloyd, R. M. (1966). Oxygen isotope enrichment of sea water by evaporation. *Geochim. Cosmochim. Acta* **30**(8), 801–814.

Lobanova, V. V. (1949). Petrography of potash deposits in the eastern Carpathian region (in Russian). *Dokl. Akad. Nauk SSSR* **66**(6), 1161–1164.

Lobanova, V. V. (1953). The genesis of langbeinite in potassium-bearing formations of the Carpathian foothills (in Russian). *Dokl. Akad. Nauk SSSR* **88**(2), 145–147.

Lobanova, V. V. (1960). Petrographic characteristics of salt beds in the western Azgir upland (in Russian). *Tr. Vses. Nauchno–Issled. Proektn. Inst. Galurgii* **40,** 116–126.

Lobanova, V. V. (1963). A new mineral tatarskite (in Russian). *Zap. Vses. Mineral. O–va.* **92**(6), 697–702.

Lobanova, V. V. (1974). Petrography of Borislav potassium deposit (in Russian). *Tr. Vses. Nauchno–Issled. Proekt. Inst. Galurgii* **68,** 52–64.

Lo Cicero, G., and Catalano, R. (1976). Facies and petrography of some Messinian evaporites of the Ciminna Basin, Sicily. *Mem. Soc. Geol. Ital.* **16,** 63–81.

Loeffler, J. (1960a). Die Carnallitgesteine von Aschersleben–Schierstedt. *Freiberger Forschungsh.*, [C] **87,** 1–63.

Loeffler, J. (1960b). Primäre Sedimentationsunterschiede im Zechstein 2 und 3. *Geologie* **9**(7), 768–777.

Loeffler, J. (1962). Zur Genese der Augensalze im Zechstein der Deutschen Demokratischen Republik. *Z. Angew. Geol.* **8**(11), 583–589.

Loeffler, J. (1963). Zur Verbreitung des Kalisalzflözes Stassfurt (K-2) im östlichen Mitteleuropa. *Z. Angew. Geol.* **9**(10), 523–527.

Loewe, F. (1953). Some remarks concerning the seasonal storage of heat in the surface layers of the earth. *Notos (Pretoria),* **2,** 270–272.

Logvinenko, A. T., and Savinkina, M. A. (1967a). Effects of organic impurities on the phase composition and properties of gypsum calcination products (in Russian). *Izv. Akad. Nauk SSSR, Sib. Otd., Ser. Khim. Nauk* No. 5, pp. 88–93.

Logvinenko, A. T., and Savinkina, M. A., (1967b). Effects of organic impurities on the phase composition and properties of gypsum calcination products (in Russian). *Tr. Vses. Nauchno–Issled. Proektn. Inst. Galurgii* **68,** 52–64.

Lohse, H. H. (1963). Die Struktur des Könenits und seiner Umwandlungsprodukte. *Fortschr. Mineral.* **41**(2), 153.

Lomonosov, N. F. (1972). Origin of one type of sinuous sulfate aggregates. *Byull. Mosk. O–va. Ispyt. Prir., Otd. Geol.* **46**(6), 96–101; *Int. Geol. Rev. (Engl. Transl.)* **14**(12), 1365–1368.

Lomova, O. S. (1979). Palygorskites and sepiolites as indicators of geologic conditions (in Russian). *Tr. Geol. Inst. Akad. Nauk SSSR* **336,** 180 pp.

Longinelli, A., and Ricchiuto, T. E. (1979). Il ruolo delle acque meteoriche durante la crisi di salinità del Messiniano. *Bull. Soc. Geol. Ital.* **96**(3), 423–428.

Longinelli, A., Dongarra, G., and Ricchiuto, T. E. (1978). Geochimica isotopica di due formazioni evaporitiche della Sicilia. *Rend. Soc. Ital. Mineral. Petrol.* **34**(1), 153–160.

Longworth, L. G. (1957). The temperature dependence of the Soret coefficient of aqueous potassium chloride. *J. Phys. Chem.* **61,** 1557–1562.

Lopez, R. C., and Borrell, S. (1944). Analysis of a saline deposit from the Sahara. *An. Fiz. Quim.* **40,** 1291–1295.

Lopez-Rubio, F. B., and de la Rubia-Pacheco, J. (1950). Some physico–chemical considerations on the constitution of native alkali metal salts. *Ion* **10,** 586–588, 592.

Lotze, F. (1957). "Steinsalz und Kalisalze." 2nd ed., vol. 1, Gebr. Borntraeger, Berlin, 466 pp.

Lounsbury, R. W. (1963). Clay mineralogy of the Salina Formation, Detroit, Michigan. *In* "Sympos. on Salt'' (A. C. Bersticker, K. E., Hoekstra, and J. F. Hall, eds.), pp. 56–63. N. Ohio Geol. Soc., Cleveland, Ohio.

Love, L. G. (1958). Micro-organisms and the presence of syngenetic pyrite. *Qu. J. Geol. Soc. London* **113**(452), 429–438.

Love, L. G. (1962). Biogenic primary sulfide of the Permian Kupferschiefer and Marl Slate. *Econ. Geol.* **57**(3), 350–366.

Lucas, J., Chaabani, F., and Prevot, L. (1979). Phosphorites et evaporites: deux formations de milieux sédimentaires voisins étudiés dans la coupe du Paléogène de Foum Selja (Metlaoui, Tunisia). *Bull. Sci. Géol.* **32**(1–2), 7–19.

Lucia, F. J. (1972). Recognition of evaporite–carbonate shoreline sedimentation. In ''Recognition of Ancient Sedimentary Environments'' (J. K. Rigby and W. K. Hamblin, eds.), *Soc. Econ. Paleont. Mineral. Spec. Publ.* **16,** 160–191.

Ludewig, R., and Reuther, F. (1923). Untersuchung der durch Radiumstrahlen hervorgebrachten Farbänderung von Kristallen mit Hilfe des Ostwaldschen Farbmessungsverfahrens. *Z. Phys.* **18,** 183–198 and **26,** 45–53 (1924).

Ludlum, J. C. (1959). Rock salt, rhythmic bedding and salt-crystal impressions in the Upper Silurian limestones of West Virginia. *Southwest Geol.* **1**(1), 22–31.

Lueck, H. (1913). Beitrag zur Kenntnis des älteren Salzgebirges im Berlepsch Bergwerk bei Stassfurt, nebst Bemerkungen über die Pollenführung des Salztones. Ph.D. Dissertation, Univ. Leipzig.

Lupinovich, Yu. I., and Kislik, V. Z. (1965). On ripple marks in rock salt of the second potassium horizon of Starobinsk deposits (in Russian). *Litol. Polez. Iskop.* No. 2, pp. 174–176.

Lupinovich, Yu., and Kislik, V. Z. (1971). Characteristics of structural and textural features and the composition of rocks in the Byeolorussian potash deposits (in Russian). *In* ''Flotation of Soluble Salts'' (V. A. Glembotskiy, ed.). Izd. Nauka i Tekhnika, Minsk, pp. 20–26.

Lyman, J., and Fleming, R. H. (1940). Composition of sea water. *J. Mar. Res.* **3**(2), 134–146.

Mabbutt, J. A. (1977). ''Desert Landforms.'' MIT Press, Cambridge, Massachusetts, 340 pp.

McAdie, H. G. (1964). Effect of water vapour upon the dehydration of $CaSO_4 \cdot 2H_2O$. *Can. J. Chem.* **42**(4), 792–801.

MacArthur, C. G. (1916). Solubility of oxygen in salt solutions and the hydrates of these salts. *J. Phys. Chem.* **20,** 495–502.

McClay, K. R., and Carlyle, D. G. (1978). Mid-Proterozoic sulfate evaporites at Mount Isa mine, Queensland, Australia. *Nature (London)* **274**(5668), 240–241.

MacDonald, G. J. F. (1953). Anhydrite–gypsum equilibrium relations. *Am. J. Sci.* **251**(12), 884–898.

MacFadyen, W. A. (1950). Sandy gypsum crystals from Berbera, British Somaliland. *Geol. Mag.* **87**(6), 409–420.

McGregor, D. J. (1954). Gypsum and anhydrite deposits in southwest Indiana. *Prog. Rep. Indiana Geol. Surv.* **8,** 1–24.

McIntire, W. L. (1968). Effect of temperature on the partition of rubidium between sylvite crystals and aqueous solutions. *In* ''Saline Deposits'' (R. B. Mattox *et al.*, eds.). *Spec. Paper Geol. Soc. Am.* **88,** 505–524.

McIntosh, R. A., and Wardlaw, N. C. (1968). Barren halite bodies in the sylvinite mining zone at Esterhazy, Saskatchewan. *Can. J. Earth Sci.* **5**(5), 1221–1238.

McIver, R. D. (1973). Geochemical significance of gas and gasoline range hydrocarbons and another organic matter in a Miocene sample from site 134 Balearic Abyssal Plain. *In* Intl. Rep. Deep-Sea Drlg. Proj., vol. 13 (Ryan, W. B. F., Hsu, K. J. *et al.*, eds.), Pt. 2, pp. 813–816. U.S. Govt. Print. Off., Washington, D.C.

McKelvey, J. G., and Milne, I. H. (1962). The flow of salt solutions through compacted clay. *Clays Clay Miner.* **9,** 248–259.

McKenzie, J. A., Hsu, K. J., and Schneider, F. (1980). Movement of subsurface waters in the sabkha, Abu Dhabi, UAE, and its relation to evaporative dolomite genesis. *In* "Concepts and Models of Dolomitization" (D. H. Zenger, J. B. Dunham, and R. L. Etherington, eds.), *Spec. Publ. Soc. Econ. Paleontol. Mineral.* **28,** 11–30.

McLaughlin, D. H., Jr. (1972). Evaporite deposits of Bogota area, Cordillera Oriental, Colombia. *Bull. Am. Assoc. Pet. Geol.* **56**(11), 2240–2259.

MacNamara, E. E., and Usselman, T. (1972). Salt minerals in soil profiles and as surficial crusts and efflorescences, Coastal Enderby Land, Antarctica. *Bull. Geol. Soc. Am.* **83**(10), 3145–3150.

Macquar, J. C. (1978). Les concentrations barytiques stratiformes de l'Hettangien supérieur des Causses. Une minéralisation en milieu margino–littoral dans un environnement évaporitique. *Sci. Terre* **22**(2), 113–126.

Madigan, C. T. (1930). Lake Eyre, Australia. *Geogr. J.* **76**(3), 215–240.

Madison, R. J. (1970). Effect of a causeway on the chemistry of a brine in Great Salt Lake, Utah. *Water-Resour. Bull. (Utah Geol. Mineral. Surv.)* No. 14, pp. 1–52.

Magaritz, M., and Turner, P. (1982). Carbon cycle changes of the Zechstein Sea: Isotopic transition zone in the Marl Slate. *Nature (London)* **297**(5865), 389–390.

Maglione, G. (1974a). Présence de northupite dans une dépression interdunaire du Kanem (littoral septentrional du lac Tchad); ses implications géochimiques. *C. R. Hebd. Séances Acad. Sci. Paris,* [D] **279**(5), 377–379.

Maglione, G. (1974b). Géochimie et mécanisme de mise en place actuelle des évaporites dans le bassin tchadien. *Bull. Liaison, Assoc. Sénégal. Étude Quat. (Ouest Afr.)* No. 42–43, pp. 33–44.

Maglione, G. (1976). Géochimie des évaporites et silicates néoformés en milieu continental confinés; les dépressions interdunaires du Tchad. *Trav. Doc., Off. Rech. Sci. Tech. Outre–Mer* **50,** 1–335.

Maiklem, W. R., Bebout, D. G., and Glaister, R. F. (1969). Classification of anhydrite—a practical approach. *Bull. Can. Pet. Geol.* **17**(2), 194–233.

Makhnach, A. A. (1981). Halocatagenesis: Specific bios of superimposed postdiagenetic processes (in Russian). *Izv. Akad. Nauk SSSR, Ser. Geol.,* No. 10, 141–145.

Maley, J. (1981). Études palynologiques dans le bassin du Tchad et paléoclimatologie de l'Afrique nord–tropicale de 30000 ans à l'époque actuelle. *Trav. Doc., Off. Rech. Scient. Tech. Outre–Mer (Paris)* **129,** 604 p.

Malikova, E. (1967). "Regularities in the Distribution of Rubidium, Thallium and Bromine in Potash Salt Deposits" (in Russian). Izd. Nauka, Sib. Otd., Novosibirsk, 148 pp.

Manecki, A., and Pawlikowski, M. (1975). Mineralogical and petrographic study of selected rock salt sequences from the Wieliczka salt deposit (in Polish). *Pr. Mineral. Pol. Akad. Nauk, Oddzial Krakowie, Kom. Nauk Mineral.* **40,** 41–64.

Manegold, E., and Hofmann, R. (1930). Über die spezifische Durchlässigkeit von Kolloid–Membranen für molekular zerstreute Lösungen. *Kolloid–Z.* **50,** 207–217.

Manheim, F. T., and Horn, M. K. (1968). Composition of deeper subsurface waters along the Atlantic continental margin. *Southeast. Geol.* **9**(4), 215–236.

Mannion, L. E. (1963). Virgin Valley salt deposits, Clark County, Nevada. In "Symp. on Salt" (A. C. Bersticker, K. E. Hoekstra, and J. F. Hall, eds.), pp. 166–175. N. Ohio Geol. Soc., Cleveland, Ohio.

Mansfield, C. F. (1979). Possible biogenic origin for some sedimentary dolomite (abstr.). *Bull. Am. Assoc. Pet. Geol.* **63**(3), 490.

Mansfield, C. F. (1980). A urolith of biogenic dolomite; another clue in the dolomite mystery. *Geochim. Cosmochim. Acta* **44**(6), 829–840.

Manzhurnet, V. V. (1939). Dehydration of gypsum. *Keramika* No. 9, pp. 41–47.

Marr, U. (1957a). Zur Verteilung der Eisengehalte in Salzgesteinen des Stassfurt–Zyklus. *Geologie* **6**(1), 41–70; (2), 148–169.

Marr, U. (1957b). Über die Steinsalzleitbänke und die Unstrutbänke sowie die Möglichkeit beide Schemata zu parallelisieren. *Geologie* **6**(4), 423–444.

Marr, U. (1959). Die Bildung des Stassfurt–Flözes unter Berücksichtigung geochemischer Untersuchungen. *Freiberger Forschungsh.*, [A] **123,** 70–82.

Mars, P. (1963). Les faunes et la stratigraphie du Quaternaire Méditerranéen. *Rec. Trav. St. Mar. Endoume, Bull.* **24**(43), 61–97.

Marsal, D. (1952). Der Einfluss des Druckes auf das System $CaSO_4$–H_2O. *Beitr. Mineral. Petrogr.* **3**(4), 289–296.

Marschner, H. (1969). Hydrocalcite ($CaCO_3 \cdot H_2O$) and nesquehonite ($MgCO_3 \cdot 3H_2O$) in carbonate scales. *Science* **165**(3898), 1119–1121.

Marshall, W. L., and Slusher, R. (1966). Thermodynamics of calcium sulfate dihydrate in aqueous sodium chloride solutions 0–110°C. *J. Phys. Chem.* **70,** 4015–4027.

Marshall, W. L., and Slusher, R. (1968). Aqueous systems at high temperature. XIX. Solubility to 200°C of calcium sulfate and its hydrates in sea water and saline water concentrates, and temperature–concentration limits. *J. Chem. Eng. Data* **13,** 83–93.

Martin, H. (1963). A suggested theory for the origin and a brief description of some gypsum deposits of South West Africa. *Trans. Geol. Soc. S. W. Afr.* **66,** 345–349.

Martina, E. (1974). Un deposito salino nel Messiniano della Piana di Formia (Lazio Meridionale). *Boll. Soc. Geol. Ital.* **93**(3), 705–707.

Marzela, C. (1981). Klimazeugen aus dem jüngsten Präkambrium und Unterkambrium in Süd–Marokko. *Geol. Rundsch.* **81**(2), 473–479.

Mascarenhas Neto, M. G. N. (1961). As bacias sedimentares de Benguela e Mocamedes. *Bol. Serv. Geol. Minas Angola* No. 3, pp. 63–93.

Maslenitskiy, I. N. (1939). Hydration of natural gypsum (in Russian). *Byull. Vses. Nauchno–Issled. Proektn. Inst. Galurgii* No. 10–11, pp. 3–17.

Masson, P. H. (1955). An occurrence of gypsum in Southwest Texas. *J. Sediment. Petrol.* **25**(1), 72–77.

Matsubaya, O., Sakai, O., Torii, T., Burton, H., and Kerry, K. (1979). Antarctic saline lakes: Stable isotope ratios, chemical composition and evolution. *Geochim. Cosmochim. Acta* **43**(1), 7–26.

Matsui, S., and Fukawa, E. (1955). Hydrogen sulfide production of bacteria in a differential medium (in Japanese). *Japan. J. Bacteriol.* **10,** 761–763.

Matsuoka, K., Iwahisa, T., and Miyamoto, M. (1978). Effects of organic compounds on the formation of dolomite under hydrothermal conditions. *Rep. Res. Lab. Hydrotherm. Chem. (Kochi, Jpn)* **2**(9–15), 42–43. *Chem. Abstr.* **93,** 124,871.

Matthews, R. D., and Egleson, G. C. (1974). Origin and implications of a mid-basin potash facies in the Salina salt of Michigan. *In* "Symp on Salt, 4th." (A. H. Coogan, ed.), vol. 1, pp. 15–34. N. Ohio Geol. Soc., Cleveland, Ohio.

Maunaye, M., Anoun, H., and LeGuyadev, M. (1981). Synthèse de la dolomite. *C. R. Hebd. Séances Acad. Sci. Paris,* [2] **293**(10), 753–756.

Maus, H., Guntlach, H., and Poderfal, P. (1979). Über den Sellait (MgF_2) der Grube Clara, Oberwolf, Mittlerer Schwarzwald. *Neues Jahrb. Mineral., Abh.* **136**(1), 10–25.

Maxim, A. T. (1929). Contributiuni la explicaria fenomenului de incalzire al apelor lacurilor sarate din Transilvania. Kontribution zur Erklärung des Erwärmungsprozesses des Wassers der Salzteiche von Transilvanien. I. Lacurile Sovata, Die heissen Seen von Sovata (in Romanian with German Abstr.). *Rev. Muz. Geol. Mineral. Univ. Cluj* **3**(1), 49–86.

Maxim, A. T. (1930). Contributiuni la explicaria fenomenului de incalzire al apelor lacurilor sarate din Transilvania. Kontribution zur Erklärung des Erwärmungsprozesses des Wassers der Salzteiche von Transilvanien. II. Lacurile de la Ocna–Sibului. Die Teiche von Ocna–Sibului (in Romanian with German Abstr.). *Rev. Muz. Geol. Mineral. Univ. Cluj* **4**(1), 47–111.

Maxim, A. T. (1936). Contributiuni la explicaria fenomenului de incalzire al apelor lacurilor sarate din Transilvania. Kontribution zur Erklärung des Erwärmungsprozesses des Wassers der Salzteiche von Transilvanien. III. Lacurile Sarate dela Turda (in Romanian with German Abstr.). *Rev. Muz. Geol. Mineral. Univ. Cluj,* **6**(1–2), 209–320.

May, I., Schnepfe, M., and Naeser, C. R. (1961). Interaction of anhydrite with solutions of strontium and cesium. *U.S. Geol. Surv. Prof. Paper* **424D,** 336–338.

Mayer, L. M., Macko, S. A., Mook, W. H., and Murray, A. (1981). The distribution of bromine in coastal sediments and its use as a source indicator for organic matter. *Org. Geochem.* **3**(1–2), 37–42.

Mayrhofer, H. (1953). Beiträge zur Kenntnis des alpinen Salzgebirges. *Z. Dtsch. Geol. Ges.* **105**(4), 752–775.

Mayrhofer, H. (1955). Über ein Vorkommen von Langbeinit und Kainit in den Salzlagerstätten von Ischl. *Karinthin* **30,** 94–99.

Mazor, E., and Mantel, M. (1966). Epsomite efflorescence and the composition of shallow groundwaters in the southern Negev, Israel. *Isr. J. Earth Sci.* **15**(2), 71–75.

Medwenitsch, W. (1963). Probleme der alpinen Salzlagerstatten. *Z. Dtsch. Geol. Ges.* **115**(2–3), 863–866.

Medwenitsch, W. (1972). Geology and tectonics of Alpine evaporite formations of Austria. *In* "Geology of Saline Deposits" (G. Richter-Bernburg, ed.), *U.N.E.S.C.O. Earth Sci. Ser.* **7,** 271–272.

Meerwein, H. (1930). Über Änderungen in den Eigenschaften chemischer Verbindungen durch Komplexbildung. *Sitzungsber. Ges. Förder. Gesamt. Naturwiss. Marburg* **64,** 119–135.

Mehl, J. W., and Schmidt, C. L. A. (1937). Diffusion of certain amino acids in aqueous solution. *Publ. Physiol. Univ. Calif. (Berkeley)* **8,** 165–188.

Mehta, S. K. (1976). Some studies in the system calcium sulfate water. *Trans. Indian Ceram. Soc.* **34**(4), 65–71.

Meier, R. (1969a). Petrographic differentiation in the Stassfurt potash horizon of the German Zechstein and its causes (in Russian). *Vestn. Mosk. Univ., Ser.* 4: *Geol.* **24**(4), 71–75.

Meier, R. (1969b). Beitrag zur Geologie des Kaliflözes Stassfurt (Zechstein 2). *Geologie, Beih.* **65,** 1–99.

Meier, R. (1975). Zu einigen Sedimentgefügen der Werra-Sulfate am Osthang der Eichsfeld-Schwelle. *Z. Geol. Wiss.* **3**(10), 1333–1347.

Meier, R. (1977). Turbidite und Olisthostrome—Sedimentationsphänomene des Werra–Sulfats (Zechstein 1) am Osthang der Eichsfeld–Schwelle im Gebiet des Südharzes. *Veröff. Akad. Wiss. D.D.R., Zentralinst. Physik der Erde, Potsdam* **50,** 1–45.

Meier, R., Bartmann, W., Valyashko, G., and Kusayeva, N. S. (1971). Die Rubidiumverteilung in Hartsalz/Carnallit Übergängen des Kaliflözes Stassfurt in der D.D.R. *Z. Angew. Geol.* **17**(11), 449–453.

Meinhold, R. (1980). Die Rolle frühdiagenetischer Erdöl—und Erdgasbildung. *Z. Angew. Geol.* **26**(7), 363–369.

Meister, E. M., and Aurich, N. (1972). Geologic outline and oilfields of Sergipe Basin, Brazil. *Bull. Am. Assoc. Pet. Geol.* **56**(6), 1034–1047.

Mel'nikova, Z. M., and Moshkina, I. A. (1973). Solubility of anhydrite and gypsum in an aqueous system composed of chlorides and sulfates of sodium, magnesium and calcium at 25°C (in Russian). *Izv. Akad. Nauk SSSR Ser. Khim. Nauk* No. 2, 17–26.

Mel'nikova, Z. M., Moshkina, I. A., and Kolosov, A. S. (1971). The solubility of gypsum and anhydrite in watery solutions of $CaCl_2$ at 25° and 50°C (in Russian). *Izv. Akad. Nauk SSSR, Sib. Otd., Ser. Khim. Nauk* **14**(6), 15–20.

Mel'nikova, Z. M., Moshkina, I. A., and Kolosov, A. S. (1977). Physicochemical studies of the conditions of anhydrite and gypsum formation (in Russian). *In* "Problems of Salt Accumulation" (A. L. Yanshin and M. A. Zharkov, eds.), vol. 1, pp. 128–134. Izd. Nauka Sib. Otd., Novosibirsk.

Merritt, C. A. (1936). "Castellated" dolomite from Major County, Oklahoma. *Am. Mineral.* **21**(9), 604–607.

Metler, A. V., and Ostroff, A. G. (1967). The proximate calculation of the solubility of gypsum in natural brines from 28°C to 70°C. *Environ. Sci. Technol.* **1**(10), 815–819.

Meyer, T. A., Prutton, C. F., Lightfoot, W. J. (1949). Equilibria in saturated salt solutions. V. The quinary system $CaCl_2$–$MgCl_2$–KCl–NaCl–H_2O at 35°C. *J. Am. Chem. Soc.* **71**(4), 1236–1237.

Michel, M. (1955). Étude calorimétrique de la déshydratation du gypse. *C. R. Hebd. Séances Acad. Sci. Paris* **241**(21), 1462–1464.

Mikhailov, O. V. (1965). Relief of the Mediterranean sea floor (in Russian). *In* "Basic Features of the Geologic Structure of the Hydrogeologic Regime and the Biology of the Mediterranean Sea" (L. M. Fomin, ed.), pp. 10–19. Izd. Nauka, Moscow.

Mileshina, A. G., and Komissarova, I. N. (1972). Change in the composition of crude oils moving through halite (in Russian). *Geol. Nefti Gaza* **16**(6), 51–55.

Miller, A. D., and Fisher, E. I. (1974). Dissolution of gold in ferric chloride solutions and its deposition on pyrite (laboratory model) (in Russian). *Geokhimiya* No. 3, pp. 411–417.

Millot, G. (1970). "Geology of Clays." Springer-Verlag, New York, 429 pp.

Milne, J. K., Saunders, M. J., and Woods, P. J. E. (1977). Iron–boracite from the English Zechstein. *Mineral. Mag.*, **41**(319), 404–405.

Milone, M., and Ferrero, F. (1947). The relation between surface tension and crystalline habit. *Gazz. Chim. Ital.* **77,** 348–352. *Chem. Abstr.* **42,** 2,832.

Miropol'skiy, L. M. (1935). Characteristics of the mineral complex and the fundamental geochemical processes in Permian deposits near the town of Syukeyev in the Tatarian republic (in Russian). *Uch. Zap., Kaz. Gos. Univ.* **95**(3–4), 5–6.

Miropol'skiy, L. M. (1941). Microcrystalline dolomites, their origin and pseudomorphoses of anhydrite, and gypsum after dolomite rhombohedra in the Lower Permian deposits of Tataria (in Russian). *Dokl. Akad. Nauk SSSR* **32**(8), 572–574.

Miropol'skiy, L. M., and Kovyazin, N. M. (1950). Gypsum in Hauterivian sediments of Tatariya (in Russian). *Dokl. Akad. Nauk SSSR* **70,** 481–484.

Mishin, G. T. (1971). Nature of radioactivity of potassic salts in the Upper Kama and Starobinsk deposits (in Russian). *Tr. Vses. Nauchno–Issled. Proektn. Inst. Galurgii* **55,** 10–15.

Mizutaní, S. (1977). Progressive ordering of cristobalite silica in the early stages of diagenesis. *Contrib. Mineral. Petrol.* **61**(2), 129–140.

Mizutani, Y. (1962). Volcanic sublimates and encrustations from Showashinzan. *J. Earth Sci., Nagoya Univ.* **10,** 149–164.

Moberg, E. G., Greenberg, D. M., Revelle, R., and Allen, E. C. (1934). The buffer mechanism of sea water. *Bull. Scripps Inst. Oceanogr., Tech. Ser.* **3,** 231–278.

Moetzing, R. (1968). Auftreten und Bedeutung von Pseudomorphosen nach Langbeinit im Kalisalzlager "Flöz Stassfurt". *Bergakademie* **20**(6), 324–328; (12), 714–717.

Moetzing, R. (1978). D'Ansit in Zechstein-2. *Z. Angew. Geol.* **24**(12), 513–518.

Moiola, R. J., and Glover, E. D. (1965). Recent anhydrite from Clayton Playa, Nevada. *Am. Mineral.* **50**(11–12), 2063–2069.

Mokhnach, M. F. (1978). The mechanism of formation of sylvite layers in the Upper Kama deposit of potash salts (in Russian). *In* "Studies in Mathematical Geology" (M. A. Romanova and N. A. Sapogov, eds.), pp. 55–62. Izd. Nauka, Leningrad.

Montoriol-Pouss, J. (1965). A new type of gypsum in the lava tubes of Lanzarote (Canary Islands). *Bol. R. Soc. Esp. Hist. Nat., Sect. Geol.* **63**(1), 77–85. *Chem. Abstr.* **64,** 13,948.

Monty, C. (1980). Considerations on Cyanophyceae in hypersaline settings. *Bull. Cent. Rech. Explor.–Prod. Elf–Aquitaine* **4**(1), 371.

Monzhalei, N. D. (1938). The solubility of aluminum hydroxide in normal potassium chloride solutions as affected by pH (in Russian). *Tr. Nauchno–Issled. Inst. Udobren. Agrotekhn. i Agropochvovedeniya,* Otd. *Leningrad* **2,** 307–316.

Moody, J. D. (1959). Discussion of relationship of primary evaporites to oil accumulation (by L. L. Sloss). *Proc. World Pet. Congr., 5th., Sect. 1,* pp. 134–157.

Moore, G. W. (1960). Origin and chemical composition of evaporite deposits. Ph.D. Dissertation, Yale Univ., New Haven, Connecticut, 170 pp.

Moore, G. W. (1971). Geologic significance of the minor element composition of marine salt deposits. *Econ. Geol.* **66**(1), 187–191.

Moore, J., and Runkles, J. R. (1968). The evaporation from brine solutions under controlled laboratory conditions. *Rep. Tex. Water Dev. Board* **77,** 1–69.

Moore, L. R. (1969). Geomicrobiology and geomicrobiological attac on sedimentary organic matter. *In* "Organic Geochemistry" (G. Eglinton and M. T. J. Murphy, eds.), Pt. 2, pp. 265–303. Springer-Verlag, Berlin.

Morachevskiy, Yu. V. (1938). Gases in salt deposits (in Russian). *Byull. Vses. Nauchno–Issled. Proektn. Inst. Galurgii* No. 5, 1–14.

Morachevskiy, Yu. V. (1940). Geochemical investigations of the Upper Kama salt deposits (in Russian). *Uch. Zap. Leningr. Gos. Univ., Ser. Khim. Nauk* **5**(54), 175–189.

Morachevskiy, Yu. V., Samartseva, A. G., and Cherepennikov, A. A. (1937). Gas occlusions in the potash salt deposits of the Upper Kama district (in Russian). *Kalii* **6**(7), 24–31.

Morita, R. Y. (1980). Calcite precipitation by marine bacteria. *Geomicrobiol. J.* **2**(1), 63–82.

Mormil', S. I. (1974). Geochemistry of cesium in the halogene formations of the western Ural region. *Litol. Polezn. Iskop.,* No. 5, pp. 62–71; *Lithol. Mineral. Resour. (Engl. Transl.)* **9**(5), 561–568.

Morozova, A. I., and Firsova, G. N. (1956). The solubility of calcium sulfate in salt mixtures that are ordinarily present in natural waters. *Nauchn. Tr. Novocherk. Politekh. Inst.* **27,** 151–165. *Chem. Abstr.* **52,** 16,841.

Mortensen, H. (1933). Die Salzsprengung und ihre Bedeutung fur die regionale klimatische Gliederung der Wüste. *Petermann's Geogr. Mitt.,* pp. 130–135.

Mossman, D. J., Delabio, R. N., Mackintosh, A. D. (1982). Mineralogy of clay marker seams in some Saskatchewan potash mines. *Can. J. Earth Sci.* **19**(11), 2126–2140.

Mossop, G. D., and Shearman, D. J. (1973). Origin of secondary gypsum rocks. *Trans. Inst. Min. Metall., Sect. B* **82,** 147–154.

Muegge, O. (1913). Über die Minerale im Rückstand des roten Carnallits von Stassfurt und des schwarzen Carnallits von der Hildesia. *Kali (Halle)* **7**(1), 1–3.

Mueller, A., and Schwartz, W. (1953). Über das Vorkommen von Mikroorganismen in Salzlagerstätten. *Z. Dtsch. Geol. Ges.* **105**(4), 789–802.

Mueller, George (1958). A theory of concentration of nitrates in northern Chilean salt deposits by means of capillary migration (in Sapnish). *An. Fac. Ing., Univ. Concepcion* **7,** 41–45. *Chem. Abstr.* **54,** 18,222.

Mueller, George (1959). Theory explaining the high nitrate concentration in the mines of northern Chile (in Spanish). *Bol. Soc. Chilena Quim.* **9,** 45–48.

Mueller, Gerhard (1964). Ein Beitrag zur Geochemie des Strontiums in Ca–Sulfat–Gesteinen. Ph.D. Dissertation, Univ. Saarbrücken, 166 pp.

Mueller, Gerhard (1969). Zum Vorkommen von Mirabilit (Glaubersalz) und Ulexit bei Ihn, Kreis Saarlouis. *Neues Jahrb. Mineral., Abh.* **110**(2), 188–198.

Mueller, German (1962). Zur Geochemie des Strontiums in ozeanischen Evaporiten unter besonderer Berücksichtigung der sedimentären Cölestinlagerstätte von Hemmelte-West (Süd-Oldenburg). *Geologie, Beih.* **11,** 1–90.

Mueller, German (1968). Genetic histories of nitrate deposits from Antarctica and Chile. *Nature (London)* **219**(5159), 1131–1134.

Mueller, German (1971). Aragonite: Inorganic precipitation in a freshwater lake. *Nature (London) Phys. Sci.* **229**(1), 18.

Mueller, German, and Irion, G. (1969a). "Salt biscuits"—a special growth structure of NaCl in salt sediments of the Tuz Gölü ("Salt Lake"), Turkey. *J. Sediment. Petrol.* **39**(4), 1604–1607.

Mueller, German, and Irion, G. (1969b). Subaerial cementation and subsequent dolomitization of lacustrine carbonate muds and sands from Paleo-Tuz Gölü ("Salt Lake"), Turkey. *Sedimentology* **12**(3–4), 193–204.

Mueller, J., and Fabricius, F. H. (1978). Luneburgite [$Mg_3(PO_4)_2 \cdot B_2O(OH)_4 \cdot 6H_2O$] in Upper Miocene sediments of the eastern Mediterranean Sea. *In* Intl. Rep. Deep–Sea Drlg. Proj., vol. 42A (K. J. Hsu, L. Montadert *et al.*, eds.), pp. 661–664. U.S. Govt. Print. Off., Washington, D.C.

Mueller, P., and Heymel, W. (1956). Die Bestimmung der Gaskonzentration in den gas–enthaltenden Salzen der Kaligruben im Südharz und im Werrabezirk. *Bergbautechnik* **6,** 313–319.

Muir, J. L. (1934). Anhydrite–gypsum problem of Blaine Formation, Oklahoma. *Bull. Am. Assoc. Pet. Geol.* **18**(10), 1297–1312.

Muir, M., Lock, D., and von der Borch, C. (1980). The Coorong model for penecontemporaneous dolomite formation in the Middle Proterozoic McArthur Group, Northern Territory, Australia. *In* "Concepts and Models of Dolomitization" (D. H. Zenger, J. B. Dunham, and R. L. Etherington, eds.), *Spec. Publ. Soc. Econ. Paleontol. Mineral.* **28,** 51–67.

Mumindzhanova, M. A., and Osichkina, R. G. (1982). Experimental determination of coefficients of rubidium and cesium distribution between sylvite crystals and a solution in potassium chloride–water and sodium chloride–potassium chloride–water systems (in Russian). *Uzb. Khim. Zh.* No. 1, pp. 11–15.

Murakami, K., Shimamura, Y., and Tanaka, H. (1957). Dehydration temperature of gypsum (in Japanese). *Gypsum Lime (Jpn)* **1,** 1522–1528.

Murat, M. (1971). Sur l'effet exothermique associé à la transformation de l'anhydrite hexagonale en anhydrite orthorhombique. *J. Thermal Anal.* **3**(3), 259–264.

Murat, M. (1977). Structure, cristallochimie et réactivité des sulfates de calcium. *In* "Sulfates de calcium et matériaux derivés (Calcium Sulfates and Derived Materials)" (M. Murat and M. Foucault eds.), pp. 59–172. C. R. Colloq. Int. Saint-Remy-les-Chevreuse, Réunion Int. Lab. d'Essais Rech. Matér. Constr., Paris.

Murat, M., and Comel, C. (1970). Particularités présentées par la déshydratation du gypse en formation de l'origine du minéral. *C. R. Hebd. Séances Acad. Sci. Paris,* [D] **270**(15), 1849–1852.

Murata, K. J., and Smith, R. L. (1946). Manganese and lead as coactivators of red fluorescence in halite. *Am. Mineral.* **31**(11–12), 527–538.

Murin, A., and Popov, D. (1953). Additivity of Soret coefficients (in Russian). *Dokl. Akad. Nauk SSSR* **88,** 879–882.

Murray, R. C. (1964). Origin and diagenesis of gypsum and anhydrite. *J. Sediment. Petrol.* **34**(3), 512–523.

Murray, R. C. (1969). Hydrology of South Bonaire, N. A.—a rock-selective dolomitization model. *J. Sediment. Petrol.* **39**(3), 1007–1013.

Murzayev, P. M. (1941). Saltspar and its origin (in Russian). *Dokl. Akad. Nauk SSSR* **33,** 306–307.

Muza, J. P., and Wise, S. W., Jr. (1983). An authigenic gypsum, pyrite, and glauconite association in a Miocene deep sea biogenic ooze from the Falkland Plateau, Southwest Atlantic Ocean. *In* Intl. Rep. Deep-Sea Drlg. Proj., vol. 71 (W. J. Ludwig, V. A. Krasheninikov *et al.*, eds.), Pt. 1, 361–375. U.S. Govt. Print. Off., Washington, D.C.

Myagkov, V. F., and Burmistrov, D. V. (1964). Bromine distribution in carnallite of the Upper Kama deposit (in Russian). *Geokhimiya,* No. 7, 684–686.

Nabiev, M. N., and Osichkina, R. G. (1965). "Potash Salts of Tyubegatan" (in Russian). Izd. Nauka Uzb. SSR, Tashkent, 126 pp.

Nachsel, G. (1966). Quarz als Faziesindikator. *Z. Angew. Geologie* **12**(6), 322–326.

Nacken, R., and Fill, K. (1931). Über die Umwandlung von Gips–Halbhydrat zu Anhydrit in feuchter Luft bei normalem Luftdruck. *Tonind.-Ztg.* **55,** 1194–1196, 1222–1224.

Nadler, A., and Magaritz, M. (1980). Studies of marine solution basins—isotopical and compositional changes during evaporation. *In* "Hypersaline Brines and Evaporitic Environments" (A. Nissenbaum, ed.), pp. 115–130. Elsevier, Amsterdam.

Nagai, S., Sekiya, M., and Ikeda, T. (1952). Effects of amino acids on the crystallization of gypsum (in Japanese). *Gypsum (Jpn)* **1,** 292–300.

Nakayama, F. S. (1971). Calcium complexing and the enhanced solubility of gypsum in concentrated sodium solutions. *Proc. Soil Sci. Am.* **35,** 881–883.

Namyslowski, B. (1913). Über unbekannte halophile Mikroorganismen aus dem Innern des Salzbergwerkes Wieliczka. *Bull. Acad. Sci. Krakow,* [B], No. 3 and 4; *Kali (Halle)* **7** (20), 520.

Narkelyun, L. F., and Bezrudykh, Yu. P. (1966). Ore mineralization in the Lena cupriferous sandstones. *Litol. Polezn. Iskop.* No. 6, pp. 88–92; *Lithol. Mineral. Resour.* **1**(6), 752–755.

Nash, C. R. (1972). Primary anhydrite in Precambrian gneisses from the Swakopmund district, Southwest Africa. *Contrib. Mineral. Petrol.* **36**(1), 27–32.

Nazina, T. N., Rozanova, E. P., and Kalininskaya, T. A. (1979). Fixation of molecular nitrogen by sulfate reducing bacteria from oilfields (in Russian). *Mikrobiologiya* **48**(1), 133–136.

Neal, J. T. (1975). Playa surface features as indicators of environment. *In* "Playas and Dried Lakes, Occurrence and Development" (J. T. Neal, ed.), pp. 363–388. Dowden, Hutchinson & Ross, Inc., Stroudsburg, Pennsylvania.

Neal, J. T., and Motts, W. S. (1967). Recent geomorphic changes in playas of the southwestern United States. *J. Geol.* **75**(5), 511–525.

Neev, D. (1979). Deep-water gypsum deposits as indicated by the Neogene geological history of the central coastal plain of Israel. *Sediment. Geol.* **23**(1–4), 127–136.

Neev, D., and Emery, K. O. (1967). The Dead Sea, depositional processes and environments of evaporites. *Bull. Isr. Geol. Surv.* **41,** 147 pp.

Neev, D., Almagor, G., Arad, A., Ginsburg, A., and Hall, J. K. (1976). The geology of the southeastern Mediterranean Sea. *Bull. Isr. Geol. Surv.* **68,** 51 pp.

Nesmelova, Z. N. (1959). Gasses in potassium salts of the Bereznikovsk mine. *Tr. Vses. Nauchno-Issled. Proektn. Inst. Galurgii* **35,** 206–243.

Nesmelova, Z. N. (1961). Geochemical features of gases in salt-bearing rocks (in Russian). *Tr. Vses. Nauchno–Issled. Neft. Geologorazved. Inst.* **174,** 177–185.

Nesteroff, W. D. (1973a). Mineralogy, petrography, distribution and origin of the Messinian Mediterranean evaporites. *In* Intl. Rep. Deep-Sea Drlg. Proj., vol. 13 (W. B. F. Ryan, K. J. Hsu *et al.*, eds.), Pt. 2, pp. 673–694. Govt. Print. Off., Washington, D.C.

Nesteroff, W. D. (1973b). Un modèle pour les évaporites messiniennes en Méditerranée des bassins peu profonds avec dépôt d'évoporites lagunaires. *In* "Messinian Events in the Mediterranean" (C. W. Drooger *et al.*, eds.), pp. 68–81. North–Holland, Amsterdam.

Newell, N. D., Rigby, J. K., Fischer, A., Whiteman, A. J., Hickox, J. E., and Bradley, J. S. (1953). "The Permian Reef Complex of the Guadelupe Mountains Region, Texas and New Mexico: A study in Paleoecology." Freeman, San Francisco, California, 236 pp.

Newell, R. E., Herman, G. F., Gould-Stewart, S., and Tanaka, M. (1975). Decreased global rainfall during the past Ice Age. *Nature (London)* **253**(5486), 33–34.

Newman, E. S., and Wells, L. S. (1938). Heats of hydration and transition of calcium sulfate. *J. Res. Natl. Bur. Stand. (U.S.)* **20,** 825–836.

Nguyen, T. C., Lobanova, V. V., and Frank-Kamenetskiy, V. A. (1973). First find of vanthoffite in evaporite formations in the eastern Ciscarpathians (in Russian). *Zap. Vses. Mineral. O–va.* **102**(2), 192–193.

Nicholson, S. E., and Flohn, H. (1980). African environmental and climatic changes and the general atmospheric circulation in late Pleistocene and Holocene. *Clim. Change* **2**(4), 313–348.

Nicol. H. (1942). Microbiology of petroleum products. *Petroleum (London)* **5,** 205.

Nicol, W. (1828). Observations on the fluids contained in crystallized minerals. *Edinburgh New Philos. J.* No. 5, pp. 94–96.

Niemann, H. (1960). Untersuchungen am grauen Salzton der Grube Königshall-Hindenburg, Reyershausen bei Göttingen. *Beitr. Mineral. Petrogr.* **7,** 137–165.

Niggli, P. (1952). Gesteine und Minerallagerstätten, v. 2. Exogene Gesteine und Minerallagerstätten. Birkhäuser, Basel, 557 pp.

Nikolayev, V. I., and Fradkina, Kh. B. (1949). Origin and industrial use of sulfate deposits in the Aral region (in Russian). *Izv. Sekt. Fiz. Khim. Anal., Inst. Obshch. Neorg. Khim. Akad. Nauk SSSR* **17,** 383–395.

Nikolayev, V. I., and Ravich, M. I. (1931). Singular folds in the ternary system: $Na_2O–HBr–H_2O$ (in Russian). *Zh. Obhch. Khim.* **1,** 785–791.

Nikolić, D., and Poharć, V. (1977). Epsomite from the gypsum ore body of Lipnica in Serbia (in Serbian). *Zap. Srp. Geol. Druš.,* (1975–76), pp. 51–56.

Nikol'skaya, Yu. P., and Gordeyeva, G. I. (1967). Synthesis of dolomite (in Russian). *Izv. Sib. Otd. Akad. Nauk SSSR, Ser. Khim, Nauk* No. 5, pp. 80–87.

Nilsson, L. Y. (1968). Short-time variations of the ground water and its reasons. *In* "Ground Water Problems" (E. Eriksson, Y. Gustafsson, and K. Nilsson, eds.), pp. 57–72. Pergamon Press, New York.

Niskiewicz, J. (1980). Metasomatic phenomena in Zechstein copper ores of Lower Silesia (in Polish). *Geol. Sudetica* **15**(2), 7–82.

Niyazov, A. N., and Dzhumayev, R. (1983). Some results of a study of organic compounds in brines in Kara Bogaz Gol Bay (in Russian). *Izv. Akad. Nauk Turkm. SSR, Ser- Fiz.-Tekh., Khim. Geol. Nauk* No. 2, pp. 72–75.

Noll, W. (1934). Zur Genesis porphyrischer Struktur in Gipsgesteinen. *Chem. Erde* **9**(1), 1–21.

Nordenskjoeld, O. (1914). Einige Züge der physischen Geographie und der Entwicklungsgeschichte Süd-Grönlands. *Geogr. Z.* **20**(9–10), 505–524.

Norman, R. J. (1935). Interaction of amino acids and salts. *J. Biol. Chem.* **111,** 479–499.

Norton, F. J., and Johnston, J. (1926). Transition temperature and solubility of sodium sulfate in the presence of chloride or sodium bromide. *Am. J. Sci.* **12,** 477–483.

Nurmi, R. D., and Friedman, G. M. (1977). Sedimentology and depositional environment of basin-center evaporites, lower Salina Group (Upper Silurian), Michigan Basin. *In* "Reefs and Evaporites" (J. H. Fisher, ed.). *Am. Assoc. Pet. Geol. Stud. Geol.* **5,** 23–52.

Nury, D. (1968). De l'influence marine dans les dépôts à gypse d'Aix. *C. R. Hebd. Séances Acad. Sci. Paris,* [D] **267**(19), 1489–1491.

Ochsenius, K. (1876). Über die Salzbildung der Egelnschen Mulde. *Z. Dtsch. Geol. Ges.* **28**(4), 654–667.

Ochsenius, K. (1877). "Die Bildung der Steinsalzlager und ihrer Mutterlaugensalze unter spezieller Berücksichtigung der Flöze von Douglashall in der Egeln'schen Mulde." Halle/Saale, C.E.M. Pfeffer Verlag, 172 pp.

Ochsenius, K. (1878). Beiträge zur Erklärung der Bildung von Steinsalzlagern und ihrer Mutterlaugensalze. *Nova Acta, Kais. Leopoldisch-Carolinisch. Dtsch. Akad. Naturforsch.* **40**(7), 121–166.

Ochsenius, K. (1888). On the formation of rock salt beds and mother liquor salts. *Proc. Philadelphia Acad. Sci.* Pt. 2, pp. 181–187.

Ochsenius, K. (1892). Die Bildung von Kohlenflözen. *Z. Dtsch. Geol. Ges.* **44,** 84–98.

Ochsenius, K. (1893). Bedeutung des orographischen Elementes "Barre" in Hinsicht auf Bildungen und Veränderungen von Lagerstätten und Gesteinen. *Z. Prakt. Geol.* **1**(5), 189–201; (6), 217–233.

Ochsenius, K. (1897). Unsere Mutterlaugen—(Kali–)Salze. *"Industrie," Fachztg. Kohlen– Kalibergbau, Suppl.,* 7 pp.; ref. also in *Z. Prakt. Geol.* (1898), p. 112.

Ochsenius, K. (1898). Zur Erdölbildung. *Prometheus* **9**(421), 69–71.

Oelsner, O. (1959). Bemerkungen zur Herkunft der Metalle im Kupferschiefer. *Freiberg. Forschungsh.,* [C] **58,** 106–113.

Oesterle, F. P., and Lippolt, A. J. (1975a). Isotopische Datierung der Langbeinit–Bildung in der Kalisalzlagerstätte des Fulda–Beckens. *Kali Steinsalz* **6**(11), 391–398.

Oesterle, F. P., and Lippolt, A. J. (1975b). Alter der Langbeinit–Bildung in den Kalilagern des Fulda– und Werra–Beckens. *Fortschr. Mineral., Beih.* **53**(1), 62.

Oesterlind, S. (1950). Inorganic carbon sources of green algae. *Physiol. Plant.* **3,** 353–360.

Ogienko, V. S. (1959). Distribution of bromine in the rock salt of the Angara–Lena salt basin and the possibility of finding potassium salts. *Geokhimiya* No. 8, pp. 721–726; *Geochem. Int. (Engl. Transl.)* pp. 893–900.

Ogniben, L. (1957). Petrografia della Serie Solfifera siciliana e condizioni geologische relative. *Mem. Descritt., Carta Geol. d'Italia,* **23,** 275 p.

Ogorelec, B., Mišić, M., Serčelj, A., Cimerman, F., Faganeli, J., and Stegnar, P. (1981). Sediments of the Sečovlje salt marsh (in Slovenian). *Geologija* **24**(2), 179–216.

Ohde, S., and Kitano, Y. (1981). Dissolution and transformation of natural magnesian calcite in brine. *Jpn. Geochem. J.* **15**(1), 39–45.

Oka, S. (1938). Van der Waal forces and the Debye–Hueckel theory of strong electrolytes. *Proc. Phys.–Math. Soc. Jpn.* **20,** 11–14.

Oka, S. (1944). Composition of bittern produced in solar salt garden. II. Effect of diluting bittern. *J. Soc. Chem. Ind. Jpn.* **47,** 317–319.

Oka, S., and Hishikari, S. (1944). Change in composition of sea water on isothermal evaporation at 25°C. *J. Soc. Chem. Ind. (Jpn.)* **47,** 314–316.

Oka, S., and Inagaki, H. (1942a). The composition of sea water bittern of solar salt garden. *J. Soc. Chem. Ind. (Jpn.)* **45,** 677–682.

Oka, S., and Inagaki, H. (1942b). The composition of the sea water bittern and the heterogeneous equilibrium of sea salt solution. *J. Soc. Chem. Ind. (Jpn)* **45,** 791–794.

Oka, S., and Kaneko, K. (1942). The removal of sulfate ion from the sea water bittern with calcium chloride. *J. Soc. Chem. Ind. Jpn.* **45,** 288–291.

Okada, A., and Shima, M. (1973). Authigenic minerals in the sediments of the Sea of Japan. Framboidal pyrite, glauconite and gypsum (in Japanese). *Sci. Rep. Inst. Phys. Chem. Res. (Jpn.)* **67**(3), 148–154.

Okamura, T., and Matsushita, T. (1967). Fluorite in common salt (in Japanese). *Food Nutrition (Jpn.)* **19**(5), 331–333.

Olausson, E. (1961). Studies of deep-sea cores. *Rep. Swedish Deep-Sea Exped. 1948–1949* **3,** pp. 285–334; **4,** pp. 335–391.

Olszewski, W. (1973). Carnallite decomposition in the presence of urea (in Polish). *Pr. Nauk Inst. Technol. Nieorg. Nawozow Miner. Politech. Wroclaw* No. 5, pp. 89–98.

Omori, K. (1960). Consideration of the origin of some saline minerals (concepts of astronomical geology I). *Sci. Rep. Tohoku Univ.,* [3] **6,** 405–407.

Ongley, E. D., Bynoe, M. C., and Percival, J. B. (1981). Physical and geochemical characteristics of suspended solids, Wilton Creek, Ontario. *Can. J. Earth Sci.* **18**(8), 1365–1379.

Ordonez, S., Lopez Aguayo, F., Garcia del Cura, M. A. (1976). Conceptual models of intramontane basins of the Tajo, Duero and Jucar Rivers (Spain). *Int. Geol. Congr., 25th,* 1976, *Sydney, N.S.W., Abstr.,* v. 1, [6A], pp. 275–276.

Ordonneau, J. (1950). Hygroscopicity of salts: Potassium chloride. *Ann. Mines* **139,** 35–63. *Chem. Abstr.* **44,** 6,587.

Orekhova, A. I., Savinkova, E. I., and Pavlova, N. V. (1974). Thermal decomposition of bischofite containing sodium chloride (in Russian). *Izv. Vyssh. Uchebn. Zaved., Tsvetn. Metall.* **17**(6), 56–59.

Orekhova, A. I., Savinkova, E. I., and Pavlova, N. V. (1975). Determination of the equilibrium pressure of water vapor over magnesium chloride crystal hydrates containing sodium chloride (in Russian). *Izv. Vyssh. Uchebn. Zaved., Tsvetn. Metall.* **18**(4) 53–57.

Orr, W. L., and Gaines, A. G., Jr. (1974). Observations on rate of sulphate reduction and organic matter oxidation in the bottom waters of an estuarine basin: The upper basin of the Pettaquamscutt River (Rhode Island). *Proc. Int. Meet. Adv. Org. Geochem., 6th (1973),* p. 791–812.

Orti Cabo, F., and Pueyo Mur, J. J. (1977). Asociacion halita bandeada anhidrita nodular del yacimiento de Remolinos, Zaragoza (sector central de la Cuenca del Ebro), nota petrogenetica. *Publ. Inst. Invest. Geol., Diputacion Barcelona* **32,** 167–202.

Orti Cabo, F., and Shearman, D. J. (1977). Estructuras y fabricas deposicionales en las evaporitas del mioceno superior (Messiniense) de San Miguel de Salinas (Alicante, España). *Publ. Inst. Invest. Geol., Diputacion Barcelona* **32,** 5–54.

Osaka, J. (1965). Volcanic sublimates (in Japanese). *Kazan (Jpn.)* **10,** 205–213.

Oshakpayev, T. A. (1977). On the conditions of formation of evaporite deposits in the Caspian Lowlands (in Russian). *Tr. Inst. Geol. Akad. Nauk Kaz. SSR* **36,** 54–61.

Osichkina, R. G. (1978a). Geochemical features and development conditions of salt deposits in Upper Jurassic Halogen Formation of South Central Asia. *Litol. Polezn. Iskop.* No. 4, pp. 102–111. *Lithol. Mineral. Resour. (Engl. Transl.)* **13**(4), 467–474.

Osichkina, R. G. (1978b). Rubidium and cesium in salt deposits of an Upper Jurassic evaporite formation in southern Central Asia as an index of the conditions of their formation (in Russian). *Geokhimiya* No. 6, 921–925; *Geochem. Int. (Engl. Transl.)* **15**(3), 168–173.

Osichkina, R. G. (1978c). Characteristics of bromine geochemistry in the Upper Jurassic halide association of south-central Asia. *Geokhimiya* No. 10, pp. 1530–1536; *Geochem. Int. (Engl. Transl.)* **15**(5), 157–162.

Osichkina, R. G. (1980). Physicochemical characteristics of evaporite sedimentation in an Upper Jurassic basin in the southern part of Central Asia (in Russian). *Uzb. Khim. Zh.* No. 3, pp. 3–9.

Ostroff, A. G. (1964). Conversion of gypsum to anhydrite in aqueous salt solution. *Geochim. Cosmochim. Acta* **28**(9), 1363–1372.

Ostroff, A. G., and Metler, A. V. (1966). Solubility of calcium sulfate dihydrate in the system $NaCl–MgCl_2–H_2O$ from 28°C to 70°C. *J. Chem. Eng. Data* **11**(3), 346–350.

Otsuki, A., and Hanya, T. (1972). Production of dissolved organic matter from dead green algal cells. I. Aerobic microbial decomposition. *Limnol. Oceanogr.* **17**(2), 248–257.

Ottemann, J. (1950). Über Lösung und Hydratation des Anhydrits; ein Beitrag zu Grundlagenforschung des Anhydrits als Bindebaustoff. *Geol. Landesamt Berlin, Abh.* **219,** 1–16.

Ovcharenko, F. D., and Tretinnik, V. Yu. (1962). The effect of electrolytes on the structural–mechanical and filtration properties of aqueous suspensions of clays (in Russian). *Dopov. Akad. Nauk Ukr. SSR,* pp. 1210–1212.

Pagnier, H. (1978). Depth of deposition of Messinian selenitic gypsum in the basin of Sorbas (S. E. Spain). *Mem. Soc. Geol. Ital.* **16,** 363–367.

Palacas, P. G., Swanson, V. E., Lova, A. H. (1968). Organic geochemistry of recent sediments in the Choctawhatchee Bay area, Florida. A preliminary report. *U.S. Geol. Surv. Prof. Paper* **600-C,** 97–106.

Paliwal, B. S. (1977). The source of salt in Rajasthan—an investigation of the salt lake of Didwara. *Ann. Arid Zone* **6**(2), 221–230.

Pamić, J. (1955). An occurrence of gypsum and anhydrite in the upper reaches of the Rama and Doljanka Rivers. *Geol. Glasn. (Sarajevo),* No. 2, pp. 187–197.

Pannekoek, A. J. (1965). Shallow-water and deep-water evaporite deposition. *Am. J. Sci.* **263**(3), 284–285.

Panno, S. V., Harbottle, G., Sayre, E. V., and Hood, W. C. (1981). Genetic implications of halogen host rock aureoles surrounding a Mississippi Valley–type ore deposit. *Geol. Soc. Am., Abstr. Prog.* **13**(7), 525.

Parea, G. C., and Ricci-Lucchi, F. (1972). Resedimented evaporites in the Periadriatic Trough (Upper Miocene, Italy). *Isr. J. Earth Sci.* **21**(3–4), 125–141.

Parfenov, S. I. (1967). Special features in anhydrite gypsification. *Litol. Polezn. Iskop.* No. 3, 117–127; *Lithol. Mineral. Resour. (Engl. Transl.)* **1**(3), 371–379.

Parkin, D. W. (1976). Solar constant during glaciation. *Nature (London)* **260**(5546), 28–31.

Parkin, D. W., and Shackleton, N. J. (1973). Trade Wind and temperature correlation down a deep-sea core off the Saharan coast. *Nature (London)* **245**(5426), 455–457.

Parmenter, C., and Folger, D. W. (1974). Eolian biogenic detritus in deep-sea sediments: A possible index of equatorial ice age aridity. *Science* **185**(4152), 695–698.

Parsons, A. L. (1927). Dehydration of gypsum. *Univ. Toronto Stud. Geol., Contrib. Can. Mineral.* No. 24, pp. 24–27.

Partridge, E. P., and White, A. H. (1929). Solubility of calcium sulfate from 0–200°C. *J. Am. Chem. Soc.* **51,** 360–370.

Pastukhova, M. V. (1965). Authigenic minerals in chemogenic-terrigenic rocks of the Tuztag salt sequence (in Russian). *Litol. Polezn. Iskop.* No. 1, pp. 31–52.

Pasztor, A. J., and Snover, J. S. (1983). How to treat metal contamination from heavy clear brines. *Oil Gas J.* **81**(29), 141–146.

Patel, K. P., and Seshadri, K. (1975). Phase transformation of carnallite to kainite. *Salt Res. Ind. (India)* **11**(1), 47.

Patterson, G. W. (1967). The effect of culture conditions on the hydrocarbon content of *Chlorella vulgaris. J. Phycol.* **3**(1), 22–23.

Patterson, R. J. (1972). Relation between the hydrology, fluid chemistry, and diagenetic mineral formation in coastal areas of the Persian Gulf. *Proc. Symp. Fundam. Transp. Phenom. Porous Media, 2nd 1972* vol. 2, pp. 683–698.

Patterson, R. J., and Kinsman, D. J. J. (1977). Marine and continental sources in a Persian Gulf coastal sabkha. *In* "Reefs and Related Carbonates; Ecology and Sedimentology" (S. H. Frost, M. P. Weiss, and J. B. Saunders, eds.), *Am. Assoc. Pet. Geol. Stud. Geol.* **4,** 381–397.

Patterson, R. J., and Kinsman, D. J. J. (1982). Formation of diagenetic dolomite in coastal sabkha along Arabian (Persian) Gulf. *Bull. Am. Assoc. Pet. Geol.* **66**(1), 28–43.

Pauca, M. (1968). Beiträge zur Kenntnis der miozänen Salzlagerstätten Rumäniens. *Geol. Rundsch.* **57**(2), 514–531.

Paul, J. (1980). Upper Permian algal stromatolitic reefs; Harz Mountains, (F. R. Germany). *Contrib. Sedimentol.* **9,** 253–268.

Pauly, H. (1963a). Ikaite, a new mineral from Greenland (in Danish). *Nat. Verden* **46,** 2–12.

Pauly, H. (1963b). Ikaite, a new mineral from Greenland. *Arctic* **16,** 263–264.

Pavlov, D. I., and Dvorov, V. I. (1977). Formation of increased metal content in the Cheleken thermal brines and problems in the genesis of stratiform lead–zinc ore mineralization (in Russian). *In* "Processes of Sedimentary and Volcano–Sedimentary Accumulation of Base

Metals (Siberia and the Far East)'' (Yu. P. Kazanskiy and L. F. Narkelyun, eds.), pp. 108–113. Izd. Nauka, Sib. Otd., Novosibirsk.

Pavlyuchenko, M. M. (1967). Über das Haften von Salzteilchen an Luftblasen. *In* Int. Kalisymp., 3rd. 1965, vol. 1, pp. 99–121. VEB Dtsch. Verlag Grundstoffind., Leipzig.

Pavlyuchenko, M. M., Akulovich, V. M., Filonov, B. O., Dubovik, V. V., and Pikulik, V. A. (1961). Accessory elements in potassium salts of the Starobino deposit (in Russian). *Tr. Soveshch. Ispol'z. Obogashch. Kaliinykh Solei Byelorussii* (Proc. Conf. on Exploitation and Enrichment of Potash Salts in Byelorussia, 1960), pp. 31–40.

Pavlyuchenko, M. M., Medvedova, A. P., and Mazal, A. I. (1961). The bromine content of the Starobino potassium salts (in Russian). *Tr. Soveshch. Ispol'z. Obogashch. Kaliinykh Solei Byelorussii* (Conf. on Exploitation and Enrichment of Potash Salts in Byelorussia, 1960), pp. 41–47.

Pawlikowski, M. (1978). Petrographic studies of the Wieliczka salt deposit (in Polish). *Pr. Mineral., Pol. Akad. Nauk, Oddzial Krakow., Kom. Mineral. Nauk* **58,** 65–124.

Pawlikowski, M., and Ksiazek, E. (1975). Floor–nucleated crystals from Wieliczka. *Mineral. Pol.* **6**(2), 99–107.

Pawlikowski, M., and Stasik, I. (1980). Mineralogical study of zuber–salt from Inowroclaw (in Polish). *Pr. Mineral., Pol. Akad. Nauk, Oddzial Krakow., Kom. Mineral. Nauk* **66,** 35–44.

Pawlowska, K. O. (1962). Gypsum and native sulphur in rocks beneath the gypsum in the Holy Cross Mountains (in Polish). *In* ''Ksiega pamiatkowa prof. J. Samsonowicza'' (E. Passendorfer, ed.), pp. 16–81. Pol. Akad. Nauk, Warsaw.

Peach, P. A. (1949). Liquid inclusions in geothermometry. *Am. Mineral.* **34**(5–6), 460–461.

Pearson, W. J. (1963). Salt deposits of Canada. *In* ''Symp. on Salt'' (A. C. Bersticker, K. E. Hoekstra, and J. F. Hall, eds.), pp. 197–239. N. Ohio Geol. Soc., Cleveland, Ohio.

Peirce, G. J. (1914). The behavior of certain microorganisms living in brine. *In* ''The Salton Sea'' (D. T. MacDougal *et al.*, eds.), pp. 49–69. Carnegie Inst., Washington, D.C.

Pelikan, A. (1891). Das Tetrakishexaeder (102) am Steinsalz von Starunia. *Tschermak's Mineral. Petrogr. Mitt.* **12**(6), 483–486.

Pelikan, A. (1900). Pseudomorphose von Edelopal nach Gips. *Tschermak's Mineral. Petrogr. Mitt.* **19**(4), 339–340.

Pelišek, J. (1941). Distribution of soluble salts in the profiles of Southern Moravian salt soils (in Czech). *Sb. Akad. Zemědělská* **16,** 301–304.

Pel'sh, A. D. (1940). The simultaneous crystallization of mirabilite and rock salt from the Kara Bogaz brines saturated with NaCl (in Russian). *Byull. Vses. Nauchno–Issled. Proektn. Inst. Galurgii* **3,** 11–24.

Pel'sh, A. D. (1953). Diagram of 0, 2.5, and 5° isotherms of the reciprocal aqueous system $Na_2SO_4 + MgCl_2 \leftrightarrows 2NaCl + MgSO_4$ (in Russian). *Byull. Vses. Nauchno-Issled. Proektn. Inst. Galurgii* **27,** 17–33.

Perel'man, A. I. (1967). ''Geochemistry of Epigenesis.'' Plenum Press, New York, 266 pp.

Pering, K. L. (1973). Bitumen associated with lead, zinc and fluorite ore minerals in north Derbyshire, England. *Geochim. Cosmochim. Acta* **37**(3), 401–417.

Perova, A. P. (1961). Decomposition of polyhalite in aqueous solutions of potassium and magnesium chlorides and of potassium sulfate (in Russian). *Zh. Neorg. Khim.* **6,** 1713–1717.

Perrier, C. (1915). Artificial crystals of gypsum. *Atti Accad. Naz. Lincei, Cl. Sci. Fis., Mat., Nat., Rend.*, [5] **24**(1), 159–163.

Perrott, K. W. (1981). Effect of pH and aluminosilicate composition on potassium–magnesium exchange selectivity of amorphous aluminosilicates. *Geoderma* **26**(4), 311–322.

Perthuisot, J. P. (1971). Recent polyhalite from Sebkha El Melah (Tunisia). *Nature (London) Phys. Sci.* **232**(35), 186–187.

Perthuisot, J. P. (1974). Les dépôts salins de la sebkha El Melah de Zarzis: Conditions et modalités de la sédimentation évaporitique. *Rev. Géogr. Phys. Géol. Dynam.*, [2] **16**(2), 177–188.

Perthuisot, J. P. (1975). La sebkha El Melah de Zarzis. Génèse et évolution d'un bassin salin paralique. *Trav. Lab. Geol., École Norm. Supér., Paris* **9,** 1–252.

Perthuisot, J. P. (1976). Une sebkha sulfatée sodique en pays sédimentaire. La sebkha Oum el Krilate (Sund tunisien). *Geol. Mediterr.* **3**(4), 265–274.

Perthuisot, J. P. (1977). La sebkha de Doukhane (Qatar) et la transformation: Gypse—anhydrite + eau. *Bull. Soc. Géol. Fr.*, [7] **19**(5), 1145–1449.

Perthuisot, J. P. (1980). Sites et processes de la formation d'évaporites dans la nature actuelle. *Bull. Cent. Rech.-Explor. Elf-Aquitaine* **4**(3), 207–233.

Perthuisot, J. P., and Jauzein, A. (1978). Le Khour el Aadid, lagune sursalée de l'émirat de Qatar (Khor el Odeid, a hypersaline lagoon in Qatar). *Rev. Géogr. Phys. Géol. Dynam.*, [2] **20**(4), 347–358.

Perthuisot, J. P., Floridia, S., and Jauzein, A. (1972). Un modèle récent de bassin côtier à sédimentation saline: La sebkha El Melah (Zarzis, Tunisie). *Rev. Géogr. Phys. Géol. Dynam.*, [2] **14**(1), 67–84.

Perthuisot, V., Guilhaumou, N., and Touray, J. C. (1978). Les inclusions fluides hypersalines et gazeuses des quartz et dolomites du Trias évaporitique nord-tunisien; essai d'interprétation géodynamique. *Bull. Soc. Geol. Fr.*, [7] **20**(2), 145–155.

Peterson, I. (1982). Keeping radiation waste out of sight. *Sci. News* **121**(1), 9–11.

Peterson, J. A., and Hite, R. J. (1969). Pennsylvanian evaporite–carbonate cycles and their relation to petroleum occurrence, southern Rocky Mountains. *Bull. Am. Assoc. Pet. Geol.* **53**(4), 884–908.

Petraschek, W. (1947). Bitumen und Erdgas im Haselgebirge des alpinen Salzbergbaues. *Berg-Hüttenm. Monatsh.* **92,** 106–109.

Petrichenko, O. I. (1973). "Methods of Studying Inclusions in Minerals of Saline Rocks" (in Ukrainian). Izd. Naukova Dumka, Kiev, 91 pp.

Petrichenko, O. I., and Slivko, Y. P. (1967). Accessory disseminated alkaline elements in evaporite deposits of the Ukraine (in Russian). *Geokhimiya* No. 4, pp. 461–466.

Petrichenko, O. I., and Slivko, Y. P. (1973). Conditions of mineral formation during the formation of the salt deposits in the Donets Basin (in Ukrainian). *Mineral. Sb. (L'viv)* **27**(3), 263–274.

Petrichenko, O. I., Slivko, E. P., and Shaidetskaya, V. S. (1973a). Composition of intercrystalline solutions of Devonian rock salt in the Dnieper–Donets region (in Russian). *In* "Problems of Pore Fluids" (G. V. Bogomolov, ed.), pp. 49–50. Izd. Nauka i Tekhnika, Minsk.

Petrichenko, O. I., Slivko, E. P., and Shaidetskaya, V. S. (1973b). Geochemical features of the formation of an evaporite bed in the Transcarpathian Solotvinsk depression (in Ukrainian). *Geol. Geokhim. Goryuch. Iskop.* **35,** 37–41.

Petrichenko, O. I., Kovalevich, V. M., and Chalyi, V. N. (1974). Geochemical conditions of salt formation in the Tortonian evaporite basin of northwestern Ciscarpathia (in Ukrainian). *Geol. Geokhim. Goryuch. Iskop.* **41,** 74–80.

Petrov, M. P. (1976). "Deserts of the World." Wiley, New York, 447 pp.

Petrov, N. P., and Chistyakov, P. A. (1964). "Lithology of Mesozoic Salt and Redbed Deposits in the Southwestern Spurs of the Gissar" (in Russian). Izd. Nauka, Tashkent, 222 pp.

Petrov, N. P., and Chistyakov, P. A. (1972). Genetic and paragenetic relations between the evaporite and red bed formations in Uzbekistan (in Russian). *Zap. Vses. Miner. O–va. Uzbek. Otd.* **25,** 53–58.

Petrova, N. S. (1973). Partition of Rb between carnallite and solution in the $NaCl–MgCl_2–H_2O$ system at 25°C. *Geokhimiya,* No. 6, pp. 919–924; *Geochem. Int. (Engl. Transl.)* **10**(3), 709–714.

Petukhov, A. V., Zvereva, O. V., and Tikhomirova, E. S. (1981). Montmorillonitization of clay by hydrocarbons (in Russian). *Dokl. Akad. Nauk SSSR* **258**(6), 1450–1453.

Pfeifer, H. R. (1977). A model for fluids in metamorphosed ultramafic rocks. Observations at surface and subsurface conditions (high pH spring waters). *Schweiz. Mineral. Petrogr. Mitt.* **57**(3), 361–396.

Pfennig, H. (1962). Dampfdruck einer gesättigten wässrigen $MgCl_2$–Lösung in Abhängigkeit von der Temperatur. *Naturwissenschaften* **49**(4), 81.

Phleger, F. B. (1969). A modern evaporite deposit in Mexico. *Bull. Am. Assoc. Pet. Geol.* **53**(4), 824–829.

Pick, H., and Weber, H. (1950). Dichte–Änderung von Kalichlorid–Kristallen durch Einbau zweiwertiger Ionen. *Z. Phys.* **128**(3), 409–413.

Pierre, C., and Catalano, R. (1976). Stable isotopes ^{18}O, ^{13}C and ^{2}H in the evaporitic sequence of the Ciminna Basin (Sicily). *Mem. Soc. Geol. Ital.,* **16,** 55–62.

Pierre, C., Ortlieb, L., and Person, A. (1981). Formation actuelle de dolomite supralittorale dans des sables quartzofeldspathiques: Un exemple au sud de la lagune Ojo de Liebre (Basse Californie, Mexique). *C. R. Hebd. Séances Acad. Sci. Paris,* [2] **293**(1), 73–78.

Pieszczek, E. (1905). Über die Natur des blauen Steinsalzes. *Pharm. Ztg.* **50,** 929–931; **51,** 700–703.

Pilot, J., and Blank, P. (1967). K–Ar Bestimmungen von Salzgesteinen des Zechsteins. *Z. Angew. Geol.* **13**(11–12), 661–662.

Pilot, J., and Roesler, H. V. (1967). Altersbestimmung von Kalisalzmineralen. *Naturwissenschaften* **54**(18), 490.

Pilot, J., Roesler, H. V., and Mueller, P. (1972). Zur geochemischen Entwicklung des Meereswassers und mariner Sedimente im Phanerozoikum mittels Untersuchungen von S-, O- und C-Isotopen. *Neue Bergbautech.* **2,** 161–168.

Pisarchik, Ya. K., Golubchina, M. N., and Toksubaya, A. I. (1977). Isotopic composition of sulfur in calcium sulfates from the Cambrian Upper Lena Formation (Siberian Platform). *Geokhimiya* No. 4, pp. 623–626; *Geochem. Int. (Engl. Transl.)* **14**(2), 182–185.

Plaziat, J. C., and Desprairies, A. (1969). Les pseudomorphoses de cristaux de sel gemme du keuper inférieur de Lorraine; mode de formation et répartition paléogeographique. *Bull. Soc. Géol. Fr.,* [7] **11**(3), 400–406.

Plet-Lajoux, C., Monnier, G., and Pedro, G. (1971). Étude expérimentale sur la génèse et la mise en place des encroûtements gypseux. *C. R. Hebd. Séances Acad. Sci. Paris Sér. D* **272**(24), 3017–3020.

Ploss, R. S. (1964). NaCl: Modification of crystal habit by chemical agents. *Science* **144**(3615), 169–170.

Plummer, P. S., and Gostin, V. A. (1981). Shrinkage cracks: Desiccation or synaeresis. *J. Sediment. Petrol.* **31**(4), 1147–1156.

Poborski, J. (1959). Gases in Polish salt mines (in Polish). *Przegl. Gorn.* **15**(1), 51–53.

Poborski, J. W. (1970). The Upper Permian Zechstein in the Eastern Province of Central Europe. *In* "Symp. on Salt, 3rd." (J. L. Rau and L. F. Dellwig, eds.), vol. 1, pp. 24–29. N. Ohio Geol. Soc., Cleveland, Ohio.

Pochon, J., and Jaton, C. (1967). Causes of the deterioration of building materials: The role of microbial agencies in the deterioration of stone. *J. Chem. Ind.* **38,** 1587–1589.

Pochon, J., Rose, A., Tchan, Y. T., and Augier, J. (1949). Formation de gypse par voie biologique, dans certaines altérations des pierres des monuments. *C. R. Hebd. Séances Acad. Sci. Paris* **228,** 438–439.

Podemski, M. (1972). The Zechstein rock and potash salts of the cyclothems Z2 and Z1 around Nowa Sol' (in Polish). *Biul. Pol. Inst. Geol. Warsaw* **260**(11), 5–64.

Podemski, M. (1975). The Zechstein salts in the area of the Rybak structure (in Polish). *Biul. Pol. Inst. Geol. Warsaw* **286**(3), 5–63.

Polivanova, A. I. (1977). Relationship between salt accumulation and gas formation (in Russian). *In* "Physicochemical Processes in Facies" (A. L. Yanshin and M. A. Zharkov, eds.), pp. 146–153. Izd. Nauka, Sib. Otd., Novosibirsk.

Polivanova, A. I. (1982). The role of density and composition in solution migration (on the basis of experimental data) (in Russian). *In* "New Data on the Geology and Geochemistry of Subsurface Waters and of Economic Deposits in Evaporite Basins" (A. L. Yanshin and M. A. Zharkov, eds.), vol. 2, pp. 16–18. Izd. Nauka, Sib. Otd., Novosibirsk.

Polyakov, L. F., Krasnokutskiy, N. P., and Zhevzhik, E. P. (1971). Gas bearing capacity of the third sylvinite stratum in the Soligorsk potassium mines (in Russian). *Gorn. Zh.* **147**(9), 68–71.

Pompeckij, F. J. (1914). Das Meer des Kupferschiefers, *Branca Festschr.*, pp. 444–494, Leipzig.

Ponahlo, J. (1979). The chemical behaviour of powders of calcite, wollastonite, and quartz mixtures of them with naphthenic acids. *Neues Jahrb. Mineral., Abh.* **136**(1), 77–92.

Ponizovskiy, A. M., Meleshko, E. P., and Globina, N. I. (1953). Viscosity and specific heat of sea water and natural brines (in Russian). *Tr. Krym. Fil. Akad. Nauk Ukr. RSR* **4**(1), 75–80.

Ponizovskiy, A. M., Chernova, Z. S., Sultanyants, V. I., and Shenker, M. A. (1974). Solubility of gypsum dihydrate and hemihydrate in solutions of magnesium and calcium chlorides (in Russian). *In* "Industrial Magnesium Production Technology" (B. A. Shoikhat and V. I. Kuznetsov, eds.), pp. 42–47. Izd. Nauka, Moscow. *Chem. Abstr.* **87,** 91,527.

Popov, V. I., and Vorobiev, A. L. (1947). Polyhydrite (hemihydrite) in continental desert deposits of Central Asia (in Russian). *Zap. Vseross. Miner. O–va., [2]* **76,** 268–270.

Popov, V. S. (1968a). Conditions of formation of potash facies (as illustrated by the Upper Jurassic evaporite formation of Central Asia) (in Russian). *In* "Physico-Chemical Facies Processes" (A. G. Kossokovskaya, ed.), pp. 146–153. Izd. Nauka, Moscow.

Popov, V. S. (1968b). Upper Jurassic evaporites of intracontinental marine saline basins of central Asia. *Litol. Polezn. Iskop.*, No. 1, pp. 56–69; *Lithol. Mineral. Resour. (Engl. Transl.)*, No. 1, pp. 45–55.

Popov, V. S. (1975). Geochemistry of gold in the Upper Jurassic evaporite formation of southern Central Asia. *Dokl. Akad. Nauk SSSR,* **224**(4), 929–932; *Dokl. Acad. Sci. USSR, Earth Sci. Sect. (Engl. Transl. Am. Geol. Inst.)* **224**(1–6), 204–207.

Popov, V. S. (1977). Distribution of bromine in salt of central Asiatic Upper Jurassic evaporite formation (in Russian). *Zap. Uzb. Otd. Vses. Mineral. O–va.* **30,** 173–176.

Popov, V. S., and Osichkina, R. G. (1973). On the behavior of bromine during evaporite formation in the Upper Jurassic salt basin of Central Asia. *Geokhimiya,* No. 3, pp. 404–415; *Geochem. Int. (Engl. Transl.)* **10**(2), 283–293.

Popov, V. S., and Sadykov, T. S. (1970). Geochemistry of bromine in an Upper Jurassic evaporite formation of the southern Tadzhik depression (in Russian). *Dokl. Akad. Nauk Uzb. SSR* **27**(11), 49–51.

Porch, E. L., Jr. (1917). The Rustler Springs sulfur deposit. *Bull. Univ. Tex.* No. 1722, pp. 1–71.

Poroshin, V. D. (1981). Formation of highly mineralized calcium chloride brines (in Russian). *Litol. Polezn. Iskop.*, No. 6, pp. 55–61.

Porter, A. W. (1927). Note on the Soret effect. *Trans. Faraday Soc.* **23,** 314–316.

Pošepný, F. (1877). Genese der Salzablagerungen besonders in Nordamerika. *Sitzungsber. k.u.k. Akad. Wiss. Wien, Math.–Naturwiss. Kl.*, [1] **76,** 179–212.

Posnjak, E. (1938). The system $CaSO_4$–H_2O. *Am. J. Sci.* **35A,** 247–272.

Posnjak, E. (1940). Deposition of calcium sulfate from seawater. *Am. J. Sci.* **238**(8), 559–568.

Posokhov, E. V. (1949). The "calcium chloride" lakes of central Kazakhstan (in Russian). *Dokl. Akad. Nauk SSSR* **66,** 421–423.

Post, F. J. (1977). The microbial ecology of the Great Salt Lake. *Microbiol. Ecol.* **3,** 143–165.

Post, F. J. (1980). Oxygen-rich gas domes of microbial origin in the salt crust of the Great Salt Lake, Utah. *Geomicrobiol. J.* **2**(2), 127–140.

Postma, H. (1965). Water circulation and suspended matter in Baja California lagoons. *Neth. J. Sea Res.* **2**(4), 566–604.

Potonié, R. (1928). "Petrographie der Ölschiefer und ihrer Verwandten." Gebr. Borntraeger, Berlin, 173 pp.

Pouget, M. (1968). Contribution à l'étude des croûtes et encroûtements gypseux de nappe dans le Sud–Tunesien. *Cah. Off. Rech. Sci. Tech. Outre–Mer, Sér. Pédol.* **6**(3–4), 309–365.

Powell, D. A., and Way, S. J. (1962). Transformation of calcium sulfate hemihydrate to insoluble anhydrite. *Aust. J. Chem.* **15,** 386–389.

Power, W. H., Fabuss, B. M., and Satterfield, C. N. (1966). Transient solute concentrations and phase changes of calcium sulfate in aqueous sodium chloride. *J. Chem. Eng. Data* **11**(2), 149–154.

Powers, D. W., Lambert, S. J., Shaffer, S. E., Hill, L. R., and Weart, W. D. (1978). "Geological Characterization Report, Waste Isolation Pilot Plant (W.I.P.P.) Site, Southeastern New Mexico." Sandra Laboratories, Albuquerque, New Mexico, var. pagin.

Poyarkov, N. F. (1957). Formation of two horizons of halite-bloedite deposits in salt lakes of Central Asia (in Russian). *Izv. Akad. Nauk Uzb. SSR, Ser. Khim. Nauk,* No. 1, pp. 15–21.

Pratt, A. R., Heylman, E. B., Cohenour, R. E. (1966). Salt deposits of Sevier Valley, Utah. *In* "Symp. on Salt, 2nd." (J. L. Rau, ed.), vol. 1, pp. 48–58. N. Ohio Geol. Soc., Cleveland, Ohio.

Pratt, L. B. (1962). The Runn of Cutch. *J. Sediment. Petrol.* **32**(1), 92–98.

Precht, H. (1898). Sekundäre Salzbildungen im Kalisalzlager. *Z. Ver. Dtsch. Ing.* **42,** 677–710.

Precht, H. (1907). Über das Vorkommen von Erdöl in dem Kalibergwerk Desdemona bei Alfeld a. Leine. *Kali (Halle)* **1**(4), 63.

Predtechenskiy, P. P. (1950). The dynamics of climate in relation to changes in solar activity (in Russian). *Tr. Gl. Geofiz. Obs.* [*N.S.*] **19**(81), 193–208.

Price, N. J. (1975). Fluids in the crust of the earth. *Sci. Prog. (Oxford)* **62**(245), 59–87.

Price, R. H. (1982). Effects of anhydrite and pressure on the mechanical behavior of synthetic rock salt. *Geophys. Res. Lett.* **9**(9), 1029–1032.

Prikryl, J. (1966). Swelling of clay formations and chemical treatment of drilling muds for deep drilling. *Pr. Výsk, Úst. Čs. Naft. Dolů* **23,** 178–179.

Prinz, W. (1908). Studien an farblosem und blauem Steinsalz. *Bull. Soc. Belge Géol. Paléont. Hydrol.* **22,** 63–82.

Privalova, L. A. (1971). Petrographic characteristics of rock salt in the Ilek deposit (in Russian). *Tr. Vses. Nauchno–Issled. Inst. Solyanoi Prom–sti.* **19,** 12–20.

Protopopov, A. L. (1969). Secondary textures in saline rocks from the third potassium horizon in the Starobinsk deposit (in Russian). *In* "Geology and Petrography of Potash Salts in Byeolorussia" (A. S. Makhnach, ed.), pp. 348–355. Izd. Nauka i Tekhnika, Minsk.

Przibram, K. (1926). Eine Erklärung der Farbänderung in Salzen. *Sitzungsber. Akad. Wiss. Wien, Math.–Naturwiss. Kl.,* [2a] **135,** 213–216.

Przibram, K. (1927a). Über die Farbänderung von Steinsalz unter Druck. *Sitzungsber. Akad. Wiss. Wien, Math.–Naturwiss. Kl.,* [2a] **136,** 43–56, 345–346.

Przibram, K. (1927b). Das Rätsel des blauen Steinsalzes. *Kali (Halle)* **21**(17), 253–255.

Przibram, K. (1929). Über natürliches blaues Steinsalz. *Sitzungsber. Akad. Wiss. Wien, Math.–Naturwiss. Kl.,* [2a] **138,** 353–362, 483–495, 781–797; **141,** 567–569 (1932); **143,** 489–498 (1934).

Przibram, K. (1931). Färbung und Lumineszenz durch Becquerel–Strahlung. *Z. Phys.* **68**(3), 403–422.

Przibram, K. (1932). Umkristallisierung und Färbung. *Sitzungsber. Akad. Wiss. Wien, Math.–Naturwiss. Kl.*, [2a] **141,** 639–643.

Przibram, K. (1936). Das Rätsel des blauen Steinsalzes. II. *Kali, Verw. Salze, Erdöl* **30**(7), 61–63.

Przibram, K. (1947). Genügen die bekannten Strahlungsquellen um die Farbe von natürlichem blauem Steinsalz und von ähnlichen Mineralverfärbungen zu erklären? *Acta Phys. Austriaca* **1,** 131–136.

Przibram, K. (1950). Über das blaue Steinsalz. *Tschermak's Mineral. Petrogr. Mitt.*, [3] **2**(1), 124–129.

Przibram, K. (1951). Über die Rolle eines optimalen Störungsgrades und der Diffusion in der Verteilung gewisser Farben in Mineralen. *Anz. Österr. Akad. Wiss., Math.–Nat. Kl.* **88,** 1–6.

Przibram, K., and Belar, M. (1924). Über die Verfärbung durch Becquerel–Strahlung und die Frage des blauen Steinsalzes. *Sitzungsber. Akad. Wiss. Wien, Math.–Naturwiss. Kl.*, [2a] **132,** 261–277.

Puchelt, H., Lutz, F., and Schock, H. H. (1972). Verteilung von Bromid zwischen Lösungen und chloridischen Salzmineralen. *Naturwissenschaften* **59**(1), 34–35.

Pueyo Mur, J. J. (1977). El bromo y el rubidio como indicatores geneticos en las evaporitas de la cuenca potasica catalana. *Publ. Inst. Invest. Geol., Diputacion Barcelona* **32,** 77–85.

Pusch, R. (1973). The influence of salinity and organic matter on the formation of clay microstructures. *Proc. Int. Symp. Soil Struct., Göteborg 1973,* pp. 161–173.

Pustyl'nikov, A. M. (1975). Origin of blue color in halite in Cambrian salt deposits of the Siberian Platform. *Lithol. Mineral. Resour. (Engl. Transl.)* **10**(3), 388–389.

Pytkowicz, R. M. (1965). Rates of inorganic calcium carbonate nucleation. *J. Geol.* **73**(1), 196–199.

Qu, Y., Han, W., and Cai, K. (1975). The mineralogical study of d'ansite, a rare saline mineral. First find in China (in Chinese). *Acta Geol. Sin.* No. 2, 180–186.

Quaide, W. L. (1958). Clay minerals from salt concentration ponds. *Am. J. Sci.* **256**(6), 431–437.

Quievreux, F. (1935). Esquisse du monde vivant sur les rives de la lagune potassique. *Bull. Soc. Ind. Mulhouse* **101,** 161–187.

Rabochev, I. (1949). The influence of gypsum seams in the soils of the Gobi steppe on the effectiveness of flushing the solonchak (in Russian). *Pochvovedeniye,* pp. 377–386.

Rabotnov, V. T., Gudzenko, V. T., Murogova, R. N., and Krayevskiy, V. I. (1980). Composition of gases in the Upper Precambrian and Lower Cambrian of the Siberian Platform (in Russian). *Geol. Nefti Gaza,* No. 3, pp. 25–28.

Rachid, M. A., and Leonard, J. D. (1973). Modification in the solubility and precipitation behavior of various metals as a result of their interaction with sedimentary humic acid. *Chem. Geol.* **11**(1), 89–97.

Radczewski, O. E. (1968). Feine Teilchen (Staube) in Natur und Technik, ihre Bestimmung als Verunreinigungen und der Nachweis von Fluorverbindungen in der Luft. *Ber. Dtsch. Keram. Ges.* **45,** 550–556.

Ragotskie, R. A., and Likens, G. E. (1964). The heat balance of the Antarctic lakes. *Limnol. Oceanogr.* **9**(3), 412–425.

Ramsay, W. (1876). On the influence of various substances in accelerating the precipitation of clays suspended in water. *Qu. J. Geol. Soc. London* **32,** 129–132.

Ramsdell, L. S., and Partridge, E. P. (1929). The crystal form of calcium sulfate. *Am. Mineral.* **14**(2), 59–74.

Ramser, J. H. (1957). Theory of thermal diffusion under linear fluid shear. *J. Ind. Eng. Chem.* **49,** 155–158.

Randazzo, A. F., and Hickey, E. W. (1978). Dolomitization in the Florida aquifer. *Am. J. Sci.* **278**(8), 1177–1184.

Rao, I. R., and Rao, C. S. (1936). Dissociation of strong electrolytes in concentrated solutions. *Nature (London)* **137**(3466), 580.

Rasmussen, R. A. (1974). Emission of biogenic hydrogen sulfide. *Tellus* **26,** 254–260.

Rassonskaya, I. S., Semendyaeva, N. K., and Kudinov, I. B. (1968). Thermal analysis of salt mixtures and crystal hydrates (in Russian). *Proc. All-Union Conf. Experim. Tech. Mineral. Petrogr., 8th,* (V. V. Lapin, ed.), pp. 102–107.

Raucq, P. (1954). La dépression de la Pande et ses gîsements de gypse (Haute Katanga). *Ann. Soc. Belge Géol. Paléont. Hydrol.,* [B] **77**(7–9), 315–334.

Raup, O. (1970). Brine mixing: An additional mechanism for formation of basin evaporites. *Bull. Am. Assoc. Pet. Geol.* **54**(12), 2246–2259.

Raup, O. B. (1982). Gypsum precipitation by mixing seawater brines. *Bull. Am. Assoc. Pet. Geol.* **66**(3), 363–367.

Raup, O. B., Hite R. J., and Groves, H. L., Jr. (1970). Bromine distribution in the Paradox Basin. *In* "Symp. on Salt, 3rd." (J. L. Rau and L. F. Dellwig, eds.), vol. 1, pp. 40–47. N. Ohio Geol. Soc., Cleveland, Ohio.

Rayevskiy, V. I. (1967). On the origin of variegated sylvinites in the Upper Kama deposit (in Russian). *Geokhimiya,* No. 3, pp. 378–380.

Raymond, L. R. (1953). Some geological results from the exploration for potash in north-east Yorkshire. *Qu. J. Geol. Soc. London* **108**(431), 283–310.

Reed, H. G. (1963). Salt deposits of the Williston Basin, United States portion. *In* "Symp. on Salt" (A. C. Bersticker, K. E. Hoekstra, and J. F. Hall, eds.), pp. 147–165. N. Ohio Geol. Soc., Cleveland, Ohio.

Reeder, R. J., and Wenk, H. R. (1979). Microstructures in low temperature dolomites. *Geophys. Res. Lett.* **6**(2), 77–80.

Reeves, C. C., Jr. (1968). "Introduction to Paleolimnology." Elsevier, Amsterdam, 228 pp.

Register, J. K., and Brookins, D. G. (1980). Rb–Sr isochron age of evaporite minerals from the Salado Formation (Late Permian), southeastern New Mexico. *N. M. Bur. Mines Mineral. Resour., Isochron/West,* No. 29, p. 39.

Reichenbach, W. (1970). Die lithologische Gliederung der regressiven Folge von Zechstein 2—5 in ihrer Beckenausbildung. *Ber. Dtsch. Ges. Geol. Wiss.,* [A] **15**(4), 555–563.

Reichenbach, W., and Boehm, H. (1976). Distribution of bromine in the Leine series (Zechstein 3) and its significance for the genesis of the potash salt deposit "Ronnenberg seam" of the Calvoerde block, German Democratic Republic (in Russian). *In* "The Bromium in Brines as a Geochemical Indicator of the Genesis of Salt Deposits" (A. P. Vinogradov, ed.), pp. 232–302. Moscow Univ. Press, Moscow. Shorter version in *Neue Bergbautech.* **9**(5), 247–253.

Reimer, T. O., and Utter, T. (1979). Kapillarit: Eine ungewöhnliche natürliche Erscheinungsform des Steinsalzes. *Neues Jahrb. Mineral., Monatsh.,* No. 3, pp. 93–100.

Reinold, P. (1965). Beitrag zur Geochemie der ostalpinen Salzlagerstätten. *Tschermak's Mineral. Petrogr. Mitt.,* [3] **10**(1–4), 505–527.

Reiser, R., and Tasch, P. (1960). Investigation of the viability of osmophilic bacteria of great geologic age. *Trans. Kan. Acad. Sci.* **63**(1), 31–34.

Reitemeyer, R. F., and Buehrer, T. F. (1940). The inhibiting action of sodium hexametaphosphate on calcium carbonate from ammoniacal solutions. *J. Phys. Chem.* **44**(1), 535–537.

Renard, M. (1975). Étude géochimique de la fraction carbonatée d'un faciès de bordure de dépôt gypseux (exemple du gypse ludien du bassin de Paris). *Sediment. Geol.* **13**(3), 191–231.

Renfro, A. (1974). Genesis of evaporite-associated stratiform metalliferous deposits—a sabkha process. *Econ. Geol.* **69**(1), 33–45.

Rentsch, J. (1964). Kenntnisstand über die Metall- und Erzmineralverteilung im Kupferschiefer. *Z. Angew. Geologie* **10,** 281–288.

Repshe, I. K. (1936). Gold colloids in sylvite crystals (in Russian). *Acta Physicochim. Akad. Nauk SSSR* **5,** 173–188.

Reuss, A., (1867). Die fossile Fauna der Steinsalzablagerungen von Wieliczka in Galizien. *Sitzungsber. k.u.k. Akad. Wiss. Wien, Math.–Naturwiss. Kl.*, [1], pp. 17–182.

Reventos, M. M., Fenoll-Castillo, A. L., and Amigo, J. M. (1974). Estequiometria y tipos estructurales que presentan los haluros minerales. *Bol. Soc. Esp. Hist. Nat., Secc. Geol.* **72**(1–4), 161–171.

Reynolds, T. D., and Gloyna, E. F. (1961). Creep measurements in salt mines. *Bull. Pennsylvania State Univ. Mineral. Industries Expt. Sta., (Proc. Symp. on Rock Mech., 4th.)* **76,** 11–17.

Reznikov, V. A., and Bel'dy, M. P. (1974). Kinetics of underground dissolution of bischofite (in Russian). *Tr. Vses. Nauchno–Issled. Proektn. Inst. Galurgii* **66,** 106–115.

Ricci-Lucchi, F. (1969). Composizione e morfometria di un conglomerato risedimentato nel flysch miocenico romagnolo (Fontanelice, Bologna). *Giorn. Geol.*, [2a] **36**(1), 1–47 (1968).

Ricci-Lucchi, F. (1973a). Depositional cycles in two turbidite formations of northern Apennines (Italy). *J. Sediment. Petrol.* **45**(1), 3–43.

Ricci-Lucchi, F. (1973b). Resedimented evaporites: Indicators of slope instability and deep-basin conditions in Periadriatic Messinian (Apennines Foredeep, Italy). *In* "Messinian Events in the Mediterranean" (C. W. Drooger *et al.*, eds.), pp. 142–149. North-Holland, Amsterdam.

Rich, J. L. (1951). Three critical environments of deposition and criteria for recognition of rock deposited in each of them. *Bull. Geol. Soc. Am.* **62**(1), 1–20.

Richards, F. A. (1970). The enhanced preservation of organic matter in anoxic marine environments. *In* "Symp. on Organic Matter in Natural Waters" (D. W. Hood, ed.). *Occas. Publ. Univ. Alaska, Inst. Mar. Sci.* **1,** 399–411.

Richardson, L. T., and Wells, R. C. (1931). Heat of solution of some potash minerals. *J. Wash. Acad. Sci.* **21**(11), 243–248.

Richardson, W. A. (1920). The fibrous gypsum of Nottinghamshire. *Mineral. Mag.* **19**(91), 77–95.

Richter, A. 1962. Die Rotfärbung in den Salzen der deutschen Zechsteinlagerstätten—I. *Chem. Erde* **22**(6), 508–546.

Richter, A., 1964. Die Rotfärbung in den Salzen der deutschen Zechsteinlagerstätten—II. *Chem. Erde* **23**(3), 179–203.

Richter, R. (1941). Risse durch Innenschrumpfung und Risse durch Lufttrocknung. *Senckenbergiana* **23**(1–3), 165–167.

Richter, W. (1953). Untersuchungen zum Auffinden von CO_2–Nestern im Kalibergbau mit gaselektrischen Methoden. *Freiberger Forschungsh.*, [C] **7,** 87–103.

Richter-Bernburg, G. (1950). Zur Frage der absoluten Geschwindigkeit geologischer Vorgänge. *Naturwissenschaften* **37**(1), 1–6.

Richter-Bernburg, G. (1953). Über salinare Sedimentation. *Z. Dtsch. Geol. Ges.* **105,** 593–645.

Richter-Bernburg, G. (1957a). Zur Paläogeographie des Zechsteins. *Atti Conv. Milano* "I Giacimenti Gessiferi dell'Europa Occidentale." *Atti Accad. Naz. Lincei Ente Naz. Idrocarburi* **1,** 87–99.

Richter-Bernburg, G. (1957b). Isochrone Warven im Anhydrit des Zechstein–2. *Geol. Landesanst., Geol. Jahrb.* **74,** 601–610; **75,** 629–639.

Richter-Bernburg, G. (1960). Zeitmessung geologischer Vorgänge nach Warven–Korrelationen im Zechstein. *Geol. Rundsch.* **49**(1), 132–148.

Richter-Bernburg, G. (1964). Solar cycle and other climatic periods and varved evaporites. *In* "Problems in Palaeoclimatology" (A. E. M. Nairn, ed.), pp. 510–532. Wiley (Interscience), New York.

Richter-Bernburg, G. (1977). Einflüsse progressiver and regressiver Salinität auf Entstehung und Strukturformen von Salzgesteinen—eine Problematik des Muschelkalksalzes. *Jahresber. Mitt. Oberrhein. Geol. Ver., N.S.* **59,** 273–301.

Richter-Bernburg, G. (1979). Eodiagenetische Vorgänge bei der Bildung von Salinargesteinen. *Geol. Rundsch.* **68**(3), 1055–1065.

Richter-Bernburg, G. (1980). Aberrant vertical structures in well bedded halite deposits. *In* "Symp on Salt, 5th." (A. H. Coogan and L. Hauber, eds.), vol. 1, pp. 159–166. N. Ohio Geol. Soc., Cleveland, Ohio.

Richter-Bernburg, G. (1981). Some peculiarities in the sedimentation of the German Permian. *Proc. Int. Symp. Cent. Eur. Permian, Jablonna, 1978,* pp. 48–55.

Rickard, D. T., Willden, M., Marde, Y., and Ryhage, R. (1975). Hydrocarbons associated with lead–zinc ores at Laiswell, Sweden. *Nature (London)* **255**(5504), 131–133.

Rickard, L. V. (1970). Gamma-ray logs and the origin of salt. *In* "Symp. on Salt, 3rd." (J. L. Rau and L. F. Dellwig, eds.), vol. 1, pp. 34–39. N. Ohio Geol. Soc., Cleveland, Ohio.

Ricour, J., Bourcart, J., and Leveque, P. (1958). Répartition and origine des sulfates du Trias rencontre par les sondages profonds du Bassin de Paris. *C. R. Hebd. Séances Acad. Sci. Paris,* **247**(21), 1882–1885.

Ridge, M. J., and Surkevicius, H. (1961). Variations in the kinetics of setting of calcined gypsum. *J. Appl. Chem.* **11,** 420–434.

Rieke, H. H., III, and Chilingar, G. V. (1962). pH of brines. *Sedimentology* **1**(1), 75–79.

Riemann, C. (1907). Die Entstehung der Salzlager. *Kali (Halle)* **1**(1), 2–7; (2), 21–23.

Riley, C. M., and Byrne, J. V. (1961). Genesis of primary structures in anhydrite. *J. Sediment. Petrol.* **31**(4), 553–559.

Rios, J. M. (1968). Saline deposits of Spain. *In* "Saline Deposits" (R. B. Mattox *et al.*, eds.). *Geol. Soc. Am. Spec. Paper* **88,** 59–74.

Ripun, M. B. (1961). The petrography of the gypsum–anhydrite stratum in the Carpathian syncline (in Ukrainian). *Geol. Zh. (Ukr. Ed.)* **21**(5), 76–82.

Ripun, M. B. (1965). Organic substances in an anhydrite veinlet in Jurassic dolomites in the northern Medinychi area (in Ukrainian). *Sb. Resp. Mezhredomstv. Akad. Nauk Ukr. RSR,* **1,** 51–55. *Chem. Abstr.* **65,** 1,997.

Risacher, F. (1978). Génèse d'une croûte de gypse dans un bassin de l'Altiplano bolivien. *In* "Evolution récente des haute plateaux Andins en Bolivie" *Cah. Off. Rech. Sci. Tech. Outre-Mer. Ser. Geol.* **10**(1), 91–100.

Robert, C., and Chamley, H. (1974). Gypse et sapropels profonds de la Méditerranée orientale. *C. R. Hebd. Séances Acad. Sci. Paris,* [D] **278,** 843–846.

Roberts, N., Erol, O., de Meester, T., and Uerpmann, H. P. (1979). Radiocarbon chronology of the Pleistocene of late Pleistocene Konya Lake, Turkey. *Nature (London)* **281**(5733), 662–664.

Robinson, R. A., and Stokes, R. H. (1949). The role of hydration in the Debye–Hueckel theory. *Ann. New York Acad. Sci.* **51,** 593–604.

Rockel, W., and Ziegenhardt, W. (1979). Strukturelle Kriterien der Laugenbildung im tieferen Zechstein im Raum südlich Berlin. *Z. Geol. Wiss.* **7**(7), 847–860.

Roda, C. (1970). Rock salt, potash beds, and sulfur in Italy. *Boll. Sedute Accad. Gioenia Sci. Nat. Catania* **10**(5), 385–395. *Chem. Abstr.* **78,** 32,544.

Rode, K. P. (1944). The submarine volcanic origin of rock salt deposits. *Proc. Indian Acad. Sci.,* [*B*] **20,** 130–142.

Roedder, E. (1968). Temperature, salinity, and origin of the ore forming fluids at Pine Point, Northwest Territories, Canada. *Econ. Geol.* **63**(5), 439–450.

Roedder, E. (1976). Fluid inclusion evidence on the genesis of ores in sedimentary and volcanic rocks. *In* "Handbook of Strata-Bound and Stratiform Ore Deposits" (K. H. Wolf, ed.), vol. 2, pp. 67–110. Elsevier, Amsterdam.

Roedder, E., and Belkin, H. E. (1980). Thermal gradient migration of fluid inclusions in single crystals of salt from the Waste Isolation Pilot Plant (WIPP) site. *Sci. Basis Nucl. Waste Manage.,* vol. 2, pp. 453–464. *Chem. Abstr.* **94,** 197,166.

Rognon, P. (1980). Une extension des déserts (Sahara et Moyen–Orient) au cours de Tardiglaciaire (18,000–10,000 ans B.P.). *Rev. Géol. Dyn. Géogr. Phys.* **22**(4–5), 313–328.

Rogozovskaya, M. Z., Luk'yanova, N. K., and Kononchuk, T. I. (1977). On the properties of the calcium sulfate residue in the purification of a brine for chlorine production (in Russian). *Khim. Prom-st' (Moscow)* No. 10, pp. 770–772.

Rogozovskaya, M. Z., Kononchuk, T. I., and Luk'yanova, N. K. (1979). Influence of pH on the formation of gypsum incrustations (in Russian). *Zh. Prikl. Khim.* **52**(3), 563–566.

Roller, J. (1931). Chemical activity and particle size. *J. Phys. Chem.* **35**(4), 1133–1142; **36**(4), 1202–1231 (1932).

Romanenk, V. I., Peres Eiris, M., Kudryavtsev, V. M., and Aurora Pubienes, M. (1976). Microbiological processes in meromictic lake Vae-de-San Juan, Cuba. *Microbiology (Engl. Transl.)* **45**(3), 466–472.

Rooney, L. F., and French, R. R. (1968). Allogenic quartz and the origin of penemosaic texture in evaporites of the Detroit River Formation (Middle Devonian) in northern Indiana. *J. Sediment. Petrol.* **38**(3), 755–765.

Rooth, J. (1965). The flamingos on Bonaire (Netherlands Antilles). Habit, diet and reproduction of *Phaenicopterus ruber ruber*. *Natuurwet. Studiekr. Suriname Ned. Antillen* **41,** 13–151.

Rose, A. W. (1976). The effect of cuprous chloride complexes on the origin of red-bed cooper and related deposits. *Econ. Geol.* **71**(6), 1036–1048.

Rose, H., (1839). Über das Knistersalz von Wieliczka. *Ann. Phys. Chem.* **48**(2), 353–361.

Rosenqvist, I. T. (1955). Investigations in the clay–electrolyte–water system. *Publ. Norw. Geotech. Inst.* **9,** 1–125.

Rosenqvist, I. T. (1966). Norwegian research in the properties of quick clay—a review. *Eng. Geol.* **1**(6), 445–450.

Rosset, C. (1965). Inclusions fluides salines dans les cristaux du gypse oligocène à foraminifères, du Portel (Aude). *C. R. Hebd. Séances Acad. Sci. Paris* **260**(8), 2267–2269.

Roth, R. (1937). Custer Formation of Texas. *Bull. Am. Assoc. Pet. Geol.* **21**(4), 421–424.

Rothbaum, H. P. (1958). Vapor pressure of sea water concentrates. *Ind. Eng. Chem., Chem. Eng. Data Ser.* **3,** 50–52.

Rouchy, J. M. (1980). La génèse des évaporites messiniennes de Méditerranée: un bilan. *Bull. Cent. Rech. Pau* **4**(1), 511–545.

Rouchy, J. M. (1981). La génèse des évaporites messiniennes de Méditerranée. *Mus. Natl. Hist. Nat., Paris,* Ph.D. Dissertation, 295 pp.

Rouchy, J. M. (1982). La crise évaporitique messinienne de Méditerranée: nouvelles propositions pour une interprétation génétique. *Bull. Mus. Natl. Hist. Nat.,* [C] **4**(3–4), 107–136.

Rouchy, J. M., and Monty, C. (1981). Stromatolites and cryptalgal laminites associated with Mesum of Cyprus. *In* ''Phanerozoic Stromatolites'' (C. Monty, ed.), pp. 155–180. Springer-Verlag, New York.

Roulston, B. V., and Waugh, D. C. E. (1981). A borate mineral assemblage from the Penobsquia and Salt Springs evaporite deposits of southern New Brunswick. *Can. Mineral.* **19**(2), 291–301.

Rouse, J. E., and Sherif, N. (1980). Major evaporite deposition from remobilized salts. *Nature (London)* **285**(5765), 470–472.

Rowlands, D. L. G., and Webster, R. K. (1971). Precipitation of vaterite in lake water. *Nature (London) Phys. Sci.* **229**(5), 158.

Rozen, B. Ya. (1970). ''Geochemistry of Bromine and Iodine'' (in Russian). Izd. Nedra, Moscow, 143 pp.

Rozen, Z. (1926). Pénétration lamelleuse de sylvine et de sel gemme. *Kosmos* (Lwów) **51**(3), 505–516.

Rozsa, M. (1911). "Neuere Daten zur Kenntnis der warmen Salzseen." Friedlaender, Berlin, 32 pp.

Rozsa, M. (1913a). Über die periodische Entstehung doppelter Temperaturmaxima in den warmen Salzseen. *Ann. Hydrogr. Marit. Meteorol.* **41,** 511–513.

Rozsa, M. (1913b). Daten zur Kenntnis des organischen Aufbaus der Stassfurter Salzablagerungen. *Kali (Halle)* **7**(10), 242–247.

Rozsa, M. (1915). Die physikalischen Bedingungen der Akkumulation von Sonnenwärme in den Salzseen. *Physik. Z.* **16**(6), 108–111.

Rozsa, M. (1920). Petrogenesis und petroklimatologische Beziehungen der Salzablagerungen im Tertiär des Oberelsass. *Kali (Halle)* **14**(4), 61–70.

Rubanov, V. I. (1964). Some geochemical characteristics of salt accumulation in the northern part of the Lyavlyakan group of lakes in the Kyzylkum (in Russian). *Uzb. Geol. Zh.* **8**(5), 7–14.

Rubanov, V. I. (1966). Geology and mineralogical–geochemical characteristics of salt deposits in Lake Dengizkul' (in Russian). *In* "Lithology and Sedimentary Mineral Resources of Uzbekistan" (P. A. Chistyakov, ed.), pp. 117–135. Izd. FAN, Tashkent.

Rubanov, V. I., Mirakhmedov, M., and Sharipova, A. (1964). Anhydrite in Recent salt sediments of the Sarikamsk lakes (in Russian). *Dokl. Akad. Nauk SSSR* **158**(3), 622–624.

Rudolph, H., Weyrich, H., and Zimmermann, K. D. (1966). Ein neues Vorkommen von Cölestin im norddeutschen Zechsteinsalinar. *Chem. Erde* **25**(2), 169–171.

Ruggieri, G., and Sproveri, R. (1978). The "desiccation theory" and its evidence in Italy and in Sicily. *Mem. Geol. Soc. Ital.* **16,** 165–169.

Rutherford, R. L. (1936). Geologic age of potash deposits. *Bull. Geol. Soc. Am.* **47**(8), 1207–1215.

Rutskov, A. P. (1961). Comparative appraisal of the lyotropic effect of ions on the properties of aqueous electrolyte solutions (in Russian). *Zh. Fiz. Khim.* **35**(1), 3–8.

Rutten, M. G. (1954). Continental origin of fossil salt layers. *Geol. Mijnbouw* **16**(3), 61–88.

Rza-Zade, P. F., and Rustamov, P. I. (1961). Isothermal system $MgCl_2$–$CaCl_2$–$CaSO_4$–H_2O at 25°C (in Russian). *Azerb. Khim. Zh.* No. 4, pp. 127–132.

Sabouraud, C. (1974). Étude des liquides inclus dans les cristaux du Gypse Parisien. *Bull. Inf. Assoc. Geol. Bassin Paris* **41,** 19–21.

Sabouraud-Rosset, C. (1974). Determination par activation neutronique des rapports Cl/Br des inclusions fluids de divers gypses. Correlation avec les données de la microcryoscopie et interprétations génétiques. *Sedimentology* **21**(3), 415–432.

Safonova, G. I., and Mileshina, A. G. (1972). Significance of the composition of n-paraffins in crude oils for typical migration paths (in Russian). *Tr. Vses. Nauchno–Issled. Neft. Geologorazved. Inst.* **119,** 181–192.

Sahama, T. G. (1945). Abundance relation of fluorite and sellaite in rocks. *Ann. Acad. Sci. Fenn.,* [A] No. 9, p. 121.

Sahores, J. (1962). Contribution à l'étude des phénomènes méchaniques qui accompagnient l'hydration de l'anhydrite. Construct. Trav. Publ., Rev. Matér., No. 558, pp. 65–72; No. 559, pp. 97–102; No. 560, pp. 131–141; No. 561, pp. 157–167; No. 562/563, pp. 210–223; No. 564, pp. 257–267; No. 565, pp. 285–298; No. 566, pp. 325–329; No. 567, pp. 357–365.

Sahores, M. N. (1955). Sur les efforts exercis par la plâtre et l'anhydrite aucours de leur hydration. *C. R. Hebd. Séances Acad. Sci. Paris* **241**(2), 223–225.

Sakai, H. (1957). Fractionation of sulphur isotopes in nature. *Geochim. Cosmochim. Acta* **12**(1–2), 150–169.

Sakai, H. (1972). Oxygen isotope ratios of some evaporites from Precambrian to Recent ages. *Earth Planet. Sci. Lett.* **15**(2), 201–205.

Sakai, W. (1943). The solubility change of gypsum in water by addition of urea. *J. Soc. Chem. Ind. Jpn.* **46,** 157–158.

Salvan, H. M. (1968). L'évolution du problème des évaporites et ses conséquences sur l'interprétation des gîsements marocaines. *Mines Géol. (Maroc)* **11**(27), 5–30.

Salvan, H. M. (1974). Les séries salifères du Trias marocain, caractères généraux et possibilités d'interprétation. *Bull. Soc. Géol. Fr.* **16**(6), 724–731.

Salvinien, J., and Brun, B. (1964). Mésure des coefficients de self-diffusion des constituants (ions et solvant) de solutions salines entre de larges limites de concentration. *C. R. Hebd. Séances Acad. Sci. Paris* **259**(3), 565–567.

Samama, J. C., Miguel de Sa, L. C., and Rey, M. (1978). Les minéralizations plombo–zincifères du trias ardechois en tant que pre-évaporites: le problème des relations a diverses echelles. *Sci. Terre* **22**(2), 167–175.

Sannemann, D., Zimdars, J., and Plein, E. (1978). Der basale Zechstein (A2–T1) zwischen Weser und Ems. *Z. Dtsch. Geol. Ges.* **129,** 33–69.

Sarg, J. F. (1981). Petrology of the carbonate–evaporite facies transition of the Seven Rivers Formation (Guadelupian, Permian), southeastern New Mexico. *J. Sediment. Petrol.* **51**(1), 73–96.

Sasaki, A. (1970). Seawater sulfate as a possible determinant for sulfur isotope compositions of some strata-bound sulfide ores (in Japanese). *Geochem. J. (Jpn.)* **4**(1), 41–51.

Sassi, S. (1969). Contribution à l'étude de la sebkha Tegdimane et du Chott el Guettar. *Notes Tunis. Géol. Surv.* **24,** 1–107.

Sastri, G. G. K. (1962). Origin of the desert gypsum in Rajasthan. *Rec. Geol. Surv. India* **87,** 781–786.

Šatava, V. (1977). A study of gypsum dehydration in aqueous medium. *In* "Sulfates de calcium et matériaux dérivés" (Calcium Sulfates and Derived Materials) (M. Murat and M. Foucault, eds.), pp. 181–196. C. R. Colloq. Int. Saint–Remy–les–Chevreuse, Réunion Int. Lab. d'Essais Rech. Matér. Constr., Paris.

Šatava, V., and Prokop, R. (1963). Kinetics of dehydration of gypsum in solutions of calcium chloride (in Czech). Silikáty (Prague) **7**(2), 118–124.

Šatava, V., and Zbůžek, B. (1971). Effect of water pressure on the conversion of gypsum to hemihydrate (in Czech). Silikáty (Prague) **15**(2), 127–135.

Sato, T., and Hayashi, K. (1961). The viscosity of concentrated aqueous solutions of strong electrolytes (in Japanese). *Bull. Chem. Soc. Jpn.* **34,** 1260–1264.

Sauthein, M. (1978). Sand deserts during glacial maximum and climatic optimum. *Nature (London)* **272**(5648), 43–46.

Savelli, D., and Wezel, F. C. (1978). Schema geologico del Messiniano del Pesarese. *Boll. Soc. Geol. Ital.* **97**(1–2), 165–188.

Savinkova, E. I., Rudakov, V. A., Orekhova, A. I., Sudarkina, N. V., Zharova, A. P., and Nikitina, S. A. (1976). Effect of the composition of a magnesium chloride solution on conversion of potassium chloride to carnallite (in Russian). *Tr. Vses. Inst. Nauchno–Tekhn. Inform., Dep. Doc.* No. 1371–76, 11 pp.; *Chem. Abstr.* **88,** 39479.

Savostyanova, M. (1930). Über die kolloidale Natur der färbenden Substanz im verfärbten Steinsalz. *Z. Phys.* **64,** 262–278.

Saxena, G. M., and Seshadri, T. R. (1956). Origin of salt in Rajasthan. *J. Sci. Ind. Res. (India)* **15A,** 505–508.

Scavnicar, B., and Scavnicar, S. (1980). Authigenic celestite in Lower Triassic limestones and Permo–Triassic evaporite deposits in Dalmatia, southern Croatia. *Geol. Vjesn.* **31,** 279–286.

Schaaffs, W. (1954). Rhythmic crystallization of several inorganic salts. *Kolloid–Z.* **137**(1), 12–14.

Schachl, E. (1954). Das Muschelkalksalz in Südwestdeutschland. Neues Jahrb. Geol. Paläontol., Abh. **98**(3), 309–394.

Schaller, W. T., and Henderson, E. P. (1932). Mineralogy of drill cores from the potash field of New Mexico and Texas. *Bull. U.S. Geol. Surv.* **833,** 124 pp.

Schauberger, O. (1953). Zur Genese des alpinen Haselgebirges. *Z. Dtsch. Geol. Ges.* **105**(4), 736–751.

Schauberger, O. (1960). Über das Auftreten von Naturgasen in alpinen Salinen. *Erdöl–Ztg* **76,** 226–233.

Schauberger, O. (1972). The problem of the primary development and distribution of the Alpine salt formation with reference to the most recent test drillings in the Salzkammergut. *In* ''Geology of Saline Deposits'' (G. Richter-Bernburg, ed.), U.N.E.S.C.O., Paris, Earth Sci. Ser. **7,** 217–221.

Schauberger, O., and Kuehn, R. (1969). Über die Entstehung des alpinen Augensalzes. *Neues Jahrb. Geol. Paläont., Monatsh.* No. 6, pp. 247–259.

Schedling, J. A., and Wein, J. (1955). Differentialthermoanalytische Untersuchungen an $CaSO_4 \cdot 2H_2O$ und seinen durch Entwässerung entstehenden Folgeprodukten. *Sitzungsber. Österr. Akad. Wiss., Math.–Naturwiss. Kl.,* [2] **164,** 175–187.

Schenk, P. (1969). Carbonate–sulfate–redbed facies and cyclic sedimentation of the Windsorian stage (Middle Carboniferous), Maritime Provinces. *J. Can. Earth Sci.* **6**(5), 1037–1066.

Scherreiks, R. (1970). Cölestin-Versteinerungen im Hauptdolomit der östlichen Lechtaler Alpen. *Naturwissenschaften* **57**(7), 353–354.

Schettler, H. (1972). The stratigraphical significance of idiomorphic quartz crystals in the saline formations of the Weser–Ems area, north-western Germany (with discussion). *In* ''Geology of Saline Deposits'' (G. Richter–Bernburg, ed.), U.N.E.S.C.O., Paris, Earth Sci. Ser. **7,** 111–127.

Schidlowski, M., Junge, C. E., and Pietnik, H. (1977). Sulfate isotope variation in marine sulfate evaporites and the Phanerozoic oxygen budget. *J. Geophys. Res.* **82**(18), 2557–2565.

Schilling, J. H. (1973). K–Ar dates on Permian potash minerals from southeastern New Mexico. *N. M. Bur. Mines Mineral. Resour., Isochron/West* No. 6, pp. 37–38.

Schlager, W., and Bolz, H. (1977). Clastic accumulation of sulphate evaporites in deep water. *J. Sediment. Petrol.* **47**(2), 600–609.

Schlanger, S. O. (1965). Dolomite–evaporite relations on Pacific islands. *Sci. Rep. Tohoku Univ., Ser. 2: Geol.,* **37**(1), 15–29.

Schlenker, B. (1971). Petrographische Untersuchungen am Gipskeuper und Lettenkeuper von Stuttgart. *Abh. Oberrhein. Geol.* **20**(1–2), 69–102.

Schmalz, R. (1969). Deep-water evaporite deposition: A genetic model. *Bull. Am. Assoc. Pet. Geol.* **53**(4), 798–823.

Schmalz, R. F. (1971). Evaporites and petroleum: A reply. *Bull. Am. Assoc. Pet. Geol.* **55**(11), 2033–2061.

Schmidt, E. (1960). Über den Kohlendioxidgehalt von Mineralsalzen ozeanischen Ursprungs. *Bergakademie* **12,** 693–697.

Schneider, J. (1979). Stromatolithische Milieus in Salinen der Nord–Adria. *In* ''Cyanobakterien Bakterien oder Algen?'' Oldenburg Symp. on Cyanobacteria, 1st. (W. E. Krumbein, ed.), pp. 93–106. Univ. Oldenburg.

Schneider, T. R. (1965). Zwei eigenartige Sulfatvorkommen. *Schweiz. Mineral. Petrogr. Mitt.* **45**(1), 153–165.

Schneiderhoehn, H. (1923). Chalkographische Untersuchung des Mansfelder Kupferschiefers. *Neues Jahrb. Mineral. Geol. Paläontol., Beilageband* **47,** 1–3.

Schneiderhoehn, H. (1926). Erzführung und Gefüge des Mansfelder Kupferschiefers. *Metall Erz* **23,** 143–146.

Schnepfe, M. M. (1972). Determination of total iodine and iodate in sea water and in various evaporites. *Anal. Chim. Acta* **58**(1), 83–89.

Schnerb, J., and Bloch, M. R. (1958). The influence of metal salts on the crystallization of potassium chloride from supersaturated solutions. *Bull. Res. Counc. Isr., Sect. A,* **7,** 179–185.

Schock, H. H., and Puchelt, H. (1970). Distribution of rubidium and cesium in potassium salts—experimental and analytical investigations. *In* "Symp. on Salt, 3rd." (J. L. Rau and L. F. Dellwig, ed.), vol. 1, pp. 232–238. N. Ohio Geol. Soc., Cleveland, Ohio.

Schock, H. H., and Puchelt, H. (1971). Rubidium and cesium distribution in salt minerals. I. Experimental investigation. *Geochem. Cosmochim. Acta* **35**(3), 307–317.

Scholz, H. (1968). Die Böden der Wüste Namib, Südwestafrika. *Z. Pflanzenernähr. Bodenkd.* **119,** 91–107.

Schott, J. (1973). Sur la mésure des coefficients de Soret par la méthode thermogravitationelle. Conséquences sur l'étude de la sélectivité des crystallizations naturelles. *C. R. Hebd. Séances Acad. Sci. Paris,* [C] **276**(6), 459–462.

Schrader, H. J., and Gersonde, R. (1978). The late Messinian Mediterranean brackish to freshwater environment, diatom floral evidence. *In* Intl. Rep. Deep-Sea Drlg. Proj., vol. 42A (K. J. Hsu, L. Montadert *et al.,* eds.), pp. 761–775. U.S. Govt. Print. Off., Washington, D.C.

Schramm, L. L., and Kwak, J. C. T. (1984). Hydrolysis of alkali and alkali earth forms of montmorillonite in dilute solutions. *Soil Sci.* **137**(1), 1–6.

Schrauzer, G. H., Strampad, N., Hui, L. N., and Palmer, M. R. (1983). Nitrogen photoreduction on desert sands under sterile conditions. *Proc. Natl. Acad. Sci. U.S.A.* **80**(12), 3873–3876.

Schreiber, B. C. (1973). Survey of the physical features of Messinian sediments. *In* "Messinian Events in the Mediterranean" (R. W. Drooger *et al.,* eds.), pp. 101–110. North-Holland, Amsterdam.

Schreiber, B. C., and Decima, A. (1976). Sedimentary facies produced under evaporitic environments: A review. *Mem. Soc. Geol. Ital.* **16,** 111–126.

Schreiber, B. C., and Friedman, G. M. (1976). Depositional environments of Upper Miocene (Messinian) evaporites of Sicily as determined from analysis of intercalated carbonates. *Sedimentology* **23**(2), 255–270.

Schreiber, B. C., and Schreiber, E. (1977). The salt that was. *Geology* **5**(9), 527–528.

Schreiber, B. C., Friedman, G. M., Decima, A., and Schreiber, E. (1976). Depositional environments of Upper Miocene (Messinian) evaporite deposits of the Sicilian Basin. *Sedimentology* **23,** 729–760.

Schreiber, B. C., Roth, M. S., and Helman, M. L. (1982). Recognition of primary facies characteristics of evaporites and the differential of these forms from diagenetic overprints. *In* "Depositional and Diagenetic Spectra of Evaporites" (C. R. Handford, R. G. Loucks, and G. R. Davies, eds.). *Soc. Econ. Paleont. Mineral. Core Workshop* **3,** 1–32.

Schroll, E. (1961). Ein Vorkommen von Magnesit im alpinen Steinsalzlager. *Radex Rundsch.* No. 6, pp. 704–771.

Schubel, J. R. (1968). Suspended sediments of the northern Chesapeake Bay. *Tech. Rep. Chesapeake Bay Inst.* **35,** 264 pp., Johns Hopkins Univ., Baltimore, Maryland.

Schuchert, C. (1915). The conditions of black shale deposition as illustrated by the Kupferschiefer and Lias of Germany. *Trans. Am. Phil. Soc.* **54**(218), 259–269.

Schulze, G. (1958). Beitrag zur Stratigraphie und Genese der Steinsalzserien I–IV des mitteldeutschen Zechsteins unter besonderer Berücksichtigung der Bromverteilung. *Freiberger Forschungsh.,* [A] **123,** 175–196.

Schulze, G. (1960a). Beitrag zur Genese des Polyhalites im deutschen Zechsteinsalinar. *Z. Angew. Geol.* **6**(7), 310–317.

Schulze, G. (1960b). Stratigraphische und genetische Deutung der Bromverteilung in den mitteldeutschen Steinsalzlagern des Zechsteins. *Freiberger Forschungsh.,* [C] **83,** 1–114.

Schulze, G., Greulich, C. and Seifert, H. (1962). Untersuchungen am grauen Salzton im Spaltendiapir des oberen Allertales. *Z. Angew. Geol.* **8**(1), 1–9.

Schulze, G., and Seifert, H. (1959). Der Bromgehalt an der Basis des Leinesteinsalzes und seine Beziehungen zu Faziesdifferenzierungen im Flöz Stassfurt. *Z. Angew. Geol.* **5**(2), 62–69.

Schwabe, K. (1964). Säuremessungen in konzentrierten Salzlösungen und organischen Lösungsmitteln. *Österr. Chemiker-Ztg.* **65**(11), 339–357.

Schwabe, K., and Ferse, A. (1962). Der Einfluss fremder Ionen auf die pH–Bestimmung. *Zuckererzeugung* **6**(8), 194–198.

Schwartz, W. (1972). Mikrobiologie des Erdöls von der Entstehung bis zur Lagerstätte. *Naturwissenschaften* **59**(8), 356–360.

Schwerdtner, W. M. (1962). Untersuchungen an bunten Augensalzen in der Ronnenberg–Gruppe (Zechstein 3) von Hannover. *Kali Steinsalz* **1**(8), 265–275.

Schwerdtner, W. M. (1964). Genesis of potash rocks in Middle Devonian Prairie Evaporite Formation of Saskatchewan. *Bull. Am. Assoc. Pet. Geol.* **48**(7), 1108–1115.

Scroggin, M. P. (1978). Cationic flotation of sylvite. *Circ. N. M. Bur. Mines Mineral. Resour.* **159,** 65–67.

Scruton, R. C. (1953). Deposition of evaporites. *Bull. Am. Assoc. Pet. Geol.* **37**(11), 2498–2512.

Sedel'nikov, G. S. (1958). Über die hydrochemischen Verhältnisse bei der Salzbildung in der Karabugas–Bucht. *Freiberger Forschungsh.*, [A] **123,** 166–174.

Sedletskiy, V. I., and Derevyagin, V. S. (1969). Distribution of bromine in Mesozoic evaporites of Central Asia and its significance in geochemical correlation. *Dokl. Akad. Nauk SSSR,* **185**(1), 185–187; *Dokl. Acad. Sci. USSR, Earth Sci. Sect. (Engl. Transl. Am. Geol. Inst.)* **185**(1–6), 155–158.

Sedletskiy, V. I., and Derevyagin, V. S. (1971). Distribution of bromine in the salt deposits of the southern part of Central Asia. *Geokhimiya* Nr. 1, pp. 92–100; *Geochem. Int. (Engl. Transl.)* **8**(1–2), 55–64.

Sedletskiy, V. I., and Mel'nikova, Ye. M. (1970). Typomorphic characteristics of carbonates in different stages of halogenesis. *Litol. Polez. Iskop.* **5**(1), 89–93; *Lithol. Mineral. Resour. (Engl. Transl.)* No. 5, pp. 110–114.

Sedletskiy, V. I., Trufanov, V. N., and Mel'nikova, Ye. M. (1971). Some features of authigenic quartz formation in evaporite basins (in Russian). *Geol. Geofiz.* Nr. 5, pp. 72–77.

Seidel, A. (1917). "Solubilities of Inorganic and Organic Substances." 2nd ed., Van Nostrand-Reinhold, Princeton, New Jersey, 367 pp.

Seidel, G. (1966). Die paläogeographischen Ursachen des faziellen Überganges von der kieseritischen Hartsalzfazies des Unstrut–Saale–Bezirkes in die anhydritische des Südharzbezirkes. *Geologie* **15**(4–5), 434–442.

Seifert, H. (1952). Habitusmodifikation wachsender Kristalle durch Verunreinigungen: Ein Adsorptionsproblem. *Z. Elektrochem.* **56,** 331–338.

Seitz, F. (1946). Color centers in alkali halide crystals. *Mod. Phys. Rev.* **18,** 384–408.

Seitz, F. (1951). Color centers in additively colored alkali halide crystals containing alkaline earth ions. *Phys. Rev.* **83,** 134–140.

Seliber, G. L. (1950). Degradation and synthesis in decomposition of fats by sulfate-reducing bacteria (in Russian). *Mikrobiologiya* **19,** 294–298.

Selli, R., and Fabbri, R. (1971). Tyrrhenian: A Pliocene deep sea. *Atti Accad. Naz. Lincei, Cl. Sci. Fiz. Mat. Nat., Rend.,* [8] **50**(5), 580–592.

Serra, A. (1949). Coloring principles in colored minerals. *Ric. Sci.* **18,** 1059–1060. *Chem. Abstr.* **43,** 6116.

Serratosa, J. M. (1979). Surface properties of fibrous clay minerals (palygorskite and sepiolite). *In* "Proceedings of the International Clay Conference 1978" (M. M. Mortland and V. C. Farmer, eds.), pp. 99–109. Elsevier, Amsterdam.

Sha, Q., Pan, Zh., and Wang, Y. (1979). Recent dedolomitization in the vadose zone (in Chinese). *Sci. Geol. Sin.* No. 4, pp. 378–383.

Shaffer, L. H. (1966). An investigation of the solubility of calcium sulfate in seawater and seawater concentrates at temperatures from ambient to 65°. Part I. Gypsum. *Abstr. 152nd Annu. Meet. Am. Chem. Soc.*, p. N–21.

Shaposhnikov, A. A., and Khetchikov, L. N. (1970). Effect of solution concentration on the homogenization temperature of gas–liquid inclusions in quartz (in Russian). *Tr. Vses. Nauchno–Issled. Inst. Sint. Miner. Syr'ya* No. 12, pp. 101–103.

Shcherbina, V. N. (1948). Glauberite in Tertiary saline deposits (in Russian). *Dokl. Akad. Nauk SSSR* **63,** 441–443.

Shcherbina, V. N. (1949). Genesis of mirabilite in salt deposits of northern Kirgisia (in Russian). *Dokl. Akad. Nauk SSSR* **67,** 357–359.

Shcherbina, V. N. (1962). Elements of long-term rhythmicity in mineralogic–petrographic sylvinite zones of the salt horizons in the Pripyat salt basin (in Russian). *Dokl. Akad. Navuk BSSR* **6**(8), 510–513.

Shcherbina, V. N., and Shirokikh, I. N. (1971). Thermodynamic and experimental data on stability of gypsum, hemihydrate and anhydrite under hydrothermal conditions. *Geol. Geofiz.* No. 9, pp. 102–105; *Int. Geol. Rev. (Engl. Transl.)* **13**(11), 1671–1673.

Shearman, D. J. (1966). Origin of marine evaporites by diagenesis. *Trans. Inst. Min. Metall., Sect. B* **75,** 208–215.

Shearman, D. J. (1970). Recent halite rock, Baja California, Mexico. *Trans. Inst. Min. Metall., Sect. B* **79,** 155–162.

Shearman, D. J., and Orti Cabo, F. (1976). Upper Miocene gypsum: San Miguel de Salinas, S. E. Spain. *Mem. Soc. Geol. Ital.* **16,** 327–329.

Shearman, D. J., and Fuller, J. G. C. M. (1969). Anhydrite diagenesis, calcitization, and organic laminites, Winnipegosis Formation, Middle Devonian, Saskatchewan. *Bull. Can. Pet. Geol.* **17**(4), 496–525.

Shearman, D. J., and Shirmohammadi, N. H. (1969). Distribution of strontium in dedolomites from the French Jura. *Nature (London)* **223**(5206), 606–608.

Shell Development Co. (1973). K–Ar dates on Permian potash minerals from southeastern New Mexico. *N. M. Bur. Mines Mineral. Resour., Isochron/West* **6,** 37.

Sheng, G. Y., Fan, S. F., Liu, D. H., Su, N. X., Zhou, H. M. (1981). Characteristics of n-alkanes with even–odd predominance (in Chinese). *Chin. J. Oil Nat. Gas Geol.* **2**(1), 57–65. *Chem. Abstr.* **95,** 135,309.

Sherlock, R. L. (1921). Rock salt and brine. *Geol. Surv. G. B., Mem., Spec. Rep. Mineral. Resour.* No. 18. pp. 1–121.

Sherlock, R. L., Smith, B., and Hollingworth, S. E. (1938). Gypsum and anhydrite, celestine and strontianite. *Geol. Surv. G. B., Mem., Spec. Rep. Mineral. Resour.* No. 3, pp. 1–57.

Shields, L. M., Mitchell, C., and Drouet, F. (1957). Alga- and lichen-stabilized surface crusts as soil nitrogen sources. *Am. J. Bot.* **44,** 489–498.

Shleimovich, R. E. (1976). Bromine in salt rocks of the Upper Kama deposit (in Russian). *In* "The Bromium in Brines as a Geochemical Indicator of the Genesis of Salt Deposits" (A. P. Vinogradov, ed.), pp. 48–140. Moscow Univ. Press, Moscow.

Shlezinger, N. A., Zorkin, F. P., and Larina, A. P. (1940a). The chemical composition of the Ozin salts and the outlook for their industrial utilization (in Russian). *Uch. Zap. Sarat. Gos. Univ.* **15**(1), 175–188.

Shlezinger, N. A., Zorkin, F. P., and Petrukhova, E. V. (1940b). Über die Bildungsbedingungen des Kainits. *Dokl. Akad. Nauk SSSR* **27**(6), 466–469.

Shlichta, P. J. (1968). Growth deformation and defect structure of salt crystals. *In* "Saline Deposits" (R. B. Mattox *et al.*, eds.). *Spec. Paper Geol. Soc. Am.* **88,** 597–617.

Shternina, E. B. (1948). Calculation of the solubility of calcium sulfate in aqueous solution of salt (in Russian). *Dokl. Akad. Nauk SSSR* **60**(2), 247–250.

Shternina, E. B. (1949). Solubility of gypsum in salt solutions (in Russian). *Izv. Sekt. Fiz. Khim. Anal. Inst. Obshch. Neorg. Khim. Akad. Nauk SSSR* **17,** 351–469.

Shternina, E. B. (1960). Solubility of gypsum in aqueous solutions of electrolytes. *Izv. Akad. Nauk SSSR, Otd. Khim. Nauk,* No. 1, pp. 3–8; *Int. Geol. Rev. (Engl. Transl.)* **1,** 605–616.

Shuman, A. C. (1966). Gross imperfections and habit modifications in salt crystals. *In* "Symp. on Salt, 2nd." (J. L. Rau, ed.), vol. 2, pp. 246–258. N. Ohio Geol. Soc., Cleveland, Ohio.

Shynsareva, M. N. (1969). Hydroglauberite, a new mineral in the group of hydrous sulfates (in Russian). *Zap. Vses. Miner. O–va.,* [2] **98**(1), 59–62.

Siebenthal, C. E. (1915). Origin of the lead and zinc deposits of the Joplin region. *Bull. U.S. Geol. Surv.* **606,** 283 pp.

Siedentopf, H. (1905). Ultramikroskopische Untersuchungen über Steinsalzverfärbungen. *Phys. Z.* **6**(24), 855–866.

Siedentopf, H. (1908). Über künstlichen Dichroismus von blauem Steinsalz. *Phys. Z.* **8**(23), 850–851.

Siegel, F. R. (1958). Effect of strontium on the aragonite–calcite ratios of Pleistocene corals. *Bull. Geol. Soc. Am.* **69**(12), 1643.

Siegel, F. R. (1960). Effect of strontium on the aragonite–calcite ratios of Pleistocene corals. *J. Sediment. Petrol.* **30**(2), 297–304.

Siegl, W. (1941). Zur Genesis des Kupferschiefers. *Mineral. Petrogr. Mitt.* **52,** 347–362.

Siemeister, G. (1961). Primärparagenese und Metamorphose des Ronnenberglagers nach Untersuchungen im Grubenfeld Salzdetfurth. Ph.D. Dissertation, Bergakademie, Clausthal, 208 pp.; republished in *Bundesanstalt Bodenforsch., Geol. Jahrb.* **62,** 122 pp. (1969).

Siemens, G. (1961). Primärparagenese und Metamorphose des Ronnenberglagers nach Untersuchungen im Grubenfeld Salzdetfurth. Ph.D. Dissertation, Bergakademie, Clausthal, 207 pp.

Siesser, W. G. (1978). Petrography and geochemistry of pyrite and marcasite in DSDP Leg 40 sediments. Intl. Rep. Deep-Sea Drlg. Proj., Suppl. to v. 38–41, pp. 767–775.

Siesser, W. G., and Rogers, J. (1976a). Authigenic pyrite and gypsum in South West African continental slope sediments. *Tech. Rep., Mar. Geol. Prog., Dept. Geol., Cape Town Univ.* **8,** 86–91.

Siesser, W. G., and Rogers, J. (1976b). Authigenic pyrite and gypsum in South West African continental slope sediments. *Sedimentology* **23**(4), 567–577.

Sigl, W., Chamley, H., Fabricius, F., d'Argoud, G. G., and Mueller, J. (1978). Sedimentology and environmental conditions of sapropels. *In* Intl. Rep. Deep-Sea Drlg. Proj., vol. 42A (K. J. Hsu, L. Montadert *et al.,* eds.), pp. 445–466. U.S. Govt. Print. Off., Washington, D.C.

Simon, C. (1981). Quakes in the East. *Sci. News* **120**(15), 232–234.

Simon, J. (1929). Petrogenetische Studie der Salzlagerstätten der Gewerkshaften Volkenrode und Poethen im Südharzgebiet. *Kali (Halle)* **23,** 118–120.

Simon, P. (1967). Feinstratigraphische und paläogeorgraphisch–fazielle Untersuchungen des Stassfurt–Steinsalzes (Zechstein 2) im Kalisalzbergwerk "Königshall–Hindenburg." *Bundesanst. Bodenforsch., Geol. Jahrb.* **84,** 341–364.

Simon, P. (1972). Stratigraphie und Bromgehalt des Stassfurt–Steinsalzes (Zechstein 2) im hannoverischen Kalisalzebergbaugebiet. *Bundesanst. Geowiss., Geol. Jahrb.* **90,** 67–118.

Simpson, D. R. (1965). Experiments on the precipitation of salts from sea water. *Proc. Pennsylvania Acad. Sci.* **38**(2), 168–169.

Singh, D. R. (1972). Topotaktische Phanomene bei der Calcinierung, Sulfatisierung und Chloridisierung einiger Karbonat–Einzelkristalle. *Neues Jahrb. Mineral., Monatsh.* No. 1, pp. 12–22.

Singh, G., Joshi, R. D., and Singh, A. B. (1972). Stratigraphic and radiocarbon evidence for age and development of three salt lake deposits in Rajasthan, India. *Quat. Res.* **2**(4), 496–505.

Sisler, F. D., and Zobell, C. E. (1951). Hydrogen utilization by some marine sulphate–reducing bacteria. *J. Bacteriol.* **62,** 117–127.

Sjoegren, S. A. H., (1893). Om Vaetskeinneslutningar i gips fran Sicilien. *Geol. Foren. Forh.* **15**(3/150), 136–139.

Skempton, A. W. (1953). Soil mechanics in relation to geology. *Proc. Yorkshire Geol. Soc.* **29**(1), 33–62.

Skrotskiy, S. S. (1971). New data on oil and gas shows in the western part of the Caspian Basin (in Russian). *In* "Geologic Structure and Mineral Resources of the Kalmyk ASSR," vol. 1, pp. 39–42.

Skrotskiy, S. S. (1974a). Evidence of vertical hydrocarbon migration through salt in the western part of the Caspian Basin. *Dokl. Akad. Nauk SSSR,* **217**(4), 929–930. *Dokl. Acad. Sci. USSR, Earth Sci. Sect. (Engl. Transl. Am. Geol. Inst.)* **217**(1–6), 201–202.

Skrotskiy, S. S. (1974b). Pulsating nature of the filling of some evaporite basins (in Russian). *In* "Geochemical Questions in the Caspian Lowland" (S. S. Kumeyev, ed.), pp. 59–64. Kalmyk Gos. Univ., Elista.

Slaughter, M., and Milne, I. (1958). The formation of chlorite-like structures from montmorillonite. *Proc. Natl. Conf. Clay Miner.* **7,** 114–124.

Slivko, E. P., Slivko, M. M., and Petrichenko, O. I. (1973). Geochemistry of trace elements in salt deposits of the Carpathian region (in Ukrainian). *Vopr. Litol. Petrogr.* **2,** 129–136.

Sloss, L. L. (1969). Evaporite deposition from layered solutions. *Bull. Am. Assoc. Pet. Geol.* **53**(4), 776–789.

Smekal. A. (1927). Über weitere Untersuchungen an deformierten Steinsalzkristallen. *Anz. Akad. Wiss. Wien, Math.–Naturwiss. Kl.* **64,** 22–24.

Smith, D. B. (1970). The paleogeography of the British Zechstein. *In* "Symp. On Salt, 3rd." (J. L. Rau and L. F. Dellwig, eds.), vol. 1, pp. 20–23. N. Ohio Geol. Soc., Cleveland, Ohio.

Smith, D. B. (1971). Possible displacive halite in the Permian Upper Evaporite Group of northeast Yorkshire. *Sedimentology* **17**(3–4), 221–232.

Smith, D. B. (1973). The origin of the Permian middle and upper potash deposits of Yorkshire: An alternative hypothesis. *Proc. Yorkshire Geol. Soc.* **39**(3), 327–346.

Smith, D. B. (1974). Possible displacive halite in the Permian Upper Evaporite Group of northeastern Yorkshire. *Sedimentology* **17**(3–4), 221–232.

Smith, D. B. (1980). The evolution of the English Zechstein basin. *Contrib. Sedimentol.* **9,** 7–34.

Smith, D. B. (1981). The evolution of the English Zechstein Basin. *Proc. Int. Symp. Cent. Eur. Permian, Jablonna, 1978,* pp. 9–47.

Smith, D. B., and Crosby, A. (1979). The regional and stratigraphic context of Zechstein 3 and 4 potash deposits in the British sector of the southern North Sea and adjoining land areas. *Econ. Geol.* **74**(2), 397–408.

Smith, G. I., Friedman, I., and Matsuo, S. (1970). Salt crystallization temperatures in Searles Lake, California. *Mineral. Soc. Am. Spec. Paper* **3,** 257–259.

Snarsky, A. N. (1962). Die primäre Migration des Erdöls. *Freiberger Forschungsh.*, [C] **123,** 63–73.

Snowdon, P. N., and Turner, J. C. R. (1960). Concentration dependence of the Soret effect. *Trans. Faraday Soc.* **56,** 1812–1819.

Soerensen, H., Leonardsen, E. S., and Petersen, O. V. (1970a). Trona and thermonatrite from the Ilimaussaq alkaline intrusion, South Greenland. *Grønl. Geol. Unders., Misc. Publ.* No. 85, 19–37.

Soerensen, H., Leonardsen, E. S., and Petersen, O. V. (1970b). Trona and thermonatrite from the Ilimaussaq alkaline intrusion, South Greenland. *Bull. Geol. Soc. Denm.* **20**(1), 1–19.

Soerensen, J., Christensen, D., and Joergensen, B. B. (1981). Volatile fatty acids and hydrogen as substrate for sulfate-reducing bacteria in anaerobic marine sediment. *Appl. Environ. Microbiol.* **42**(1), 5–11.

Sokolov, A. S. (1965). On the genesis of native sulfur deposits (in Russian). *Litol. Polezn. Iskop.* No. 2, pp. 50–59.

Sokolov, G. A., and Pavlov, D. E. (1964). The source and the role of chlorine in magmatogene ore deposits. *Int. Geol. Congr., 22nd. 1964, New Delhi, India, Rep., Abstr.*, pp. 89–90; *Dokl. Soviet. Geol., Problem* No. 5, pp. 79–83.

Sokolov, N. A. (1957). Gas possibilities of the carnallite layers of the Solikamsk mine and its influence on the conduct of mining operations (in Russian). *Tr. Nauchno–Issled. Inst. Gorno-Khim. Syr'ya* No. 3, pp. 5–14.

Sokolov, P. N. (1969). Pyroclastic material in evaporite formations of the Irkutsk amphitheater (in Russian). *Tr. Sib. Nauchno–Issled. Inst. Geol. Geofiz. Mineral. Syr'ya* **91,** 85–91.

Sokolov, P. N. (1971). Stages in salt accumulation in the Cambrian of the Irkutsk amphitheatre (in Russian). *Geol. Geofiz.* No. 9, pp. 30–39.

Sokolova, M. N., Kapelyushnikova, L. M., and Mironov, S. I. (1962). The nature of the nitrogen compounds in Devonian bitumina in the Minusinsk region (in Russian). *In* "Geochemistry of Combustible Rocks and their Deposits," Gos. Kom. Sov. Min. SSSR po Toplivin. Prom., Inst. Geol. Razrabotki Goryuch. Iskop., pp. 5–35.

Solé, A. (1954). Die rhythmischen Kristallisationen im Influenztagogramm. *Kolloid–Z.* **137**(1), 15–19.

Solowjow, B. A. (1978). Allgemeine Gesetzmässigkeiten der Verteilung von Erdöl und Erdgas in den Salzbildungsbecken Europas. *Z. Angew. Geol.* **24**(6), 243–249.

Sonnenfeld, P. (1964). Dolomites and dolomitization—a review. *Bull. Can. Pet. Geol.* **12**(1), 101–132.

Sonnenfeld, P. (1974). The Upper Miocene evaporite basins in the Mediterranean region—a study in paleo-oceanography. *Geol. Rundsch.* **63**(3), 1133–1172.

Sonnenfeld, P. (1975). The significance of Upper Miocene (Messinian) evaporites in the Mediterranean Sea. *J. Geol.* **83**(3), 287–311.

Sonnenfeld, P. (1977). Origins of Messinian sediments in the Mediterranean region—some constraints on their interpretation. *Ann. Geol. Pays Hell.* **27,** 160–189.

Sonnenfeld, P. (1978). Effects of a variable sun at the beginning of the Cenozoic era. *Clim. Change* **1,** 355–382.

Sonnenfeld, P. (1979). Brines and evaporites. *Geosci. Can.* **6**(2), 83–90.

Sonnenfeld, P. (1980). Postulates for massive evaporite formation. *Rev. Géol. Méditerr.* **7**(1), 103–113.

Sonnenfeld, P. (1981). The Phanerozoic Tethys Sea. *In* "Tethys—the Ancestral Mediterranean" (P. Sonnenfeld, ed.), pp. 18–53. Hutchinson Ross, Inc., Stroudsburg, Pennsylvania.

Sonnenfeld, P., and Hudec, P. P. (1977). Stratified brines in restricted basins as a source of oil and oilfield brines. *Proc. Symp. on Water–Rock Interact.*, 2nd., Strasbourg, 1977, vol. 2, pp. 42–49.

Sonnenfeld, P., and Hudec, P. P. (1978). Geochemistry of a meromictic brine lake. *Geol. Mijnbouw* **57**(2), 333–337.

Sonnenfeld, P., and Hudec, P. P. (1980). Heliothermal lakes. *In* "Hypersaline Brines and Evaporitic Environments" (A. Nissenbaum, ed.), pp. 93–100. Elsevier, Amsterdam.

Sonnenfeld, P., Hudec, P. P., and Turek, A. (1976). Stratified heliothermal brines as metal concentrators. *Acta Cient. Venez.* **27,** 190–195.

Sonnenfeld, P., Hudec, P. P., Davis, M. W., and Dermitsakis, M. (1978). Messinian clastics in the Mediterranean region. *Ann. Géol. Pays Hell.* **29,** 595–602.

Sonnenfeld, P., Hudec, P. P., Turek, A., and Boon, J. A. (1977). Base metal concentration in a density stratified evaporite pan. *In* "Reefs and Evaporites—Concepts and Depositional Models" (J. H. Fisher, ed.), *Am. Assoc. Pet. Geol. Stud. Geol.* **5,** 181–186.

Sonnenfeld, P., Hudec, P. P., and Dermitsakis, M. (1983). Matrix provenance of Upper Miocene gypsiferous conglomeratic sandstones. *Rapp. Proc.-v. Comm. Int. Mer Méditerr.* **28**(4), 47–49.

Soriano, J., Marfil, R., de la Peña, J. A. (1977). Sedimentacion salina actual en las lagunas del norte de Alcazar de San Juan (Ciudad Real). *Estud. Geol. (Madrid)* **33**(2), 123–130.

Sorokin, J. I. (1970). Interrelations between sulphur and carbon cycles in meromictic lakes. *Arch. Hydrobiol.* **66**(4) 391–446.

Sorokin, J. I. (Yu. I.) (1972). The bacterial population and the processes of hydrogen sulphide oxidation in the Black Sea. *J. Cons. Int. Explor. Mer* **34**(3), 423–454.

Sorokin, J. I. (1975). Sulfide formation and chemical composition of bottom sediments of some Italian lakes. *Hydrobiologia* **47**(2), 231–240.

Sorokin, J. I., and Donats, N. (1975). Carbon and sulfur metabolism in meromictic lake Faro (Sicily). *Hydrobiologia* **47**(2), 241–252.

Southard, J. C. (1940). Heat of hydration of calcium sulfate. *J. Ind. Eng. Chem.* **32,** 442–445.

Sozanskiy, V. I. (1971). Geologic aspects of evaporite formation (in Russian). *In* "Problems of the Inorganic Origin of Petroleum" (V. B. Porfir'ev, ed.), pp. 151–172. Izd. Naukova Dumka, Kiev.

Sozanskiy, V. I. (1973). "Geology and Genesis of Salt-Bearing Formations" (in Ukrainian). Izd. Naukova Dumka, Kiev, 199 pp.

Sozinov, N. A., Nechipornenko, G. O., and Bondarenko, G. F. (1982). Experimental data on the effect of organic matter during formation of high-magnesium calcite ("protodolomite"). *Dokl. Akad. Nauk SSSR* **264**(3), 708–711; *Dokl. Acad. Sci. USSR, Earth Sci. Sect. (Engl. Transl. Am. Geol. Inst.)* **264**(1–6), 192–195.

Spezia, G. (1888). Über die Druckwirkung bei der Anhydritbildung. *Z. Kristallogr. Mineral.* **13,** 302.

Spezia, G. (1909). Metallic sodium as the alleged cause of the natural blue color of rock salt. *Centralbl. Mineral., Geol., Paläont.*, pp. 398–404. *Chem. Abstr.* **3,** 2106.

Spiro, B. (1977). Bacterial sulfate reduction and calcite precipitation in hypersaline deposition of bituminous shales. *Nature (London)* **269**(5625), 235–237.

Srebrodol'skiy, B. I. (1972). Supergene transformation of gypsum into anhydrite. *Dokl. Akad. Nauk SSSR,* **204**(4), 964–965; *Dokl. Acad. Sci. USSR, Earth Sci. Sect. (Engl. Transl. Am. Geol. Inst.)* **204**(1–6), 208–209.

Srebrodol'skiy, B. I. (1973). Dehydration of minerals from sulfur ore under desert conditions. *Dokl. Akad. Nauk SSSR,* **213**(6), 1403–1405; *Dokl. Acad. Sci. USSR, Earth Sci. Sect. (Engl. Transl. Am. Geol. Inst.)* **213**(1–6), 153–155.

Srivastava, K. K. (1970). Gypsum. *Bull. Geol. Surv. India,* [A] **32,** 1–109.

Stadnikoff, G. (1930). "Die Entstehung von Kohle und Erdöl. Die Umwandlung organischer Substanz im Laufe geologischer Zeitperioden." Enke, Stuttgart, 254 pp.

Stadtbaeumer, F. J. (1976). The influence of inorganic salts on some soil mechanics parameters of clays. *Bull. Int. Assoc. Eng. Geol.,* No. 14, pp. 65–69.

Stadtbaeumer, F. J. (1977). Die Beeinflussbarkeit physikalischer Eigenschaften von Tonen durch anorganische Salze. *Muenstersche Forsch. Geol. Palaeont.* **46,** 135 pp.

Steiger, G. (1910). A reconnaissance of the gypsum deposits in California. *Bull. U.S. Geol. Surv.* **413,** 1–37.

Steinbrecher, B. (1959). Die Subrosion des Zechsteins im östlichen und nördlichen Harzvorland, besonders in der Edderitz–Synklinale. *Geologie* **8,** 489–522.

Steinike, U. (1962). Lösungs- und Wachstumsbehinderung im System KCl–H_2O durch Blockierung mit einer Deckschicht komplexer Cyanide. *Z. Anorg. Allg. Chem.,* [B] **317**(3–4), 186–203.

Steinmetz, H. (1932). Die blaue Farbe des Steinsalzes als eine Begleiterscheinung der Funkenentladung in Salzkristallen. Neues *Jahrb. Mineral. Geol., Beilageband,* [A] **65,** 119–128.

Stephens, D. W., and Gillespie, D. M. (1976). Phytoplankton production in the Great Salt Lake, Utah, and a laboratory study of algal response to enrichment. *Limnol. Oceanogr.* **21,** 74–87.

Stern, M. E. (1980). Mixing in stratified fluids. *In* "Hypersaline Brines and Evaporitic Environments" (A. Nissenbaum ed.), pp. 9–32. Elsevier, Amsterdam.

Sterne, E. J., Reynolds, R. C., Jr., and Zantop, H. (1982). Natural ammonium illites from black shales hosting a stratiform base metal deposit, Delong Mountains, northern Alaska. *Clays, Clay Miner.* **30**(3), 161–166.

Stewart, A. J. (1979). A barred-basin evaporite in the Upper Proterozoic of the Amadeus Basin, central Australia. *Sedimentology* **26**(1), 33–62.

Stewart, F. H. (1949). The petrology of the evaporites of the Eksdale No. 2 boring, east Yorkshire; part I. The lower evaporite bed. *Mineral. Mag.* **28**(206), 621–675.

Stewart, F. H. (1951). The petrology of the evaporites of the Eksdale No. 2 boring, east Yorkshire; part II. The middle evaporite bed. *Mineral. Mag.* **29**(212), 445–475.

Stewart, F. H. (1953). Early gypsum in the Permian evaporites of northeastern England. *Proc. Geol. Assoc. London* **64**(1), 33–39.

Stewart, F. H. (1954). Permian evaporites and associated rocks in Texas and New Mexico compared with those of northern England. *Proc. Yorkshire Geol. Soc.* **29,** 185–235.

Stewart, F. H. (1956). Replacements involving early carnallite in the potassium-bearing evaporites of Yorkshire. *Mineral. Mag.* **31**(233), 127–135.

Stewart, F. H. (1963a). Marine evaporites. *U.S. Geol. Surv. Prof. Paper* **440-Y,** 53 pp.

Stewart, F. H. (1963b). The Permian lower evaporites of Fordon in Yorkshire. *Proc. Yorkshire Geol. Soc.* **34**(1), 1–4.

Stewart, F. H. (1965). Mineralogy of the British Permian evaporites. *Mineral. Mag.* **34**(268), 460–470.

Stewart, F. H., and Vincent, E. A. (1951). The petrology of the evaporites of the Eksdale No. 2 boring, east Yorkshire; part III. The upper evaporite bed. *Mineral. Mag.* **29**(213), 557–572.

Stieglitz, J. (1908). The relationship of equilibrium between the carbon dioxide of the atmosphere and calcium sulphate, calcium carbonate and calcium bicarbonate in water solutions in contact with it. *Publ. Carnegie Inst. Washington, D.C.* **107,** 235–264.

Stockdale, P. B. (1936). Rare stylolites. *Am. J. Sci.* **32**(188), 129–133.

Stoecklin, J. (1968). Salt deposits of the Middle East. *In* "Saline Deposits" (R. B. Mattox *et al.*, eds.). *Spec. Paper Geol. Soc. Am.* **88,** 157–182.

Stoertz, G. E., and Ericksen, G. E. (1974). Geology of salars in northern Chile. *U.S. Geol. Surv. Prof. Paper* **811,** 1–65.

Stoessell, R. K., and Byrne, P. A. (1982). Salting-out of methane in single-salt solutions at 25°C and below 800 psia. *Geochim. Cosmochim. Acta* **46**(8), 1327–1332.

Stoffers, P., and Fishbeck, R. (1974). Monohydrocalcite in the sediments of Lake Kivu, Africa; with discussion by W. E. Krumbein. *Sedimentology* **21**(4), 163–170; **22**(4), 631–636.

Stoffers, P., and Kuehn, R. (1974a). Red Sea evaporites. A petrographic and geochemical study. *In* Intl. Rep. Deep-Sea Drlg. Proj., vol. 23 (R. B. Whitmarsh, D. E. Weser, D. A. Ross, *et al.*, eds.), pp. 821–847. U.S. Govt. Print. Off., Washington, D.C.

Stoffers, P., and Kuehn, R., (1974b). Geochemische und petrographische Untersuchungen an Evaporit–Bohrkernen aus dem Roten Meer. *Kali Steinsalz* **6**(8), 290–299.

Stoiber, R. E., and Rose, W. I., Jr. (1974). Fumarole encrustations at active central American volcanoes. *Geochim. Cosmochim. Acta* **38**(4), 495–516.

Stokes, R. H., and Robinson, R. A. (1948). Ion hydration and activity in electrolyte solutions. *J. Am. Chem. Soc.* **70,** 1870–1878.

Stokes, R. H., and Robinson, R. A. (1949). Standard solutions for humidity control at 25°C. *Ind. Eng. Chem.* **41,** 2013.

Stommel, H., and Farmer, H. G. (1952). Abrupt changes in width in two layer open channel flow. *J. Mar. Res.* **11**(2), 205–214.

Stoops, G. J., and Zavaleta, A. (1978). Micromorphological evidence of barite neoformation in soils. *Geoderma* **20**(1), 63–70.

Storck, U. (1953). Die Entstehung der Vertaubungen in den Kalisalzlagern im Zusammenhang mit regelmässigen Begleiterscheinungen auf dem Kaliwerk Königshall-Hindenburg. *Z. Dtsch. Geol. Ges.* **105**(4), 685.

Storck, U. (1954). Die Entstehung der Vertaubungen und des Hartsalzes im Flöz Stassfurt im Zusammenhang mit regelmässigen Begleiterscheinungen auf dem Kaliwerk Königshall–Hindenburg. *Kali Steinsalz* **1**(6), 21–31.

Storck, U. (1964). Die Entstehung und Metamorphose ozeaner Salzlagerstätten. *Kali Steinsalz* **12**(2), 61–63.

Stoughton, R. W., and Lietzke, M. H. (1965). Calculations of some thermodynamic properties of sea salt solutions at elevated temperatures from data on NaCl solutions. *J. Chem. Eng. Data* **10**(3), 254–260.

Strakhov, N. M. (1962). ''Principles of Lithogenesis.'' Izd. Akad. Nauk SSSR, Moscow (transl.: Consultants Bureau, New York, 1967–1970, 3 vols.).

Street, F. A., and Grove, A. T. (1976). Environmental and climatic implications of late Quaternary lake-level fluctuations in Africa. *Nature (London)* **261**(5559), 385–390.

Stroitelev, S. A. (1960). The effect of cosolutes on the crystallization of some sulfates (in Russian). *Uch. Zap. Tomsk. Gos. Pedagog. Inst.*, No. 36, pp. 158–170.

Strong, M. W. (1956). Marine iron bacteria as rock forming organisms. *Adv. Sci.* **12**(49), 583–585.

Strutt, R. J. (1909). Helium in saline minerals and its probable connection with potassium. *Proc. R. Soc. London,* [4] **81,** 278–279.

Stuart, L. S. (1940). The growth of halophilic bacteria in concentration of sodium chloride above three molar. *J. Agric. Res. (Washington, D.C.)* **61,** 259–265.

Stuart, L. S., and James, L. H. (1937). The effect of Eh and sodium chloride concentration on the physiology of halophilic bacteria. *J. Bacteriol.* **35**(4), 381–396.

Stuart, L. S., and James, L. H. (1938). The effect of sodium chloride on the Eh of protogenous media. *J. Bacteriol.* **35,** 369–380.

Stuart, L. S., and Swenson, T. L. (1934). Morphological and physiological observations on salt-tolerant bacteria. *J. Am. Leather Chem. Assoc.* **29,** 142–158.

Stumm, W., and Morgan, J. J. (1970). ''Aquatic Chemistry.'' Wiley (Interscience), New York, 583 pp.

Sturmfels, E. (1943). Das Kalisalzlager von Buggingen (Südbaden). *Neues Jahrb. Mineral., Geol., Paläontol., Abh., Beil.,* **78A,** 131–216.

Stutzer, O. (1933). Über die im Mikroskop sichtbare Bitumenführung des Hauptdolomits von Volkenrode und des ''Stinkschiefers'' von Mansfeld. *Kali (Halle)* **27**(4), 43–45; (5), 53–56; (6), 73–75.

Suess, E. (1970). Interaction of organic compounds with calcium carbonate. *Geochim. Cosmochim. Acta* **34**(2), 157–168; **37**(11), 2435–2437.

Suess, E. (1982). Authigenic barite in varved clays: Result of marine transgression over freshwater deposits associated changes in interstitial water chemistry. *In* ''The Dynamic Environment of the Ocean Floor'' (K. A. Fanning and F. T. Manheim, eds.), pp. 339–355. Lexington Books, Lexington, Massachusetts.

Sugden, W. (1963). The hydrology of the Persian Gulf and its significance in respect to evaporite deposition. *Am. J. Sci.* **261**(8), 741–755.

Sun, M. S. (1962). Oriented growths of cryptomelane in sylvite, Carlsbad, New Mexico. *Am. Mineral.* **47**(1–2), 152–155.

Sun, T. P. (1974). Origin of Recent potash deposits in a certain lake in China (in Chinese). *Geochemistry (China)* No. 4, pp. 230–248.

Surdam, R. C., and Parker, R. D. (1972). Authigenic aluminosilicate minerals in the tuffaceous rocks of the Green River Formation, Wyoming. *Bull. Geol. Soc. Am.* **83**(3), 689–700.

Suzuki, R. (1943). The dehydration of gypsum (in Japanese). *Jpn. Assoc. Mineral., Petrol., Econ. Geol.* **30**(1), 11–32.

Svalgaard, L. (1973). Solar activity and the weather. *In* "Correlated Interplanetary and Magnetospheric Observations" (D. E. Page, ed.), pp. 627–639. Reidel, Dordrecht, The Netherlands.

Sweeney, J. F. (1976). Evolution of the Sverdrup Basin, Arctic Canada. *Tectonophysics* **36**(1–3), 181–196.

Sweeney, J. F. (1977). Subsidence of the Sverdrup Basin, Canadian Arctic Islands. *Bull. Geol. Soc. Am.* **88**(1), 41–48.

Swineford, A., and Runnels, R. T. (1953). Identification of polyhalite (a potash mineral) in Kansas Permian salt. *Trans. Kans. Acad. Sci.* **56,** 364–370.

Swoboda, A. R., and Thomas, G. W. (1965). The movement of sulfate salts in soils. *Proc. Soil Sci. Soc. Am.* **29**(5), 540–544.

Sydow, W. (1959). Die Ausbildung des Ronnenberg-Lagers unter besonderer Berücksichtigung des petrographischen Aufbaus und seiner sekundären Veränderung. *Kali Steinsalz* **2**(12), 406–418.

Szatmari, P., 1980. The origin of oil deposits; a model based on evaporites. *Proc. Congr. Brasil. Geol. Balnear. Camboria, 31st., Soc. Brasil. Geol.*, pp. 455–499.

Szatmari, P., Carvalho, R. S., and Simoes, I. A. (1979). A comparison of evaporite facies in the Late Paleozoic Amazon and the Middle Cretaceous South Atlantic salt basins. *Econ. Geol.* **74**(2), 432–447.

Taft, W. H. (1961). Authigenic dolomite in modern carbonate sediments along the south coast of Florida. *Science* **134**(3478), 561–562.

Tagaeva, N. (1935). Über die Entstehung der Erdölwasser. *Pet. Z.* **31**(32), 15–32.

Tagaki, T. (1931). Biologische Untersuchungen über die Durchlässigkeit von Gips und Glimmer fur Ultraviolet. *Strahlentherapie* **40,** 189–192.

Tammann, G. (1931). Umkristallisierung. *Z. Elektrochem.* **37,** 429–436.

Tanaka, T., and Sugimoto, M. (1965). Über die Zersetzung von Gips durch Erhitzung. *Zem.–Kalk–Gips* **18**(12), 636–640.

Tanaka, Y., Nakamura, K., and Hara, R. (1931). On the calcium sulphate in sea water. *J. Soc. Chem. Ind. Jpn.* **34,** Suppl., 284–287.

Taneda, S. (1949). The anhydrite from a Himeshima lava (in Japanese). *J. Jpn. Assoc. Mineral., Petrol., Econ. Geol.* **33**(3), 69–73.

Tanner, C. C. (1927). The Soret effect. *Trans. Faraday Soc.* **23,** 75–95.

Tanner, C. C. (1952). The Soret effect. *Nature (London)* **170**(4314), 34–35.

Tarakanova, Ye. I. (1973). Authigenic minerals of sulfur in recent peat bogs of the Urals. *Litol. Polezn. Iskop.* **8**(4), 78–82; *Lithol. Mineral. Resour. (Engl. Transl.)* **8**(4), 462–465.

Tarasevich, S. I., Tsakhnovskiy, M. A., Chechel, E. I., Mashevich, Ya. G., and Vasilevskiy, A. F. (1971). New data on the absolute age of rock salt from the Siberian Platform. *Dokl. Akad. Nauk SSSR* **199**(4), 905–908; *Dokl. Acad. Sci. USSR, Earth Sci. Sect. (Engl. Transl. Am. Geol. Inst.)* **199**(1–6), 69–71.

Tarasov, N. I. (1939). Low pH values in natural waters and the biology of hypersaline brines (in Russian). *Priroda,* No. 10, pp. 76–77.

Tardy, Y., Krempp, G., and Trauth, N. (1972). Le lithium dans les minéraux argileux des sédiments et des sols. *Geochim. Cosmochim. Acta* **36**(4), 397–412.

Tarr, W. A. (1929). Doubly terminated quartz crystals occurring in gypsum. *Am. Mineral.* **14**(1), 19–25.

Tasch, P. (1963). Fossil content of salt and association evaporites. *In* "Symp. on Salt" (A. C. Bersticker, K. E. Hoekstra, and J. F. Hall, eds.), pp. 96–102. N. Ohio Geol. Soc., Cleveland, Ohio.

Tasch, P. (1970). Biochemical and geochemical aspects of the White Salt Pan—Bonaire, N. A. *In* "Symp. on Salt, 3rd." (J. L. Rau and L. F. Dellwig, eds.), vol. 1, pp. 204–210. N. Ohio Geol. Soc., Cleveland, Ohio.

Tatarskiy, V. B. (1949). Dedolomitization of rocks (in Russian). *Dokl. Akad. Nauk SSSR* **69,** 849–851.

Taylor, G. R. (1982). A mechanism for framboid formation as illustrated by a volcanic exhalative sediment. *Mineral. Deposita* **17**(1), 23–36.

Taylor, J. C. M. (1980). Origin of the Werraanhydrit in the U.K. Southern North Sea—a reappraisal. *Contrib. Sedimentol.* **9,** 91–113.

Taylor, J. C. M., and Coulter, V. S. (1975). Zechstein of the English Sector of the Southern North Sea Basin. *In* "Petroleum and the Continental Shelf of North-West Europe" (A. W. Woodland, ed.), vol. 1, pp. 249–263. Wiley, New York.

Taylor, R. E. (1937). Water insoluble residue in rock salt of Louisiana salt plugs. *Bull. Am. Assoc. Pet. Geol.* **21**(10), 1268–1310; (11), 1496; (12), 1594.

Taylor, R. E. (1972). Caprock genesis and occurrence of sulphur deposits. *In* "Geology of Saline Deposits" (G. Richter-Bernburg, ed.), U.N.E.S.C.O., Paris, Earth Sci. Ser. **7,** 253–254.

Teichmueller, R. (1958). Ein Querschnitt durch den Südteil des niederrheinischen Zechsteinbeckens. *Bundesanst. Bodenforsch., Geol. Jahrb.* **73,** 39–50.

Tennyson, C. (1963). Eine Systematik der Borate auf Kristallchemischer Grundlage. *Fortschr. Mineral.* **41**(1), 64–91.

Teodorovich, G. I. (1958). "Authigenic Minerals in Sedimentary Rocks." Akad. Nauk SSSR Moscow (transl. Consultants Bureau, New York, 1961, 120 pp.).

Teraoka, I., and Ito, S. (1958). Decomposition of syngenite (in Japanese). *Gypsum Lime (Jpn.)* **1,** 1622–1626.

Teruggi, M. E., Dalla Salda, L. H., and Dangaus, N. V. (1974). La presencia de yeso en la Laguna de las Barrancas, Provincia de Buenos Aires. *An., Lab. Ensayo Mater. Invest. Tecnol. Prov. Buenos Aires,* [2], No. 2, pp. 121–131.

Tew, R. W. (1980). Halotolerant Ectothiorhodospira survival in mirabilite: Experiments with a model of chemical stratification by hydrate deposition in saline lakes. *Geomicrobiol. J.* **2**(1), 13–20.

Tezuka, Y., Takii, S., and Kitamura, H. (1963). Anaerobic decomposition of organic matter by the microflora of polluted river waters. Organic acid fermentation coupled with sulfate reduction. *Jpn. J. Ecol.* **13,** 188–196.

Theilig, F., and Pensold, G. (1964). Über das Vorkommen von Cölestin im Hauptdolomit des norddeutschen Zechsteins. *Chem. Erde* **23**(3), 215–218.

Thiede, D. S., and Cameron, E. N. (1978). Concentration of heavy metals in the Elk Point evaporite sequence, Saskatchewan. *Econ. Geol.* **73**(3), 405–415.

Thompson, E. F., and Gilson, H. C. (1937). Chemical and physical investigations: Introduction. *Sci. Rep., John Murray Exped. (1934)* **2**(2), 15–81.

Thompson, T. G., and Nelson, K. H. (1956). Concentration of brines and deposition of salts from seawater under frigid conditions. *Am. J. Sci.* **254**(4), 227–238; (12), 758–760.

Tilden, J. E. (1930). Phycological examination of fossil red salt from three localities in southern states. *Am. J. Sci.* **19,** 297–304.

Til'mans, Yu. Ya. (1951). Dendritic crystallization of salts from aqueous solutions. *Dokl. Akad. Nauk SSSR* **78**(1), 83–86.

Til'mans, Yu. Ya. (1952). Dendritic crystallization of salts from aqueous solutions. *Zh. Obshch. Khim.* **22,** 385–396; *J. Gen. Chem. USSR (Engl. Transl.)* **22,** 461–467.

Titov, A. V. (1939). Effect of temperature on the composition of complexes obtained by the interaction of chloride solutions (in Russian). *Tr. Ivanov. Khim.-Technol. Inst.* **2,** 12–24.

Titov, A. V. (1949). Complex compounds formed by magnesium chloride in aqueous solution (in Russian). *Zh. Obshch. Khim.* **19,** 458–461.

Tittel, M. (1959). Thermodynamische Berechnungen zu Untersuchungen über eine einstufige Carnallitentwässerung. *Freiberger Forschungsh.*, [A] **123,** 457–463.

Tokarska, K. (1971). Geochemical description of bituminous substances in the Zechstein copper shales (in Polish). *Kwart. Geol.* **15**(1), 67–76.

Tokarski, A. (1965). Stratigraphy of the salt-bearing Roet in the Fore–Sudetic monocline (in Polish). *Acta Geol. Pol.* **15**(2), 105–129.

Tolchel'nikov, Yu. S. (1962). Neoformation of calcium sulfates and carbonates in sandy soils of deserts (in Russian). *Pochvovedenie* No. 6, pp. 88–96.

Tollert, H. (1950a). Über den Nachweis von Molekülen höherer Ordnung in Mischungen verdünnter wässriger Elektrolyt–Lösungen und über einen neuen Trenneffekt. *Z. Phys. Chem.* **195,** 237–243.

Tollert, H. (1950b). Über die Verteilung der Temperatur, der Konzentration und der Strömungsgeschwindigkeit in dynamisch–polythermen Systemen. *Z. Phys. Chem.* **195,** 281–294.

Tollert, H. (1956). Die kinetische und stationare Bestimmung von Lösungsgewichten leichtlöslicher Salze und deren thermodynamische Grundlagen zur Deutung des metastabilen Sättigungszustandes mit Hilfe der Hydratationsenthalpien. *Z. Phys. Chem.* [*N.S.*] **6,** 242–260.

Tollert, H. (1964). Beitrag zur Porosität von Salzgesteinen. *Kali Steinsalz* **4**(2), 55–60.

Tolloczko, S. (1910). The solution velocity of different faces of gypsum. *Bull. Int. Acad. Sci. Krakow,* pp. 209–218.

Tone, K. (1934). A study on gypsum. *Rep. Imp. Ind. Res. Inst. Osaka Prefect.* **15**(6), 1–56.

Tooms, J. S. (1970). Review of knowledge of metalliferous brines and related deposits. *Trans. Inst. Min. Metall., Sect. B* **79,** 116–126.

Torii, T., Murata, S., and Yamagata, N. (1981). Geochemistry of the Dry Valley Lakes. *J. R. Soc. N. Z.* **11**(4), 387–399.

Toriumi, T., and Hara, R. (1938). On the transition point of calcium sulphate in water and concentrated sea water. *Technol. Rep. Tohoku Imp. Univ.* **12**(4), 572–590.

Touray, J. C. (1970). Analyse thermo-optique des familles d'inclusions à dépôts salins (principalement halite). *Schweiz. Mineral. Petrogr. Mitt.* **50**(1), 67–79.

Touray, J. C., and Yajima, J. (1967). Hydrocarbures liquides inclus dans des fluorites tunisiennes. *Mineral. Deposita* **2,** 286–290.

Tovbin, M. V., and Boyevudskaya, Z. L. (1956). Self–adsorption I. surface tension of salt solution (in Ukrainian). *Ukr. Khim. Zh.* **22,** 173–179.

Tovbin, M. V., and Krasnova, S. I. (1955). The stability of supersaturated solutions of slightly soluble salts (in Ukrainian). *Ukr. Khim. Zh.* **21,** 32–38.

Trashliev, S. (1971). Bassanite and anhydrous calcium sulfate in the gypsum horizon of northwestern Bulgaria (in Bulgarian). *Izv. Geol. Inst. Bulg. Akad. Nauk, Ser. Rud. Nerud. Polezn. Iskop.* No. 19–20, pp. 147–158.

Trauth, N. (1977). Argiles évaporitiques dans la sédimentation carbonatée continentale et epicontinentale tertiaire; bassins de Paris, de Mormoiron et de Sabinelles (France), Jbel Gassoul (Maroc). *Mém. Soc. Géol. Strasbourg.* **49,** 162 pp.

Traveria Cros, A., Amigo, J. M., and Montoriol-Pous, J. (1971). Nota sobre una masa de "rosas del desierto" recogida en el Gran Erg oriental (Argelia). *Acta Geol. Hisp.* **6**(2), 49–52.

Treesh, M. I., and Friedman, G. M. (1974). Sabkha deposition of the Salina Group (Upper Silurian) of New York State. *In* "Symp. on Salt, 4th." (A. H. Coogan, ed.), vol. 1, pp. 35–46. N. Ohio Geol. Soc., Cleveland, Ohio.

Tret'yakov, Yu. A. (1974). Zones of impoverishment in the Upper Kama potash deposit. *Litol. Polezn. Iskop.* **9**(1), 75–85; *Lithol. Mineral. Resour. (Engl. Transl.)* **9**(1), 59–67.

Trichet, J. (1963). Description d'une forme d'accumulation de gypse par voie éolienne dans le sud Tunisien. *Bull. Soc. Géol. Fr.*, [7] **5**(4), 617–621.

Trofimuk, P. I. (1970). Bromine chlorine ratios in the Cambrian evaporite beds of the Angara–Lena basin and their interpretation during prospecting for potash salt deposits (in Russian). *Tr. Inst. Geol. Geofiz. Akad. Nauk SSSR, Sib. Otd.*, **116,** 83–99.

Truc, G. (1980). Evaporites in a subsident continental basin (Ludian and Stampian of Mormoiron–Pernes in southeastern France). Sequential aspects of deposition, primary facies and their diagenetic evolution. *In* "Evaporite Deposits" (G. Busson, ed.), pp. 61–71. Ed. Technip, Paris.

Truesdell, A. H., and Jones, B. F. (1969). Ion association in natural brines. *Chem. Geol.* **4**(1), 51–62.

Trusheim, F. (1971). Zur Bildung der Salzlager im Rotliegenden und Mesozoikum Europas. *Bundesanst. Bodenforsch., Geol. Jahrb., Beih.* **112,** 1–51.

Tschoertner, U.S. (1969). The influence of artificial physical changes on the ecology of Rondevlei, Cape Province. *Invest. Rep. Cape Good Hope Provinc. Admin., Dep. Nat. Conserv.* **13,** 1–12.

Tseng Cheng, P., *et al.* (1965). Separation of carnallite from sea salt bittern (in Chinese). *J. Chem. Eng. (China),* No. 2, pp. 69–80. *Chem. Abstr.* **65,** 16,554.

Tucker, M. E. (1978). Gypsum crusts (gypcrete) and patterned ground from northern Iraq. *Z. Geomorph.*, [*N.S.*] **22**(1), 89–100.

Turk, L. J. (1974). Leakage from solar evaporation ponds associated with sediment–brine interactions. *In* "Symp. on Salt, 4th." (A. H. Coogan, ed.), vol. 2, pp. 403–407. N. Ohio Geol. Soc., Cleveland, Ohio.

Twenhofel, W. H. (1923). Development of shrinkage cracks in sediments without exposure to the atmosphere (abstr.). *Bull. Geol. Soc. Am.* **34**(1), 64.

Udden, J. A. (1924). Laminated anhydrite in Texas. *Bull. Geol. Soc. Am.* **35**(2), 347–354.

Ujueta, L. G. (1969). Salt in the eastern Cordillera of Colombia. *Bull. Geol. Soc. Am.* **80**(11), 2317–2320.

Ul'masova, M. I. (1971). Varieties of halite rocks in the Khodzhaikan deposit (in Russian). *Nauchn. Tr. Tashk. Gos. Univ.* **403,** 8–13. *Chem. Abstr.* **78,** 60,632.

Urazov, G. G. (1930). Sequence of formation of the Solikamsk salt deposits from the viewpoint of a study of crystallization of the system: KCl–NaCl–$MgCl_2$–H_2O. *Ann. Inst. Anal. Phys.–Khim. (Leningrad)* **4**(2), 41–83.

Urazov, G. G. (1932). Sequence of salt deposition in the Solikamsk potash deposit (in Russian). *Tr. Vses. Geologorazved. Ob'edin.* **43,** 41–83.

Urry, W. D. (1937). Helium ratio of Florida anhydrite. *Am. Mineral.* **22**(3), 212.

Usdowski, R., and Usdowski, H. E. (1970). Use of minerals of marine evaporites as humidity indicators. *Contrib. Mineral. Petrol.* **29**(2), 135–144.

Usiglio, J. (1849). Analyse de l'eau de la Méditerranée sur les côtes de France. *Ann. Chim. Phys.*, [3] **27**(2), 92–107; (3), 177–191.

Vai, G. B., and Ricci-Lucchi, F. (1977). Algal crusts, autochthonous and clastic gypsum in a cannibalistic evaporite basin: A case history from the Messinian of Northern Appennines. *Sedimentology* **24,** 211–244.

Vai, G. B., and Ricci-Lucchi, F. (1978). The Vera del Gesso in northern Appennines: Growth and mechanical breakdown of gypsified algal crust. *Mem. Soc. Geol. Ital.* **16,** 217–247.

Vakhremeyova, V. A. (1954). On the genesis of varicolored sylvinites of the Upper Kama deposit (in Russian). *Tr. Vses. Nauchno–Issled. Proektn. Inst. Galurgii* **29,** 129–142.

Vaksman, E. G., and Onischenko, S. K. (1972). Gypsum-containing soils of lake–colluvial salt-accumulating conditions (in Russian). *Tr. Tadzh. Nauchno–Issled. Inst. Pochvoved.* **15**(1), 107–123.

Valentine, J. W. (1973). "Evolutionary Paleoecology." Prentice Hall, Englewood Cliffs, New Jersey, 512 pp.

Valentinov, S. (1916). Heliumgehalt im blauen Steinsalz. *Kali (Halle)* **6**(1), 1–3.

Valette, J. N. (1972). Étude minéralogique et géochimique des sédiments de mer Alboran: Resultats préliminaires. *C. R. Hebd. Séances Acad. Sci. Paris,* [D] **275**(21), 2287–2290.

Valyaev, B. M. (1970). Evaluation of the influence of basin isolation on salt accumulation (based on water–salt balance calculations) (in Russian). *Litol. Polezn. Iskop.* No. 6, pp. 83–90; Lithol. Mineral. Resour. (Engl. Transl.) No. 6, pp. 720–726.

Valyashko, M. G. (1951). Volume relations of liquid and solid phases in evaporating seawater (in Russian). *Dokl. Akad. Nauk SSSR* **77,** 1055–1058.

Valyashko, M. G. (1956). Geochemistry of bromide in the processes of salt deposition and the use of the bromide content as a genetic and prospecting tool. *Geokhimiya* No. 6, pp. 33–48; *Geochemistry USSR (Engl. Trans.)* No. 6, pp. 570–587.

Valyashko, M. G. (1958). Die wichtigsten geochemischen Parameter für die Bildung der Kalisalzlagerstätten. *Freiberger Forschungsh.,* [A] **123,** 197–235.

Valyashko, M. G. (1959). Geochemistry of bromide in the processes of salt deposition and the use of the bromide content as a genetic and prospecting tool. *Proc. Int Geol. Congr., 20th Mexico 1956* ("Symp. on Geochem. Prospect."), vol. 2, pp. 261–281.

Valyashko, M. G. (1962). "The Sequence of Formation of Potash Salt Deposits" (in Russian). Izd. Mosk. Gos. Univ., Moscow, 398 pp.

Valyashko, M. G. (1968). Physicochemical analysis of the leaching of complex potash rocks. *Litol. Polezn. Iskop.* No. 1, pp. 70–80; *Lithol. Mineral. Resour. (Engl. Transl.)* **2**(1), 56–63.

Valyashko, M. G. (1972). Playa Lakes: Necessary stage in the development of a salt bearing basin. *In* "Geology of Saline Deposits" (G. Richter-Bernburg, ed.), U.N.E.S.C.O., Paris, Earth Sci. Ser. **7,** 41–51.

Valyashko, M. G., and Lavrova, A. N. (1976). Some new possibilities of using bromine–chlorine relations for studying the conditions of formation of evaporite deposits (in Russian). *In* "The Bromium in Brines as a Geochemical Indicator of the Genesis of Salt Deposits" (A. P. Vinogradov, ed.), pp. 343–353. Moscow Univ. Press, Moscow.

Valyashko, M. G., and Solov'eva, E. F. (1949). Metastable equilibria in the system $2Na^{+}$–$2K^{+}$–Mg^{++}–SO_4^{--}–$2Cl^{-}$–H_2O (in Russian). *Tr. Vses. Nauchno–Issled. Proektn. Inst. Galurgii* **21,** 197–217.

Valyashko, M. G., and Solov'eva, E. F. (1953). Crystallization of sylvite upon evaporation of seawater (in Russian). *Tr. Vses. Nauchno–Issled. Proektn. Inst. Galurgii* **27,** 159–170.

Valyashko, M. G., and Vlasova, N. K. (1975). Stability of magnesium salts in solutions of marine origin and their geochemical significance (in Russian). *Vestn. Mosk. Univ., Ser. 4: Geol.* No. 4, pp. 16–27.

Valyashko, M. G., Zherebtsova, I. K., Lavrova, A., and U Bi Hao (1976a). Distribution of bromine between salt and solutions of different composition and concentration (in Russian). *In* "The Bromium in Brines as a Geochemical Indicator of the Genesis of Salt Deposits" (A. P. Vinogradov, ed.), pp. 381–404. Moscow Univ. Press, Moscow.

Valyashko, M. G., Zherebtsova, I. K., Grebennikov, N. P., and Yermakov, V. A. (1976b). Genesis of potash salts and bischofite in evaporites of the Volga region monocline (in Russian). *In* "The Bromium in Brines as a Geochemical Indicator of the Genesis of Salt Deposits" (A. P. Vinogradov, ed.), pp. 436–453. Moscow Univ. Press, Moscow.

Valyashko, M. G., Zherebtsova, I. K., Yermakov, V. A., and Grebennikov, N. P. (1979). Geochemische Analyse der Endstadien der Halogenese am Beispiel der Bischofit–Ablagerungen der Wolga-Monoklinale. *Z. Geol. Wiss.* **7**(8), 931–943.

van de Poll, H. W. (1978). Paleoclimatic control and stratigraphic limits of synsedimentary mineral occurrences in Mississippian–Early Pennsylvanian strata of eastern Canada. *Econ. Geol.* **73**(6), 1069–1081.

van der Hammen, T. (1972). Changes in vegetation and climate in the Amazon Basin and surrounding areas during the Pleistocene. *Geol. Mijnbouw* **51**(6), 641–643.

van der Hammen, T. (1974). Pleistocene changes in vegetation and climate in tropical South America. *J. Biogeogr.* **1,** 3–26.

van der Meer Mohr, C. G. (1972). Recent carbonates from the Netherlands Antilles: Hypersaline to open marine environments. *Ann. Soc. Belge Géol. Paléont. Hydrol.* **95,** 407–412.

van der Zwaan, G. J. (1982). Paleoecology of Late Miocene Mediterranean foraminifera. *Utrecht Micropaleontol. Bull.* **25,** 202 pp.

van Doesburg, J. D. J., Vergouwen, L., and van der Plas, L. (1982). Konyaite, $Na_2Mg(SO_4)_2 \cdot 5H_2O$, a new mineral from the Great Konya Basin, Turkey. *Am. Mineral.* **67,** 1035–1038.

van Rosmalen, G. M., Marchee, W. G. J., and Bennema, P. (1976). A comparison of gypsum crystals grown in silica gel and agar in the presence of additives. *J. Cryst. Growth* **35**(2), 167–176.

van Ruyven, B. H. (1953). An investigation of particles in concentrated solutions of electrolytes with respect to the hydration of ions. *Recl. Trav. Chim. Pays-Bas* **72,** 739–762.

van Tassel, R. (1965). Crystal particularities in dolomite synthesis. *Am. Mineral.* **50,** 503–504.

van t'Hoff, J. H., Armstrong, E. F., Hinricheson, W., Weigert, F., Just, G. (1903). Gips und Anhydrit. *Z. Phys. Chem.* **45,** 257–306.

van t'Hoff, J. H., *et al.* (1912). "Untersuchungen über die Bildungsverhältnisse der ozeanen Salzablagerungen insbesondere des Stassfurter Salzlagers." Akad. Verlagsanst., Leipzig, 374 pp.

van Werveke, L. (1924). Versuch einer Erklärung des seitlichen Übergangs der Salzlagerstätten im Lothringischen Keuper in eine Anhydritbildung. *Kali (Halle)* **18,** 265–268; 317–318.

van Zeist, W., and Wright, H. E. (1963). Preliminary pollen studies at Lake Zeribar, Zagros Mountains, southwestern Iran. *Science* **140**(3562), 65–67.

van Zeist, W., and Wright, H. E. (1967). Late Quaternary vegetation history of western Iran. *Rev. Palaeobot. Palynol.* **2**(1–4), 301–311.

Vardukadze, A. (1959). The influence of exchangeable cations on the value of the zeta-potential of clay suspensions (in Russian). *Tr. Gos. Univ. Tbilisi* **74,** 203–209.

Varlamov, A. A., Kozlov, S. S., Gemp, S. D., and Fomina, V. D. (1974). Some postsedimentary alterations of potash horizons of the Starobinsk deposit (in Russian). *Tr. Vses. Nauchno–Issled. Proektn. Inst. Galurgii* **71,** 70–84.

Vater, H. (1900). Einige Versuche über die Bildung des marinen Anhydrites. *Sitzungsber. Kgl. Preuss. Akad. Wiss.* No. 1, 269–295.

Vdovykin, G. P., and Shorokhov, N. R. (1980). Gases of Lower Cambrian rock salt of Mirnyi, Kuyumbinsk, Ust–Kama, and Vendian rock salt of the Upper Vilyuchan areas (East Siberia) (in Russian). *Geol. Geofiz.* No. 6, 135–138.

Veber, V. V., and Gorbunova, L. I. (1969). Role of redbed facies in oil formation (in Russian). *Sov. Geol.* **12**(9), 3–15.

Veitch, J. D., and McLeroy, D. G. (1972). Organic mobilization of ore metals in low-temperature carbonate environments. *Geol. Soc. Am., Abstr. Prog.* **4**(7), 110–111.

Veizer, J., and Compton, W. (1974). $^{87}Sr/^{86}Sr$ composition of seawater during the Phanerozoic. *Geochim. Cosmochim. Acta* **38**(9), 1461–1484.

Velde, B. (1977). "Clays and Clay Minerals in Natural and Synthetic Systems." Elsevier, Amsterdam, 218 pp.

Vennum, W. R. (1979). Evaporite encrustations and yellow and green surficial salts from Orville Coast and eastern Ellsworth Island. *Antarctic J. U.S.* **14**(5), 22–24.

Vergnaud-Grazzini, G., and Bartolini, G. (1970). Évolution paléoclimatique des sédiments wurmiens et post-wurmiens en mer d'Alboran. *Rev. Géogr. Phys. Géol. Dynam.*, [2] **12**(4), 325–334.

Vergouwen, L. (1981). Eugsterite, a new salt mineral. *Am. Mineral.* **66**(5–6), 632–636.

Verner, A. R., and Orlovskiy, N. V. (1948). The role of sulfate-reducing bacteria in the salt balance of Baraba soils (in Russian). *Pochvovedenie* **9,** 553–560.

Verwey, J. (1931). The depth of coral reefs in relation to their oxygen consumption and the penetration of light in the water. *Treubia* **13**(2), 169–216.

Vetter, F. (1911). Beitrag zu unserer Kenntnis des Kalzium Karbonatniederschlags von einer Bikarbonatlösung. *Z. Kristall. Mineral.* **48,** 45–109.

Vetter, O. J. (1972). An evaluation of scale inhibitors. *J. Pet. Technol.* **24**(8), 997–1006.

Vicq, G., Fornies–Marquina, J. M., and Nguyen, B. C. (1977). Contribution a l'étude de la déshydratation de quelques composés cristallins par adsorption dipolaire Debye aux hyperfréquences. *J. Chim. Phys. Phys.–Chim. Biol.* **74**(2), 249–252.

Vil'denburg, E. V. (1975). Content and chemical composition of natural gases in evaporite strata of the Permian formations in sections of the Orenburg region (in Russian). *Mater. Geol. Polezn. Iskop. Orenb. Obl.* **5,** 132–138.

Viluyanskiy, Ya. E., and Menshikova, E. M. (1933). Speed of solution of potassium minerals (in Russian). *Kalii* No. 1, pp. 17–26.

Vinogradov, B. V. (1955). Macropolygonal structure of clay plains. *Dokl. Akad. Nauk SSSR* **104,** 118–120; transl. *in* "Playas and Dried Lakes. Occurrence and Development" (J. T. Neal, ed.), pp. 323–326. Dowden, Hutchinson & Ross, Inc., Stroudsburg, Pennsylvania, 1975.

Vinokurov, V. M. (1958). Blue rock salt from the Solikamsk district (in Russian). *Zap. Vses. Mineral. O–va.* **87,** 504–507.

Visse, L. (1947). Le gypse des argiles à lignites du Laonnais (Aisne). *Ann. Soc. Géol. Nord* **67,** 706–715.

Vlasov, N. A., and Gorodkova, M. M. (1961). Physico–chemical analysis of annual cycles in the Selenga sulfate lake (in Russian). *Izv. Fiz.–Khim. Nauchno–Issled. Inst. Irkutsk. Univ.* **5**(2), 110–128.

Vochten, F. C. R., and Stoops, G. (1978). Gypsum crystals in Rupelian clay of Betekom (Prov. Antwerpen, Belgium). *Ann. Soc. Belge Géol. Paléont. Hydrol.* **101,** 79–81.

Vol'fkovich, S. I., and Margolis, F. G. (1943). Catalysts for the production of K_2SO_4 or Na_2SO_4 from KCl or NaCl by reaction with SO_2 (in Russian). *Dokl. Akad. Nauk SSSR* **41**(1), 23–25.

von der Borch, C. C., and Lock, D. (1979). Geological significance of Coorong dolomites. *Sedimentology* **26**(6), 813–824.

von der Borch, C. C., Bolton, B., and Warren, J. K. (1977). Environmental setting and microstructure of subfossil lithified stromatolites associated with evaporites, Marion Lake, South Australia. *Sedimentology* **24**(5), 693–708.

von der Borch, C. C., Lock, D., and Schwebel, D. (1975). Groundwater-formation of dolomite in the Coorong region of South Australia. *Geology* **3**(5), 283–285.

Vonder Haar, S. P. (1976). Evaporites and algal mats at Laguna Mormona, Pacific Coast, Baja California, Mexico. Ph.D. Dissertation, University of Southern California, Los Angeles.

von Gottesmann, W. (1963). Eine häufig auftretende Struktur des Halits im Kaliflöz Stassfurt. *Geologie* **12,** 576–581.

von Kaleczinszky, A. (1901). Über die ungarischen heissen und warmen Kochsalzseen als natürliche Wärme-Accumulatoren. *Földt. Közl.* **31,** 409–431.

Voronova, M. L. (1960). Petrographic characteristics of the Lower Cambrian salt-bearing deposits in the southeastern part of the Siberian Platform (in Russian). *Tr. Vses. Nauchno–Issled. Proektn. Inst. Galurgii* **40,** 70–100.

Voronova, M. L. (1962). Kalistrontite, a new mineral of potassium and strontium sulfate (in Russian). *Zap. Vses. Mineral. O–va.* **91,** 712–717.

Voronova, M. L. (1968). Petrographic features of the salt-bearing sequence in the Erevan Basin (Razdan area) (in Russian). *In* ''Geology of Salt and Potash Deposits'' (S. M. Korenevskiy, ed.). *Tr. Vses. Nauchno–Issled. Geol. Inst.* **161,** 138–151.

Vovk, I. F. (1979). On the source of hydrogen in potassium deposits. *Geokhimiya* No. 1, pp. 122–127; *Geochem. Int. (Engl. Transl.)* **15**(1), 86–90.

Vyalov, O. S., and Flerov, K. K. (1953). New finds of vertebrates in the Ciscarpathian Dobrotov beds (in Russian). *Dokl. Akad. Nauk SSSR* **90**(3), 1–16.

Vysotskiy, E. A., Kislik, V. Z., Protasevich, B. A., and Avkhimovich, V. I. (1980). Characteristics of the structure and facies of the halite member in the northwestern part of the Pripyat depression (in Russian). *Dokl. Akad. Navuk BSSR* **24**(1), 67–70.

Wagner, W. (1953). Die tertiären Salzlagerstätten im Oberrheintal–Graben. *Z. Dtsch. Geol. Ges.* **105**(4), 706–728.

Waldschmidt, W. A. (1958). Halite as cementing mineral in sandstones. *Bull. Am. Assoc. Pet. Geol.* **42**(4), 871–892.

Walker, D. (1978). Quaternary climates of the Australian region. *In* ''Climatic Change and Variability'' (A. B. Pittock *et al.*, eds.), pp. 82–97. Cambridge Univ. Press, London.

Walker, R. N., Muir, M. D., Diver, W. L., Williams, N., and Wilkins, N. (1977). Evidence of major sulfate-evaporite deposits in the Proterozoic McArthur group, Northern Territory, Australia. *Nature (London)* **265**(5594), 526–529.

Walter, J. C., Jr. (1953). Paleontology of Rustler Formation, Culberson County, Texas. *Paleontol.* **27**(5), 697–702.

Walter-Levy, L., and Maarten de Wolff, P. (1949). Contribution à l'étude du ciment Sorel. *C. R. Hebd. Séances Acad. Sci. Paris* **229**(21), 1077–1079.

Walther, J. (1903). Die Entstehung von Salz und Gips durch topographische oder klimatische Ursachen. *Centralbl. Mineral. Geol. Paläont.* No. 2, pp. 211–217.

Walther, J. (1924). ''Das Gesetz der Wüstenbildung in Gegenwart und Vorzeit.'' 4th ed., Quelle und Meyer, Leipzig, 421 pp.

Walton, A. W. (1978). Evaporites, relative humidity control of mineral facies: A discussion. *J. Sediment. Petrol* **48**(4), 1357–1359.

Ward, W. C., Folk, R. L., and Wilson, J. L. (1970). Blackening of eolianite and caliche adjacent to saline lakes, Islas Mujeres, Quintana Roo, Mex. *J. Sediment. Petrol.* **40**(2), 548–555.

Wardlaw, N. C. (1968). Carnallite–sylvite relationships in the Middle Devonian Prairie Evaporite Formation, Saskatchewan, Canada. *Bull. Geol. Soc. Am.* **79**(10), 1273–1294.

Wardlaw, N. C. (1970). Effects of fusion, rates of crystallization and leaching on bromide and rubidium solid solutions in halite, sylvite and carnallite. *In* ''Symp. on Salt, 3rd.'' (J. L. Rau and F. L. Dellwig, eds.), vol. 1, pp. 223–231. N. Ohio Geol. Soc., Cleveland, Ohio.

Wardlaw, N. C. (1972a). Unusual marine evaporites with salts of calcium and magnesium chloride in Cretaceous basins of Sergipe, Brazil. *Econ. Geol.* **67**(2), 156–168.

Wardlaw, N. C. (1972b). Synsedimentary folds and associated structures in Cretaceous salt deposits of Sergipe, Brazil. *J. Sediment. Petrol.* **42**(3), 572–577.

Wardlaw, N. C., and Christie, D. L. (1975). Sulfates of submarine origin in Pennsylvanian Otto Fiord Formation of Canadian Arctic. *Bull. Can. Pet. Geol.* **23**(1), 149–171.

Wardlaw, N. C., and Hartzell, W. G. (1963). Geothermometry of halite from the Middle Devonian Prairie Evaporite Formation, Saskatchewan. *Bull. Can. Inst. Min. Metall.* **56**(610), 155.

Wardlaw, N. C., and Reinson, G. (1971). Carbonate and evaporite deposition and diagenesis,

Middle Devonian Winnipegosis and Prairie Evaporite Formations of Saskatchewan. *Bull. Am. Assoc. Pet. Geol.* **55**(10), 1759–1786.

Wardlaw, N. C., and Schwerdtner, W. (1963). Koenenite from Saskatchewan, Canada. *Neues Jahrb. Geol. Paläont., Monatsh.* No. 2, 76–77.

Wardlaw, N. C., and Schwerdtner, W. M. (1966). Halite-anhydrite seasonal layers in the Middle Devonian Prairie Evaporite Formation, Saskatchewan, Canada. *Bull. Geol. Soc. Am.* **77**(4), 331–342.

Wardlaw, N. C., and Watson, D. W. (1966). Middle Devonian salt formations and their bromide content, Elk Point area, Alberta. *Can. J. Earth Sci.* **3,** 263–275.

Warrington, G. (1974). Les évaporites du Trias britannique. *Bull. Soc. Géol. Fr.* **7**(16), 708–723.

Watson, A. (1979). Gypsum crusts in deserts. *J. Arid Environ.* **2**(1), 3–20.

Wattenberg, H. (1936). Kohlensäure und Kalziumkarbonat im Meere. *Fortschr. Mineral. Kristallogr. Petrogr.* **20,** 168–195.

Wattenberg, H., and Timmermann, E. (1936). Über die Sättigung des Seewassers an $CaCO_3$ und die anorgane Bildung von Kalksedimenten. *Ann. Hydrogr. Marit. Meteorol.* **64,** 23–31.

Weaver, C. E., and Pollard, L. D. (1973). "The Chemistry of Clay Minerals." Elsevier, Amsterdam, 213 pp.

Weber, K. (1931). Geologisch–petrographische Untersuchungen am Stassfurt–Egelner Sattel unter besonderer Berücksichtigung der Genese der Polyhalit– und Kieseritregion. *Kali (Halle)* **25**(2), 17–23; (3), 33–38; (4), 49–55; (5), 65–71; (6), 82–88; (7), 97–104; (8), 122–123.

Weber, K. (1961). Untersuchungen über die Faziesdifferenzierungen, Bildungs- und Umbildungserscheinungen in den beiden Kalilagern des Werra–Fulda–Gebietes unter besonderer Berücksichtigung der Vertaubungen. PhD Dissertation, Bergakademie, Clausthal, 164 pp.

Wedepohl, K. H. (1964). Untersuchungen am Kupferschiefer in Nordwest-Deutschland. *Geochim. Cosmochim. Acta* **28**(3), 305–364.

Wedepohl, K. H. (1971). "Kupferschiefer" as a prototype of syngenetic sedimentary ore deposits. *Soc. Min. Geol. Jpn, Spec. Issue* No. 3, pp. 268–273.

Weeks, L. G. (1958). "The Habitat of Oil," pp. 1–101. Am. Assoc. Pet. Geol., Tulsa, Oklahoma.

Weete, J. D. (1972). Aliphatic hydrocarbons of the fungi. *Phytochemistry* **11**(4), 1201–1205.

Weiler, Y., Sass, E., and Zak, I. (1974). Halite oolites and ripples in the Dead Sea, Israel. *Sedimentology* **21,** 623–632.

Weisman, R. N., and Brutsaert, W. (1973). Evaporation and cooling of a lake under unstable atmospheric conditions. *Water Resour. Res.* **9,** 1242–1257.

Wellman, H. W., and Wilson, A. T. (1963). Salt weathering, a neglected geological erosive agent in coastal and arid environments. *Nature (London)* **205**(4976), 1097–1098.

Wells, A. J. (1962). Recent dolomite in the Persian Gulf. *Nature (London)* **194**(4825), 274–275.

Wells, A. J., and Illing, L. V. (1964). Present–day precipitation of calcium carbonate in the Persian Gulf. *In* "Deltaic and Shallow Marine Deposits" (L.M.J.U. van Straaten, ed.), pp. 429–435. Elsevier, Amsterdam.

Wemelsfelder, R. J. (1970). Sea level observation as a fact and as an illusion. *In* Rep. on the Symp. "Coastal Geodesy" (R. Sigl, ed.), pp. 65–80. Tech. Univ., Munich.

Wendling, E., von Hodenberg, R., and Kuehn, R. (1972). Congolit, der trigonale Eisenboracit. *Kali Steinsalz* **6**(1), 1–3.

Werner, D. H. (1938). Die Entstehung von Grossfluten in der Perm- und Triaszeit. *Kali (Halle)* **32**(7), 61–66.

West, I. M. (1975). Evaporites and associated sediments of the basal Purbeck Formation (Upper Jurassic) of Dorset. *Proc. Geol. Assoc. London* **86,** 205–225.

West, I. M., Ali, Y. A., and Hilmy, M. E. (1979). Primary gypsum nodules in a modern sabkha on the Mediterranean coast of Egypt. *Geology* **7**(7), 354–358.

Wetzel, R. G. (1975). "Limnology." W. B. Saunders, Philadelphia, Pennsylvania, 743 pp.

Wetzel, W. (1928). Salzbildungen der chilenischen Wüste. *Chem. Erde* **3**(3–4), 375–436.

Wetzel, W. (1932). Nitrat. *In* "Die wichtigsten Lagerstätten der Nichterze" (O. Stutzer, ed.), vol. 4, pp. 297–372. Gebr. Borntraeger, Berlin.

Wetzel, W. (1939). Sedimentpetrographische Untersuchungen an deutschen Salzgesteinen. *Jahresber., Niedersächs. Geol. Ver.* **29,** 89–100.

Wexler, A., and Hasegawa, S. (1954). Relative humidity–temperature relationships of some saturated salt solutions in the temperature range 0 to 50°C. *J. Res. Natl. Bur. Stand. (U.S.)* **53,** 19–26.

Weyrich, H. (1961). Veränderungerscheinungen am Anhydrit. *Z. Angew. Geol.* **7**(2), 96.

Wezel, F. C., Savelli, D., Bellagamba, M., Tramontana, M., and Bartole, R. (1981). Plio–Quaternary depositional style of sedimentary basins along insular Tyrrhenian margins. *In* "Sedimentary Basins of Mediterranean Margins" (F. C. Wezel, ed.), pp. 239–269, Tecnoprint, Bologna.

Whelan, J. A. (1972). Ochsenius bar theory of saline deposition supported by quantitative data, Great Salt Lake, Utah. *Rep. Int. Geol. Congr., 24th, Montreal 1972, Sess. 10,* pp. 296–303.

White, A. F. (1977). Sodium and potassium coprecipitation in aragonite. *Geochim. Cosmochim. Acta* **41**(5), 613–625.

White, D. E. (1968). Environments of generation of some base-metal ore deposits. *Econ. Geol.* **63**(4), 301–335.

White, D. E., Hem, J. D., and Waring, G. A. (1963). Chemical composition of subsurface waters. *U.S. Geol. Surv. Prof. Paper* **440–F,** 67 pp.

White, W. A. (1961). Colloid phenomena in sedimentation of argillaceous rocks. *J. Sediment. Petrol.* **31**(4), 560–570.

Whitman, W. G., Russel, R. P., and Davis, G. H. B. (1925). The solubility of ferrous hydroxide and its effect on corrosion. *J. Am. Chem. Soc.* **47**(1), 70–79.

Whittemore, D. O., Basel, C. L., Galle, O. K., and Waugh, T. C. (1981). Identification of salt sources contaminating an alluvial aquifer in central Kansas. *Geol. Soc. Am., Abstr. Prog.* **13**(7), 580.

Wicke, E., and Eigen, M. (1953). Über die thermodynamischen Eigenschaften von mässig konzentrierten wässrigen Elektrolyt–Lösungen. *Z. Elektrochem.* **57,** 219–330.

Wied, J. I., and Syrojezkina, J. W. (1965). Effect of surface active substances on the crystal habit of $CaSo_4 \cdot 2H_2O$. *Cem.–Wapno–Gips* **1,** 181–185.

Wieninger, L. (1950). Verfärbung gedruckter Steinsalzkristalle durch Bestrahlung mit alpha–Teilchen von Radium. *Sitzungsber. Österr. Akad. Wiss., Math.–Naturwiss. Kl.,* [2a] **159,** 381–393.

Wieninger, L. (1951). Über die Natur und Herkunft der violetten oder blauen Farbung natürlicher Steinsalzkristalle. *Sitzungsber. Österr. Akad. Wiss., Math.–Naturwiss. Kl.,* [2a] **160,** 147–180.

Wieninger, L., and Adler, N. (1950). Der Einfluss von Erwärmung auf das Absorptionsspektrum von Steinsalz, das verfärbt ist durch alpha–Teilchen von Radium. *Sitzungsber. Österr. Akad. Wiss., Math.–Naturwiss. Kl.,* [2a] **159,** 395–406.

Wigley, T. M. L., and Plummer, L. N. (1976). Mixing of carbonate waters. *Geochim. Cosmochim. Acta* **40**(9), 989–995.

Wilde, W. (1961). Über die Bildung von Schaumspat und Vaterit. *Z. Angew. Geol.* **7,** 539–541.

Wilfarth, M. (1933). Sedimentationsprobleme in der germanischen Senke zur Perm- und Triaszeit. *Geol. Rundsch.* **24**(6), 349–377.

Wilfarth, M. (1934). Strömungserscheinungen im Wellenkalkmeer. *Z. Dtsch. Geol. Ges.* **86**(5), 265–285.

Wilhelm, H. G., and Ackermann, W. (1972). Altersbestimmung nach der K–Ca Methode am Sylvin des Oberen Zechsteins des Werragebietes. *Z. Naturforsch.*, [A] **27**(8–9), 1256–1259.

Wilkansky, B. (1936). Life in the Dead Sea. *Nature (London)* **138**(3489), 467.

Williams, R. E. (1968). Groundwater flow systems and the origin of evaporite deposits. *Pam. Idaho Bur. Mines Geol.* No. 141, pp. 1–15.

Wills, L. J. (1970). The Triassic succession in the coastal Midlands in its regional setting. *Qu. J. Geol. Soc. London* **126,** 225–283.

Wilser, B. (1926). Zur Frage der Entstehung und Herkunft der alttertiären Salzablagerungen im Rheintalgraben. *Kali (Halle)* **20**(6), 86–88.

Wink, W. A., and Sears, G. A. (1950). Equilibrium of relative humidities above saturated salt solutions at various temperatures. *Tappi* **33**(9), 96–99. *Chem. Abstr.* **44,** 10,416.

Winkler, E., and Singer, P. C. (1972). Crystallization pressure of salts in stone and concrete. *Bull. Geol. Soc. Am.* **83**(11), 3509–3513.

Winkler, E. M., and Wilhelm, E. J. (1970). Salt burst by hydration pressures in architectural stone in urban atmosphere. *Bull. Geol. Soc. Am.* **81**(2), 567–572.

Winnock, E. (1981). Structure du bloc pelagien. *In* "Sedimentary Basins of Mediterranean Margins" (F. C. Wezel, ed.), pp. 445–464. Tecnoprint, Bologna.

Winogradsky, S. (1887). Über Schwefelbakterien. *Bot. Ztg.* **45,** 489–600.

Winogradsky, S. (1888). Über Eisenbakterien. *Bot. Ztg.* **46,** 261–276.

Winters, K., Parker, P. L., and van Baalen, C. (1969). Hydrocarbons of bluegreen algae: Geochemical significance. *Science* **163**(3866), 467–468.

Wirth, H. E. (1937). The partial molal volume of potassium chloride, potassium bromide, and potassium sulfate in sodium chloride solutions. *J. Am. Chem. Soc.* **59**(12), 2549–2554.

Withington, C. F. (1961). Origin of mottled structure in bedded calcium sulfate. *U.S. Geol. Surv. Prof. Paper* **424D,** 342–344.

Wollam, J. S., and Wallace, W. E. (1956). A comparison of pycnometric and x-ray densities for the sodium chloride–sodium bromide–potassium bromide systems. *J. Phys. Chem.* **60,** 1654–1656.

Wood, G. V., and Wolfe, M. J. (1969). Sabkha cycles in the Arab/Darb formation off the Trucial Coast of Arabia. *Sedimentology* **12**(3–4), 165–191.

Wood, J. A. (1977). Mineralogical and petrological study of the low-temperature minerals in carbonaceous chondrites. *Rep. Natl. Aeronaut. Space Admin., Contract* CR–149, 811, 6 pp.

Wood, J. R. (1975). Thermodynamics of brine–salt equilibria. I. The system NaCl–KCl–$MgCl_2$–$CaCl_2$–H_2O and NaCl–$MgSO_4$–H_2O at 25°C. *Geochim. Cosmochim. Acta* **39**(8), 1147–1163.

Woods, P. J. E. (1973). Potash exploration in Yorkshire: Boulby Mine pilot borehole. *Trans. Inst. Min. Metall., Sect. B* **82,** 99–106.

Woods, P. J. E. (1979). The geology of Boulby mine. *Econ. Geol.* **74**(2), 409–418.

Woolnough, W. G. (1937). Sedimentation in barred basins and source rocks of oil. *Bull. Am. Assoc. Pet. Geol.* **21**(9), 1101–1157; (10), 1350–1351.

Worsley, N., and Fuzesy, A. (1979). The potash-bearing members of the Devonian Prairie Evaporite of southeastern Saskatchewan south of the mining area. *Econ. Geol.* **74**(2); 377–388.

Wright, P. B. (1978). The southern oscillation. *In* "Climatic Change and Variability" (A. B. Pittock *et al.*, eds.), pp. 180–190. Cambridge Univ. Press, London.

Wu, B., Qi, Z., Wang, M., and Qu, J. (1980). Geochemistry of salt-bearing deposits in Basin Q, Hubei Province, China (in Chinese). *Acta Geol. Sin.* **54**(4), 324–333.

Wu, J. (1977). A note on the slope of a density interface between two stably stratified fluids under wind. *J. Fluid Mech.* **81**(2), 335–339.

Wuest, G. (1934). Salzgehalt und Wasserbewegung im Suezkanal. *Naturwissenschaften* **22**(26), 447–450.

Wuest, J., and Lange, I. (1925). Lösungs– und Verdünnungswärmen von Salzen von der äussersten Verdünnung bis zur Sättigung. I. Alkalihalogenide KCl, KBr, KJ, NaCl, NaBr, NaJ. *Z. Phys. Chem. Stöchiom. Verwandtschaftl.* **116,** 161–214.

Xavier, A., and Klemm, D. D. (1979). Authigenic gypsum in deep-sea manganese nodules. *Sedimentology* **26**(2), 307–310.

Xu, G. (1982). The dynamic equilibrium calculation of phase-transformation effects of the intercrystalline brine in a salt lake. *Bull. Soc. Geol. Sin.* No. 1, pp. 96–102.

Yaalon, D. H. (1965). Downward movement and distribution of anions in soil profiles with limited wetting. *In* "Experimental Pedology" (E. G. Hallworth and D. V. Crawford, eds.), pp. 157–164. Butterworth, London.

Yamamoto, T. (1938). Growth of crystals. IX. Relation between concentration of cations added to solution and their amount in the crystal formed. *Bull. Inst. Phys. Chem. Res. Tokyo* **17** 1278–1291.

Yamasaki, N., Sawada, M., Mitsushio, H., and Matsuoka, K. (1981). Effects of organic compounds on hydrothermal formation of calcite (in Japanese). *Rep. Res. Lab. Hydrotherm. Chem. Kochi (Jpn)* **4**(197), 11–13; *Chem. Abstr.* **98,** 74,941.

Yanat'eva, O. K., and Orlova, V. T. (1958). Crystallization of schoenite in the marine system K, Na, Mg ‖ Cl, SO_4–H_2O at 10°C (in Russian). *Zh. Neorg. Khim.* **3,** 2408–2413.

Yang, Q. (1982). Sediment formation of potash deposits in the Qarhan inland salt lake (in Chinese). *Sci. Geol. Sin.* **56**(3), 281–292; *Chem. Abstr.* **97,** 185,623.

Yanshin, A. L. (1961). On the depth of salt-bearing basins and some problems of forming thick salt series (in Russian). *Tr. Inst. Geol. Geofiz. Akad. Nauk SSSR, Sib. Otd.* No. 1, pp. 3–15.

Yarzhemskiy, Ya. Ya. (1949a). Secondary quartz in halite rocks (in Russian). *Dokl. Akad. Nauk SSSR* **66**(5), 915–918.

Yarzhemskiy, Ya. Ya. (1949b). Polyhalite in potash salt deposits (in Russian). *Dokl. Akad. Nauk SSSR* **66**(6), 1157–1160.

Yarzhemskiy, Ya. Ya. (1950). The schoenitization of langbeinite in water vapor (in Russian). *Dokl. Akad. Nauk SSSR* **74**(5), 1015–1017.

Yarzhemskiy, Ya. Ya. (1954). The question of polyhalite genesis in potash deposits (in Russian). *Tr. Vses. Nauchno–Issled. Proektn. Inst. Galurgii* **29,** 223–259.

Yarzhemskiy, Ya. Ya. (1955). The role of dolomite and magnesite in salt deposits (in Russian). *Dokl. Akad. Nauk SSSR* **104,** 622–625.

Yarzhemskiy, Ya. Ya. (1966). Reasons for the occurrence of bischofite in sulfatic evaporite basins of the marine type. *Dokl. Akad. Nauk SSSR* **167**(6), 1373–1375; *Dokl. Acad. Sci. USSR, Earth Sci. Sect. (Engl. Transl. Am. Geol. Inst.)* **167**(1–6), 158–159.

Yarzhemskiy, Ya. Ya. (1967). "Potash Rocks and Potash Bearing Evaporites" (in Russian). Izd. Nauka, Sib. Otd., Novosibirsk, 133 pp.

Yashkichev, V. I. (1963). Surface tension of aqueous solutions of salts and influence of ions in the water structure (in Russian). *Zh. Strukt. Khim.* **4**(6), 837–843.

Yefremova, A. G. (1978). Effect of sulfate reduction on the formation and distribution of recent gas and bitumen producing deposits (in Russian). *Nauchno–Tekh. Obz., Ser. Geol. Razved. Gazov. Gazokondens. Mestorozhd.* No. 2, pp. 13–21. *Chem. Abstr.* **89,** 62,047.

Yermakov, V. A., and Grebennikov, N. P. (1977). Structural regularities of bischofite deposits in evaporite beds of the lower Volga River region and paleogeographic conditions of their accumulation (in Russian). *In* "Problems of Salt Accumulation" (A. L. Yanshin and M. A. Zharkov, eds.), vol. 2, pp. 40–45. Izd. Nauka, Sib. Otd., Novosibirsk.

Yesair, J. (1930). Bacterial content of salt. *Cann. Trade* **52**(27), 112–115.

Yeventov, Yu. S., Mileshina, A. G., and Komissarova, I. N. (1971). Oil permeability of mineral salt (in the Caspian Basin) (in Russian). *Geol. Nefti Gaza* No. 4, pp. 37–42; *German Transl.: Z. Angew. Geol.* **19**(2), 67–71 (1973).

Yeventov, J. S., Mileschina, A. G., and Komissarova, I. N. (1973). Über die Erdöldurchlassigkeit fossiler Salze. *Z. Angew. Geol.* **19**(2), 67–71.

Young, D. M. (1952). Adsorption of argon on octahedral potassium chloride. *Trans. Faraday Soc.* **48,** 548–561.

Youngblood, W. W., Bloomer, M., Guillard, R. L., and Fiore, F. (1971). Saturated and unsaturated hydrocarbons in marine benthic algae. *Mar. Biol.* **8**(3), 190–191.

Yurk, Y. Y., and Lebedeva, A. D. (1960). Gypsum crystals from Pliocene sediments of the Tarutino region (in Ukrainian). *Tr. Inst. Geol. Nauk Akad. Nauk Ukr. RSR, Ser. Petrogr. Mineral. Geokhim.* No. 6, pp. 178–184.

Zablocki, J. (1928). Tertiäre Flora der Salzlager von Wieliczka I. *Acta Soc. Bot. Pol.* **5**(2), 174–208.

Zablocki, J. (1930). Tertiäre Flora der Salzlager von Wieliczka II. *Acta Soc. Bot. Pol.* **7**(2), 215–230.

Zabrodin, N. P. (1977). Present-day evolution of Kara Bogaz Gol: Problems connected with its industrial utilization (in Russian). *In* "Problems of Salt Accumulation" (A. L. Yanshin and M. A. Zharkov, eds.), vol. 1, pp. 235–237. Izd. Nauka, Sib. Otd., Novosibirsk.

Zaenker, G. (1970). Über einige sedimentäre und diagenetische Gefügemerkmale in chloridischen Salzablagerungen. *Ber. Dtsch. Ges. Geol. Wiss.,* [A] **15**(4), 495–502.

Zaenker, G. (1979). Zur Frage der Beckentiefe und synsedimentären Beckenabsetzungen im höheren Zechstein 2 des Südharzebietes. *Z. Geol. Wiss.* **7**(7), 861–869.

Zafirov, S. I. (1959). The origin of gypsum in the gypsiferous district of Radnevo (in Bulgarian). *God. Geol. Geogr. Fak. Sofii. Univ., Mineral. Geol. Inst.* **5,** 153–165.

Zaghloul, Z. M., El-Ayouty, M. K., and El Sawy, M. M. (1977). The subsurface Miocene evaporites in the Gulf of Suez region and their genetic relation with petroleum. *Egypt. J. Geol.* **18**(2), 77–86.

Zajic, J. E. (1969). "Microbial Biogeochemistry." Academic Press, New York, 345 pp.

Zak, I. (1974). Sedimentology and bromine chemistry of marine and continental evaporites in the Dead Sea basin. *In* "Symp. on Salt, 4th." (A. H. Coogan, ed.), vol. 1, pp. 349–361. N. Ohio Geol. Soc., Cleveland, Ohio.

Zak, I., and Bentor, Y. K. (1972). Some new data on the salt deposits of the Dead Sea area, Israel. *In* "Geology of Saline Deposits" (G. Richter-Bernburg, ed.), U.N.E.S.C.O., Paris, Earth Sci. Ser. **7,** 137–146.

Zaritskiy, P. V. (1960). Celestite from Lower Permian sediments of the Don Basin (in Russian). *Dokl. Akad. Nauk SSSR* **133**(3), 446–449.

Zaslavskiy, A., and Harzestein, N. (1930). Über die Einwirkung gewisser Salze auf obligat halophile Schwefelbakterien. *Zentralbl. Bakteriol. Parasitenkd. Infektionskr. Hyg.,* [2] **80,** 165–169.

Zdanovskiy, A. B. (1946). The role of the interphase solution in the kinetics of the solution of salts (in Russian). *Zh. Fiz. Khim.* **20,** 869–880.

Zednicek, W. (1954). Gipskristalle aus dem Russ des Karawankentunnels. *Karinthin* **144,** 29–32.

Zekert, B. (1927). Die Färbung von Steinsalz und Kunzit durch Becquerelstrahlen. *Sitzungsber. Akad. Wiss.* Wien, *Math.-Naturwiss. Kl.,* [2a] **136,** 337–355.

Zelizna, S. T., and Roskosh, Ya. T. (1973). Mechanism of carbonate metasomatism of sulfate rocks (in Ukrainian). *Geol. Geokhim. Goryuch. Kopalin* No. 35, pp. 94–99.

Zemmels, I., and Cook, H. E. (1973). X-ray mineralogy studies of selected deep-sea samples from the sea floor of the Northeast Atlantic and Mediterranean Sea. *In* Intl. Rep. Deep-Sea Drlg. Proj., vol. 13 (Ryan, W. B. F., Hsu, K. J., *et al.*), Pt. 2, pp. 605–665. Govt. Print. Off., Washington, D.C.

Zen, E. (1960). Early stages of evaporite deposition. *U.S. Geol. Surv. Prof. Paper* **400–B,** 458–461.

Zen, E. (1965). Solubility measurements in the system $CaSO_4$–NaCl–H_2O. *J. Petrol.* **6**(1), 124–164.

Zharkov, M. A. (1978). "History of Paleozoic Salt Accumulation." *Tr. Inst. Geol. Geofiz. Akad. Nauk SSSR, Sib. Otd.* **354,** 270 pp.; transl.: Springer-Verlag, New York, (1981), 308 pp.

Zharkov, M. A., Zharkova, T. M., and Merzlyakov, G. A. (1979). Die Volumenverhältnisse halogener Gesteine im paläozoischen Salinarbecken und das Problem der Stoffbestandsentwicklung des Meerwassers. *Z. Geol. Wiss.* **7**(7), 827–841.

Zherebtsova, I. K. (1970). The behavior of bromine in the eutonic stage of concentration of sea water (in Russian). *Tr. Inst. Geol. Geofiz. Akad. Nauk SSSR* **116,** 272–278.

Zherebtsova, I. K., and Volkova, N. N. (1966). Experimental study of the behavior of microelements in the process of natural solar evaporation of Black Sea water and brine of Lake Sasyk–Sivash (in Russian). *Geokhimiya* No. 7, pp. 832–844.

Zhuze, N. G. (1972). Bituminosity of layers interbedding Frasnian salt deposits in the central areas of the Dnieper–Donets syncline (in Russian). *Neftegazov. Geol. Geofiz.* No. 6, pp. 21–28.

Zieglar, D. L. (1963). Pre-Piper post-Minnekahta red beds in the Williston Basin. *In* "Symp. on Salt" (A. C. Bersticker, K. E. Hoekstra, and J. F. Hall, eds.), pp. 170–178. N. Ohio Geol. Soc., Cleveland, Ohio.

Ziegler, G. (1898). An den Herausgeber des Prometheus: Absonderliche Temperaturverhältnisse in einem Solbehalter. *Prometheus* **9,** 79, and discussion p. 325.

Ziehr, H., Matzke, K., Ott, G., and Voutsidis, V. (1980). Ein stratiformes Fluoritvorkommen im Zechstein–Dolomit bei Eschwege und Sontra in Hessen. *Geol. Rundsch.* **69**(2), 325–348.

Zimmermann, E. (1907). Über den Pegmatitanhydrit und den mit ihm verbundenen Roten Salzton im jüngeren Steinsalz des Zechsteins vom Stassfurter Typus und über Pseudomorphosen nach Gips in diesem Salzton. *Z. Dtsch. Geol. Ges., Monatsber.* **59,** 136–143.

Zimmermann, E. (1908). Über Wellenfurchen im Steinsalz. *Kali (Halle)* **2**(10), 214.

Zimmermann, E. (1914). Der thüringische Plattendolomit und seine Vertreter im Stassfurter Zechsteinprofil sowie eine Bemerkung zur Frage der Jahresringe. *Z. Dtsch. Geol. Ges., Monatsber.* **65,** 357–372.

ZoBell, C. E. (1946). Studies on redox potential of marine sediments. *Bull. Am. Assoc. Pet. Geol.* **30**(4), 477–513.

Zoikin, L. M., Voitov, G. I., and Dorogokupets, T. I. (1975). An evaluation of the potential use of gases in salts for a prognosis of petroleum and gas content in formations beneath the salt (in Russian). *Tr. Vses. Nauchno–Issled. Inst. Yad. Geofiz. Geokhim.* **22,** 32–39.

Zolotov, V. A. (1957). Intercrystalline layer in gypsum. *Dokl. Akad. Nauk SSSR* **115,** 534–536; *Dokl. Acad. Sci. USSR, Phys. Chem. Sect. (Engl. Transl.)* **115,** 499–500.

Zolotov, V. A., and Lavrov, M. N. (1960). Effect of plastic deformation of gypsum crystals on the subsequent dehydration process (in Russian). *Uch. Zap. Arzamas. Gos. Pedagog. Inst.* **4,** 104–115. *Chem. Abstr.* **57,** 4326.

Zolotukhin, V. V. (1954). Two varieties of gypsum from Zaleshchiki (in Ukrainian). *Mineral. Sb. (L'viv)* **8,** 253–260.

Zonneveld, J. I. S. (1968). Quaternary climatic changes in the Caribbean and northern South America. *Eiszeitalter Ggw.* **19,** 203–208.

Zor'kin, L. M., Kozlov, V. G., Stadnik, E. V. (1979). Geochemistry of subsurface waters of evaporites of the eastern Russian Platform (in Russian). *In* "Problems in Theory and Regional Hydrochemistry," pp. 61–63. Izd. Nauka, Moscow; Chem. Abstr. **92,** 183,982.

Zubakhina, Z. K., and Gerasimenko, E. I. (1974). Effect of temperature and composition on the pH level of liquids of a mixer and distiller for soda production (in Russian). *Tr. Gos. Nauchno–Issled. Proektn. Inst. Osnovn. Khim.* **34,** 58–63.

Zwanzig, H. (1958). Die Problematik des Tachhydritgehaltes der mitteldeutschen Kalilagerstätten. *Bergbautechnik* **8,** 93–94.

Zwanzig, W. (1928). Die Zechsteinsalzlagerstätte im oberen Allertal bei Wefensleben-Belsdorf. *Kali, Verw. Salze, Erdöl* **22**(4), 45–49; (5) 62–66; (6) 76–79; (7) 92–94; (9) 113–115.

Index

C

D

M

Q

T

U

X

Z